Excursions in
Modern Mathematics

Excursions in Modern Mathematics

FIFTH EDITION

PETER TANNENBAUM

CALIFORNIA STATE UNIVERSITY—FRESNO

PEARSON EDUCATION, INC.
Upper Saddle River, NJ 07458

Library of Congress Cataloging-in-Publication Data
Tannenbaum, Peter (date)
  Excursions in modern mathematics.—5th ed. / Peter Tannenbaum.
    p. cm.
  Includes bibliographical references and index.
  ISBN 0-13-100191-4
  1. Mathematics. I. Title.

QA36.T35 2004
510—dc21   2003040569

Editor in Chief/Acquisitions Editor: *Sally Yagan*
Project Manager: *Jacquelyn Riotto*
Vice President/Director of Production and Manufacturing: *David W. Riccardi*
Executive Managing Editor: *Kathleen Schiaparelli*
Senior Managing Editor: *Linda Mihatov Behrens*
Production Editor: *Barbara Mack*
Assistant Managing Editor, Math Media Production: *John Matthews*
Media Production Editor: *Donna Crilly*
Manufacturing Buyer: *Michael Bell*
Manufacturing Manager: *Trudy Pisciotti*
Marketing Manager: *Krista M. Bettino*
Editorial Assistant/Print Supplements Editor: *Joanne Wendelken*
Art Director: *Heather Scott*
Interior Designer: *Circa '86/Brian Molloy*
Cover Designer: *John Christiana*
Art Editor: *Thomas Benfatti*
Managing Editor, Audio/Video Assets: *Patricia Burns*
Creative Director: *Carole Anson*
Director of Creative Services: *Paul Belfanti*
Director, Image Resource Center: *Melinda Reo*
Manager, Rights and Permissions: *Zina Arabia*
Interior Image Specialist: *Beth Boyd-Brenzel*
Cover Image Specialist: *Karen Sanatar*
Image Permission Coordinator: *Charles Morris*
Photo Researcher: *Teri Stratford*
Cover Photo: *Antelope Canyon, Arizona/R. Gerth/Masterfile Corporation*
Art Studio: *Scientific Illustrators*
Compositor: *Lithokraft II*

PEARSON
Prentice
Hall
© 2004, 2001, 1998, 1995, 1992 by Pearson Education, Inc.
Pearson Education, Inc.
Upper Saddle River, New Jersey 07458

Printed in the United States of America
10 9 8 7 6 5 4 3 2

Reprinted with corrections August, 2003

ISBN 0-13-100191-4

Pearson Education LTD., *London*
Pearson Education Australia PTY, Limited, *Sydney*
Pearson Education Singapore, Pte. Ltd
Pearson Education North Asia Ltd, *Hong Kong*
Pearson Education Canada, Ltd., *Toronto*
Pearson Educación de Mexico, *S.A. de C.V.*
Pearson Education—Japan, *Tokyo*
Pearson Education Malaysia, Pte. Ltd

*To Sally, Nicholas, David, Paul, and Kathryn,
and in loving memory of my mother, Anna*

CONTENTS

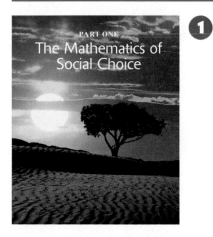

Fair Division   86
The Mathematics of Sharing

The Mathematics of Apportionment   136
Making the Rounds

PART 2   Management Science

8   **The Mathematics of Scheduling   314**
Directed Graphs and Critical Paths

PART 3   Growth and Symmetry

9   **Spiral Growth in Nature   358**
Fibonacci Numbers and the Golden Ratio

PART 4 Statistics

Normal Distributions   626

Everything Is Back to Normal (Almost)

PREFACE

To most outsiders, modern mathematics is unknown territory. Its borders are protected by dense thickets of technical terms; its landscapes are a mass of indecipherable equations and incomprehensible concepts. Few realize that the world of modern mathematics is rich with vivid images and provocative ideas.

Ivars Peterson, The Mathematical Tourist

Excursions in Modern Mathematics is, as we hope the title might suggest, a collection of "trips" into that vast and alien frontier that many people perceive mathematics to be. While the purpose of this book is quite conventional—it is intended to serve as a textbook for a college-level liberal arts mathematics course—its contents are not. By design, the topics in this book are chosen with the purpose of showing the reader a different view of mathematics from the one presented in a traditional general education mathematics curriculum. The notion that general education mathematics must be dull, unrelated to the real world, highly technical, and deal mostly with concepts that are historically ancient is totally unfounded.

The "excursions" in this book represent a collection of topics chosen to meet a few simple criteria.

- **Applicability.** The connection between the mathematics presented here and down-to-earth, concrete real-life problems is direct and immediate. The often heard question, "What is this stuff good for?" is a legitimate one and deserves to be met head on. The often heard answer, "Well, you need to learn the material in Math 101 so that you can understand Math 102 which you will need to know if you plan to take Math 201 which will teach you the real applications," is less than persuasive and in many cases reinforces students' convictions that mathematics is remote, labyrinthine, and ultimately useless to them.

- **Accessibility.** Interesting mathematics need not always be highly technical and built on layers upon layers of concepts. As a general rule, the choice of topics in this book is such that a heavy mathematical infrastructure is not needed—by and large, Intermediate Algebra is an appropriate and sufficient prerequisite. (In the few instances in which more advanced concepts are unavoidable, an effort has been made to provide enough background to make the material self-contained.) A word of caution—this does not mean that the material is easy! In mathematics, as in many other walks of life, simple and straightforward is not synonymous with easy and superficial.

- **Age.** Much of the mathematics in this book has been discovered within the last 100 years; some as recently as 20 years ago. Modern mathematical discoveries do not have to be only within the grasp of experts.

- **Aesthetics.** The notion that there is such a thing as beauty in mathematics is surprising to most casual observers. There is an important aesthetic component in mathematics and, just as in art and music (which mathematics very much resembles), it often surfaces in the simplest ideas. A fundamental objective of this book is to develop an appreciation for the aesthetic elements of mathematics. Hopefully, every open-minded reader will find some topics about which they can say, "I really enjoyed learning this stuff!"

Outline

The material in the book is divided into four independent parts. Each of these parts in turn contains four chapters dealing with interrelated topics.

- **Part 1 (Chapters 1 through 4).** **The Mathematics of Social Choice.** This part deals with mathematical applications in social science. How do groups make decisions? How are elections decided? What is power? How can power be measured? What is fairness? How are competing claims on property resolved in a fair and equitable way? How are seats apportioned in the House of Representatives?

- **Part 2 (Chapters 5 through 8).** **Management Science.** This part deals with methods for solving problems involving the organization and management of complex activities—that is, activities involving either a large number of steps and/or a large number of variables (routing the delivery of packages, landing a spaceship on Mars, organizing a banquet, scheduling classrooms at a big university, etc.). Efficiency is the name of the game in all these problems. Some limited or precious resource (time, money, raw materials) must be managed in such a way that waste is minimized. We deal with problems of this type (consciously or unconsciously) every day of our lives.

- **Part 3 (Chapters 9 through 12).** **Growth and Symmetry.** This part deals with nontraditional geometric ideas. How do sunflowers and seashells grow? How do animal populations grow? What are the symmetries of a snowflake? What is the true pattern behind that wallpaper pattern? What is the geometry of a mountain range? What kind of symmetry lies hidden in our circulatory system?

- **Part 4 (Chapters 13 through 16).** **Statistics.** In one way or another, statistics affects all of our lives. Government policy, insurance rates, our health, our diet, and public opinion are all governed by statistical laws. This part deals with some of the most basic aspects of statistics. How should statistical data be collected? How is data summarized so that it is intelligible? How should statistical data be interpreted? How can we measure the inherent uncertainty built into statistical data? How can we draw meaningful conclusions from statistical information? How can we use statistical knowledge to predict patterns in future events?

Exercises and Projects

An important goal for this book is that it be flexible enough to appeal to a wide range of readers in a variety of settings. The exercises, in particular, have been designed to convey the depth of the subject matter by addressing a broad spectrum of levels of difficulty—from the routine drill to the ultimate challenge. For convenience (but with some trepidation) the exercises are classified into three levels of difficulty:

- **Walking.** These exercises are meant to test a basic understanding of the main concepts, and they are intended to be within the capabilities of students at all levels.

- **Jogging.** These are exercises that can no longer be considered as routine—either because they use basic concepts at a higher level of complexity, or they require slightly higher order critical thinking skills, or both.

- **Running.** This is an umbrella category for problems that range from slightly unusual or slightly above average in difficulty to problems that can be a real challenge to even the most talented of students.

Traditional exercises sometimes are not sufficient to convey the depth and richness of a topic. A new feature in this edition is the addition of a **Projects and Papers** section following the exercise sets at the end of each chapter. One of the nice things about the "excursions" in this book is that they often are just a starting point for further exploration and investigation. This section offers some potential topics and ideas for some of these explorations, often accompanied with suggested readings and leads for getting started. In most cases, the projects are well suited for group work, be it a handful of students or an entire small class.

What Is New in This Edition?

The two most visible additions to this edition are the **Projects and Papers** section discussed above and a **biographical profile** at the end of each chapter (in the chapter on Apportionment, a historical section detailing the checkered story of apportionment in the U.S. House of Representatives was added instead). Each biographical profile features a scientist (they are not always mathematicians) who made a significant contribution to the subject covered in the chapter. In keeping with the spirit of modernity, most are contemporary and in many cases still alive.

Other changes in this edition worth mentioning are:

- In Chapter 2 the European Union is introduced as another important example of a weighted voting system. Both the Banzhaf and the Shapley-Shubik power distribution of the member nations in the EU are given. The power distribution of the Electoral College has been updated to reflect the 2000 Census data.

- In Chapter 4 I expanded the discussion of how to use trial and error to find divisors for Jefferson's and Adams's methods. The new explanations are illustrated with flowcharts. An historical section on apportionment (which was an appendix in earlier editions) has been expanded and moved to the end of the chapter. I added a new appendix (Appendix 2), showing the apportionments in the U.S. House of Representatives for each state under each of the methods discussed in the chapter.

- In Chapter 5 I added a brief discussion on algorithms in general.

- In Chapter 10 several new examples have been added to give more realistic illustrations of the use of exponential growth models.

- In Chapter 11 several new tables have been added to better clarify the classification of symmetry types. I also included a flowchart for the classification of wallpaper patterns.

- In Chapter 14 a new section on computing pth percentiles in general has been added. The computations of the median and the quartiles now follow as special cases of the general case.

Teaching Extras Available with the Fifth Edition

- **Companion Web Site (*www.prenhall.com/tannenbaum*)** Features a syllabus manager, online quizzes. Internet projects, graphing calculator help, PowerPoint downloads, and dozens of additional resource links.
- **Instructor's Solutions Manual 0-13-100557-X** Contains solutions to all the exercises in the text. Also includes a brief guide to give an overview of the text and how it may be successfully used.
- **Printed Test Bank 0-13-100559-6** Printed version of TestGen.
- **TestGen win/mac CD 0-13-100556-1** Test-generating software that creates randomized tests and offers an onscreen LAN-based testing environment, complete with Instructor Gradebook.
- **Student Solutions Manual 0-13-177485-9** Contains worked out solutions to odd-numbered problems from the text. Also contains a glossary of terms for each chapter.

A Final Word

This book grew out of the conviction that a liberal arts mathematics course should teach students more than just a collection of facts and procedures. The ultimate purpose of this book is to instill in the reader an overall appreciation of mathematics as a discipline and an exposure to the subtlety and variety of its many facets: problems, ideas, methods, and solutions. Last, but not least, I have tried to show that mathematics can be interesting and fun.

Acknowledgments

Special thanks to my good friend Bob Arnold for co-authoring the first four editions of this book with me. His talents, insights, and hard work will be missed.

Dale Buske of St. Cloud State contributed greatly to this edition—most of the new exercises and many of the projects are his. He also read (and reread) the manuscript and made many suggestions for improvements. I am greatly indebted to Dale for both the quantity and quality of his contributions.

The following mathematicians reviewed this edition and previous editions of the book and made many invaluable suggestions:

Teri Anderson, *Northern Wyoming Community College, Gillette Campus*
Carmen Artino, *College of Saint Rose*
Donald Beaton, *Norwich University*
Terry L. Cleveland, *New Mexico Military Institute*
Leslie Cobar, *University of New Orleans*
Crista Lynn Coles, *Elon University*
Ronald Czochor, *Rowan College of New Jersey*
Nancy Eaton, *University of Rhode Island*
Lily Eidswick, *The University of Montana, Missoula*

Kathryn E. Fink, *Moorpark College*
Stephen I. Gendler, *Clarion University*
Marc Goldstein, *Goucher College*
Josephine Guglielmi, *Meredith College*
William S. Hamilton, *Community College of Rhode Island*
Harold Jacobs, *East Stroudsburg University of Pennsylvania*
Tom Kiley, *George Mason University*
Jean Krichbaum, *Broome Community College*
Kim L. Luna, *Eastern New Mexico University*

Mike Martin, *Johnson County Community College*

Thomas O'Bryan, *University of Wisconsin—Milwaukee*

Daniel E. Otero, *Xavier University*

Philip J. Owens, *Austin Community College*

Matthew Pickard, *University of Puget Sound*

Lana Rhoads, *William Baptist College*

David E. Rush, *University of California at Riverside*

Kathleen C. Salter, *Eastern New Mexico University*

Theresa M. Sandifer, *Southern Connecticut State University*

Paul Schembari, *East Stroudsburg University of Pennsylvania*

Marguerite V. Smith, *Merced College*

William W. Smith, *University of North Carolina at Chapel Hill*

David Stacy, *Bellevue Community College*

Zoran Sunik, *University of Nebraska, Lincoln*

John Watson, *Arkansas Tech University*

Sarah N. Ziesler, *Dominican University*

I was extremely fortunate to have Barbara Mack again as my production editor. Barbara is, by far, the best production editor I have ever worked with.

Last, but not least, the person most responsible for the success of this book is Sally Yagan. There is an editor behind every book, but few that can match her vision, "can-do" attitude, and leadership.

PART ONE

The Mathematics of Social Choice

Mumble.
Grumble.
Complain.
Wallow.
Hope.
Despair.
Worry.

Vote.

Just a reminder: the one on the bottom changes things a lot faster.

Call 1-800-345-VOTE to register.

CHAPTER 1

The Mathematics of Voting

The Paradoxes of Democracy

Vote! In a democracy, the rights and duties of citizenship are captured in that simple one-word mantra. And, by and large, we do vote. We vote in presidential elections, gubernatorial elections, local elections, school bonds, stadium bonds, and initiatives large and small. The paradox is that the more opportunities we have to vote, the less we seem to appreciate the meaning of voting. We wonder if our vote really counts, and if so, how?

We can best answer these questions once we understand the full story behind an election. The reason we have elections is that we don't all think alike. Since we cannot all have things our way, we vote. But *voting* is only the first half of the story, the one we are most familiar with. As playwright Tom Stoppard suggests, it's the second half of the story—the *counting*—

It's not the voting that's democracy; it's the counting.
Tom Stoppard

that is at the very heart of the democratic process. How does it work, this process of sifting through the many voices of individual voters to find the collective voice of the group? And even more importantly, how well does it work? Is the process always fair? Answering these questions and explaining a few of the many intricacies and subtleties of **voting theory** are the purpose of this chapter.

But wait just a second! Voting theory? Why do we need a fancy theory to figure out how to count the votes? It all sounds pretty simple: We have an election; we count the ballots. Based on that count, we decide the outcome of the election in a manner that is consistent and fair. Surely, there must be a reasonable way to accomplish all of this! Surprisingly, there isn't.

In the late 1940s, the mathematical economist Kenneth Arrow discovered a remarkable fact: For elections involving three or more candidates, there is no consistently fair democratic method for choosing a winner. In fact, Arrow demonstrated that *a method for determining election results that is democratic and always fair is a mathematical impossibility*. This, the most famous fact in voting theory, is known as **Arrow's impossibility theorem**. In 1972, Kenneth Arrow was awarded the Nobel Prize in Economics for his pioneering work in what is now known as social-choice theory. (For more on Arrow and his work, see the biographical profile on p. 27.)

In this chapter we will explore a few of the more common **voting methods** used in elections—how they work, what their implications are, and how they stack up when we put them to some basic tests of fairness. In so doing, we will also gain some insight into the meaning and significance of Arrow's impossibility theorem.

1.1 Preference Ballots and Preference Schedules

We kick off our discussion of voting theory with a simple but important example, which we will revisit several times throughout the chapter. You may want to think of this example as a mathematical parable—its importance being not in the story itself, but in what lies hidden behind it.

EXAMPLE 1.1 The Math Club Election

The Math Appreciation Society (MAS) is a student organization dedicated to an unsung but worthy cause—that of fostering the enjoyment and appreciation of mathematics among college students. The Tasmania State University chapter of MAS is holding its annual election for president. There are four candidates running for president: Alisha, Boris, Carmen, and Dave (A, B, C, and D for short). Each of the 37 members of the club votes by means of a ballot indicating his or her first, second, third, and fourth choice. The 37 ballots submitted are shown in Fig. 1-1. Once the ballots are in, it's decision time. Who should be the *winner* of the election? Why?

Ballot	Ballot	Ballot	Ballot	Ballot	Ballot	Ballot	Ballot	Ballot	Ballot	Ballot
1st A	1st B	1st A	1st C	1st B	1st C	1st A	1st B	1st C	1st A	1st C
2nd B	2nd D	2nd B	2nd B	2nd D	2nd D	2nd B	2nd D	2nd B	2nd B	2nd B
3rd C	3rd C	3rd C	3rd D	3rd C	3rd D	3rd C	3rd C	3rd D	3rd C	3rd D
4th D	4th A	4th D	4th A	4th A	4th A	4th D	4th A	4th A	4th D	4th A

Ballot	Ballot	Ballot	Ballot	Ballot	Ballot	Ballot	Ballot	Ballot	Ballot	Ballot	Ballot	Ballot
1st D	1st A	1st A	1st C	1st A	1st C	1st D	1st C	1st A	1st D	1st D	1st C	1st C
2nd C	2nd B	2nd B	2nd B	2nd B	2nd B	2nd C	2nd B	2nd B	2nd C	2nd C	2nd B	2nd B
3rd B	3rd C	3rd C	3rd D	3rd C	3rd D	3rd B	3rd D	3rd C	3rd B	3rd B	3rd D	3rd D
4th A	4th D	4th D	4th A	4th D	4th A	4th A	4th A	4th D	4th A	4th A	4th A	4th A

Ballot	Ballot	Ballot	Ballot	Ballot	Ballot	Ballot	Ballot	Ballot	Ballot	Ballot	Ballot	Ballot
1st D	1st A	1st D	1st C	1st A	1st D	1st B	1st A	1st C	1st A	1st A	1st D	1st A
2nd C	2nd B	2nd C	2nd B	2nd B	2nd C	2nd D	2nd B	2nd D	2nd B	2nd B	2nd C	2nd B
3rd B	3rd C	3rd B	3rd D	3rd C	3rd B	3rd C	3rd C	3rd B	3rd C	3rd C	3rd B	3rd C
4th A	4th D	4th A	4th A	4th D	4th A	4th A	4th D	4th A	4th D	4th D	4th A	4th D

FIGURE 1-1

The 37 anonymous ballots for the MAS election

Before we try to answer these two deceptively simple questions, let's introduce a bit of terminology. The essential ingredients in every election are a set of *voters* (in our example the members of MAS), and a set of *candidates* or *choices* (in our example A, B, C, D). Typically, the word *candidate* is used when electing people, and the word *choice* is associated with nonhuman alternatives (cities, colleges, pizza toppings, etc.), but in this chapter we will use the two words interchangeably.

Individual voters are asked to express their opinions through *ballots*, which can come in many forms. In the MAS election, the ballots asked the voters to rank each of the candidates in order of preference, and ties were not allowed. A ballot in which the voters are asked to rank the candidates in order of preference

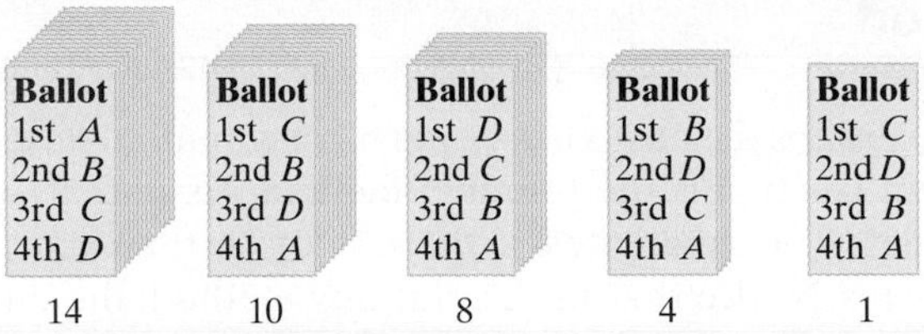

Ballot
1st A
2nd B
3rd C
4th D

14

Ballot
1st C
2nd B
3rd D
4th A

10

Ballot
1st D
2nd C
3rd B
4th A

8

Ballot
1st B
2nd D
3rd C
4th A

4

Ballot
1st C
2nd D
3rd B
4th A

1

FIGURE 1-2
The 37 MAS election ballots organized into piles

TABLE 1-1   Preference Schedule for the MAS Election

Number of voters	14	10	8	4	1
1st choice	A	C	D	B	C
2nd choice	B	B	C	D	D
3rd choice	C	D	B	C	B
4th choice	D	A	A	A	A

is called a **preference ballot**. A ballot in which ties are not allowed is called a **linear** ballot.

In this chapter we will illustrate all of our examples using linear preference ballots as the preferred (no pun intended) format for voting. While it is true that the preference ballot is not the most typical way we cast our vote (in most elections for public office, for example, the ballot asks for just the top choice), it is also true that it is one of the best ways to vote since it allows us to express our opinion on the relative merit of *all* the candidates.

A quick look at Fig. 1-1 is all it takes to see that there are many repeats among the ballots submitted, reflecting the fact that different voters have ranked the candidates exactly the same way. Thus, a logical way to organize the ballots is to group together identical ballots (Fig. 1-2), and this leads in a rather obvious way to Table 1-1, which is called the **preference schedule** for the election. The preference schedule is the simplest and most compact way to summarize the voting in an election based on preference ballots.

Transitivity and Elimination of Candidates

There are two important facts that we need to keep in mind when we work with preference ballots. The first is *the transitivity of individual preferences*, which is a fancy way of saying that if a voter prefers A to B and B to C, then it follows automatically that this voter must prefer A to C. A useful consequence of this observation is this: *If we need to know which candidate a voter would vote for if it came down to a choice between just two candidates, all we have to do is look at which candidate was placed higher on that voter's ballot.* We will use this fact throughout the chapter.

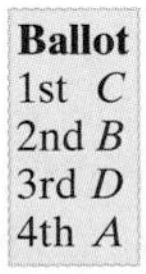

Ballot
1st C
2nd B
3rd D
4th A

FIGURE 1-3

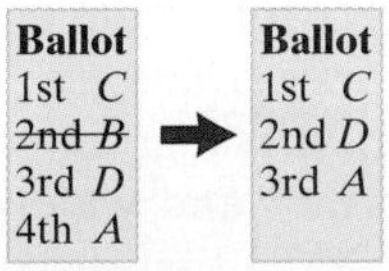

The other important fact is that the relative preferences of a voter are not affected by the elimination of one or more of the candidates. Take, for example, the ballot shown in Fig. 1-3 and pretend that candidate B drops out of the race right before the ballots are submitted. How would this voter now rank the remaining three candidates? As Fig. 1-4 shows, the relative positions of the remaining candidates are unaffected: C remains the first choice, D moves up to the second choice, and A moves up to the third choice.

Ballot
1st C
2nd B
3rd D
4th A

➡

Ballot
1st C
2nd D
3rd A

FIGURE 1-4

Let's now return to the business of deciding the outcome of elections in general and the MAS election (Example 1.1) in particular.

1.2 The Plurality Method

Perhaps the best known and most commonly used method for finding a winner in an election is the **plurality method**. Essentially this method says that the candidate (or candidates, if there is more than one) with the *most* first-place votes wins. Notice that in the plurality method, the only information that we use from the ballots are the votes for first place—nothing else matters. In practice, preference ballots are not used with the plurality method—all one asks of the voters is to vote for their first choice.

When we apply the plurality method to the Math Appreciation Society election, this is what we get:

The Math Appreciation Society News
ALISHA ELECTED PRESIDENT OF MAS!

A gets 14 first-place votes.

B gets 4 first-place votes.

C gets 11 first-place votes.

D gets 8 first-place votes.

In this case, the results of the election are clear—the winner is A (Alisha).

The popularity of the plurality method stems not only from its simplicity but also from the fact it is a natural extension of the principle of **majority rule**: In a democratic election between *two* candidates, the one with the majority (more than half) of the votes wins.

When there are three or more candidates, the majority rule cannot always be applied: In the MAS election, 19 first-place votes (out of 37) are needed for a majority, but none of the candidates received 19 first-place votes, so no one has the required majority. Alisha, with 14 first-place votes, has more than anyone else, so she has a *plurality*.

The Majority Criterion

While a plurality does not imply a majority, a majority does imply a plurality: A candidate that has more than half of the first-place votes must automatically have more first-place votes than any other candidate. Thus, a candidate that has a majority of the first-place votes is automatically the winner under the plurality method.

The notion that having a majority of the first-place votes should automatically guarantee a victory makes good sense and is an important requirement for a fair and democratic election. In fact, it is important enough to have a name: the **majority criterion**.

> **The Majority Criterion.** If a choice receives a majority of the first-place votes in an election, then that choice should be the winner of the election.

If a choice receives a majority of the first-place votes in an election but does not win the election, we have a *violation* of the majority criterion. When there is no majority choice in an election, the majority criterion does not come into play, and we certainly would not want to say that it is being violated.

Since a candidate having a majority of first-place votes is guaranteed to win under the plurality method the plurality method can never violate the majority criterion. We say that *the plurality method satisfies the majority criterion.*

In a democracy we tend to think of the majority criterion as a given. We will soon see that this need not be the case. There are important and widely used voting methods where a candidate could have a majority of the first-place votes and still lose the election.

What's Wrong with the Plurality Method?

In spite of its widespread use, the plurality method has many flaws and is usually a poor method for choosing the winner of an election when there are more than two candidates. Its principal weakness is that it fails to take into consideration the voters' preferences other than first choice, and in so doing can lead to some very bad election results.

To underscore the point, consider the following example.

EXAMPLE 1.2

Tasmania State University has a superb marching band. They are so good that this coming New Year they have been invited to march at five different bowl games: The Rose Bowl (*R*), the Hula Bowl (*H*), the Cotton Bowl (*C*), the Orange Bowl (*O*), and the Sugar Bowl (*S*). An election is held among the 100 members of the band to decide in which of the five bowl games they will march. A preference schedule giving the results of the election is shown in Table 1-2.

TABLE 1-2	Preference Schedule for the Band Election		
Number of voters	**49**	**48**	**3**
1st choice	R	H	C
2nd choice	H	S	H
3rd choice	C	O	S
4th choice	O	C	O
5th choice	S	R	R

If the plurality method is used, the winner of the election is the Rose Bowl, with 49 first-place votes. Note, however, that the Hula Bowl (*H*), which has 48 first-place votes, also has 52 second-place votes. Simple common sense tells us that the Hula Bowl is a far better choice to represent the wishes of the entire band. In fact, we can make the following persuasive argument in favor of the Hula Bowl: If we compare the Hula Bowl to any other bowl on a *head-to-head* basis, the Hula Bowl is always the preferred choice. Take, for example, a comparison between the Hula Bowl and the Rose Bowl. There are 51 votes for the Hula Bowl (48 from the second column plus the 3 votes in the last column) versus 49 votes for the Rose Bowl. Likewise, a comparison between the Hula Bowl and the Cotton Bowl would result in 97 votes for the Hula Bowl (first and second columns) and 3 votes for the Cotton Bowl. And when the Hula Bowl is compared to either the Orange Bowl or the Sugar Bowl, it gets all 100 votes.

We can now summarize the problem with Example 1.2 as follows: Although *H* wins in a head-to-head comparison between it and any other choice, the plurality method fails to choose *H* as the winner. In the language of voting theory, we say that the plurality method *violates* a basic requirement of fairness called the **Condorcet[1] criterion**.

Marie Jean Antoine Nicolas Caritat, Marquis de Condorcet (1743–1794)

> **The Condorcet Criterion.** If there is a choice that in head-to-head comparisons is preferred by the voters over each of the other choices, then that choice should be the winner of the election.

A candidate who wins in every head-to-head comparison against each of the other candidates is called the **Condorcet candidate**. The Condorcet criterion simply says that when there is a Condorcet candidate, then that candidate should be the winner. When there is no Condorcet candidate, the Condorcet criterion does not apply.

The plurality method violates the Condorcet criterion because we can find examples of elections where there is a Condorcet candidate that does not win the election. The band election is one such example. We should not conclude that this happens every time there is a Condorcet candidate—it doesn't. The problem with the plurality method is that violations of the Condorcet criterion can happen at all.

We will return to the idea of head-to-head comparisons between the candidates shortly. In the meantime, we conclude this section by discussing another important weakness of the plurality method: the ease with which election results can be manipulated by a voter or a block of voters through **insincere voting**. (A voter who changes the true order of his or her preferences in the ballot in an effort to influence the outcome of the election against a certain candidate is said to vote *insincerely*.) As an example, consider once again Table 1-2. The last column of the preference schedule represents the ballots of three specific band members, who, let's imagine, are dead set against the Rose Bowl (allergies!). Assuming that they have some idea of how the election is likely to turn out and that their first choice (the Cotton Bowl) has no chance of winning the election, their best strategy is to manipulate the election by voting insincerely, moving their second choice (the Hula Bowl) to first.

In real-world elections insincere voting can have serious and unexpected consequences and almost all voting methods are manipulable, but the plurality method is the one most impacted (see Project D). Take, for example, the overwhelming tendency for a two-party system in American politics. Why is the two-party system so entrenched in the United States? Partly, it is because the plurality method encourages insincere voting. It is well known that third-party candidates have a hard time getting their just share of the votes. Many voters who actually prefer the third-party candidate end up reluctantly voting for one of the two major-party candidates for fear of "wasting" their vote. Allegedly, this occurred in the 1992 presidential election when many voters who were inclined to vote for Ross Perot actually voted (insincerely) for Bill Clinton or George Bush.

1.3 The Borda Count Method

An entirely different approach to finding the winner in an election is the **Borda count method**.[2] In this method each place on a ballot is assigned points. In an election with N candidates we give 1 point for last place, 2 points for second from last place ..., and N points for first place. The points are tallied for each candidate separately, and the candidate with the highest total is the winner.

---

[2]This method is named after the Frenchman Jean-Charles de Borda (1733–1799). Borda was a military man—a cavalry officer and naval captain—who wrote on such diverse subjects as mathematics, physics, the design of scientific instruments, and voting theory.

Let's use the Borda count method to choose the winner of the Math Appreciation Society election. Table 1-3 shows the point values under each column based on first place worth 4 points, second place worth 3 points, third place worth 2 points, and fourth place worth 1 point.

TABLE 1-3   Borda Points for the MAS Election

Number of voters	14	10	8	4	1
1st choice: 4 points	A: 56 pts	C: 40 pts	D: 32 pts	B: 16 pts	C: 4 pts
2nd choice: 3 points	B: 42 pts	B: 30 pts	C: 24 pts	D: 12 pts	D: 3 pts
3rd choice: 2 points	C: 28 pts	D: 20 pts	B: 16 pts	C: 8 pts	B: 2 pts
4th choice: 1 point	D: 14 pts	A: 10 pts	A: 8 pts	A: 4 pts	A: 1 pt

Now we tally the points:

A gets $56 + 10 + 8 + 4 + 1 = 79$ points;

B gets $42 + 30 + 16 + 16 + 2 = 106$ points;

C gets $28 + 40 + 24 + 8 + 4 = 104$ points;

D gets $14 + 20 + 32 + 12 + 3 = 81$ points;

and we find that the winner is Boris!

What's Wrong with the Borda Count Method?

In contrast to the plurality method, the Borda count method takes into account *all* the information provided by the voters' preferences and produces as a winner the *best compromise candidate*. This is good! The real problem with the Borda count method is that it *violates the majority criterion!* In other words, a candidate with a majority of first-place votes can lose the election. The next example illustrates how this can happen.

EXAMPLE 1.3

The last principal at George Washington Elementary School has just retired and the School Board must hire a new principal. The four finalists for the job are Mrs. Amaro, Mr. Burr, Mr. Castro, and Mrs. Dunbar (A, B, C, and D, respectively). After interviewing the four finalists, the eleven members of the school board vote by ranking the four candidates and then use the Borda count method to decide the winner. The results of the voting are shown in Table 1-4.

TABLE 1-4   Preference Schedule for Example 1.3

Number of voters	6	2	3
1st choice	A	B	C
2nd choice	B	C	D
3rd choice	C	D	B
4th choice	D	A	A

It is a simple matter of arithmetic (which we leave to the reader to verify) that under the Borda count method, Mr. Burr gets the principal's job, with a total of 32 points. This happens in spite of the fact that Mrs. Amaro has 6 out of the 11 first-place votes and therefore a majority. What we have here is a violation of the majority criterion.

2002 Heisman Trophy winner
Carson Palmer of USC.

Here is another problem with the Borda count method: Since any violation of the majority criterion is an automatic violation of the Condorcet criterion as well (see Exercises 19 and 20), we can conclude from Example 1.3 that the Borda count method also *violates the Condorcet criterion*.

In spite of these drawbacks, the Borda count method is widely used in a variety of important real-world elections, especially when there is a large number of candidates. The winner of the Heisman award; the American and National Baseball Leagues MVP's (see Exercise 63); the winners of various music awards (Country Music Vocalist of the Year, etc.); the hiring of school principals and university presidents; and a host of other jobs, awards, and distinctions are decided using the Borda count method or some simple variation of it.

1.4 The Plurality-with-Elimination Method (Instant Runoff Voting)

In most municipal and local elections, a candidate needs a majority of the first-place votes to get elected. When there are three or more candidates running, it is often the case that no candidate gets a majority. Typically, the candidate or candidates with the fewest first-place votes are eliminated, and a runoff election is held. Since runoff elections are expensive to both the candidates and the municipality, this is an inefficient and cumbersome method for choosing a mayor or a county supervisor.

A much more efficient way to implement the same process without needing a separate runoff election is to use preference ballots. The information in the preference ballots tells us not only which candidate each voter wants to win but also which candidate the voter would want to win in a runoff between any pair of candidates, thus eliminating the need for an actual runoff. The method has become increasingly popular in the last few years and is fashionably referred to as **instant runoff voting** (IRV). In March of 2002, the voters in San Francisco passed Proposition A, which made instant runoff voting the official voting method for all San Francisco municipal elections. Alaska, Vermont, and New Mexico are all considering adopting some form of instant runoff voting as well (see Project C).

Before instant runoff voting became fashionable, the method of eliminating the candidates with the fewest first-place votes one at a time until one of them gets a majority had been known by various names, including the **plurality-with-elimination method** and the **Hare method**. We will call it the *plurality-with-elimination method*—it is the most descriptive of the three names.

A formal description of the plurality-with-elimination method goes like this:

- **Round 1.** Count the first-place votes for each candidate, just as you would in the plurality method. If a candidate has a majority of first-place votes, that candidate is the winner. Otherwise, eliminate the candidate (or candidates if there is a tie) with the fewest first-place votes.

- **Round 2.** Cross out the name(s) of the candidates eliminated from the preference schedule and recount the first-place votes. (Remember that when a candidate is eliminated from the preference schedule, in each column the candidates below it move up a spot.) If a candidate has a majority of first-place votes, declare that candidate the winner. Otherwise, eliminate the candidate with the fewest first-place votes.

- **Rounds 3, 4, etc.** Repeat the process, each time eliminating one or more candidates, until there finally is a candidate with a majority of first-place votes, which is then declared the winner.

Let's apply the plurality-with-elimination method to the Math Appreciation Society election. For the reader's convenience Table 1-5 shows the preference schedule again—it is exactly the same as Table 1-1.

TABLE 1-5	Preference Schedule for the MAS Election				
Number of voters	**14**	**10**	**8**	**4**	**1**
1st choice	A	C	D	B	C
2nd choice	B	B	C	D	D
3rd choice	C	D	B	C	B
4th choice	D	A	A	A	A

■ **Round 1.**

Candidate	A	B	C	D
Number of first-place votes	14	4	11	8

Since B has the fewest first-place votes, he is eliminated first.

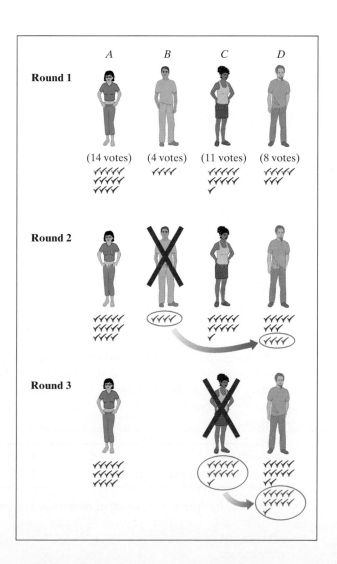

FIGURE 1-5

- **Round 2.** Once B is eliminated, the four votes that originally went to B in round 1 will now go to D, the next-best candidate in the opinion of these four voters (see Fig. 1-5). The new tally is shown below:

Candidate	A	B	C	D
Number of first-place votes	14		11	12

In this round C has the fewest first-place votes and is eliminated.

- **Round 3.** The 11 votes that went to C in round 2 now go to D (just check the relative positions of D and A in the second and fifth columns of Table 1-5). This gives the following:

Candidate	A	B	C	D
Number of first-place votes	14			23

The Math Appreciation Society News
DAVE ELECTED PRESIDENT OF MAS!

We now have a winner, and lo and behold, it's neither Alisha nor Boris. The winner of the election, with 23 first-place votes, is Dave!

Applying the Plurality-with-Elimination Method

The next two examples are intended primarily to illustrate some subtleties that can come up when applying the plurality-with-elimination method. To speed things up, we describe each election by simply showing the preference schedule.

EXAMPLE 1.4

Table 1-6 shows the preference schedule for an election between five candidates A, B, C, D, and E.

TABLE 1-6	Preference Schedule for Example 1.4					
Number of voters	**10**	**5**	**2**	**1**	**4**	**4**
1st choice	A	B	C	C	D	E
2nd choice	B	D	A	E	C	D
3rd choice	C	E	E	B	A	C
4th choice	D	C	B	A	E	A
5th choice	E	A	D	D	B	B

From the preference schedule, we can determine that the number of voters is $10 + 5 + 2 + 1 + 4 + 4 = 26$, and therefore 14 or more votes are needed for a majority. Let's use the plurality-with-elimination method to find a winner.

- **Round 1.**

Candidate	A	B	C	D	E
Number of first-place votes	10	5	3	4	4

Here C has the fewest number of first-place votes and is eliminated first.

■ **Round 2.** Of the three votes originally going to C, now two go to A (look at the third column of the preference schedule) and one goes to E (from the fourth column of the preference schedule).

Candidate	A	B	C	D	E
Number of first-place votes	12	5		4	5

In this round D has the fewest first-place votes and is eliminated.

■ **Round 3.** The four votes originally going to D would next go to C (look at the fifth column of the preference schedule). Because C is out of the picture at this point, however, we dip further down into the column. Thus, the four votes go to the next top candidate in that column, candidate A.

Candidate	A	B	C	D	E
Number of first-place votes	16	5			5

At this point we can stop, as there is no need to go on! Candidate A has a majority of the first-place votes and is the winner of the election. ■

EXAMPLE 1.5

Table 1–7 shows the result of an election among the four candidates $W, X, Y,$ and Z.

TABLE 1-7	Preference Schedule for Example 1.5			
Number of voters	**8**	**6**	**2**	**19**
1st choice	W	X	Y	Z
2nd choice	X	Z	Z	X
3rd choice	Z	Y	W	W
4th choice	Y	W	X	Y

The number of voters in this election is $8 + 6 + 2 + 19 = 35$, so it takes 18 or more votes for a majority. But notice that candidate Z has 19 first-place votes right out of the gate. This means that we are done—Z is automatically the winner! ■

There is a simple but important lesson to be learned from Example 1.5: *The plurality-with-elimination method satisfies the majority criterion.*

What's Wrong with the Plurality-with-Elimination Method?

The main problem with the plurality-with-elimination method is quite subtle and is illustrated by the next example.

EXAMPLE 1.6

Three cities, Athens (A), Babylon (B), and Carthage (C), are competing to host the next Summer Olympic Games. The final decision is made by a secret vote of the 29 members of the Executive Council of the International Olympic Committee, and the winner is chosen using the plurality-with-elimination method. Two days before the actual election is to be held, a straw vote is conducted by the Executive Council just to see how things stand. The results of the straw poll are shown in Table 1–8.

TABLE 1-8	**Preference Schedule in Straw Vote Two Days Before the Actual Election**			
Number of voters	7	8	10	4
1st choice	A	B	C	A
2nd choice	B	C	A	C
3rd choice	C	A	B	B

The results of the straw vote are as follows: In the first round Athens has 11 votes, Babylon has 8, and Carthage has 10, which means that Babylon is eliminated first. In the second round, Babylon's 8 votes go to Carthage (see the second column of Table 1–8), so Carthage ends up with 18 votes, more than enough to lock up the election.

Although the results of the straw poll are supposed to be secret, the word gets out that unless some of the voters turn against Carthage, Carthage is going to win the election. Because everybody loves a winner, what ends up happening in the actual election is that even more first-place votes are cast for Carthage than in the straw poll. Specifically, the four voters in the last column of Table 1–8 decide as a block to switch their first-place votes from Athens to Carthage. Surely, this is just frosting on the cake for Carthage, but to be sure, we recheck the results of the election.

Table 1–9 shows the preference schedule for the actual election. [Table 1–9 is the result of switching A and C in the last column of Table 1–8 and combining columns 3 and 4 (they are now the same) into a single column.]

TABLE 1-9	**Preference Schedule for the Actual Election**		
Number of voters	7	8	14
1st choice	A	B	C
2nd choice	B	C	A
3rd choice	C	A	B

When we apply the plurality-with-elimination method to Table 1–9, Athens (with 7 first-place votes) is eliminated first, and the 7 votes originally going to Athens now go to Babylon, giving it 15 votes *and the win!* How could this happen? How could Carthage lose an election it had locked up simply because some voters moved Carthage from second to first choice? To the people of Carthage this was surely the result of an evil Babylonian plot, but double-checking the figures makes it clear that everything is on the up and up—Carthage is just the victim of a quirk in the plurality-with-elimination method: the possibility that you can actually do worse by doing better! In the language of voting theory this is known as a *violation of the monotonicity criterion.* ■

> **The Monotonicity Criterion.** If choice X is a winner of an election and, in a reelection, the only changes in the ballots are changes that only favor X, then X should remain a winner of the election.

We now know that the plurality-with-elimination method *violates the monotonicity criterion*. We leave it as an exercise for the reader to verify that plurality with elimination also *violates the Condorcet criterion* (see Exercises 33 and 34).

In spite of its flaws, the plurality-with-elimination method is used in many real-world situations, usually in elections in which there are few candidates (typically three or four, rarely more than six). While Example 1.6 was just a simple dramatization, it is a fact that when choosing which city gets to host the Olympic Games, the International Olympic Committee uses the plurality-with-elimination method. (For details as to how the 2000 Summer Olympics were awarded to Sydney, the reader is referred to Appendix 2 at the end of this chapter.)

A simple variation of the plurality-with-elimination method commonly known as *plurality with a runoff* is used in elections for local political office (city councils, county boards of supervisors, school boards, etc.). Plurality with a runoff works just like plurality with elimination except that *all* candidates except the top two get eliminated in the first round (see Exercise 56).

1.5 The Method of Pairwise Comparisons (Copeland's Method)

So far, all three voting methods we have discussed violate the Condorcet criterion, but this is not an insurmountable problem. It is reasonably easy to come up with a voting method that satisfies the Condorcet criterion. Our next method does this. It is sometimes known as the **method of pairwise comparisons** and other times as **Copeland's method**.

The method of pairwise comparisons is like a round-robin tournament in which every candidate is matched *one-on-one* with every other candidate. Each of these one-on-one matchups is called a *pairwise comparison*. In a pairwise comparison between candidates X and Y each vote is assigned to either X or Y, *the vote going to whichever of the two candidates is higher on the ballot*. The winner of the pairwise comparison is the one with the most votes, and as in an ordinary tournament, a win is worth 1 point (a loss is worth nothing!). In case of a tie each candidate gets $\frac{1}{2}$ point. The winner of the election is the candidate with the most points after all the pairwise comparisons are tabulated. In case of a tie, which is common under this method, we can either have more than one winner or use a predetermined tie-breaking procedure if multiple winners are not permitted.

Once again, we will illustrate the method of pairwise comparisons using the Math Appreciation Society election.

Let's start with a pairwise comparison between A and B. Looking at Table 1–10, we can see that there are 14 voters that prefer A to B (first column) and 23 voters that prefer B to A (last four columns). Consequently, the winner of the pairwise comparison between A and B is B. We summarize this result as follows:

A versus *B*: 14 votes to 23 votes (*B* wins). *B* gets 1 point.

TABLE 1-10 Comparing Candidates A and B

Number of voters	14	10	8	4	1
1st choice	$\widehat{A}$	C	D	$\widehat{B}$	C
2nd choice	B	$\widehat{B}$	C	D	D
3rd choice	C	D	$\widehat{B}$	C	$\widehat{B}$
4th choice	D	A	A	A	A

Let's next look at the pairwise comparison between C and D (Table 1–11). In this one there are 25 voters that prefer C to D (first, second, and last columns) and only 12 voters that prefer D to C (third and fourth columns). This point goes to C.

C versus D: 25 votes to 12 votes (C wins). C gets 1 point.

TABLE 1-11 Comparing Candidates C and D

Number of voters	14	10	8	4	1
1st choice	A	$\widehat{C}$	$\widehat{D}$	B	$\widehat{C}$
2nd choice	B	B	C	$\widehat{D}$	D
3rd choice	$\widehat{C}$	D	B	C	B
4th choice	D	A	A	A	A

If we continue in this manner, comparing in all possible ways two candidates at a time, we end up with the following scoreboard:

A versus B: 14 votes to 23 votes (B wins). B gets 1 point.

A versus C: 14 votes to 23 votes (C wins). C gets 1 point.

A versus D: 14 votes to 23 votes (D wins). D gets 1 point.

B versus C: 18 votes to 19 votes (C wins). C gets 1 point.

B versus D: 28 votes to 9 votes (B wins). B gets 1 point.

C versus D: 25 votes to 12 votes (C wins). C gets 1 point.

The final tally produces 0 points for A, 2 points for B, 3 points for C, and 1 point for D. Can it really be true? Yes! The winner of the election under the method of pairwise comparisons is Carmen!

For the reader who likes straight answers to simple questions, what's been happening with the Math Appreciation Society election may be somewhat disconcerting. The question, "Who is the winner of the election?" is totally ambiguous. The answer seems to depend as much on the counting as it does on the voting.

Voting method	Winner
Plurality	Alisha
Borda count	Boris
Plurality with elimination	Dave
Pairwise comparisons	Carmen

It is easy to see that the method of pairwise comparisons *satisfies the Condorcet criterion*—a Condorcet candidate wins every pairwise comparison and thus gets the highest number of point under this method.

It is also true that the method of pairwise comparisons *satisfies both the majority criterion and the monotonicity criterion* (see Exercises 60 and 67). Hmmm ... this is beginning to look promising.

So, What's Wrong with the Method of Pairwise Comparisons?

Unfortunately, the method of pairwise comparisons is not without flaws. The next example illustrates the most serious one.

EXAMPLE 1.7

As the newest expansion team in the NFL, the Los Angeles Web Surfers will be getting the number-one choice in the upcoming draft of college football players. After narrowing the list of candidates to five players (Allen, Byers, Castillo, Dixon, and Evans), the coaches and team executives meet to discuss the candidates and eventually have a vote, a decision of major importance to both the team and the chosen player. According to team rules, the final decision must be made using the method of pairwise comparisons. Table 1–12 shows the preference schedule after all the ballots are turned in.

TABLE 1-12   Preference Schedule for LA's Draft Choices

Number of voters	2	6	4	1	1	4	4
1st choice	A	B	B	C	C	D	E
2nd choice	D	A	A	B	D	A	C
3rd choice	C	C	D	A	A	E	D
4th choice	B	D	E	D	B	C	B
5th choice	E	E	C	E	E	B	A

We leave it to the reader to verify that the results of the 10 possible pairwise comparisons are as follows:

A versus *B*: 7 votes to 15 votes. *B* gets 1 point.

A versus *C*: 16 votes to 6 votes. *A* gets 1 point.

A versus *D*: 13 votes to 9 votes. *A* gets 1 point.

A versus *E*: 18 votes to 4 votes. *A* gets 1 point.

B versus *C*: 10 votes to 12 votes. *C* gets 1 point.

B versus *D*: 11 votes to 11 votes. *B* gets $\frac{1}{2}$ point, *D* gets $\frac{1}{2}$ point.

B versus *E*: 14 votes to 8 votes. *B* gets 1 point.

C versus *D*: 12 votes to 10 votes. *C* gets 1 point.

C versus *E*: 10 votes to 12 votes. *E* gets 1 point.

D versus *E*: 18 votes to 4 votes. *D* gets 1 point.

The final tally produces 3 points for *A*, $2\frac{1}{2}$ points for *B*, 2 points for *C*, $1\frac{1}{2}$ points for *D*, and 1 point for *E*. It looks as if Allen (*A*) is the lucky young man who will make millions of dollars playing for the Los Angeles Web Surfers.

The interesting twist to the story surfaces when it is discovered right before the draft that one of the other players (Castillo) had accepted a scholarship to go to medical school and will not be playing professional football. Since Castillo was not the top choice, this fact should have no effect on the choice of Allen as the draft choice. Or should it?

Suppose we were to eliminate Castillo from the original election, which we can easily do by crossing C from the preference schedule shown in Table 1–12 and thus getting the preference schedule shown in Table 1–13.

TABLE 1-13	Preference Schedule for LA's Draft Choices After C Is Eliminated						
Number of voters	2	6	4	1	1	4	4
1st choice	A	B	B	B	D	D	E
2nd choice	D	A	A	A	A	A	D
3rd choice	B	D	D	D	B	E	B
4th choice	E	E	E	E	E	B	A

The results of the six possible pairwise comparisons between the four remaining candidates would now be as follows:

A versus B: 7 votes to 15 votes. B gets 1 point.

A versus D: 13 votes to 9 votes. A gets 1 point.

A versus E: 18 votes to 4 votes. A gets 1 point.

B versus D: 11 votes to 11 votes. B gets $\frac{1}{2}$ point, D gets $\frac{1}{2}$ point.

B versus E: 14 votes to 8 votes. B gets 1 point.

D versus E: 18 votes to 4 votes. D gets 1 point.

In this new scenario A would have 2 points, B would have $2\frac{1}{2}$ points, D would have $1\frac{1}{2}$ points, and E would have 0 points, and the winner would be Byers. In other words, if the election had been conducted with the knowledge that Castillo was not really a candidate, then Byers, and not Allen, would have been the winner, and the millions of dollars that are going to go to Allen would have gone instead to Byers! On its surface, the original outcome seems grossly unfair to Byers. ■

The strange happenings in Example 1.7 help illustrate an important fact: The method of pairwise comparisons may satisfy all of our previous fairness criteria, but unfortunately, it violates another requirement of fairness known as *the independence-of-irrelevant-alternatives criterion.*

The Independence-of-Irrelevant-Alternatives Criterion. If choice X is a winner of an election and one (or more) of the other choices is disqualified and the ballots recounted, then X should still be a winner of the election.

A second problem with the method of pairwise comparisons is that sometimes it can produce an outcome in which everyone is a winner.

EXAMPLE 1.8

The Icelandia State University varsity hockey team is on a road trip. An important decision needs to be made: Where to go for dinner? In the past, this has led to some heated arguments, so this time they decide to hold an election. The choices boil down to three restaurants: Hunan (H), Pizza Palace (P), and Danny's (D). The decision is to be made using the method of pairwise comparisons. Table 1–14 shows the results of the voting by the 11 players on the squad.

TABLE 1-14	Preference Schedule for Example 1.8		
Number of voters	**4**	**2**	**5**
1st choice	H	P	D
2nd choice	P	D	H
3rd choice	D	H	P

Here *H* beats *P* (9 to 2), *P* beats *D* (6 to 5), and *D* beats *H* (7 to 4). This results in a three-way tie for first place. What now? In this particular example it is unrealistic to declare the result of the election a three-way tie and have everybody go to the restaurant of their choice. Here, as in most situations, it becomes necessary to break the tie.

In general, there is no set way to break a tie, and in practice, it is important to establish the rules as to how ties are to be broken ahead of time. Otherwise, consider what might happen. Those who want to eat at Danny's could argue, not unreasonably, that the tie should be broken by counting first-place votes. In this case, Danny's would win. On the other hand, those who want to eat at the Hunan could make an equally persuasive argument that the tie should be broken by counting total points (Borda count). In this case the Hunan would get 24 points and Danny's 23, so the Hunan would win. As the reader can see, it would have been smart to think about these things before the election. For a more detailed discussion of ties and how to break them, you are encouraged to look at Appendix 1 at the end of this chapter. ■

How Many Pairwise Comparisons?

One practical difficulty with the method of pairwise comparisons has to do with the amount of work required to come up with a winner. You may have noticed that there seem to be a lot of pairwise comparisons to check out. Exactly how many? Since comparisons are made between two candidates at a time, the answer obviously depends on the number of candidates. We already saw that with *four* candidates there are *six* pairwise comparisons possible, and in Example 1.7 we saw that with *five* candidates there are *ten* possible pairwise comparisons. How many pairwise comparisons are there between six, seven, ..., *N* candidates? Suppose we have an election with 12 candidates, and let's try to count the comparisons systematically, making sure that we don't count any comparison twice.

■ We compare the first candidate with each of the other 11 candidates—*11 pairwise comparisons.*

■ We compare the second candidate with each of the other candidates except the first one, since that comparison has already been made—*10 pairwise comparisons.*

■ We compare the third candidate with each of the other candidates except the first and second candidates, since those comparisons have already been made—*9 pairwise comparisons.*

.
.
.

■ We compare the eleventh candidate with each of the other candidates except the first 10 candidates, since those comparisons have already been made. In other words, we compare the eleventh candidate with the twelfth candidate—*1 pairwise comparison.*

We see that the total number of pairwise comparisons is

$$1 + 2 + 3 + 4 + 5 + 6 + 7 + 8 + 9 + 10 + 11 = 66.$$

How many pairwise comparisons are there in an election with 100 candidates? Well, using an argument similar to the preceding one, we find that the total number of pairwise comparisons is

$$1 + 2 + 3 + 4 + \cdots + 99.$$

In general, if there are N candidates, the total number of pairwise comparisons is

$$1 + 2 + 3 + 4 + \cdots + (N - 1).$$

Note that the last number added is one less than the number of candidates.

Next, we are going to learn about a very useful mathematical formula. Let's go back to the case of 100 candidates. How much is $1 + 2 + 3 + \cdots + 99$? Although we could just add up these numbers, there is a much better way.

Suppose that before a pairwise comparison between two candidates takes place, each candidate gives the other one his or her business card. Then, clearly, each candidate would end up with the business card of every other candidate, and there would be a total of $99 \times 100 = 9900$ cards handed out (each of the 100 candidates would hand out 99 cards, one to each of the other candidates). But since each comparison resulted in two cards being handed out, the total number of comparisons must be half as many as the number of cards. Consequently,

$$1 + 2 + 3 + 4 + \cdots + 99 = \frac{99 \times 100}{2} = 4950.$$

Similar arguments show that if there are N candidates, the number of pairwise comparisons needed is

$$1 + 2 + 3 + 4 + \cdots + (N - 1) = \frac{(N - 1)N}{2}.$$

1.6 Rankings

Quite often it is important not only to know who wins the election but also to know who comes in second, third, etc. Let's consider once again the Math Appreciation Society election. Suppose now that instead of electing just the president we need to elect a board of directors consisting of a president, a vice president, and a treasurer. The club's bylaws state that rather than having separate elections for each office, the winner of the election gets to be the president, the second-place candidate gets to be the vice president, and the third-place candidate gets to be the treasurer. In a situation like this, we need a voting method that gives us not just a winner but also a second place, a third place, etc.—in other words, a **ranking** of the candidates.

Extended Ranking Methods

Each of the four voting methods we discussed earlier in this chapter has a natural extension that can be used to produce a ranking of the candidates.

Let's start with the plurality method and see how we might extend it to produce a ranking of the four candidates in the Math Appreciation Society election. For the reader's convenience, the preference schedule is shown again in Table 1-15.

TABLE 1-15	Preference Schedule for the MAS Election				
Number of voters	14	10	8	4	1
1st choice	A	C	D	B	C
2nd choice	B	B	C	D	D
3rd choice	C	D	B	C	B
4th choice	D	A	A	A	A

The count of first-place votes is as follows:

A: 14 first-place votes

B: 4 first-place votes

C: 11 first-place votes

D: 8 first-place votes

We know that, using the plurality method, A is the winner. Who should be second? The answer seems obvious: C has the second most first-place votes (11), so we declare C to come in second. By the same token, we declare D to come in third (8 votes) and B last. In short, the *extended plurality method* gives us a complete ranking of the candidates, shown in Table 1-16.

TABLE 1-16	Ranking the Candidates in the MAS Election Using the Extended Plurality Method		
Office	Place	Candidate	First-place votes
President	1st	A	14
Vice president	2nd	C	11
Treasurer	3rd	D	8
	4th	B	4

Ranking the candidates using the extended Borda count method is equally simple. In the MAS election, for example, the point totals under the Borda count method were

A: 79 Borda points

B: 106 Borda points

C: 104 Borda points

D: 81 Borda points

The resulting ranking, based on the *extended Borda count method*, is shown in Table 1-17.

TABLE 1-17	Ranking the Candidates in the MAS Election Using the Extended Borda Count Method		
Office	Place	Candidate	Borda points
President	1st	B	106
Vice president	2nd	C	104
Treasurer	3rd	D	81
	4th	A	79

Ranking the candidates using the *extended plurality-with-elimination method* is a bit more subtle. We rank them in reverse order of elimination: The first candidate eliminated is ranked last, the second candidate eliminated is ranked next to last, etc. In cases where a candidate gets a majority of first-place votes before the ranking of all the candidates is complete, we continue the process of elimination to rank the remaining candidates.

Table 1-18 shows the results of ranking the candidates in the MAS election using the *extended plurality-with-elimination method*.

TABLE 1-18	Ranking the Candidates in the MAS Election Using the Extended Plurality-with-Elimination Method		
Office	**Place**	**Candidate**	**Eliminated in**
President	1st	*D*	
Vice president	2nd	*A*	3rd round
Treasurer	3rd	*C*	2nd round
	4th	*B*	1st round

Last, we can rank the candidates using the *extended method of pairwise comparisons* according to the number of pairwise comparisons won (recall that we count a tie as $\frac{1}{2}$ point). In the case of the MAS election, *C* won 3 pairwise comparisons, *B* won 2 pairwise comparisons, *D* won 1 pairwise comparison, and *A* won none. The results of ranking the candidates under the *extended method of pairwise comparisons* are shown in Table 1-19.

TABLE 1-19	Ranking the Candidates in the MAS Election Using the Extended Method of Pairwise Comparisons		
Office	**Place**	**Candidate**	**Points**
President	1st	*C*	3
Vice president	2nd	*B*	2
Treasurer	3rd	*D*	1
	4th	*A*	0

A summary of the results of the MAS election using the different extended ranking methods is shown in Table 1-20.

TABLE 1-20	Ranking the Candidates in the MAS Election: A Tale of Four Methods			
	Ranking			
Method	**1st**	**2nd**	**3rd**	**4th**
Extended plurality	*A*	*C*	*D*	*B*
Extended Borda count	*B*	*C*	*D*	*A*
Extended plurality with elimination	*D*	*A*	*C*	*B*
Extended pairwise comparisons	*C*	*B*	*D*	*A*

The most striking thing about Table 1-20 is the wide discrepancy of results. While it is somewhat frustrating to see this much equivocation, it is important to keep things in context: This is the exception rather than the rule. One purpose of the MAS example is to illustrate how crazy things can get in some elections, but in most real-life elections there tends to be much more consistency among the various methods.

Recursive Ranking Methods

We will now discuss a different, somewhat more involved strategy for ranking the candidates, which we will call the **recursive** approach. The basic strategy here is the same regardless of which voting method we choose—only the details are different.

Let's say we are going to use some voting method X and the recursive approach to rank the candidates in an election. We first use method X to find the winner of the election. So far, so good. We then remove the name of the winner on the preference schedule and obtain a new, modified preference schedule with one less candidate on it. We apply method X once again to find the "winner" based on this new preference schedule, and this candidate is ranked second. (This makes a certain amount of sense: What we're saying is that, after the winner is removed, we run a brand-new race, and the best candidate in that race is the second-best candidate overall.) We repeat the process again (cross out the name of the last winner, calculate the new preference schedule, and apply method X to find the next winner, who is then placed next in line in the ranking) until we have ranked as many of the candidates as we want.

We will illustrate the basic idea of recursive ranking with a couple of examples, both based on the Math Appreciation Society election.

EXAMPLE 1.9

Suppose we want to rank the four candidates in the MAS election using the *recursive plurality method*. The preference schedule, once again, is given in Table 1-21.

TABLE 1-21   Preference Schedule for the MAS Election

Number of voters	14	10	8	4	1
1st choice	A	C	D	B	C
2nd choice	B	B	C	D	D
3rd choice	C	D	B	C	B
4th choice	D	A	A	A	A

- **Step 1.** (Choose the winner using plurality.) We already know the winner is A with 14 first-place votes.

- **Step 2.** (Choose second place.) First we remove A from the original schedule—this give us a "new" preference schedule to work with (Table 1-22).

TABLE 1-22 Preference Schedules for the MAS Election

The original election						After A has been removed					
Number of voters	14	10	8	4	1	Number of voters	14	10	8	4	1
1st choice	A	C	D	B	C	1st choice	B	C	D	B	C
2nd choice	B	B	C	D	D	2nd choice	C	B	C	D	D
3rd choice	C	D	B	C	B	3rd choice	D	D	B	C	B
4th choice	D	A	A	A	A						

In this schedule the winner using plurality is B, with 18 first-place votes. Thus, *second place goes to B*.

- **Step 3.** (Choosing third place.) We now remove B from the preceding schedule. The resulting schedule is shown on the right in Table 1-23.

TABLE 1-23 Preference Schedules for the MAS Election

After A has been removed						After A and B have been removed		
Number of voters	14	10	8	4	1	Number of voters	25	12
1st choice	B	C	D	B	C	1st choice	C	D
2nd choice	C	B	C	D	D	2nd choice	D	C
3rd choice	D	D	B	C	B			

Using plurality, the winner for this schedule is C with 25 first-place votes. This means that third place goes to C and last place goes to D. The final ranking of the candidates under the *recursive plurality method* is shown in Table 1-24.

TABLE 1-24 Ranking the Candidates in the MAS Election Using the Recursive Plurality Method

Office	Place	Candidate
President	1st	A
Vice president	2nd	B
Treasurer	3rd	C
	4th	D

It is worth noting how different this ranking is from the ranking obtained using the *extended plurality method*. In fact, except for first place (which will always be the same), all the other positions turned out to be different. ∎

EXAMPLE 1.10

For our last example, we will apply the *recursive plurality-with-elimination method* to rank the candidates in the MAS election.

To help the reader understand how this method works, we will make a semantic distinction. In running the plurality-with-elimination method, candidates are "eliminated" until there is a winner left. Having locked up a place in the ranking, this winner is then "removed" so that the election can be rerun for the next place in the ranking. Here is how it works:

- **Step 1.** We apply the plurality-with-elimination method to the original preference schedule. After eliminating B, C, and D in that order, we get a winner: D. (See pp. 11–12 for details.)
- **Step 2.** We now remove the winner D from the preference schedule (Table 1-25).

TABLE 1-25 Preference Schedules for the MAS Election

The original election						After the winner D has been removed					
Number of voters	14	10	8	4	1	Number of voters	14	10	8	4	1
1st choice	A	C	D	B	C	1st choice	A	C	C	B	C
2nd choice	B	B	C	D	D	2nd choice	B	B	B	C	B
3rd choice	C	D	B	C	B	3rd choice	C	A	A	A	A
4th choice	D	A	A	A	A						

In the new election (right half of Table 1-25), C is the automatic winner by virtue of having 19 first-place votes (a majority). This means that second place in the original election goes to C.

- **Step 3.** We now remove C from the last preference schedule (Table 1-26).

TABLE 1-26 Preference Schedules for the MAS Election

After the winner D has been removed						After D and C have been removed					
Number of voters	14	10	8	4	1	Number of voters	14	10	8	4	1
1st choice	A	C	C	B	C	1st choice	A	B	B	B	B
2nd choice	B	B	B	C	B	2nd choice	B	A	A	A	A
3rd choice	C	A	A	A	A						

The winner of this new election under plurality with elimination is B. This means that third place goes to B.

The final ranking of the candidates under the *recursive plurality with elimination method* is shown in Table 1-27.

TABLE 1-27 Ranking the Candidates in the MAS Election
Using the Recursive Plurality-with-Elimination Method

Office	Place	Candidate
President	1st	D
Vice president	2nd	C
Treasurer	3rd	B
	4th	A

While somewhat more complicated than the extended ranking methods, the recursive ranking methods are an interesting example of an important idea in mathematics: the concept of a recursive process. We will discuss recursive processes in greater detail in Chapters 9, 10, and 12.

Conclusion

Fairness and Arrow's Impossibility Theorem

When is a voting method fair? Throughout this chapter we have introduced several standards of fairness known as *fairness criteria*.[3] Let's review what they are.

- **Majority Criterion.** If there is a choice that has a majority of the first-place votes, then that choice should be the winner of the election.

- **Condorcet Criterion.** If there is a choice that is preferred by the voters over each of the other choices, then that choice should be the winner of the election.

- **Monotonicity Criterion.** If choice X is a winner of an election and, in a re-election, all the changes in the ballots are changes favorable only to X, then X should still be a winner of the election.

- **Independence-of-Irrelevant-Alternatives Criterion.** If choice X is a winner of an election, and one (or more) of the other choices is disqualified and the ballots recounted, then X should still be a winner of the election.

Each of the preceding four criteria represents a basic standard of fairness, and it is reasonable to expect that a fair voting method ought to satisfy all of them. Surprisingly, none of the four voting methods we discussed in this chapter does. The question remains: Is there a democratic voting method that satisfies all of the fairness criteria—if you will, a perfectly fair voting method? For elections involving more than two alternatives the answer is No! *No perfectly fair voting method exists.*

At first glance, this fact seems a little surprising. Given the obvious importance of elections in a democracy and given the collective intelligence and imagination of social scientists and mathematicians, how is it possible that no one has come up with a voting method satisfying all of the fairness criteria? Up until the late 1940s this was one of the most challenging questions in social-choice theory. Finally, in 1949, Kenneth Arrow (see profile on opposite page) demonstrated the now famous **Arrow's impossibility theorem**: *It is mathematically impossible for a democratic voting method to satisfy all of the fairness criteria.* No matter how hard we look for it, there can be no perfectly fair voting method. Ironically, total and consistent fairness is inherently impossible in a democracy.

The search of the great minds of recorded history for the perfect democracy, it turns out, is the search for a chimera, a logical self-contradiction.

Paul Samuelson

---

[3]Singular: criterion; plural: criteria.

Kenneth J. Arrow (1921–)

Kenneth Arrow is one of the best known and most versatile mathematical economists of the 20th century. Born and raised in New York City, Arrow completed his undergraduate studies at City College of New York, where at the age of 19 he received a Bachelor's degree in Social Science and Mathematics. He continued his graduate studies at Columbia University, where he received a Master of Arts (M.A.) in Mathematics in 1941. At this time he became interested in mathematical economics, and by 1942 he completed all the necessary course work for a Ph.D. in Economics at Columbia. His graduate studies at Columbia were interrupted by World War II, and between 1942 and 1946 he was assigned to serve as a weather officer in the U.S. Army Air Corps. As a result of his weather work for the Army, Arrow was able to publish his first mathematical research paper—*On the Optimal Use of Winds for Flight Planning*, the first demonstration of his versatility and his ability to find creative uses for mathematics in just about any setting. Arrow's versatility and ability to see significant mathematical ideas in seemingly nonmathematical subjects became one of the hallmarks of his work.

In 1946 Arrow returned to Columbia to write his doctoral thesis, which he completed in 1949. (By then he was already working as a research associate and Assistant Professor of Economics at the University of Chicago.) Arrow's doctoral thesis, entitled *Social Choice and Individual Values*, became a landmark work in mathematical economics and would eventually lead to his being awarded the 1972 Nobel Prize in Economics. Using the mathematical principles of game theory, Arrow proved his famous *impossibility theorem*, which in essence says that it is impossible for a democratic society to take the individual opinions of its voters (the "individual values") and always make a collective decision (the "social choice") that is fair. As one might imagine for a doctoral thesis and Nobel Prize–quality work, the details are quite technical, but the mathematics itself was not difficult. Arrow's genius was to find a way to redefine the idea of "fairness" in a way that lends itself to mathematical treatment.

In 1949 Arrow was appointed Assistant Professor of Economics and Statistics at Stanford University, eventually becoming Professor of Economics, Statistics, and Operations Research. Between 1968 and 1979 he was a Professor of Economics at Harvard University, but in 1979 he returned to Stanford, where he is currently Emeritus Professor of Economics and still active in research. (His faculty Web page lists his current research interests as *economics of information and organization, collective decision making, general equilibrium theory, environment, and growth.*

Arrow's impossibility theorem
Borda count method
Condorcet candidate
Condorcet criterion
extended rankings methods
independence-of-irrelevant-
  alternatives criterion
insincere voting
linear ballot

majority criterion
method of pairwise comparisons
monotonicity criterion
plurality method
plurality-with-elimination method
preference ballot
preference schedule
rankings
recursive ranking methods

WALKING

A. Ballots and Preference Schedules

1. The management of the XYZ Corporation has decided to treat their office staff to dinner. The choice of restaurants is The Atrium (*A*), Blair's Kitchen (*B*), The Country Cookery (*C*), and Dino's Steak House (*D*). Each of the 12 staff members is asked to submit a preference ballot listing his or her first, second, third, and fourth choices among these restaurants. The resulting preference ballots are as follows:

Ballot	Ballot	Ballot	Ballot	Ballot	Ballot	Ballot	Ballot	Ballot	Ballot	Ballot	Ballot
1st *A*	1st *C*	1st *B*	1st *C*	1st *C*	1st *C*	1st *A*	1st *C*	1st *A*	1st *A*	1st *C*	1st *A*
2nd *B*	2nd *B*	2nd *D*	2nd *B*	2nd *B*	2nd *B*	2nd *B*	2nd *B*	2nd *B*	2nd *B*	2nd *B*	2nd *B*
3rd *C*	3rd *D*	3rd *C*	3rd *A*	3rd *A*	3rd *D*	3rd *C*	3rd *A*	3rd *C*	3rd *C*	3rd *D*	3rd *C*
4th *D*	4th *A*	4th *A*	4th *D*	4th *D*	4th *A*	4th *D*	4th *D*	4th *D*	4th *D*	4th *A*	4th *D*

   (a) How many first-place votes are needed for a majority?

   (b) Which restaurant has the most first-place votes? Is it a majority or a plurality?

   (c) Write out the preference schedule for this election.

2. The Latin Club is holding an election to choose its president. There are three candidates, Arsenio, Beatrice, and Carlos (*A*, *B*, and *C* for short). Following are the votes of the 11 members of the club that voted.

Voter	Sue	Bill	Tom	Pat	Tina	Mary	Alan	Chris	Paul	Kate	Ron
1st choice	*C*	*A*	*C*	*A*	*B*	*C*	*A*	*A*	*C*	*B*	*A*
2nd choice	*A*	*C*	*B*	*B*	*C*	*B*	*C*	*C*	*B*	*C*	*B*
3rd choice	*B*	*B*	*A*	*C*	*A*	*A*	*B*	*B*	*A*	*A*	*C*

   (a) How many first-place votes are needed for a majority?

   (b) Which candidate has the most first-place votes? Is it a majority or a plurality?

   (c) Write out the preference schedule for this election.

3. An election is held to choose the Chair of the Mathematics Department at Tasmania State University. The candidates are Professors Argand, Brandt, Chavez, Dietz, and Epstein. The preference schedule for the election is as follows:

Number of voters	5	3	5	3	2	3
1st choice	*A*	*A*	*C*	*D*	*D*	*B*
2nd choice	*B*	*D*	*E*	*C*	*C*	*E*
3rd choice	*C*	*B*	*D*	*B*	*B*	*A*
4th choice	*D*	*C*	*A*	*E*	*A*	*C*
5th choice	*E*	*E*	*B*	*A*	*E*	*D*

(a) How many people voted in this election?

(b) How many first-place votes are needed for a majority?

(c) Which candidate had the most first-place votes?

(d) Which candidate had the least first-place votes?

(e) Which candidate had the least last-place votes?

(f) Which candidate had the most last-place votes?

4. A math class is asked by the instructor to vote among four possible times for the final exam—A (December 15, 8:00 A.M.), B (December 20, 9:00 P.M.), C (December 21, 7:00 A.M.), and D (December 23, 11:00 A.M.). The following is the class preference schedule.

Number of voters	3	4	9	9	2	5	8	3	12
1st choice	A	A	A	B	B	B	C	C	D
2nd choice	B	B	C	C	A	C	D	A	C
3rd choice	C	D	B	D	C	A	B	D	A
4th choice	D	C	D	A	D	D	A	B	B

(a) How many students in the class voted?

(b) How many first-place votes are needed for a majority?

(c) Which alternative(s) had the most first-place votes?

(d) Which alternative(s) had the least first-place votes?

(e) Which alternative(s) had the least last-place votes?

(f) Which alternative(s) had the most last-place votes?

5. This exercise refers to the election for Mathematics Department Chair discussed in Exercise 3. Suppose that the election rules are that when there is a candidate with a majority of the first-place votes, he/she is the winner. Otherwise, all candidates with 20% or less of the first-place votes are eliminated and the ballots are recounted.

(a) Which candidates are eliminated in this election?

(b) Find the preference schedule for the recount.

(c) Which candidate is the majority winner after the recount?

6. The student body at Eureka High School is having an election for Homecoming Queen. The candidates are Alicia, Brandy, Cleo, and Dionne. The preference schedule for the election is as follows:

Number of voters	153	102	55	202	108	20	110	160	175	155
1st choice	A	A	A	B	B	B	C	C	D	D
2nd choice	C	B	D	D	C	C	A	B	A	B
3rd choice	B	D	C	A	D	A	D	A	C	C
4th choice	D	C	B	C	A	D	B	D	B	A

Suppose that the election rules are that when there is a candidate with a majority of the first-place votes, she is the winner. Otherwise, all candidates with 25% or less of the first-place votes are eliminated and the ballots are recounted.

(a) Which candidates are eliminated in this election?

(b) Find the preference schedule for the recount.

(c) Which candidate is the majority winner after the recount?

In this chapter we used preference ballots that list ranks (1st choice, 2nd choice, etc.) and ask the voter to put the name of a candidate or choice next to each rank. Exercises 7 and 8 refer to an alternative format for preference ballots in which the names of candidates appear in some order and the voter is asked to put a rank (1, 2, 3, etc.) next to each name.

7. Rewrite the following preference schedule in the conventional format used in the book.

Number of voters	47	36	24	13	5
A	3	1	2	4	3
B	1	2	1	2	5
C	4	4	5	3	1
D	5	3	3	5	4
E	2	5	4	1	2

8. Rewrite the following conventional preference schedule in the alternative format. Assume the candidates are listed in alphabetical order on the ballots.

Number of voters	47	36	24	13	5
1st choice	*A*	*B*	*D*	*C*	*B*
2nd choice	*C*	*A*	*B*	*A*	*D*
3rd choice	*B*	*D*	*C*	*E*	*E*
4th choice	*E*	*C*	*E*	*B*	*A*
5th choice	*D*	*E*	*A*	*D*	*C*

B. Plurality Method

9. This exercise refers to the election for Homecoming Queen at Eureka High School discussed in Exercise 6.

(a) Find the winner(s) of the election under the plurality method.

(b) Suppose that in case of a tie, the winner is decided by choosing the candidate with the fewest last-place votes. In this case, which candidate would win the election?

10. This exercise refers to the election discussed in Exercise 4.

(a) Which alternative would win under the plurality method?

(b) Suppose that ties are broken by choosing the alternative with the fewest last-place votes. In this case, when would the final exam be given?

11. This exercise refers to the election for Homecoming Queen at Eureka High School discussed in Exercises 6 and 9. Imagine that one of the students at Eureka High School, whom we'll call Miss Insincere, likes Cleo a lot and is extremely jealous of Dionne.

    (a) Describe Miss Insincere's original ballot in the election.

    (b) Describe how Miss Insincere could vote insincerely and affect the results of the election under the plurality method.

12. This exercise refers to the election discussed in Exercises 4 and 10. The preference schedule is repeated in the following table.

Number of voters	3	4	9	9	2	5	8	3	12
1st choice	A	A	A	B	B	B	C	C	D
2nd choice	B	B	C	C	A	C	D	A	C
3rd choice	C	D	B	D	C	A	B	D	A
4th choice	D	C	D	A	D	D	A	B	B

Imagine that there is a student in the class who has another final on December 20 and thus is dead set against option *B* and, at the same time, would like to catch a ride home on December 22.

    (a) Describe this student's ballot in the election.

    (b) Describe how this student could vote insincerely and affect the results of the election under the plurality method.

13. An election with 4 candidates (*A, B, C, D*) and 150 voters is to be decided using the plurality method. After 120 ballots have been recorded, *A* has 26 votes, *B* has 18 votes, *C* has 42 votes, and *D* has 34 votes.

    (a) What is the smallest number of the remaining 30 votes that *A* must receive to guarantee a win for *A*? Explain.

    (b) What is the smallest number of the remaining 30 votes that *C* must receive to guarantee a win for *C*? Explain.

14. An election with 4 candidates (*A, B, C, D*) and 150 voters is to be decided using the plurality method. After 120 ballots have been recorded, *A* has 26 votes, *B* has 18 votes, *C* has 42 votes, and *D* has 34 votes.

    (a) What is the smallest number of the remaining 30 votes that *B* must receive to guarantee a win for *B*? Explain.

    (b) What is the smallest number of the remaining 30 votes that *D* must receive to guarantee a win for *D*? Explain.

15. An election is to be decided using the plurality method. There are 5 candidates and 407 voters.

    (a) What is the smallest number of votes that a winning candidate can have?

    (b) What is the smallest number of votes that a winning candidate can have if there can be no ties for the winner?

16. An election is to be decided using the plurality method. There are 4 candidates and 306 voters.

    (a) What is the smallest number of votes that a winning candidate can have?

    (b) What is the smallest number of votes that a winning candidate can have if there can be no ties for the winner?

C. Borda Count Method

17. This exercise refers to the election for Mathematics Department Chair discussed in Exercise 3. The candidates are Professors Argand, Brandt, Chavez, Dietz, and Epstein, and the preference schedule is repeated in the following table.

Number of voters	5	3	5	3	2	3
1st choice	A	A	C	D	D	B
2nd choice	B	D	E	C	C	E
3rd choice	C	B	D	B	B	A
4th choice	D	C	A	E	A	C
5th choice	E	E	B	A	E	D

Suppose that the election is to be decided under the Borda count method.

(a) Find the winner of the election.

(b) Suppose that before the votes are counted, Professor Epstein withdraws from the race. Find the preference schedule for the new election and the winner under the Borda count method.

18. This exercise refers to the election for Homecoming Queen at Eureka High School discussed in Exercises 6 and 9. The candidates are Alicia, Brandy, Cleo, and Dionne, and the preference schedule for the election is repeated in the following table.

Number of voters	153	102	55	202	108	20	110	160	175	155
1st choice	A	A	A	B	B	B	C	C	D	D
2nd choice	C	B	D	D	C	C	A	B	A	B
3rd choice	B	D	C	A	D	A	D	A	C	C
4th choice	D	C	B	C	A	D	B	D	B	A

(a) Find the winner of the election under the Borda count method. (You will probably want to use a calculator to do the arithmetic.)

(b) Suppose that before the votes are counted, Cleo is found to be ineligible because of her grades. Find the preference schedule for the new election and the winner under the Borda count method.

19. The editorial board of *Gourmet* magazine is having an election to choose the "Restaurant of the Year." The candidates are Andre's, Borrelli, Casablanca, Dante, and Escargot. The preference schedule for the election is given in the following table.

Number of voters	8	7	6	2	1
1st choice	A	D	D	C	E
2nd choice	B	B	B	A	A
3rd choice	C	A	E	B	D
4th choice	D	C	C	D	B
5th choice	E	E	A	E	C

(a) Find the winner under the Borda count method.

(b) Explain why this election illustrates a violation of the majority criterion.

(c) Explain why this election illustrates a violation of the Condorcet criterion.

20. The members of the Tasmania State University soccer team are having an election to choose the captain of the team from among the four seniors— Anderson, Bergman, Chou, and Delgado. The preference schedule for the election is given in the following table.

Number of voters	4	1	9	8	5
1st choice	A	B	C	A	C
2nd choice	B	A	D	D	D
3rd choice	D	D	A	B	B
4th choice	C	C	B	C	A

(a) Find the winner under the Borda count method.

(b) Explain why this election illustrates a violation of the majority criterion.

(c) Explain why this election illustrates a violation of the Condorcet criterion.

21. An election is held among four candidates (A, B, C, D). Each column in the following preference schedule shows the percentage of voters voting that way.

Percentage of voters	40%	25%	20%	15%
1st choice	A	C	B	B
2nd choice	D	B	D	A
3rd choice	B	D	A	D
4th choice	C	A	C	C

(a) Assuming there are 100 voters, find the number of voters for each column in the preference schedule and then find the winner of the election under the Borda count method.

(b) Assuming the number of voters is $100N$, where N is a positive integer, find the number of voters for each column in the preference schedule (in terms of N) and then find the winner of the election under the Borda count method.

(c) Does your answer in part (b) depend on N? Explain.

22. An election is held among four candidates (A, B, C, D). Each column in the following preference schedule shows the percentage of voters voting that way.

Percentage of voters	48%	24%	16%	12%
1st choice	A	C	B	B
2nd choice	D	B	D	A
3rd choice	B	D	A	D
4th choice	C	A	C	C

(a) Assuming there are 100 voters, find the number of voters for each column in the preference schedule and then find the winner of the election under the Borda count method.

(b) Assuming the number of voters is $100N$, where N is a positive integer, find the number of voters for each column in the preference schedule (in terms of N) and then find the winner of the election under the Borda count method.

(c) Does your answer in part (b) depend on N? Explain.

23. An election with four candidates and 50 voters is to be determined using the Borda count method.

(a) What is the maximum number of points a candidate can receive?

(b) What is the minimum number of points a candidate can receive?

24. An election with three candidates and 100 voters is to be determined using the Borda count method.

(a) What is the maximum number of points a candidate can receive?

(b) What is the minimum number of points a candidate can receive?

25. An election is to be decided using the Borda count method. There are four candidates (A, B, C, D) in this election.

(a) How many points are given out by one ballot?

(b) If there are 110 voters in the election, what is the total number of points given out to the candidates?

(c) If candidate A gets 320 points, candidate B gets 290 points, and candidate C gets 180 points, how many points did candidate D get?

26. An election is to be decided using the Borda count method. There are five candidates (A, B, C, D, E) and 40 voters. If candidate A gets 139 points, candidate B gets 121 points, candidate C gets 80 points, and candidate D gets 113 points, who is the winner of the election? (If you have trouble with this exercise, try Exercise 25 first.)

D. Plurality-with-Elimination Method

27. This exercise refers to the election for Mathematics Department Chair discussed in Exercises 3 and 17. The preference schedule is repeated in the following table.

Number of voters	5	3	5	3	2	3
1st choice	A	A	C	D	D	B
2nd choice	B	D	E	C	C	E
3rd choice	C	B	D	B	B	A
4th choice	D	C	A	E	A	C
5th choice	E	E	B	A	E	D

Find the winner of the election under the plurality-with-elimination method.

28. This exercise refers to the Homecoming Queen election discussed in Exercises 6 and 18. The preference schedule is repeated in the following table.

Number of voters	153	102	55	202	108	20	110	160	175	155
1st choice	A	A	A	B	B	B	C	C	D	D
2nd choice	C	B	D	D	C	C	A	B	A	B
3rd choice	B	D	C	A	D	A	D	A	C	C
4th choice	D	C	B	C	A	D	B	D	B	A

Find the winner of the election under the plurality-with-elimination method.

29. This exercise refers to the "Restaurant of the Year" election discussed in Exercise 19. The preference schedule is repeated in the following table.

Number of voters	8	7	6	2	1
1st choice	A	D	D	C	E
2nd choice	B	B	B	A	A
3rd choice	C	A	E	B	D
4th choice	D	C	C	D	B
5th choice	E	E	A	E	C

(a) Find the winner of the election under the plurality-with-elimination method.

(b) Explain why the winner in (a) can be determined in the first round.

30. This exercise refers to the Tasmania State University soccer captain election discussed in Exercise 20.

(a) Find the winner of the election under the plurality-with-elimination method.

(b) Explain why the winner in (a) can be determined in the first round.

31. This exercise refers to the preference schedule shown in Exercise 21. Find the winner of the election under the plurality-with-elimination method.

32. This exercise refers to the preference schedule shown in Exercise 22. Find the winner of the election under the plurality-with-elimination method.

33. An election is held by the 27 members of the National Football League Executive Committee to choose the host city for Super Bowl XL. The finalists are Atlanta, Boston, Chicago, and Denver. The preference schedule for the election is given in the following table.

Number of voters	10	6	5	4	2
1st choice	A	B	B	C	D
2nd choice	C	D	C	A	C
3rd choice	B	C	A	D	B
4th choice	D	A	D	B	A

(a) Find the winner of the election under the plurality-with-elimination method.

(b) There is a Condorcet candidate in this election. Find it.

(c) Explain why the plurality-with-elimination method violates the Condorcet criterion.

34. The Board of Directors of the XYZ Corporation is holding an election to choose a new Chairman of the Board. The candidates are Allen, Beckman, Cole, Dent, and Emery. The preference schedule for the election is given in the following table.

Number of voters	10	8	5	4	3
1st choice	A	D	B	C	E
2nd choice	C	C	C	B	A
3rd choice	B	B	D	D	C
4th choice	D	E	A	E	B
5th choice	E	A	E	A	D

(a) Find the winner of the election under the plurality-with-elimination method.

(b) There is a Condorcet candidate in this election. Find it.

(c) Explain why the plurality-with-elimination method violates the Condorcet criterion.

E. Pairwise Comparisons Method

35. Find the winner of the election given by the following preference schedule under the method of pairwise comparisons.

Number of voters	8	6	5	5	2
1st choice	B	A	C	D	D
2nd choice	A	D	D	A	A
3rd choice	C	B	B	C	B
4th choice	D	C	A	B	C

36. Find the winner of the election given by the following preference schedule under the method of pairwise comparisons. (Note: This is the final exam election discussed in Exercises 4, 10, and 12.)

Number of voters	3	4	9	9	2	5	8	3	12
1st choice	A	A	A	B	B	B	C	C	D
2nd choice	B	B	C	C	A	C	D	A	C
3rd choice	C	D	B	D	C	A	B	D	A
4th choice	D	C	D	A	D	D	A	B	B

37. Find the winner of the election given by the following preference schedule under the method of pairwise comparisons. (Note: This is the Tasmania State University Math Department election discussed in Exercises 3, 17, and 27.)

Number of voters	5	3	5	3	2	3
1st choice	A	A	C	D	D	B
2nd choice	B	D	E	C	C	E
3rd choice	C	B	D	B	B	A
4th choice	D	C	A	E	A	C
5th choice	E	E	B	A	E	D

38. Find the winner of the election given by the following preference schedule under the method of pairwise comparisons. (Note: This is the "Restaurant of the Year" election discussed in Exercises 19 and 29.)

Number of voters	8	7	6	2	1
1st choice	A	D	D	C	E
2nd choice	B	B	B	A	A
3rd choice	C	A	E	B	D
4th choice	D	C	C	D	B
5th choice	E	E	A	E	C

39. An election with five candidates A, B, C, D, and E is held under the method of pairwise comparisons. Partial results of the pairwise comparisons are as follows: A wins two pairwise comparisons, B wins two and ties one, C wins one, and D wins one and ties one. Find the winner of the election.

40. An election with six candidates A, B, C, D, E, and F is held under the method of pairwise comparisons. Partial results of the pairwise comparisons are as follows: A wins three pairwise comparisons, B and C both win two, D and E both win two and tie one.

(a) Find the winner of the election.

(b) Give the result of the pairwise comparison between D and E.

F. Ranking Methods

41. For the election given by the following preference schedule,

Number of voters	4	1	9	8	5
1st choice	A	B	C	A	D
2nd choice	C	A	D	B	C
3rd choice	B	D	A	D	B
4th choice	D	C	B	C	A

(a) rank the candidates using the extended plurality method.

(b) rank the candidates using the extended Borda count method.

(c) rank the candidates using the extended plurality-with-elimination method.

(d) rank the candidates using the extended pairwise comparisons method.

42. For the election given by the following preference schedule,

Number of voters	14	10	8	4	1
1st choice	A	C	D	B	C
2nd choice	B	B	C	D	D
3rd choice	C	D	B	C	B
4th choice	D	A	A	A	A

(a) rank the candidates using the extended plurality method.

(b) rank the candidates using the extended Borda count method.

(c) rank the candidates using the extended plurality-with-elimination method.

(d) rank the candidates using the extended pairwise comparisons method.

43. For the election given by the following preference schedule,

Number of voters	4	1	9	8	5
1st choice	A	B	C	A	D
2nd choice	C	A	D	B	C
3rd choice	B	D	A	D	B
4th choice	D	C	B	C	A

(a) rank the candidates using the recursive plurality method.

(b) rank the candidates using the recursive Borda count method.

(c) rank the candidates using the recursive plurality-with-elimination method.

(d) rank the candidates using the recursive pairwise comparisons method.

44. For the election given by the following preference schedule,

Number of voters	14	10	8	4	1
1st choice	A	C	D	B	C
2nd choice	B	B	C	D	D
3rd choice	C	D	B	C	B
4th choice	D	A	A	A	A

(a) rank the candidates using the recursive plurality method.

(b) rank the candidates using the recursive Borda count method.

(c) rank the candidates using the recursive plurality-with-elimination method.

(d) rank the candidates using the recursive pairwise comparisons method.

G. Miscellaneous

Find the sums in Exercises 45–48.

45. $1 + 2 + 3 + \cdots + 498 + 499 + 500$

46. $1 + 2 + 3 + \cdots + 3218 + 3219 + 3220$

47. $501 + 502 + 503 + \cdots + 3218 + 3219 + 3220$
(*Hint:* Do Exercises 45 and 46 first.)

48. $1801 + 1802 + 1803 + \cdots + 8843 + 8844 + 8845$
(*Hint:* Do Exercise 47 first.)

49. In an election with 15 candidates,

(a) how many pairwise comparisons are there?

(b) if it takes one minute to calculate a pairwise comparison, approximately how long would it take to calculate the results of the election using the method of pairwise comparisons?

50. Suppose that 21 players sign up for a round-robin Ping-Pong tournament. (In a round-robin tournament, everyone plays everyone else once.)

(a) How many matches must be scheduled for the entire tournament?

(b) If six matches can be scheduled per hour and the tournament hall is available for 12 hours each day, for how many days would the tournament hall have to be reserved?

51. Consider the election given by the following preference schedule.

Number of voters	7	4	2
1st choice	A	B	D
2nd choice	B	D	A
3rd choice	C	C	C
4th choice	D	A	B

(a) Find the Condorcet candidate in this election.

(b) Find the winner of this election under the Borda count method.

(c) Suppose that candidate C drops out of the race. Who among the remaining candidates wins the election under the Borda count method?

(d) Based on (a) through (c), which of the four fairness criteria are violated in this election? Explain.

52. Consider the election given by the following preference schedule.

Number of voters	10	6	5	4	2
1st choice	A	B	C	D	D
2nd choice	C	D	B	C	A
3rd choice	D	C	D	B	B
4th choice	B	A	A	A	C

(a) Find the Condorcet candidate in this election.

(b) Find the winner of this election under the plurality-with-elimination method.

(c) Suppose that candidate D drops out of the race. Who among the remaining candidates wins the election under the plurality-with-elimination method?

(d) Based on (a) through (c), which of the four fairness criteria are violated in this election? Explain.

53. Consider the election, to be determined by the Borda count method, given by the following preference schedule.

Number of voters	7	5	3
1st choice	A	B	C
2nd choice	B	A	A
3rd choice	C	C	B

(a) Does this election have a Condorcet candidate?

(b) Does this election have a candidate with the majority of first-place votes?

(c) Does this election violate the Condorcet criterion?

(d) Does this election violate the majority criterion?

54. Consider the election, to be determined by the plurality-with-elimination method, given by the following preference schedule.

Number of voters	7	5	3
1st choice	A	B	C
2nd choice	B	C	A
3rd choice	C	A	B

(a) Does this election have a Condorcet candidate?

(b) Does this election have a candidate with the majority of first-place votes?

(c) Does this election violate the Condorcet criterion?

(d) Does this election violate the majority criterion?

JOGGING

55. **Two candidate elections.** Explain why when there are only two candidates, the four voting methods we discussed in this chapter give the same winner and the winner is determined by straight majority.

56. **Plurality with a runoff.** This is a simple variation of the plurality-with-elimination method. Here, if a candidate has a majority of the first-place votes, then that candidate wins the election; otherwise we eliminate *all* candidates except the two with the most first-place votes. The winner is then chosen between these two by recounting the votes in the usual way.

    (a) Use the MAS election to show that plurality with a runoff can produce a different outcome than plurality with elimination.

    (b) Give an example that shows that plurality with a runoff violates the monotonicity criterion.

    (c) Give an example that shows that plurality with a runoff violates the Condorcet criterion.

57. **Equivalent Borda count (Variation 1).** The following simple variation of the Borda count method described in the chapter is sometimes used: A first place is worth $N - 1$ points, second place is worth $N - 2$ points, ..., last place is worth 0 points (where N is the number of candidates). The candidate with the most points is the winner.

    (a) Suppose that a candidate gets p points using the Borda count as originally described in the chapter and q points under this variation. Explain why if k is the number of voters, then $p = q + k$.

    (b) Explain why this variation is equivalent to the original Borda count described in the chapter (i.e., it produces exactly the same election results).

58. **Equivalent Borda count (Variation 2).** Another commonly used variation of the Borda count method described in the chapter is the following: A first place is worth 1 point, second place is worth 2 points, ..., last place is worth N points (where N is the number of candidates). The candidate with the fewest points is the winner, second fewest points is second, etc.

    (a) Suppose that a candidate gets p points using the Borda count as originally described in the chapter and r points under this variation. Explain why if k is the number of voters, then $p + r = k(N + 1)$.

    (b) Explain why this variation is equivalent to the original Borda count described in the chapter (i.e., it produces exactly the same election results).

59. Give an example of an election with four candidates (A, B, C, and D) satisfying the following: (i) No candidate has a majority of the first-place votes; (ii) C is a Condorcet candidate but has no first-place votes; (iii) B is the winner under the Borda count method; (iv) A is the winner under the plurality method.

60. Explain why the method of pairwise comparisons satisfies the majority criterion.

61. Explain why the plurality method satisfies the monotonicity criterion.

62. The following table shows the three top-ranked college football teams at the end of the 1993 football season, according to the CNN/*USA Today* coaches' poll. This poll is based on the votes of 62 coaches, each one of whom ranks the top 25 teams. (The remaining 22 teams are not shown because they are irrelevant to this exercise.) A team gets 25 points for each first-place vote, 24 points for each second-place vote, 23 points for each third-place vote, etc.

Team	Points	Number of first-place votes
1. Florida State	1523	36
2. Notre Dame	1494	25
3. Nebraska	1447	1
⋮	⋮	⋮

(a) Based on the information given in the table it is possible to conclude that all 62 coaches had Florida State, Notre Dame, and Nebraska in some order as their top three choices. Explain why this is true.

(b) Find the number of second- and third-place votes for each of the 3 teams.

63. The AL MVP. Each year, the Most Valuable Player of the American League is chosen by a group of 28 sports writers using a variation of the Borda count method. This is one of the most important individual awards in professional baseball, not only because of the honor, but also because there is a large cash prize that goes with it. The winners and their votes for the years 1997–1999 are as follows.

Year	Winner	Total Points	Votes
1997	Ken Griffey Jr.	392	1st place: 28
1998	Juan Gonzales	357	1st place: 21; 2nd place: 7
1999	Ivan Rodriguez	252	1st place: 7; 2nd place: 6; 3rd place: 7; 5th place: 5; 6th place: 2; 7th place: 1

(a) Based on the information above, determine the rules for the AL MVP election. (How many points are given for 1st place, 2nd place, 3rd place, ..., 10th place?) You can assume that the number of points for each of the first ten places in the ballot is a different positive integer.

(b) Give a fictitious example of how this voting method can produce a violation of the majority criterion.

64. The Coombs method. This method is just like the plurality-with-elimination method except that in each round we eliminate the candidate with the *largest number of last-place votes* (instead of the one with the fewest first-place votes).

(a) Find the winner of the MAS election using the Coombs method.

(b) Give an example showing that the Coombs method violates the Condorcet criterion.

(c) Give an example showing that the Coombs method violates the monotonicity criterion.

65. An election with four candidates and 50 voters is to be determined using the Borda count method. What is the minimum number of points a candidate can receive and be the only winner of the election?

RUNNING

66. Show that if, in an election with an odd number of voters, there is no Condorcet candidate, then any ranking of the candidates based on the extended pairwise comparisons method must result in at least two candidates ending up tied in the rankings. Explain why the result does not have to be true with an even number of voters.

67. Show that the method of pairwise comparisons satisfies the monotonicity criterion.

68. **The Pareto criterion.** The following fairness criterion was proposed by the Italian economist Vilfredo Pareto (1848–1923): *If every voter prefers alternative X over alternative Y, then a voting method should not choose Y as the winner.* Show that all four voting methods discussed in the chapter satisfy the Pareto criterion. (A separate analysis is needed for each of the four methods.)

69. Suppose the following was proposed as a fairness criterion: *If a majority of the voters prefer alternative X to alternative Y, then the voting method should rank X above Y.* Give an example to show that all four of the extended voting methods discussed in the chapter can violate this criterion. (*Hint:* Consider an example with no Condorcet candidate.)

70. Consider the following fairness criterion: *If a majority of the voters prefer every alternative over alternative X, then a voting method should not choose alternative X as the winner.*

(a) Give an example to show that the plurality method violates this criterion.

(b) Give an example to show that the plurality-with-elimination method violates this criterion.

(c) Explain why the method of pairwise comparisons satisfies this criterion.

(d) Explain why the Borda count method satisfies this criterion.

71. **The Condorcet loser criterion.** *If there is an alternative that loses in a one-to-one comparison to each of the other alternatives, then that alternative should not be the winner of the election.* (This fairness criterion is a sort of mirror image of the regular Condorcet criterion.)

(a) Give an example that shows that the plurality method violates the Condorcet loser criterion.

(b) Explain why the plurality-with-elimination method violates the Condorcet loser criterion.

(c) Explain why the Borda count method satisfies the Condorcet loser criterion.

72. Consider a variation of the Borda count method in which a first-place vote in an election with N candidates is worth F points (where $F > N$) and all other places in the ballot are the same as in the ordinary Borda count: $N - 1$ points for second place, $N - 2$ points for third place, $\ldots$, 1 point for last place. By choosing F large enough, we can make this variation of the Borda count method satisfy the majority criterion. Find the smallest value of F (expressed in terms if N) for which this happens.

PROJECTS AND PAPERS

A. Ballots, Ballots, Ballots!

In this chapter we discussed elections in which the voters cast their votes by means of linear preference ballots. There are many other types of ballots used in real-life elections, ranging from the simple (winner only) to the exotic (each voter has a fixed number of points to divide among the candidates any way he or she sees fit). In this project you are to research other types of ballots; how, where, and when they are used; and what are the arguments for and against their use.

B. Sequential Voting

Sequential voting is a voting method used by legislative bodies and committees to choose one among a list of alternatives. In sequential voting the alternatives are presented in some order *A, B, C, D*, etc. The voters then choose between *A* and *B*, the winner is then matched against *C*, the winner of that vote against *D*, and so on. In sequential voting, the *agenda* (the order in which the options are presented) can have a critical impact on the outcome of the vote. Write a research paper on sequential voting, its history, and its use, paying particular attention to the issue of the agenda and its impact on the outcome. Illustrate your points with examples that *you* have made up.

C. Instant Runoff Voting

Imagine you are a political activist in your community. The city council is having hearings to decide if the *method of instant runoff voting* should be adopted in your city. (For a brief discussion of instant runoff voting, see Section 1.4.) Stake out a position for or against instant runoff voting, and prepare a brief to present to the city council that justifies that position. To make an effective case, your argument should include mathematical, economic, political, and social considerations. (Remember that your city council members are not as well versed as you are on the mathematical aspects of elections. Part of your job is to educate them.)

D. Manipulability of an Election

A voter is said to *manipulate* the results of an election if he or she is able to produce a more desirable outcome by voting insincerely than by voting according to his or her true preferences. We touched briefly on *insincere voting* and *manipulability* in our discussion of the plurality method, but this is a big subject. In this project you should research the issue of manipulability of elections and its political and social costs. Your research should include the mathematical aspects of manipulability, culminating with a description and interpretation of the *Gibbard-Satterthwaite manipulability theorem*.

Note: A suggested starting point is to read the article "The Manipulability of Voting Systems," by Alan Taylor in *The American Mathematical Monthly* (v. 109, April 2002).

E. The 2000 Presidential Election and the Florida Vote

The unusual circumstances surrounding the 2000 presidential election and the Florida vote are a low point in American electoral history. Write an analysis paper on the 2000 Florida vote, paying particular attention to what went wrong and how a similar situation can be prevented in the future. You should touch on technology issues (outdated and inaccurate vote tallying methods and equipment, poorly designed ballots, etc.), political issues (the two-party system, the Electoral College, etc.), and, as much as possible, on issues related to concepts from this chapter (can presidential elections be improved by changing to preference ballots, using a different voting method, etc.).

F. Short Story

Write a fictional short story using an election as the backdrop. Weave into the dramatic structure of the story elements and themes from this chapter (fairness, manipulation, monotonicity, and independence of irrelevant alternatives all lend themselves to good drama). Be creative and have fun.

| APPENDIX 1 | Breaking Ties |

By and large, most of the examples given in the chapter were carefully chosen to avoid tied winners, but of course in the real world ties are bound to occur.

In this appendix we will discuss very briefly the problem of how to break ties when necessary. Tie-breaking methods can raise some fairly complex issues, and our purpose here is not to study such methods in great detail but rather to make the reader aware of the problem and give some inkling as to possible ways to deal with it.

For starters, consider the election with preference schedule shown in Table A-1. If we look at this preference schedule carefully, we can see that there is complete *symmetry* in the positions of the three candidates. Essentially, this means that we could interchange the names of the candidates and the preference schedule would not change. Given the complete symmetry of the preference schedule, it is clear that no rational voting method could choose one candidate as the winner over the other two. In this situation, a tie is inevitable regardless of the voting method used. We call this kind of tie an **essential tie.** We cannot break essential ties using a rational tie-breaking procedure, and must instead rely on some sort of outside intervention such as chance (flip a coin, draw straws, etc.), a third party (the judge, mom, etc.), or even some outside factor (experience, age, etc.).

TABLE A-1	A Three-Way Essential Tie		
Number of voters	7	7	7
1st choice	A	B	C
2nd choice	B	C	A
3rd choice	C	A	B

TABLE A-2	A Tie That Could Be Broken					
Number of voters	**5**	**3**	**5**	**3**	**2**	**4**
1st choice	A	A	C	D	D	B
2nd choice	B	B	E	C	C	E
3rd choice	C	D	D	B	B	A
4th choice	D	C	A	E	A	C
5th choice	E	E	B	A	E	D

Most ties are not essential ties, and we can often break them in more rational ways: either by implementing some tie-breaking rule or by using a different voting method to break the tie. To illustrate some of these ideas, let's consider as an example the election with preference schedule shown in Table A-2.

If we decide this election using the method of pairwise comparisons, we have the following:

A versus B: 13 votes to 9 votes. A gets 1 point.

A versus C: 12 votes to 10 votes. A gets 1 point.

A versus D: 12 votes to 10 votes. A gets 1 point.

A versus E: 10 votes to 12 votes. E gets 1 point.

B versus C: 12 votes to 10 votes. B gets 1 point.

B versus D: 12 votes to 10 votes. B gets 1 point.

B versus E: 17 votes to 5 votes. B gets 1 point.

C versus D: 14 votes to 8 votes. C gets 1 point.

C versus E: 18 votes to 4 votes. C gets 1 point.

D versus E: 13 votes to 9 votes. D gets 1 point.

In this election A and B, with three wins each, tie for first place. How could we break this tie? Here is just a sampler of the many possible ways:

1. Use the results of a pairwise comparison between the winners. In the preceding example, since A beats B 13 votes to 9, the tie would be broken in favor of A.

2. Use the total point differentials. For example, since A beats B 13 to 9, the point differential for A is $+4$, and since A lost to E 10 to 12, the point differential for A is -2. Computing the total point differentials for A gives $4 + 2 + 2 - 2 = 6$. Likewise, the total point differential for B is $2 + 2 + 12 - 4 = 12$. In this case the point differentials favor B, so B would be declared the winner.

3. Use first-place votes. In the example, A has 8 and B has 4. With this method, the winner would be A.

4. Use Borda count points to choose between the two winners. Here

   A has $(5 \times 8) + (3 \times 4) + (2 \times 7) + (1 \times 3) = 69$ points;

   B has $(5 \times 4) + (4 \times 8) + (3 \times 5) + (1 \times 5) = 72$ points;

and the tie would be broken in favor of B.

By now we should not be at all surprised that different tie-breaking methods produce different winners and that there is no single *right* method for breaking ties. In retrospect, flipping a coin might not be such a bad idea!

APPENDIX 2	A Sampler of Elections in the Real World

Olympic Venues. The selection of the city that gets to host the Olympic Games has tremendous economic and political impact for the cities involved, and it goes without saying that it always generates a fair amount of controversy. The selection process is carried out by means of an election very much like some of the ones we studied in this chapter (see Example 6). The voters are the members of the International Olympic Committee, and the actual voting method used to select the winner is the plurality-with-elimination method with a minor twist: Instead of indicating their preferences all at once, the voters let their preferences be known one round at a time. Here are the actual details of how Sydney, Australia, was chosen to host the 2000 Summer Olympic Games.

On September 23, 1993, the 89 members of the International Olympic Committee met in Monte Carlo, Monaco, to vote on the selection of the site for the 2000 Summer Olympics. Five cities made bids: Beijing (China), Berlin (Germany), Istanbul (Turkey), Manchester (England), and Sydney (Australia). In each round, the delegates voted for just one city, and the city with the fewest votes was eliminated. The voting went as follows:

- **Round 1.**

City	Beijing	Berlin	Istanbul	Manchester	Sydney
Votes	32	9	7	11	30

Istanbul was eliminated in round 1.

- **Round 2.**

City	Beijing	Berlin	Manchester	Sydney
Votes	37	9	13	30

Berlin was eliminated in round 2.

- **Round 3.**

City	Beijing	Manchester	Sydney	Abstentions
Votes	40	11	37	1

Manchester was eliminated in round 3.

- **Round 4.**

City	Beijing	Sydney	Abstentions
Votes	43	45	1

Beijing was eliminated in round 4; Sydney gets the 2000 Summer Olympics!

The Academy Awards. The Academy of Motion Picture Arts and Sciences gives its annual Academy Awards ("Oscars") for various achievements in connection with motion pictures (best picture, best director, best actress, etc.). Eligible members of the Academy elect a winner in each category. The election process varies slightly from award to award and is quite complicated. For the sake of brevity we will describe the election process for best picture. (The process is almost identical for each of the major awards.) The election takes place in two stages: (1) the nomination stage, in which the five top pictures are nominated, and (2) the final balloting for the winner.

We describe the second stage first because it is so simple: Once the five top pictures are nominated, each eligible member of the Academy votes for one candidate, and the winner is chosen by simple plurality. Because the number of voters is large (somewhere between 4000 and 5000), ties are not likely to occur, but if they do, they are not broken. Thus, it is possible for two candidates to share an award, as in 1968 when Katherine Hepburn and Barbra Streisand shared the award for Best Actress.

The process for selecting the five nominations is considerably more complicated and is based on a voting method called **single transferable voting**. Each eligible member of the Academy submits a preference ballot with the names of his or her top five choices ranked from first to fifth. Based on the total number of valid ballots submitted, the minimum number of votes needed to get a nomination (called the **quota**) is established, and any picture with enough first-place votes to make the quota is automatically nominated.

The quota is always chosen to be a number that is over one-sixth (16.66%) but not more than one-fifth (20%) of the total number of valid ballots cast. (Setting the quota this way ensures that it is impossible for six or more pictures to get automatic nominations.) While in theory it is possible for five pictures to make the quota right off the bat and get an automatic nomination (in which case the nomination process is over), this has never happened in practice. In fact, what usually happens is that there are no pictures that make the quota automatically. Then the picture with the fewest first-place votes (say X) is eliminated, and on all the ballots that originally had X as the first choice, X's name is crossed off the top and all the other pictures are moved up one spot. The ballots are then counted again. If there are still no pictures that make the quota, the process of elimination is repeated. Eventually, there will be one or more pictures that make the quota and are nominated.

The moment that one or more pictures are nominated, there is a new twist: Nominated pictures "give back" to the other pictures still in the running (not nominated but not eliminated either) their "surplus" votes. This process of giving back votes (called a **transfer**) is best illustrated with an imaginary example. Suppose that the quota is 400 (a nice, round number) and at some point a picture (say Z) gets 500 first-place votes, enough to get itself nominated. The surplus for Z is $500 - 400 = 100$ votes, and these are votes that Z doesn't really need. For this reason the 100 surplus votes are taken away from Z and divided fairly among the second-place choices on the 500 ballots cast for Z. The way this is done may seem a little bizarre, but it makes perfectly good sense. Since there are 100 surplus votes to be divided into 500 equal shares, each second-place vote on the 500 ballots cast for Z is worth $\frac{100}{500} = \frac{1}{5}$ vote. While one-fifth of a vote may not seem like much, enough of these fractional votes can make a difference and help some other picture or pictures make the quota. If that's the case, then once again the surplus or surpluses are transferred back to the remaining pictures following the procedure described above; otherwise, the process of elimination is started up again. Eventually, after several possible cycles of eliminations and transfers, five pictures get enough votes to make the quota and be nominated, and the process is over.

The method of single transferable voting is not unique to the Academy Awards. It is used to elect officers in various professional societies as well as the members of the Irish Senate.

Corporate Boards of Directors. In most corporations and professional societies, the members of the Board of Directors are elected by a method called **approval voting**. In approval voting, a voter does not cast a preferential ballot but rather votes for as many candidates as he or she wants. Each of these votes is simply a yes vote for the candidate, and it means that the voter approves of that candidate. The candidate with the most approval votes wins the election.

Table A-3 shows an example of a hypothetical election based on approval voting.

The results of this election are as follows: Winner, A (6 approval votes); second place, C (4 approval votes); last place, B (3 approval votes). Note that a voter can cast anywhere

TABLE A-3 An Election Based on Approval Voting

Candidates	Sue	Bill	Tito	Prince	Tina	Van	Devon	Ike
A	Yes		Yes	Yes	Yes		Yes	Yes
B			Yes		Yes	Yes		
C	Yes				Yes	Yes	Yes	

from no approval votes at all (such as Bill did in Table A-3) to approval votes for all the candidates (such as Tina did in Table A-3). It is somewhat ironic that the effect of Tina's vote is exactly the same as that of Bill's.

In the last few years a strong case has been made suggesting that for political elections, approval voting is a big improvement over the more traditional voting methods. In particular, approval voting encourages voter turnout. The reason for this is psychological: Voters are more likely to vote when they feel they can make intelligent decisions, and unquestionably it is easier for a voter to give an intelligent answer to the question, Do you approve of this candidate—yes or no? than it is to the question, Which candidate is your first choice, second choice, etc.? The latter requires a much deeper knowledge of the candidates, and in today's complex political world it is a knowledge that very few voters have.

REFERENCES AND FURTHER READINGS

1. Arrow, Kenneth J., *Social Choice and Individual Values*. New York: John Wiley & Sons, Inc., 1963.

2. Baker, Keith M., *Condorcet: From Natural Philosophy to Social Mathematics*. Chicago: University of Chicago Press, 1975.

3. Brams, Steven J., and Peter C. Fishburn, *Approval Voting*. Boston: Birkhäuser, 1982.

4. Dummett, M., *Voting Procedures*. New York: Oxford University Press, 1984.

5. Farquharson, Robin, *Theory of Voting*. New Haven, CT: Yale University Press, 1969.

6. Fishburn, Peter C., and Steven J. Brams, "Paradoxes of Preferential Voting," *Mathematics Magazine*, 56 (1983), 207–214.

7. Gardner, Martin, "Mathematical Games (From Counting Votes to Making Votes Count: The Mathematics of Elections)," *Scientific American*, 243 (October 1980), 16–26.

8. Kelly, J., *Arrow Impossibility Theorems*. New York: Academic Press, 1978.

9. Merrill, S., *Making Multicandidate Elections More Democratic*. Princeton, NJ: Princeton University Press, 1988.

10. Niemi, Richard G., and William H. Riker, "The Choice of Voting Systems," *Scientific American*, 234 (June 1976), 21–27.

11. Nurmi, H., *Comparing Voting Systems*. Dordretch, Holland: D. Reidel, 1987.

12. Saari, Donald G., *Basic Geometry of Voting*. New York: Springer-Verlag, 1995.

13. Saari, Donald G., *Chaotic Elections: A Mathematician Looks at Voting.* Providence, R.I.: American Mathematical Society, 2001.

14. Saari, Donald G., and F. Valognes "Geometry, Voting, and Paradoxes," *Mathematics Magazine*, 71 (Oct. 1998), 243–259.

15. Straffin, Philip D., Jr., *Topics in the Theory of Voting*, UMAP Expository Monograph. Boston: Birkhäuser, 1980.

16. Taylor, Alan, *Mathematics and Politics: Strategy, Voting, Power and Proof.* New York: Springer-Verlag, 1995.

17. Taylor, Alan, "The Manipulability of Voting Systems," *The American Mathematical Monthly,* 109 (April 2002), 321–337.

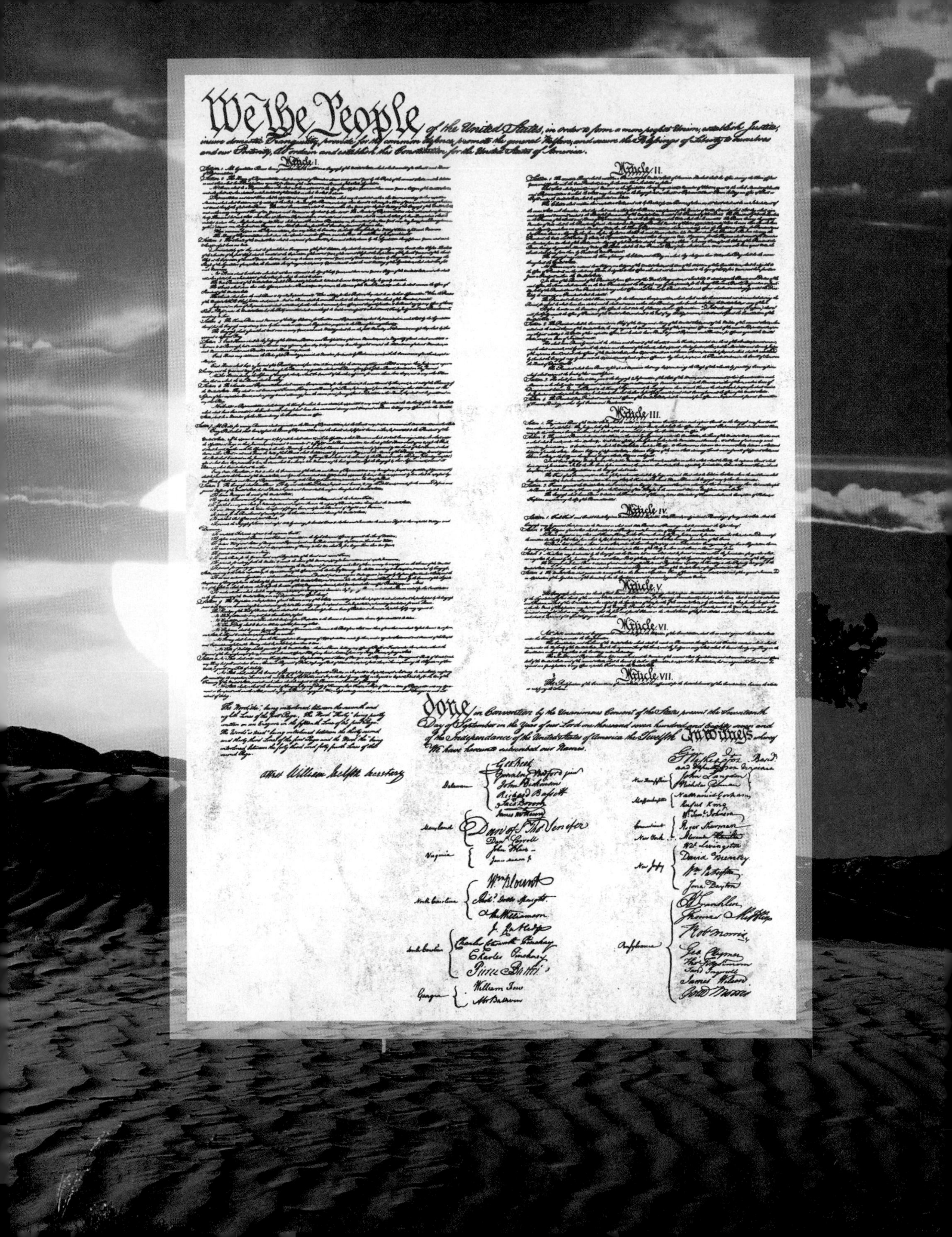

We the People of the United States, in Order to form a more perfect Union, establish Justice, insure domestic Tranquility, provide for the common defence, promote the general Welfare, and secure the Blessings of Liberty to ourselves and our Posterity, do ordain and establish this Constitution for the United States of America.

Article. I.

Article. II.

Article. III.

Article. IV.

Article. V.

Article. VI.

Article. VII.

done in Convention by the Unanimous Consent of the States present the Seventeenth Day of September in the Year of our Lord one thousand seven hundred and Eighty seven and of the Independance of the United States of America the Twelfth In witness whereof We have hereunto subscribed our Names.

Weighted Voting Systems

The Power Game

In a democracy we take many things for granted, not the least of which is the idea that we are all equal. When it comes to voting rights, the democratic ideal of equality translates into the principle of *one person–one vote*. But is the principle of *one person–one vote* always justified? Must it apply when the *voters* are something other than individuals, such as organizations, states, and even countries?

In a diverse society, it is in the very nature of things that voters—be they individuals or institutions—are not equal, and sometimes it is actually desirable to recognize their differences by giving them different amounts of say over the outcome of an election. What we are talking about here is the exact opposite of the principle of *one voter–one vote*, a principle best described as *one voter–x votes*, and more formally known as **weighted voting**.

One of the best known and most controversial examples of weighted voting is the Electoral College, that uniquely American institution bequeathed to the nation by the Founding Fathers through Article II, Section 1 of the U.S. Constitution. In the Electoral College, each of the 50 states controls a number of votes equal to its number of Representatives plus Senators in Congress (in addition, the District of Columbia controls 3 votes). At one end of the spectrum is California, an electoral heavyweight with 55 electoral votes; at the other end of the spectrum are the little guys (Alaska, Montana, North Dakota, etc.) with a meager 3 electoral votes each. The other states fall somewhere in between. (See the appendix at the end of this chapter for full details.) The 2000 presidential election brought to the surface, in a very dramatic way, the vagaries and complexities of the Electoral College system, and, in particular, the critical role that a few voters in a single state (in this case Florida) can end up having in the final outcome of the election.

The question of power in an election—who has it and how much do they have—caught our attention during the 2000 presidential election because of the unusual circumstances surrounding the Florida vote, but it is a question of fundamental importance in every weighted voting situation. (In one person–one vote situations, the question of power is moot, since everyone is equal.) It turns out that power in weighted voting can be measured in a systematic way using basic mathematical ideas, and these are the ideas we will discuss in this chapter.

Examples of weighted voting are more common than one would imagine. In addition to the Electoral College, in this chapter we will discuss how power is divided in the United Nations Security Council, in the European Union, in some county boards of supervisors, and in corporations. All of these are examples of situations that involve some form of weighted voting.

2.1 Weighted Voting Systems

We will start by introducing some terminology. Any formal voting arrangement in which the voters are not necessarily equal in terms of the number of votes they control is called a **weighted voting system**. To keep things simple, we will only consider voting between two candidates or alternatives. A vote involving only two choices can always be thought of as a *yes-no* vote and is generally referred to as a **motion**.

Every weighted voting system is characterized by three elements: the *players*, the *weights* of the players, and the *quota*. The **players** are just the voters themselves. (From now on we will stick to the usual convention of using "voters" when we are dealing with a *one person–one vote* situation as in Chapter 1, and "players" in the case of a weighted voting system.) We will use the letter N to represent the number of players and the symbols $P_1, P_2, \ldots, P_N$ to represent the names of the players—it is a little less personal but a lot more convenient than using Archie, Betty, Jughead, etc. In a weighted voting system, each player controls a certain number of votes, and this number is called the player's **weight**. We will use the symbols $w_1, w_2, \ldots, w_N$ to represent the weights of $P_1, P_2, \ldots, P_N$, respectively. Finally, there is the **quota**, the minimum number of votes needed to pass a motion. We will use the letter q to denote the quota.

It is important to note that the quota q can be something other than a strict majority of the votes. There are many voting situations in which a majority of the votes is not enough to pass a motion—the rules may stipulate a different definition of what is needed for passing. Take, for example, the rules in the U.S. Senate. To pass an ordinary law, a simple majority of the votes is sufficient, but when the Senate is attempting to override a presidential veto, the Constitution requires a quota of two-thirds of the votes. In other organizations the rules may stipulate that three-fourths (75%) of the votes are needed or four-fifths (80%) or even unanimity (100%). In fact, any number larger than half the total number of votes but not more than the total number of votes can be a reasonable choice for the quota q. To put it somewhat more formally,

$$\frac{w_1 + w_2 + \cdots + w_N}{2} < q \leq w_1 + w_2 + \cdots + w_N.$$

Notation and Examples

A convenient way to describe a weighted voting system is

$$[q: w_1, w_2, \ldots, w_N].$$

The quota is always given first, followed by a colon and then the respective weights of the individual players. It is customary to write the weights in numerical order, starting with the highest, and we will adhere to this convention throughout the chapter.

EXAMPLE 2.1

Consider a corporation with four shareholders P_1, P_2, P_3, and P_4. P_1 has 8 votes, P_2 has 6 votes, P_3 has 5 votes, and P_4 has 1 vote. The bylaws of the corporation specify that two-thirds of the 20 votes are needed to pass a motion. Using our new, simplified notation, this corporation can be described as the weighted voting system $[14:8, 6, 5, 1]$. Note that the two-thirds requirement to pass a motion translates in this case to the quota $q = 14$, the first whole number larger than two-thirds of 20. ∎

EXAMPLE 2.2

Consider again a corporation with four shareholders P_1, P_2, P_3, and P_4. In this case, P_1 has 5 votes, P_2 and P_3 both have 4 votes, and P_4 has 2 votes. Now suppose that the quota to pass a motion were 7 votes. We might be tempted to think of this example as a weighted voting system $[7: 5,4,4,2]$, but what we have here is a real mess. Imagine that there is a motion on the table to sell off the corporation. If P_1 and P_4 vote yes and P_2 and P_3 vote no, both groups would win. This is a mathematical version of anarchy and the reason why we insist that *in a legal weighted voting system the quota must be more that half the total number of votes*. ∎

EXAMPLE 2.3

Consider the same four shareholders described in Example 2.2 (P_1 has 5 votes, P_2 and P_3 both have 4 votes, and P_4 has 2 votes), with a quota of 17 votes to pass a motion. Given that all the votes combined add up to 15, this is a ridiculous situation where every motion is doomed to fail. *In a legal weighted voting system, the quota cannot be more than the total number of votes*! ∎

EXAMPLE 2.4

Consider the weighted voting system $[11: 4, 4, 4, 4, 4]$.

In this weighted voting system all 5 players are equal. To pass a motion at least 3 out of the 5 players are needed. [Note that if the quota ($q = 11$) were changed to 12, the situation would still remain the same—at least 3 out of the 5 players would be needed.] What we really have here, somewhat in disguise, is a *one person–one vote situation with simple majority needed for passing a motion*.

In terms of how it works, this weighted voting system is equivalent to the weighted voting system $[3: 1, 1, 1, 1, 1]$. ∎

EXAMPLE 2.5

Consider the weighted voting system $[15: 5, 4, 3, 2, 1]$.

Here we have 5 players with a total of 15 votes. Since the quota is 15, *the only way a motion can pass is by unanimous consent of the players*. How does this voting system differ from the voting system $[5: 1, 1, 1, 1, 1]$? Well, the latter also has 5 players, and the only way a motion can pass is by unanimous consent of the players. So, in terms of how they work, $[15: 5, 4, 3, 2, 1]$ and $[5: 1, 1, 1, 1, 1]$ are equivalent weighted voting systems. ∎

The surprising conclusion of Example 2.5 is that the weighted voting system $[15: 5, 4, 3, 2, 1]$ describes a one person–one vote situation in disguise. This seems like a contradiction only if we think of a *one person–one vote* situation as implying that all players have an *equal number of votes rather than an equal say in the outcome of the election*. Apparently, these two things are not the same! As the example makes abundantly clear, just looking at the number of votes a player controls can be very deceptive.

Power; More Terminology; More Examples

Let's look at a few more examples of weighted voting systems and start to informally focus on the notion of power.

EXAMPLE 2.6

Consider the weighted voting system [11: 12, 5, 4].

Here is a situation in which a single player (P_1) controls enough votes to pass any measure single-handedly. Such a player has all the power, and, not surprisingly, we call such a player a *dictator*. ■

In general, we will say that a player is a **dictator** if the player's weight is bigger than or equal to the quota. Notice that whenever there is a dictator, all the other players, regardless of their weights, have absolutely no power. A player with no power is called a **dummy**.

EXAMPLE 2.7

Consider the weighted voting system [12: 9, 5, 4, 2].

Here we have a situation in which player P_1, while not a dictator, has the power to obstruct by preventing any motion from passing. This happens because even if all the remaining players were to vote together, they wouldn't have the votes to pass a motion against the will of P_1. ■

A player that is not a dictator, but that can single-handedly prevent the rest of the players from passing a motion, is said to have **veto power**.

EXAMPLE 2.8

The *XYZ* Corporation has an interesting power structure. The founder, Mr. *X*, has 99 votes. The founder's only daughter, Mrs. *Y*, has 98 votes. Ms. *Z*, a distant relative, has a measly 3 votes. Decisions are made by majority vote, which in this case requires 101 out of the total 200 votes. In essence, the XYZ Corporation operates as a weighted voting system [101: 99, 98, 3].

The distribution of power in this example is surprising. At first glance, we would guess that Mr. *X* and Mrs. *Y* share most of the power and that Ms. *Z* has very little power, if any. On closer inspection, however, we notice that it takes two players to pass a motion, and in fact, a motion will pass if *any* two of the three players vote for it. Remarkably, this weighted voting system operates exactly as if each shareholder had one vote with two votes required to pass a motion. Hard as it is to believe, despite the great difference in their votes, all three players have equal power in this weighted voting system. ■

2.2 The Banzhaf Power Index

We are almost ready to formally introduce our first mathematical definition of power for weighted voting systems. This particular definition of power was suggested by John Banzhaf[1] in 1965.

Let's consider once again the weighted voting system [101: 99, 98, 3]. We will use this weighted voting system to introduce some important concepts.

Which sets of players can combine their votes and carry a motion? Looking at the numbers, we can see that there are four such sets:

John Banzhaf (1940–   ).

---

[1]John F. Banzhaf III (1940–   ) is a law professor at George Washington University and the founder and executive director of Action on Smoking and Health, a national antismoking organization. When he proposed his original idea for the Banzhaf power index in an article entitled "Weighted Voting Doesn't Work," Banzhaf was interested in issues of equity and fair representation in state and local governing bodies.

- P_1 and P_2 (this pair controls 197 votes)
- P_1 and P_3 (this pair controls 102 votes)
- P_2 and P_3 (this pair controls 101 votes, just enough to win)
- P_1, P_2, and P_3 (this group controls all the votes)

From now on we will use the standard terminology of voting theory and call any set of players that might join forces to vote together a **coalition**. (We use the word *coalition* in a rather generous way and will allow for even single-player coalitions.) The total number of votes controlled by a coalition is called the **weight of the coalition**. Of course, some coalitions have enough votes to win and some don't. Quite naturally, we call the former **winning coalitions** and the latter **losing coalitions**. The coalition consisting of all the players is called the **grand coalition**. Since a grand coalition controls all the votes, it is always a winning coalition.

Since coalitions are just sets of players, the most convenient way to describe coalitions mathematically is to use set notation. For example, the coalition consisting of players P_1 and P_2 can be written as the set $\{P_1, P_2\}$, the coalition consisting of just player P_2 by itself can be written as the set $\{P_2\}$, and so on.

Table 2-1 summarizes the facts regarding the coalitions for the weighted voting system [101: 99, 98, 3]. If we analyze the winning coalitions in Table 2-1, we notice that in coalitions 4, 5, and 6 both players are *critical* for the win—if either player were to leave the coalition, the coalition would no longer have the votes to carry a motion—while in coalition 7 no *single* player is critical to the win—if a player were to leave the coalition, the coalition would have enough votes to carry a motion.

TABLE 2-1 The Seven Possible Coalitions for [101: 99, 98, 3]

	Coalition	Coalition weight	Win or lose	Critical players
1	$\{P_1\}$	99	Lose	Not applicable
2	$\{P_2\}$	98	Lose	Not applicable
3	$\{P_3\}$	3	Lose	Not applicable
4	$\{P_1, P_2\}$	197	Win	P_1 and P_2
5	$\{P_1, P_3\}$	102	Win	P_1 and P_3
6	$\{P_2, P_3\}$	101	Win	P_2 and P_3
7	$\{P_1, P_2, P_3\}$	200	Win	None

We will look for players whose desertion turns a winning coalition into a losing coalition, and we will call such a player a **critical player** for the coalition. Notice that a winning coalition can have more than one critical player, and occasionally a winning coalition has no critical players. Losing coalitions never have critical players.

The critical-player concept is the basis for the definition of the **Banzhaf power index**. Banzhaf's key idea is that a player's power is proportional to the number of coalitions for which that player is critical, so that the more often the player is critical, the more power he or she holds.

We know from Table 2-1 that in the weighted voting system [101: 99, 98, 3] each player is critical twice, so they all have equal power. Since there are three players, we can say that each player holds one-third of the power or, more formally, that each has a Banzhaf power index equal to $\frac{1}{3}$.

We will now describe a general method for finding the Banzhaf power index of a player in a weighted voting system with N players.

FINDING THE BANZHAF POWER INDEX OF PLAYER P

- **Step 1.** Make a list of all possible coalitions.
- **Step 2.** Determine which of the above are winning coalitions.
- **Step 3.** In each winning coalition, determine which of the players are *critical* players.
- **Step 4.** Count the total number of times player P is critical. (Let's call this number B.)
- **Step 5.** Count the total number of times all players are critical. (Let's call this number T.)

The Banzhaf power index of player P is then given by the fraction B/T. It represents the proportion of times that player P is critical out of all the times that players are critical.

EXAMPLE 2.9

Foreman & Sons is a family-owned corporation. Three generations of Foremans (George I, George II, and George III) are involved in its management, but, their names notwithstanding, the Foremans are not all the same. When it comes to making final decisions, George I has 3 votes, George II has 2 votes, and George III has 1 vote. A majority of 4 (out of the 6 possible votes) is needed to carry a motion. How is the power divided among the three Georges?

What we have here is the weighted voting system [4: 3, 2, 1]. To find the Banzhaf power index of each player we follow the five steps described in the preceding box. (For consistency, we will use P_1 for George I, P_2 for George II, and P_3 for George III.)

- **Step 1.** There are 7 possible coalitions. They are $\{P_1\}$, $\{P_2\}$, $\{P_3\}$, $\{P_1, P_2\}$, $\{P_1, P_3\}$, $\{P_2, P_3\}$, $\{P_1, P_2, P_3\}$.
- **Step 2.** The winning coalitions are $\{P_1, P_2\}$, $\{P_1, P_3\}$, and $\{P_1, P_2, P_3\}$.
- **Step 3.**

Winning coalitions	Critical players
$\{P_1, P_2\}$	P_1 and P_2
$\{P_1, P_3\}$	P_1 and P_3
$\{P_1, P_2, P_3\}$	P_1 only

- **Step 4.**

  P_1 is critical three times.

  P_2 is critical one time.

  P_3 is critical one time.

- **Step 5.** Adding the numbers in Step 4, we get the total number of times the players are critical, which is five.

The Banzhaf power index of each of the players is

$P_1: \frac{3}{5}$, $P_2: \frac{1}{5}$, $P_3: \frac{1}{5}$

Surprisingly, P_2 and P_3 have the same amount of power! ■

We will refer to the complete listing of the Banzhaf power indexes as the **Banzhaf power distribution** of a weighted voting system. It is a common practice to write power indexes as percentages, rather than fractions. Percentagewise, the Banzhaf power distribution of the weighted voting system in Example 2.9 is

$$P_1: 60\%, \quad P_2: 20\%, \quad P_3: 20\%$$

EXAMPLE 2.10

Among the most important decisions a professional basketball team must make is the drafting of college players. In many cases the decision as to whether to draft a specific player is made through weighted voting. Take, for example, the case of the Akron Flyers. In their system, the head coach (HC) has 4 votes, the general manager (GM) has 3 votes, the director of scouting operations (DS) has 2 votes, and the team psychiatrist (TP) has 1 vote. Of the 10 votes cast, a simple majority of 6 votes is required for a yes vote on a player to be drafted. In essence, the Akron Flyers operate as the weighted voting system $[6: 4, 3, 2, 1]$.

We will now find the Banzhaf power distribution of this weighted voting system. Table 2-2 shows the 15 possible coalitions, which ones are winning and which are losing coalitions, and, for each winning coalition, the critical players (underlined).

TABLE 2-2 **The 15 Coalitions for the Akron Flyers Management Team with Critical Players Underlined**

Coalition	Weight	Win or lose
$\{HC\}$	4	Lose
$\{GM\}$	3	Lose
$\{DS\}$	2	Lose
$\{TP\}$	1	Lose
$\{\underline{HC}, \underline{GM}\}$	7	Win
$\{\underline{HC}, \underline{DS}\}$	6	Win
$\{HC, TP\}$	5	Lose
$\{GM, DS\}$	5	Lose
$\{GM, TP\}$	4	Lose
$\{DS, TP\}$	3	Lose
$\{\underline{HC}, \underline{GM}, DS\}$	9	Win
$\{\underline{HC}, \underline{GM}, TP\}$	8	Win
$\{\underline{HC}, \underline{DS}, TP\}$	7	Win
$\{\underline{GM}, \underline{DS}, \underline{TP}\}$	6	Win
$\{HC, GM, DS, TP\}$	10	Win

All we have to do now is count the number of times each player is underlined and divide by the total number of underlines. The Banzhaf power distribution is

$$HC: \tfrac{5}{12} = 41\tfrac{2}{3}\%$$
$$GM: \tfrac{3}{12} = 25\%$$
$$DS: \tfrac{3}{12} = 25\%$$
$$TP: \tfrac{1}{12} = 8\tfrac{1}{3}\%$$

Note that the power indexes always add up to 1. This fact provides a useful check on your calculations.

How Many Coalitions?

Before we go on to the next example, let's take a brief detour and consider the following mathematical question: For a given number of players, how many different coalitions are possible? Here, our identification of coalitions with sets will come in particularly handy. Except for the empty subset { }, we know that every other subset of the set of players can be identified with a different coalition. This means that we can count the total number of coalitions by counting the number of subsets and subtracting one. So, how many subsets does a set have?

A careful look at Table 2-3 shows us that each time we add a new element we are doubling the number of subsets—the same subsets we had before we added the element plus an equal number consisting of each of these subsets but with the new element thrown in.

TABLE 2-3   **The Subsets of a Set**

Set	$\{P_1, P_2\}$	$\{P_1, P_2, P_3\}$		$\{P_1, P_2, P_3, P_4\}$		$\{P_1, P_2, P_3, P_4, P_5\}$
Number of subsets	4	8		16		32
Subsets	{ }	{ }	$\{P_3\}$	{ }	$\{P_4\}$	The 16 subsets from the previous column along with each of these with P_5 thrown in.
	$\{P_1\}$	$\{P_1\}$	$\{P_1, P_3\}$	$\{P_1\}$	$\{P_1, P_4\}$	
	$\{P_2\}$	$\{P_2\}$	$\{P_2, P_3\}$	$\{P_2\}$	$\{P_2, P_4\}$	
	$\{P_1, P_2\}$	$\{P_1, P_2\}$	$\{P_1, P_2, P_3\}$	$\{P_1, P_2\}$	$\{P_1, P_2, P_4\}$	
				$\{P_3\}$	$\{P_3, P_4\}$	
				$\{P_1, P_3\}$	$\{P_1, P_3, P_4\}$	
				$\{P_2, P_3\}$	$\{P_2, P_3, P_4\}$	
				$\{P_1, P_2, P_3\}$	$\{P_1, P_2, P_3, P_4\}$	

Since each time we add a new player we are doubling the number of subsets, we will find it convenient to think in terms of powers of 2. Table 2-4 summarizes what we have learned.

TABLE 2-4   **The Number of Possible Coalitions**

Players	Number of subsets	Number of coalitions
P_1, P_2	$4 = 2^2$	$2^2 - 1 = 3$
P_1, P_2, P_3	$8 = 2^3$	$2^3 - 1 = 7$
P_1, P_2, P_3, P_4	$16 = 2^4$	$2^4 - 1 = 15$
P_1, P_2, P_3, P_4, P_5	$32 = 2^5$	$2^5 - 1 = 31$
$\vdots$	$\vdots$	$\vdots$
$P_1, P_2, \ldots, P_N$	2^N	$2^N - 1$

EXAMPLE 2.11

The disciplinary committee at George Washington High School has five members: the principal (P_1), the vice principal (P_2), and three teachers $(P_3, P_4,$ and $P_5)$. When voting on a specific disciplinary action the principal has three votes, the vice principal has two votes, and each of the teachers has one vote. A total of five votes are needed for a motion to carry. We can describe the disciplinary committee as the voting system $[5: 3, 2, 1, 1, 1]$.

We now know that with five players there are 31 possible coalitions. Rather than plow straight ahead and list them all, we can sometimes save ourselves a lot of work by figuring out directly which are the winning coalitions. Table 2–5 shows the winning coalitions only, with the critical players in each coalition underlined. We leave it to the reader to verify the details.

TABLE 2-5   Winning Coalitions for Example 2.11 with Critical Players Underlined

Winning coalitions	Comments
$\{\underline{P_1}, \underline{P_2}\}$	Only possible winning two-player coalition.
$\{\underline{P_1}, \underline{P_2}, P_3\}$ $\{\underline{P_1}, \underline{P_2}, P_4\}$ $\{\underline{P_1}, \underline{P_2}, P_5\}$ $\{\underline{P_1}, \underline{P_3}, \underline{P_4}\}$ $\{\underline{P_1}, \underline{P_3}, \underline{P_5}\}$ $\{\underline{P_1}, \underline{P_4}, \underline{P_5}\}$	Winning three-player coalitions must contain P_1 plus any two other players.
$\{\underline{P_1}, P_2, P_3, P_4\}$ $\{\underline{P_1}, P_2, P_3, P_5\}$ $\{\underline{P_1}, P_2, P_4, P_5\}$ $\{\underline{P_1}, P_3, P_4, P_5\}$ $\{\underline{P_2}, \underline{P_3}, \underline{P_4}, \underline{P_5}\}$	All four-player coalitions are winning coalitions.
$\{P_1, P_2, P_3, P_4, P_5\}$	The grand coalition always wins.

The Banzhaf power distribution of the disciplinary committee is

Principal(P_1): $\qquad\qquad\qquad \frac{11}{25} = 44\%$

Vice principal(P_2): $\qquad\qquad \frac{5}{25} = 20\%$

Teachers$(P_3, P_4,$ and $P_5)$: $\qquad \frac{3}{25} = 12\%$ each   ■

EXAMPLE 2.12

The Tasmania State University Promotion and Tenure committee consists of five members: the dean (D) and four other faculty members of equal standing $(F_1, F_2, F_3,$ and $F_4)$. In this committee motions are carried by strict majority, but the dean never votes except to break a 2-2 tie. How is power distributed in this voting system?

While in this example we are not given the weights of the various players, we can still proceed in the usual manner. In the coalitions with three players (three faculty or two faculty plus the dean) each of the players is critical. In the only other possible winning coalition (four faculty) none of the players is critical. Table 2-6 shows the winning coalitions with the critical players underlined.

TABLE 2-6	Winning Coalitions for Example 2.12 with Critical Players Underlined
Winning coalitions without the dean	**Winning coalitions with the dean**
$\{\underline{F_1}, \underline{F_2}, \underline{F_3}\}$	$\{\underline{D}, \underline{F_1}, \underline{F_2}\}$
$\{\underline{F_1}, \underline{F_2}, \underline{F_4}\}$	$\{\underline{D}, \underline{F_1}, \underline{F_3}\}$
$\{\underline{F_1}, \underline{F_3}, \underline{F_4}\}$	$\{\underline{D}, \underline{F_1}, \underline{F_4}\}$
$\{\underline{F_2}, \underline{F_3}, \underline{F_4}\}$	$\{\underline{D}, \underline{F_2}, \underline{F_3}\}$
$\{F_1, F_2, F_3, F_4\}$	$\{\underline{D}, \underline{F_2}, \underline{F_4}\}$
	$\{\underline{D}, \underline{F_3}, \underline{F_4}\}$

The Banzhaf power distribution in this committee is

$D: \frac{6}{30} = 20\%$

$F_1: \frac{6}{30} = 20\%$

$F_2: \frac{6}{30} = 20\%$

$F_3: \frac{6}{30} = 20\%$

$F_4: \frac{6}{30} = 20\%$

Surprise! All the members (including the dean) have the same amount of power.

An interesting variation of Example 2.12 occurs in the U.S. Senate, where the vice president of the United States votes only to break a tie. An analysis similar to the one in Example 2.12 would show that, assuming all 100 senators are voting, the vice president has exactly the same amount of power as any other member of the senate.

2.3 Applications of the Banzhaf Power Index

The Nassau County Board of Supervisors, New York. John Banzhaf first introduced the Banzhaf power index in 1965 in an analysis of how power was distributed in the Board of Supervisors of Nassau County, New York. Although Banzhaf was a lawyer, it was his mathematical analysis of power in the Nassau County Board that provided the legal basis for a series of lawsuits involving the mathematics of weighted voting systems and its implications regarding the "equal protection" guarantee of the Fourteenth Amendment (see Project D).

Nassau County is divided into six different districts, and, based on 1964 population figures, a total of 115 votes were allocated to the districts, with 58 votes needed to pass a motion. Table 2–7 shows the names of the districts and their allocation of votes.

TABLE 2-7	**Nassau County Board (1964)**
District	**Votes in 1964**
Hempstead #1	31
Hempstead #2	31
Oyster Bay	28
North Hempstead	21
Long Beach	2
Glen Cove	2

In effect, the Nassau County Board of Supervisors operated as the weighted voting system [58: 31, 31, 28, 21, 2, 2]. So far, so good, but what about the power of each district? In his lawsuit, Banzhaf argued that in this instance, all the power in the County Board was concentrated in the hands of the top three districts—Hempstead #1, Hempstead #2, and Oyster Bay. After a moment's reflection we can see why this was so: No winning coalition was possible without two of the top three players in it, and since any two of the top three already formed a winning coalition, none of the last three players could have ever been critical players. (We leave it to the reader to verify all the details—see Exercise 21.) The long and the short of it was that, as Banzhaf successfully argued, this County Board was in practice a three-member board, with Hempstead #1, Hempstead #2, and Oyster Bay each having one-third of the power, and North Hempstead, Glen Cove, and Long Beach having absolutely no power at all!

Based on Banzhaf's analysis, the number of votes allocated to each district was changed several times since 1965. Since 1994, the Nassau County Board has operated as the weighted voting system [65: 30, 28, 22, 15, 7, 6] (see Exercise 22).

The United Nations Security Council. The main body responsible for maintaining the international peace and security of nations is the United Nations Security Council. The Security Council is a classic example of a weighted voting system. It consists of fifteen voting nations—five of them are the *permanent* members (Britain, China, France, Russia, and the United States); the other ten nations are *nonpermanent* members appointed for a two-year period on a rotating basis. To pass a motion in the Security Council requires a yes vote from each of the permanent members (in effect giving each permanent member *veto power*) plus additional yes votes from at least four of the ten nonpermanent members. Thus, the winning coalitions consist of all five of the permanent members plus four or more nonpermanent members. There is a total of 848 such coalitions (Exercise 63). In each of these winning coalitions, each permanent member is critical. The nonpermanent members are only critical in the minimal winning coalitions (five permanent and exactly four nonpermanent members). There are 210 of these (Exercise 63). The total number of times all players are critical is 5080 (Exercise 63). With all this information, we can conclude that the

The U.N. Security Council: a weighted voting system in which five players have most of the power and the rest have hardly any power.

Banzhaf power index of each permanent member is 848/5080, or approximately 16.7%; and the Banzhaf power index of each nonpermanent member is 84/5080, or approximately 1.65%. Notice the discrepancy in power between the permanent and nonpermanent members: A permanent member has more than ten times as much power as a nonpermanent member. One has to wonder if this was the original intent of the United Nations charter or perhaps a miscalculation based on a less than clear understanding of the mathematics of weighted voting.

The European Union. Since 1993, the principal nations of Europe have joined forces in a political and economic confederation called the European Union (EU). Currently, the EU consists of 15 member nations (with plans to expand soon to 21 nations and eventually to 26 nations). The main decision-making body of the EU is the Council of Ministers, a unique legislative body in which member nations are not all equal in terms of votes. The 15 member nations and their respective weights in the Council are France, Germany, Italy, and the United Kingdom (10 votes each); Spain (8 votes); Belgium, Greece, Netherlands, and Portugal (5 votes each); Austria and Sweden (4 votes each); Denmark, Finland, and Ireland (3 votes each); Luxembourg (2 votes). The total number of votes in the Council is 87, and the quota is $q = 62$.

The EU Council is a classic example of a weighted voting system—in fact, as a legislative body, it is nothing more than the weighted voting system [62: 10, 10, 10, 10, 8, 5, 5, 5, 5, 4, 4, 3, 3, 3, 2]. In theory, we could calculate the Banzhaf power distribution of the Council using pencil and paper and the methods described in this section, but that might not be such a hot idea—there is a total of 2549 winning coalitions, and a total of 16,565 critical players lurking in these coalitions. Fortunately, it is not hard to program a computer to do these calculations, and the results are given in Table 2–8.

We can determine how well a weighted voting system accomplishes its intended purpose by matching votes and power. The easiest way to do this is to compare percentages of votes and percentages of power. When we look at the two percentage columns in Table 2-8, we see that there is an extremely close match, an indication that as a weighted voting system, the European Union Council works remarkably well.

TABLE 2-8	Banzhaf Power in the European Union ($q = 62$)			
	Votes		Banzhaf Power	
Country	Number	Percent	Index	Percent
France Germany Italy U.K.	10	11.49%	$\dfrac{1849}{16{,}565}$	11.16%
Spain	8	9.20%	$\dfrac{1531}{16{,}565}$	9.24%
Belgium Greece Netherlands Portugal	5	5.75%	$\dfrac{973}{16{,}565}$	5.87%
Austria Sweden	4	4.60%	$\dfrac{793}{16{,}565}$	4.79%
Denmark Finland Ireland	3	3.45%	$\dfrac{595}{16{,}565}$	3.59%
Luxembourg	2	2.30%	$\dfrac{375}{16{,}565}$	2.26%

2.4 The Shapley-Shubik Power Index

In this section we will discuss a different approach to measuring power, first proposed jointly by Lloyd Shapley and Martin Shubik[2] in 1954. The key difference between the Shapley-Shubik interpretation of power and Banzhaf's centers around the concept of a **sequential coalition**. In the Shapley-Shubik method, coalitions are assumed to be formed sequentially: Every coalition starts with a first player, who may then be joined by a second player, then a third, and so on. Thus, to an already complicated situation we are adding one more wrinkle—the question of the order in which the players joined the coalition.

Let's illustrate the difference with a simple example. According to the Banzhaf interpretation of power, a coalition such as $\{P_1, P_2, P_3\}$ means that P_1, P_2, and P_3 have joined forces and will vote together. We don't care who joined the coalition when. According to the Shapley-Shubik interpretation of power, the

[2]Lloyd Shapley and Martin Shubik became lifelong friends and collaborators while graduate students at Princeton University. Shapley went on to become one of the founders of the mathematical theory of games while at the Rand Corporation (for more on Shapley, see the biographical profile on p. 71). Shubik is a Professor of Economics at Yale University.

same three players can form six different sequential coalitions: $\langle P_1, P_2, P_3 \rangle$ (this means that P_1 started the coalition, then P_2 joined in, and last came P_3); $\langle P_1, P_3, P_2 \rangle$; $\langle P_2, P_1, P_3 \rangle$; $\langle P_2, P_3, P_1 \rangle$; $\langle P_3, P_1, P_2 \rangle$; $\langle P_3, P_2, P_1 \rangle$. The six sequential coalitions of three players are graphically illustrated in Fig. 2-1.

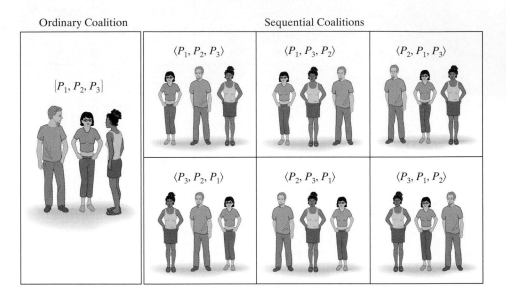

Ordinary Coalition   Sequential Coalitions

$\{P_1, P_2, P_3\}$

$\langle P_1, P_2, P_3 \rangle$ $\langle P_1, P_3, P_2 \rangle$ $\langle P_2, P_1, P_3 \rangle$

$\langle P_3, P_2, P_1 \rangle$ $\langle P_2, P_3, P_1 \rangle$ $\langle P_3, P_1, P_2 \rangle$

FIGURE 2-1

Note the change in notation: From now on the notation $\langle \rangle$ will indicate that we are dealing with a sequential coalition (that is, we care about the order in which the players are listed).

Factorials

It is now time to consider another one of those *How many?* questions. For a given number of players *N, how many sequential coalitions containing the N players are there*? We have just seen that with three players there are six sequential coalitions. What happens if we have four players? We could try to write down all of the sequential coalitions, a somewhat tedious task. Instead, let's argue as follows: To fill the first slot in a coalition we have 4 choices (any one of the 4 players); to fill the second slot we have 3 choices (any one of the players except the one in the first slot); to fill the third slot we have only 2 choices; and to fill the last slot we have only 1 choice. We can now combine these choices by multiplying them. Thus, the total number of possible sequential coalitions with four players turns out to be $4 \times 3 \times 2 \times 1 = 24$.

The one question that may still remain is, Why did we multiply? The answer lies in a basic rule of mathematics called the **multiplication rule**: *If there are m different ways to do X and n different ways to do Y, then X and Y together can be done in m × n different ways.* For example, if an ice cream shop offers 2 different types of cones and 3 different flavors of ice cream, then according to the multiplication rule there are $2 \times 3 = 6$ different cone/flavor combinations. Figure 2-2 shows why this is so. We will discuss the multiplication rule and its uses in greater detail in Chapter 15.

If we have 5 players, following up on our previous argument, we can count on a total of $5 \times 4 \times 3 \times 2 \times 1 = 120$ sequential coalitions. In general, the number of sequential coalitions with *N* players is $1 \times 2 \times 3 \times \cdots \times N$.

3 flavors

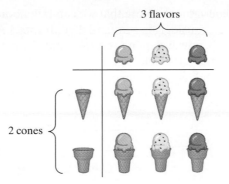

2 cones

FIGURE 2-2

Numbers of the form $1 \times 2 \times 3 \times \cdots \times N$ are among the most important numbers in mathematics and will show up several times in this book. The number $1 \times 2 \times 3 \times \cdots \times N$ is called the **factorial** of N and is written in the shorthand form $N!$. The factorial of 5, for example, is written $5!$ and equals $1 \times 2 \times 3 \times 4 \times 5 = 120$, while $10! = 3,628,800$ (check it out!).

> The number of sequential coalitions with N players is
> $$N! = 1 \times 2 \times 3 \times \cdots \times N.$$

Back to the Shapley-Shubik Power Index

Suppose that we have a weighted voting system with N players. We know from the preceding discussion that there is a total of $N!$ different sequential coalitions containing *all* the players. In each of these coalitions there is one player that tips the scales—the moment that player joins the coalition, the coalition changes from a losing to a winning coalition (see Fig. 2-3). We call such a player a **pivotal player** for the sequential coalition. The underlying principle of the Shapley-Shubik approach is that the pivotal player deserves special recognition. After all, the players who came before the pivotal player did not have enough votes to carry a motion, and the players who came after the pivotal player are a bunch of

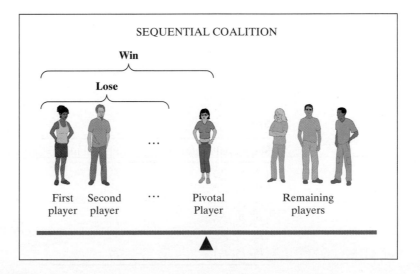

FIGURE 2-3
The pivotal player tips the scales.

Johnny-come-latelies. (Note that we can talk about "before" and "after" only because we are considering sequential coalitions.) According to Shapley and Shubik, a player's power depends on the total number of times that player is pivotal in relation to all other players.

The formal description of the procedure for finding the **Shapley-Shubik power index** of any player in a generic weighted voting system with N players is as follows:

FINDING THE SHAPLEY-SHUBIK POWER INDEX OF PLAYER P

- **Step 1.** Make a list of all sequential coalitions containing all N players. (There are $N!$ of them.)
- **Step 2.** In each sequential coalition determine *the* pivotal player. (There is one in each sequential coalition.)
- **Step 3.** Count the total number of times P is pivotal and call this number S.

    The Shapley-Shubik power index of P is then given by the fraction $S/N!$.

A listing of the Shapley-Shubik power indexes of all the players gives the **Shapley-Shubik power distribution** of the weighted voting system.

We will illustrate the procedure for computing Shapley-Shubik power by revisiting Example 2.9.

EXAMPLE 2.13

Let's consider, once again, an analysis of power at Foreman & Sons, a family company that operates as the weighted voting system [4: 3, 2, 1]. This time we will use the Shapley-Shubik interpretation of power.

- **Step 1.** There are $3! = 6$ sequential coalitions of the three players. They are

$\langle P_1, P_2, P_3 \rangle$
$\langle P_1, P_3, P_2 \rangle$
$\langle P_2, P_1, P_3 \rangle$
$\langle P_2, P_3, P_1 \rangle$
$\langle P_3, P_1, P_2 \rangle$
$\langle P_3, P_2, P_1 \rangle$

- **Step 2.**

Sequential coalition	Pivotal player
$\langle P_1, P_2, P_3 \rangle$	P_2
$\langle P_1, P_3, P_2 \rangle$	P_3
$\langle P_2, P_1, P_3 \rangle$	P_1
$\langle P_2, P_3, P_1 \rangle$	P_1
$\langle P_3, P_1, P_2 \rangle$	P_1
$\langle P_3, P_2, P_1 \rangle$	P_1

■ **Step 3.**

P_1 is pivotal four times.

P_2 is pivotal one time.

P_3 is pivotal one time.

The Shapley-Shubik power distribution is

$$P_1: \frac{4}{6} = 66\frac{2}{3}\%, \quad P_2: \frac{1}{6} = 16\frac{2}{3}\%, \quad P_3: \frac{1}{6} = 16\frac{2}{3}\%$$

Note that the power distribution is different from the Banzhaf power distribution (P_1: 60%; P_2: 20%; P_3: 20%) obtained in Example 2.9. Under the Shapley-Shubik interpretation of power, P_1 has even more power—his son and grandson each have a little less. ■

EXAMPLE 2.14

We will now reconsider Example 2.10, the one about the Akron Flyers system for picking players in the draft. The weighted voting system in this example is [6: 4, 3, 2, 1], and we will now find its Shapley-Shubik power distribution.

There are 24 different sequential coalitions involving the 4 players. They are listed in Table 2-9 with the pivotal player underlined.

TABLE 2-9   **The 24 Sequential Coalitions for Example 2.14 with Pivotal Players Underlined**

$\langle HC, \underline{GM}, DS, TP \rangle$	$\langle GM, \underline{HC}, DS, TP \rangle$	$\langle DS, \underline{HC}, GM, TP \rangle$	$\langle TP, HC, \underline{GM}, DS \rangle$
$\langle HC, \underline{GM}, TP, DS \rangle$	$\langle GM, \underline{HC}, TP, DS \rangle$	$\langle DS, \underline{HC}, TP, GM \rangle$	$\langle TP, HC, \underline{DS}, GM \rangle$
$\langle HC, \underline{DS}, GM, TP \rangle$	$\langle GM, DS, \underline{HC}, TP \rangle$	$\langle DS, GM, \underline{HC}, TP \rangle$	$\langle TP, GM, \underline{HC}, DS \rangle$
$\langle HC, \underline{DS}, TP, GM \rangle$	$\langle GM, DS, \underline{TP}, HC \rangle$	$\langle DS, GM, \underline{TP}, HC \rangle$	$\langle TP, GM, \underline{DS}, HC \rangle$
$\langle HC, TP, \underline{GM}, DS \rangle$	$\langle GM, TP, \underline{HC}, DS \rangle$	$\langle DS, TP, \underline{HC}, GM \rangle$	$\langle TP, DS, \underline{HC}, GM \rangle$
$\langle HC, TP, DS, GM \rangle$	$\langle GM, TP, \underline{DS}, HC \rangle$	$\langle DS, TP, \underline{GM}, HC \rangle$	$\langle TP, DS, \underline{GM}, HC \rangle$

The Shapley-Shubik power distribution is

$$HC: \frac{10}{24} = 41\frac{2}{3}\%$$
$$GM: \frac{6}{24} = 25\%$$
$$DS: \frac{6}{24} = 25\%$$
$$TP: \frac{2}{24} = 8\frac{1}{3}\%$$
■

It is worth mentioning that in the preceding example the Shapley-Shubik power distribution turns out to be exactly the same as the Banzhaf power distribution (see Example 2.10). If nothing else, this shows that it is not impossible for these power distributions to agree. In general, however, for randomly chosen real-life situations, it is very unlikely that the Banzhaf and Shapley-Shubik methods will give the same answer.

EXAMPLE 2.15

The city of Cleansburg operates under what is called a "strong-mayor" system. The strong-mayor system in Cleansburg works like this: There are five council members, namely the mayor and four "ordinary" council members. A motion can pass only if the mayor and at least two other council members vote for it, or alternatively, if all four of the ordinary council members vote for it. (This situation is usually described by saying that the *mayor has veto power but a unanimous vote of the other council members can override the mayor's veto*.)

Common sense tells us that under these rules, the four ordinary council members have the same amount of power but the mayor has more. We will now use the Shapley-Shubik interpretation of power to determine exactly how much more.

Since there are 5 players in this voting system, there are 5! = 120 sequential coalitions to consider. Obviously, we will want to find some kind of a shortcut. We will first try to find the Shapley-Shubik power index of the mayor. In what position does the mayor have to be in order to be the pivotal player in a sequential coalition?

Does the mayor have to be in first place? No way! No player can be pivotal in the first position unless he or she is a dictator. Second? No—an ordinary council member plus the mayor are not enough to carry a motion. Third place? Yes! If the mayor is in third place, he is the pivotal player in that sequential coalition (see Fig. 2-4a). Likewise, if the mayor is in fourth place, he is the pivotal player in that sequential coalition, because the three preceding ordinary members are not enough to carry a motion (see Fig. 2-4b). Finally, when the mayor is in the last (fifth) place in a sequential coalition, he is not the pivotal player—the four ordinary members preceding him do have enough votes to carry the motion (see Fig. 2-4c).

Now comes a critical question: In how many (of these 120) sequential coalitions is the mayor in first place? second place? ... fifth place? The symmetry of the positions tells us that there should be just as many sequential coalitions in which the mayor is in first place as in any other place. It follows that the 120 sequential coalitions can be divided into 5 groups of 24—24 with the mayor in first place, 24 with the mayor in second place, etc.

We finally have a handle on the mayor. The mayor is the pivotal player in all sequential coalitions in which he is either in the third or fourth position, and there are 24 of each. Thus, the Shapley-Shubik power index of the mayor is

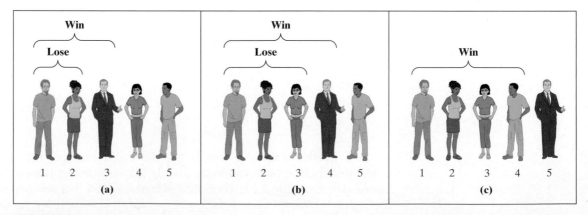

FIGURE 2-4
The mayor is the pivotal player when he is in third or fourth position only.

48/120 = 2/5 = 40%. Since the four ordinary council members must share the remaining 60% of the power equally, it follows that each of them must have a Shapley-Shubik power index of 15%. We are done! ■

For the purposes of comparison, the reader is encouraged to calculate the Banzhaf power distribution of the Cleansburg city council (see Exercise 64).

2.5 Applications of the Shapley-Shubik Power Index

The Electoral College. Calculating the Shapley-Shubik power index of the states in the electoral college is no easy task. There are 51! sequential coalitions, a number so large (67 digits long) we don't even have a name for it. Checking every possible sequential coalition would take literally thousands of years, so a direct approach is out of the question. There are, however, some sophisticated mathematical shortcuts, which, when coupled with a computer and the right kind of software, allow the calculations to be done quite efficiently. The appendix at the end of this chapter shows both the Banzhaf and the Shapley-Shubik power indexes for each state. Comparing the Banzhaf and the Shapley-Shubik power indexes shows that there is a very small difference between the two. This example illustrates the fact that in some situations the Banzhaf and Shapley-Shubik power indexes give essentially the same answer. The next example illustrates a very different situation.

The United Nations Security Council. As mentioned earlier, the United Nations Security Council consists of fifteen member nations—five are permanent members and ten are nonpermanent members appointed on a rotating basis. The voting rules are that a motion can pass only if it has the support of each of the five permanent members (they have *veto power*) plus at least four of the ten nonpermanent members. This arrangement makes the Security Council a weighted voting system. In fact, the voting rules are equivalent to giving each permanent member 7 votes, each nonpermanent member 1 vote, and making the quota equal to 39 votes (see Exercise 63). We will sketch a rough outline of how the Shapley-Shubik power distribution of the Security Council can be calculated. (The details, while not terribly difficult, go beyond the scope of this book.) First, there are 15! sequential coalitions involving the 15 members. This is about *1.3 trillion sequential coalitions*. Second, a nonpermanent member can be pivotal in one of these sequential coalitions *only if it is the 9th player in the coalition, preceded by all five of the permanent members and three nonpermanent members*. There are approximately 2.44 billion sequential coalitions in which this happens. It follows that any one of the nonpermanent members is pivotal in approximately 2.44 billion of the 1.3 trillion sequential coalitions, giving it a *Shapley-Shubik power index of 0.19%* (2.44 billion/1.3 trillion ≈ 0.0019 = 0.19%). This implies that the Shapley-Shubik power of the ten nonpermanent members together adds up to less than 2% of the power. The remaining 98% of the power is divided equally between the five permanent members, giving *each a Shapley-Shubik power index of 19.6%, roughly 100 times the power of a nonpermanent member!*

The European Union. We discussed in the previous section the background of the European Union and the fact that its Council of Ministers operates as the weighted voting system [62: 10, 10, 10, 10, 8, 5, 5, 5, 5, 4, 4, 3, 3, 3, 2]. We already saw that the Banzhaf power of each member nation matches remarkably well the percentage of votes that it holds (see Table 2–8). What about the Shapley-Shubik power? Once again, there isn't much future in trying to calculate the power indexes by hand—the number of sequential coalitions is, as we know, 15!, which doesn't impress very much until we find out that 15! = 1,307,674,368,000 (yes—*one trillion, three hundred and seven billion, six hundred and seventy four million and some spare change*).

In spite of the enormous number of sequential coalitions that need to be considered, the Shapley-Shubik power distribution of the EU Council can be computed using some sophisticated software (a version of the program is available at this book's Web site: *www.prenhall.com/tannenbaum*). The power distribution is summarized in Table 2-10:

TABLE 2-10

Country	Votes	Shapley-Shubik power
France, Germany, Italy, U.K.	10	11.67%
Spain	8	9.55%
Belgium, Greece, Netherlands, Portugal	5	5.52%
Austria, Sweden	4	4.54%
Denmark, Finland, Ireland	3	3.53%
Luxembourg	2	2.07%

Conclusion

In any society, no matter how democratic, some individuals and groups have more power than others. This is simply a consequence of the fact that individuals and groups are not all equal. Diversity is the inherent reason why the concept of power exists.

Power itself comes in many different forms. We often hear cliches such as "In strength lies power" or "Money is power" (and the newer cyber version, "Information is power"). In this chapter we discussed the notion of power as it applies to formal voting situations called *weighted voting systems* and saw how mathematical methods allow us to measure the power of an individual or group by means of a *power index*. In particular, we looked at two different kinds of power indexes: the *Banzhaf power index* and the *Shapley-Shubik power index*.

These indexes provide two different ways to measure power, and while they occasionally agree, they often differ significantly. Of the two, which one is closer to reality?

Unfortunately, there is no simple answer. Both of them are useful, and in some sense the choice is subjective. Perhaps the best way to evaluate them is to think of them as being based on a slightly different set of assumptions. The idea behind the Banzhaf interpretation of power is that players are free to come and go, negotiating their allegiance for power (somewhat like professional athletes since the advent of free agency). Underlying the Shapley-Shubik interpretation of power is the assumption that when a player joins a coalition, he or she is making a commitment to stay. In the latter case a player's power is generated by his ability to be in the right place at the right time.

In practice, the choice of which method to use for measuring power is based on which of the assumptions better fits the specifics of the situation. Contrary to what we've often come to expect, mathematics does not give us the answer, just the tools that might help us make an informed decision.

Lloyd S. Shapley (1923–   )

Lloyd Shapley is one of the giants of modern *game theory*—which, in a nutshell, is the mathematical study of *games*. (The word *game* is used here as a metaphor for any situation involving competition and cooperation among individuals or groups each trying to further their own separate goals. This broad interpretation encompasses not only real games like poker and Monopoly, but also economic, political, and military "games," thus making game theory one of the most important branches of modern applied mathematics.)

Lloyd Stowell Shapley was born in Cambridge, Massachusetts to a prominent scientific family. His father, Harlow Shapley, was a renowned astronomer at Harvard University and director of the Harvard Observatory. Shapley entered Harvard in 1942, but he interrupted his studies to join the Army during World War II. From 1943 to 1945 he served in the Army Air Corps in China, where he was awarded a Bronze Star for breaking the Japanese weather code. In 1945 he returned to Harvard, receiving a Bachelor of Arts in Mathematics in 1948.

Upon graduation from Harvard, Shapley joined the Rand Corporation, a famous think tank in Santa Monica, California, where much of the pioneering research in game theory was being conducted. After a year at Rand, and with a reputation as a budding young star in game theory in tow, Shapley went to Princeton University as a graduate student in mathematics. At Princeton he joined a circle that included some of the most brilliant mathematical minds of his time, including John von Neumann, the father of game theory, and fellow graduate student John Nash, the mathematical genius whose life is portrayed in the Academy Award–winning movie *A Beautiful Mind* (with Russell Crowe playing John Nash).

As a graduate student, Shapley had a reputation as a fierce competitor in everything he took on, including card games and board games, some of his own invention. One of the games Shapley invented with a couple of other graduate students (one of which was Nash and the other was Martin Shubik, who at the time was an economics graduate student as well as Shapley's roommate) was a fiendish board game called *So Long Sucker,* in which players formed coalitions to gang up and double-cross each other. Some of the experiences and observations that Shapley and Shubik (who became life-long friends) derived from their graduate student game-playing days eventually led to their development of the Shapley-Shubik index for measuring power in a weighted voting system (a weighted voting system, after all, is just a simple *game* in which players form coalitions in order to pass or block a specific action by the group).

After receiving his Ph.D. in Mathematics from Princeton in 1953, Shapley returned to the Rand Corporation, where he worked as a research mathematician until 1981. During this time he made many major contributions to game theory, including the invention of *convex* and *stochastic* games, the *Shapley value* of a game, and the *Shapley-Shubik power index*. The citation when he was awarded the Von Neumann Prize in Game Theory in 1981 partly read, "His individual work and his joint research with Martin Shubik has helped build bridges between game theory, economics, political science, and practice."

In 1981 Shapley joined the Mathematics department at UCLA, where he is currently Professor Emeritus of Mathematics and Economics.

KEY CONCEPTS

Banzhaf power distribution	multiplication rule
Banzhaf power index	pivotal player
coalition	player
coalition weight	quota
critical player	sequential coalition
dictator	Shapley-Shubik power distribution
dummy	Shapley-Shubik power index
factorial	veto power
grand coalition	weighted voting system
losing coalition	weight
motion	winning coalition

EXERCISES

WALKING

A. Weighted Voting Systems

1. In the weighted voting system $[13: 7, 4, 3, 3, 2, 1]$, find

    (a) the total number of players

    (b) the total number of votes

    (c) the weight of P_2

    (d) the minimum percentage of the votes needed to pass a motion (rounded to the next whole percent)

2. In the weighted voting system $[31: 12, 8, 6, 5, 5, 5, 2]$, find

    (a) the total number of players

    (b) the total number of votes

    (c) the weight of P_3

    (d) the minimum percentage of the votes needed to pass a motion (rounded to the next whole percent)

3. A committee has four members (P_1, P_2, P_3, and P_4). In this committee P_1 has twice as many votes as P_2; P_2 has twice as many votes as P_3; P_3 has twice as many votes as P_4. Describe the committee as a weighted voting system when the requirements to pass a motion are

    (a) at least two-thirds of the votes

    (b) more than two-thirds of the votes

    (c) at least 80% of the votes

    (d) more than 80% of the votes

4. A committee has six members (P_1, P_2, P_3, P_4, P_5, and P_6). In this committee P_1 has twice as many votes as P_2; P_2 and P_3 have the same number of votes, which is twice as many as P_4; P_4 has twice as many votes as P_5; P_5 and P_6 have the same number of votes. Describe the committee as a weighted voting system when the requirements to pass a motion are

    (a) a simple majority of the votes

    (b) at least three-fourths of the votes

    (c) more than three-fourths of the votes

    (d) at least two-thirds of the votes

    (e) more than two-thirds of the votes

5. Consider the weighted voting system $[q: 10, 6, 5, 4, 2]$.

    (a) What is the smallest value that the quota q can take?

    (b) What is the largest value that the quota q can take?

6. Consider the weighted voting system $[q: 5, 3, 2, 2, 1, 1]$.

    (a) What is the smallest value that the quota q can take?

    (b) What is the largest value that the quota q can take?

7. In each of the following weighted voting systems, determine which players, if any, (i) are dictators; (ii) have veto power; (iii) are dummies.

    (a) $[6: 4, 2, 1]$

    (b) $[6: 7, 3, 1]$

    (c) $[10: 9, 9, 1]$

8. In each of the following weighted voting systems, determine which players, if any, (i) are dictators; (ii) have veto power; (iii) are dummies.

    (a) $[95: 95, 80, 10, 2]$

    (b) $[95: 65, 35, 30, 25]$

    (c) $[48: 32, 16, 8, 4, 2, 1]$

9. In each of the following weighted voting systems, determine which players, if any, (i) are dictators; (ii) have veto power; (iii) are dummies.

    (a) $[19: 9, 7, 5, 3, 1]$

    (b) $[15: 16, 8, 4, 1]$

    (c) $[17: 13, 5, 2, 1]$

    (d) $[25: 12, 8, 4, 2]$

10. In each of the following weighted voting systems, determine which players, if any, (i) are dictators; (ii) have veto power; (iii) are dummies.

    (a) $[27: 12, 10, 4, 2]$

    (b) $[22: 10, 8, 7, 2, 1]$

    (c) $[21: 23, 10, 5, 2]$

    (d) $[15: 11, 5, 2, 1]$

B. Banzhaf Power

11. Consider the weighted voting system $[10: 6, 5, 4, 2]$.

    (a) What is the weight of the coalition formed by P_1 and P_3?

    (b) Write down all winning coalitions.

    (c) Which players are critical in the coalition $\{P_1, P_2, P_3\}$?

    (d) Find the Banzhaf power distribution of this weighted voting system.

12. Consider the weighted voting system $[5: 3, 2, 1, 1]$.

    (a) What is the weight of the coalition formed by P_1 and P_3?

    (b) Which players are critical in the coalition $\{P_1, P_2, P_3\}$?

    (c) Which players are critical in the coalition $\{P_1, P_3, P_4\}$?

    (d) Write down all winning coalitions.

    (e) Find the Banzhaf power distribution of this weighted voting system.

13. **(a)** Find the Banzhaf power distribution of the weighted voting system $[6: 5, 2, 1]$.

    (b) Find the Banzhaf power distribution of the weighted voting system $[3: 2, 1, 1]$. Compare your answers in (a) and (b).

14. **(a)** Find the Banzhaf power distribution of the weighted voting system $[7: 5, 2, 1]$.

    (b) Find the Banzhaf power distribution of the weighted voting system $[5: 3, 2, 1]$. Compare your answers in (a) and (b).

15. **(a)** Find the Banzhaf power distribution of the weighted voting system $[10: 5, 4, 3, 2, 1]$. (If possible, do it without writing down all coalitions—just the winning ones.)

    (b) Find the Banzhaf power distribution of the weighted voting system $[11: 5, 4, 3, 2, 1]$. [*Hint*: Note that the only change from (a) is in the quota, and use this fact to your advantage.]

16. **(a)** Find the Banzhaf power distribution of the weighted voting system $[9: 5, 5, 4, 2, 1]$.

    (b) Find the Banzhaf power distribution of the weighted voting system $[9: 5, 5, 3, 2, 1]$.

17. Consider the weighted voting system $[q: 8, 4, 2, 1]$. Find the Banzhaf power distribution of this weighted voting system when

    (a) $q = 8$

    (b) $q = 9$

    (c) $q = 10$

    (d) $q = 12$

    (e) $q = 14$

18. Consider the weighted voting system $[q: 5, 3, 1]$. Find the Banzhaf power distribution of this weighted voting system when

    (a) $q = 5$

    (b) $q = 6$

    (c) $q = 7$

    (d) $q = 8$

    (e) $q = 9$

19. A business firm is owned by 4 partners, *A, B, C,* and *D.* When making decisions, each partner has one vote and the majority rules, except in the case of a 2-2 tie. Then, the coalition that contains *D* (the partner with the least seniority) loses. What is the Banzhaf power distribution in this partnership?

20. Consider the weighted voting system $[8: 5, 3, 1, 1, 1]$.

    (a) Make a list of all winning coalitions. (*Hint*: There aren't too many!)

    (b) Using (a), find the Banzhaf power distribution of this weighted voting system.

    (c) Suppose that P_1, with 5 votes, sells one of her votes to P_2, resulting in the weighted voting system $[8: 4, 4, 1, 1, 1]$. Find the Banzhaf power distribution of this system.

    (d) Compare the power index of P_1 in (b) and (c). Describe the paradox that occurred.

Exercises 21 and 22 refer to the Nassau County (N.Y.) Board of Supervisors, as discussed in this chapter.

21. In 1964, the Nassau County Board of Supervisors operated as the weighted voting system [58: 31, 31, 28, 21, 2, 2]. Find the Banzhaf power distribution of the 1964 board.

22. By 1994, after a series of court decisions, the votes of the six representatives of the Nassau County Board of Supervisors were changed to 30, 28, 22, 15, 7, and 6, with a quota of 60% of the votes needed to pass a motion.

   (a) Describe the 1994 Nassau County Board as a weighted voting system.

   (b) Find the Banzhaf power distribution of the 1994 Nassau County Board.

C. Shapley-Shubik Power

23. Consider the weighted voting system [16: 9, 8, 7].

   (a) Write down all the sequential coalitions involving all three players.

   (b) In each of the sequential coalitions in (a), underline the pivotal player.

   (c) Find the Shapley-Shubik power distribution of this weighted voting system.

24. Consider the weighted voting system [8: 7, 6, 2].

   (a) Write down all the sequential coalitions involving all three players.

   (b) In each of the sequential coalitions in (a), underline the pivotal player.

   (c) Find the Shapley-Shubik power distribution of this weighted voting system.

25. Find the Shapley-Shubik power distribution of the weighted voting system [5: 3, 2, 1, 1].

26. Find the Shapley-Shubik power distribution of the weighted voting system [60: 32, 31, 28, 21].

27. Find the Shapley-Shubik power distribution of each of the following weighted voting systems.

   (a) [8: 8, 5, 1]

   (b) [8: 7, 5, 2]

   (c) [8: 7, 6, 1]

   (d) [8: 6, 5, 1]

   (e) [8: 6, 5, 3]

28. Find the Shapley-Shubik power distribution of each of the following weighted voting systems.

   (a) [6: 4, 3, 2, 1]

   (b) [7: 4, 3, 2, 1]

   (c) [8: 4, 3, 2, 1]

   (d) [9: 4, 3, 2, 1]

   (e) [10: 4, 3, 2, 1]

29. Consider the weighted voting system $[q: 5, 3, 1]$. Find the Shapley-Shubik power distribution of this weighted voting system when

   (a) $q = 5$

   (b) $q = 6$

     (c) $q = 7$

     (d) $q = 8$

     (e) $q = 9$

30. Consider the weighted voting system $[q: 4, 3, 2]$. Find the Shapley-Shubik power distribution of this weighted voting system when

     (a) $q = 5$

     (b) $q = 6$

     (c) $q = 7$

     (d) $q = 8$

     (e) $q = 9$

31. Find the Shapley-Shubik power distribution of each of the following weighted voting systems.

     (a) $[51: 40, 30, 20, 10]$

     (b) $[59: 40, 30, 20, 10]$ [*Hint*: Compare this situation with the one in (a).]

     (c) $[60: 40, 30, 20, 10]$

32. Find the Shapley-Shubik power distribution of each of the following weighted voting systems.

     (a) $[41: 40, 10, 10, 10]$

     (b) $[49: 40, 10, 10, 10]$ [*Hint*: Compare this situation with the one in (a).]

     (c) $[60: 40, 10, 10, 10]$

33. A business firm is owned by four partners, *A, B, C,* and *D*. When making decisions, each partner has one vote and the majority rules. In case of a 2-2 tie, the tie is broken by going against *D* (i.e., if *D* votes yes, the decision is no, and vice versa). Find the Shapley-Shubik power distribution in this partnership.

34. A business firm is owned by four partners, *A, B, C,* and *D*. When making decisions, each partner has one vote and the majority rules. In case of a 2-2 tie, the tie is broken in favor of *A* (the senior partner). Find the Shapley-Shubik power distribution in this partnership.

D. Miscellaneous

Exercises 35 and 36 refer to the computation of factorials using a calculator. Practically all scientific and business calculators have a factorial key (either x! or n!).

35. Using a calculator, compute each of the following factorials. In cases in which the answer is not exact, give the approximate answer in scientific notation.

     (a) 13!

     (b) 18!

     (c) 24!

36. Using a calculator, compute each of the following factorials. In cases in which the answer is not exact, give the approximate answer in scientific notation.

     (a) 12!

     (b) 15!

     (c) 30!

37. In this exercise, you should do your work without using a calculator.

    (a) Find 10! given that 9! = 362,880.

    (b) Find 19! given that 20! = 2,432,902,008,176,640,000.

    (c) Find 100!/99!.

    (d) Find 9!/6!.

38. In this exercise, you should do your work without using a calculator.

    (a) Find 11! given that 9! = 362,880.

    (b) Find 14! given that 15! = 1,307,674,368,000.

    (c) Find 100!/98!.

    (d) Find 11!/8!.

39. An approximate value of 99! in scientific notation is 9.33262×10^{155}. Give a corresponding value for 100! in scientific notation.

40. An approximate value of 200! in scientific notation is 8.0×10^{374}. Give a corresponding value for 199! in scientific notation.

41. Consider the weighted voting system [18: 6, 4, 3, 3, 2, 1].

    (a) Find the total number of coalitions in this weighted voting system.

    (b) Find the number of winning coalitions in this weighted voting system.

    (c) Find the number of sequential coalitions in this weighted voting system.

42. Consider the weighted voting system [28: 10, 8, 7, 5, 1].

    (a) Find the total number of coalitions in this weighted voting system.

    (b) Find the number of winning coalitions in this weighted voting system.

    (c) Find the number of sequential coalitions in this weighted voting system.

JOGGING

43. Veto power. A player P is said to have veto power if the coalition consisting of all players other than P is a losing coalition. Explain why each of the following is true.

    (a) If P has veto power, then P is a member of every winning coalition.

    (b) If P is a critical member in every winning coalition, then P has veto power.

44. Consider the weighted voting system [21: 6, 5, 4, 3, 2, 1]. (Note that here the quota equals 100% of the votes.)

    (a) How many coalitions are there?

    (b) Write down the winning coalitions only and underline the critical players.

    (c) Find the Banzhaf power index of each player.

    (d) Explain why in any weighted voting system with N players in which the quota equals 100% of the votes, the Banzhaf power index of each player is $1/N$.

45. Consider the weighted voting system [21: 6, 5, 4, 3, 2, 1].

    (a) How many different sequential coalitions are there?

    (b) There is only one way in which a player can be pivotal in one of these sequential coalitions. Describe it.

(c) In how many sequential coalitions is P_6 pivotal?

(d) What is the Shapley-Shubik power index of P_6?

(e) What are the Shapley-Shubik power indexes of the other players?

(f) Explain why in any weighted voting system with N players in which the quota equals 100% of the votes, the Shapley-Shubik power index of each player is $1/N$.

46. The disciplinary board at Tasmania State University is composed of five members, two of which must be faculty and three of which must be students. To pass a motion requires at least three votes, and at least one of the votes must be from a faculty member.

(a) Find the Banzhaf power distribution of the disciplinary board.

(b) Describe the disciplinary board as a weighted voting system $[q: f, f, s, s, s]$.

47. A professional basketball team has four coaches, a head coach (H), and three assistant coaches (A_1, A_2, A_3). Player personnel decisions require at least three yes votes, one of which must be H's.

(a) If we use $[q: h, a, a, a]$ to describe this weighted voting system, find q, h, and a.

(b) Find the Shapley-Shubik power distribution of the weighted voting system.

48. **(a)** Consider the weighted voting system $[22: 10, 10, 10, 10, 1]$. Are there any dummies? Explain your answer.

(b) Without doing any work [but using your answer for (a)], find the Banzhaf and Shapley-Shubik power distributions of this weighted voting system.

(c) Consider the weighted voting system $[q: 10, 10, 10, 10, 1]$. Find all the possible values of q for which P_5 is not a dummy.

(d) Consider the weighted voting system $[34: 10, 10, 10, 10, w]$. Find all positive integers w which make P_5 a dummy.

49. Consider the weighted voting system $[q: 8, 4, 1]$.

(a) What are the possible values of q?

(b) Which values of q result in a dictator? (Who? Why?)

(c) Which values of q result in exactly one player with veto power? (Who? Why?)

(d) Which values of q result in more than one player with veto power? (Who? Why?)

(e) Which values of q result in one or more dummies? (Who? Why?)

50. Consider the weighted voting systems $[9: w, 5, 2, 1]$.

(a) What are the possible values of w?

(b) Which values of w result in a dictator? (Who? Why?)

(c) Which values of w result in a player with veto power? (Who? Why?)

(d) Which values of w result in one or more dummies? (Who? Why?)

51. **(a)** Verify that the weighted voting systems $[12: 7, 4, 3, 2]$ and $[24: 14, 8, 6, 4]$ result in exactly the same Banzhaf power distribution. (If you need to make calculations, do them for both systems side by side and look for patterns.)

(b) Based on your work in (a), explain why the two proportional weighted voting systems $[q: w_1, w_2, \ldots, w_N]$ and $[cq: cw_1, cw_2, \ldots, cw_N]$ always have the same Banzhaf power distribution.

52. **(a)** Verify that the weighted voting systems $[12: 7, 4, 3, 2]$ and $[24: 14, 8, 6, 4]$ result in exactly the same Shapley-Shubik power distribution. (If you need to make calculations, do them for both systems side by side and look for patterns.)

    (b) Based on your work in (a), explain why the two proportional weighted voting systems $[q: w_1, w_2, \ldots, w_N]$ and $[cq: cw_1, cw_2, \ldots, cw_N]$ always have the same Shapley-Shubik power distribution.

53. **A dummy is a dummy is a dummy.** ... This exercise shows that a player that is a dummy is a dummy regardless of which interpretation of power is used.

    (a) Explain why a player that has a Banzhaf power index of 0 (i.e., is never critical) must also have a Shapley-Shubik power index of 0 (i.e., is never pivotal).

    (b) Explain why a player that has a Shapley-Shubik power index of 0 (i.e., is never pivotal) must also have a Banzhaf power index of 0 (i.e., is never critical).

54. Consider the weighted voting system $[q: 5, 4, 3, 2, 1]$.

    (a) For what values of q is there a dummy?

    (b) For what values of q do all players have the same power?

55. The weighted voting system $[6: 4, 2, 2, 2, 1]$ represents a partnership among five people (P_1, P_2, P_3, P_4, and you!). You are the last player (the one with 1 vote), which in this case makes you a dummy! Not wanting to remain a dummy, you offer to buy one vote. Each of the other four partners is willing to sell you one of their votes, and they are all asking the same price. Which partner should you buy from in order to get as much power for your buck as possible? Use the Banzhaf power index for your calculations. Explain your answer.

56. The weighted voting system $[27: 10, 8, 6, 4, 2]$ represents a partnership among five people (P_1, P_2, P_3, P_4, and P_5). You are P_5, the one with two votes. You want to increase your power in the partnership and are prepared to buy one share (1 share = 1 vote) from any of the other partners P_1, P_2, and P_3 are each willing to sell cheap ($1000 for one share), but P_4 is not being quite as cooperative—she wants $5000 for one share. Given that you still want to buy one share, who should you buy it from? Use the Banzhaf power index for your calculations. Explain your answer.

57. The weighted voting system $[18: 10, 8, 6, 4, 2]$ represents a partnership among five people (P_1, P_2, P_3, P_4, and P_5). You are P_5, the one with two votes. You want to increase your power in the partnership and are prepared to buy shares (1 share = 1 vote) from any of the other partners.

    (a) Suppose that each partner is willing to sell one share and they are all asking the same price. Assuming that you decide to buy only one share, which partner should you buy from? Use the Banzhaf power index for your calculations.

    (b) Suppose that each partner is willing to sell two shares and they are all asking the same price. Assuming that you decide to buy two shares from a single partner, which partner should you buy from? Use the Banzhaf power index for your calculations.

(c) If you have the money and the cost per share is fixed, should you buy one share or two shares (from a single person)? Explain.

58. Sometimes in a weighted voting system, two or more players decide to merge—that is to say, to combine their votes and always vote the same way. (Notice that a merger is different from a coalition—coalitions are temporary, whereas mergers are permanent.) For example, if in the weighted voting system $[7:5, 3, 1]$ P_2 and P_3 were to merge, the weighted voting system would then become $[7:5, 4]$. In this exercise, we explore the effects of mergers on a player's power.

   (a) Consider the weighted voting system $[4:3, 2, 1]$. In Example 9 we saw that P_2 and P_3 each have a Banzhaf power index of $\frac{1}{5}$. Suppose that P_2 and P_3 merge and become a single player P^*. What is the Banzhaf power index of P^*?

   (b) Consider the weighted voting system $[5:3, 2, 1]$. Find first the Banzhaf power indexes of players P_2 and P_3 and then the Banzhaf power index of P^* (the merger of P_2 and P_3). Compare.

   (c) Rework the problem in (b) for the weighted voting system $[6:3, 2, 1]$.

   (d) What are your conclusions from (a), (b), and (c)?

59. **Decisive voting systems.** A weighted voting system is called **decisive** if for every losing coalition, the coalition consisting of the remaining players (called the *complement*) must be a winning coalition.

   (a) Show that the weighted voting system $[5:4, 3, 2]$ is decisive.

   (b) Show that the weighted voting system $[3:2, 1, 1, 1]$ is decisive.

   (c) Explain why any weighted voting system with a dictator is decisive.

   (d) Find the number of winning coalitions in a decisive voting system with N players.

60. **Equivalent voting systems.** Two weighted voting systems are **equivalent** if they have the same number of players and exactly the same winning coalitions.

   (a) Show that the weighted voting systems $[8:5, 3, 2]$ and $[2:1, 1, 0]$ are equivalent.

   (b) Show that the weighted voting systems $[7:4, 3, 2, 1]$ and $[5:3, 2, 1, 1]$ are equivalent.

   (c) Explain why equivalent weighted voting systems must have the same Banzhaf power distribution.

   (d) Explain why equivalent weighted voting systems must have the same Shapley-Shubik power distribution.

RUNNING

61. **Minimal voting systems.** A weighted voting system is called **minimal** if there is no equivalent weighted voting system with a smaller quota or with a smaller total number of votes. (For the definition of equivalent weighted voting systems, see Exercise 60.)

   (a) Show that the weighted voting system $[3:2, 1, 1]$ is minimal.

   (b) Show that the weighted voting system $[4:2, 2, 1]$ is not minimal and find an equivalent weighted voting system that is minimal.

(c) Show that the weighted voting system [8: 5, 3, 1] is not minimal and find an equivalent weighted voting system that is minimal.

(d) Given a weighted voting system with N players and a dictator, describe the minimal voting system equivalent to it.

62. **The Nassau County Board of Supervisors.** Since 1994, the Nassau County Board of Supervisors has operated as the weighted voting system [65: 30, 28, 22, 15, 7, 6] (see Exercise 22). Show that the weighted voting system [15: 7, 6, 5, 4, 2, 1] is

(a) equivalent to [65: 30, 28, 22, 15, 7, 6].

(b) minimal (see Exercise 61). (*Hint*: First show that all six weights must be positive and all must be different. Then examine possible quotas that would give the correct results for the three coalitions $\{P_1, P_2, P_5\}$, $\{P_1, P_2, P_6\}$, and $\{P_2, P_3, P_4\}$. Conclude that the players' weights *cannot* be 6, 5, 4, 3, 2, 1. Use the same coalitions to conclude that $w_3 + w_4 > w_1 + w_6$, and finally that if $w_1 = 7$, then $w_4 = 4$.)

63. **The United Nations Security Council.** The U.N. Security Council is made up of 15 member countries—5 permanent members and 10 nonpermanent members. For a motion to pass, it must have the yes vote of each of the 5 permanent members plus at least 4 of the nonpermanent members.

(a) The Banzhaf power index of each permanent member is 848/5080. Explain how both numerator and denominator come about. You may use the following two facts: (i) There are 210 coalitions consisting of five permanent members and *four* nonpermanent members, and (ii) there are 638 coalitions consisting of five permanent members and *five or more* nonpermanent members.

(b) Find the Banzhaf power distribution of the Security Council.

(c) Explain why the U.N. Security Council is equivalent to a weighted voting system in which each nonpermanent member has 1 vote, each permanent member has 7 votes, and the quota is 39 votes.

64. **The Cleansburg City Council.** Find the Banzhaf power distribution in the Cleansburg City Council. (See Example 2.15 for details.)

65. **The Fresno City Council.** In Fresno, California, the city council consists of seven members (the mayor and six other council members). A motion can be passed by the mayor and at least three other council members, or by at least five of the six ordinary council members.

(a) Describe the Fresno City Council as a weighted voting system.

(b) Find the Shapley-Shubik power distribution for the Fresno City Council. (*Hint*: See Example 2.15 for some useful ideas.)

66. Suppose that in a weighted voting system there is a player A who hates another player P so much that he will always vote the opposite way of P, regardless of the issue. We will call A the **antagonist** of P.

(a) Suppose that in the weighted voting system [8; 5, 4, 3, 2], P is the player with two votes and his antagonist A is the player with five votes. What are the possible coalitions under these circumstances? What is the Banzhaf power distribution under these circumstances?

(b) Suppose that in a generic weighted voting system with N players there is a player P who has an antagonist A. How many coalitions are there under these circumstances?

(c) Give examples of weighted voting systems where a player A can

    (i)   increase his Banzhaf power index by becoming an antagonist of another player

    (ii) decrease his Banzhaf power index by becoming an antagonist of another player

(d) Suppose that the antagonist A has more votes than his enemy P. What is a strategy that P can use to gain power at the expense of A?

67. (a) Give an example of a weighted voting system with four players and such that the Shapley-Shubik power index of P_1 is $\frac{3}{4}$.

(b) Show that in any weighted voting system with four players, a player cannot have a Shapley-Shubik power index of more than $\frac{3}{4}$ unless he or she is a dictator.

(c) Show that in any weighted voting system with N players, a player cannot have a Shapley-Shubik power index of more than $(N - 1)/N$ unless he or she is a dictator.

(d) Give an example of a weighted voting system with N players and such that P_1 has a Shapley-Shubik power index of $(N - 1)/N$.

68. (a) Give an example of a weighted voting system with three players and such that the Shapley-Shubik power index of P_3 is $\frac{1}{6}$.

(b) Explain why in any weighted voting system with three players, a player cannot have a Shapley-Shubik power index of less than $\frac{1}{6}$ unless he or she is a dummy.

(c) Give an example of a weighted voting system with four players and such that the Shapley-Shubik power index of P_4 is $\frac{1}{12}$.

(d) Explain why in any weighted voting system with four players, a player cannot have a Shapley-Shubik power index of less than $\frac{1}{12}$ unless he or she is a dummy.

69. (a) Give an example of a weighted voting system with N players having a player with veto power who has a Shapley-Shubik power index of $1/N$.

(b) Explain why in any weighted voting system with N players, a player with veto power must have a Shapley-Shubik power index of at least $1/N$.

70. (a) Give an example of a weighted voting system with N players having a player with veto power who has a Banzhaf power index of $1/N$.

(b) Explain why in any weighted voting system with N players, a player with veto power must have a Banzhaf power index of at least $1/N$.

71. An alternative way to compute Banzhaf power. As we know, the first step in computing the Banzhaf power index of player P is to compute B, the total number of times P is a critical player. The following formula gives an alternative way to compute B:

$$B = 2 \cdot W_P - W,$$

where W is the number of winning coalitions and W_P is the number of winning coalitions containing P. Explain why the preceding formula is true.

A. The Johnston Power Index

The Banzhaf and Shapley-Shubik power indexes are not the only two mathematical methods for measuring power. The Johnston power index is a subtle but rarely used variation of the Banzhaf power index in which the power of a player is based not only on how often he or she is critical in a coalition, but also on the number of other players in the coalition. Specifically, being a critical player in a coalition of 2 players contributes $\frac{1}{2}$ toward your power score; being critical in a coalition of 3 players contributes $\frac{1}{3}$ toward your power score; and being critical in a coalition of 10 contributes only $\frac{1}{10}$ toward your power score.

A players' *Johnston power score* is obtained by adding all such fractions over all coalitions in which the player is critical. The player's *Johnston power index* is his or her Johnston power score divided by the sum of all players' power scores. (A more detailed description of how to compute the Johnston power index is given in reference 17.)

Prepare a presentation on the Johnston power index. Include a mathematical description of the procedure for computing Johnston power, give examples, and compare the results with the ones obtained using the Banzhaf method. Include your own personal analysis on the merits of the Johnston method compared with the Banzhaf method.

B. The Past, Present, and Future of the Electoral College

Starting with the Constitutional Convention of 1776 and ending with the Bush-Gore presidential election of 2000, give a historical and political analysis of the Electoral College. You should address some or all of the following issues: How did the Electoral College get started? Why did some of the Founding Fathers want it? How did it evolve? What has been its impact over the years in affecting presidential elections? (Pay particular attention to the 2000 presidential election.) What does the future hold for the Electoral College? What are the prospects that it will be reformed or eliminated?

C. Mathematical Arguments in Favor of the Electoral College

As a method for electing the president, the Electoral College is widely criticized as being undemocratic. At the same time, different arguments have been made over the years to support the case that the Electoral College is not nearly as bad as it seems. Massachusetts Institute of Technology physicist Alan Natapoff has recently used mathematical ideas (many of which are connected to the material in this chapter) to make the claim that the Electoral College is a better system than a direct presidential election. (Natapoff's arguments are nicely described in reference 6.) Summarize and analyze Natapoff's mathematical arguments in support of the Electoral College.

D. Banzhaf Power and the Law

John Banzhaf was a lawyer, and he made his original arguments on behalf of his mathematical method to measure power in court cases, most of which involved the Nassau County Board of Supervisors in New York state (see pp. 60–61). Among the more significant court cases were *Graham v. Board of Supervisors* (1966); *Bechtle v. Board of Supervisors* (1981); *League of Women Voters v. Board of Supervisors* (1983); and *Jackson v. Board of Supervisors* (1991). Other important legal cases based on the Banzhaf method for measuring power but not involving Nassau County were *Ianucci v. Board of Supervisors of Washington County* and *Morris v. Board of Estimate* (U.S. Supreme Court, 1989). Choose one or two of these cases, read their background, arguments, and the court's decision, and write a brief for each.

Note: This is a good project for prelaw and political science majors, but it might require access to a good law library.

APPENDIX

Power in the Electoral College (2001–2010)

State	Electoral Votes		Power Index (%)	
	Number	**Percent**	**Shapley-Shubik**	**Banzhaf**
Alabama	9	1.673	1.639	1.640
Alaska	3	0.558	0.540	0.546
Arizona	10	1.859	1.824	1.823
Arkansas	6	1.115	1.086	1.092
California	55	10.223	11.036	11.402
Colorado	9	1.673	1.639	1.640
Connecticut	7	1.301	1.270	1.274
Delaware	3	0.558	0.540	0.546
District of Columbia	3	0.558	0.540	0.546
Florida	27	5.019	5.087	5.012
Georgia	15	2.788	2.761	2.744
Hawaii	4	0.743	0.722	0.728
Idaho	4	0.743	0.722	0.728
Illinois	21	3.903	3.910	3.865
Indiana	11	2.045	2.010	2.007
Iowa	7	1.301	1.270	1.274
Kansas	6	1.115	1.086	1.092
Kentucky	8	1.487	1.454	1.457
Louisiana	9	1.673	1.639	1.640
Maine	4	0.743	0.722	0.728
Maryland	10	1.859	1.824	1.823
Massachusetts	12	2.230	2.197	2.190
Michigan	17	3.160	3.141	3.116
Minnesota	10	1.859	1.824	1.823
Mississippi	6	1.115	1.086	1.092
Missouri	11	2.045	2.010	2.007
Montana	3	0.558	0.540	0.546
Nebraska	5	0.929	0.904	0.910
Nevada	5	0.929	0.904	0.910
New Hampshire	4	0.743	0.722	0.728
New Jersey	15	2.788	2.761	2.744
New Mexico	5	0.929	0.904	0.910
New York	31	5.762	5.888	5.795
North Carolina	15	2.788	2.761	2.744
North Dakota	3	0.558	0.540	0.546
Ohio	20	3.717	3.717	3.677
Oklahoma	7	1.301	1.270	1.274
Oregon	7	1.301	1.270	1.274
Pennsylvania	21	3.903	3.910	3.865

Rhode Island	4	0.743	0.722	0.728
South Carolina	8	1.487	1.454	1.457
South Dakota	3	0.558	0.540	0.546
Tennessee	11	2.045	2.010	2.007
Texas	34	6.320	6.499	6.393
Utah	5	0.929	0.904	0.910
Vermont	3	0.558	0.540	0.546
Virginia	13	2.416	2.384	2.375
Washington	11	2.045	2.010	2.007
West Virginia	5	0.929	0.904	0.910
Wisconsin	10	1.859	1.824	1.823
Wyoming	3	0.558	0.540	0.546

REFERENCES AND FURTHER READINGS

1. Banzhaf, John F., III, "One Man, 3.312 Votes: A Mathematical Analysis of the Electoral College," *Villanova Law Review*, 13 (1968), 304–332.
2. Banzhaf, John F., III, "Weighted Voting Doesn't Work," *Rutgers Law Review*, 19 (1965), 317–343.
3. Brams, Steven J., *Game Theory and Politics*. New York: Free Press, 1975, chap. 5.
4. Felsenthal, Dan, and Moshe Machover, *The Measurement of Voting Power: Theory and Practice, Problems and Paradoxes*. Cheltenham, England: Edward Elgar, 1998.
5. Grofman, B., "Fair Apportionment and the Banzhaf Index," *American Mathematical Monthly*, 88 (1981), 1–5.
6. Hively, Will, "Math Against Tyranny," *Discover*, November 1996, 74–85.
7. Imrie, Robert W., "The Impact of the Weighted Vote on Representation in Municipal Governing Bodies of New York State," *Annals of the New York Academy of Sciences*, 219 (November 1973), 192–199.
8. Lambert, John P., "Voting Games, Power Indices and Presidential Elections," *UMAP Journal*, 3 (1988), 213–267.
9. Merrill, Samuel, "Approximations to the Banzhaf Index of Voting Power," *American Mathematical Monthly*, 89 (1982), 108–110.
10. Meyerson, Michael I., *Political Numeracy: Mathematical Perspectives on Our Chaotic Constitution*, New York: W. W. Norton, 2002, chap. 2.
11. Riker, William H., and Peter G. Ordeshook, *An Introduction to Positive Political Theory*. Englewood Cliffs, NJ: Prentice-Hall, Inc., 1973, chap. 6.
12. Shapley, Lloyd, and Martin Shubik, "A Method for Evaluating the Distribution of Power in a Committee System," *American Political Science Review*, 48 (1954), 787–792.
13. Sickels, Robert J., "The Power Index and the Electoral College: A Challenge to Banzhaf's Analysis," *Villanova Law Review*, 14 (1968), 92–96.
14. Straffin, Philip D., Jr., "The Power of Voting Blocs: An Example," *Mathematics Magazine*, 50 (1977), 22–24.
15. Straffin, Philip D., Jr., *Topics in the Theory of Voting, UMAP Expository Monograph*. Boston: Birkhäuser, 1980, chap. 1.
16. Tannenbaum, Peter, "Power in Weighted Voting Systems," *The Mathematica Journal*, 7 (1997), 58–63.
17. Taylor, Alan, *Mathematics and Politics: Strategy, Voting, Power and Proof*. New York: Springer-Verlag, 1995, chaps. 4 and 9.

The Lion, the Fox, and the Ass

One of Aesop's Fables

The Lion, the Fox, and the Ass entered into an agreement to assist each other in the hunt. Having secured a large booty, the Lion on their return from the forest asked the Ass to allot its due portion to each of the three partners in the treaty.

The Ass carefully divided the spoil into three equal shares and modestly requested the two others to make the first choice. The Lion, bursting out into a great rage, devoured the Ass. Then he requested the Fox to do him the favor to make the division. The Fox accumulated all that they had killed into one large heap and left to himself the smallest possible morsel.

The Lion said, "Who has taught you, my very excellent fellow, the art of division? You are perfect to a fraction," to which the Fox replied, "I learned it from the Ass, by witnessing his fate."

Fair Division

The Mathematics of Sharing

We start to learn about sharing at a very young age—sharing toys, sharing treats, sharing attention. As we get older, we learn about more abstract forms of sharing such as sharing duties, responsibilities, and even blame. This business of dividing things among ourselves—be they "good" things (food, toys, love) or "bad" things (chores, responsibility, guilt)—is among the most social of all human interactions. Social animals also divide, although not always in agreeable ways (consider the fate of the poor Ass in Aesop's sobering tale). Of course, our long history of war and conquest bears witness to the fact that we humans often do it just as badly, possibly worse. But we can also do it better, quite well in fact, when we set our minds to it. Dividing things fairly using reason and logic, instead of bullying our way to a solution, is one of the great achievements of social science, and, once again, we can trace the roots of this achievement to simple mathematics.

> If you want to know the true character of a person, divide an inheritance with him.
>
> Ben Franklin

We take our first mathematical stab at *fair division* somewhere around the third or fourth grade. A typical problem goes like this: *There are 20 pieces of candy to be divided among four equally deserving children.* What is a fair solution? We know, of course, the standard answer: Give each child five pieces. The problem with this answer is that it may not produce a fair division after all. What if the pieces of candy are not all the same? Say, for example, the twenty pieces are made up of a wide variety of goodies: Snickers, Milk Duds, caramels, bubblegum, etc., with some pieces clearly more desirable than others, and with each child having a different set of preferences. Can we take these diverse opinions into account and still divide the pieces fairly? And by the way, what does *fairly* mean in this situation? These are some of the many thought-provoking questions we will discuss in this chapter.

Why are these questions important, you may wonder? After all, it's only candy and kids. Not exactly. Just like the booty in Aesop's fable, the candy is a metaphor—we could just as well be dividing an inheritance (jewelry, family heirlooms, works of art, etc.) And, on an even grander scale, the problem takes on added significance. The division of entire nations (as in the case of the former Yugoslavia in the 1990s), the division of rights of access to mine the ocean floor (as in the Convention of the Law of the Sea), and the division of responsibilities

The Judgment of Solomon, by
Nicolas Poussin (1649).

for environmental cleanup (as in the 1994 NAFTA treaties) are all issues of
broad significance, and yet, in essence, they are variations on one basic theme—
they are all *problems of fair division*.

Problems of fair division are as old as humankind. One of the best-known
and best-loved biblical stories is built around a fair-division problem: Two women,
both claiming to be mothers of the same baby, make their case to King Solomon.
As a solution, King Solomon proposes to cut the baby in half and give each
woman an equal "share," a division that is totally unacceptable to the true mother, who would rather see the baby go to the other woman than be slaughtered.
The final settlement, of course, is that the baby is returned to its rightful mother.

The basic issue in all fair-division problems can be stated in reasonably simple terms: How can something that must be shared by a set of competing parties
be divided among them in a way that ensures that each party receives a fair
share? It could be argued that a good general answer to this question would go a
long way in solving most of the problems of humankind, but unfortunately, good
answers are not easy to come by. On the other hand, under the right set of circumstances, these kinds of problems can be solved using basic mathematical
ideas. In this chapter we will learn how to identify and solve some of these types
of problems, so that at the end, while not quite as wise as King Solomon, we might
be wise enough not to try to divide a booty with a lion.

3.1 Fair-Division Games

In this section we will introduce some of the basic concepts and terminology of
fair division. Much as we did in Chapter 2 when we studied weighted voting
systems, we will think of a fair-division in the context of a game, with players,
goals, rules, and strategies. Regardless of whether the game is dividing a cake
or an expensive estate, the underlying elements of every *fair-division game* are
the same:

- A set of *goods* to be divided. These goods can be anything that has a potential value. Typically, the goods are tangible physical objects, such as candy, cake, pizza, jewelry, art, property such as cars, houses, land, etc. In more esoteric situations the goods may be intangible things such as rights (water rights, drilling rights, broadcast licenses, etc.).[1] We will use the symbol S to denote the object or objects to be divided (the "booty" if you will).

- A set of *players*, who are the parties entitled to share the set S. We will call the players $P_1, P_2, \ldots, P_N$. The players are usually persons, but they could also be countries, states, ethnic or political groups, and institutions. The one key characteristic that each of the players must have is his own *value system*—the ability to assign value to the set S as well as to any subset of S.

Given the booty S, and players $P_1, P_2, P_3, \ldots, P_N$, each with his or her own opinions of how S should be divided, the ultimate goal of the game is to end up with a **fair division** of S, that is, to divide S into shares (one share for each player) in such a way that each player gets a *fair share*. We will define precisely what we mean by a *fair share* soon.

In the remaining sections of this chapter, we will discuss different methods by which fair divisions can be accomplished. We call these **fair-division methods** (they are also known as fair-division *protocols* or fair-division *schemes*). Essentially, we can think of a *fair-division method* as the *set of rules* defining how the game is to be played. This means that a specific fair-division game involves not only a given set S and a given set of players $P_1, P_2, P_3, \ldots, P_N$, but also a specific method by which we plan to accomplish the fair division (this is very similar to Chapter 1, where a specific election was determined by three things: candidates, ballots, and voting method). There are many different fair-division methods known, but in this chapter we will only discuss a few of the classic ones.

Like most games, fair-division games are predicated on certain assumptions about the players. For the purposes of our discussion, we will make the following assumptions:

- *Cooperation*: Players are willing participants and accept the rules of the game as binding. The rules are such that after a *finite* number of moves by the players the game terminates with a division of S. (There are no outsiders such as judges or referees involved in these games—just the players and the rules.)

- *Rationality*: Players act rationally, and their value systems conform to the basic laws of arithmetic.

- *Privacy*: Players have no useful information on the other players' likes and dislikes, and thus, of what kinds of moves they are going to make in the game.[2]

- *Symmetry*: Players have *equal* rights in sharing the set S. A corollary of this assumption is that at a minimum, each player is entitled to a *proportional*

---

[1]To keep things simple we will stick to positive goods throughout our discussion, but it is also possible for the "goods" to have negative value (chores, responsibility, guilt, etc.) in which case one could call them "bads." With minor variations, all the methods discussed in this chapter can be used to divide "bads" just as well as "goods" (see Exercises 63 and 64).

[2]This assumption does not always hold in practice, when often the players are siblings or friends. In these cases, the game can still be played, but the final division may not be totally fair, as players can take advantage of their inside knowledge. Much like in a card game, a player should not be privy to another player's hand.

share of S—when there are two players, each is entitled to at least one-half of S, with three players each is entitled to at least one-third of S, and so on.[3]

When all of the preceding assumptions are satisfied, a fair-division method *guarantees to each player the opportunity to get a fair share of the set S.* This means that there is a strategy available to each player that, if followed, guarantees that he or she will receive a fair share.

What Is a Fair Share?

The last question that remains to be answered is how to define what constitutes a *fair share* of S. In this chapter, we will adopt the following convention: Given a share s of S and a player P (one of the N players in the game), we will say that s is a **fair share** to player P if, *in the opinion of P, s is worth at least* $(1/N)$th of the total value of S. (Such a share is often called a *proportional fair share*, but for simplicity we will refer to it just as a *fair share.*)

Two comments about the preceding definition are in order. First, notice that the concept of a fair share is relative—what may be a fair share to P may not necessarily be a fair share to a different player Q. Second, it is possible for a share to be a fair share to P, but not necessarily the one that P likes the best. For example, suppose that in a division among four players, Paul thinks that s_1 is worth 15% of S, s_2 is worth 35% of S, s_3 is worth 20% of S, and s_4 is worth 30% of S. If Paul were to get s_4, worth 30% of the total, then he would be getting a fair share although obviously he would rather have s_2, which, in his opinion, is the *best share*. (A fair division in which each player gets a share that he or she considers to be the best share is called an *envy-free* fair division. For more on envy-free fair division, see Project A.)

Types of Fair-Division Games

Depending on the nature of the set S, a fair-division game can be classified as one of three types: *continuous, discrete,* or *mixed.*

In a **continuous** fair-division game the set S is divisible in infinitely many ways, and shares can be increased or decreased by arbitrarily small amounts. Typical examples of continuous fair-division games involve the division of land, a cake, a pizza, etc.

A fair-division game is **discrete** when the set S is made up of objects that are indivisible like paintings, houses, cars, boats, jewelry, etc. As far as candy is concerned, yes, a piece could be chopped up into smaller and smaller pieces, but nobody would really do that (it's messy), so let's agree that throughout this chapter we will think of candy as indivisible and therefore discrete (a semantic convenience).

A *mixed* fair-division game is one in which some of the components are continuous and some are discrete. Dividing an estate consisting of jewelry, a house, and a parcel of land is a mixed fair-division game.

Fair-division methods are classified according to the nature of the problem involved. Thus, there are *discrete fair-division* methods (used when the set S is made up of indivisible, discrete objects), and there are *continuous fair-division*

---

[3] All of the fair-division methods we will discuss in this chapter can be extended to *asymmetric* situations in which different players are entitled by right to different sized shares of S. (See Project B.)

methods (used when the set S is an infinitely divisible, continuous set). Mixed fair-division games can usually be solved by dividing the continuous and discrete parts separately, so we will not study them in this chapter.

We will start our discussion with continuous fair-division methods.

3.2   Two Players: The Divider-Chooser Method

The **divider-chooser method** is undoubtedly the best known of all continuous fair-division methods. This method can be used anytime there is a continuous fair-division problem involving just two players. Most of us have unwittingly used it at some time or another, and informally it is best known as the *you cut—I choose method*. As this name suggests, one player, the *divider*, divides the *cake* (a convenient metaphor for any continuous set S) into two pieces, and the second player, the *chooser*, picks the piece he or she wants, leaving the other piece to the divider. When played properly, this method guarantees that both players will get a share they believe to be worth *at least one-half* of the total. The divider can guarantee this by dividing the cake into two halves of equal value, and the chooser is guaranteed a fair share by choosing the piece he or she likes best.

EXAMPLE 3.1

On their first date, Damian and Cleo go to the county fair. With a $2 raffle ticket they win the chocolate-strawberry cake shown in Fig. 3-1(a).

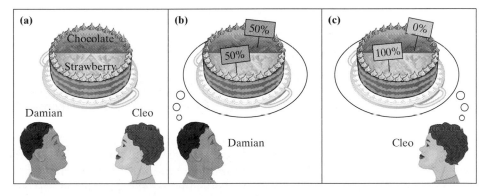

FIGURE 3-1
A chocolate-strawberry cake. The values are in the eyes of the beholder.

To Damian, chocolate and strawberry are equal in value—he has no preference for one over the other. Thus, in Damian's eyes the value of the cake is distributed evenly between the chocolate and strawberry parts [Fig.3-1(b)]. On the other hand, Cleo hates chocolate (she is allergic to it and gets sick if she eats any). Thus, in Cleo's eyes the value of the cake is concentrated entirely in the strawberry half; the chocolate half has *zero value* [Fig. 3-1(c)]. Since this is their first date, we can assume neither one of them knows anything about the other's likes and dislikes.

Let's now see how Damian and Cleo might divide this cake using the divider-chooser method. Damian volunteers to go first and be the divider. His cut is shown in Fig. 3-2(a). Note that this is a perfectly rational cut based on Damian's value system—each piece is worth one-half of the total value of the cake. It is now Cleo's turn to choose, and her choice is obvious—she will pick the piece having the largest strawberry part [Fig.3-2(b)].

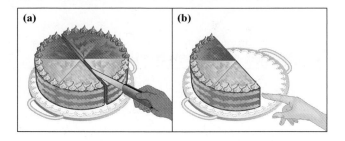

FIGURE 3-2
(a) Damian cuts (b) Cleo picks.

Notice that while Damian gets a share that (to him) is worth exactly one-half, Cleo ends up with a share that (to her) is worth much more than one-half. ■

Example 3.1 illustrates why, given a choice, it is always better to be the chooser than the divider—the divider is guaranteed a piece that is worth exactly one-half of the total, but the chooser has a chance to get a piece that is worth more than one-half. This happened in Example 3.1 and will happen any time the divider divides S into shares that the chooser does not agree are equal in value. Notice that it is the disagreement in the value systems of the players that makes the solution interesting.

Since a fair-division method should treat all players equally, both players should have an equal chance of being the chooser. The obvious way to handle this is to determine who gets to be the divider and who gets to be the chooser by flipping a fair coin.

The idea behind the divider-chooser method is an ancient one. When Lot and Abraham argued over grazing rights, the Old Testament tells us that Abraham proposed, in effect a divider-chooser approach. But what do we do when there are three players? Four? *N*? Surprisingly, extensions of the divider-chooser method that work for three or more players only came about in the 1940s, a result of the pioneering work of the Polish mathematician Hugo Steinhaus (for more on Steinhaus, see the biographical profile on p. 111).

There are three different approaches to extending the divider-chooser method when there are more than two players: the *lone-divider method*, the *lone-chooser method*, and the *last-diminisher method*. In the lone-divider method, one of the players is the divider and all the rest are choosers. In the lone-chooser method, one of the players is the chooser and all the rest are dividers. In the last-diminisher method, each player has a chance to be a divider or a chooser.

> *Let us divide the land into north and south, If you go north, I will go south; and if you go south, I will go north*
>
> *(Genesis 13:1–17)*

3.3 The Lone-Divider Method

The first important breakthrough in the mathematics of fair division came in 1943, when Steinhaus came up with a clever way to extend some of the ideas in the divider-chooser method to the case of a fair division among three players. In 1967, Steinhaus's approach was generalized by Princeton mathematician Harold Kuhn to any number of players.

We start this section with a description of Steinhaus's *lone-divider method* for the case of $N = 3$ players.

Steinhaus's Lone-Divider Method for Three Players

- **Preliminaries.** One of the three players is designated to be the divider. The other two players will be choosers. We'll call the divider D and the choosers C_1 and C_2. Since it is better to be a chooser than a divider, the decision of who is what is made by a random draw.

"... *The Ass carefully divided the spoil into three equal shares and modestly requested the two others to make their first choice* ..."

Aesop

- **Step 1. (Division).** The divider D divides the cake into three pieces (s_1, s_2, and s_3). D will get one of these pieces, but at this point does not know which one. This forces D to divide the cake in such a way that all three pieces have equal value, namely one-third of the value of the entire cake.

- **Step 2. (Bidding).** C_1 declares (usually by writing on a slip of paper) which of the three pieces cut up by the divider are fair shares to her. Independently, C_2 does the same. These are the *choosers' bids*. To preserve the privacy requirement, it is important that the bids be made independently, without the choosers seeing each other's bids. A chooser must bid for any piece that he or she values to be worth one-third or more of the cake. Thus, a chooser can bid for one, two, or even all three pieces. It is logically impossible, however, for a chooser not to bid on any piece—at least one of the three pieces must be worth one-third or more.

- **Step 3. (Distribution).** Who gets which piece? The answer, of course, depends on the bids. For convenience, we will separate the pieces into two groups: the C-pieces ("chosen" pieces that are listed in one or both of the choosers' bids), and the U-pieces (unwanted pieces that neither chooser considered fair shares). We now consider two cases, depending on the number of C-pieces.

  Case 1. There are two or more C-pieces. Here, it is possible to give each chooser one of the pieces that she bid for and to give the divider the last remaining piece. Once this is done, every player has received a fair share, and our goal of fair division has been met. Note that this method of distribution does not preclude the possibility that each chooser may like the other chooser's piece better, in which case it is perfectly reasonable to let them swap their pieces. This would make each of them happier than they already were, and who could be against that?

  Case 2. There is only one C-piece. The two other pieces are U-pieces. Now we are in trouble, because this implies that both choosers covet the same piece. Here is the way out of this impasse. We first choose one of the two U-pieces to give to the divider, to whom all pieces are equal in value. (The best way to do this is to try to get the two choosers to agree on which U-piece they would rather part with. If they can't agree, the decision can be made by flipping a coin.) Once one of the two U-pieces is given to the divider, the other U-piece is combined with the C-piece to make a single big piece which we'll call the B-piece [see Figs. 3-3(a) and (b)]. Now what? First, a simple matter of arithmetic: the B-piece plus the divider's piece make up the entire cake, and the value of the divider's piece to both choosers is less than one-third of the total (neither chooser bid for it in Step 2). This means that in the eyes of both choosers, the B-piece is worth more than two-thirds of the value of the original cake. The trick now is to divide the B-piece fairly between the two choosers. If we can do this, we are done. But in fact, we do know a way to divide a piece fairly among two players: the *divider-chooser method*. Applying the divider-chooser method gives us a way to give each of the two choosers a fair half of the B-piece, and thus, a fair share of the original cake.

FIGURE 3-3
Case 2 in the lone-divider method (three players). (a) One of the *U*-pieces goes to the divider *D*. (b) The *C*-piece and the remaining *U*-piece are recombined into the *B*-piece. (c) The *B*-piece is divided in two shares using the divider-chooser method.

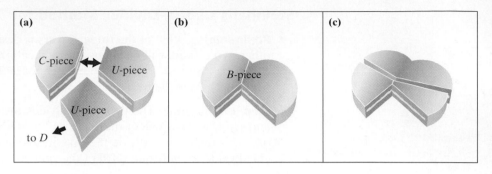

We will illustrate the lone-divider method for three players with several examples. In all of these examples, we will assume that the divider *D* has already divided the cake into three pieces s_1, s_2, and s_3. In each example, the values that each of the three players assigns to the pieces, expressed as percentages of the total value of the cake, are shown in the form of a table. The reader should remember, however, that this information is never available in full to the players—an individual player only knows the percentages on his or her row.

EXAMPLE 3.2

Table 3-1 shows the values of the three pieces in the eyes of each of the players. With three players, a fair share is any piece valued at $33\frac{1}{3}\%$ or more. Looking at the table, we can see that C_1's bid should be $\{s_1, s_3\}$, and C_2's bid should also be $\{s_1, s_3\}$. In this case the s_1, s_3 are the *C*-pieces, and there are two possible ways to complete the fair division. One division would be to give s_2 to *D*, s_1 to C_1, and s_3 to C_2. An even better division (from both choosers' point of view) would be to give s_2 to *D*, s_3 to C_1, and s_1 to C_3, but the players themselves have no way of knowing this, since the information available in the game is just the bids (and not the percentages shown on Table 3-1). After the division has been made, however, if two of the players want to swap pieces, it is perfectly permissible for them to do so. If C_1 were to get s, and C_2 got s_3, it is clear that both choosers would benefit by exchanging their pieces, and under the rationality assumption they would undoubtedly do so.

TABLE 3-1

	s_1	s_2	s_3
D	$33\frac{1}{3}\%$	$33\frac{1}{3}\%$	$33\frac{1}{3}\%$
C_1	35%	10%	55%
C_2	40%	25%	35%

EXAMPLE 3.3

Table 3-2 shows the values of the three pieces in the eyes of each of the players. In this example, C_1's bid consists of just $\{s_2\}$ and C_2's bid consists of just $\{s_1\}$. Here, the only possible distribution of the pieces is to give s_2 to C_1, s_1 to C_2, and s_3 to D.

TABLE 3-2

	s_1	s_2	s_3
D	$33\frac{1}{3}\%$	$33\frac{1}{3}\%$	$33\frac{1}{3}\%$
C_1	30%	40%	30%
C_2	60%	15%	25%

EXAMPLE 3.4

Table 3-3 shows the values of the three pieces in the eyes of each of the players. Here we are in a case 2 situation: both C_1's and C_2's bid consists of just $\{s_3\}$, so s_3 is the only C-piece. Of the U-pieces, both C_1 and C_2 like s_1 least, so they would agree that s_1 goes to D. This makes $s_2 + s_3$ the B-piece, which is to be divided between C_1 and C_2 using the divider-chooser method. Note that the B-piece is worth 80% of the original cake to C_1 and 90% of the original cake to C_2. Thus, C_1 will end up with a piece that is worth at least 40% of the original cake and C_2 will end up with a piece that is worth at least 45% of the original cake—and both end up happy as clams.

TABLE 3-3

	s_1	s_2	s_3
D	$33\frac{1}{3}\%$	$33\frac{1}{3}\%$	$33\frac{1}{3}\%$
C_1	20%	30%	50%
C_2	10%	20%	70%

The Lone-Divider Method for More Than Three Players

In 1967 Harold Kuhn, a game-theorist at Princeton University, was able to extend Steinhaus's lone-divider method to any number of players. Kuhn's method is based on some fairly sophisticated mathematical ideas and is rather difficult to describe in full detail, so in what follows we will only give a general outline of how the method works and then illustrate it with a couple of examples in the case of four players. (For a detailed description of Kuhn's lone-divider method, see reference 12.)

With N players, one of the players is randomly chosen to be the divider D, and the remaining $N - 1$ players are going to be all choosers. The divider starts by dividing the set S into N shares. Each of the choosers then makes an independent bid, writing down all the pieces that he or she considers fair shares. When the bids are opened, there are two possible scenarios that we must consider:

Case 1. There is a way to give to each chooser one of the shares listed in his or her bid. The divider, to whom all shares are supposed to be of equal value, gets the last unassigned share. At the end, players may choose to swap pieces if they want.

Case 2. There is a *standoff*. A standoff happens when there are two choosers both bidding on the same share, or three choosers bidding on just two shares, or K choosers bidding on less than K shares. This is a much more complicated case, and what follows is a rough sketch of what to do. To resolve a standoff, we first set aside the shares involved in the standoff from the remaining shares. Likewise, the players involved in the standoff are temporarily separated from the rest. Each of the remaining players (including the divider) can be assigned a fair share from among the remaining shares and sent packing. All the shares left are recombined into a new set S to be divided among the players involved in the standoff, and the process starts all over again.

The following two examples will illustrate some of the ideas behind the lone-divider method in the case of $N = 4$ players. The first example is one without a standoff; the second example involves a standoff.

EXAMPLE 3.5

We have one divider D, and three choosers C_1, C_2, and C_3. The divider D first divides S into four shares s_1, s_2, s_3, and s_4. Table 3-4 shows how each of the players values each of the four shares. Remember that the information on each row of Table 3-4 is private and only known to that player.

TABLE 3-4

	s_1	s_2	s_3	s_4
D	25%	25%	25%	25%
C_1	30%	20%	35%	15%
C_2	20%	20%	40%	20%
C_3	25%	20%	20%	35%

Each chooser bids based on his or her value system: C_1's bid is $\{s_1, s_3\}$, C_2's bid is just $\{s_3\}$, and C_3's bid is $\{s_1, s_4\}$.

It is now possible to assign to each of the choosers one of the shares they bid for. For starters, C_2 gets s_3—there is no other option. This forces C_1 to get s_1 and C_3 to get s_4. Finally, we give the last remaining piece s_2 to the divider D. This completes the fair division of S, with each player getting a fair share although not necessarily the best share. ∎

EXAMPLE 3.6

This game starts out exactly as in Example 3.5, with D dividing S into four shares s_1, s_2, s_3, and s_4. Table 3-5 shows how each of the players values each of the four shares.

TABLE 3-5

	s_1	s_2	s_3	s_4
D	25%	25%	25%	25%
C_1	20%	20%	20%	40%
C_2	15%	35%	30%	20%
C_3	22%	23%	20%	35%

We can see from Table 3-5 what is going to happen when the bids are opened: C_1's bid is $\{s_4\}$, C_2's bid is $\{s_2, s_3\}$, and C_3's bid is $\{s_4\}$. This creates a standoff involving C_1 and C_3, which are both bidding for s_4.

We proceed by first setting s_4 aside and postponing the assignments to C_1 and C_3. C_2 could get either s_2 or s_3. A coin toss is used to determine which one. Say C_2 ends up with s_3. D is taken care of next by another coin toss to randomly choose between s_1 and s_2 (s_3 is gone and s_4 has been set aside). Say D ends up with s_1. This leaves shares s_2 and s_4 and players C_1 and C_3.

The final step is to recombine s_2 and s_4 into a single share $(s_2 + s_4)$ and have C_1 and C_3 divide it using the divider-chooser method. Since $s_2 + s_4$ is worth more than 50% of S to both players (60% to C_1 and 58% to C_3—check it out in Table 3-5), they are both guaranteed more than 25% of S at the end. This completes the fair division of S. ∎

3.4 The Lone-Chooser Method

The lone-chooser method was first proposed by A. M. Fink, a mathematician at Iowa State University. Once again, we will start with a description of the method for the case of three players.

The Lone-Chooser Method for Three Players

- **Preliminaries.** We have one chooser and two dividers. Let's call the chooser C and the dividers D_1 and D_2. As usual, we decide who is what by a random draw.

- **Step 1. (Division).** D_1 and D_2 divide S [Fig. 3-4(a)] between themselves into *two* fair shares. To do this, they use the divider-chooser method. Let's say that D_1 ends up with s_1 and D_2 ends up with s_2 [Fig.3-4(b)].

- **Step 2. (Subdivision).** Each divider divides his share into three subshares. Thus, D_1 divides s_1 into three subshares, which we will call s_{1a}, s_{1b}, and s_{1c}. Likewise, D_2 divides s_2 into three subshares, which we will call s_{2a}, s_{2b}, and s_{2c}. [Fig.3-4(c)].

- **Step 3. (Selection).** The chooser C now selects one of D_1's three subshares and one of D_2's three subshares (whichever she likes best). These two subshares make up C's final share. D_1 then keeps the remaining two subshares from s_1, and D_2 keeps the remaining two subshares from s_2 [Fig.3-4(d)].

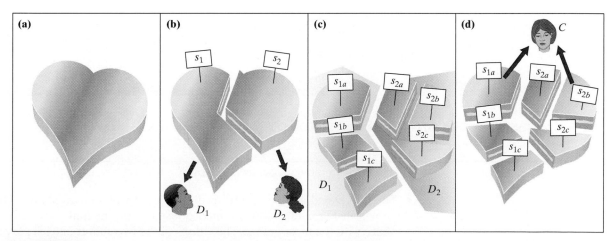

FIGURE 3-4
(a) The original cake, (b) first division, (c) second division, and (d) selection

Why is this a fair division of S? D_1 ends up with two-thirds of s_1. To D_1, s_1 was worth at least one-half of the total value of S, so two-thirds of s_1 is at least one-third—a fair share. The same argument applies to D_2. What about the chooser's share? We don't know what s_1 and s_2 are each worth to C, but it really doesn't matter. Let's say, for the sake of argument, that in C's eyes, s_1 was worth only 30% of S. This automatically implies that s_2 was worth 70%. Now C got a subshare from s_1 worth at least one-third of 30% (10%), and another subshare from s_2 worth at least one-third of 70% ($23\frac{1}{3}\%$). Between the two, C got a fair share. The argument works no matter how C values s_1 and s_2.

The following example illustrates in detail how the lone-chooser method works.

EXAMPLE 3.7

David, Dinah, and Cher are planning to divide an orange-pineapple cake valued by each of them at $27 [Fig. 3-5(a)] using the *lone-chooser method*. They draw straws and Cher gets to be the chooser, so David and Dinah first divide the cake using the *divider-chooser method*. Since David drew a shorter straw than Dinah, he will be the one to cut the cake. Now, for their value systems:

■ David likes pineapple and orange the same. To him, value is synonymous with size, so in his eyes the cake looks like Fig.3-5(b).

■ Dinah likes orange but hates pineapple. To her the entire value of the cake is concentrated in the orange half, so in her eyes the cake looks like Fig.3-5(c).

■ Cher likes pineapple twice as much as she likes orange. In her eyes the cake looks like Fig.3-5(d).

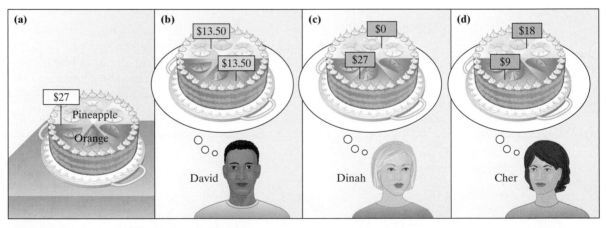

FIGURE 3-5
(a) The original cake (b) in David's eyes (c) in Dinah's eyes (d) in Cher's eyes

The division now proceeds as follows:

■ **Step 1.** David starts by cutting the cake into two equal shares. His cut is shown in Fig. 3-6(a). Since Dinah doesn't like pineapple, she will take the share with the most orange. The values of the two shares in each player's eyes are shown in Fig.3-6.

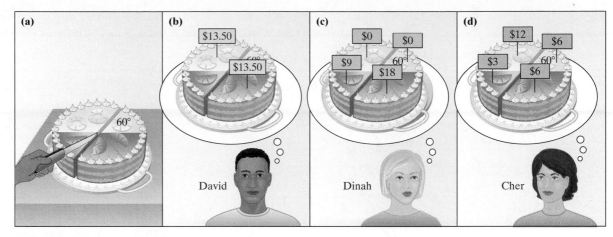

FIGURE 3-6
The first cut and the values of the shares in the eyes of each player.

- **Step 2.** David divides his share into three subshares that in his opinion are of equal value. Notice that the subshares [Fig. 3-7(a)] are all the same size. Dinah also divides her share into three smaller subshares that in her opinion are of equal value. Remember that Dinah hates pineapple. Thus, she has made her cuts in such a way as to have one-third of the orange in each of the subshares [as shown in Fig.3-7(b)].

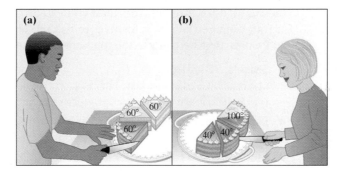

FIGURE 3-7
(a) David cuts his share.
(b) Dinah cuts her share

- **Step 3.** It's now Cher's turn to choose one subshare from David's three and one subshare from Dinah's three. Figure 3-8 shows the values of the sub-shares in Cher's eyes.

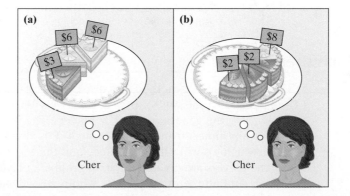

FIGURE 3-8
The values of the subshares in
Cher's eyes.

The final division of the cake is shown in Fig. 3-9. Notice that each person has received a share that is worth at least $9 and is therefore a fair share of the original cake.

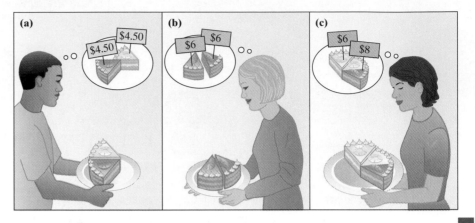

FIGURE 3-9
Each player's final fair share.

The Lone-Chooser Method for *N* Players

In the general case of N players, the lone chooser method involves $N-1$ players $D_1, D_2, \ldots, D_{N-1}$ that we call the *dividers* and one player C that is the *lone chooser*. As always, it is preferable to be a chooser than a divider, so the chooser is determined by a random draw. The method is based on an inductive strategy—if you can do it for three players then you can do it for four players, if you can do it for four then you can do it for five, and so on. Thus, when we get to N players, we can assume that we can use the lone-chooser method with $N-1$ players.

- **Step 1. (Division).** $D_1, D_2, \ldots, D_{N-1}$ divide fairly the set S among themselves, as if C didn't exist. This is a fair division among $N-1$ players, so each one gets a share they consider worth at least $1/(N-1)$th of S.

- **Step 2. (Subdivision).** Each divider subdivides his or her share into N subshares.

- **Step 3. (Selection).** The chooser C finally gets to play. C selects one subshare from each divider—one subshare from D_1, one from D_2, and so on. At the end, C ends up with $N-1$ subshares, which make up C's final share, and each divider gets to keep the remaining $N-1$ subshares in his or her subdivision.

When properly played, the lone-chooser method guarantees that everyone, dividers and chooser alike, ends up with a fair share (Exercise 66).

3.5   The Last-Diminisher Method

The last-diminisher method was discovered by the Polish mathematicians Stefan Banach and Bronislaw Knaster in the 1940s. We will describe the *last-diminisher method* for the general case of N players. The basic idea behind this method is that at any time throughout the game, S is divided into two pieces, a claimed piece, which we will call the C-piece and the rest of S, which we call the R-piece, and the players are divided into two groups, a player who is the "claimant" of the C-piece and all the other players, whom we will call the "nonclaimants." As the game progresses, each nonclaimant has an opportunity to become a claimant by

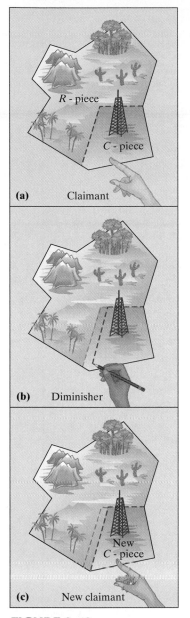

FIGURE 3-10
(a) *C*-piece claimed by claimant.
(b) Diminisher "trims" *C*-piece.
(c) Diminisher becomes new claimant.

"trimming" the *C*-piece, or remain a nonclaimant. Thus, the *C*-piece, the *R*-piece, the claimant, and the nonclaimants all can change. (This is what keeps the players honest!) Here are the details of exactly how it all works:

- **Preliminaries.** Before the game starts, the players are randomly assigned an order (P_1 first, P_2 second, ..., P_N last), and the players will play in this order throughout the game. The game is played in rounds, and at the end of each round there is one fewer player and a smaller *S* to be divided.

- **Round 1.** The first player, P_1, starts by becoming the first claimant by "cutting" for herself a share that she believes to be an *exact* (1/*N*)th of *S*. This will be the *C*-piece, claimed by P_1. Since P_1 does not know whether or not she will end up with this share, she must be careful that her claim is neither too small (in case she does) nor too large (in case someone else does). The next player, P_2, now has the right to become a claimant (*play*) or to remain a nonclaimant (*pass*) on the *C*-piece. P_2 should play only if he thinks that the *C*-piece is worth more than (1/*N*)th of *S*, otherwise he should pass and remain a nonclaimant. If P_2 plays, he must do so by trimming the *C*-piece to where, to him, *it is a fair share.* When this happens, P_2 becomes the claimant of the diminished *C*-piece, the sliver cut off from the old *C*-piece becomes a part of the *R*-piece, and P_1 happily (because the *R*-piece got bigger) goes back to the nonclaimant group. It is now P_3's turn to pass or play on the *C*-piece, regardless of whether it belongs to P_1 or P_2. If P_3 passes, then nothing changes, and we move on to the next player. If P_3 thinks that the *C*-piece is worth more than 1/*N*th of *S*, she must trim it so that it is worth exactly 1/*N*th of *S* [see Fig 3-10(b)]. The trimmed piece is added to the *R*-piece, and the previous claimant happily joins the nonclaimant group [Fig.3-10(c)]. They continue in this way until all the players in order have a chance to pass or play. The player who is the claimant at this point (*the last diminisher*) gets to keep his claimed piece, which is a fair share, and is out of the game. What happens to the remaining players (the nonclaimants)? They move on to the next round.

- **Round 2.** The *R*-piece becomes the new set *S* to be divided fairly among the $N-1$ remaining players. Since none of them thought that the old *C*-piece was a fair share, they should all be happy to be in the position of having to divide this new *S* fairly among $N-1$ players. This is done by repeating the whole process (claimants, nonclaimants, *C*-pieces, and *R*-pieces), but remembering that now, with one fewer player, the threshold for a fair share has changed. It is now 1($N-1$)th of the new *S*. At the end of this round, the last diminisher gets to keep her claimed piece and is out of the game.

- **Rounds 3, 4, etc.** Repeat the process, each time with one fewer player and a smaller *S*, until there are just two players left. At this point, divide the remaining piece between the final two players using the *divider-chooser method.*

Five sailors we will call P_1, P_2, P_3, P_4, and P_5 are marooned on a lush, deserted tropical island. Liking what they see, they decide to claim ownership of the island, divide it among themselves, and lead the good life there forever. Having learned something about fair division they decide to do the division using the last-diminisher method. Here pictures speak louder than words, so the whole story unfolds in Figs. 3-11 through 3-15.

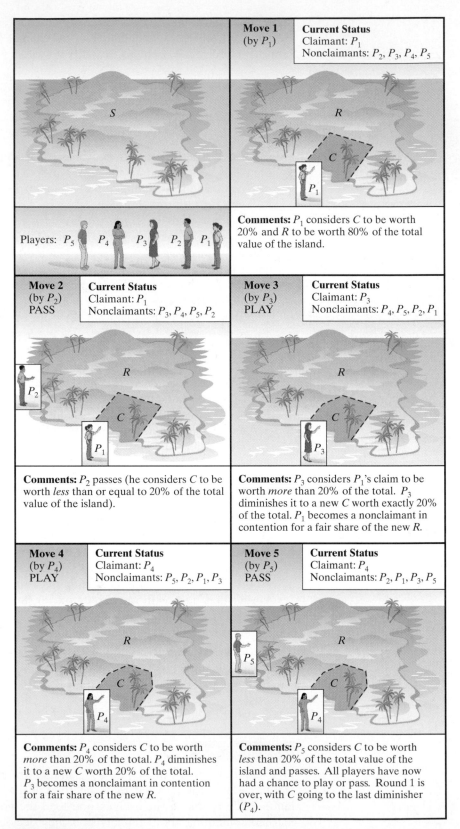

FIGURE 3-11
Example 3.8, Round 1.

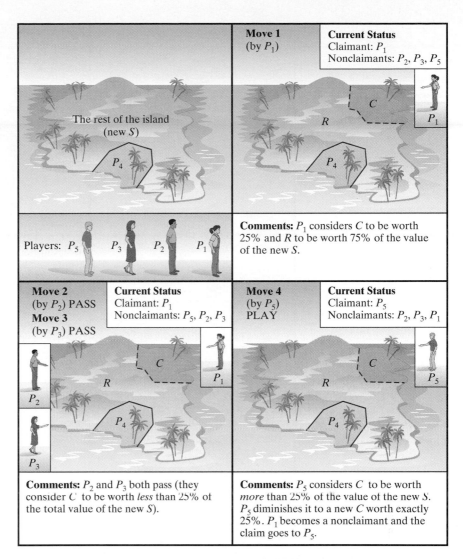

FIGURE 3-12
Example 3.8, Round 2.
(4 players left)

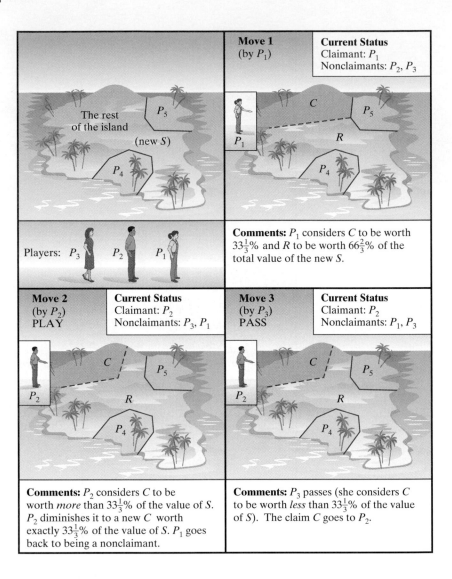

FIGURE 3-13
Example 3.8, Round 3.
(3 players left)

FIGURE 3-14
Example 3.8, last round (divider-chooser method).

FIGURE 3-15
The final division of the island.

We will now move on to *discrete* fair-division games, where the set S consists of objects that are indivisible—items such as houses, cars, paintings, candy, etc.

3.6 The Method of Sealed Bids

One of the most important discrete fair-division methods is a method discovered by Hugo Steinhaus and Bronislaw Knaster around 1948 which we will call the **method of sealed bids.** The easiest way to illustrate how this method works is by means of an example.

EXAMPLE 3.9

In her last will and testament, Grandma plays a little joke on her four grandchildren (Art, Betty, Carla, and Dave) by leaving just three valuable items—a house, a Rolls Royce, and a Picasso painting—with the stipulation that the items must remain with the grandchildren (not sold to outsiders) and must be divided fairly in equal shares among them. How can we possibly resolve this conundrum? The method of sealed bids gives an ingenious and elegant solution.

- **Step 1. (Bidding).** Each of the players is asked to make a bid for each of the items in the estate, giving his or her honest assessment of the dollar value of each item. To satisfy the privacy assumption, it is important that the bids are done independently, and no player should be privy to another player's bids before making his or her own. The easiest way to accomplish this is for each player to submit his or her bid in a sealed envelope. When all the bids are in, they are opened. Table 3–6 shows each player's bid on each item in the estate.

TABLE 3-6   The Bids

	Art	Betty	Carla	Dave
House	220,000	250,000	211,000	198,000
Rolls Royce	40,000	30,000	47,000	52,000
Picasso	280,000	240,000	234,000	190,000

- **Step 2. (Allocation).** Each item goes to the highest bidder for that item. In this example, the house goes to Betty, the Rolls Royce goes to Dave, and the Picasso painting goes to Art. Carla gets nothing. So far, this doesn't sound very fair!

- **Step 3. (Payments).** Now come the payments. Depending on what items (if any) a player gets in Step 2, he or she will owe money to or be owed money by the estate. To determine how much is owed, we first calculate how much each player believes his or her fair share is worth. This is done by adding the player's bids and dividing by the number of players. The last row of Table 3–7 shows the value of a fair share to each player. If the total value of the items that the player gets in Step 2 is more than the value of that player's fair share, the player pays the estate the difference. If the total value of the items that the player gets is less than the value of the player's fair share, the player collects the difference in cash. Here is how it works with each of our four players.

TABLE 3-7

	Art	Betty	Carla	Dave
Home	220,000	250,000	211,000	198,000
Rolls Royce	40,000	30,000	47,000	52,000
Picasso	280,000	240,000	234,000	190,000
Total	540,000	520,000	492,000	440,000
Fair share	135,000	130,000	123,000	110,000

- **Art.** By his own estimation, Art's fair share is worth $135,000 and he is getting a Picasso painting worth $280,000. This means that Art must pay the estate $145,000 ($280,000 − $135,000). Assuming Art bid honestly, the Picasso minus the $145,000 make for a fair share of the estate.

- **Betty.** Betty's fair share is worth $130,000. She is getting the house, which she values at $250,000, so she must pay the estate the difference of $120,000. If she bid honestly, Betty is now getting a fair share of the estate.

- **Carla.** Carla's fair share is, by her own estimation, $123,000. Since she is getting no items from the estate, she receives her full $123,000 in cash. Her fair share of the estate is now settled.

- **Dave.** Dave's assessment of the value of his fair share is $110,000. Dave is getting the Rolls, which he values at $52,000, so he has an additional $58,000 coming to him in cash.

At this point each of the heirs has received a fair share, and we might consider our job done, but ... if we add Art's and Betty's payments to the estate and subtract the payments made by the estate to Carla and Dave, we discover that something quite amazing is happening: There is $84,000 left over ($145,000 and $120,000 coming in; $123,000 and $58,000 going out). This leads to the next move, where everybody wins!

- **Step 4. (Dividing the Surplus).** The surplus money is divided equally among the four heirs. In our example each player's share of the $84,000 surplus is $21,000. This means that in the final settlement each player gets a fair share (in items plus or minus cash) plus an unexpected bonus of $21,000 in cash. ■

The method of sealed bids works so well because it tweaks a basic idea in economics. In most ordinary transactions there is a buyer and a seller, and the buyer knows the other party is the seller and vice versa. In a sense, this works to both parties' disadvantage. In the method of sealed bids, each player is simultaneously a buyer and a seller, without actually knowing which one until all the bids are opened. This keeps the players honest and, in the long run, works out to everyone's advantage.

The method of sealed bids works well as long as the following two important conditions are satisfied.

- Each player must have enough money to play the game. If a player is going to make honest bids on the items, he must be prepared to buy some or all of them, which means that he may have to pay the estate certain sums of money. If the player does not have this money available, he is at a definite disadvantage in playing the game.

■ Each player must accept money (if it is a sufficiently large amount) as a substitute for any item. This means that no player can consider any of the items priceless. *I want Grandma's house, and no amount of money in the world is going to make me change my mind*! is not an attitude conducive to a good resolution of the problem.

The method of sealed bids takes a particularly simple form in the case of two players and one item. Consider the following example:

EXAMPLE 3.10

Al and Betty are getting a divorce. The only common property of value is their house. Since the divorce is amicable and they are not particularly keen on going to court or hiring an attorney, they decide to divide the house using the method of sealed bids. Their bids on the house are shown in Table 3-8. Betty, being the highest bidder, gets the house but must pay the estate $71,000 because she is entitled to only half of the value of the house. Al's fair share is half of his bid, namely, $65,000. The surplus of $6000 is divided equally between Al and Betty, and the bottom line is that Betty gets the house but pays Al $68,000. (Notice that this result is equivalent to assessing the value of the house as the value halfway between the two bids ($136,000) and splitting this value equally between the two parties, with the house going to the highest bidder and the cash to the other party.) ■

TABLE 3-8

Al	Betty
$130,000	$142,000

3.7 The Method of Markers

The *method of markers* is a discrete fair-division method proposed in 1975 by William F. Lucas, a mathematician at the Claremont Graduate School. The method has the great virtue that it does not require the players to put up any of their own money. On the other hand, unlike the method of sealed bids, this method cannot be used effectively unless there are many more items to be divided than there are players and the items are reasonably close in value.

In this method, we start with the items lined up in an *array* (a fixed sequence which cannot be changed). For convenience, think of the array as a string of objects. Each player independently bids for segments of consecutive items in the array by "cutting" the string. If there are N players, then each player must cut the string into N segments, each of which represents an acceptable share of the entire set of items. Notice that to cut a string into N sections, we need $N-1$ cuts. In practice, one way to make the "cuts" is to lay markers in the places where the cuts are made. Thus, each player can make his or her bids by placing $N-1$ markers so that they divide the array into N segments. To insure privacy, no player should see the markers of another player before laying down his or her own.

What the method of markers essentially accomplishes is to guarantee that each player ends up with one of his or her bid segments (a section between two consecutive markers). The easiest way to explain how the method works is with an example.

EXAMPLE 3.11

Four children—Alice, Bianca, Carla, and Dana (*A, B, C,* and *D*)—are to divide the 20 pieces of candy shown in Fig. 3-16. Their teacher, Mrs. Jones, offers to divide the candy, but the children reply that they can do it themselves, thank you.

FIGURE 3-16
The booty.

FIGURE 3-17
The original array.

The 20 pieces are randomly arranged into the array shown in Fig. 3-17. For convenience, we will label the pieces of candy 1 through 20.

- **Step 1. (Bidding).** Each child writes down independently on a piece of paper exactly where she wants her three markers. (Remember, four players means three markers per player). The bids are opened, and the results are shown in Fig. 3-18.

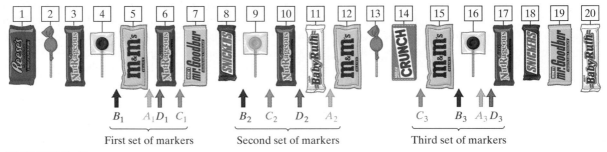

B_1 $A_1 D_1$ C_1 B_2 C_2 D_2 A_2 C_3 B_3 $A_3 D_3$

First set of markers   Second set of markers   Third set of markers

FIGURE 3-18
The results of the bidding.

- **Step 2. (Allocations).** We are now ready to allocate one segment of the array to each child. To do so we start scanning the array from left to right, until we find someone's first marker. Here the first *first marker* going from left to right is Bianca's (B_1), so we give Bianca her first segment (pieces 1 through 4, Fig. 3-19). Bianca has now received a fair share of the candy and is happily gone. At this point her markers can be removed, since they are no longer needed. We now continue scanning from left to right looking for the first *second marker*. This marker belongs to Carla (C_2). We now give to Carla her second segment, going from first marker to second marker (pieces 7 through 9, Fig. 3-20). Once Carla and her markers are out of the picture, we continue scanning from left to right until we find the first *third marker*. This is

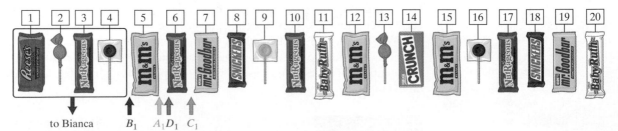

to Bianca B_1 $A_1 D_1$ C_1

FIGURE 3-19
Bianca, the owner of the first first marker, gets her first segment.

FIGURE 3-20
Carla, the owner of the first
second marker (among the
remaining players), gets
her second segment.

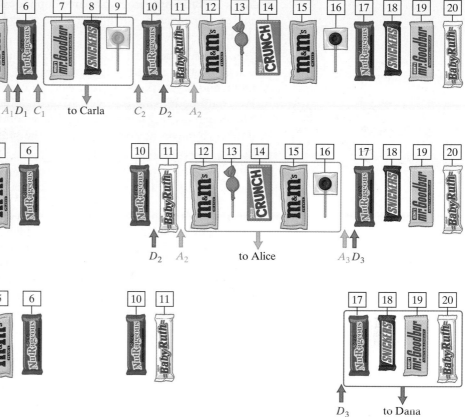

FIGURE 3-21
Alice and Dana both own
the first third marker. After
a coin toss, the third segment goes to Alice.

FIGURE 3-22
Dana is the last player left.
She gets her last segment.

a tie between Alice and Dana (A_3, D_3), and we can break the tie randomly. After a coin toss, Alice ends up with her third segment (pieces 12 through 16, Fig. 3-21), and finally we give to the last player (Dana) her last segment (pieces 17 through 20, Fig. 3-22). Now each player has gotten one of her chosen segments. The amazing part is that there is *leftover candy*!

- **Step 3. (Dividing the Leftovers).** Usually, there are just a few pieces of candy left over, not enough to play the game all over again. The simplest thing to do is randomly draw lots and let the children go in order picking one piece at a time until there is no more candy left. Here the leftover pieces are 5, 6, 10, and 11. The players now draw lots; Carla gets to choose first and takes piece 11. Dana chooses next and takes piece 5. Bianca and Alice receive pieces 6 and 10, respectively.

FIGURE 3-23
The leftovers (to be given randomly to the players one at a
time) are a bonus.

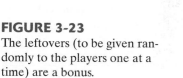

We now give the general description of the **method of markers** with N players and M items that are arranged into an array.

- **Step 1. (Bidding).** Each player independently divides the array into N fair shares by placing $N-1$ markers. The markers separate each fair share from the next.
- **Step 2. (Allocations).** Scan the array from left to right until the first *first marker* is located. The player owning that marker gets to keep his first segment, and his markers are removed. In case of a tie, break the tie randomly. We continue moving from left to right, looking for the first *second marker*. The player owning it gets to keep her second segment. Continue this process until each player has received one of the segments.
- **Step 3. (Leftovers).** The leftover items can be divided among the players by some form of lottery, and, in the rare case that there are many more items than players, the method of markers can be used again.

In spite of its simple elegance, the method of markers can be used only under some fairly restrictive conditions. In particular, the method assumes that every player is able to divide the array of items into segments in such a way that each of the segments has approximately equal value. This is usually possible when the items are of small and homogeneous value, but almost impossible to accomplish when there are expensive items involved. (Imagine, for example, trying to divide fairly a bunch of pieces of candy plus a gold coin using the method of markers.)

Conclusion

The problem of dividing an object or set of objects among the members of a group is a practical problem that comes up regularly in our daily lives. When the object is a pizza, a cake, or a bunch of candy, we don't always pay a great deal of attention to the issue of fairness, but when the object is an estate, land, jewelry, or some other valuable asset, dividing things fairly becomes a critical issue.

On the surface, problems of fairness seem far removed from the realm of mathematics. We are more likely to think of economics, political science, or law as being the proper fields for a discussion of this topic. It is surprising, therefore, that when certain basic conditions are satisfied, mathematics can provide fair-division methods that not only guarantee fairness but often do much better than that.

In this chapter we discussed the problem of fair division using concepts and terminology borrowed from game theory, and covered several important *fair-division methods*. The choice of which is the best fair-division method to use in a particular situation is not always clear, and in fact there are many situations in which a fair division is mathematically unattainable. (We will discuss an important example of this in Chapter 4.) At the same time, in a large number of every-day situations the fair-division methods we described in this chapter (or simple variations thereof) will work. Remember these methods the next time you must divide an inheritance, a piece of real estate, or even some of the chores around the house. They may serve you well.

Hugo Steinhaus (1887–1972)

Dividing things, be it the spoils of war or the fruits of peace, is an issue as old as man. Throughout most of human history, fairness in dividing things was seen as a problem far removed from the mathematical arena—the province of kings, priests, judges, and politicians. It wasn't until the 1940s that the revolutionary notion that mathematical methods could be used to tackle successfully many types of fair division problems came about. This breakthrough can be traced to the inspiration and creativity of one man—the great Polish mathematician Hugo Steinhaus.

Hugo Steinhaus was born in Jaslo, Poland. After completing his secondary education in his homeland, Steinhaus went to Germany to study mathematics at the University of Gottingen, which in the early 1900s was the most prestigious center of mathematical research in the world. At Gottingen, Steinhaus studied under many famous mathematicians, including David Hilbert, his doctoral supervisor and arguably the most famous mathematician of his time. Steinhaus was awarded a doctorate in mathematics (with distinction) in 1911.

After completing four years of military service in World War I, Steinhaus returned to Poland, where he began a distinguished academic career as a professor of mathematics, first in Krakow and then in Lvov, where he founded the famous Lvov school of mathematics. Steinhaus was a prolific and versatile mathematician, who made important contributions to many different branches of mathematics, including functional analysis, trigonometric series, and probability theory. But Steinhaus's interests went beyond conventional mathematical research—he loved to discuss and expound on simple but interesting real-life applications of mathematics and to foster an appreciation for the power and beauty of mathematics among young people. In 1937, Steinhaus wrote a classic book called *Mathematical Snapshots*, a collection of vignettes intended to appeal, in Steinhaus's own words, to the *scientist in the child and the child in the scientist*. After many editions, *Mathematical Snapshots* is still in print and widely quoted.

Sometime in the late 1930s, Steinhaus made one of his more famous mathematical discoveries, a theorem informally known as the *Ham Sandwich Theorem*. Essentially, the theorem can be paraphrased as follows: If you have a sandwich consisting of *three* ingredients (say bread, ham, and cheese), then there is a way to slice the sandwich (using a single straight cut) into two parts each of which has exactly half of the volume of bread, half of the ham and half of the cheese—you can share the sandwich with your friend and each of you gets exactly half of each of the three ingredients. (You can do this with even the funkiest of sandwiches, as long as you stick to just three ingredients. Unfortunately, if you want to divide *n* ingredients equally, you need to live in the *n*th dimension!)

The Ham Sandwich Theorem is a beautiful theoretical result but has little practical value—it says that the desired cut is possible in theory, but it doesn't give even the slightest hint as to how to do it. It promises a fair division but it cannot deliver it. The limitations of the Ham Sandwich Theorem forced Steinhaus to think about a different, truly practical approach to the problem of dividing things *equally*, and this led to the foundations of the theory of fair division as we know it today. By the late 1940s, Steinhaus, (together with Stefan Banach and Bronislaw Knaster, two of his students and collaborators) had developed many of the most important fair division methods we know today, including the *last-diminisher method*, the *method of sealed bids*, and a continuous version of the *method of markers*.

Hugo Steinhaus died in 1972, at the age of 85. His mathematical legacy includes over 170 articles, 5 books, and the invaluable insight that mathematics can help humankind solve some of its disputes.

KEY CONCEPTS

continuous fair-division game   last-diminisher method
discrete fair-division game   lone-chooser method
divider-chooser method   lone-divider method
fair division   method of markers
fair-division method   method of sealed bids
fair share

EXERCISES

WALKING

A. Fair Division Concepts

1. Alex buys a chocolate-strawberry mousse cake [shown in (i)] for $12. Alex values chocolate 3 times as much as he values strawberry.

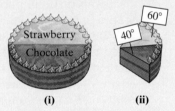

(i)   (ii)

   (a) What is the value of the chocolate half of the cake to Alex?

   (b) What is the value of the strawberry half of the cake to Alex?

   (c) A piece of the cake is cut as shown in (ii). What is the value of the piece to Alex?

2. Jody buys a chocolate-strawberry mousse cake [shown in (i)] for $13.50. Jody values strawberry 4 times as much as she values chocolate.

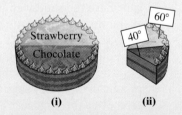

(i)   (ii)

   (a) What is the value of the chocolate half of the cake to Jody?

   (b) What is the value of the strawberry half of the cake to Jody?

   (c) A piece of the cake is cut as shown in (ii). What is the value of the piece to Jody?

3. Kala buys a chocolate-strawberry-vanilla cake [shown in (i)] for $12. Kala values strawberry twice as much as vanilla and values chocolate three times as much as vanilla.

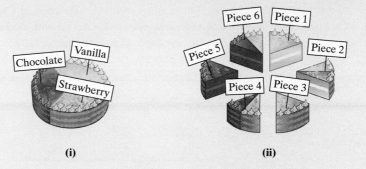

(i)   (ii)

(a) What is the value of the chocolate part of the cake to Kala?

(b) What is the value of the strawberry part of the cake to Kala?

(c) What is the value of the vanilla part of the cake to Kala?

(d) If the cake is cut into the six 60° wedges shown in (ii), find the value to Kala of each of the six pieces.

4. Malia buys a chocolate-strawberry-vanilla cake [shown in (i)] for $11.20. Malia values strawberry twice as much as chocolate and values chocolate twice as much as vanilla.

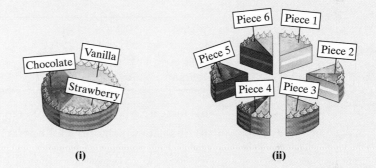

(i)   (ii)

(a) What is the value of the chocolate part of the cake to Malia?

(b) What is the value of the strawberry part of the cake to Malia?

(c) What is the value of the vanilla part of the cake to Malia?

(d) If the cake is cut into the six 60° wedges shown in (ii), find the value to Malia of each of the six pieces.

5. Three players (Ana, Ben, and Cara) must divide a cake among themselves. Suppose the cake is divided into 3 slices (s_1, s_2, and s_3). The values of the entire cake and of each of the 3 slices in the eyes of each of the players are shown in the following table.

	Whole cake	s_1	s_2	s_3
Ana	$12.00	$3.00	$5.00	$4.00
Ben	$15.00	$4.00	$4.50	$6.50
Cara	$13.50	$4.50	$4.50	$4.50

(a) Indicate which of the three slices are fair shares to Ana.

(b) Indicate which of the three slices are fair shares to Ben.

(c) Indicate which of the three slices are fair shares to Cara.

6. Three players (Alex, Betty, and Cindy) must divide a cake among themselves. Suppose the cake is divided into three slices (s_1, s_2, and s_3). The following table shows the percentage of the value of the entire cake that each slice represents to each player.

	s_1	s_2	s_3
Alex	30%	40%	30%
Betty	35%	25%	40%
Cindy	$33\frac{1}{3}\%$	50%	$16\frac{2}{3}\%$

(a) Indicate which of the three slices are fair shares to Alex.

(b) Indicate which of the three slices are fair shares to Betty.

(c) Indicate which of the three slices are fair shares to Cindy.

7. Four partners (Adams, Benson, Cagle, and Duncan) jointly own a piece of land which is subdivided into four parcels (s_1, s_2, s_3, and s_4). The following table shows the percentage of the value of the land that each parcel represents to each partner.

	s_1	s_2	s_3	s_4
Adams	30%	24%	20%	26%
Benson	35%	25%	20%	20%
Cagle	25%	15%	40%	20%
Duncan	20%	20%	20%	40%

(a) Indicate which of the four parcels are fair shares to Adams.

(b) Indicate which of the four parcels are fair shares to Benson.

(c) Indicate which of the four parcels are fair shares to Cagle.

(d) Indicate which of the four parcels are fair shares to Duncan.

(e) Assuming that the four parcels cannot be changed or further subdivided, describe a fair division of the land.

8. Four players (Abe, Betty, Cory, and Dana) must divide a cake among themselves. Suppose the cake is divided into four slices (s_1, s_2, s_3, and s_4). The values of the entire cake and of each of the four slices in the eyes of each of the players are shown in the following table.

	Whole cake	s_1	s_2	s_3	s_4
Abe	$15.00	$3.00	$5.00	$5.00	$2.00
Betty	$18.00	$4.50	$4.50	$4.50	$4.50
Cory	$12.00	$4.00	$3.50	$1.50	$3.00
Dana	$10.00	$2.75	$2.40	$2.45	$2.40

(a) Indicate which of the four slices are fair shares to Abe.

(b) Indicate which of the four slices are fair shares to Betty.

(c) Indicate which of the four slices are fair shares to Cory.

(d) Indicate which of the four slices are fair shares to Dana.

(e) Using the four given slices, describe a fair division of the cake.

B. The Divider-Chooser Method

9. Two friends (David and Paul) decide to divide the pizza shown in the accompanying figure using the divider-chooser method. David likes pepperoni, sausage, and mushrooms equally well, but hates anchovies. Paul likes anchovies, mushrooms, and pepperoni equally well, but hates sausage. Neither one knows anything about the other one's likes and dislikes (they are new friends).

   (a) Suppose that David is the divider. Which of the cuts (i) through (iv) shown below is consistent with David's value system?

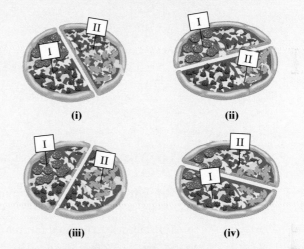

   (b) For each of the cuts consistent with David's value system, which of the two pieces is Paul's best choice?

10. Raul and Karli want to divide a chocolate-strawberry mousse cake. Raul values chocolate three times as much as he values strawberry. Karli values chocolate twice as much as she values strawberry.

   (a) If Raul is the divider, which of the following cuts are consistent with Raul's value system?

Cut 1   Cut 2   Cut 3   Cut 4   Cut 5

(b) For each of the cuts consistent with Raul's value system, indicate which of the pieces is Karli's best choice.

11. This exercise is a continuation of Exercise 9.

   (a) Suppose Paul is the divider. Draw three different cuts that are consistent with his value system.

   (b) For each of the cuts in (a), indicate which of the pieces is David's best choice.

12. This exercise is a continuation of Exercise 10.

   (a) Suppose Karli is the divider. Draw three different cuts that are consistent with her value system.

   (b) For each of the cuts in (a), indicate which of the pieces is Raul's best choice.

13. Jamie and Mo want to divide an orange-pineapple cake using the divider-chooser method. Jamie values orange four times as much as he values pineapple. Mo is the divider and cuts the cake as shown in (ii).

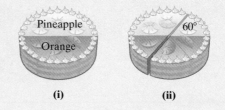

(i)   (ii)

   (a) What percent of the value of the cake is the pineapple half in Mo's eyes?

   (b) What percent of the value of the cake is each piece in (ii) in Jamie's eyes?

   (c) Describe the final fair division of the cake.

14. Susan and Veronica want to divide an orange-pineapple cake using the divider-chooser method. Susan values orange four times as much as she values pineapple. Veronica is the divider and cuts the cake as shown in (ii) in the following figure.

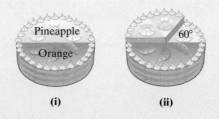

(i)   (ii)

   (a) What percent of the value of the cake is the pineapple half in Veronica's eyes?

   (b) What percent of the value of the cake is each piece in (ii) in Susan's eyes?

   (c) Describe the final fair division of the cake.

C. The Lone-Divider Method

15. Three partners (Chase, Chandra, and Divine) want to divide a plot of land fairly using the lone-divider method. Using a map, Divine divides the property into three parcels (s_1, s_2, s_3).

    (a) If the chooser declarations are

    Chase: $\{s_2, s_3\}$

    Chandra: $\{s_1, s_3\}$,

    describe a possible fair division of the land.

    (b) If the chooser declarations are

    Chase: $\{s_1, s_2, s_3\}$

    Chandra: $\{s_1\}$,

    describe a possible fair division of the land.

    (c) If the chooser declarations are

    Chase: $\{s_1\}$

    Chandra: $\{s_2\}$,

    describe a possible fair division of the land.

    (d) If the chooser declarations are

    Chase: $\{s_1\}$

    Chandra: $\{s_1\}$,

    describe how to proceed to obtain a possible fair division of the land.

16. Four partners (Childs, Choate, Chou, and DiPalma) want to divide a piece of land fairly using the lone-divider method. Using a map, DiPalma divides the land into four parcels (s_1, s_2, s_3, s_4), and the choosers make the following declarations:

    Childs: $\{s_2, s_3\}$

    Choate: $\{s_3, s_4\}$

    Chou: $\{s_4\}$.

    (a) Describe a fair division of the land.

    (b) Explain why your answer in (a) is the only possible fair division of the land using the four given parcels.

17. Four players want to divide a cake fairly using the lone-divider method. The divider cuts the cake into four slices (s_1, s_2, s_3, s_4), and the choosers make the following declarations:

    Chooser 1: $\{s_2, s_3\}$

    Chooser 2: $\{s_1, s_3\}$

    Chooser 3: $\{s_1, s_2\}$.

    (a) Describe a fair division of the cake.

    (b) Describe a fair division of the cake different from the one given in (a).

    (c) Is it possible to find a fair division of the cake such that the divider doesn't get s_4? Explain your answer.

18. Four players want to divide a cake fairly using the lone-divider method. The divider cuts the cake into 4 slices (s_1, s_2, s_3, s_4), and the choosers make the following declarations:

Chooser 1: $\{s_1, s_2\}$

Chooser 2: $\{s_1, s_2\}$

Chooser 3: $\{s_2\}$.

Describe how to proceed to obtain a possible fair division of the cake.

19. Five players want to divide a cake fairly using the lone-divider method. The divider cuts the cake into five slices $(s_1, s_2, s_3, s_4, s_5)$, and the choosers make the following declarations:

Chooser 1: $\{s_2, s_4\}$

Chooser 2: $\{s_2, s_4\}$

Chooser 3: $\{s_2, s_3, s_4\}$

Chooser 4: $\{s_2, s_3, s_5\}$.

(a) Describe a fair division of the cake.

(b) Describe a fair division of the cake different from the one given in (a).

(c) Is it possible to find a fair division of the cake such that the divider doesn't get s_1? Explain your answer.

20. Five players want to divide a cake fairly using the lone-divider method. The divider cuts the cake into 5 slices $(s_1, s_2, s_3, s_4, s_5)$, and the choosers make the following declarations:

Chooser 1: $\{s_2, s_5\}$

Chooser 2: $\{s_1, s_2, s_5\}$

Chooser 3: $\{s_1, s_4, s_5\}$

Chooser 4: $\{s_2, s_4\}$.

(a) Describe a fair division of the cake.

(b) Describe a fair division of the cake different from the one given in (a).

(c) Is it possible to find a fair division of the cake such that the divider doesn't get s_3? Explain.

21. Six players want to divide a cake fairly using the lone-divider method. The divider cuts the cake into six slices $(s_1, s_2, s_3, s_4, s_5, s_6)$, and the choosers make the following declarations:

Chooser 1: $\{s_2, s_3, s_5\}$

Chooser 2: $\{s_1, s_5, s_6\}$

Chooser 3: $\{s_3, s_5, s_6\}$

Chooser 4: $\{s_2, s_3\}$

Chooser 5: $\{s_3\}$.

(a) Describe a fair division of the cake.

(b) Explain why the answer in (a) is the only possible fair division of the cake.

22. Six players want to divide a cake fairly using the lone-divider method. The divider cuts the cake into six slices $(s_1, s_2, s_3, s_4, s_5, s_6)$, and the choosers make the following declarations:

Chooser 1: $\{s_1\}$

Chooser 2: $\{s_2, s_3\}$

Chooser 3: $\{s_4, s_5\}$

Chooser 4: $\{s_4, s_5\}$

Chooser 5: $\{s_1\}$.

Describe how to proceed to obtain a fair division of the cake.

23. Four partners want to divide a piece of land valued at $120,000 using the lone-divider method. Using a map, the divider cuts the land into four parcels (s_1, s_2, s_3, s_4) as shown in the following figure.

The value of each parcel (in thousands of dollars) in each chooser's eyes is given in the following figure.

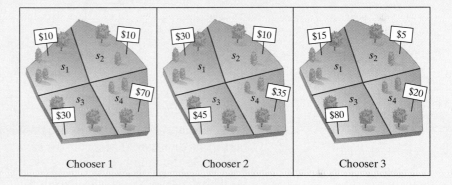

(a) What should each chooser's declarations be?

(b) Describe a possible fair division of the land.

24. Four partners want to divide a piece of land using the lone-divider method. Using a map, the divider cuts the land into four parcels (s_1, s_2, s_3, s_4) as shown in the figure.

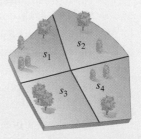

The value of each parcel (as a percentage of the total value of the land) in each chooser's eyes is given in the following figure.

(a) What should each chooser's declarations be?

(b) Describe a possible fair division of the land.

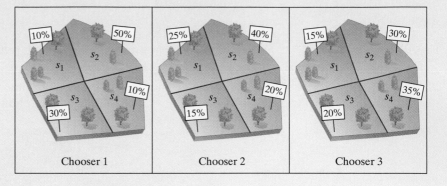

D. The Lone-Chooser Method

Exercises 25 through 28 refer to three players (Angela, Boris, and Carlos) who decide to divide a $12 vanilla-strawberry cake using the lone-chooser method. The dollar amounts of the cake in each player's eyes are given in the following figure.

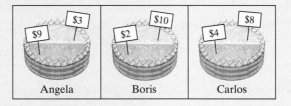

25. Suppose that Angela and Boris are the dividers and Carlos is the chooser. In the first division, Angela cuts the cake vertically through the center and Boris picks the right half.

    (a) Draw a possible second division that Angela might make of the left half of the cake.

    (b) Draw a possible second division that Boris might make of the right half of the cake.

    (c) Based on the second divisions you gave in (a) and (b), describe a possible final fair division of the cake.

    (d) For the final fair division you described in (c), find the dollar value of each share in the eyes of the player receiving it.

26. Suppose that Carlos and Angela are the dividers and Boris is the chooser. In the first division, Carlos cuts the cake vertically through the center and Angela picks the right half.

    (a) Draw a possible second division that Carlos might make of the left half of the cake.

    (b) Draw a possible second division that Angela might make of the right half of the cake.

    (c) Based on the second divisions you gave in (a) and (b), describe a possible final fair division of the cake.

    (d) For the final fair division you described in (c), find the dollar value of each share in the eyes of the player receiving it.

27. Suppose that Angela and Boris are the dividers and Carlos is the chooser. In the first division, Angela cuts the cake.

(a) Draw a possible first division by Angela other than a straight line vertical cut through the center, and indicate which of the two pieces Boris would choose.

(b) Based on the first division you gave in (a), draw a possible second division that Angela might make.

(c) Based on the first division you gave in (a), draw a possible second division that Boris might make.

(d) Based on the second divisions you gave in (b) and (c), describe a possible final fair division of the cake.

(e) For the final fair division you described in (d), find the dollar value of each share in the eyes of the player receiving it.

28. Suppose that Carlos and Angela are the dividers and Boris is the chooser. In the first division, Carlos cuts the cake.

(a) Draw a possible first division by Carlos other than a straight line vertical cut through the center, and indicate which of the two pieces Angela would choose.

(b) Based on the first division you gave in (a), draw a possible second division that Carlos might make.

(c) Based on the first division you gave in (a), draw a possible second division that Angela might make.

(d) Based on the second divisions you gave in (b) and (c), describe a possible final fair division of the cake.

(e) For the final fair division you described in (d), find the dollar value of each share in the eyes of the player receiving it.

Exercises 29 through 32 refer to three players (Arthur, Brian, and Carl) who decide to divide the cake shown in the following figure using the lone-chooser method.

The players value the different parts of the cake as follows:

Arthur likes chocolate and orange equally well, but hates strawberry and vanilla.

Brian likes chocolate and strawberry equally well, but hates orange and vanilla.

Carl likes chocolate and vanilla equally well, but hates orange and strawberry.

29. Suppose that Arthur and Brian are the dividers, with Arthur making the first cut.

(a) Draw a possible first division by Arthur, and indicate which of the two pieces Brian would choose.

(b) Based on the first division you gave in (a), draw a possible second division that Arthur might make.

(c) Based on the first division you gave in (a), draw a possible second division that Brian might make.

(d) Based on the second divisions you gave in (b) and (c), describe a possible final fair division of the cake.

(e) For the final fair division you described in (d), find the value of each share (as a percentage of the total value of the cake) in the eyes of the player receiving it.

30. Suppose that Carl and Arthur are the dividers, with Carl making the first cut.

(a) Draw a possible first division by Carl, and indicate which of the two pieces Arthur would choose.

(b) Based on the first division you gave in (a), draw a possible second division that Carl might make.

(c) Based on the first division you gave in (a), draw a possible second division that Arthur might make.

(d) Based on the second divisions you gave in (b) and (c), describe a possible final fair division of the cake.

(e) For the final fair division you described in (d), find the value of each share (as a percentage of the total value of the cake) in the eyes of the player receiving it.

31. Suppose that Brian and Carl are the dividers, with Brian making the first cut.

(a) Draw a possible first division by Brian, and indicate which of the two pieces Carl would choose.

(b) Based on the first division you gave in (a), draw a possible second division that Brian might make.

(c) Based on the first division you gave in (a), draw a possible second division that Carl might make.

(d) Based on the second divisions you gave in (b) and (c), describe a possible final fair division of the cake.

(e) For the final fair division you described in (d), find the value of each share (as a percentage of the total value of the cake) in the eyes of the player receiving it.

32. Suppose that Arthur and Carl are the dividers, with Arthur making the first cut.

(a) Draw a possible first division by Arthur, and indicate which of the two pieces Carl would choose.

(b) Based on the first division you gave in (a), draw a possible second division that Arthur might make.

(c) Based on the first division you gave in (a), draw a possible second division that Carl might make.

(d) Based on the second divisions you gave in (b) and (c), describe a possible final fair division of the cake.

(e) For the final fair division you described in (d), find the value of each share (as a percentage of the total value of the cake) in the eyes of the player receiving it.

E. The Last-Diminisher Method

33. A cake is to be divided among five players (P_1, P_2, P_3, P_4, and P_5) using the last-diminisher method. The players play in a fixed order, with P_1 first, P_2 second, etc. In round 1, P_1 cuts a piece s, and P_2 and P_4 are the only diminishers.

(a) Is it possible for P_3 to end up with any part of s in his final share? Explain.

(b) Which player gets a piece at the end of round 1?

(c) Which player cuts the piece at the beginning of round 2?

(d) Who is the last player with an opportunity to diminish the piece in round 2?

34. A cake is to be divided among four players (P_1, P_2, P_3, and P_4) using the last-diminisher method. The players play in a fixed order, with P_1 first, P_2 second, etc. In round 1, P_1 cuts a piece s, P_2 and P_3 pass, and P_4 diminishes it.

(a) Is it possible for P_2 to end up with any part of s in his final share? Explain.

(b) Which player gets a piece at the end of round 1?

(c) Which player cuts the piece at the beginning of round 2?

(d) Who is the last player who has an opportunity to diminish the piece in round 2?

35. A cake is to be divided among 12 players (P_1, P_2, P_3, ..., P_{12}) using the last-diminisher method. The players play in a fixed order, with P_1 first, P_2 second, etc. In round 1, P_1 cuts a piece, and P_3, P_7, and P_9 are the only diminishers. In round 2 the only diminisher is P_5, and in round 3 there are no diminishers.

(a) Which player gets the piece at the end of round 1?

(b) Which player cuts the piece at the beginning of round 2?

(c) Who is the last player with an opportunity to diminish the piece in round 2?

(d) Which player gets the piece at the end of round 2?

(e) Which player gets the piece at the end of round 3?

(f) Which player cuts the piece at the beginning of round 4?

(g) Who is the last player with an opportunity to diminish the piece in round 4?

36. A cake is to be divided among six players (P_1, P_2, P_3, P_4, P_5, P_6) using the last-diminisher method. The players play in a fixed order, with P_1 first, P_2 second, etc. In round 1, P_1 cuts a piece, and P_2, P_5, and P_6 are the only diminishers. In round 2 there are no diminishers. In round 3, after the first cut, each successive player is a diminisher.

(a) Which player gets the piece at the end of round 1?

(b) Which player cuts the piece at the beginning of round 2?

(c) Who is the last player with an opportunity to diminish the piece in round 2?

(d) Which player gets the piece at the end of round 2?

(e) Which player cuts the piece at the beginning of round 3?

(f) Which player gets the piece at the end of round 3?

(g) Who is the last player with an opportunity to diminish the piece in round 4?

37. An island is to be divided among seven players $(P_1, P_2, P_3, \ldots, P_7)$ using the last-diminisher method. The players play in a fixed order, with P_1 first, P_2 second, etc. P_3 gets his fair share at the end of round 1, and P_7 gets her fair share at the end of round 3. There are no diminishers in rounds 2, 4, and 5.

(a) Who is the last diminisher in round 1?

(b) Which player gets a fair share at the end of round 2?

(c) Which player cuts at the beginning of round 3?

(d) Which player gets a fair share at the end of round 4?

(e) Which player gets a fair share at the end of round 5?

(f) Which player is the chooser in the final round?

38. An island is to be divided among eight players $(P_1, P_2, P_3, \ldots, P_8)$ using the last-diminisher method. The players play in a fixed order, with P_1 first, P_2 second, etc. In rounds 1 and 5, everyone who has an opportunity to diminish does so. P_5 gets her fair share at the end of round 2. There are no diminishers in rounds 3 and 4. P_4 gets his fair share at the end of round 6.

(a) Which player gets a fair share at the end of round 1?

(b) Which player gets a fair share at the end of round 4?

(c) Which player gets a fair share at the end of round 5?

(d) How many diminishers are there in round 6?

(e) Which player is the divider in the final round?

F. The Method of Sealed Bids

39. Three sisters (Ana, Belle, and Chloe) wish to use the method of sealed bids to divide up 4 pieces of furniture they shared as children. Their bids on each of the items are given in the following table.

	Ana	Belle	Chloe
Dresser	$150	$300	$275
Desk	180	150	165
Vanity	170	200	260
Tapestry	400	250	500

Describe the final outcome of this fair-division problem.

40. Robert and Peter equally inherit their parents' old cabin and classic car. They decide to divide the two items using the method of sealed bids. Robert bids $29,200 on the car and $60,900 on the cabin. Peter bids $33,200 on the car and $65,300 on the cabin. Describe the final outcome of this fair-division problem.

41. Bob, Ann, and Jane wish to dissolve their partnership using the method of sealed bids. Bob bids $240,000 for the partnership, Ann bids $210,000, and Jane bids $225,000.

(a) Who gets the business and for how much?

(b) What do the other two partners get?

42. Three heirs (Andre, Bea, and Chad) wish to divide up an estate consisting of a house, a small farm, and a painting, using the method of sealed bids. The heirs' bids on each of the items are given in the following table.

	Andre	Bea	Chad
House	$150,000	$146,000	$175,000
Farm	430,000	425,000	428,000
Painting	50,000	59,000	57,000

Describe the final outcome of this fair-division problem.

43. Three players (A, B, and C) wish to divide up four items using the method of sealed bids. Their bids on each of the items are given in the following table.

	A	B	C
Item 1	$20,000	$18,000	$15,000
Item 2	46,000	42,000	35,000
Item 3	3,000	2,000	4,000
Item 4	201,000	190,000	180,000

Describe the final outcome of this fair-division problem.

44. Three players (A, B, and C) wish to divide up five items using the method of sealed bids. Their bids on each of the items are given in the following table.

	A	B	C
Item 1	$14,000	$12,000	$22,000
Item 2	24,000	15,000	33,000
Item 3	16,000	18,000	14,000
Item 4	16,000	16,000	18,000
Item 5	18,000	24,000	20,000

Describe the final outcome of this fair-division problem.

45. Five heirs (A, B, C, D, and E) wish to divide up an estate consisting of six items using the method of sealed bids. The heirs' bids on each of the items are given in the following table.

	A	B	C	D	E
Item 1	$352	$295	$395	$368	$324
Item 2	98	102	98	95	105
Item 3	460	449	510	501	476
Item 4	852	825	832	817	843
Item 5	513	501	505	505	491
Item 6	725	738	750	744	761

Describe the final outcome of this fair-division problem.

G. The Method of Markers

46. Three players (A, B, and C) agree to divide the 13 items shown by lining them up in order and using the method of markers. The players' bids are as indicated.

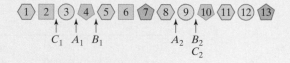

    (a) Describe the allocation of items to each player.

    (b) Which items are left over?

47. Three players (A, B, and C) agree to divide the 13 items shown by lining them up in order and using the method of markers. The players' bids are as indicated.

    (a) Describe the allocation of items to each player.

    (b) Which items are left over?

48. Two players (A and B) agree to divide the 12 items shown by lining them up in order and using the method of markers. The players' bids are as indicated.

    (a) Describe the allocation of items to each player.

    (b) Which items are left over?

49. Three players (A, B, and C) agree to divide the 12 items shown by lining them up in order and using the method of markers. The players' bids are as indicated.

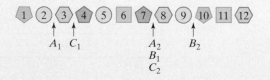

    (a) Describe the allocation of items to each player.

    (b) Which items are left over?

50. Three players (A, B, and C) agree to divide the 12 items shown by lining them up in order and using the method of markers. The players' bids are as indicated.

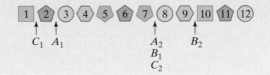

(a) Describe the allocation of items to each player.

(b) Which items are left over?

51. Five players (*A*, *B*, *C*, *D*, and *E*) agree to divide the 20 items shown by lining them up in order and using the method of markers. The players' bids are as indicated.

1 2 3 4 5 6 7 8 9 10 11 12 13 14 15 16 17 18 19 20

C_1 A_1 B_1 C_2 D_1 E_2 B_2 C_3 A_2 D_3 B_3 C_4 A_3 B_4 A_4
E_1 D_2 E_3 D_4 E_4

(a) Describe the allocation of items to each player.

(b) Which items are left over?

52. Four players (*A*, *B*, *C*, and *D*) agree to divide the 15 items shown below by lining them up in order and using the method of markers. The players' bids are as indicated.

1 2 3 4 5 6 7 8 9 10 11 12 13 14 15

C_1 B_1 C_2 A_1 A_2 B_2 D_1 D_2 A_3 C_3 B_3
                                   D_3

(a) Describe the allocation of items to each player.

(b) Which items are left over?

53. Four players (*A*, *B*, *C*, and *D*) agree to divide the 15 items shown by lining them up in order and using the method of markers. The players' bids are as indicated.

1 2 3 4 5 6 7 8 9 10 11 12 13 14 15

D_1 A_1 A_2 B_1 A_3 B_2 B_3 C_1 D_3 C_2 C_3
                        D_2

(a) Describe the allocation of items to each player.

(b) Which items are left over?

JOGGING

54. Every Friday night, Marty's Ice Cream Parlor sells "Kitchen Sink Sundacs" (Fridaes?) for $6.00 each. A KiSS consists of 12 mixed scoops of whatever flavors Marty wants to get rid of. The customer has no choice. Three friends (Abe, Babe, and Cassandra) decide to share one. Abe wants to eat half of it and pays $3.00 while Babe and Cassie pay $1.50 each. They decide to divide it by the lone-divider method. Abe spoons the sundae onto four plates (*P*, *Q*, *R*, and *S*) and says that he will be satisfied with any two of them.

(a) If both Babe and Cassie find only *Q* and *R* acceptable, discuss how to proceed.

(b) If Babe finds only *Q* and *R* acceptable, and Cassie finds only *P* and *S* acceptable, discuss how to proceed.

(c) If Babe and Cassie both find only *R* acceptable, discuss how to proceed.

55. Three friends, Peter, Paul, and Mary, each contribute $1.20 to purchase a $3.60 half-gallon brick of "Neapolitan" ice cream made of equal size bricks of strawberry, vanilla, and chocolate. To divide the ice cream they decide to use the lone-chooser method with Mary as the chooser. The value of the three flavors to each player is shown in the following figure:

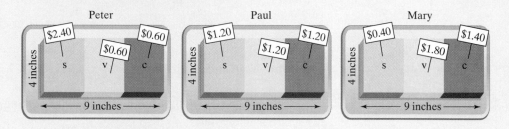

Peter starts by cutting the whole brick into two pieces as shown in the following figure.

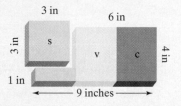

Assuming that all players play honestly and that all of the remaining cuts are horizontal, describe how the rest of the division would proceed. Who gets what, and how much is each player's share worth to the player receiving it.

56. Three players (*P*₁, *P*₂, and *P*₃) agree to divide the property shown using the last-diminisher method. The order of the players is P_1, P_2, P_3. The first player to play, P_1, makes a claim *C* as shown in the following figure.

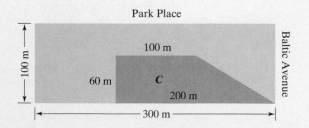

We know that both P_2's and P_3's value systems are the same and that they value the land uniformly.

(a) Give a geometric argument for why P_2 and P_3 would both pass in round 1 and P_1 would end up with C.

(b) Describe a possible cut that the divider in round 2 might make.

(c) Suppose that, after round 1 is over, P_2 and P_3 discover that the city requires that the next cut be made parallel to Park Place. Describe a possible cut that the divider in round 2 might make in this case.

(d) Repeat (c) for a cut that must be made parallel to Baltic Avenue.

57. Three players (P_1, P_2, and P_3) agree to divide the property shown using the last-diminisher method. The order of the players is P_1, P_2, P_3. The first player to play, P_1, makes a claim C as shown in the following figure.

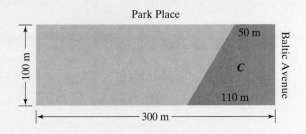

We know that both P_2's and P_3's value systems are the same and that they value the land uniformly.

(a) Give a geometric argument for why P_2 and P_3 would both pass in round 1 and P_1 would end up with C.

(b) Describe a possible cut that the divider in round 2 might make.

(c) Suppose that, after round 1 is over, P_2 and P_3 discover that the city requires that the next cut be made parallel to Baltic Avenue. Describe a possible cut that the divider in round 2 might make in this case.

58. Three players (P_1, P_2, and P_3) agree to divide the property shown using the last-diminisher method. The order of the players is P_1, P_2, P_3. The first player to play, P_1, makes a claim C as shown in the following figure.

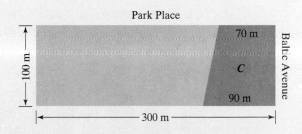

We know that both P_2's and P_3's value systems are the same and that they value the land uniformly, except for the square 20-meter-by-20-meter plot in the upper left corner of the property. This plot is contaminated by an old, underground gas station tank that will cost twice as much to remove and clean up as that square plot would otherwise be worth.

(a) Give an argument why P_2 and P_3 would both pass in round 1 and P_1 would end up with C.

(b) Suppose that after round 1 is over, P_2 and P_3 discover that the city requires that the next cut be made parallel to Baltic Avenue. Describe a possible cut that the divider in round 2 might make in this case.

59. Two players (A and B) wish to dissolve their partnership using the method of sealed bids. A bids $\$x$ and B bids $\$y$, where $x < y$.

   (a) How much are A and B's original fair shares worth?

   (b) How much is the surplus after the original allocations are made?

   (c) When all is said and done, how much must B pay A for A's half of the partnership?

60. Consider the following variation of the divider-chooser method for two players. After the divider cuts the cake into two pieces, the chooser (who is unable to see either piece) picks his piece randomly by flipping a coin. The divider, of course, gets the other piece.

   (a) Is this a fair-division scheme according to our definition? Explain your answer.

   (b) Who would you rather be—divider or chooser? Explain.

61. Four roommates (Quintin, Ramon, Stephone, and Tim) want to divide 18 small items by the method of markers. The items are lined up as shown.

❶ ❷ ❷ ❸ ❷ ❶ ❶ ❹ ❸ ❸ ❸ ❷ ❸ ❷ ❸ ❷ ❹ ❹

The (secret) values of the items by the roommates are shown in the following table.

	Quintin	Ramon	Stephone	Tim
Each ❶ is worth	$12	$9	$8	$5
Each ❷ is worth	$7	$5	$7	$4
Each ❸ is worth	$4	$5	$6	$4
Each ❹ is worth	$6	$11	$14	$7

   (a) Show where each roommate would place his markers. Use Q_1, Q_2, Q_3 for Quintin's markers, R_1, R_2, R_3 for Ramon's markers, etc.

   (b) Describe who gets what piece and which pieces are left over.

   (c) How would you divide the leftovers?

62. Three players (A, B, and C) agree to divide some candy using the method of markers. The candy consists of 3 Nestle Crunch Bars, 6 Snickers Bars, and 6 Reese's Peanut Butter Cups lined up exactly as shown in the following array.

The players' value systems are as follows.

A loves Nestle Crunch Bars but does not like Snickers Bars or Reese's Peanut Butter Cups at all.

B loves Snickers Bars and Nestle Crunch Bars equally well (i.e., 1 Snickers Bar = 1 Nestle Crunch Bar), but does not like Reese's Peanut Butter Cups at all.

C loves Snickers Bars and Reese's Peanut Butter Cups equally well (i.e., 1 Snickers Bar = 1 Reese's Peanut Butter Cup), but is allergic to Nestle Crunch Bars.

(a) What bid would *A* make to ensure that she gets her fair share (according to her value system)?

(b) What bid would *B* make to ensure that he gets his fair share (according to his value system)?

(c) What bid would *C* make to ensure that she gets her fair share (according to her value system)?

(d) Describe the allocations to each player.

(e) What items are left over?

RUNNING

Exercises 63 and 64 show how the method of sealed bids can be used when some values are negative. If you offer to pay a bid (as for a purchase), the bid is listed as a positive amount. It follows that, if you offer to receive a bid (as for labor), the bid is listed as a negative amount. Regardless, the winning bid is the highest number.

63. Three women (Ruth, Sarah, and Tamara) share a house and wish to divide the chores: bathrooms, cooking, dishes, laundry, and vacuuming. For each chore, they privately write the least they are willing to receive monthly (their negative valuation) in return for doing that chore. The results are shown in the following table.

	Ruth	Sarah	Tamara
Clean bathrooms	$ -20	$ -30	$ -40
Do cooking	$ -50	$ -10	$ -25
Wash dishes	$ -30	$ -20	$ -15
Mow the lawn	$ -30	$ -20	$ -10
Vacuum and dust	$ -20	$ -40	$ -15

Divide the chores using the method of sealed bids. Who does which chores? Who gets paid, and how much? Who pays, and how much?

64. Four roommates are going their separate ways after graduation and wish to divide up their jointly owned furniture (equal shares) and the moving chores by the method of sealed bids. Their bids (in dollars) on the items are shown in the following table.

	Quintin	Ramon	Stephone	Tim
Stereo	300	250	200	280
Couch	200	350	300	100
Table	250	200	240	80
Desk	150	150	200	220
Cleaning the rugs	-80	-70	-100	-60
Patching nail holes	-60	-30	-60	-40
Repairing the window	-60	-50	-80	-80

(a) What is each roommate's estimate of his part of the total?

(b) How much surplus cash is there?

(c) What is the final outcome?

(d) What percentage of the total value (of everything) does each roommate get using the roommate's own valuation?

(e) If Tim is dishonest and sneaks a peek at the bid lists of the other 3 roommates before filling out his own, how could he adjust his bids (in whole dollars) so as to get the same furniture as before, but no chores, and also maximize his cash receipts? Explain your reasoning.

65. Three players (A, B, and C) are going to share a cake consisting of three equal sections of chocolate, strawberry, and vanilla. The three figures below show how each player values each of the sections of cake (given as a percentage of the total value of the cake). Describe how you could cut this cake into three pieces so that each player can get a piece that he or she values at exactly 50% of the value of the cake.

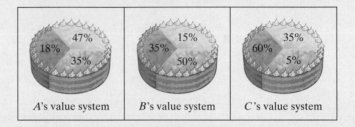

| A's value system | B's value system | C's value system |

66. (a) Suppose that four players divide a cake using the lone-chooser method. The chooser is C, the dividers are D_1, D_2, and D_3. Explain why, when properly played, the method guarantees to each player a share that is worth at least 25% of the cake.

(b) Suppose that N players divide a cake using the lone-chooser method. The chooser is C, the dividers are D_1, D_2, ..., D_{N-1}. Explain why, when properly played, the method guarantees to each player a fair share.

67. (a) Explain why, in the method of sealed bids, the surplus (after the original allocation and payments are made) is always positive or zero.

(b) Describe the circumstances under which the surplus is zero.

PROJECTS AND PAPERS

A. Envy-Free Fair Division

An *envy-free fair division* is a fair division in which each player ends up with a share that he or she feels is as good or better than that of any other player. Thus, in an envy-free fair division a player would never envy or covet another player's share. In the last decade, several important envy-free fair division methods have been developed.

Write a paper discussing the topic of envy-free fair division.

Notes: Some ideas of topics for your paper: (i) Discuss how envy-free fair division differs from the (proportional) type of fair division discussed in this chapter, (ii) describe a continuous envy-free fair division method for $N = 3$ players, (iii) give an outline of the Brams-Taylor method for continuous envy-free fair division for any number of players.

Strongly suggested readings for this paper are as follows (in this order):

* Stewart, Ian, "Division without Envy," *Scientific American*, January 1999.

* Hively, Will, "Dividing the Spoils," *Discover*, March 1995.

* Brams Steven, and Taylor, Alan, "An Envy-Free Cake Division Protocol," *American Mathematical Monthly*, January 1995.

B. Fair Divisions with Unequal Shares

All the fair-division problems we discussed in this chapter were based on the assumption of *symmetry* (i.e., all players have equal rights in the division). Sometimes, players are not all equal and are entitled to larger or smaller shares than other players. This type of fair-division problem is called an *asymmetric* fair-division (*asymmetric* means that the players are not all equal in their rights.) For example, Grandma's will may stipulate that her estate is to be divided as follows: Art is entitled to 25%, Betty is entitled to 35%, Carla is entitled to 30%, and Dave is entitled to 10%. (After all, it is her will, and if she wants to be difficult, she can!)

Write a paper discussing how some of the fair-division methods discussed in this chapter can be adapted for the case of asymmetric fair division. Discuss at least one discrete and one continuous asymmetric fair-division method.

C. The Mathematics of Forgiveness and Cooperation

It is generally acknowledged that as a species, humans are more selfless, cooperative, and forgiving to each other than any other animal species. On the surface, this appears to be a contradiction to the general Darwinian notion that only the ruthless get ahead. In recent years, scientists have been able to model the *cooperation* versus *noncooperation* question using a mathematical game called the *prisoner's dilemma*. Recent studies using the *prisoner's dilemma* have shown that (i) in the long run, cooperation makes sense mathematically, and (ii) the human impulse to cooperate may have a biochemical root.

Write a research paper describing these recent developments. This is a topic where mathematics, game theory, biology and neuroscience all interact, and you should touch on these interactions in your paper.

Notes: You should use at least the following three sources:

* Bass, Thomas, "Forgiveness Math," *Discover*, May 1993.

* Browne, Malcolm W., "Biologists Use Computer Game to Tally Generosity's Rewards," *New York Times*, April 14, 1992.

* Angier, Natalie, "Why We're So Nice: We're Wired to Cooperate," *New York Times*, July 23, 2002.

Why We're So Nice:
We're Wired to Cooperate

By NATALIE ANGIER

What feels as good as chocolate on the tongue or money in the bank but won't make you fat or risk a subpoena from the Securities and Exchange Commission?

Hard as it may be to believe in these days of infectious greed and sabers unsheathed, scientists have discovered that the small, brave act of cooperating with another person, of choosing trust over cynicism, generosity over selfishness, makes the brain light up with quiet joy.

Studying neural activity in young women who were playing a classic laboratory game called the Prisoner's Dilemma, in which participants can select from a number of greedy or cooperative strategies as they pursue financial gain, researchers found that when the women chose mutualism over "me-ism," the mental circuitry normally associated with reward-seeking behavior swelled to life.

And the longer the women engaged in a cooperative strategy, the more strongly flowed the blood to the pathways of pleasure.

The researchers, performing their work at Emory University in Atlanta, used magnetic resonance imaging to take what might be called portraits of the brain on hugs.

"The results were really surprising to us," said Dr. Gregory S. Berns, a psychiatrist and an author on the new report, which appears in the current issue of the journal Neuron. "We went in expecting the opposite."

The researchers had thought that the biggest response would occur in cases where one person cooperated and the other defected, when the cooperator might feel that she was being treated unjustly.

Instead, the brightest signals arose in cooperative alliances and in those neighborhoods of the brain already known to respond to desserts, pictures of pretty faces, money, cocaine, and any number of licit or illicit delights.

"It's reassuring," Dr. Berns said. "In some ways, it says that we're wired to cooperate with each other."

The study is among the first to use MRI technology to examine social interactions in real time, as opposed to taking brain images while subjects stared at static pictures or thought prescribed thoughts.

It is also a novel approach to exploring an ancient conundrum: Why are humans so, well, nice? Why are they willing to cooperate with people whom they barely know and to do good deeds and to play fair a surprisingly high percentage of the time?

Scientists have no trouble explaining the evolution of competitive behavior. But the depth and breadth of human altruism, the willingness to forgo immediate personal gain for the long-term common good, far exceeds behaviors seen even in other large-brained social species like chimpanzees and dolphins, and it has as such been difficult to understand.

"I've pointed out to my students how impressive it is that you can take a group of young men and women of prime reproductive age, have them come into a classroom, sit down, and be perfectly comfortable and civil to each other," said Dr. Peter J. Richerson, a professor of environmental science and policy at the University of California at Davis and an influential theorist in the field of cultural evolution. "If you put 50 male and 50 female chimpanzees that don't know each other into a lecture hall, it would be a social explosion."

Dr. Ernst Fehr of the University of Zurich and colleagues recently presented findings on the importance of punishment in maintaining cooperative behavior among humans and the willingness of people to punish those who commit crimes or violate norms, even when the chastisers take risks and gain nothing themselves while serving as ad hoc police.

In her survey of the management of so-called commons in small-scale communities where villagers have the right, for example, to graze livestock on commonly held land, Dr. Elinor Ostrom of Indiana University found that all communities have some form of monitoring to gird against cheating or using more than a fair share of the resource.

In laboratory games that mimic small-scale commons, Dr. Richerson said, 20 to 30 percent have to be coerced by a threat of punishment to cooperate.

Fear alone is not highly likely to inspire cooperative behavior to the degree observed among humans. If research like Dr. Fehr's shows the stick side of the equation, the newest findings present the neural carrot—people cooperate because it feels good to do it.

In the new findings, the researchers studied 36 women from 20 to 60 years old, many of them students at Emory and inspired to participate by the promise of monetary rewards. The scientists chose an all-female sample because so few brain-imaging studies have looked at only women. Most have been limited to men or to a mixture of men and women.

But there is a vast body of nonimaging data that rely on using the Prisoner's Dilemma.

"It's a simple and elegant model for reciprocity," said Dr. James K. Rilling, an author on the Neuron paper who is at Princeton. "It's been referred to as the E. coli of social psychology."

From past results, the researchers said, one can assume that neuroimaging studies of men playing the game would be similar to their new findings with women.

The basic structure of the trial had two women meet each other briefly ahead of time. One was placed in the scanner while the other remained outside the scanning room. The two interacted by computer, playing about 20 rounds of the game. In every round, each player pressed a button to indicate whether she would "cooperate," or "defect." Her answer would be shown on-screen to the other player.

The monetary rewards were apportioned after each round. If one player defected and the other cooperated, the defector earned $3 and the cooperator nothing. If both chose to cooperate, each earned $2. If both opted to defect, each earned $1.

Hence, mutual cooperation from start to finish was a far more profitable strategy, at $40 a woman, than complete mutual defection, which gave each $20.

The risk that a woman took each time she became greedy for a little bit more was that the cooperative strategy would fall apart and that both would emerge the poorer.

In some cases, both women were allowed to pursue any strategy that they chose. In other cases, the nonscanned woman would be a "confederate" with the researchers, instructed, unbeknown to the scanned subject, to defect after three consecutive rounds of cooperation, the better to keep things less rarefied and pretty and more lifelike and gritty.

In still other experiments, the woman in the scanner played a computer and knew that her partner was a machine. In other tests, women played a computer but thought that it was a human.

The researchers found that as a rule the freely strategizing women cooperated. Even occasional episodes of defection, whether from free strategizers or confederates, were not necessarily fatal to an alliance.

"The social bond could be reattained easily if the defector chose to cooperate in the next couple of rounds," another author of the report, Dr. Clinton D. Kilts, said, "although the one who had originally been "betrayed" might be wary from then on."

As a result of the episodic defections, the average per-experiment take for the participants was in the $30's. "Some pairs, though, got locked into mutual defection," Dr. Rilling said.

Analyzing the scans, the researchers found that in rounds of cooperation, two broad areas of the brain were activated, both rich in neurons able to respond to dopamine, the brain chemical famed for its role in addictive behaviors.

One is the anteroventral striatum in the middle of the brain right above the spinal cord. Experiments with rats have shown that when electrodes are placed in the striatum, the animals will repeatedly press a bar to stimulate the electrodes, apparently receiving such pleasurable feedback that they will starve to death rather than stop pressing the bar.

Another region activated during cooperation was the orbitofrontal cortex in the region right above the eyes. In addition to being part of the reward-processing system, Dr. Rilling said, it is also involved in impulse control.

"Every round, you're confronted with the possibility of getting an extra dollar by defecting," he said. "The choice to cooperate requires impulse control."

Significantly, the reward circuitry of the women was considerably less responsive when they knew that they were playing against a computer. The thought of a human bond, but not mere monetary gain, was the source of contentment on display.

In concert with the imaging results, the women, when asked afterward for summaries of how they felt during the games, often described feeling good when they cooperated and expressed positive feelings of camaraderie toward their playing partners.

Assuming that the urge to cooperate is to some extent innate among humans and reinforced by the brain's feel-good circuitry, the question of why it arose remains unclear. Anthropologists have speculated that it took teamwork for humanity's ancestors to hunt large game or gather difficult plant foods or rear difficult children. So the capacity to cooperate conferred a survival advantage on our forebears.

Yet as with any other trait, the willingness to abide by the golden rule and to be a good citizen and not cheat and steal from one's neighbors is not uniformly distributed.

"If we put some CEOs in here, I'd like to see how they respond," Dr. Kilts said. "Maybe they wouldn't find a positive social interaction rewarding at all."

A Prisoner's Dilemma indeed.

REFERENCES AND FURTHER READINGS

1. Brams, Steven, and Alan Taylor, "An Envy-Free Cake Division Protocol," *American Mathematical Monthly*, 102 (1995), 9–18.

2. Brams, Steven, and Alan Taylor, *Fair Division*. Cambridge, England: Cambridge University Press, 1996.

3. Brams, Steven, and Alan Taylor, *The Win-Win Solution: Guaranteeing Fair Shares to Everybody*. New York: W. W. Norton, 1999.

4. Dubins, L. E., "Group Decision Devices," *American Mathematical Monthly*, 84 (1977), 350–356.

5. Fink, A. M., "A Note on the Fair Division Problem," *Mathematics Magazine*, 37 (1964), 341–342.

6. Gardner, Martin, *aha! Insight*. New York: W. H. Freeman, 1978.

7. Hill, Theodore, "Determining a Fair Border," *American Mathematical Monthly*, 90 (1983), 438–442.

8. Hill, Theodore, "Mathematical Devices for Getting a Fair Share," *American Scientist*, 88 (2000), 325–331.

9. Hively, Will, "Dividing the Spoils," *Discover*, 16 (1995), 49–57.

10. Jones, Martin, "A Note on a Cake Cutting Algorithm of Banach and Knaster," *American Mathematical Monthly*, 104 (1997), 353–355.

11. Kaluza, Roman, *Through A Reporter's Eyes: The Life of Stefan Banach*. Boston, MA: Birkhauser, 1996.

12. Kuhn, Harold W., "On Games of Fair Division," *Essays in Mathematical Economics*, Martin Shubik, ed. Princeton, NJ: Princeton University Press, 1967, 29–37.

13. Olivastro, Dominic, "Preferred Shares," *The Sciences*, March–April 1992, 52–54.

14. Robertson, Jack, and William Webb, *Cake Cutting Algorithms: Be Fair If You Can*. Natick, MA: A. K. Peters, 1998.

15. Steinhaus, Hugo, *Mathematical Snapshots*, 3rd ed. New York: Dover, 1999.

16. Steinhaus, Hugo, "The Problem of Fair Division," *Econometrica*, 16 (1948), 101–104.

17. Stewart, Ian, "Fair Shares for All," *New Scientist*, 146 (June, 1995), 42–46.

18. Stewart, Ian, "Mathematical Recreations: Division without Envy," *Scientific American*, (1999), 110–111.

19. Stromquist, Walter, "How to Cut a Cake Fairly," *American Mathematical Monthly*, 87 (1980), 640–644.

Supreme Court Upholds Method
Used in Apportionment of House

By LINDA GREENHOUSE
Special to The New York Times

WASHINGTON, March 31 — In a decision that dashed Montana's hope of retaining two seats in the House of Representatives, the Supreme Court today upheld the constitutionality of the method Congress has used for 50 years to apportion seats among the states.

The unanimous ruling overtu decision issued

in 1941, after long study and on the basis of expert advice, meets constitutional requirements.

The Court acted with unusual speed to decide the Federal Government's appeal from the District Court ruling. The case, U.S. Department merce v. M

requires at one Representative for each state.

Absolute or Relative?

So the question in this case was whether Congress is constitutionally obliged to minimize the absolute population differences among the districts, as the District Court had concluded in rejecting the method of examining relative, rather than absolute, differences.

In his opinion for the Court today, Justice John Paul Stevens said that "neither mathematical analysis nor constitutional interpretation provides a conclusive answer" to the question of "what is the better measure of inequality."

Since there is bound to be "a significant departure from the ideal," he said, the method that Congress adopted

that had been by Alexander Hamilton and was thought to favor Northern states.

The method Congress adopted in 1941, on the recommendation of the National Academy of Sciences, is a mathematical formula known as the method of equal proportions. It minimizes the relative difference between the size of Congressional districts and between the number of Representatives per person. Perhaps most significantly, as Justice Stevens noted today, of the five formulas that Congress considered, this one incorporates the least bias toward small or large states.

Today's decision will result in Montana having the biggest single Congressional district in the country, including all the state's 804,000 people, while the ideal Congressional district would have 572,000. Had the state been divided into two districts of 402,000 each, each district would have been 171,000 smaller than the ideal.

The Mathematics of Apportionment

Making the Rounds

In the stifling heat of the Philadelphia summer of 1787, delegates from the thirteen states met to draft a Constitution for a new nation. Except for Thomas Jefferson (then minister to France) and Patrick Henry (who refused to participate), all the main names of the American Revolution were there—George Washington, Ben Franklin, Alexander Hamilton, James Madison.

Without a doubt, the most important and heated debate at the Constitutional Convention concerned the makeup of the legislature. The small states, led by New Jersey, wanted all states to have the same number of representatives. The larger states, led by Virginia, wanted some form of proportional representation. The final compromise, known as the Connecticut Plan, is all too familiar to us: a Senate, in which every state has two senators, and a House of Representatives, in which each state has a number of representatives that is a function of its population.

> *Representatives ... shall be apportioned among the several States, which may be included within this Union, according to their respective Numbers ...*
> *Article I, Section 2.*
> *U.S. Constitution*

While the Constitution is clear about the fact that seats in the House of Representatives are to be allocated to the states based on their populations (... *according to their respective Numbers* ...), it does not say how the calculations are to be done. Undoubtedly, the Founding Fathers felt that this was a relatively minor detail—a matter of simple arithmetic that could be easily figured out and agreed upon by reasonable people. Certainly it was not the kind of thing to clutter a Constitution with, or spend time arguing over in the heat of the summer. What the Founding Fathers did not realize is that Article 1, Section 2, set the Constitution of the United States into a collision course with a mathematical iceberg known today (but certainly not then) as *the apportionment problem*.

What is an apportionment? Why is it a problem? Why is the problem so complicated? Why should anyone care? These, in essence, are the questions we will answer in this chapter. In so doing, we will learn some interesting mathematics and at the same time get a glimpse of a little-known but fascinating chapter of United States history.

4.1 The Apportionment Problem

What is generally now known as the **apportionment problem** is really a special kind of *discrete fair-division problem*—a sort of dual of some of the problems we discussed in Chapter 3. As in Chapter 3, we have *indivisible objects* that we would like to divide fairly among a set of *players*. The difference is this: In Chapter 3, each player was entitled to an equal share but the objects were different; now *the objects are all going to be the same, but the players are going to be entitled to different-sized shares*.

The most important example of an apportionment problem is that of *proportional representation* in a legislative body, exactly the kind of problem faced by our Founding Fathers in 1787. Here, the identical, *indivisible objects* to be apportioned are *seats* in the legislature, and the players are the *states* (or provinces, regions, etc.). The idea of proportional representation is that each state is entitled to a number of seats that is proportional to its population. Most of our discussion for this chapter will take place in the context of this particular type of apportionment problem, but it is important to realize that apportionment problems occur in many other guises as well. The point is best illustrated with a couple of examples.

ap · pôr · tion: *to divide and assign in due and proper proportion or according to some plan*

Webster's New Twentieth Century Dictionary

EXAMPLE 4.1 Kitchen Capitalism

Mom has a total of 50 identical, indivisible pieces of caramel candy which she is going to divide among her five children. Like any good mom, she is intent on doing this fairly. Of course, the easiest thing to do would be to give each child 10 caramels, but mom has loftier goals—she wants to teach her children about the value of work and about the relationship between work and reward. This leads her to the following idea. The candy is going to be *apportioned* (i.e., *divided in due and proper proportion*) among the children based strictly on the amount of time each child spends helping with the weekly kitchen chores.

Here we are, trying to divide candy once again! But now things are quite different from the way they were in Chapter 3. We have 50 identical objects (the caramels) to be divided among 5 players (the kids), each of which is entitled to a different share of the total. How should this be done?

Table 4-1 shows the amount of work (in minutes) done by each child during the week.

TABLE 4-1   **Amount of Work (in minutes) per Child**

Child	Alan	Betty	Connie	Doug	Ellie	Total
Minutes worked	150	78	173	204	295	900

Once the figures are in, it is time to apportion the candy. According to the ground rules, Alan, who worked 150 out of a total of 900 minutes, is entitled to $16\frac{2}{3}\%$ of the 50 pieces of candy [$(150/900) = 16\frac{2}{3}\%$], or $8\frac{1}{3}$ pieces. Here comes the problem: Since the pieces of candy are indivisible, it is impossible for Alan to get the exact share he is entitled to—he can get 8 pieces (and get shorted) or he can get 9 pieces (and someone else will get shorted). A similar problem occurs with each of the other children. Betty's exact fair share should be $4\frac{1}{3}$ pieces; Connie's should be $9\frac{11}{18}$ pieces; Doug's, $11\frac{1}{3}$ pieces; and Ellie's $16\frac{7}{18}$ pieces. (We leave it to the reader to verify these figures.) Because none of these shares can be realized, an absolutely fair apportionment of the candy is going to be impossible. What should mom do? (What would you do in her place?)   ▪

Our next example shows a more classical version of an apportionment problem.

EXAMPLE 4.2 The Intergalactic Congress

It is the year 2525, and all the planets in the galaxy have finally signed a peace treaty. Five of the planets (Alanos, Betta, Conii, Dugos, and Ellisium) decide to join forces and form an Intergalactic Federation. The Federation will be ruled by an Intergalactic Congress consisting of 50 delegates, and each of the 5 planets will be entitled to a number of delegates that is proportional to its population. The population data for each of the planets (in billions) is shown in Table 4-2. How many delegates should each planet get?

TABLE 4-2	Intergalactic Federation: Population Figures (in billions) for 2525					
Planet	Alanos	Betta	Conii	Dugos	Ellisium	Total
Population	150	78	173	204	295	900

This example is not just another example of an apportionment problem—it is Example 4.1 revisited. When we compare Example 4.2 with Example 4.1, we see that the numbers are identical—it is only the setting that has changed. While the merits of the problem may be different, mathematically speaking, Examples 4.1 and 4.2 are one and the same!   ▪

Between the extremes of apportioning the seats in the Intergalactic Congress (important, but too far away!) and apportioning the caramels among the children (closer to home, but the galaxy will not come to an end if some of the kids get shorted!) fall many other applications that are both important and relevant: apportioning nurses to shifts in a hospital, apportioning telephone calls to switchboards in a network, apportioning subway cars to routes in a subway system, etc.

Our primary purpose in this chapter is to learn various **apportionment methods** for solving apportionment problems, something that sounds reasonably simple but has many subtleties and surprises. In fact, our discussions in this chapter will be somewhat reminiscent of our experiences with *voting methods* in Chapter 1.

Over the years, statesmen, politicians, and mathematicians have designed many ingenious apportionment methods, and we will study some of the best known in this chapter. Interestingly, the names associated with many of these apportionment methods—Alexander Hamilton, Thomas Jefferson, John Quincy Adams, and Daniel Webster—one would expect to find in a history book, rather than a mathematics book. This is one of the fascinating sidebars to the apportionment problem—the way history, politics, and mathematics become intertwined.

4.2   The Mathematics of Apportionment: Basic Concepts

In this section we will introduce some of the basic concepts in the mathematics of apportionment. We will motivate these concepts with the following example. (Much of our discussion throughout the chapter will be centered around this example.) While the story may be ficticious, the issues are real.

EXAMPLE 4.3 The Congress of Parador

Parador is a new republic located in Central America. It is made up of six states: Azucar, Bahia, Cafe, Diamante, Esmeralda, and Felicidad (*A, B, C, D, E,* and *F* for short). According to the new constitution of Parador, the Congress will have 250 seats, divided among the six states according to their respective populations. The population figures for each state are given in Table 4-3.

TABLE 4-3	Republic of Parador (Population Data by State)						
State	*A*	*B*	*C*	*D*	*E*	*F*	Total
Population	1,646,000	6,936,000	154,000	2,091,000	685,000	988,000	12,500,000

A natural starting point for our mathematical adventure is to calculate the ratio of national population to number of seats in Congress, a sort of national average of people per seat. Since the total population of Parador is 12,500,000 and the number of seats is 250, this average is 12,500,000/250 = 50,000.

In general, for any apportionment problem in which the total population of the country is *P* and the number of seats to be apportioned is *M,* the ratio *P/M* gives *the number of people per seat in the legislature on a national basis.* We will call the ratio *P/M* the **standard divisor.**[1]

$$\text{Standard divisor } (SD) = \frac{\text{population } (P)}{\text{total number of seats } (M)}$$

Using the standard divisor, we can calculate the *fraction of the total number of seats that each state would be entitled to if fractional seats were possible.* This number, which we will call the state's **standard quota** (sometimes known as the *exact* or *fair quota*), is obtained by dividing the state's population by the standard divisor.

$$\text{State } X\text{'s standard quota} = \frac{\text{state } X\text{'s population}}{SD}$$

Table 4-4 shows the standard quota of each state in Parador. The numbers are given to two decimal places (which is usually enough). The quotas were obtained by dividing each state's population by the standard divisor, in this case 50,000. Notice that the sum of all the standard quotas is 250, the total number of seats to be divided.

---

[1]Note that, although our definition is given in the context of Example 4.3, the concept of standard divisor applies to any apportionment problem, regardless of the context. In our original candy example, the standard divisor is *P/M* = 900/50 = 18, and it represents the average number of minutes of work needed to earn a single caramel.

TABLE 4-4	Republic of Parador: Standard Quotas for Each State ($SD = 50{,}000$)						
State	A	B	C	D	E	F	Total
Population	1,646,000	6,936,000	154,000	2,091,000	685,000	988,000	12,500,000
Standard quota	32.92	138.72	3.08	41.82	13.70	19.76	250

Associated with each state's standard quota are two whole numbers: the state's **lower quota**, the standard quota rounded down, and the state's **upper quota**, the standard quota rounded up. (For example, state A with a standard quota of 32.92 has a lower quota of 32 and an upper quota of 33.) In the unusual case that the state's standard quota is a whole number, then the lower and upper quotas are both equal to the standard quota.

The standard quotas represent each state's exact fair share of the 250 seats, and if the seats in the legislature could be chopped up into fractional parts, we would be done. Unfortunately, seats have to be given out whole, so it now becomes a question of *how* to round the standard quotas into whole numbers. At first glance, the obvious strategy would appear to be the traditional approach to rounding we learned in school, which we call *conventional rounding*: Round down if the fractional part is less than 0.5, round up otherwise. Unfortunately, this approach is not guaranteed to work. Look at what happens when we try it with the standard quotas of the states in Parador (Table 4-5).

TABLE 4-5	Conventional Rounding Doesn't Always Work!		
State	Population	Standard quota	Rounded to
A	1,646,000	32.92	33
B	6,936,000	138.72	139
C	154,000	3.08	3
D	2,091,000	41.82	42
E	685,000	13.70	14
F	988,000	19.76	20
Total	12,500,000	250.00	**251**

As the total in the last column of Table 4-5 shows, we have a slight problem: We are giving out 251 seats in Congress! Where is that extra seat going to come from?

Table 4-5 illustrates the problem with conventional rounding of the standard quotas—a seductive idea that doesn't always work. In the case of Parador's Congress, we ended up apportioning more seats than we were supposed to; other times we could end up apportioning fewer. Occasionally, by sheer luck, the numbers might work out just right.

The fact that conventional rounding of the standard quotas does not usually work is disappointing, but hardly a surprise at this point. After all, if this obvious and simple-minded approach worked all the time, the whole issue of apportionment would be mathematically trivial (and there wouldn't be a reason for this chapter to exist!). Given this, we will be forced to consider more sophisticated approaches to apportionment. Our strategy for the rest of this chapter will be to look at several important apportionment methods (important both historically and mathematically) and find out what is good and bad about each one.

4.3 Hamilton's Method and the Quota Rule

Alexander Hamilton, (1757–1804).*Alexander Hamilton, by John Trumball, c. 1804, oil on canvas, 30 1/2 x 25 1/2 inches, negative number 6242, accession number 1867.305.* © Collection of The New-York Historical Society.

While historically Hamilton's method did not come first, we will discuss it first because it is mathematically the simplest.[2]

HAMILTON'S METHOD

- **Step 1.** Calculate each state's standard quota.
- **Step 2.** Give to each state (for the time being) its *lower quota*. In other words, round each state's quota down.
- **Step 3.** Give the surplus seats (one at a time) to the states with the largest fractional parts until there are no more surplus seats.

Essentially, Hamilton's method can be described as follows: Every state gets at least its lower quota. As many states as possible get their upper quota, with the one with highest fractional part having first priority, the one with second highest fractional part second priority, and so on.

EXAMPLE 4.4 Hamilton's Method Meets Parador's Congress

Let's apply Hamilton's method to apportion Parador's Congress. Table 4-6 shows all the details and speaks for itself. (The reader is reminded that the standard quotas are found in Table 4-5.)

TABLE 4-6 Republic of Parador: Apportionment Based on Hamilton's Method

State	Population	Standard quota (Step 1)	Lower quota (Step 2)	Fractional part	Surplus seats (Step 3)	Final apportionment
A	1,646,000	32.92	32	0.92	1 (1st)	33
B	6,936,000	138.72	138	0.72	1 (4th)	139
C	154,000	3.08	3	0.08		3
D	2,091,000	41.82	41	0.82	1 (2nd)	42
E	685,000	13.70	13	0.70		13
F	988,000	19.76	19	0.76	1 (3rd)	20
Total	12,500,000	250.00	246	4.00	4	250

At first glance, Hamilton's method appears to be quite fair. However, a careful look at Example 4.4 already shows some hints of possible unfairness. Compare the fates of state *B*, with a fractional part of 0.72, and state *E*, with a fractional part of 0.70. State *B* gets the last surplus seat; state *E* gets nothing! Sure enough, 0.72 is more than 0.70, so following the rules of Hamilton's method we give priority to *B*. By the same token, *B* is a huge state, and as a percentage of

---

[2]Hamilton's method is also known as the *method of largest remainders* and sometimes as *Vinton's method*. The method is still used to apportion the legislatures of other countries, including Costa Rica and Sweden.

its population the 0.72 represents an insignificant amount, whereas state E is a relatively small state, and its fractional part of 0.70 (for which it gets nothing) represents more than 5% of its population ($0.70/13.70 \approx 0.051 = 5.1\%$). It could be reasonably argued that the method's absolute reliance on fractional parts without consideration as to what those fractional parts represent could lead to what many might consider unfair apportionments.[3]

An apportionment method is said to be *neutral* if it shows no systematic bias in favor of larger states over smaller ones, or vice versa. Under a neutral apportionment method, any state has the same opportunity for a favorable apportionment as any other state. It turns out to be the case that Hamilton's method fails to meet the neutrality criterion—it consistently favors large states over small ones. And, as we will soon see, this is just one of the many flaws of Hamilton's method.

To be totally fair, Hamilton's method has two important things going for it: (1) It is very easy to understand, and (2) it satisfies an extremely important requirement for fairness called the *quota rule*.

The Quota Rule

Suppose that the standard quota of some state X is 38.59. This number represents the benchmark for what a fair apportionment of seats to X ought to be. Since giving X exactly 38.59 seats is impossible, it is reasonable to argue that the only two fair choices are giving X its upper quota of 39 seats or its lower quota of 38 seats. Thus, if for some reason, X ended up with 40 seats, we might consider this a violation of a basic principle of fairness—after all, it is clear that X is getting much more than its fair share. On the flip side, say X ended up with 37 seats. There is something wrong with this picture as well—this an unfair apportionment because X is getting seriously shorted.

We will now formalize the preceding observations.

THE QUOTA RULE

- A state's fair apportionment should be either its upper quota or its lower quota.

- An apportionment *method* that guarantees that every state will be apportioned either its lower quota or its upper quota is said to *satisfy the quota rule*.

- Violations of the quota rule can be of two types: when a state ends up with an apportionment smaller than its lower quota, it is called a **lower-quota violation**; when a state ends up with an apportionment larger than its upper quota, it is called an **upper-quota violation**.

Hamilton's method satisfies the quota rule: It guarantees to every state at least its lower quota, and, since no state gets more than one surplus seat, no state can end up with more than its upper quota. So far, so good!

---

[3]The idea that relative rather than absolute fractional parts should determine the order in which the surplus seats are handed out is the basis for an apportionment method known as *Lowndes's method*. For details, the reader is referred to Exercise 44.

4.4 The Alabama Paradox

The most serious (in fact, the fatal) flaw of Hamilton's method is commonly known as the **Alabama paradox**. In essence, the Alabama paradox occurs when an *increase in the total number of seats being apportioned, in and of itself, forces a state to lose one of its seats*. The best way to understand what this means is to look carefully at the following example.

EXAMPLE 4.5

A small country consists of three states: *A*, *B*, and *C*. Table 4-7 shows the apportionment under Hamilton's method when there are $M = 200$ seats to be apportioned. Table 4-8 shows the apportionment under Hamilton's method when there are $M = 201$ seats to be apportioned. The reader is encouraged to verify the necessary calculations. (Here is some help: The standard divisor when $M = 200$ is 100; the standard divisor when $M = 201$ is approximately 99.5.)

TABLE 4-7   Apportionment Under Hamilton's Method for $M = 200$

State	Population	Standard quota when $M = 200$	Apportionment under Hamilton's method
A	940	9.4	10
B	9030	90.3	90
C	10,030	100.3	100
Total	20,000	200.0	200

Using Hamilton's method to apportion the seats, we can see that when there are $M = 200$ seats to be apportioned, *A* gets the only surplus seat and the final apportionment gives 10 seats to *A*, 90 seats to *B*, and 100 seats to *C* (Table 4-7).

What happens when the number of seats to be apportioned increases to $M = 201$? Now there are 2 surplus seats but they go to *B* and *C*, so that *A* ends up with just 9 seats (Table 4-8).

TABLE 4-8   Apportionment Under Hamilton's Method for $M = 201$

State	Population	Standard quota when $M = 201$	Apportionment under Hamilton's method
A	940	9.45	9
B	9030	90.75	91
C	10,030	100.80	101
Total	20,000	201.00	201

The shocking conclusion of Example 4.5 is that under Hamilton's method it is possible for a state to receive a smaller apportionment with a larger legislature than with a smaller one. Undoubtedly, this is a very unfair situation. The first serious instance of this problem occurred in 1882, when it was noted that if the House of Representatives were to have 299 seats, Alabama would get 8 seats, but if the House of Representatives were to have 300 seats, Alabama would end up with 7 (see Table 4-9). This is how the name *Alabama paradox* came about.

TABLE 4-9   Hamilton's Method and the Alabama Paradox, 1882

State	Standard quota with $M = 299$	Apportionment with $M = 299$	Standard quota with $M = 300$	Apportionment with $M = 300$
Alabama	7.646	8	7.671	7
Texas	9.64	9	9.672	10
Illinois	18.64	18	18.702	19

Mathematically, the Alabama paradox is the result of some quirks of basic arithmetic. When we increase the number of seats to be apportioned, each state's standard quota goes up, but not by the same amount. Thus, the *priority order* for surplus seats used by Hamilton's method can become scrambled, with some states moving from the front of the priority order to the back, and vice versa. This can result in some state or states losing seats they already had. This is exactly what happened in Example 4.5 and in the apportionment of 1882.

In the the U.S. House of Representatives, the final kiss of death for Hamilton's method came during the 1901 apportionment debate. As part of the debate, House sizes of anywhere between $M = 350$ and $M = 400$ seats were considered. Within that range, the apportionment for the state of Maine went up and down like a roller coaster: For M between 350 and 356, Maine would get 4 seats but for $M = 357$ Maine's apportionment goes down to 3 seats. Then up again to 4 seats for M between 358 and 381, then back down to 3 for $M = 382$, and so on. When a bill with $M = 357$ came out of the Census committee, all hell broke loose on the House floor. Fortunately, cooler heads prevailed and the bill never passed. Hamilton's method was never to be used again. (For more details, see the historical section at the end of this chapter.)

> *In Maine comes and out Maine goes—God help the State of Maine when mathematics reach for her to strike her down.*
>
> Representative Charles E. Littlefield of Maine, 1901

4.5 The Population and New-States Paradoxes

In addition to the Alabama paradox, Hamilton's method suffers from two other systemic problems called the *population paradox* and the *new-states paradox*. We will discuss these two paradoxes briefly in this section. These paradoxes underscore the fact that in apportionment, appearances can be deceiving—a seemingly simple and fair method can sometimes produce surprising and bizarre results.

The Population Paradox

Sometime in the early 1900s, it was discovered that under Hamilton's method too much growth in its population can end up costing a state seats.

Specifically, the **population paradox** occurs when state X loses a seat to state Y even though X's population grew at a higher rate than Y's.

Too weird to be true you say? Check out the next example.

EXAMPLE 4.6

Back to the future. We are going to revisit the Intergalactic Federation of 2525 (Example 4.2). Table 4-10 shows the population figures once again.

TABLE 4-10	Intergalactic Federation: Population Figures (in billions) for 2525					
Planet	Alanos	Betta	Conii	Dugos	Ellisium	Total
Population	150	78	173	204	295	900

The total population for the 5 planets comes to 900 billion. If we divide this number by 50, we get a standard divisor of 18 billion. Using this standard divisor, we can obtain the standard quotas (column 3 of Table 4-11) and then carry out steps 2 and 3 of Hamilton's method, as shown in columns 4 and 5 of Table 4-11, respectively. The final apportionment is shown in the last column. We call the reader's attention to two planets that play a key role in this story: Betta (4 delegates) and Ellisium (17 delegates).

TABLE 4-11	Intergalactic Federation: Apportionment of 2525 (Hamilton's Method)			
Planet	**Population (in billions)**	**Standard quota (Population ÷ 18)**	**Lower quota (Step 2)**	**Surplus seats (Step 3)**	**Final apportionment**
Alanos	150	$8.\overline{3}$	8		8
Betta	78	$4.\overline{3}$	4		4
Conii	173	$9.6\overline{1}$	9	1	10
Dugos	204	$11.\overline{3}$	11		11
Ellisium	295	$16.3\overline{8}$	16	1	17
Total	900	50.00	48	2	50

Intergalactic Federation. Part II. Ten years have gone by, and it is time to reapportion the Intergalactic Congress. Actually not much has changed (population-wise) within the Federation. Conii's population has increased by 8 billion and Ellisium's population has increased by 1 billion. All other planets have stayed exactly the same (Table 4-12).

TABLE 4-12	Intergalactic Federation: Population Figures (in billions) for 2535					
Planet	Alanos	Betta	Conii	Dugos	Ellisium	Total
Population	150	78	181	204	296	909

Since the total population is now 909 billion and the number of delegates is still 50, the standard divisor changes to 909/50 = 18.18. Table 4-13 shows the

TABLE 4-13	Intergalactic Federation: Apportionment of 2535 (Hamilton's Method)			
Planet	**Population (in billions)**	**Standard quota (Population ÷ 18.18)**	**Lower quota (Step 2)**	**Surplus seats (Step 3)**	**Final apportionment**
Alanos	150	8.25	8		8
Betta	78	4.29	4	1	5
Conii	181	9.96	9	1	10
Dugos	204	11.22	11		11
Ellisium	296	16.28	16		16
Total	909	50.00	48	2	50

steps for Hamilton's method based on this new standard divisor. Once again, the final apportionment is shown in the last column. Do you notice something terribly wrong with this apportionment? Ellisium, whose population went up by 1 billion, is losing a delegate to Betta, whose population did not go up at all! ■■

The New-States Paradox

A third type of paradox was discovered when Oklahoma became a state in 1907. At the time Oklahoma joined the Union, the House of Representatives had 386 seats. Based on its population, Oklahoma was entitled to 5 seats, so the size of the House of Representatives was changed from 386 to 391. The obvious intent in adding the extra 5 seats was to leave all the other states' apportionments unchanged. However, when the total population figure was adjusted to include Oklahoma's population, and the apportionments recalculated under Hamilton's method, something truly bizarre took place: Maine's apportionment went up (from 3 to 4 seats) and New York's went down (from 38 to 37 seats). The mere addition of Oklahoma (with its fair share of seats) to the Union would force New York to give a seat to Maine! The perplexing fact that *the addition of a new state with its fair share of seats can, in and of itself, affect the apportionments of other states*, is called the **new-states paradox**.

The following example gives a simple illustration of the new-states paradox. For a change of pace, we will discuss something other than legislatures.

EXAMPLE 4.7

Central School District has two high schools, North High with an enrollment of 1045 students, and South High with an enrollment of 8955. The school district has a counseling staff of 100 counselors, who are to be apportioned between the two schools using Hamilton's method. This results in an apportionment of 10 counselors to North High and 90 counselors to South High. The computation is summarized in Table 4-14.

TABLE 4-14 Apportionment of Counselors to the Two High Schools Based on Hamilton's Method

School	Enrollment	Standard quota (Standard divisor = 100)	Apportionment
North High	1045	10.45	10
South High	8955	89.55	90
Total	10,000	100.00	100

Suppose now that a new high school (New High) is added to the district. New High has an enrollment of 525 students, so the district (using the standard divisor of 100 students per counselor) decides to hire 5 new counselors and assign them to New High. After this is done, someone has the bright idea of having the entire apportionment recalculated. The calculations are shown in detail in Table 4-15. The surprising result is that one of South's counselors is heading North!

TABLE 4-15	Apportionment of Counselors to the Three High Schools Based on Hamilton's Method		
School	**Enrollment**	**Standard quota** (Standard divisor $\approx$ 100.24)	**Apportionment**
North High	1045	10.42	11
South High	8955	89.34	89
New High	525	5.24	5
Total	10,525	105	105

4.6 Jefferson's Method

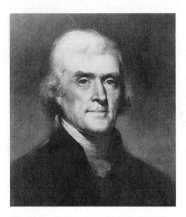

Thomas Jefferson (1743–1826).

We are now ready to study **Jefferson's method**,[4] another apportionment method of both historical and mathematical importance (see the historical section at the end of the chapter). Ironically, we will explain the idea behind Jefferson's method by taking one more look at Hamilton's method.

Recall that under Hamilton's method we start by dividing every state's population by a fixed number (the standard divisor). This gives us the standard quotas. Step 2 is then to round *every* state's standard quota down. Notice that up to this point Hamilton's method uses a uniform policy for all states—every state is treated in exactly the same way. If you are looking for fairness, this is obviously good! But now comes the bad part (step 3). We have some leftover seats which we need to distribute, but not enough for every state. Thus, we are forced to choose some states over others for preferential treatment. No matter how fair we try to be about it, there is no getting around the fact that some states get that extra seat and others don't. From the fairness point of view, this is the major weakness of Hamilton's method.

Wouldn't it be nice if we could eliminate step 3 in Hamilton's method? Or, to put it another way, wouldn't it be nice if we could rig things up so that after dividing every state's population by the same number (step 1) and then rounding the resulting quotas down (step 2) we were *left with no surplus seats*?

How could we work such magic? In theory, the answer is simple. We need to use a modified divisor (different from the standard divisor) that will give us new *modified quotas* that, when rounded down, will total the exact number of seats to be apportioned. In essence, we have just described *Jefferson's method*. Before we give a detailed description of the method, it's time to look at an example.

EXAMPLE 4.8 Jefferson's Method Meets Parador's Congress

Once again we will use the Parador example. (Recall that the standard divisor in this example is 50,000.) Table 4-16 shows the calculations based on the standard divisor. We are already familiar with these calculations—they are exactly what we used in steps 1 and 2 of Hamilton's method.

---

[4]Jefferson's method is also known as the *method of greatest divisors*, and in Europe as *d'Hondt's method*. The method is still used to apportion the legislature in many countries, including Austria, Brazil, Finland, Germany, and the Netherlands.

TABLE 4-16 Republic of Parador: Calculations Using *SD* = 50,000

State	Population	Standard quota (Population ÷ 50,000)	Lower quota (Step 2)
A	1,646,000	32.92	32
B	6,936,000	138.72	138
C	154,000	3.08	3
D	2,091,000	41.82	41
E	685,000	13.70	13
F	988,000	19.76	19
Total	12,500,000	250.00	246

Table 4-17 shows us similar calculations based on a *modified divisor* of *D* = 49,500. Let's not worry right now about where this number came from. The important thing is that with this smaller divisor, all the modified quotas are higher than the standard quotas, and the modified lower quotas add up to 250— exactly the right total. The last column of Table 4-17 shows the apportionment produced by Jefferson's method.

TABLE 4-17 Republic of Parador: Calculations Using *D* = 49,500

State	Population	Standard quota	Modified quota (Population ÷ 49,500)	Modified lower quota
A	1,646,000	32.92	33.25	33
B	6,936,000	138.72	140.12	140
C	154,000	3.08	3.11	3
D	2,091,000	41.82	42.24	42
E	685,000	13.70	13.84	13
F	988,000	19.76	19.96	19
Total	12,500,000	250	252.52	250

Before continuing with our discussion of Jefferson's method, we will make official some of the terminology we have already used: We call the number *D* used in step 1 the **modified divisor**, and the result of dividing the state's population by the modified divisor *D* the state's **modified quota**. We are now ready for a formal description of Jefferson's method.

JEFFERSON'S METHOD

- **Step 1.** Find a modified divisor *D* such that when each state's *modified quota* (state's population divided by *D*) is rounded *downward* (*modified lower quota*), the total is the exact number of seats to be apportioned.
- **Step 2.** Apportion to each state its modified lower quota.

There is one important issue we still haven't addressed. How does one go about finding this "magic" divisor D that makes Jefferson's method work? For example, how does one come up with $D = 49,500$ in Example 4.8? One way is through educated trial and error. With a calculator (or better yet, a spreadsheet) the trial-and-error method usually works rather well. Let's start with the fact that the modified divisor we are looking for has to be smaller than the standard divisor. (Remember, we want the modified quotas to be bigger than the standard quotas, so we must divide by a smaller amount.) So we pick a number D that we hope will work. We now carry out the calculations required by Jefferson's method: (1) Divide the population by D; (2) round the results downward; (3) add up the total. If we are lucky, the total is exactly right and we are finished. Otherwise we change our guess (make it higher if the total is too high, lower if the total is too low) and try again (see Fig. 4-1). In most cases, it takes at most two or three guesses before we find a divisor D that works; usually there is more than one.

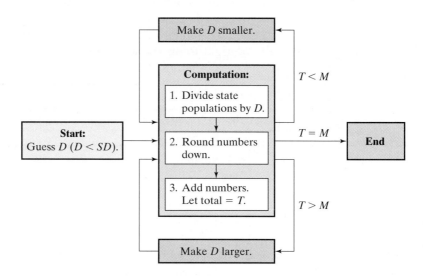

FIGURE 4-1

Flow chart for trial-and-error implementation of Jefferson's method.

Let's go through the paces using Example 4.8. We know we are looking for a modified divisor D that is less than 50,000. Let's start with a guess of $D = 49,000$. It turns out that this divisor doesn't work—it gives us a total of 252, which is too high (we leave it to the reader to verify the details). This means that we need to make D larger (thereby lowering the modified quotas), so we try $D = 49,500$. Bingo! Note that the divisor $D = 49,450$ also works, as do many others.

Jefferson's Method and the Quota Rule

If we look carefully at Table 4-17 in Example 4.8, we notice a problem: State B, with a standard quota of 138.72, is going to end up with 140 seats. This, of course, is an upper-quota violation, and an illustration of Jefferson's method main flaw—it violates the quota rule.

Jefferson's method was the very first apportionment method used by the U.S. House of Representatives. When adopted, in 1791, no one realized that violations of the quota rule were possible, but it didn't take long for the problem to rear its ugly head. In the apportionment of 1832, New York, with a standard quota of 38.59, received 40 seats. This horrified practically everyone except the New York delegation. In one of his more famous speeches, Daniel Webster argued that such an apportionment was unconstitutional:

The House is to consist of 240 members. Now, the precise portion of power, out of the whole mass presented by the number of 240, to which New York would be entitled according to her population, is 38.59; that is to say, she would be entitled to thirty-eight members, and would have a residuum or fraction; and even if a member were given her for that fraction, she would still have but thirty-nine. But the bill gives her forty ... for what is such a fortieth member given? Not for her absolute numbers, for her absolute numbers do not entitle her to thirty-nine. Not for the sake of apportioning her members to her numbers as near as may be because thirty-nine is a nearer apportionment of members to numbers than forty. But it is given, say the advocates of the bill, because the process [Jefferson's method] which has been adopted gives it. The answer is, no such process is enjoined by the Constitution.[5]

The apportionment of 1832 was to be the last time the House of Representatives used Jefferson's method. It was clear that something new had to be tried, and the search for an apportionment method that did not violate the quota rule was on.

4.7 Adams's Method

John Quincy Adams
(1767–1848).

At about the time that Jefferson's method was falling into disfavor because of its violation of the quota rule, John Quincy Adams was proposing a method that was a mirror image of it. It was based on the same idea, but instead of being based on the modified lower quotas, it was based on the *modified upper quotas*.[6]

ADAMS'S METHOD

- **Step 1.** Find a modified divisor D such that when each state's *modified quota* (state's population divided by D) is rounded *upward* (*modified upper quota*), the total is the exact number of seats to be apportioned.
- **Step 2.** Apportion to each state its modified upper quota.

Undoubtedly, Adams thought that by doing this he could avoid upper-quota violations, the big weakness of Jefferson's method. He was only partly right.

EXAMPLE 4.9 Adams's Method Meets Parador's Congress

We will use educated trial and error to find the right divisor D. We start by guessing a possible divisor D that we hope will work. We know that D will have to be bigger than 50,000 so that the modified quotas will be smaller than the standard quotas. Remembering that 49,500 worked for Jefferson's method, we suspect a good guess might be $D = 50,500$.

Table 4-18 shows the calculations based upon $D = 50,500$.

---

[5] Daniel Webster, *The Writings and Speeches of Daniel Webster, Vol. VI* (Boston: Little, Brown and Company, 1903).
[6] Adams's method is also known as the *method of smallest divisors*.

TABLE 4-18   Republic of Parador: Calculations for Adams's Method Based on *D* = 50,500

State	Population	Standard quota	Modified quota (Population ÷ 50,500)	Modified upper quota
A	1,646,000	32.92	32.59	33
B	6,936,000	138.72	137.35	138
C	154,000	3.08	3.05	4
D	2,091,000	41.82	41.41	42
E	685,000	13.70	13.56	14
F	988,000	19.76	19.56	20
Total	12,500,000	250.00	247.52	251 ← Too high!

Since the total is too high, we need to try a higher divisor, say *D* = 50,700. Table 4-19 shows the calculations based on *D* = 50,700. Now it works! The last column of Table 4-19 shows the apportionment produced by Adams's method.

TABLE 4-19   Republic of Parador: Calculations for Adams's Method Based on *D* = 50,700

State	Population	Standard quota	Modified quota (Population ÷ 50,700)	Modified upper quota
A	1,646,000	32.92	32.47	33
B	6,936,000	138.72	136.80	137
C	154,000	3.08	3.04	4
D	2,091,000	41.82	41.24	42
E	685,000	13.70	13.51	14
F	988,000	19.76	19.49	20
Total	12,500,000	250.00	246.55	250 ← That's it!

Are there any problems with Adams's method? You bet! Look at state *B*'s apportionment of 137 seats and compare it with *B*'s standard quota of 138.72—a deficit of 1.72 seats! Much like Jefferson's method, Adams's method violates the quota rule. The difference is that in Jefferson's method, the violations are all upper-quota violations whereas in Adams's method the violations are all lower-quota violations (see Exercise 47).

4.8 Webster's Method

Daniel Webster (1782–1852).

It is clear that both Jefferson's method and Adams's method share the same philosophy: Treat all states exactly the same way. (The only difference is that whereas Jefferson's method rounds the modified quotas down, Adams's method rounds the modified quotas up.) For a while this sounded like a good idea, but as we now know both methods have a major flaw: They violate the quota rule.

In 1832, Daniel Webster proposed an obvious compromise between Jefferson's method and Adams's method: Round the quotas to the nearest integer, the way we round decimals in practically every other walk of life—down if the fractional part is less than 0.5, up otherwise. But, an alert reader would argue, we have tried this idea before, and there is no guarantee it will work (see Table 4-5)! There is, however, a new twist. In our first attempt, we were married to the notion that we had to use the standard quotas. Webster's method[7] is based on the use of modified quotas chosen specifically so that conventional rounding does work.

WEBSTER'S METHOD

- **Step 1.** Find a modified divisor D such that when each state's *modified quota* (state's population divided by D) is rounded the *conventional way* (to the nearest integer) the total is the exact number of seats to be apportioned.
- **Step 2.** Apportion to each state its modified quota rounded the conventional way.

EXAMPLE 4.10 Webster's Method Meets Parador's Congress.

We will now apportion Parador's Congress using Webster's method. Our first decision is to make a guess at the divisor D. Should it be more than the standard divisor (50,000) or should it be less? Here we will use the standard quotas as a starting point. When we round off the standard quotas to the nearest integer (see Table 4-5), we get a total of 251. This number is too high, which tells us that we should try a divisor D larger than the standard divisor. Let us try $D = 50,100$. Table 4-20 shows the calculations based on $D = 50,100$. When we round the modified quotas to the nearest integer we see that it works! The last column of Table 4-20 shows the apportionment under Webster's method.

TABLE 4-20 Republic of Parador: Calculations for Webster's Method Based on $D = 50,100$

State	Population	Modified quota (Population ÷ 50,100)	Rounded to
A	1,646,000	32.85	33
B	6,936,000	138.44	138
C	154,000	3.07	3
D	2,091,000	41.74	42
E	685,000	13.67	14
F	988,000	19.72	20
Total	12,500,000	249.49	250 ← It worked!

The flow chart in Fig. 4-2 shows how to implement Webster's method using educated trial and error.

The most significant difference when we use trial and error to implement Webster's method as opposed to Jefferson's method (compare Figs. 4-1 and 4-2) is the starting value of D. With Webster's method we always try the standard divisor SD first. If we are lucky and SD happens to work, we are done. In this case we

---

[7]Webster's method is sometimes known as the *Webster-Willcox method* as well as the *method of major fractions*.

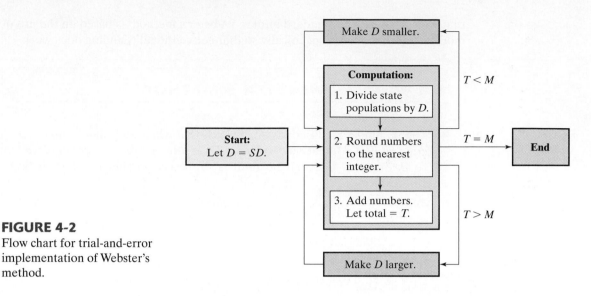

FIGURE 4-2
Flow chart for trial-and-error implementation of Webster's method.

will get as good an apportionment as one can hope for, since every state gets a number of seats equal to its standard quota rounded to the nearest integer. This means that every state gets an apportionment that is within 0.5 of its standard quota.

Under less fortunate circumstances (when the standard divisor *SD* doesn't quite do the job), we proceed with the trial-and-error approach until we find a divisor that works. In this case there will be at least one state with an apportionment that differs by more than 0.5 from its standard quota. With very bad luck, it may happen that some state gets an apportionment that differs by more than 1 from its standard quota. When this happens, we are witnessing a violation of the quota rule (see Exercise 48). Under Webster's method, this is possible but rare.

Of all the apportionment methods we discussed in this chapter, Webster's method is the one that comes closest to consistently satisfying the main requirements for fairness—it does not suffer from any paradoxes, and it shows no favoritism between small and large states. Its one flaw is that violations of the quota rule are possible in theory, although they are rare in practice.[8]

Conclusion

Balinski and Young's Impossibility Theorem

In this chapter we introduced four different *apportionment methods*. Table 4-21 summarizes the results of apportioning Parador's Congress (Example 4.3) under each of the four methods.

Note that in this example each of the four methods produced a different apportionment. This, of course, need not always happen, and often one can get the same results from different methods.

---

[8]There have been 22 U.S. Censuses between 1790 and 2000, and if Webster's method had been used with all of these, not a single violation of the quota rule would have taken place.

TABLE 4-21   Parador's Congress: A Tale of Four Methods

State	Population	Standard quota	Hamilton	Jefferson	Adams	Webster
A	1,646,000	32.92	33	33	33	33
B	6,936,000	138.72	139	140	137	138
C	154,000	3.08	3	3	4	3
D	2,091,000	41.82	42	42	42	42
E	685,000	13.70	13	13	14	14
F	988,000	19.76	20	19	20	20
Total	12,500,000	250.00	250	250	250	250

Of the four methods we discussed, one (Hamilton's) is based on a strict adherence to the standard divisor, whereas the other three (Jefferson's, Adams's, and Webster's) are examples of divisor methods—methods based on the philosophy that quotas can be conveniently modified by the appropriate choice of divisor. While some of the methods are clearly better than others, none of them is perfect. Each either *violates the quota rule* or *produces paradoxes*. Table 4-22 summarizes the characteristics of the four methods.

TABLE 4-22   How the Four Methods Stack Up

	Hamilton	Jefferson	Adams	Webster
Violates quota rule?	No	Yes (upper quota only)	Yes (lower quota only)	Yes (either)
Alabama paradox?	Yes	No	No	No
Population paradox?	Yes	No	No	No
New-states paradox?	Yes	No	No	No
Favors	Large states	Large states	Small states	Neutral

For many years, the ultimate hope held by scholars interested in the apportionment problem, both inside and outside Congress, was that mathematicians would eventually come up with an *ideal* apportionment method—one that never violates the quota rule, does not produce any paradoxes, and is neutral in its treatment of large and small states. As Congressman Ernest Gibson of Vermont stated in 1929, "The apportionment of Representatives to the population is a mathematical problem. Then why not use a method that will stand the test [of fairness] under a correct mathematical formula?"[9]

Indeed, why not? The answer was provided in 1980 by a surprising discovery made by two mathematicians—Michel L. Balinski of the State University of New York at Stony Brook and H. Peyton Young of The Johns Hopkins University—and is known as **Balinski and Young's impossibility theorem**: *There cannot be a perfect apportionment method. Any apportionment method that does not violate the quota rule must produce paradoxes, and any apportionment method that does not produce paradoxes must violate the quota rule.*

Once again, we reach an eerily familiar conclusion in a slightly different setting: Fairness and proportional representation are inherently incompatible.

No invasions of the constitution are so fundamentally dangerous as the tricks played on their own numbers, apportionment.
Thomas Jefferson

[9][*Congressional Record*, 70th Congress, 2d Session, 70 (1929), p. 1500].

HISTORICAL NOTE

A Brief History of Apportionment in the United States

The apportionment of the U.S. House of Representatives has an intriguing and convoluted history. As mandated by the Constitution, apportionments of the House are to take place every ten years, following each census of the population. The real problem was that the Constitution left the method of apportionment and the number of seats to be apportioned essentially up to Congress.[10]

Following the 1790 Census, and after considerable and heated debate, Congress passed the first "act of apportionment" in 1792. The bill, sponsored by Secretary of the Treasury Alexander Hamilton, established a House of Representatives with $M = 120$ seats, to be apportioned using *Hamilton's method*. In April of 1792, at the urging of Thomas Jefferson, then Secretary of State, George Washington vetoed the bill—the first use of the presidential veto power in United States history. Jefferson convinced Washington to support a completely different apportionment bill, using *Jefferson's method* and a House with $M = 105$ seats (see Project E).

Unable to override the president's veto and facing a damaging political stalemate, Congress finally adopted Jefferson's plan. This is how the first House of Representatives came to be constituted.

George Washington and his Cabinet. From left to right: Washington, Henry Knox, Alexander Hamilton, Thomas Jefferson, and Edmund Randolph.

---

[10]The only constitutional restrictions are that *The number of Representatives shall not exceed one for every thirty thousand but each State shall have at least one Representative* (Article I, Section 2, United States Constitution).

Jefferson's method remained in use for five decades, up to and including the apportionment of 1832. The great controversy during the 1832 apportionment debate centered around New York's apportionment. Under Jefferson's method, New York was getting 40 seats, but its *standard quota* was only 38.59—an egregious violation of the *quota rule*. Two alternative apportionment bills were considered, one proposed by John Quincy Adams that would apportion the House using *Adams's method*, and a second one, sponsored by Daniel Webster, that would do the same using *Webster's method*. Both of these proposals were defeated, and the original apportionment bill passed, but the need for change became obvious. The apportionment of 1832 was to be the last based on Jefferson's method.

Webster's method was adopted for the 1842 apportionment, but in 1852 Congress passed a law making Hamilton's method the "official" apportionment method for the House of Representatives. Since it is not unusual for Hamilton's method and Webster's method to produce exactly the same apportionment, an "unofficial" compromise was also adopted in 1852: Choose the number of seats in the House so that the apportionment is the same under either method. This was done again with the apportionment bills of 1852 and 1862.

In 1872, as a result of a power grab among states, an apportionment bill was passed that can only be described as a total mess—it was based on no particular method and produced an apportionment that was inconsistent with both Hamilton's method and Webster's method. The apportionment of 1872 was in violation of both the constitution (which requires that some method be used) and the 1852 law (which designated Hamilton's method as the method of choice.)

In 1876 Rutherford B. Hayes defeated Samuel L. Tilden in one of the most controversial and disputed presidential elections in United States history. Hayes won in the Electoral College (despite having lost the popular vote) after Congress awarded him the disputed electoral votes from three southern states—Florida, Louisiana, and South Carolina. One of the many dark sidebars of the

The election of 1876. The botched apportionment of 1872 resulted in the election of Rutherford B. Hayes over Samuel Tilden.

1876 election was that, had the House of Representatives been legally apportioned, Tilden would have been the clear-cut winner in the Electoral College.

In 1882, the *Alabama paradox* first surfaced. In looking at possible apportionments for different House sizes, it was discovered that for $M = 299$ Alabama would get 8 seats, but if the House size was increased to $M = 300$ Alabama's apportionment would decrease to 7 seats. So how did Congress deal with this disturbing discovery? It essentially glossed it over, choosing a House with $M = 325$ seats, a number for which Hamilton's method and Webster's method would give the same apportionment. The same strategy was adopted in the apportionment bill of 1892.

In 1901, the *Alabama paradox* finally caught up with Congress. When the Census Bureau presented to Congress tables showing the possible apportionments under Hamilton's method for all House sizes between 350 and 400 seats, it was pointed out that two states—Maine and Colorado—were impacted by the Alabama paradox: For most values of M starting with $M = 350$, Maine would get 4 seats, but for $M = 357, 382, 386, 389$, and 390, Maine would go down to 3 seats. Colorado would get 3 seats for all possible values of M except $M = 357$, for which it would only get 2 seats. For $M = 357$, both Maine and Colorado lose, and coincidentally, this just happened to be the House size that was proposed for the 1901 bill. Faster than you can say *we are being robbed*, the debate in Congress escalated into a frenzy of name-calling and accusations, with the end result being that the bill was defeated and Hamilton's method was scratched for good. The final apportionment of 1901 used Webster's method and a House with $M = 386$ seats.

Webster's method remained in use for the apportionments of 1901, 1911, and 1931 (no apportionment bill was passed following the 1920 Census, in direct violation of the Constitution).

In 1941, Congress passed a law that established a fixed size for the House of Representatives (435 seats) and a permanent method of apportionment—a method called the *Huntington-Hill method* or *method of equal proportions* (a detailed description of the Huntington-Hill method is given in Appendix 1 at the end of this chapter). The 1941 law [*Public Law 291, H.R. 2665, 55 Stat 761: An Act to Provide for Apportioning Representatives in Congress among the Several States by the Equal Proportions Method*] represented a realization by Congress that politics should be taken out of the apportionment debate, and that the apportionment of the House of Representatives should be purely a mathematical issue.

Are apportionment controversies then over? Not a chance. With a fixed-size House, one state's gain has to be another state's loss. In the 1990 apportionment, Montana was facing the prospect of losing one of its two seats—seats it had held in the House for 80 years. Not liking the message, Montana tried to kill the messenger. In 1991 Montana filed a lawsuit in Federal District Court (*Montana v. United States Department of Commerce*) in which it argued that the Huntington-Hill method is unconstitutional, and that either *Adams's method* or *Dean's method* (see Project A) should be used. (Coincidentally, under either one of these methods Montana would keep its 2 seats.) A panel of three federal judges ruled by a 2 to 1 vote in favor of Montana. The case then went on appeal to the Supreme Court, which overturned the decision of the lower federal court and upheld the constitutionality of the Huntington-Hill method (for more details see the story on the opposite page and Project D).

Supreme Court Upholds Method Used in Apportionment of House

By LINDA GREENHOUSE
Special to The New York Times

WASHINGTON, March 31 — In a decision that dashed Montana's hope of retaining two seats in the House of Representatives, the Supreme Court today upheld the constitutionality of the method Congress has used for 50 years to apportion seats among the states.

The unanimous ruling overturned a decision issued last fall by a special three-judge Federal District Court in Montana. That court, ruling in a lawsuit brought by the state, had ordered the use of a different method under which Montana would have kept the two House seats it has had since 1910 and the State of Washington would have lost one of its nine.

While the decision affected only those two states, it was of much broader interest because it was the Court's

Ruling ends Montana's hope to keep two seats.

first look at Congressional apportionment in light of the strict one-person, one-vote requirement the Court now applies to legislative districting.

The Court insists on virtual mathematical equality for Congressional districts in a state. But such equality is not possible for districts in different states, because districts may not cross state lines and the Constitution requires at least one Representative for each state.

Absolute or Relative?

So the question in this case was whether Congress is constitutionally obliged to minimize the absolute population differences among the districts, as the District Court had concluded in rejecting the method of examining relative, rather than absolute, differences.

In his opinion for the Court today, Justice John Paul Stevens said that "neither mathematical analysis nor constitutional interpretation provides a conclusive answer" to the question of "what is the better measure of inequality."

Since there is bound to be "a significant departure from the ideal," he said, the method that Congress adopted in 1941, after long study and on the basis of expert advice, meets constitutional requirements.

The Court acted with unusual speed to decide the Federal Government's appeal from the District Court ruling. The case, U.S. Department of Commerce v. Montana, No. 91-860, was argued just four weeks ago, on March 4.

Backing Judgment of Congress

The Solicitor General, Kenneth W. Starr, urged the Court to accept the judgment of Congress rather than to "transplant into alien soil a concept that doesn't apply," that of mathematical equality.

The Government also argued that the allocation of Congressional seats was a political question to be left to the discretion of Congress without interference by Federal judges.

The Court rejected that argument today, finding the dispute to be one that is appropriate for judicial resolution. Justice Stevens said that while the Court had "respect for a coordinate branch of Government," it nonetheless had to exercise its jurisdiction to decide whether Congress had acted "within the limits dictated by the Constitution."

Congressional apportionment was the subject of heated political warfare from the earliest years of the country until 1941, when Congress placed into law the formula the Court upheld today and made its use automatic after every census. President George Washington employed his first veto against an early apportionment bill that had been endorsed by Alexander Hamilton and was thought to favor Northern states.

The method Congress adopted in 1941, on the recommendation of the National Academy of Sciences, is a mathematical formula known as the method of equal proportions. It minimizes the relative difference between the size of Congressional districts and between the number of Representatives per person. Perhaps most significantly, as Justice Stevens noted today, of the five formulas that Congress considered, this one incorporates the least bias toward small or large states.

Today's decision will result in Montana having the biggest single Congressional district in the country, including all the state's 804,000 people, while the ideal Congressional district would have 572,000. Had the state been divided into two districts of 402,000 each, each district would have been 171,000 smaller than the ideal.

Adams's method	modified divisor
Alabama paradox	modified quota
apportionment method	new-states paradox
apportionment problem	population paradox
Balinski and Young's impossibility theorem	quota rule
	standard divisor
Hamilton's method	standard quota
Jefferson's method	upper quota
lower quota	upper-quota violation
lower-quota violation	Webster's method

EXERCISES

WALKING

A. Standard Divisors and Quotas

1. The Bandana Republic is a small country consisting of four states (Apure, Barinas, Carabobo, and Dolores). The populations of each state (in millions) are given in the following table.

State	Apure	Barinas	Carabobo	Dolores
Population (in millions)	3.31	2.67	1.33	0.69

   (a) Find the standard divisor when the number of seats in the Bandana Republic legislature is $M = 160$.

   (b) Using the standard divisor you found in (a), find each state's standard quota.

   (c) Using the standard quotas you found in (b), find each state's lower and upper quotas.

2. Use the same Bandana Republic population figures given in Exercise 1.

   (a) Find the standard divisor when the number of seats in the Bandana Republic legislature is $M = 200$.

   (b) Using the standard divisor you found in (a), find each state's standard quota.

   (c) Using the standard quotas you found in (b), find each state's lower and upper quotas.

3. The Scotia Metropolitan Area Rapid Transit Service (SMARTS) operates 6 bus routes (*A, B, C, D, E,* and *F*) and 130 buses. The buses are apportioned among the routes based on the average number of daily passengers per route, which is given in the following table.

Route	*A*	*B*	*C*	*D*	*E*	*F*
Average number of passengers	45,300	31,070	20,490	14,160	10,260	8,720

   (a) Who are the "players" in this problem?

(b) Find the standard divisor. Explain what the standard divisor represents in this problem.

(c) Find the standard quota for each bus route.

(d) Find the lower and upper quotas for each bus route.

4. The Placerville General Hospital has a nursing staff of 225 nurses working in four shifts: *A* (7:00 A.M. to 1:00 P.M.), *B* (1:00 P.M. to 7:00 P.M.), *C* (7:00 P.M. to 1:00 A.M.), and *D* (1:00 A.M. to 7:00 A.M.). The number of nurses apportioned to each shift is based on the average number of patients per shift, given in the following table.

Shift	A	B	C	D
Average number of patients	871	1029	610	190

(a) Who are the "players" in this problem?

(b) Find the standard divisor. Explain what the standard divisor represents in this problem.

(c) Find the standard quota for each shift.

(d) Find the lower and upper quotas for each shift.

5. The Republic of Tropicana is a small country consisting of five states (*A, B, C, D*, and *E*). The total population of Tropicana is 23.8 million. According to the Tropicana constitution the seats in the legislature are apportioned to the states according to their populations. The standard quota of each state is given in the following table.

State	A	B	C	D	E
Standard quota	40.50	29.70	23.65	14.60	10.55

(a) Find the number of seats in the Tropicana legislature.

(b) Find the standard divisor.

(c) Find the population of each state.

6. Tasmania State University is made up of five different schools: Agriculture, Business, Education, Humanities, and Science. The total number of students at TSU is 12,500. The faculty positions at TSU are apportioned to the various schools based on the schools' respective enrollments. The standard quota for each school is given in the following table.

School	Agriculture	Business	Education	Humanities	Science
Standard quota	32.92	15.24	41.62	21.32	138.90

(a) Find the number of faculty positions at TSU.

(b) Find the standard divisor. What does the standard divisor represent in this problem?

(c) Find the number of students enrolled in each school.

B. Hamilton's Method

7. Use Hamilton's method to apportion the Bandana Republic legislature discussed in Exercise 1 when the number of seats is $M = 160$.

8. Use Hamilton's method to apportion the Bandana Republic legislature discussed in Exercise 2 when the number of seats is $M = 200$.

9. Use Hamilton's method to apportion the buses among the routes in the Scotia Metropolitan Area Rapid Transit Service as discussed in Exercise 3.

10. Use Hamilton's method to apportion the nurses among the shifts at the Placerville General Hospital discussed in Exercise 4.

11. Use Hamilton's method to apportion the Republic of Tropicana legislature discussed in Exercise 5.

12. Use Hamilton's method to apportion the faculty among the schools at Tasmania State University as discussed in Exercise 6.

13. A mother wishes to distribute 11 pieces of candy among her 3 children based on the number of minutes each child spends studying, as shown in the following table.

Child	Bob	Peter	Ron
Minutes studied	54	243	703

   (a) Find each child's apportionment using Hamilton's method.

   (b) Suppose that before mom has time to sit down and do the actual calculations, the children decide to do a little more studying. Say Bob studies an additional 2 minutes, Peter an additional 12 minutes, and Ron an additional 86 minutes. Find each child's apportionment using Hamilton's method based on the new total time studied.

   (c) Did anything paradoxical occur? Explain.

14. A mother wishes to distribute 10 pieces of candy among her three children based on the number of minutes each child spends studying, as shown in the following table.

Child	Bob	Peter	Ron
Minutes studied	54	243	703

   (a) Find each child's apportionment using Hamilton's method.

   (b) Suppose that, just prior to actually handing over the candy, mom finds another piece of candy and includes it in the distribution. Find each child's apportionment using Hamilton's method and 11 pieces of candy.

   (c) Did anything paradoxical occur? (What's the name of this paradox?)

15. A mother wishes to apportion 11 pieces of candy among three of her children based on the number of minutes each child spends studying, as shown in the following table.

Child	Bob	Peter	Ron
Minutes Studied	54	243	703

   (a) Find each child's apportionment using Hamilton's method [same as Exercise 13(a)].

   (b) Suppose that, just prior to handing over the candy, a fourth son Jim claims that he spent 580 minutes studying and so mom decides that he too should receive his fair share—an additional six pieces of candy. Find each child's apportionment using Hamilton's method had mom started with the four kids and 17 pieces of candy.

(c) Explain the paradox that took place. What's the name of this paradox?

16. A mother wishes to apportion 15 pieces of candy among her three daughters, Katie, Lilly, and Jaime, based on the number of minutes each child spent studying. The only information we have is that mom will use Hamilton's method, and that Katie's standard quota is 6.53.

(a) Explain why it is impossible for all three daughters to end up with five pieces of candy each.

(b) Explain why it is impossible for Katie to end up with nine pieces of candy.

(c) Explain why it is impossible for Lilly to end up with nine pieces of candy.

C. Jefferson's Method

17. Use Jefferson's method to apportion the Bandana Republic legislature discussed in Exercise 1 when the number of seats is $M = 160$.

18. Use Jefferson's method to apportion the Bandana Republic legislature discussed in Exercise 2 when the number of seats is $M = 200$.

19. Use Jefferson's method to apportion the buses among the routes in the Scotia Metropolitan Area Rapid Transit Service as discussed in Exercise 3.

20. Use Jefferson's method to apportion the nurses among the shifts at the Placerville General Hospital discussed in Exercise 4. (*Hint:* Divisors don't have to be whole numbers.)

21. Use Jefferson's method to apportion the Republic of Tropicana legislature discussed in Exercise 5.

22. Use Jefferson's method to apportion the faculty among the schools at Tasmania State University as discussed in Exercise 6.

23. If Jefferson's method were to be applied to the 2000 Census data, it would give California 55 seats in the House of Representatives, even though California's current standard quota is only 52.45. What does this fact illustrate about Jefferson's method?

24. According to the 2000 U.S. Census, the population of the United States was 281,424,177. Thus, each of the 435 members of the House of Representatives represents approximately 646,952 people. In 2000, Puerto Rico had a population of 3,808,610.

(a) Assuming Puerto Rico was to become a state, estimate Puerto Rico's "fair share" of representatives.

(b) Suppose that the current 435 member House of Representatives adopted Jefferson's method. If Puerto Rico were to then become a new state and if the House were to be expanded by Puerto Rico's "fair share," would it be possible for Alabama to lose one of its seven seats? Explain.

D. Adams's Method

25. Use Adams's method to apportion the Bandana Republic legislature discussed in Exercise 1 when the number of seats is $M = 160$.

26. Use Adams's method to apportion the Bandana Republic legislature discussed in Exercise 2 when the number of seats is $M = 200$.

27. Use Adams's method to apportion the buses among the routes in the Scotia Metropolitan Area Rapid Transit Service as discussed in Exercise 3.

28. Use Adams's method to apportion the nurses among the shifts at the Placerville General Hospital discussed in Exercise 4. (*Hint:* Divisors don't have to be whole numbers.)

29. Use Adams's method to apportion the Republic of Tropicana legislature discussed in Exercise 5.

30. Use Adams's method to apportion the faculty among the schools at Tasmania State University as discussed in Exercise 6.

31. If Adams's method were to be applied to the 2000 Census data, it would give California 50 seats in the House of Representatives, even though California's current standard quota is 52.45. What does this fact illustrate about Adams's method?

32. Suppose that the current 435-member House of Representatives were to adopt Adams's method. If the size of the House were then increased to 436 members, could both New York and Texas gain representation? Explain.

E. Webster's Method

33. Use Webster's method to apportion the Bandana Republic legislature discussed in Exercise 1 when the number of seats is $M = 160$.

34. Use Webster's method to apportion the Bandana Republic legislature discussed in Exercise 2 when the number of seats is $M = 200$.

35. Use Webster's method to apportion the buses among the routes in the Scotia Metropolitan Area Rapid Transit Service as discussed in Exercise 3.

36. Use Webster's method to apportion the nurses among the shifts at the Placerville General Hospital discussed in Exercise 4. (*Hint:* Divisors don't have to be whole numbers.)

37. Use Webster's method to apportion the Republic of Tropicana legislature discussed in Exercise 5.

38. Use Webster's method to apportion the faculty among the schools at Tasmania State University as discussed in Exercise 6.

F. Miscellaneous

Exercises 39 through 42 are based on the following facts: A small country consists of four states (A, B, C, and D). The 125 seats in the legislature are to be apportioned among the four states based on their populations. The following table gives the population of each state as a percentage of the national population.

State	A	B	C	D
Percent of population	6.24%	26.16%	28.48%	39.12%

39. (a) Find the standard divisor as a percentage of the national population.
    (b) Find each state's standard quota.
    (c) Find each state's apportionment under Hamilton's method.

40. (a) Find each state's modified quotas based on a modified divisor D_1 that is 0.79% of the national population.
    (b) Find each state's apportionment under Jefferson's method.

41. **(a)** Find each state's modified quotas based on a modified divisor D_2 that is 0.814% of the national population.

    (b) Find each state's apportionment under Adams's method.

42. **(a)** Find each state's modified quotas based on a modified divisor D_3 that is 0.804% of the national population.

    (b) Find each state's apportionment under Webster's method.

43. Draw a flow chart illustrating a trial-and-error implementation of Adams's method. (See Fig. 4-1.)

JOGGING

*Exercises 44 and 45 refer to a variation of Hamilton's method known as **Lowndes's method**. (This method was first proposed by South Carolina Representative William Lowndes in 1822.) The basic difference between Hamilton's and Lowndes's methods is that, in the latter method, after each state is assigned the lower quota, the surplus seats are handed out in order of **relative fractional parts**. (The relative fractional part of a number is the fractional part divided by the integer part. For example, the relative fractional part of 41.82 is 0.82/41 = 0.02, and the relative fractional part of 3.08 is 0.08/3 = 0.027. Notice that while 41.82 would have priority over 3.08 under Hamilton's method, 3.08 has priority over 41.82 under Lowndes's method because 0.027 is greater than 0.02.)*

44. **(a)** Find the apportionment of Parador's Congress (Example 3) under Lowndes's method.

    (b) Verify that the resulting apportionment is different from each of the apportionments shown in Table 4-21. In particular, list which states do better under Lowndes's method than under Hamilton's method.

45. Consider an apportionment problem with only two states, A and B. Suppose that state A has standard quota q_1 and state B has standard quota q_2, neither of which is a whole number. (Of course, $q_1 + q_2 = M$ must be a whole number.) Let f_1 represent the fractional part of q_1 and f_2 the fractional part of q_2.

    (a) Find values q_1 and q_2 such that Lowndes's method and Hamilton's method result in the same apportionment.

    (b) Find values q_1 and q_2 such that Lowndes's method and Hamilton's method result in different apportionments.

    (c) Write an inequality involving q_1, q_2, f_1, and f_2 that would guarantee that Lowndes's method and Hamilton's method result in different apportionments.

46. Consider an apportionment problem with only two states, A and B. Suppose that state A has standard quota q_1 and state B has standard quota q_2, neither of which is a whole number. (Of course, $q_1 + q_2 = M$ must be a whole number.) Let f_1 represent the fractional part of q_1 and f_2 the fractional part of q_2.

    (a) Explain why one of the fractional parts is bigger than or equal to 0.5 and the other is smaller than or equal to 0.5.

    (b) Assuming neither fractional part is equal to 0.5, explain why Hamilton's method and Webster's method must result in the same apportionment.

(c) Explain why, in any apportionment problem involving only 2 states, Hamilton's method can never produce the Alabama paradox or the population paradox.

(d) Explain why, in the above situation, Webster's method can never violate the quota rule.

47. (a) Explain why, when Jefferson's method is used, any violations of the quota rule must be upper-quota violations.

(b) Explain why, when Adams's method is used, any violations of the quota rule must be lower-quota violations.

(c) Use parts (a) and (b) to justify why, in the case of an apportionment problem with just two states, neither Jefferson's nor Adams's method can possibly violate the quota rule.

48. For an arbitrary state X, let q represent its standard quota and s represent the number of seats apportioned to X under some unspecified apportionment method. Interpret in words the meaning of each of the following mathematical statements:

(a) $s - q \geq 1$

(b) $q - s \geq 1$

(c) $|s - q| \leq 0.5$

(d) $0.5 < |s - q| < 1$

49. This exercise is based on actual data taken from the 1880 census. In 1880, the population of Alabama was given at 1,262,505. With a House of Representatives consisting of $M = 300$ seats, the standard quota for Alabama was 7.671.

(a) Find the 1880 census population for the United States (rounded to the nearest person).

(b) Given that the standard quota for Texas was 9.672, find the population of Texas (to the nearest person).

50. The purpose of this exercise is to show that under rare circumstances, Jefferson's method may not work. A small country consists of four states with populations given as follows.

State	A	B	C	D
Population	500	1000	1500	2000

There are $M = 49$ seats in the House of Representatives.

(a) Find each state's apportionment using Adams's method.

(b) Attempt to apportion the seats using Jefferson's method with the modified divisor $D = 100$. What happens if $D < 100$? What happens if $D > 100$?

(c) Explain why Jefferson's method will not work for this example.

51. The purpose of this exercise is to show that under rare circumstances, Adams's method may not work. A small country consists of four states with populations given as follows.

State	A	B	C	D
Population	500	1000	1500	2000

There are $M = 51$ seats in the House of Representatives.

(a) Find each state's apportionment using Jefferson's method.

(b) Attempt to apportion the seats using Adams's method with the modified divisor $D = 100$. What happens if $D < 100$? What happens if $D > 100$?

(c) Explain why Adams's method will not work for this example.

RUNNING

52. Explain why Jefferson's method cannot produce

(a) the Alabama paradox

(b) the new-states paradox

53. Explain why Adams's method cannot produce

(a) the Alabama paradox

(b) the new-states paradox

54. Explain why Webster's method cannot produce

(a) the Alabama paradox

(b) the new-states paradox

Exercises 55 through 58 refer to the Huntington-Hill method described in Appendix 1. (These exercises should not be attempted without first reading and understanding the material in Appendix 1.)

55. Use the Huntington-Hill method to find the apportionments of each state for a small country that consists of 5 states. The total population of the country is 24.8 million. The standard quotas of each state are as follows.

State	A	B	C	D	E
Standard quota	25.26	18.32	2.58	37.16	40.68

56. (a) Use the Huntington-Hill method to apportion Parador's Congress (Example 4.3).

(b) Compare your answer in (a) with the apportionment produced by Webster's method. What's your conclusion?

57. A country consists of six states with populations as follows.

State	Population
A	344,970
B	408,700
C	219,200
D	587,210
E	154,920
F	285,000
Total	2,000,000

There are 200 seats in the legislature.

(a) Find the apportionment under Webster's method.

(b) Find the apportionment under the Huntington-Hill method.

(c) Compare the divisors used in (a) and (b).

(d) Compare the apportionments found in (a) and (b).

58. A country consists of six states with populations as follows.

State	Population
A	344,970
B	204,950
C	515,100
D	84,860
E	154,960
F	695,160
Total	2,000,000

There are 200 seats in the legislature.

(a) Find the apportionment under Webster's method.

(b) Find the apportionment under the Huntington-Hill method.

(c) Compare the divisors used in (a) and (b).

(d) Compare the apportionments found in (a) and (b).

59. The Geometric Mean. If a and b are two positive numbers, the *geometric mean of a and b* is defined as the number $\sqrt{a \cdot b}$.

(a) Without using a calculator, find the geometric mean of 10 and 1000.

(b) Find the geometric mean of a and $a^3 (a > 0)$.

(c) Show that the *geometric mean* of a and b is always smaller or equal to the *average* of a and b.

(d) Use part (c) to explain why the cutoff for rounding under the Huntington-Hill method is always less than the cutoff for rounding under Webster's method.

PROJECTS AND PAPERS

A. Dean's Apportionment Method

This method was first proposed in 1832 by James Dean, a former professor of Daniel Webster's at Dartmouth. Dean's method is a divisor method based on the concept of the *harmonic mean* of two numbers (it is also known as the *method of harmonic means*).

In this project, pretend you are to prepare a presentation on Dean's method. Describe in detail what you would do.

Friendly suggestions: (i) Do your research first. (A search on the Web will provide plenty of information.) (ii) Start with a careful description of Dean's method. (As part of your description you should include a brief explanation of the concept of the **harmonic mean** of two numbers, including some examples.) (iii) Explain how Dean's method can be implemented using educated trial and error (you may want to include a flow chart if you think it helps). Illustrate using one of the examples from the chapter (such as Parador's Congress). (iv) Compare Dean's method with Webster's method and give an example showing that they can produce different apportionments. (v) If you know how to work with an Excel spreadsheet, you may want to calculate what the 2000 apportionment would be under Dean's method, and compare the results with the results shown in the table in Appendix 2.

B. Apportionment Methods and the 2000 Presidential Election

After tremendous controversy over hanging chads and missed votes, the 2000 presidential election was decided in the Supreme Court, with George W. Bush getting the disputed electoral votes from Florida and thus beating Al Gore by a margin of four electoral votes. Ignored in all the controversy was the significant role that the choice of apportionment method plays in a close presidential election. Would things have turned out differently if the House of Representatives had been apportioned under a different method?

In this project, you are asked to analyze and speculate how the election would have turned out had the House of Representatives (and thus the Electoral College) been apportioned under (i) Hamilton's method, (ii) Jefferson's method, and (iii) Webster's method. (Remember to add 2 to the seats in the House to get each state's Electoral College votes!)

Notes: (1) The 2000 Electoral College was based on the 1990 Census. You can go to *www.census.gov/main/www/cen1990.html* for the 1990 state population figures. (2) You can find an "apportionment calculator" that will do the calculations for you at this book's Web site: *http://cwx.prenhall.com/bookbind/pubbooks/ tannenbaum/chapter4*. (3) Give Florida's votes to Bush.

C. Apportionment Calculations via Spreadsheets

The power of a spreadsheet is its ability to be *dynamic*. That is, any and all calculations will automatically update after a change is made to the value in any given cell. This is particularly useful in implementing apportionment calculations that use educated trial and error, as well as when trying to rig up examples of apportionment problems that meet specific requirements.

This project has two parts.

Part 1. Implement each of the four apportionment methods discussed in the chapter using a spreadsheet. (Jefferson's, Adams's, and Webster's methods all fall under a similar template, so once you can implement one, the others are quite similar. Hamilton's method is a little different and, ironically, a bit harder to implement with a spreadsheet.) To check that everything works, do the different apportionments of Parador's Congress via the spreadsheets.

Part 2. Use your spreadsheets to create examples of the following situations:
  (i) an apportionment problem in which Hamilton's method and Jefferson's method produce exactly the same apportionments
  (ii) an apportionment problem in which Hamilton's method and Adams's method produce exactly the same apportionments
  (iii) an apportionment problem in which Hamilton's method and Webster's method produce exactly the same apportionments
  (iv) an apportionment problem in which Hamilton's method, Jefferson's method, Adams's method, and Webster's method all produce different apportionments
  (v) an apportionment problem in which Webster's method violates the quota rule

Notes: To do this project you should be reasonably familiar with Excel or a similar spreadsheet program. (Some Excel tips can be found at this book's Web site: *http://cwx.prenhall.com/bookbind/pubbooks/tannenbaum/chapter4*.)

D. Montana v. U.S. Department of Commerce (U.S. District Court, 1991); U.S. Department of Commerce v. Montana (U.S. Supreme Court, 1992)

Write a paper discussing these two important legal cases dealing with the constitutionality of our current method of apportionment of the House of Representatives (see Appendix 1). In this paper you should (i) present the background preceding Montana's challenge to the constitutionality of the Huntington-Hill method, (ii) summarize the arguments presented by Montana and the government in both cases, and (iii) summarize the arguments given by the District Court in ruling for Montana and the arguments given by the Supreme Court in unanimously overturning the District Court ruling. Conclude the paper with your own thoughts on the issue, incorporating as much as possible mathematical ideas from this chapter.

E. The First Apportionment of the House of Representatives

The following table shows the populations of the 13 original states in the United States obtained from the 1790 Census.

State	Population	State	Population
Connecticut	236,841	New York	331,589
Delaware	55,540	North Carolina	353,523
Georgia	70,835	Pennsylvania	432,879
Kentucky	68,705	Rhode Island	68,446
Maryland	278,514	South Carolina	206,236
Massachusetts	475,327	Vermont	85,533
New Hampshire	141,822	Virginia	630,560
New Jersey	179,570	Total	3,615,920

For the first apportionment bill of 1792, two competing proposals were considered: a proposal by Alexander Hamilton to apportion $M = 120$ seats using Hamilton's method, and an alternative proposal by Thomas Jefferson to apportion $M = 105$ seats using Jefferson's method. The original apportionment bill passed by Congress was based on Hamilton's proposal, but Jefferson was able to persuade Washington to veto the bill and support his proposal, which eventually was passed.

This project has three parts:

Part 1. Historical Research. Summarize the arguments presented by Hamilton and Jefferson on behalf of their respective proposals. Among other sources, these can be found in *The Papers of Alexander Hamilton, Vol. XI,* Harold C. Syrett (editor), and *The Works of Thomas Jefferson, Vol. VI,* Paul Leicester Ford (editor).

Part 2. Mathematics. Calculate the apportionments for the House of Representatives under both proposals.

Part 3. Analysis. It has been argued by some scholars that there was more than mathematical merit behind Jefferson's thinking and Washington's support of it. Looking at the apportionments obtained in Part 2, explain why one could be suspicious of Jefferson's and Washington's motives. (*Hint:* Jefferson and Washington were both from the same state.)

APPENDIX 1	The Huntington-Hill Method

The method currently used to apportion the U.S. House of Representatives is known as the **Huntington-Hill method**, and more commonly as the **method of equal proportions**.

Let's start with some historical background. The method was developed sometime around 1911, by Joseph A. Hill, chief statistician of the Bureau of Census, and Edward V. Huntington, professor of mechanics and mathematics at Harvard University. In 1929, the Huntington-Hill method was endorsed by a distinguished panel of mathematicians. The panel, commissioned by the National Academy of Sciences at the formal request of the Speaker of the House, investigated many different apportionment methods and recommended the Huntington-Hill method as the best possible one.

On November 15, 1941, President Franklin D. Roosevelt signed "An Act to Provide for Apportioning Representatives in Congress among the Several States by the Equal Proportions Method." This law set the Huntington-Hill method as the permanent method for apportionment of the House of Representatives, and fixed the size of the House at 435 seats. This act still stands today, but political, legal, and mathematical challenges to it have come up periodically.

There are several ways to describe how the Huntington-Hill method works. For the purposes of explanation, the method is most conveniently described by comparison to Webster's method. In fact, the two methods are almost identical. Just as in Webster's method, we will find modified quotas, and we will round some of them upward and some of them downward. The difference between the two methods is in the cutoff point for rounding up or down. Take, for example, a state with a modified quota of 3.48. Under Webster's method we know that we must round this quota downward, because the cutoff point for rounding is 3.5. It may seem like overkill, but we can put it this way: The cutoff point for rounding quotas under Webster's method is exactly halfway between the modified lower quota (L) and the modified upper quota ($L + 1$). That is,

$$\text{cutoff for Webster's method} = \frac{L + (L + 1)}{2}.$$

(Thus, for a state with a modified quota of 3.48, the cutoff point is 3.5.)

Under the Huntington-Hill method the cutoff point for rounding quotas is computed using a different formula.[11]

$$\text{cutoff for the Huntington-Hill method} = \sqrt{L \cdot (L + 1)}.$$

Thus, if a state has a modified quota of 3.48, the cutoff for rounding this quota under the Huntington-Hill method would be $\sqrt{3 \times 4} = \sqrt{12} = 3.464$. Since the modified quota 3.48 is above this cutoff, under the Huntington-Hill method this state would get 4 seats.

HUNTINGTON-HILL ROUNDING RULES

If the quota falls between L and $L + 1$, the Huntington-Hill cutoff point for rounding is $H = \sqrt{L \cdot (L + 1)}$. If the quota is below H, we round down; otherwise we round up.

---

[11]There is a handy mathematical name for the Huntington-Hill cutoffs. For any two positive numbers a and b, $\sqrt{a \times b}$ is called the *geometric mean* of a and b (see Exercise 59). Thus, we can describe each Huntington-Hill cutoff as the geometric mean of the modified lower and upper quotas.

HUNTINGTON-HILL METHOD

■ **Step 1.** Find a divisor D such that when each state's modified quota (state's population divided by D) is rounded according to the Huntington-Hill rounding rules, the total is the exact number of seats to be apportioned.

■ **Step 2.** Apportion to each state its modified quota, rounded using the Huntington-Hill rules.

Table 4-23 is convenient to have handy when working with the Huntington-Hill method.

TABLE 4-23

Modified quota between	Cutoff point for rounding under Webster's method	Cutoff point for rounding under Huntington-Hill method
1 and 2	1.5	$\sqrt{2} \approx 1.414$
2 and 3	2.5	$\sqrt{6} \approx 2.449$
3 and 4	3.5	$\sqrt{12} \approx 3.464$
4 and 5	4.5	$\sqrt{20} \approx 4.472$
5 and 6	5.5	$\sqrt{30} \approx 5.477$
6 and 7	6.5	$\sqrt{42} \approx 6.481$
7 and 8	7.5	$\sqrt{56} \approx 7.483$
8 and 9	8.5	$\sqrt{72} \approx 8.485$
9 and 10	9.5	$\sqrt{90} \approx 9.487$
10 and 11	10.5	$\sqrt{110} \approx 10.488$

We will conclude this appendix with a very simple example that shows that the Huntington-Hill method can produce an apportionment that differs from Webster's method.

EXAMPLE A1

A small country consists of three states. We want to apportion the 100 seats in its legislature to the three states according to the population figures shown in Table 4-24.

TABLE 4-24

State	A	B	C	Total
Population	3480	46,010	50,510	100,000

We will use Webster's method first and then the Huntington-Hill method.
We start by computing the standard quotas. Since the standard divisor is $100,000/100 = 1000$, this is really easy, as shown in Table 4-25.

TABLE 4-25

State	A	B	C	Total
Standard quota	3.48	46.01	50.51	100

It so happens that rounding the standard quotas the conventional way gives a total of 100, so the standard quotas work for Webster's method (Table 4-26).

TABLE 4-26

State	Population	Standard quota	Webster's apportionment
A	3,480	3.48	3
B	46,010	46.01	46
C	50,510	50.51	51
Total	100,000	100.00	100

Next, we notice that our old friend 3.48 has made an appearance. We know that under the Huntington-Hill method 3.48 is past the cutoff point of 3.464, so it has to be rounded upward (to 4). The other two standard quotas are not affected and are still rounded as before (Table 4-27).

TABLE 4-27

State	Population	Standard quota (Standard divisor = 1000)	Rounded under Huntington-Hill rules
A	3,480	3.48	4
B	46,010	46.01	46
C	50,510	50.51	51
Total	100,000	100.00	101

Since the total comes to 101, these quotas don't work—we can see they are a bit too high. But when we try a divisor just a tad bigger ($D = 1001$), the totals do come out right (Table 4-28). The last column of Table 4-28 shows the way the 100 seats would be apportioned under the Huntington-Hill method. Note that the apportionment is different from the one produced by Webster's method (see also Exercise 58).

TABLE 4-28

State	Population	Standard quota (Divisor = 1001)	Rounded under Huntington-Hill rules
A	3,480	3.477	4
B	46,010	45.964	46
C	50,510	50.460	50
Total	100,000	100.000	100

APPENDIX 2	Apportionments Based on the 2000 Census

TABLE 4-29 Apportionments Under Five Different Methods Using the 2000 Census

State	Population	Hamilton	Jefferson	Adams	Webster	Huntington-Hill*
Alabama	4,461,130	7	7	7	7	7
Alaska	628,933	1	1	1	1	1
Arizona	5,140,683	8	8	8	8	8
Arkansas	2,679,733	4	4	4	4	4
California	33,930,798	52	55	50	53	53
Colorado	4,311,882	7	7	7	7	7
Connecticut	3,409,535	5	5	6	5	5
Delaware	785,068	1	1	2	1	1
Florida	16,028,890	25	26	24	25	25
Georgia	8,206,975	13	13	13	13	13
Hawaii	1,216,642	2	1	2	2	2
Idaho	1,297,274	2	2	2	2	2
Illinois	12,439,042	19	20	19	19	19
Indiana	6,090,782	9	9	9	9	9
Iowa	2,931,923	5	4	5	5	5
Kansas	2,693,824	4	4	4	4	4
Kentucky	4,049,431	6	6	6	6	6
Louisiana	4,480,271	7	7	7	7	7
Maine	1,277,731	2	2	2	2	2
Maryland	5,307,886	8	8	8	8	8
Massachusetts	6,355,568	10	10	10	10	10
Michigan	9,955,829	15	16	15	15	15
Minnesota	4,925,670	8	7	8	8	8
Mississippi	2,852,927	4	4	5	4	4
Missouri	5,606,260	9	9	9	9	9
Montana	905,316	1	1	2	1	1
Nebraska	1,715,369	3	2	3	3	3
Nevada	2,002,032	3	3	3	3	3
New Hampshire	1,238,415	2	2	2	2	2
New Jersey	8,424,354	13	13	13	13	13
New Mexico	1,823,821	3	2	3	3	3
New York	19,004,973	29	30	28	29	29
North Carolina	8,067,673	13	13	12	13	13
North Dakota	643,756	1	1	1	1	1
Ohio	11,374,540	18	18	17	18	18
Oklahoma	3,458,819	5	5	6	5	5
Oregon	3,428,543	5	5	6	5	5
Pennsylvania	12,300,670	19	19	19	19	19
Rhode Island	1,049,662	2	1	2	2	2
South Carolina	4,025,061	6	6	6	6	6
South Dakota	756,874	1	1	2	1	1
Tennessee	5,700,037	9	9	9	9	9
Texas	20,903,994	32	33	31	32	32
Utah	2,236,714	4	3	4	3	3
Vermont	609,890	1	1	1	1	1
Virginia	7,100,702	11	11	11	11	11
Washington	5,908,684	9	9	9	9	9
West Virginia	1,813,077	3	2	3	3	3
Wisconsin	5,371,210	8	8	8	8	8
Wyoming	495,304	1	1	1	1	1
Total	**281,424,177**					

*Actual apportionment.

REFERENCES AND FURTHER READINGS

1. Balinski, Michel L., and H. Peyton Young, "The Apportionment of Representation," *Fair Allocation: Proceedings of Symposia on Applied Mathematics*, 33 (1985), 1–29.

2. Balinski, Michel L., and H. Peyton Young, *Fair Representation; Meeting the Ideal of One Man, One Vote*. New Haven, CT: Yale University Press, 1982.

3. Balinski, Michel L., and H. Peyton Young, "The Quota Method of Apportionment," *American Mathematical Monthly*, 82 (1975), 701–730.

4. Brams, Steven, and Philip Straffin, Sr., "The Apportionment Problem," *Science*, 217 (1982), 437–438.

5. Census Bureau Web site, *http://www.census.gov.*

6. Eisner, Milton, *Methods of Congressional Apportionment*, COMAP Module #620.

7. Ford, Paul L. (editor), *The Works of Thomas Jefferson*, vol. VI. New York: G. P. Putnam & Sons, 1904.

8. Hoffman, Paul, *Archimedes' Revenge: The Joys and Perils of Mathematics*. New York: W. W. Norton & Co., 1988, chap. 13.

9. Huntington, E. V., "The Apportionment of Representatives in Congress." *Transactions of the American Mathematical Society*, 30 (1928), 85–110.

10. Huntington, E.V., "The Mathematical Theory of the Apportionment of Representatives," *Proceedings of the National Academy of Sciences*, U.S.A., 7 (1921), 123–127.

11. Meder, Albert E., Jr., *Legislative Apportionment*. Boston: Houghton Mifflin Co., 1966.

12. Saari, D. G., "Apportionment Methods and the House of Representatives," *American Mathematical Monthly*, 85 (1978), 792–802.

13. Schmeckebier, L. F., *Congressional Apportionment*. Washington, DC: The Brookings Institution, 1941.

14. Steen, Lynn A., "The Arithmetic of Apportionment," *Science News*, 121 (May 8, 1982), 317–318.

15. Syrett, Harold C. (editor), *The Papers of Alexander Hamilton*, vol. XI. New York: Columbia University Press, 1966.

16. Webster, Daniel, *The Writings and Speeches of Daniel Webster, Vol. VI, National Edition*. Boston: Little, Brown, and Company, 1903.

17. Woodall, D.R., "How Proportional Is Proportional Representation?," *The Mathematical Intelligencer*, 8 (1986), 36–46.

PART TWO

Management Science

KONINGSBERGA

A. Das Schloß.   E. Sagheimsche Kirch.   I. Das Closter.
B. Alt Stewer Kirch.   F. Die Domkirch.   K. Haberbergische Kirch.
C. S. Niclaus.   G. Das Colleguim.   L. Haber kruck.
D. S. Barbara.   H. Rahthaus im Kneiphoff. M. Hospital.

Der new

Euler Circuits

The Circuit Comes to Town

Sometimes great discoveries arise from the humblest and most unexpected of origins. Such is the case with the main idea we will explore in this chapter—the mathematical study of how things are interconnected. Our story begins more than 250 years ago in the medieval town of Königsberg, in Eastern Europe. Königsberg was divided by the river Pregel into four separate land areas which were connected to each other by seven bridges. A map of Königsberg drawn by the cartographer Martin Zeiller (opposite page) shows the layout of the old town in 1736, the year a brilliant young mathematician named Leonhard Euler came passing through. (For more on Euler see the biographical profile on p. 202.)

While in Königsberg, Euler heard of an innocent little puzzle of disarming simplicity: Is it possible for a stroller to take a walk around the old town crossing each of the seven bridges once but only once? The locals had tried, repeatedly and without success. Could Euler prove mathematically that it could not be done?

Euler, perhaps sensing that something important lay behind the frivolity of the question, proceeded to solve it by demonstrating that indeed such a walk was impossible. But he actually did much more. In solving the puzzle of the Königsberg bridges Euler laid the foundations for what was at the time a totally new type of geometry, which he called *geometris situs* ("the geometry of location"). From these modest beginnings, the basic ideas set forth by Euler eventually developed and matured into one of the most important and practical branches of modern mathematics, now known as *graph theory*. Modern applications of graph theory span practically every area of science and technology—from chemistry, biology, and computer science to psychology, sociology, and management science.

Over the next four chapters we will learn how graph theory is used to solve many important and unique problems in real life. For starters, in this chapter we will become acquainted with the basic notion of a *graph*, and how graphs can be used to model certain types of real-world problems. Along the way, we will also learn how Euler solved the Königsberg bridge puzzle.

> In theory, there is no difference between theory and practice. In practice, there is.
>
> *Yogi Berra*

5.1 Routing Problems

How important to you is the work of your garbage collector? Hardly anyone thinks much about this except when the collectors go on strike, at which time our appreciation for their services grows significantly. Similar things can be said about the mail carrier and the policeman on the beat. What these people have in common is that they are *providers* of a service that is *delivered* to us (usually at our homes), as opposed to a service we must go out and get (haircuts, the movies, etc.). Usually, these types of services can be delivered economically only when they are delivered to many customers, and doing this properly requires planning. In this chapter we will study, among other things, the mathematics behind the proper planning and design of delivery routes. These kinds of mathematics problems fall under the generic title of **routing problems**.

What is a *routing problem*? To put it in the most general way, routing problems are concerned with finding ways to route the delivery of *goods* and/or *services* to an assortment of *destinations*. Examples of the goods in question are packages, mail, newspapers, raw materials; examples of services are police protection, garbage collection, Internet access; examples of destinations are houses, warehouses, computer terminals, towns. In addition, *proper routes* must satisfy what we will call the *rules of the road*: (i) if there is a "direction of traffic" (as in one-way streets, pipeline flows, and communication protocols), then the direction of traffic must be followed, and (ii) if there is no direct way to get from destination *X* to destination *Y*, then a proper route cannot go directly from point *X* to point *Y*. (This seems self-evident, but in some situations it is easy to forget this rule.)

Two fundamental questions can come up in a routing problem.

1. Is there a proper route for the particular problem?
2. If there are many possible routes, which one is the *best* (where best is a function of some predetermined variable such as *cost, distance*, or *time*)?

Question 1 calls for a yes/no answer, which is often (but not always) easy enough to provide. Question 2 tends to be a little more involved and in some situations (as we will see in Chapter 6) can actually be quite difficult to answer. In this chapter we will learn how to answer both types of questions for a special category of routing problems called **Euler circuit problems** (sometimes also known as *traversability* problems).

The routing of a garbage truck, a mail truck, or a patrol car through the streets of a city is a typical example of this type of problem. Other examples might involve routing water and electric meter readers, census takers, newspaper deliverers, tour buses, etc. Whatever the case may be, the common thread in all Euler circuit problems is the need to *traverse all* the streets (roads, lanes, bridges, etc.) within a designated area—be it a whole town or a section of it.

To clarify the concept of an Euler circuit problem, we will introduce several examples of such problems (just the problems for now—their solutions will come later in the chapter).

EXAMPLE 5.1 The Walking Patrolman

After a rash of burglaries, a private security guard is hired to patrol on foot the streets of the small neighborhood shown in Fig. 5-1. He parks his car at the corner across from the school playground (*S* in Fig. 5-1). The security guard is being paid for just one walk-through and is anxious to get the job done and go home. Being mathematically inclined, he has two questions he would like answered: (1) Is there a route that allows him to walk through every block just once (with the walk starting and ending at the corner *S* where he parked his car)? (2) If not, what is the most *efficient* possible way to walk the neighborhood, once again starting and ending at *S*? Here, efficiency is measured in total number of blocks walked.

FIGURE 5-1

EXAMPLE 5.2 The Walking Mail Carrier

Consider now the problem of a mail carrier, who has exactly the same neighborhood as the security guard (Fig. 5-1) as her designated mail delivery area. The big difference is that for those blocks in which there are homes on both sides of the street the mail carrier must walk through the block *twice* (she does each side of the street separately). Also, the mail carrier needs to start and end her trip at the local Post Office (P.O. in Fig. 5-1). The mail carrier asks two similar questions, since she is also interested in doing her route with the least amount of walking: (1) Starting at the Post Office, can she cover every sidewalk along which there are homes once and only once, ending her walk back at the Post Office? (2) If that can't be done, what is the most efficient way to deliver the mail throughout the neighborhood? Again, efficiency is measured in total number of blocks walked.

EXAMPLE 5.3 The Seven Bridges of Königsberg

Basically, this is the true story with which we opened the chapter—with a little embellishment: A prize (seven gold coins) is offered to the first person who can find a way to walk across each one of the seven bridges of Königsberg without recrossing any and return to the original starting point. (For the reader's convenience we modernized the area map, now shown as Fig. 5-2.) A smaller prize (five gold coins) is offered for anyone who can cross each of the seven bridges exactly once without necessarily returning to the original starting point. So far, no one has collected on either prize. How come?

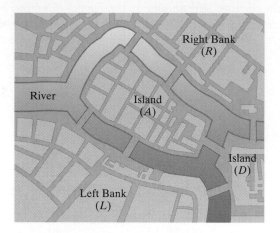

FIGURE 5-2

EXAMPLE 5.4 The Bridges of Madison County

This is a more modern version of Example 5.3. Madison County is a quaint old place, famous for its quaint old bridges. A beautiful river runs through the county, and there are four islands (*A, B, C,* and *D*) and 11 bridges joining the islands to both banks of the river (*R* and *L*) and one another (Fig. 5-3). A famous photographer is hired to take pictures of each of the 11 bridges for a national magazine. The photographer needs to drive across each bridge once for the photo shoot. Moreover, since there is a $25 toll (the locals call it a "maintenance tax") every time an out-of-town visitor drives across a bridge, the photographer wants to minimize the total cost of his trip and to recross bridges only if it is absolutely necessary. What is the best (cheapest) route for him to follow?

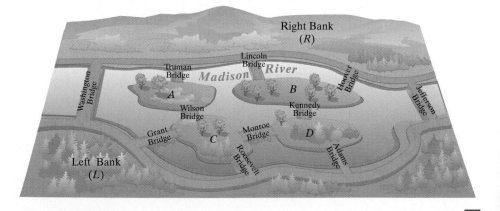

FIGURE 5-3

EXAMPLE 5.5 Child's Play?

Figure 5-4 shows some simple line drawings. Can we trace each drawing without lifting the pencil or retracing any of the lines, and end in the same place we started? What if we are not required to end back at the starting place? These kinds of tracings are called **unicursal tracings** (*closed* unicursal tracings if we have to end back where we started, *open* unicursal tracings if we don't). Many of us played such games in our childhood (those were the good old days before video games) and may actually know or can quickly figure out the answers in the case of Figs. 5-4(a), (b), and (c).[1] But what about more complicated shapes, such as the one in Fig. 5-4(d)? Can we trace this figure without lifting the pencil or re-tracing any of the lines? If so, how? In general, any unicursal tracing problem is just an abstract example of an Euler circuit problem.

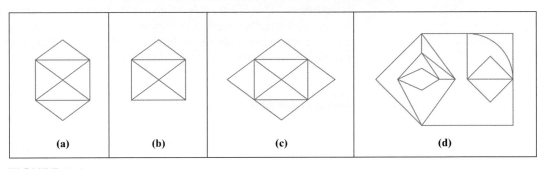

FIGURE 5-4

As the preceding examples illustrate, Euler circuit problems can come in a variety of forms. Fortunately, the same basic mathematical theory is used to solve any Euler circuit problem. In the next several sections we will develop the basic elements of this theory. In the last section we will use our newly acquired knowl-edge to come back and solve some of these examples.

5.2 Graphs

The unifying mathematical concept that will allow us to solve any Euler circuit problem is the concept of a *graph*.

For starters, let's say that a **graph**[2] is a picture consisting of dots, called **vertices**, and lines, called **edges**. The edges do not have to be straight lines, but they always have to connect two vertices. When an edge connects a vertex back with itself (which is also allowed), then it is called a **loop**.

The foregoing is not to be taken as a precise definition of a graph, but rather as an informal description that will help us get by for the time being. To get a feel for what a graph is, let's look at a few examples.

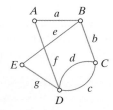

FIGURE 5-5

EXAMPLE 5.6

Figure 5-5 shows an example of a graph. This graph has five vertices called *A*, *B*, *C*, *D*, and *E* and seven edges called *a*, *b*, *c*, *d*, *e*, *f*, and *g*. (As much as possible, we will try to be consistent and use uppercase letters for vertices and lowercase let-ters for edges, but this is not mandatory.) A couple of comments about this graph: First, note that the point where edges *e* and *f* cross is *not* a vertex—it is just the

---

[1]Figure 5-4(a) has a closed unicursal tracing. Any point can be chosen as the starting and ending point. Figure 5-4(b) does not have a closed unicursal tracing but it does have an open unicursal tracing. The starting point must be one of the two bottom corners. The ending point will be the bottom corner op-posite the starting point. Figure 5-4(c) has no possible unicursal tracing.

[2]In this context, the word *graph* has no connection with the graphs of functions studied in algebra and calculus.

crossing point of two edges. Second, note that there is no rule against having more than one edge connecting the same two vertices, as is the case with vertices *D* and *C*. These are called **multiple edges**. ∎

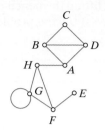

FIGURE 5-6

EXAMPLE 5.7

Figure 5-6 shows a graph with 8 vertices (*A, B, C, D, E, F, G*, and *H*) and 11 unlabeled edges. (There is no rule that says that we have to give names to the edges.) We can still specify an edge by naming the 2 vertices that are its end points. For example, we can talk about the edge *AH*, the edge *BD*, and so on. Note that there is a loop in this graph—it is the edge *GG*. ∎

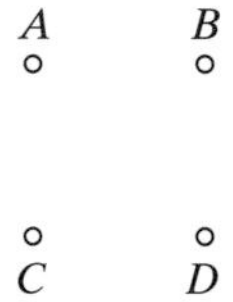

A ○   B ○

C ○   D ○

FIGURE 5-7

EXAMPLE 5.8

Does Fig. 5-7 show a graph? Yes! It is a graph with four vertices and no edges. While it does not make for a particularly interesting graph, there is nothing illegal about it. Graphs without any edges are permissible. On the other hand, we cannot have a graph without vertices, since then there can be no edges, and without vertices or edges we have nothing! ∎

EXAMPLE 5.9

Does Fig. 5-8 show a graph? We have vertices and we have edges, so the answer is yes! The vertices have funny names, but so what? Note that the graph is made up of 2 separate, disconnected pieces. Graphs of this type are said to be *disconnected*, and the individual pieces are called the *components* of the graph. (The graph in Fig. 5-7, for example, is disconnected and has four components.)

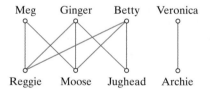

FIGURE 5-8

What might the graph in Fig. 5-8 represent? Let's suppose that Meg, Ginger, Betty, Veronica, Reggie, Moose, Jughead, and Archie are friends who went together to a party. A graph such as the one in Fig. 5-8 might be a pictorial description of who danced with whom at the party. We can learn a few things from such a picture (such as the fact that Veronica and Archie only danced with each other), but most importantly, we should appreciate the fact that the picture provides such a crisp and convenient way to describe the evening's dancing arrangements. ∎

EXAMPLE 5.10

We are now going to present a graph without a picture. This graph has four vertices (*A, D, L*, and *R*) and seven edges (*AR, AR, AD, AL, AL, DR*, and *DL*). This information completely specifies the graph. The reader is encouraged to draw a picture of this graph. Where should the vertices be placed? (It doesn't matter!) What shape should the edges have—straight, curved, wiggly? (It doesn't matter either!) Figures 5-9(a) and (b) show two different representations of this same graph.

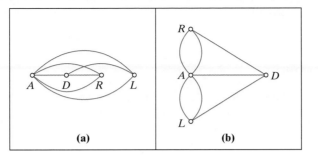

FIGURE 5-9

(a)   (b)

Example 5.10 illustrates an important point. A graph can be drawn in infinitely many different ways, and it is not the shape of the graph that matters, but rather *how* the vertices are connected to each other.

With all of the preceding examples under our belt, we might be ready for a more formal definition of a graph. A graph is a *structure for describing relationships* within a set of objects (the vertices). The story of which objects are related to each other is told by the edges. That is all the information we get out of a graph—it doesn't seem like much, but it is.

It follows that any time we have a relationship between objects, whatever that relationship might be (love, kinship, dance partner, etc.), *we can describe such a relationship by means of a graph.* This simple idea is the key reason for the tremendous usefulness of graph theory.

EXAMPLE 5.11

On any particular week of the baseball season one can look up the schedule for that week in a good newspaper or a television guide. Here is one week's schedule for the National League East pretty much as it would be reported in the newspaper.

- Monday: Pittsburgh versus Montreal, New York versus Philadelphia, Chicago versus St. Louis.
- Tuesday: Pittsburgh versus Montreal.
- Wednesday: New York versus St. Louis, Philadelphia versus Chicago.
- Thursday: Pittsburgh versus St. Louis, New York versus Montreal, Philadelphia versus Chicago.
- Friday: Philadelphia versus Montreal, Chicago versus Pittsburgh.
- Saturday: Philadelphia versus Pittsburgh, New York versus Chicago, Montreal versus St. Louis.
- Sunday: Philadelphia versus Pittsburgh.

A different way to describe the schedule is by means of a graph. Here the vertices are the teams, and each game played during that week is described by an edge between two teams, as in Fig. 5-10. (Insofar as the description of the schedule is concerned, geography is not an issue, and note that the position of the vertices has nothing to do with the geographic location of the cities.) The main point of this example is to illustrate the convenience of the graph as a way to describe the schedule. Do you want to know how many games Pittsburgh is scheduled to play during the week? Do you want to know if New York plays Pittsburgh during the week? Where would you rather look—the list or the graph? The answer is obvious.

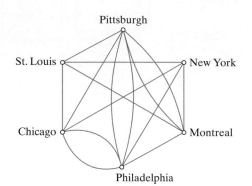

FIGURE 5-10

5.3 Graph Concepts and Terminology

Every branch of mathematics has its own peculiar jargon, and the theory of graphs has more than its share. In this section we will introduce a few essential concepts and terms that we will need in the chapter.

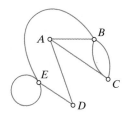

FIGURE 5-11

- **Adjacent vertices.** Two vertices are said to be **adjacent** if there is an edge joining them. (In this context, adjacent vertices do not have to be physically next to each other.) In the graph shown in Fig. 5-11 vertices E and B are adjacent; D and C are not (even though in the picture they are near each other). Also, because of the loop at E, we can say that vertex E is adjacent to itself.

- **Adjacent edges.** Two edges are **adjacent** if they share a common vertex. In Fig. 5-11, edges AB and AD are adjacent; edges AB and DE are not.

- **Degree of a vertex.** The **degree of a vertex** is the number of edges at that vertex. (A loop contributes twice toward the degree.) An **odd vertex** is a vertex of odd degree; an **even vertex** is a vertex of even degree.

  In the graph shown in Fig. 5-11, vertex A has degree 3 [which we can write as $\deg(A) = 3$], vertex B has degree 4 [$\deg(B) = 4$], $\deg(C) = 3$, $\deg(D) = 2$, and $\deg(E) = 4$ (because of the loop). All together, the graph has 2 odd vertices and 3 even vertices.

- **Paths.** A **path** is a sequence of vertices with the property that each vertex in the sequence is *adjacent* to the next one. Thus, a path can also be thought of as describing a sequence of adjacent edges—a *trip*, if you will, along the edges of the graph. Whereas a vertex can appear on the path more than once, an edge can be part of a path *only once*. The number of edges in the path is called the *length* of the path.

  The graph in Fig. 5-11 has many paths—here are just a few examples.

  - A, B, E, D. This is a path from vertex A to vertex D, consisting of edges AB, BE, and ED. The length of this path is 3.

- *A, B, C, A, D, E.* This is a path of length 5 from *A* to *E.* The path visits vertex *A* twice, but no edge is repeated.
- *A, B, C, B, E.* This is also a path from *A* to *E.* This path is possible because there are two edges connecting *B* and *C.*
- *A, C, B, E, E, D.* This path of length 5 is possible because of the loop at *E.*

  The following *are not* paths:

- *A, C, D, E.* There is no edge connecting *C* and *D.*
- *A, B, C, B, A, D.* The edge *AB* appears twice, so this is not a path.
- *A, B, C, B, E, E, D, A, C, B.* In this long string of vertices, everything is OK until the very end, when the edge *CB* appears for a third time. The first two instances are fine, because there are two edges connecting *B* and *C.* The third time, though, is one too many. One of the two edges would have to be retraveled.

- **Circuits.** A **circuit** has the same definition as a path, with the additional requirement that it must start and end at the same vertex. The following are some of the circuits in the graph in Fig. 5-11.

  - *A, B, C, A.* Note that this same circuit can also be written as *B, C, A, B* or *C, A, B, C.* This is a circuit of length 3. A circuit—like a bead necklace—has really no specified start or end. We use the words *starting vertex* and *ending vertex* only when we write circuits in linear form—which we do primarily as a matter of convenience.
  - *B, C, B.* A perfectly legitimate circuit. It could also be written as *C, B, C.*
  - *E, E.* Yes, a loop in and of itself can be considered a circuit of length 1.

    There are several other circuits in Fig. 5-11, including one of length 6. Can you find it?

- **Connected graphs.** A graph is **connected** if any two of its vertices can be joined by a path. This essentially means that it is possible to travel from any vertex to any other vertex along consecutive edges of the graph. If a graph is not connected, it is said to be **disconnected**. A graph that is disconnected is made up of pieces that are by themselves connected. Such pieces are called the *components* of the graph.

  The graph in Fig. 5-12(a) is connected. The graphs in Figs. 5-12(b) and (c) are disconnected. The one in Fig. 5-12(b) has two components; the one in Fig. 5-12(c) has three.

FIGURE 5-12
(a) This graph is connected.
(b) This graph is not connected. It has two components. (c) This graph is not connected. It has three components, one of them the isolated vertex *E.*

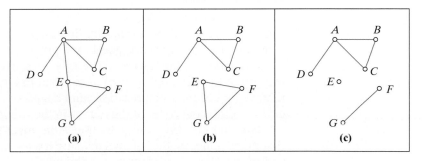

(a) (b) (c)

- **Bridges.** Sometimes in a connected graph there is an edge such that if we were to erase it, the graph would become disconnected. Not surprisingly, such an edge is called a **bridge**.

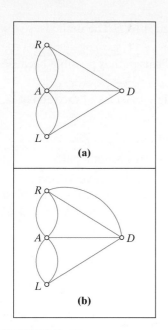

(a)

(b)

FIGURE 5-13
(a) This graph has no Euler paths; (b) This graph has several Euler paths.

In Fig. 5-12(a), the edge *AE* is a bridge because when we remove it, the graph becomes disconnected. [There is another bridge in Fig. 5-12(a). We leave it to the reader to find it.]

- **Euler paths.** An **Euler path** is a path that travels through *every* edge of a graph. Since it is a path, edges can only be traveled once. Thus, in an Euler path we travel along *every* edge of the graph *once and only once*. By definition, the length of an Euler path is the number of edges in the graph. Note that, since every edge of the graph must be included in an Euler path, only a connected graph can have such a path.

  The definition of an Euler path should ring a bell—it sounds almost the same as the concept of a unicursal tracing. In fact, they are the same idea. The former is couched in the context of graphs, the latter in the context of ordinary line drawings. Not every graph has an Euler path—it is often the case that a graph cannot be traced. The graph shown in Fig. 5-13(a) does not have an Euler path. On the other hand, the graph shown in Fig. 5-13(b) has several Euler paths. One of them is *L, A, R, D, A, R, D, L, A.* The reader is encouraged to find at least one more.

- **Euler circuits.** An **Euler circuit** is a circuit that travels through *every* edge of a graph. Thus, we have the same requirements as for an Euler path, but in addition, we require that the starting and ending vertex be the same. Note than an Euler circuit is essentially the same as a closed unicursal tracing of the graph.

5.4 Graph Models

One of Euler's most important ideas was the observation that certain types of problems can be conveniently rephrased as graph problems, and that, in fact, graphs offer the perfect model for describing many real-life situations. The notion of using a mathematical concept to describe and solve a real-life problem is one of the oldest and grandest traditions in mathematics. It is called *modeling*. Unwittingly, we have all done simple forms of modeling before, all the way back to elementary school. Every time we turn a word problem into an arithmetic calculation, an algebraic equation, or a geometric picture, we are modeling. We can now add to our repertoire one more tool for modeling: graph models.

EXAMPLE 5.12 The Seven Bridges of Königsberg: Act 2

Remember that the Königsberg bridges problem as described in Example 5.3 was to find a walk through the city that crossed each of the bridges once and returned to the starting place (good for a prize of seven gold coins) or do the same without ending at the starting place (good for five gold coins). A stylized map of the city of Königsberg is shown once again in Fig. 5-14(a). The reader is warned that we moved the exact positions and angles of some of the bridges and in general smoothed out some of the details in the original map.

Isn't this cheating? Actually not! A moment's reflection should convince us that many things on the original map are irrelevant to the problem: the shape and size of the islands and river banks, the lengths of the bridges, and even the exact location of the bridges, as long as they are still joining the same two sections of the city. Aha! We have just stumbled upon the key observation. *The only thing that truly matters in this problem is the relationship between land masses and bridges*: which land masses are connected to each other and by how many bridges

[Fig. 5-14(b)]. Thus, when we strip the map of all its superfluous information, we end up with the graph model shown in Fig. 5-14(c), where the vertices represent the four separate land masses and the edges represent the seven bridges. In this new interpretation of the puzzle a stroll around the town that crosses each bridge once and ends back at the starting point would be equivalent to an *Euler circuit* of the graph in Fig. 5–14(c). A stroll that crosses each bridge once but does not return to the starting point would be equivalent to an *Euler path* in Fig. 5–14(c).

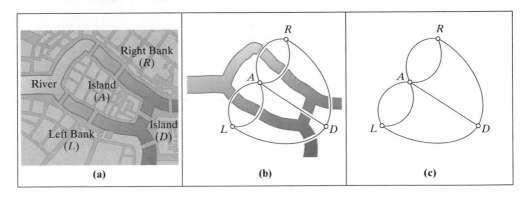

FIGURE 5-14

As big moments go, this one may not seem like much, but the reader is encouraged to take stock of what we have accomplished in Example 5.12 (and, if necessary, reread the example carefully). We have actually taken a big step—we made a connection between theory and reality.

EXAMPLE 5.13 The Walking Patrolman: Act 2

In Example 5.1, we discussed the problem of a security guard who needs to walk the streets of the neighborhood shown in Fig. 5-15(a). A graph model of this problem is given in Fig. 5-15(b), with each block of the neighborhood represented by an edge in the graph, and each intersection represented by a vertex of the graph.

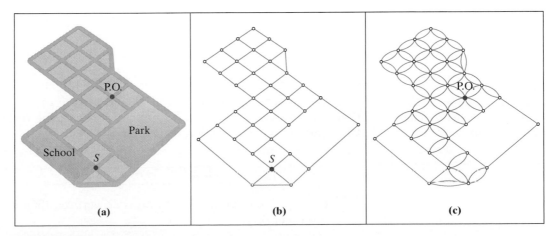

FIGURE 5-15
(a) The original neighborhood. (b) A graph model for the security guard (one pass per block). (c) A graph model for the mail carrier (one pass per sidewalk).

EXAMPLE 5.14 The Walking Mail Carrier: Act 2

As the reader may recall, the mail carrier's problem of Example 5.2 differed from the security guard's problem in that the mail carrier must deliver mail on both sides of every street (except for the blocks facing the school and the park). This means she has to walk most streets twice, once for each side. Consequently, an appropriate graph model for this problem should have one edge for every side of the street to which the mail has to be delivered, as shown in Fig. 5-15(c). ■

5.5 Euler's Theorems

The Königsberg bridges problem, as modeled in Example 5.12, was equivalent to finding an Euler circuit or an Euler path in the corresponding graph. Euler's solution to the problem was to demonstrate that neither an Euler circuit nor an Euler path is possible.

Let's start with the Euler circuit argument. Why is such a circuit impossible? Let's say for the sake of argument that the starting vertex of the desired circuit is *L*. (Since we are looking for a circuit, it makes no difference which vertex we pick for the starting point.) Somewhere along the way we will have to go through vertex *A*, and in fact, we will have to do so more than once. Let's count exactly how many times. The first visit to *A* will use up two edges, one getting in and a different one getting out. The second visit to *A* will use up two other edges, and the third visit to *A* will use up two more. Oops! There are only five edges to get in and out of *A*. Two visits to *A* won't do because there would be an untraveled edge, and three visits are too many because we would have to recross one of the edges to get out. It follows that the walk is impossible! It's the odd number of edges at *A* (or at any other vertex) that causes the problem. The argument can be extended and made general in a very natural way. We present it without any further ado.

EULER'S THEOREM 1

(a) If a graph has *any* odd vertices, then it *cannot* have an Euler circuit.

(b) If a graph is *connected* and *every* vertex is an even vertex, then it has at least one Euler circuit (and usually more).

Note that just having all even vertices does not guarantee an Euler circuit unless the graph is also connected (see, for example, Fig. 5-16).

The requirements for a graph to have an Euler path are similar. All the vertices except for the *starting* and *ending* vertices of the path must be even. The starting vertex requires *one edge to get out at the start* and two more for each visit through that vertex, so it must be *odd*. Likewise the *ending* vertex must be *odd* (two edges for every visit plus one more to come into the vertex at the end of the trip). Thus, we have the following theorem.

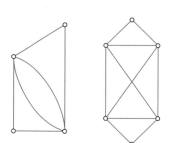

FIGURE 5-16
Every vertex is even but the graph has no Euler circuit because it is disconnected.

EULER'S THEOREM 2

(a) If a graph has *more than two* odd vertices, then it cannot have an Euler path.

(b) If a graph is connected and has exactly *two* odd vertices, then it has at least one Euler path (and usually more). Any such path must start at one of the odd vertices and end at the other one.

EXAMPLE 5.15 The Bridges of Königsberg: Conclusion

Consider the graph that models the Königsberg bridges problem [Fig. 5-17(b)]. This graph has four odd vertices, and thus neither an Euler circuit nor an Euler path can exist. *There is no possible way anyone can walk across all of the bridges without having to recross some of them!* How many bridges will need to be recrossed? With a little planning we can find a walk that recrosses just one bridge, if we can start and end the walk at different locations, or just two bridges, if we must start and end at the same location. (See Exercise 57.)

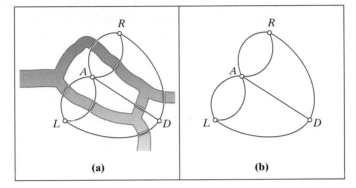

FIGURE 5-17

EXAMPLE 5.16 Unicursal Tracings Revisited

Figure 5-18 shows four graphs. These graphs are equivalent to the line drawings in Example 5.5 (Fig. 5-4). We can now easily apply Euler's theorems. The graph in Fig. 5-18(a) has an Euler circuit (and thus a closed unicursal tracing) because all vertices are even. The graph in Fig. 5-18(b) has an Euler path (open unicursal tracing) which must start at D and end at C (or vice versa) because D and C are the only two odd vertices. The graph in Fig 5-18(c) has neither an Euler path nor an Euler circuit because there are too many odd vertices. Finally, the complicated graph in Fig. 5-18(d) does have a closed unicursal tracing since, as we can verify with a quick check, every vertex is even.

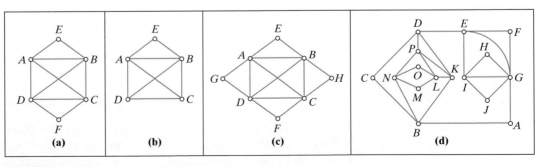

FIGURE 5-18

The careful reader may have noticed that there is an apparent gap in Euler's theorems 1 and 2. The two theorems together cover the cases of graphs with no odd vertices [Theorem 1(b)], *two* odd vertices [Theorem 2(b)], and *more than two* odd vertices [Theorem 2(a)]. What happens if a graph has *just one* odd vertex?

Didn't Euler consider this possibility? It turns out that he did, but found that it is impossible for a graph to have *just one* odd *vertex*.

The key observation that Euler made was that when *the degrees of all the vertices of a graph are added, the total is exactly twice the number of edges in the graph*. Think about it: An edge—let's say *XY*—contributes one to the degree of vertex *X* and one to the degree of vertex *Y*, so, in all, that edge contributes two to the sum of the degrees. Thus when the degrees of all the vertices of a graph are added, *the total must be an even number*, which means that it is impossible to have only one odd vertex (the sum of all the degrees would then be an odd number). In fact, we can push the logic one step further, and argue just as well that it is impossible for a graph to have 3, 5, 7, ... odd vertices. We summarize the preceding observations into a theorem.

EULER'S THEOREM 3

(a) The sum of the degrees of all the vertices of a graph equals twice the number of edges (and therefore is an even number).

(b) A graph always has an even number of *odd* vertices.

Table 5-1 is a summary of Euler's theorems 1, 2, and 3. It shows the relationship between the number of odd vertices in a connected graph and the existence of Euler paths or Euler circuits. (The assumption that the graph is connected is essential—a disconnected graph cannot have Euler paths or circuits regardless of what else is going on.)

TABLE 5-1

Number of odd vertices	Conclusion
0	Graph has Euler circuit(s)
2	Graph has Euler path(s) but no Euler circuit
4, 6, 8, ...	Graph has no Euler path and no Euler circuit
1, 3, 5, ...	Impossible (there are no graphs with an odd number of odd vertices!)

5.6 Fleury's Algorithm

Euler's theorems give us an easy way to determine if a graph has an Euler circuit or an Euler path. Unfortunately, Euler's theorems are of no help in finding an actual Euler circuit or path, if there is one. Of course, for simple graphs such as the ones shown in Figs. 5-18(a) and (b), we can find an Euler path (or circuit) by simple trial and error. But what about the graph in Fig. 5-18(d)? Or an even more complicated graph, with hundreds of vertices and edges? Do we really want to use trial and error to find an Euler circuit or an Euler path? Of course not. A trial-and-error approach for a large graph is a crapshoot—we could get lucky and find the solution right away, or we could spend hours chasing up dead ends. What we really need here is an *algorithm*.

Algorithms

There are many types of problems that can be solved by simply following a set of procedural rules—very specific rules like *when you get to this point do this, ... after you finish this, do that*, and so on. Given a specific problem *X*, a set of *procedural rules* that, when followed, always lead us to some sort of "solution"[3] to *X* is called an **algorithm** for solving *X*. The problem *X* need not be a mathematics problem—algorithms are used, sometimes unwittingly, in all walks of life: directions to find someone's house, the instructions for assembling a new bike, or a recipe for baking an apple pie, are all examples of real-life algorithms. A useful analogy is to think of the problem as a *dish* we want to prepare, and the algorithm as a *recipe* for preparing that dish.

In many cases, there are several different algorithms for solving the same problem (there is more than one way to bake an apple pie); in other cases, the problem does not lend itself to an algorithmic solution. In mathematics, algorithms are either *formula* driven (you just apply the formula or formulas to the appropriate inputs) or *directive* driven (you must follow a specific set of directives). In this section of the book (Chapters 5 through 8) we will discuss many important algorithms of the latter type.

Algorithms may be complicated, but never difficult (there is a big difference!). You don't have to be a brilliant and creative thinker to implement an algorithm—you just have to learn how to follow instructions carefully and methodically. For most of the algorithms we will discuss in this and the next three chapters, the key to success is simple: practice, practice, practice!

Fleury's Algorithm

We will now turn our attention to an algorithm that will allow us to find an *Euler circuit* in a connected graph with no odd vertices, or, alternatively, an *Euler path* in a connected graph with two odd vertices. (Technically speaking, there are two separate algorithms, but in essence they are identical, so they can be described as one.) The algorithm is attributed to a mysterious Frenchman by the name of M. Fleury, who is supposed to have published a note describing the algorithm in 1885.[4]

We will begin by describing Fleury's algorithm informally, and then work out a couple of examples and conclude with the formal description.

The idea behind Fleury's algorithm can be paraphrased by that old piece of folk wisdom: *Don't burn your bridges behind you*. In the case of a connected graph, the word *bridge* has a very specific meaning—it is the only edge connecting two separate sections (call them *A* and *B*) of the graph as illustrated in Fig. 5-19. If you are in *A*, you can only get to *B* by crossing the bridge. Moreover, if you want to get back to *A*, the only way to do it is recrossing that same bridge. Thus, you would only want to cross that bridge if you know that all edges in *A* have been traveled.

FIGURE 5-19
The bridge separates the two sections of the graph. The only way to get from *A* to *B*, or vice versa, is by crossing the bridge.

---

[3]We use the word *solution* in a rather broad sense here, and allow for approximate or imperfect "solutions" within this meaning. In the real world, sometimes that is the best we can do.
[4]The first known attribution of the algorithm to Fleury appears in a French book, *Recreations Mathematiques,* by Edouard Lucas, published in 1891.

In trying to find an Euler circuit or an Euler path, bridges are the last edges we want to cross! Simple enough, but there is a rub: The graph whose bridges we are supposed to avoid is not necessarily the original graph of the problem. Instead it is that part of the original graph which has yet to be traveled. The point is this: Once we travel along an edge, we are done with it! We will never cross it again, so from that point on, as far as we are concerned, it is as if that edge never existed. Our concerns lie only on how we are going to get around the yet-to-be-traveled part of the graph. Thus, when we talk about bridges that we want to leave as a last resort, we are really referring to *bridges of the to-be-traveled part of the graph*.

Since each time we traverse an edge the *untraveled part of the graph* changes (and consequently so do its bridges), Fleury's algorithm requires some careful bookkeeping. This does not make the algorithm difficult; it just means that we must take extra pains in separating what we have already done from what we yet need to do.

While there are many different ways to accomplish this (and readers are certainly encouraged to invent their own), a fairly reliable way goes like this: We start with two separate copies of the graph, copy 1 for making decisions and copy 2 for record keeping. Every time we traverse another edge, we erase it from copy 1 but mark it (say in red) and label it with the appropriate number on copy 2. As we progress along our Euler circuit, copy 1 gets smaller and copy 2 gets redder. Copy 1 helps us decide where to go next; copy 2 helps us reconstruct our trip (just in case we are asked to demonstrate how we did it!). Let's try a couple of examples.

EXAMPLE 5.17

The graph in Fig. 5-20 has Euler circuits—we know this is so because every vertex has even degree. Let's use Fleury's algorithm to find one of these Euler circuits. Although this is a very simple graph which could be done easily by trial and error, the real purpose of this example is to help us understand how the algorithm works.

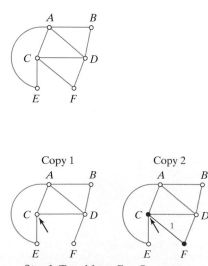

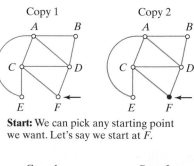

Start: We can pick any starting point we want. Let's say we start at *F*.

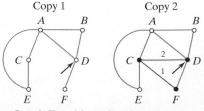

Step 1: Travel from *F* to *C*.
(Could have also gone from *F* to *D*.)

Step 2: Travel from *C* to *D*.
(Could have also gone to *A* or to *E*.)

FIGURE 5-20

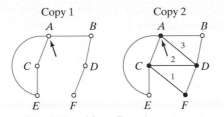

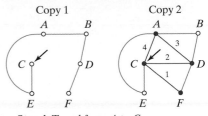

Step 3: Travel from D to A.
(Could have also gone to B but not to F — DF is a bridge!)

Step 4: Travel from A to C.
(Could have also gone to E but not to B — AB is a bridge!)

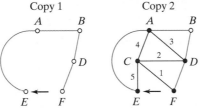

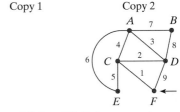

FIGURE 5-20
(continued)

Step 5: Travel from C to E.
(There is no choice!)

Steps 6, 7, 8, and 9: Only one way to go at each step.

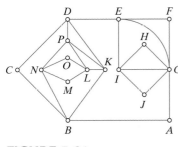

FIGURE 5-21

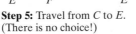

EXAMPLE 5.18

We have already discussed the fact that the graph in Fig. 5-21 has Euler circuits. Since it would be a little impractical to show each step of Fleury's algorithm with a separate picture as we did in Example 5.17, we ask the reader to do some of the work. If you haven't already done so, then, get a pencil, an eraser, and some paper. Next, make two copies of the graph. Ready? Let's go!

- **Start:** Pick an arbitrary starting point, say J.
- **Step 1:** From J we can go to either I or G. Since neither JI nor JG is a bridge, we can choose either one. Say we choose JI. (We can now erase JI on copy 1; mark and label it with a 1 on copy 2.)
- **Step 2:** From I we can go to E, H, or G. Any of these choices is OK. Say we choose IH. (Now erase IH from copy 1 and mark and label it on copy 2.)
- **Step 3:** From H there is only one way to go, and that's to G. [Erase edge HG as well as vertex H (we won't be coming back to it) from copy 1 and mark and label it on copy 2.]
- **Step 4:** From G we have several choices (to A, to F, to E, to I, or to J). We should not go to J—GJ is a bridge in copy 1. Any of the other choices is OK. Say we choose GF. (Erase edge GF from copy 1, etc.)
- **Step 5:** There is only one way to go from F (to E). (Erase edge FE as well as vertex F from copy 1, etc.)
- **Step 6:** From E we have three choices, all of which are OK. Say we choose ED. (You know what to do with copy 1 and copy 2. To speed things up, from here on we will omit this part.)
- **Step 7:** Three choices at D. All of them are OK. Say we choose DP.
- **Step 8:** Several choices at P. All of them are OK. Say we choose PK.
- **Step 9:** Several choices at K. All of them are OK. Say we choose KB.

- **Step 10:** Several choices at B. One of them is not OK (edge BA is a bridge in copy 1). Say we choose BN.
- **Step 11:** Several choices at N. All of them are OK. Say we choose NO.
- **Step 12:** Only one way to go (to L).
- **Step 13:** Several ways to go, but one of them (LK) is a bridge in copy 1. Say we choose LP.
- **Steps 14–22:** Only one way to go in each step. From P to N to M to L to K to D to C to B to A to G.
- **Step 23:** We do have some choices at G, but one of them (GJ) is a bridge in copy 1. Say we choose GI.
- **Steps 24–26:** From I to E to G to J.

We are finished! The Euler circuit we found is

$J, I, H, G, F, E, D, P, K, B, N, O, L, P, N, M, L, K, D, C, B, A, G, I, E, G, J.$

Notice that this is just one of many possible Euler circuits—making different choices along the way would lead to different Euler circuits. ■

Here is a formal description of the basic rules for Fleury's algorithm.

> ### FLEURY'S ALGORITHM FOR FINDING AN EULER CIRCUIT
>
> - First make sure that the graph is connected and all the vertices have even degree.
> - Pick any vertex as the starting point.
> - When you have a choice, always choose to travel along an edge that is not a bridge of the yet-to-be-traveled part of the graph.
> - Label the edges in the order in which you travel them.
> - When you can't travel any more, *stop.* You are done!

When a connected graph has exactly two vertices of odd degree, then we know that the graph does not have an Euler circuit, but it does have an Euler path, and we can find such a path using Fleury's algorithm with one minor change: *The starting point must be one of the two vertices of odd degree.* Other than that, the rest of the procedure is exactly the same. When they are followed properly, the trip is guaranteed to end at *the other vertex of odd degree.*

5.7 Eulerizing Graphs

We now know that when a graph has no vertices of odd degree or two vertices of odd degree, then it has an Euler circuit or an Euler path, respectively, and that when a graph has more than two vertices of odd degree, then there is no Euler circuit or Euler path. In this case there is no possible way that we can travel along all the edges of the graph without having to recross some of them.

We will now discuss a new question. How do we go about finding a route that covers all the edges of the graph while revisiting the least possible number of edges? This is important because, in many real-world routing problems, the cost is proportional to the amount of travel. Thus, the most efficient routes are those with the least amount of wasted travel (usually called *deadhead travel*), which in this case means the least amount of duplication of edges.

When we are required to travel along *all* the edges of a graph the following simple rule summarizes the situation.

Total cost of route = cost of traveling original edges in the graph
+ cost of deadhead travel.

EXAMPLE 5.19

Consider the graph in Fig. 5-22(a). The graph has eight odd vertices (*B, C, E, F, H, I, K,* and *L)*, shown as red vertices. It follows that the graph has no Euler circuit or Euler path. By adding a duplicate copy of edges *BC, EF, HI,* and *KL*, we get the graph in Fig. 5-22(b), a close cousin to the original graph. The main difference between the two graphs is that 5-22(b) has all even vertices, and thus, Euler circuits. Figure 5-22(c) shows one such Euler circuit (just travel the edges as numbered). The Euler circuit in Fig. 5-22(c) can be reinterpreted as a trip along the edges of our original graph as shown in Fig 5-22(d). In this trip we are traveling along all of the edges of the graph, but we are retracing four of them (*BC, EF, HI,* and *KL*). While this is not an Euler circuit for the original graph, it is a circuit describing the most *efficient* trip (meaning a trip with the least amount of duplication) that covers all of the edges—an *optimal* such circuit, if you will.

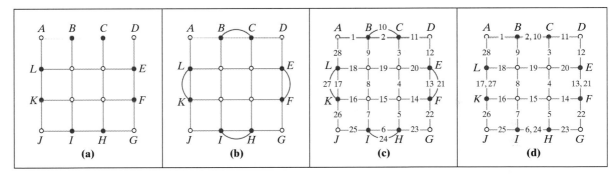

FIGURE 5-22
(a) The original graph. (b) An eulerized version of the graph in (a). (c) An Euler circuit for the graph in (b).
(d) The Euler circuit shown in (c) applied to the original graph.

In Example 5.19 we started with a given graph [Fig. 5-22(a)] and modified it by adding extra edges in such a way that the odd vertices become even vertices [Fig. 5-22(b)]. (In other words, we neutralized the "bad guys.") This process of changing a graph by adding additional edges so that the odd vertices are eliminated is called **eulerizing** the graph. There is one thing we must be careful about. The edges that we add *must be duplicates of edges that already exist*. (Remember that the point of all of this is to cover the edges of the original graph in the best possible way without creating any new edges.) Our next example clarifies this point.

EXAMPLE 5.20

Consider the graph in Fig. 5-23(a). This graph has 12 odd vertices, as shown (in red) in the figure. If we want to travel along all the edges of this graph and come back to our starting point, we know we are going to have to double up (deadhead) on some of the edges. Which ones? The answer is provided by first eulerizing the graph, so let's discuss ways in which we can do this.

Figure 5-23(b) shows how *not to do it!* Adding the edges *DF* and *NL* is not allowed, since those edges were not in the original graph.

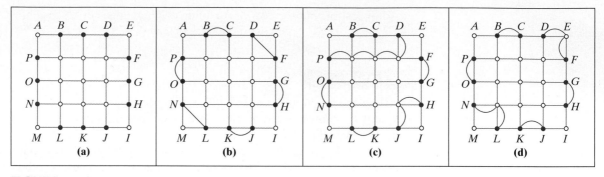

FIGURE 5-23
(a) The original graph. Odd vertices shown in red. (b) Illegal eulerization of the graph in (a). Edges *DF* and *NL* were not part of the original graph. (c) An inefficient eulerization of the graph in (a). (d) An optimal eulerization of the graph in (a).

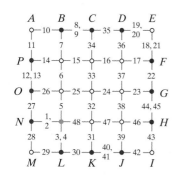

FIGURE 5-24

Figure 5-23(c) shows a legal, but wasteful, eulerization of the original graph. It is legal because we have eliminated all the vertices of odd degree by adding duplicate edges, but it is wasteful because it is obvious that we could have accomplished the same thing by adding fewer duplicate edges. Since there is a cost to traveling edges, we don't want to duplicate any more edges than is absolutely necessary!

Figure 5-23(d) shows an *optimal eulerization* of the original graph—one of several possible. This eulerization is optimal because it has the fewest possible duplicate edges (8). An optimal eulerization gives us the blueprint for an optimal round trip along the edges of the original graph. In this case we know that we are going to have to retrace 8 of the edges, and in fact we know exactly which ones. Figure 5-24 shows an actual example of an optimal trip (just follow the numbers) obtained using Fleury's algorithm on the graph in Fig. 5-23(d), and we can clearly see exactly which edges are being retraced.   ■

EXAMPLE 5.21

We will now consider a simple variation of the problem in Example 5.20. The graph shown in Fig. 5-25(a) is exactly the same as in Fig. 5-23(a). Once again, we want to travel the edges of this graph while duplicating the fewest possible edges, but this time we are not required to start and end in the same place. In this case we do what is called a **semi-eulerization** of the graph. That is, we duplicate as many edges as needed to eliminate all the odd vertices *except for* two, which we allow to remain odd. We then use these vertices as the starting and ending points of our travels. Figure 5-25(b) shows an optimal semi-eulerization of the graph in 5-25(a), with vertices *D* and *F* being the two vertices that remain odd. All the other vertices that were originally odd (*B, C, G, H, J, K, L, N, O,* and *P*) are now even. The semi-eulerization in Fig. 5-25(b) tells us that it is possible to travel all the edges of the graph in Fig. 5-25(a) by starting at *D* and ending at *F* (or vice versa) and duplicating only six of the edges (*BC, GH, JK, LM, MN,* and *OP*). Of course, there are many other ways to accomplish this, but none that do it with fewer than six duplicate edges. The reader is encouraged to find a different semi-eulerization of the graph in Fig. 5-25(a) (see Exercise 58).

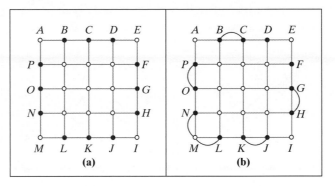

FIGURE 5-25
(a) The original graph. Odd vertices shown in red. (b) A semi-eulerization of the graph in (a). Vertices *D* and *F* remain odd.

EXAMPLE 5.22 The Bridges of Madison County: Conclusion

A graph model for this problem (first introduced in Example 5.4) is shown in Fig. 5-26, with each island and bank a vertex and each bridge an edge. The graph has four odd vertices (*R, L, B,* and *D*), so some bridges are definitely going to have to be recrossed. The photographer does not need to start and end in the same spot—when finished with the shoot he plans to move on. Thus, the ideal route involves an optimal semi-eulerization of the graph, which leaves vertices *R* and *L* as odd vertices (they will be the starting and ending points). This can be easily accomplished by duplicating the edge *BD*. The final solution to the problem is to start a trip at *R*, cross each of the bridges once, except for the Kennedy Bridge, which will have to be crossed twice, and end the trip at *L* (see Exercise 64). The total cost of this trip (in bridge tolls) will be $300 (12 bridge crossings at $25 each).

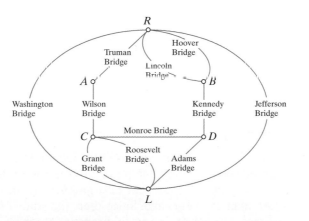

FIGURE 5-26
A graph model for the Madison bridge problem.

EXAMPLE 5.23 The Walking Patrolman: Conclusion

We have already discussed the graph model for this problem (Example 5.13). Figure 5-27(a) shows the graph, with the 18 odd vertices shown in red. Given that there are so many odd vertices, the security guard is going to have to retrace a fair number of his steps. To determine the best possible routing that starts and ends at *S* we first find an optimal eulerization of the graph. This is shown in Fig. 5-27(b).

The figure now tells us that the most efficient possible route will require that the patrolman double up on the nine blocks where an extra red edge has been added. An optimal route (there are several) can be obtained using Fleury's algorithm or just trial and error. One such route is shown in Fig. 5-27(c).

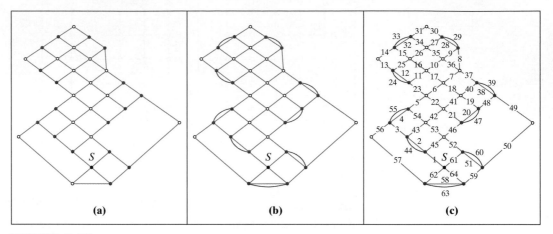

FIGURE 5-27
(a) Graph model of the neighborhood. (b) An optimal eulerization of the graph in (a). (c) An optimal route for the security guard (follow the numbers).

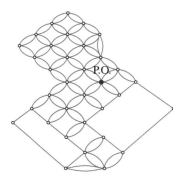

FIGURE 5-28

EXAMPLE 5.24 The Walking Mail Carrier: Conclusion

We have already found that the graph model for the mail carrier's problem is the one shown in Fig. 5-28 (see Example 5.14). All the vertices in this graph are of even degree, which means that the graph has an Euler circuit. This implies that the mail carrier will not have to waste any steps if she chooses her route carefully. The actual route, which must start and end at the Post Office, can be found using Fleury's algorithm, or just common sense and trial and error (see Exercise 59).

Conclusion

In this chapter we got our first introduction to three fundamental ideas. First, we learned about a simple but powerful concept for describing relationships within a set of objects—the concept of a *graph*. This idea can be traced back to Euler, more than 250 years ago. Since then, the study of graphs has grown into one of the most important and useful branches of modern mathematics.

The second important idea of this chapter is the concept of a *graph model*. Every time we take a real-life problem and turn it into a mathematical problem, we are, in effect, modeling. Unwittingly, we have all done some form of mathematical modeling at one time or another: using arithmetic in elementary school, and later using equations and functions to describe real-life situations. In this chapter we learned about a new type of modeling called graph modeling, in which we use graphs and the mathematical theory of graphs to solve certain types of routing problems.

By necessity, the routing problems that we solved in this chapter were fairly simplistic—the Königsberg bridge problem and some Euler circuit problems, such as the patrolman and the mail-carrier problems—but we should not be deceived by the simplicity of our examples—variations on these themes have significant practical uses. In many big cities, where the efficient routing of municipal services (police patrols, garbage collection, etc.) is a major issue, the same theory that we developed in this chapter is being used on a large scale, the only difference being that many of the more tedious details are mechanized and carried out by a computer. (In New York City, for example, garbage collection, curb sweeping, snow removal, and other municipal services have been scheduled and organized using graph models since the 1970s, and the improved efficiency has yielded savings estimated in the tens of millions of dollars a year.)

The third important concept we encountered in this chapter is that of an *algorithm*—a set of procedural rules that, when followed, provide solutions to certain types of problems. Perhaps without even realizing it, we had our first exposure to algorithms in elementary school, when we learned how to add, multiply, and divide numbers following precise and exacting procedural rules. In this chapter we learned about *Fleury's algorithm*, which helps us find an Euler circuit or an Euler path in a graph. In the next few chapters we will learn many other *graph algorithms*, some quite simple, others a bit more complicated. When it comes to algorithms of any kind, be they for doing arithmetic calculations or for finding circuits in graphs, there is one standard piece of advice that always applies: *Practice makes perfect.*

Leonhard Euler (1707–1783)

Leonhard Euler (pronounced "oiler") is universally recognized as one of the great, if not the greatest, mathematical geniuses in history. In the words of one of his biographers, *Euler was the Shakespeare of mathematics—universal, richly detailed and inexhaustible.* In terms of sheer volume, Euler's mathematical production is staggering—his *Opera Omnia* (collected works) fill over 80 volumes and run over 25,000 pages. No other mathematician—in fact, no other scientist—in the history of humankind can come close to matching Euler in terms of creative output. Euler produced groundbreaking discoveries in practically every area of mathematics—analysis, number theory, algebra, geometry, and topology (which he started). Euler's theorems in graph theory (see Section 5.5 in this chapter) and his solution of the Königsberg bridge problem are just one note in the mathematical symphony that is his work.

Leonhard Euler was born in Basel, Switzerland, the son of a Protestant minister. From an early age, Euler exhibited two traits that set him apart from mere mortals—a prodigious memory and the ability to perform incredibly complicated calculations in his head. At the age of 14 Euler entered the University of Basel to study theology and follow in the footsteps of his father. The young man's incredible mathematical gifts soon came to the attention of Johann Bernouilli, professor of mathematics at Basel and one of the most famous and influential mathematicians of his generation. Under Bernouilli's mentorship Euler eventually dropped theology and decided to concentrate on the study of mathematics. At the age of 19, Euler graduated from the University of Basel, having completed a doctoral thesis in which he analyzed the works of Descartes and Newton.

By this time Euler was already a mathematician of some fame, and was offered an academic position at the St. Petersburg Academy of Sciences in Russia—a remarkable and unprecedented offer for someone his age. For the next 14 years (1727 to 1741), Euler worked at the St. Petersburg Academy under the patronage of Catherine I. In addition to his groundbreaking work in pure mathematics, Euler was a practical man who worked on many important applied problems in physics, engineering, astronomy, and cartography. In 1733, at the age of 27, he was promoted to Professor of Mathematics at the St. Petersburg Academy. His first book, *Mechanica*, a two-volume text published in 1736, became the definitive work in mechanics for the next 50 years.

In 1738, due to an infection, Euler lost the sight in his right eye. Twenty-eight years later, he would lose the sight in his left eye, leaving him totally blind. Remarkably, neither event had a significant impact on his ability to produce copious amounts of mathematical work—he just dictated to his assistants and did most calculations in his head. In 1741 Euler accepted an offer from Frederick the Great of Prussia to become mathematics director of the Berlin Academy of Sciences. Euler worked in Berlin for the next 25 years (1741–1766), where he produced some of his greatest work, including his most famous book—a multi-volume textbook in elementary science called *Letters to a German Princess*. In 1766, mostly for political reasons, Euler returned to the St. Petersburg Academy of Sciences, where, despite his progressive blindness, he carried on with his prolific research for another 17 years.

The day he died, September 18, 1783, Euler worked in the morning on mathematical questions related to balloon flights, and in the afternoon he performed important calculations concerning the orbit of the planet Uranus—all just another day's work for him. In the late afternoon, while playing with one of his grandchildren, he suffered a massive stroke and died. His death marked the passing of one of the most creative men in the history of humankind.

adjacent edges
adjacent vertices
algorithm
bridge
circuit
connected graph
degree of a vertex
disconnected graph
edge
Euler circuit
Euler circuit problem
eulerizing a graph
Euler path
Euler's theorems

even vertex
Fleury's algorithm
graph
graph model
length of a path or circuit
loop
multiple edges
odd vertex
path
routing problems
semi-eulerization
unicursal tracing
vertex

EXERCISES

WALKING

A. Graphs: Basic Concepts

1. For each of the following graphs, list the vertices and edges and find the degree of each vertex.

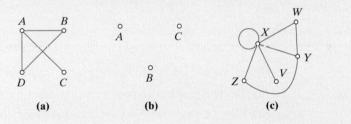

    (a)   (b)   (c)

2. For each of the following graphs, list the vertices and edges and find the degree of each vertex.

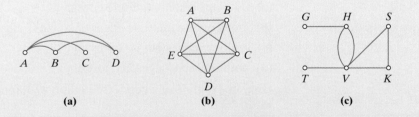

    (a)   (b)   (c)

3. For each of the following, draw two different pictures of the graph.
  (a) Vertices: *A, B, C, D*
     Edges: *AB, BC, BD, CD*
  (b) Vertices: *K, R, S, T, W*
     Edges: *RS, RT, TT, TS, SW, WW, WS*

4. For each of the following, draw two different pictures of the graph.

   (a) Vertices: *L, M, N, P*

      Edges: *LP, MM, PN, MN, PM*

   (b) Vertices: *A, B, C, D, E*

      Edges: *A* is adjacent to *C* and *E*; *B* is adjacent to *D* and *E*; *C* is adjacent to *A, D,* and *E*; *D* is adjacent to *B, C,* and *E*; *E* is adjacent to *A, B, C,* and *D*

5. (a) Explain why the following figures represent the same graph.

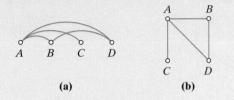

   (a)   (b)

   (b) Draw a third figure that represents the same graph.

6. (a) Explain why the following figures represent the same graph.

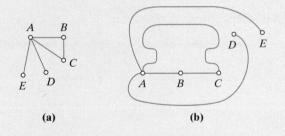

   (a)   (b)

   (b) Draw a third figure that represents the same graph.

7. (a) Draw a graph with four vertices such that each vertex has degree 2.

   (b) Draw a graph with six vertices such that each vertex has degree 3.

8. (a) Draw a graph with four vertices such that each vertex has degree 1.

   (b) Draw a graph with eight vertices such that each vertex has degree 3.

9. Give an example of a graph with four vertices, each of degree 3 with

   (a) no loops and no multiple edges.

   (b) loops but no multiple edges.

   (c) multiple edges but no loops.

   (d) both multiple edges and loops.

10. Give an example of a connected graph with five vertices, each of degree 4 with

   (a) no loops and no multiple edges.

   (b) loops but no multiple edges.

   (c) multiple edges but no loops.

   (d) both multiple edges and loops.

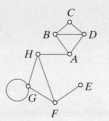

Exercises 11 through 14 refer to the graph shown in the margin.

11. **(a)** Find a path from C to F passing through vertex B but not through vertex D.

(b) Find a path from C to F passing through both vertex B and vertex D.

(c) Find a path of length 4 from C to F.

(d) Find a path of length 7 from C to F.

(e) How many paths are there from C to A?

(f) How many paths are there from H to F?

(g) How many paths are there from C to F?

12. **(a)** Find a path from D to E passing through vertex G only once.

(b) Find a path from D to E passing through vertex G twice.

(c) Find a path of length 4 from D to E.

(d) Find a path of length 8 from D to E.

(e) How many paths are there from D to A?

(f) How many paths are there from H to E?

(g) How many paths are there from D to E?

13. **(a)** Find a circuit of length 3 passing through vertex D.

(b) Find a circuit of length 4 passing through vertex G.

(c) Which edges in the graph are bridges?

14. **(a)** Find a circuit of length 3 passing through vertex H.

(b) Find a circuit of length 1 in the graph.

(c) Which edge must be added to this graph so that the resulting graph has no bridges?

B. Graph Models

15. An elementary school teacher wishes to make a seating chart for one of her reading groups. She wants to minimize the visiting among the students by separating friends as much as possible. The students in the reading group are Lynn, Jordan, Marie, Eric, Mark, Helen, Sally, and Jacob. Jordan is friends with everyone but Helen. Helen is friends with Lynn, Marie, Sally, and Mark. Eric is friends with Jordan, Mark, Jacob, and Lynn. Draw a graph that the teacher might use to represent the friendship relationships among the students in the reading group.

16. The Kangaroo Lodge of Madison County has 10 members (let's call them A, B, C, D, E, F, G, H, I, and J). The club has five working committees: the Rules Committee (A, C, D, E, I, and J), the Public Relations Committee (B, C, D, H, I, and J), the Guest Speaker Committee (A, D, E, F, and H), the New Year Eve's Committee (D, F, G, H, and I), and the Fund Raising Committee (B, D, F, H, and J).

(a) Suppose we are interested in knowing which pairs of members are on the same committee. Draw a graph that models this situation. (*Hint*: Let the vertices of the graph represent the members.)

(b) Suppose we are interested in knowing which committees have members in common. Draw a graph that models this situation. (*Hint*: Let the vertices of the graph represent the committees.)

Exercises 17 and 18 refer to the problem of routing the garbage collection trucks along the streets of the Buena Vista subdivision shown in the following figure. (All the streets are two-way streets.)

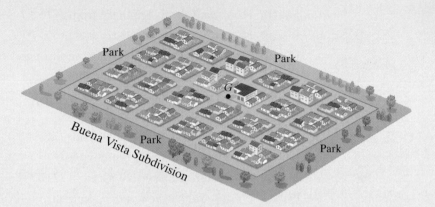

17. On weekdays the garbage is collected on each side of the street on separate passes except for the streets along the park which require only one pass. Draw a graph that models this situation.

18. On weekends, the garbage is picked up on both sides of the street on a single pass. Draw a graph that models this situation.

Exercises 19 and 20 refer to the Green Hills subdivision described by the following street map.

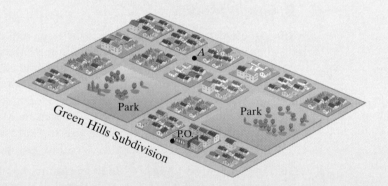

19. A night watchman must walk the streets of the Green Hills subdivision and start and end the walk at the corner labeled *A*. The night watchman needs to walk only once along each block. Draw a graph that models this situation.

20. A mail carrier must deliver mail on foot along the streets of the Green Hills subdivision. The mail carrier must walk along each block twice (once for each side of the street) except for blocks facing one of the parks, where only one pass is needed. Draw a graph that models this situation.

21. The following is a map of downtown Kingsburg, showing the Kings River running through the downtown area and the three islands (*A*, *B*, and *C*) connected to each other and both banks by seven bridges.

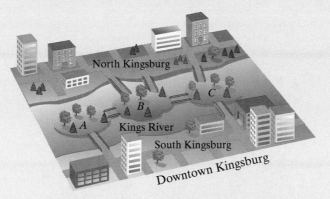

The Chamber of Commerce wants to design a walking tour that crosses all the bridges. Draw a graph that models the layout of Kingsburg.

22. The following is a map of downtown Royalton, showing the Royalton River running through the downtown area and the three islands (*A*, *B*, and *C*) connected to each other and both banks by eight bridges.

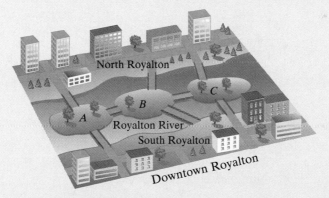

The Chamber of Commerce wants to design a walking tour that crosses all the bridges. Draw a graph that models the layout of Royalton.

C. Euler's Theorems

23. For each of the following, determine whether the graph has an Euler circuit, an Euler path, or neither of these. Explain your answer, but do not find the actual path or circuit.

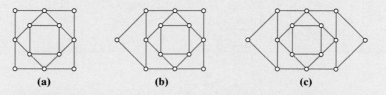

24. For each of the following, determine whether the graph has an Euler circuit, an Euler path, or neither of these. Explain your answer, but do not find the actual path or circuit.

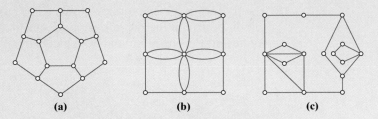

(a) (b) (c)

25. For each of the following, determine whether the graph has an Euler circuit, an Euler path, or neither of these. Explain your answer, but do not find the actual path or circuit.

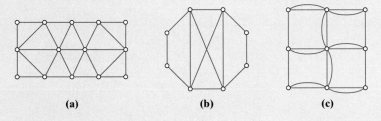

(a) (b) (c)

26. For each of the following, determine whether the graph has an Euler circuit, an Euler path, or neither of these. Explain your answer, but do not find the actual path or circuit.

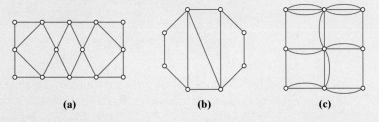

(a) (b) (c)

27. For each of the following, determine whether the graph has an Euler circuit, an Euler path, or neither of these. Explain your answer, but do not find the actual path or circuit.

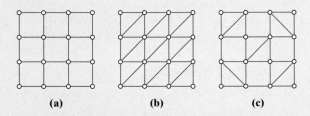

(a) (b) (c)

28. For each of the following, determine whether the graph has an Euler circuit, an Euler path, or neither of these. Explain your answer, but do not find the actual path or circuit.

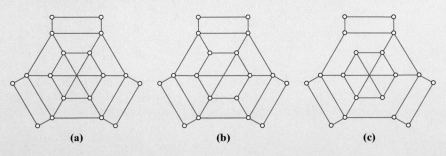

(a) (b) (c)

D. Finding Euler Circuits and Euler Paths

In Exercises 29 and 30 find an Euler circuit for the given graph. Show your answer by labeling the edges 1, 2, 3, etc. in the order in which they can be traveled.

29.

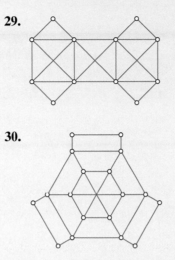

30.

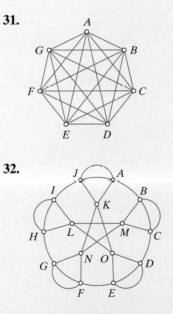

In Exercises 31 and 32 find an Euler circuit for the given graph. Show your answer by listing the vertices in the Euler circuit.

31.

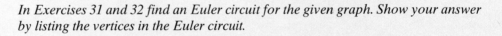

32.

In Exercises 33 and 34 find an Euler path for the given graph that starts at X and ends at Y. Show your answer by labeling the edges 1, 2, 3, etc. in the order in which they can be traveled.

33.

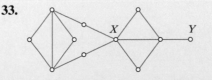

34.

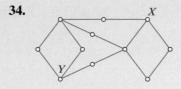

In Exercises 35 and 36 find an Euler path for the given graph. Show your answer by labeling the edges 1, 2, 3, etc. in the order in which they can be traveled.

35.

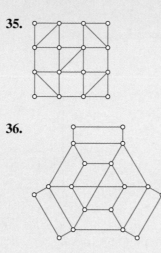

36.

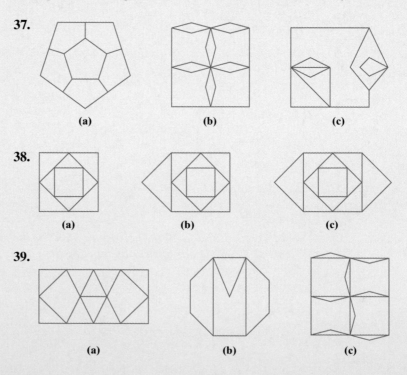

E. Unicursal Tracings

In Exercises 37 through 40, indicate in each case whether the drawing has an open unicursal tracing, a closed unicursal tracing, or neither. (If it does have a unicursal tracing, label the edges 1, 2, 3, etc. in the order in which they can be traced.)

37.

(a)   (b)   (c)

38.

(a)   (b)   (c)

39.

(a)   (b)   (c)

40.

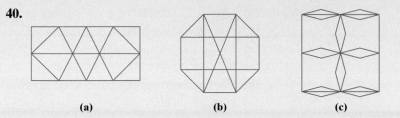

(a) (b) (c)

F. Eulerizations and Semi-eulerizations

41. Find an optimal eulerization for each of the following graphs.

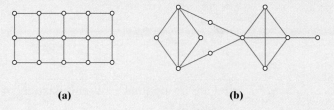

(a) (b)

42. Find an optimal eulerization for each of the following graphs.

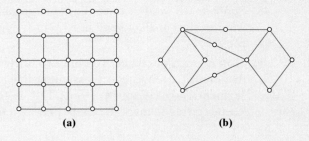

(a) (b)

43. Find an optimal semi-eulerization for each of the following graphs.

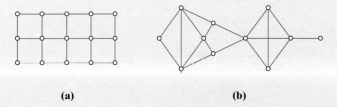

(a) (b)

44. Find an optimal semi-eulerization for each of the following graphs.

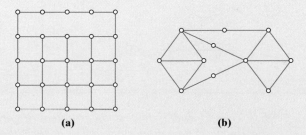

(a) (b)

G. Miscellaneous

45. (a) In the Königsberg bridge problem (Example 5.3), which of the real bridges are bridges in the graph-theory sense?

(b) Give an example of a connected graph with four vertices in which every edge is a bridge.

46. (a) In the Bridges of Madison County problem (Example 5.4), which of the real bridges are bridges in the graph-theory sense?

(b) Give an example of a connected graph with six vertices in which every edge is a bridge.

47. Suppose we want to trace the following graph with the fewest possible duplicate edges.

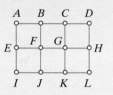

(a) Find an optimal semi-eulerization of the graph that starts at E and ends at H.

(b) Which edges of the graph will have to be retraced?

48. Suppose we want to trace the same graph as in Exercise 47, but now we want to start at B and end at K. Which edges of the graph will have to be retraced?

49. How many times would you have to lift your pencil to trace the following diagram of a tennis court, assuming that you do not trace over any line segment twice? Explain. (*Note*: This is the problem facing a grounds-keeper trying to mark the chalk lines of a tennis court.)

50. How many times would you have to lift your pencil to trace the following diagram, assuming that you do not trace over any line segment twice? Explain.

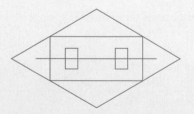

Exercises 51 and 52 refer to the problem of garbage collection along the streets of the Buena Vista subdivision shown in the following figure (see Exercises 17 and 18). All the streets are two way streets and all truck routes must start and end at the municipal garage G.

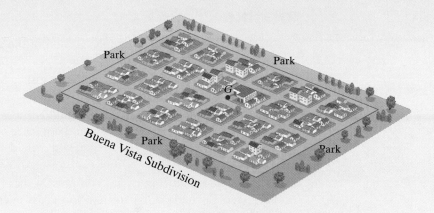

51. On weekdays the garbage is collected on each side of the street on separate passes except for the streets along the park which require only one pass. Find an optimal route for the garbage truck. Describe the route by labeling the edges 1, 2, 3, ... in the order in which they are traveled.

52. On weekends, the garbage is picked up on both sides of the street on a single pass. Find an optimal route for the garbage truck. Describe the route by labeling the edges 1, 2, 3, ... in the order in which they are traveled.

Exercises 53 and 54 refer to the Green Hills subdivision described by the following street map.

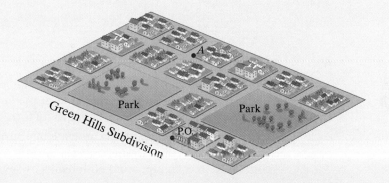

53. A night watchman must walk the streets of the Green Hills subdivision and start and end the walk at the corner labeled A. The night watchman needs to walk only once along each block (see Exercise 19). Find an optimal route for the night watchman. Describe the route by labeling the edges 1, 2, 3, ... in the order in which they are traveled.

54. A mail carrier must deliver mail on foot along the streets of the Green Hills subdivision and start and end the walk at the Post Office (P.O.). The mail carrier must walk along each block twice (once for each side of the street) except for blocks facing one of the parks, where only one pass is needed (see Exercise 20). Find an optimal route for the mail carrier. Describe the route by labeling the edges 1, 2, 3, ... in the order in which they are traveled.

JOGGING

55. (a) Explain why in every graph the sum of the degrees of all the vertices equals twice the number of edges.

(b) Explain why every graph must have either zero or an even number of vertices of odd degree.

56. Suppose a connected graph G has k odd vertices and you want to trace all of its edges. Assuming that you would not trace over any edges twice, what is the least number of times that you would have to lift your pencil? Explain.

57. Consider the following game. You must walk around the city of Königsberg (see Example 5.15) so that you cross each bridge at least once. It costs $1 each time you cross a bridge.

(a) Describe the cheapest possible walk you can make if you must start and end at the left bank (L).

(b) Describe the cheapest possible walk you can make if you are allowed to start and end at different places.

58. An optimal semi-eulerization of the following graph is given in Example 5.21. Find a different optimal semi-eulerization of the graph.

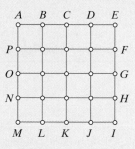

59. Describe an actual optimal route for the mail carrier's problem in Examples 5.14 and 5.24 that starts and ends at the Post Office. (Label the edges 1, 2, 3, etc. in the order in which the route is traveled.)

60. The following is a map of downtown Kingsburg, showing the Kings River running through the downtown area and the three islands (A, B, and C) connected to each other and both banks by seven bridges.

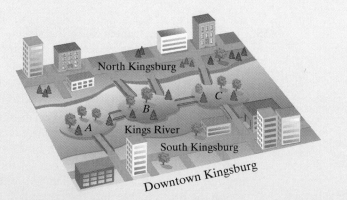

The Chamber of Commerce wants to design a walking tour that crosses all the bridges. Is it possible to take a walk such that you cross each bridge exactly once? If so, show how. If not, explain why not.

61. A policeman has to patrol along the streets of the subdivision represented by the following graph.

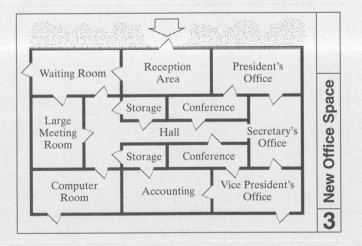

The policeman wants to start his trip at the police station (located at X) and end the trip at his home (located at Y). He needs to cover each block of the subdivision at least once and at the same time he wants to duplicate the fewest possible number of blocks.

(a) How many blocks will he have to duplicate in an optimal trip through the subdivision?

(b) Describe an optimal trip through the subdivision. Label the edges 1, 2, 3, . . . in the order the policeman would travel them.

62. (a) Give an example of a graph with 15 vertices and no multiple edges that has an Euler circuit.

(b) Give an example of a graph with 15 vertices and no multiple edges that has an Euler path but no Euler circuit.

(c) Give an example of a graph with 15 vertices and no multiple edges that has neither an Euler circuit nor an Euler path.

63. The following figure is the floor plan of an office complex.

(a) Show that it is impossible to start outside the complex and walk through each door of the complex exactly once and end up outside.

(b) Show that it is possible to walk through every door of the complex exactly once (if you start and end at the right places).

(c) Show that by removing exactly one door, it would be possible to start outside the complex, walk through each door of the complex exactly once, and end up outside.

Exercises 64 and 65 refer to the problem of the bridges of Madison County discussed in Examples 5.4 and 5.22.

64. Describe an optimal route for the photographer planning a photo shoot of all the bridges of Madison County assuming

   (a) he decides to start the shoot at the Adams Bridge.

   (b) he decides to end the shoot at the Grant Bridge.

65. Describe an optimal route for the photographer planning a photo shoot of all the bridges of Madison County assuming

   (a) he wants to start and end the route on the Right Bank.

   (b) he wants to start the shoot at the Adams Bridge and end the shoot at the Grant Bridge.

66. In his original paper on the Königsberg bridge problem (see reference 7), Euler posed the following question:

   Let us take an example of two islands with four rivers forming the surrounding water. There are fifteen bridges marked a, b, c, d, etc., across the water around the islands and the adjoining rivers. The question is whether a journey can be arranged that will pass over all the bridges but not over any of them more than once.

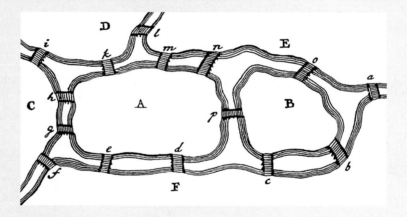

Give an answer to Euler's question. If the journey is possible, describe it. If it isn't, explain why not.

RUNNING

67. (a) Can a graph that has an Euler circuit have any bridges? If so, demonstrate it by showing an example. If not, explain why not.

   (b) Can a graph that has an Euler path have any bridges? If so, how many? Explain your answer.

68. Suppose G and H are two graphs that have no common vertices and such that each graph has an Euler circuit. Let J be a (single) graph consisting of the graphs G, H, and one additional edge joining one of the vertices of G to one of the vertices of H. Explain why the graph J has no Euler circuit but does have an Euler path.

69. Explain why in any graph in which the degree of each vertex is at least 2, there must be a circuit.

70. Suppose we have a graph with two or more vertices and without loops or multiple edges. Explain why the graph must have at least two vertices with the same degree.

71. Suppose we have a disconnected graph with exactly two vertices of odd degree. Explain why the two vertices of odd degree must be in the same component of the graph.

72. Consider the following game. You are given N vertices and required to build a graph by adding edges connecting these vertices. Each time you add an edge you must pay $1. You can stop when the graph is connected.

(a) Describe the strategy that will cost you the least money.

(b) What is the minimum amount of money needed to build the graph? (Give your answer in terms of N.)

73. Consider the following game. You are given N vertices and allowed to build a graph by adding edges connecting these vertices. For each edge you can add, you make $1. You are not allowed to add loops or multiple edges, and you must stop before the graph is connected (i.e., the graph you end up with must be disconnected).

(a) Describe the strategy that will give you the most money.

(b) What is the most money you can make building the graph? (Give your answer in terms of N.)

PROJECTS AND PAPERS

A. Original Sources

Whenever possible, it is instructive to read about a great discovery from the original source. Euler's landmark paper in graph theory with his solution to the Königsberg bridge problem was published in 1736. Luckily, the paper was written in a very readable style and the full English translation is quite accessible—it appears as an article in *Scientific American* (reference 7) and in Newman's book *The World of Mathematics* (reference 9).

Write a summary/analysis of Euler's original paper. Include (i) a description of how Euler originally tackled the Konigsberg bridge problem, (ii) a discussion of Euler's general conclusions (describe Theorems 1, 2, and 3 given in this chapter in Euler's own words), and (iii) a discussion of Euler's approach toward finding an Euler circuit/path.

B. Computer Representation of Graphs

In many real-life routing problems we have to deal with very large graphs—a graph could have thousands of vertices and tens of thousands of edges. In these cases, algorithms such as Fleury's algorithm (as well as others we will study in later chapters) are done by computer. Unlike humans, computers are not very good at interpreting pictures, so the first step in using computers to perform computations with graphs is to describe the graph in a way that the computer can understand it. The two most common ways to do so are by means of matrices.

In this project you are asked to write a short research paper describing the use of matrices to represent graphs. Explain (i) what is a **matrix**, (ii) what is the **adjacency matrix** of a graph, and (iii) what is the **incidence matrix** of a graph. Illustrate some of the graph concepts from this chapter (degrees of vertices, multiple edges, loops, etc.) in matrix terms. Include plenty of examples. You can find definitions and information on adjacency and incidence matrices of graphs in many graph theory books (see, for example, references 5 and 14).

C. The Chinese Postman Problem

A **weighted graph** is a graph in which the edges are assigned positive numbers called weights. The weights represent distances, times, or costs. The problem of finding a trip that covers *all* the edges of a *weighted graph* and that is optimal (least amount of deadhead travel) goes by the general name of the **Chinese postman problem**. These problems are a generalization of *Euler circuit* problems and, as a general rule, are much harder to solve, but most of the concepts developed in this chapter still apply.

In this project you are asked to prepare a presentation on the Chinese postman problem for your class.

Some suggestions: (i) Give several examples to illustrate Chinese postman problems and how they differ from corresponding Euler circuit problems. (ii) Describe some possible real-life applications of Chinese postman problems. (iii) Discuss how to solve a Chinese postman problem in the simplest cases when the weighted graph has no odd vertices or has only two odd vertices (these cases can be solved using techniques learned in this chapter). (iv) Give a rough outline of how one might attempt to solve a Chinese postman problem for a graph with four odd vertices.

REFERENCES AND FURTHER READINGS

1. Beltrami, E., *Models for Public Systems Analysis*. New York: Academic Press, Inc., 1977.
2. Beltrami, E., and L. Bodin, "Networks and Vehicle Routing for Municipal Waste Collection," *Networks*, 4 (1974), 65– 94.
3. Bogomolny, A., "Graphs," *http://www.cut-the-knot.com/do_you_know/graphs.shtml*.
4. Chartrand, Gary, *Graphs as Mathematical Models*. Belmont, CA: Wadsworth Publishing Co., 1977.
5. Chartrand, Gary, and Oellerman, O. R., *Applied and Algorithmic Graph Theory*, New York: McGraw-Hill, 1993.
6. Dunham, William, *Euler: The Master of Us All*. Washington, DC: The Mathematical Association of America, 1999.
7. Euler, Leonhard, "The Königsberg Bridges," trans. James Newman, *Scientific American*, 189 (1953), 66– 70.
8. Minieka, E., *Optimization Algorithms for Networks and Graphs*. New York: Marcel Dekker, Inc., 1978.

9. Newman, James R., *The World of Mathematics*, vol. 1. New York: Simon & Schuster, 1956.

10. Roberts, Fred S., "Graph Theory and Its Applications to Problems of Society," *CBMS-NSF Monograph No. 29*. Philadelphia: Society for Industrial and Applied Mathematics, 1978, chap. 8.

11. Tucker, A. C., and L. Bodin, "A Model for Municipal Street-Sweeping Operations," in *Modules in Applied Mathematics*, Vol. 3, eds. W. Lucas, F. Roberts, and R. M. Thrall. New York: Springer-Verlag, 1983, 76– 111.

12. Stein, S.K., *Mathematics: The Man-Made Universe* (3rd ed.) New York: Dover, 2000.

13. Steinhaus, H., *Mathematical Snapshots* (3rd ed.) New York: Dover, 1999.

14. West, Douglas, *Introduction to Graph Theory*. Upper Saddle River, NJ: Prentice Hall, 1996.

EARTH & MARS

AS DIFFERENT AS THEY ARE ALIKE

BOTH EARTH
AND MARS
HAVE CANYONS,
VALLEYS, CRATERS,
VOLCANOES, ICE
CAPS, STORMS,
AND SEASONS.

MARS IS HALF
THE DIAMETER
OF EARTH, BUT
BOTH PLANETS
HAVE THE SAME
AMOUNT OF DRY
LAND.

EARTH'S
ATMOSPHERE IS
77% NITROGEN
AND 21% OXY-
GEN. MARS' AT-
MOSPHERE IS 95%
CARBON DIOXIDE.

MARS HAS FOUR
SEASONS, JUST
LIKE EARTH, BUT
EACH ONE LASTS
TWICE AS LONG.

EXPLORE MARS
http://mars.jpl.nasa.gov

National Aeronautics and
Space Administration

Jet Propulsion Laboratory
California Institute of Technology
Pasadena, California

JPL 400-990 365401
EW-2001-02-009-JPL

The Traveling-Salesman Problem

Hamilton Joins the Circuit

To those of us who get a thrill out of planning for a big trip, the 21st century holds a truly special treat: We humans are going on the mother of all excursions—a trip to Mars. Why do we want to go to Mars, and what are we going to do when we get there?

For starters, Mars is the next great frontier for human exploration. Its promise lies in the fact that it is the most earthlike of all the planets in the solar system, the Martian soil is rich in chemicals and minerals and, most important of all, there is strong evidence that there may be water just below the surface.

The ultimate attraction, however, is the hope of finding life on Mars. Of all the planets in our solar system, Mars is the most likely place to show some evidence of life—probably primitive bacterial forms buried inside Martian rocks or under the Martian surface. But finding these tiny Martians—assuming they exist—raises many technical and logistical questions. What are the best places in Mars to explore? How do we get the equipment there? How long will it take to do the job? And above all, how much will it cost? Once again, lurking behind the complexities of Mars exploration is an interesting and important mathematical problem. We will discuss the problem and its many variations in this chapter.

Here are the details in a nutshell. Based on geological data already collected from earlier orbiting missions, NASA has identified a set of sites on the surface of Mars where the likelihood of finding either bacterial fossils or actual life forms is the highest (see Fig. 6-1). A main goal for the next exploration stage is to get an unmanned rover to each and every one of these locations to collect soil samples and perform experiments. One approach being planned for the near future is a *robotic sample-return mission*, in which a lander would land at one of the designated sites (probably the Ares Vallis) and release an unmanned rover controlled from Earth. The rover would then travel to each of the other sites, collecting soil samples and performing experiments. After all the sites have been visited, the rover would return to the landing site where a return rocket would bring the samples back to Earth. As of the writing of this edition, the first of these sample-return missions is scheduled to be launched in 2014.

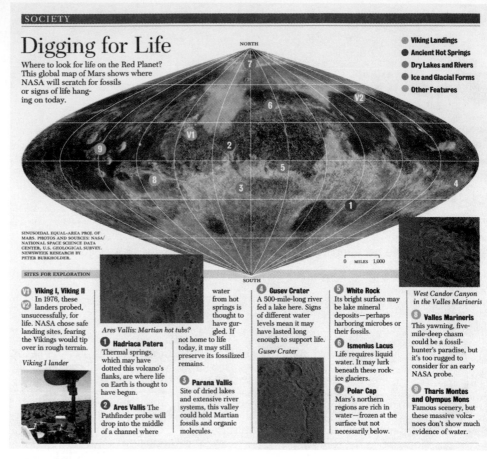

SOCIETY

Digging for Life

Where to look for life on the Red Planet? This global map of Mars shows where NASA will scratch for fossils or signs of life hanging on today.

- ● Viking Landings
- ● Ancient Hot Springs
- ● Dry Lakes and Rivers
- ● Ice and Glacial Forms
- ● Other Features

NORTH

SOUTH

SINUSOIDAL EQUAL-AREA PROJ. OF MARS. PHOTOS AND SOURCES: NASA/ NATIONAL SPACE SCIENCE DATA CENTER, U.S. GEOLOGICAL SURVEY. NEWSWEEK RESEARCH BY PETER BURKHOLDER.

0   MILES   1,000

SITES FOR EXPLORATION

V1 **Viking I, Viking II**
V2 In 1976, these landers probed, unsuccessfully, for life. NASA chose safe landing sites, fearing the Vikings would tip over in rough terrain.

Viking I lander

1 Hadriaca Patera
Thermal springs, which may have dotted this volcano's flanks, are where life on Earth is thought to have begun.

2 Ares Vallis The Pathfinder probe will drop into the middle of a channel where

Ares Vallis: Martian hot tubs?

water from hot springs is thought to have gurgled. If not home to life today, it may still preserve its fossilized remains.

3 Parana Vallis
Site of dried lakes and extensive river systems, this valley could hold Martian fossils and organic molecules.

4 Gusev Crater
A 500-mile-long river fed a lake here. Signs of different water levels mean it may have lasted long enough to support life.

Gusev Crater

5 White Rock
Its bright surface may be lake mineral deposits—perhaps harboring microbes or their fossils.

6 Ismenius Lacus
Life requires liquid water. It may lurk beneath these rock-ice glaciers.

7 Polar Cap
Mars's northern regions are rich in water—frozen at the surface but not necessarily below.

West Candor Canyon in the Valles Marineris

8 Valles Marineris
This yawning, five-mile-deep chasm could be a fossil-hunter's paradise, but it's too rugged to consider for an early NASA probe.

9 Tharis Montes and Olympus Mons
Famous scenery, but these massive volcanoes don't show much evidence of water.

FIGURE 6-1

Source: Newsweek, NASA/National Space Science Data Center, U.S. Geological Survey. *Newsweek* research by Peter Burkholder. Copyright September 23, 1996, Newsweek, Inc. All rights reserved. Reprinted by permission.

FIGURE 6-2
Mars unmanned rover.

Figure 6-3 shows one of the many possible routes that the rover might take as it travels to each of the sites to look for life and collect soil samples. And there are hundreds of other possible ways to route the rover. Which one is the best?

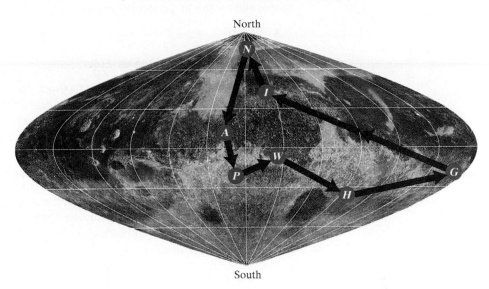

FIGURE 6-3

A: Ares Vallis (starting and ending point); *P*: Parana Vallis; *W*: White Rock; *H*: Hadriaca Patera; *G*: Gusev Crater; *I*: Ismenius Lacus; *N*: North Polar Cap.

Finding an optimal route for a Martian rover is an exotic illustration of a class of important routing problems that go by the generic name of TSPs. The acronym TSP stands for "traveling-salesman problem," so called because the classic version of this problem is that of a traveling salesman who must call on customers in several cities and wants to find the cheapest route that visits each of the cities once, returning at the end to his home town. The name has stuck as a generic name given to all sorts of similar problems, even if they have nothing to do with traveling salesmen. It is best to think of "traveling-salesman" as just a metaphor: It is a rover searching for the best route on Mars, a UPS driver trying to find the best way to deliver packages around town, or any of us, trying to plan the best route by which to run a bunch of errands on a Saturday morning.

The remarkable thing about TSPs is their deceptive simplicity. It just doesn't seem that finding an optimal route should be all that hard! It is always a surprise to find out, as we soon will, that TSPs represent one of the most interesting, important, and complex problems in graph theory, if not all of mathematics. Understanding TSPs and what it means to "solve" them will be the main purpose of this chapter.

6.1 Hamilton Circuits and Hamilton Paths

From a mathematical point of view, the main concept that will concern us in this chapter is that of a *Hamilton circuit*. Let's think back to the last chapter, where we studied Euler circuits and Euler paths. In these types of circuits (or paths) the name of the game is to visit each *edge* of the graph once and only once. But what about a circuit that must visit each *vertex* of the graph once and only once (except at the end, where it returns to the starting vertex)? This type of circuit is called a **Hamilton circuit**, after the great Irish mathematician William Rowan Hamilton (for more on Hamilton, see the biographical profile on p. 246). Likewise, a path that passes through every vertex of the graph is called a **Hamilton path**.

The difference between a Hamilton circuit and an Euler circuit (or, for that matter, a Hamilton path and an Euler path) boils down to just one word (substitute *vertex* for *edge*), but my, what a difference that word makes! As the next example shows, the two concepts are essentially unrelated.

EXAMPLE 6.1 Hamilton vs. Euler

The main purpose of this example is to show that when it comes to Hamilton circuits and Euler circuits, a graph can have one or the other, or both, or neither—anything goes.

Consider the four graphs in Fig. 6-4. The graph in Fig. 6-4(a) has many Hamilton circuits (*A, B, C, D, E, A* is one of them; *A, D, C, E, B, A* is another—there are plenty more). It also has many Hamilton paths (for example, *A, B, C, D, E* and *A, D, C, E, B*)—after all, any Hamilton circuit can be shortened into a Hamilton path by the removal of the last edge. On the other hand, since it has four vertices of odd degree, it has no Euler circuits or Euler paths.

The graph in Fig. 6-4(b) has Euler circuits because every vertex has even degree. On the other hand, it has no Hamilton circuits because whatever the starting point, we are going to have to pass through vertex *E* more than once to close the circuit. (Notice, however, that this graph does have Hamilton paths such as *A, B, E, C, D* or *C, D, E, A, B*.)

The graph in Fig. 6-4(c) has Euler circuits because every vertex has even degree and it also has Hamilton circuits such as *A, F, B, E, C, G, D, A* and many others.

Finally, the graph in Fig. 6-4(d) has no Euler circuits, no Euler paths, no Hamilton circuits, and no Hamilton paths.

FIGURE 6-4

(a) A graph with a Hamilton circuit but no Euler circuit. (b) A graph with an Euler circuit but no Hamilton circuit. (c) A graph with both a Hamilton circuit and an Euler circuit. (d) A graph with neither a Hamilton circuit nor an Euler circuit.

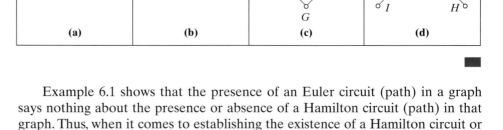

(a)   (b)   (c)   (d)

Example 6.1 shows that the presence of an Euler circuit (path) in a graph says nothing about the presence or absence of a Hamilton circuit (path) in that graph. Thus, when it comes to establishing the existence of a Hamilton circuit or path, Euler's theorems are useless.

Given an arbitrary graph, how do we tell if it has a Hamilton circuit or path? Unfortunately, there is no known simple answer to this question. Even for a small graph, such as the one in Fig. 6-5, it is not easy to determine whether it has a Hamilton circuit. Appearances to the contrary, this graph doesn't (see Exercise 65). For large graphs with hundreds of vertices, it can be quite difficult to determine if the graph has a Hamilton circuit or a Hamilton path.

The flip side of a graph that has no Hamilton circuits is a graph in which every possible sequence of the vertices turns out to produce a Hamilton circuit. We will discuss these graphs next.

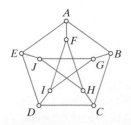

FIGURE 6-5

6.2 Complete Graphs

Figure 6-6 shows four graphs, having three, four, five, and six vertices, respectively. These graphs have one characteristic in common: There is one edge connecting every pair of vertices. A graph with N vertices in which *every* pair of vertices is joined by exactly one edge is called the **complete graph** (on N vertices) and denoted by the symbol K_N. In the complete graph on N vertices, each vertex is joined to each of the other $N-1$ vertices, so each vertex has degree $N-1$. From this it follows that the total number of edges in the complete graph with N vertices is $N(N-1)/2$ (see Exercise 59). In K_6, for example, each vertex has degree 5 and the number of edges is 15.

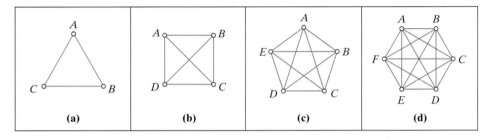

FIGURE 6-6

A complete graph has a complete repertoire of Hamilton circuits. We can write the vertices in any order we want, repeat the first vertex at the end, and presto, we have a Hamilton circuit! Take K_4, the complete graph with four vertices shown in Fig. 6-6(b). Let's randomly choose the 4 vertices in any order—say C, A, D, B. Now repeat the first vertex (C) at the end. The circuit C, A, D, B, C is indeed a Hamilton circuit of the graph. If we repeat the process with a different sequence, once again we will end up with a Hamilton circuit, although not necessarily a different one! It is important to remember that different sequences of letters can produce the same Hamilton circuit. For example, the circuit C, A, D, B, C is the same as the circuit A, D, B, C, A—we only changed the *reference point*. In the first case, we used vertex C as the reference point; in the second one we used vertex A. There are two more sequences that give this same Hamilton circuit—D, B, C, A, D (with reference point D) and B, C, A, D, B (with reference point B).

A complete listing of all the Hamilton circuits of K_4 is shown in Table 6-1. There are six altogether.

TABLE 6-1   The six Hamilton circuits of K_4. Each circuit can be written in four ways.

	Reference point is A	Reference point is B	Reference point is C	Reference point is D
1	A, B, C, D, A	B, C, D, A, B	C, D, A, B, C	D, A, B, C, D
2	A, B, D, C, A	B, D, C, A, B	C, A, B, D, C	D, C, A, B, D
3	A, C, B, D, A	B, D, A, C, B	C, B, D, A, C	D, A, C, B, D
4	A, C, D, B, A	B, A, C, D, B	C, D, B, A, C	D, B, A, C, D
5	A, D, B, C, A	B, C, A, D, B	C, A, D, B, C	D, B, C, A, D
6	A, D, C, B, A	B, A, D, C, B	C, B, A, D, C	D, C, B, A, D

The complete list of all the Hamilton circuits of K_5 is shown in Table 6-2. There are 24 of them. Notice that this time we listed them using a consistent vertex (A) as the reference point—this helps with the bookkeeping. We also paired each Hamilton circuit with its *mirror-image circuit* (the circuit traveled in reverse order). Please note that a circuit and its mirror-image circuit are not considered equal—though they are close relatives.

TABLE 6-2 The 24 Hamilton circuits of K_5 (using A as the reference point). Opposite each circuit is its mirror-image circuit.

1	A, B, C, D, E, A	13	A, E, D, C, B, A
2	A, B, C, E, D, A	14	A, D, E, C, B, A
3	A, B, D, C, E, A	15	A, E, C, D, B, A
4	A, B, D, E, C, A	16	A, C, E, D, B, A
5	A, B, E, C, D, A	17	A, D, C, E, B, A
6	A, B, E, D, C, A	18	A, C, D, E, B, A
7	A, C, B, D, E, A	19	A, E, D, B, C, A
8	A, C, B, E, D, A	20	A, D, E, B, C, A
9	A, C, D, B, E, A	21	A, E, B, D, C, A
10	A, C, E, B, D, A	22	A, D, B, E, C, A
11	A, D, B, C, E, A	23	A, E, C, B, D, A
12	A, D, C, B, E, A	24	A, E, B, C, D, A

The number of Hamilton circuits in a complete graph can be computed using *factorials*, a concept we first came across in Chapter 2. Recall that if N is any positive integer, the number $N! = 1 \times 2 \times 3 \times \cdots \times (N-1) \times N$ is called the **factorial** of N. Table 6-3 shows the first 25 values of $N!$.

TABLE 6-3 The first 25 factorials

$1! = 1$
$2! = 2$
$3! = 6$
$4! = 24$
$5! = 120$
$6! = 720$
$7! = 5040$
$8! = 40,320$
$9! = 362,880$
$10! = 3,628,800$
$11! = 39,916,800$
$12! = 479,001,600$
$13! = 6,227,020,800$
$14! = 87,178,291,200$
$15! = 1,307,674,368,000$
$16! = 20,922,789,888,000$
$17! = 355,687,428,096,000$
$18! = 6,402,373,705,728,000$
$19! = 121,645,100,408,832,000$
$20! = 2,432,902,008,176,640,000$
$21! = 51,090,942,171,709,440,000$
$22! = 1,124,000,727,777,607,680,000$
$23! = 25,852,016,738,884,976,640,000$
$24! = 620,448,401,733,239,439,360,000$
$25! = 15,511,210,043,330,985,984,000,000$

Notice that $3! = 6$ (the number of Hamilton circuits of K_4) and that $4! = 24$ (the number of Hamilton circuits of K_5). Coincidence? Not at all. As we now know, in a complete graph we can list the vertices in any order and get a Hamilton circuit. If we choose a specified vertex as the reference point, every possible ordering of the remaining $(N-1)$ vertices will result in a Hamilton circuit, and the total number of ways of ordering $(N-1)$ things is $(N-1)!$.

> The complete graph with N vertices has $(N-1)!$ Hamilton circuits.

The most important thing to notice about Table 6-3 is how quickly factorials grow and, consequently, how quickly the number of Hamilton circuits of a complete graph grows relative to the number of vertices. A modest-size graph such as the complete graph with just a dozen vertices (K_{12}) has almost *40 million Hamilton circuits* $(11! = 39,916,800)$. Double the size of the graph to K_{24} and the number of Hamilton circuits grows to an astronomical 23!, which is more than *25 trillion billions*, a number so big that it defies ordinary human comprehension.

The primary point of this discussion is this: When it comes to Hamilton circuits in complete graphs, we are facing an embarrassment of riches. The main question is no longer "Are there any?" but rather "How are we going to deal with so many?".

6.3 Traveling-Salesman Problems

What kind of a problem is a TSP? Here are a few examples of which only the first is self-evident.

EXAMPLE 6.2 A Tale of Five Cities

Meet Willy Loman, a traveling salesman. Willy has customers in five cities, which for the sake of brevity we will call *A, B, C, D,* and *E*, and he is planning an upcoming sales trip to visit each of them. Willy needs to start and end the trip at his home town of *A*. Other than that, there are no particular restrictions as to the order in which he should visit the other four cities.

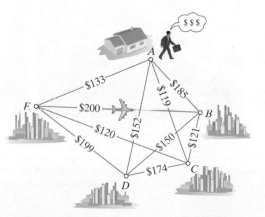

FIGURE 6-7
A five-city TSP.

The graph in Fig. 6-7 shows the cost of a *one-way*[1] airline ticket between each pair of cities. Naturally, Willy wants to cut down on his travel expenses as much as possible. What is the cheapest possible sequence in which to visit the five cities? We will return to this question soon. ■

EXAMPLE 6.3 Probing the Outer Reaches of Our Solar System

It is the year 2020. An expedition to explore the outer planetary moons in our solar system is about to be launched from planet Earth. The expedition is scheduled to visit Callisto, Ganymede, Io, Mimas, and Titan (the first three are moons of Jupiter; the last two, of Saturn), collect rock samples at each, and then return to Earth with the loot.

Figure 6-8 shows the mission time (in years) between any two moons. What is the best way to route the spaceship so that the entire trip takes the least amount of time?

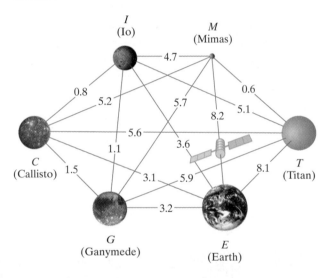

FIGURE 6-8
A six-vertex TSP.

Jupiter's moons. *Left:* Callisto, *Center:* Ganymede, *Right:* Io.

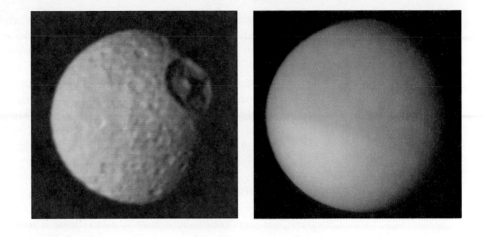

Saturn's moons. *Left:* Mimas, *Right:* Titan.

EXAMPLE 6.4 Searching for Martian Life

Figure 6-9 shows seven locations on Mars where NASA scientists believe there is a good chance of finding evidence of life. Imagine that you are in charge of planning a *sample-return* mission. First, you must land an unmanned rover in the Ares Vallis (*A*). Then you must direct the rover to travel to each site and collect and analyze soil samples. Finally, you must instruct the rover to return to the Ares Vallis landing site, where a return rocket will bring the best samples back to Earth. A trip like this will take several years and cost several billion dollars, so good planning is critical.

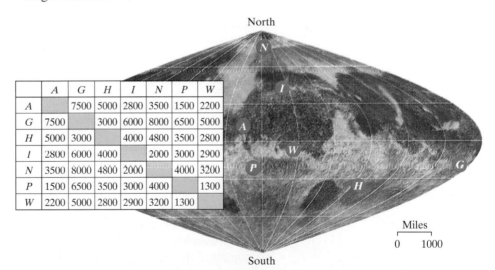

	A	*G*	*H*	*I*	*N*	*P*	*W*
A		7500	5000	2800	3500	1500	2200
G	7500		3000	6000	8000	6500	5000
H	5000	3000		4000	4800	3500	2800
I	2800	6000	4000		2000	3000	2900
N	3500	8000	4800	2000		4000	3200
P	1500	6500	3500	3000	4000		1300
W	2200	5000	2800	2900	3200	1300	

FIGURE 6-9
Approximate distances (in miles) between locations on Mars.

The table shows the estimated distances (in miles) that a rover would have to travel to get from one Martian site to another. What is the optimal sequence in which the rover should visit the different sites so that the total distance it has to travel is minimized?

Examples 6.2, 6.3, and 6.4 are variations on a single theme. In each case, we are presented with a complete graph whose edges have numbers attached to them. (For Example 6.4 it was easier to put the numbers in a table.) Any graph whose edges have numbers attached to them is called a **weighted graph**, and the numbers are called the **weights**[2] of the edges. The graphs in Examples 6.2, 6.3, and 6.4 are called **complete weighted graphs**. In each example the weights of the graph represent a different variable. In Example 6.2 the weights represent *cost*, in Example 6.3 they represent *time*, and in Example 6.4 they represent *distance*. Most important of all, in each example the problem we want to solve is the same: *to find an optimal Hamilton circuit—that is, a Hamilton circuit with least total weight—for the complete weighted graph.* These kinds of problems are known generically as **traveling-salesman problems**, or TSPs for short.

Many important real-life problems can be formulated as TSPs. The following are just a few general examples.

- **Routing School Buses.** A school bus picks up children in the morning and drops them off at the end of the day at designated stops. On a typical school bus route there may be 20 to 30 such stops, which can be represented by the vertices of a graph. The weight of the edge connecting bus-stop X to bus-stop Y represents the time of travel between the two locations (with school buses, time of travel is always the most important variable). Since the bus repeats its route every day during the school year, finding the optimal route is crucial. This is an example of a moderately sized TSP (20 to 30 vertices), which is too large to do by traditional methods but can be solved in a reasonable time with currently available computer software.

- **Package Deliveries.** Companies such as United Parcel Service (UPS) and Federal Express deal with this situation daily. Each truck has packages to deliver to a list of destinations. The travel time between any two delivery locations is known or can be estimated. The object is to deliver the packages to each of the delivery locations and return to the starting point in the least amount of time—clearly an example of a TSP. On a typical day, a UPS truck delivers packages to somewhere between 100 and 200 locations, so the graph for this TSP would have that many vertices.

- **Fabricating Circuit Boards.** In the process of fabricating integrated-circuit boards, tens of thousands of tiny holes must be drilled in each board. This is done by using a stationary laser beam and rotating the board. Efficiency considerations require that the order in which the holes are drilled be such that the entire drilling sequence be completed in the least amount of time. This is an example of a TSP, in which the vertices of the graph represent the holes on the circuit board and the weight of the edge connecting vertices X and Y represents the time needed to rotate the board from drilling position X to drilling position Y.

- **Scheduling Jobs on a Machine.** In many industries there are machines that perform multiple jobs. Think of the jobs as the vertices of a graph. After performing job X the machine needs to be set up to perform another job. The amount of time required to reset the machine to perform job Y is the weight of the edge connecting vertices X and Y. The problem is to schedule the machine to run through all the jobs in a cycle such that the total amount of time is minimized. This is another example of a TSP.

- **Running Errands Around Town.** When we have a lot of errands to run, we like to follow the route that will take us to each of our destinations and then finally home in the shortest amount of time. This is an example of a TSP. (See Exercise 43.)

---

[2]This is a different usage of the word *weight* from that in Chapter 2.

6.4 Simple Strategies for Solving TSPs

EXAMPLE 6.5 A Tale of Five Cities: Part II

At the end of Example 6.2 we left Willy the traveling salesman pondering his upcoming sales trip, dollar signs running through his head. (For the reader's convenience the graph showing the cost of one-way travel between any two cities is given again in Fig. 6-10.) Imagine now that Willy, unwilling or unable to work out the problem himself, decides to offer a reward of $20 to anyone who can find the optimal Hamilton circuit for this graph.

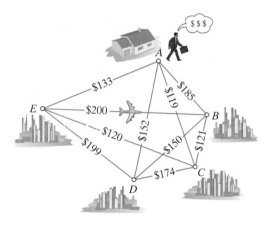

FIGURE 6-10

Would it be worth $20 to you to work out this problem? If so, how would you do it? We encourage you at this point to take a break from your reading, get a pencil and a piece of paper, and try to work out the problem on your own. (Pretend you are doing it for the money.) Give yourself about 15 to 20 minutes.

. . .

If you are like most people, you probably followed one of two standard strategies in looking for an optimal Hamilton circuit:

- **Method 1.** *(a) You made a list of all possible Hamilton circuits. (b) You calculated the total cost for each circuit. (c) You selected a circuit with the least total cost for the answer.*

Table 6-4 shows the worked-out solution in all its glory. (a) There are 24 possible Hamilton circuits in a complete graph with 5 vertices. (Since the one-way airfares are the same in either direction, a circuit and its mirror-image circuit result in the same total cost and are shown on the same row of the table. This observation saves quite a bit of work.) (b) The total cost of each circuit can be easily calculated and is shown in the middle column of the table. (c) The optimal circuits, with a total cost of $676, are shown in the second to last row (*A, D, B, C, E, A* and its mirror-image circuit *A, E, C, B, D, A*). We can use either one as a solution, shown graphically in Fig. 6-11.

TABLE 6-4		The 24 possible Hamilton circuits and their total costs.	
	Hamilton Circuit	Total Cost	Mirror-Image Circuit
1	A, B, C, D, E, A	185 + 121 + 174 + 199 + 133 = 812	A, E, D, C, B, A
2	A, B, C, E, D, A	185 + 121 + 120 + 199 + 152 = 777	A, D, E, C, B, A
3	A, B, D, C, E, A	185 + 150 + 174 + 120 + 133 = 762	A, E, C, D, B, A
4	A, B, D, E, C, A	185 + 150 + 199 + 120 + 119 = 773	A, C, E, D, B, A
5	A, B, E, C, D, A	185 + 200 + 120 + 174 + 152 = 831	A, D, C, E, B, A
6	A, B, E, D, C, A	185 + 200 + 199 + 174 + 119 = 877	A, C, D, E, B, A
7	A, C, B, D, E, A	119 + 121 + 150 + 199 + 133 = 722	A, E, D, B, C, A
8	A, C, B, E, D, A	119 + 121 + 200 + 199 + 152 = 791	A, D, E, B, C, A
9	A, C, D, B, E, A	119 + 174 + 150 + 200 + 133 = 776	A, E, B, D, C, A
10	A, C, E, B, D, A	119 + 120 + 200 + 150 + 152 = 741	A, D, B, E, C, A
11	A, D, B, C, E, A	152 + 150 + 121 + 120 + 133 = 676	A, E, C, B, D, A
12	A, D, C, B, E, A	152 + 174 + 121 + 200 + 133 = 780	A, E, B, C, D, A

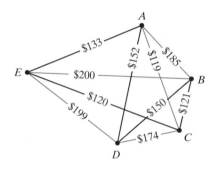

FIGURE 6-11
The optimal Hamilton circuit for the five-city TSP. Total cost: $676.

All of the above can be reasonably done in somewhere between 10 and 20 minutes—not a bad way to earn $20!

■ **Method 2.** *You started at home (A). From there you went to the city to which the cost of travel is the cheapest. Then from there you went to the next city to which the cost of travel is the cheapest, and so on. From the last city, you returned to A.*

Here are the worked-out details when we follow Method 2. We start at A. Looking at the graph, we see that the cheapest city to go to from A is C (cost: $119). From C, the cheapest city to go to (other than A) is E (cost: $120). From E, the cheapest remaining city to go to is D (cost $199), and from D we have little choice—the only remaining city to visit is B (cost: $150). From B we close the circuit and return home to A (cost: $185). Using this strategy, we get the circuit A, C, E, D, B, A, shown in Fig. 6-12. The total cost of this circuit is $773.

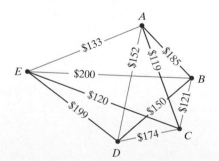

FIGURE 6-12
Circuit obtained following Method 2. Total cost: $773.

Using this last method, we can work out the problem in just a couple of minutes, which is nice, but there is a hitch. We are not going to collect any money from Willy. Justifiably, Willy is not pleased with this answer, which is $97 higher than the optimal answer obtained under Method 1. This second approach looks like a bust! (But be patient, there is more to come.) ■

EXAMPLE 6.6 Willy Expands His Territory

Let's imagine now that Willy, who has done very well with his business, has expanded his sales territory to ten cities called *A* through *K* (Fig. 6-13). Willy wants us to help him once again find an optimal Hamilton circuit that starts at *A* and goes to each of the other nine cities. Flush with success and generosity, he is offering a whopping $100 as a reward for a solution to this problem. Should we accept the challenge?

The one-way cost of travel between any two cities is shown by means of Table 6-5.[3] We now know that a foolproof method for tackling this problem is Method 1. But before we plunge into it, let's think of what we learned earlier in this chapter about factorials. The number of Hamilton circuits that we would have to check is 9!, which is 362,880. Still thinking about it? Here are some numbers that may help you to make up your mind. If you could do two circuits per minute—and that's working fast!—it would take about 3000 hours to do all 362,880 possible circuits. Shortcuts? Say you are clever and cut the work in half. That's still 1500 hours of work (a couple of month's worth if you worked nonstop, 24 hours a day, 7 days a week).

TABLE 6-5   Cost of travel between any two cities

	A	**B**	**C**	**D**	**E**	**F**	**G**	**H**	**J**	**K**
A	*	185	119	152	133	321	297	277	412	381
B	185	*	121	150	200	404	458	492	379	427
C	119	121	*	174	120	332	439	348	245	443
D	152	150	174	*	199	495	480	500	454	489
E	133	200	120	199	*	315	463	204	396	487
F	321	404	332	495	315	*	356	211	369	222
G	297	458	439	480	463	356	*	471	241	235
H	277	492	348	500	204	211	471	*	283	478
J	412	379	245	454	396	369	241	283	*	304
K	381	427	443	489	487	222	235	478	304	*

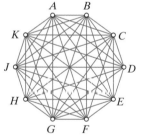

FIGURE 6-13

Let's now go back to Method 2, the one that turned out to be a pretty bad idea in Example 6.5. We start at *A*. From *A*, we would travel to *C*—at a cost of $119 it is the cheapest place to go to. Continuing with our strategy of always choosing the cheapest new city available we go from *C* to *E*, to *D*, to *B*, to *J*, to *G*, to *K*, to *F*, to *H*, and then finally back to *A*. (We leave it to the reader to verify the details.) It takes just a few minutes to use this method, and the Hamilton circuit obtained this way is *A, C, E, D, B, J, G, K, F, H, A*, with a total cost of $2153. Well, it's something, but is it a correct answer? Willy, for one, is not convinced and refuses to pay us the $100. Are we right in asking for the money? Is this an optimal circuit? What if it isn't—is it at least close? What do you think? ■

---

[3]With this many vertices and edges it is a little easier to put the weights in a table. The graph would get pretty cluttered if we tried to include the weights on the graph itself!

We will return to this example after a brief detour to formalize some of the ideas we have just discussed.

6.5 The Brute-Force and Nearest-Neighbor Algorithms

We touched on the subject of algorithms in Chapter 5, and we now revisit the concept in a little more detail. Recall that an **algorithm** is a set of procedural rules that, when properly followed, always lead to some "solution" to a problem. Both of the intuitive strategies for finding optimal Hamilton circuits that we discussed in the preceding section are examples of *graph algorithms*. Method 1 goes by the descriptive name of the **brute-force algorithm**, Method 2 has an equally descriptive name—the **nearest-neighbor algorithm**.

ALGORITHM 1: THE BRUTE-FORCE ALGORITHM

- Make a list of all the possible Hamilton circuits of the graph.
- For each Hamilton circuit calculate its *total weight* (by adding the weights of all the edges in the circuit).
- Find the circuits (there is always more than one) with the least total weight. Any one of these can be chosen as an optimal Hamilton circuit for the graph.

ALGORITHM 2: THE NEAREST-NEIGHBOR ALGORITHM

- Pick a vertex as the starting point.
- From the starting vertex go to its *nearest neighbor*—the vertex for which the corresponding edge has the smallest weight. (If there is more than one nearest neighbor, choose among them at random.)
- Continue building the circuit, one vertex at a time, by always going from a vertex to its nearest neighbor *from among the vertices that haven't been visited yet*. (Whenever there is a tie, choose at random.) Keep doing this until all the vertices have been visited.
- From the last vertex return to the starting point.

Based on what we learned in Examples 6.5 and 6.6, we know that in a general sense there are some problems with both of these algorithms.

Let's start with the brute-force algorithm. Checking through all possible Hamilton circuits to find the optimal one sounds like a great idea in theory, but, as we saw in Example 6.6, there is a practical difficulty in trying to use this approach. The difficulty resides in the fantastic growth of the number of Hamilton circuits that need to be checked. In fact, if we are doing things by hand, it would be quite foolhardy to try to use this algorithm except for graphs with a very small number of vertices.

A possible way to get around this, one might think, is to recruit a fast helper, such as a powerful computer. It seems that a computer would be exactly the right kind of helper, because the brute-force algorithm is essentially a mindless exercise in arithmetic with a little bookkeeping thrown in, and both of these are things that computers are very good at. Unfortunately, even the world's most powerful computer won't take us very far.

Let's imagine, for the sake of argument, that we have been given free access to the fastest supercomputer on the planet, one that can compute *ten billion* (that's 10^{10}) circuits per second. (That's more that any current computer could do, but since we are just fantasizing, let's think big!) Now there are 31,536,000 seconds in a year, which, roughly speaking, is 3×10^7. Altogether, this means that our supercomputer can compute about 3×10^{17} Hamilton circuits in one year. For graphs of up to 15 vertices, our helper can run through all the Hamilton circuits in a matter of seconds (or less). Things get more interesting when we start moving beyond 15 vertices. Table 6-6 illustrates what happens then.

TABLE 6-6   Solving the TSP using the brute-force algorithm with a supercomputer

Number of vertices N	Number of Hamilton circuits $(N-1)!$	Amount of time to check them all with a supercomputer
16	1,307,674,368,000	$\approx$2 minutes
17	$\approx 2.1 \times 10^{13}$	$\approx$35 minutes
18	$\approx 3.6 \times 10^{14}$	$\approx$10 hours
19	$\approx 6.4 \times 10^{15}$	$\approx 7\frac{1}{2}$ days
20	$\approx 1.2 \times 10^{17}$	$\approx$140 days
21	$\approx 2.4 \times 10^{18}$	$\approx 7\frac{1}{2}$ years
22	$\approx 5.1 \times 10^{19}$	$\approx$160 years
23	$\approx 1.1 \times 10^{21}$	$\approx$3,500 years
24	$\approx 2.6 \times 10^{22}$	$\approx$82,000 years
25	$\approx 6.2 \times 10^{23}$	$\approx$2 million years

Table 6-6 illustrates how extraordinarily fast the computational burden grows with the brute-force algorithm. Each time we increase the number of vertices of the graph by one, *the amount of work required to carry out the algorithm increases by a factor that is equal to the number of vertices in the graph*. For example, it takes 5 times as much work to go from 5 vertices to 6, 10 times more work to go from 10 vertices to 11, and 100 times as much work to go from 100 vertices to 101. Bad news!

The brute-force algorithm is a classic example of what is formally known as an **inefficient algorithm**—an algorithm for which the number of steps needed to carry it out grows disproportionately with the size of the problem. The trouble with inefficient algorithms is that they are of limited practical use—they can realistically be carried out only when the problem is small.

Fortunately, not all algorithms are inefficient. Let's discuss now the nearest-neighbor algorithm, in which we hop from vertex to vertex using a simple criterion: Where is the next "nearest" place to go to? For a graph with 5 vertices, we have to take 5 steps.[4] What happens when we double the number of vertices to ten? We now have to take 10 steps. Essentially, the amount of work doubled when the problem doubled. Could we use the nearest-neighbor algorithm in a complete graph with 100 vertices? You bet. It would take a little longer than in the case of 10 vertices (maybe an hour) but for a nice reward (say $200) it would certainly be worth our trouble.

An algorithm for which the number of steps needed to carry it out grows in proportion to the size of the input to the problem is called an **efficient algorithm**. As a practical matter, efficient algorithms are the only kind of algorithms that we can realistically use on a consistent basis to solve a graph problem.

The nearest-neighbor algorithm is an efficient algorithm, which is good! The problem with it is that, as we saw in Example 6.5, it doesn't give us what we are asking for—an optimal Hamilton circuit. So why should we even consider an algorithm that doesn't give us the optimal answer? As we will find out next, sometimes we have to take what we can get.

6.6 Approximate Algorithms

The ultimate goal in finding a general method for solving TSPs is to find an algorithm that is *efficient*, like the nearest-neighbor algorithm, and **optimal**, meaning that it guarantees us an optimal answer at all times, as the brute-force algorithm does. Unfortunately, nobody knows of such an algorithm. Moreover, *we don't even know why we don't know*. Is it because such an algorithm is actually a mathematical impossibility? Or is it because no one has yet been clever enough to find one?

Despite the efforts of some of the best mathematicians of our time, the answers to these questions have remained quite elusive. So far, no one has been able to come up with an efficient optimal algorithm for solving TSPs or, alternatively, to prove that such an algorithm is an impossibility. Because this question has profound implications in an area of computer science called complexity theory, it has become one of the most famous unsolved problems in modern mathematics.[5]

In the meantime, we are faced with a quandary. In many real-world applications, it is necessary to find some sort of solution for TSPs involving graphs with hundreds and even thousands of vertices, and to do so in real time. Since the brute-force algorithm is out of the question, and since no efficient algorithm that guarantees an optimal solution is known, the only practical strategy to fall back on is to compromise. We give up on the expectation of having an optimal solution and accept a solution that may not be optimal. In exchange, we ask for quick results. Nowadays, this is the way that TSPs are "solved."

We will use the term **approximate algorithm** to describe any algorithm that produces solutions[6] that are, most of the time, reasonably close to the optimal solution. Sounds good, but what does "most of the time" mean? And how about

---

[4]Actually, the last step (go back to the starting vertex) is automatic—but for the sake of simplicity we'll still call it a step.

[5]For an excellent account of this famous problem, see Reference 9.

[6]Note that in this context the word *solution* no longer means the "best answer," but simply "an answer."

"reasonably close"? Unfortunately, to answer these questions properly would take us beyond the scope of this book. We will have to accept the fact that in the area of analyzing algorithms we will be dealing with informal ideas rather than precise definitions. (An excellent introduction to the analysis of algorithms can be found in references 8 and 10.)

This is the end of our brief detour, and we now return to Example 6.6.

EXAMPLE 6.6 (continued)

When we left this example a few pages back, we had (a) decided not to try the brute-force algorithm (way too much work for just a $100 reward), and (b) used the nearest-neighbor algorithm to come up with the circuit *A, C, E, D, B, J, G, K, F, H, A* (total cost: $2153). We are still hoping to collect the $100, but Willy refuses to pay. It turns out that the optimal Hamilton circuit for this problem is *A, D, B, C, J, G, K, F, H, E, A* (total cost: $1914). To find this optimal circuit using the brute-force algorithm would have required spending hundreds of hours slaving over numbers and circuits. (To be truthful, we found it using a fast computer and some special software.) The net savings between the optimal solution and the quick and easy one we got using the nearest-neighbor algorithm is $239. Are the savings worth the extra effort? This is an interesting and deep question. In this case the answer is probably not. In other situations the answer may be different. ∎

The important point to take home from Example 6.6 is that approximate algorithms are not necessarily bad, and that sometimes we may be better off settling for a quick approximate answer rather than insisting on finding the optimal answer. It is often the case that approximate algorithms are the ones that give the most "bang for the buck."

In the next two sections, we will discuss a couple of new algorithms for "solving" TSPs: the *repetitive nearest-neighbor algorithm* and the *cheapest-link algorithm*. Both of these are *approximate algorithms*.

6.7 The Repetitive Nearest-Neighbor Algorithm

As one might guess, the *repetitive nearest-neighbor algorithm* is a variation of the nearest-neighbor algorithm in which we repeat several times the entire nearest-neighbor circuit-building process. Why would we want to do this? The reason is that the Hamilton circuit one gets when applying the nearest-neighbor algorithm depends on the choice of the starting vertex. If we change the starting vertex, it is likely that the Hamilton circuit we get will be different, and, if we are lucky, better. Since finding a Hamilton circuit using the nearest-neighbor algorithm is an efficient process, it is not an unreasonable burden to do it several times, each time starting at a different vertex of the graph. In this way, we can obtain several different "nearest-neighbor solutions," from which we can then pick the best.

But what do we do with a Hamilton circuit that starts somewhere other than the vertex we really want to start at? That's not a problem. Remember that once we have a circuit, we can start the circuit anywhere we want. In fact, in an abstract sense, a circuit has no starting or ending point.

To illustrate how the repetitive nearest-neighbor algorithm works, let's return to the original five-city problem we last discussed in Example 6.5.

EXAMPLE 6.7 A Tale of Five Cities: Part III

Once again, we are going to look at the TSP given by the complete graph in Fig. 6–14, representing Willy's original sales territory. (We already know that the optimal Hamilton circuit is given by A, D, B, C, E, A, so the main purpose of this example is just to illustrate how the repetitive nearest-neighbor algorithm works.)

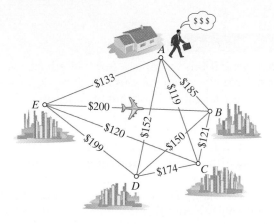

FIGURE 6-14

When we used the nearest-neighbor algorithm with A as the starting point, we got the Hamilton circuit A, C, E, D, B, A with a total cost of $773. Let's now try it with B as the starting point. We leave it to the reader to verify [see Exercise 25(a)] that now the nearest-neighbor algorithm yields the Hamilton circuit B, C, A, E, D, B with a total cost of $722. Well, that is certainly an improvement! Can a person such as Willy who must start and end his trip at A take advantage of this $722 circuit? Why not? All he has to do is rewrite the circuit B, C, A, E, D, B in its equivalent form A, E, D, B, C, A.

Having done so well so far, we might as well try the nearest-neighbor algorithm with C, D, and E as the starting points. We leave it to the reader to verify [Exercise 25(b), (c), (d)] that when the starting point is C, we get C, A, E, D, B, C with a total cost of $722; when the starting point is D, we get D, B, C, A, E, D, also with a total cost of $722; and finally, when the starting point is E, we get E, C, A, D, B, E with a total cost of $741. None of these improves on the circuit we found when we started at B (although starting at C and D actually gave us the same circuit). Thus, the best solution the repetitive nearest-neighbor algorithm gives us is the circuit A, E, D, B, C, A with a total cost of $722 (Fig. 6-15). This circuit is a nice improvement over the original $773 circuit that we got when we started at A, but is still short of the optimal circuit found in Example 6.5.

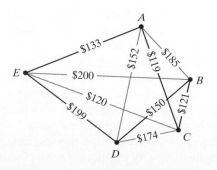

FIGURE 6-15
Hamilton circuit obtained using the repetitive nearest-neighbor algorithm. Total cost: $722.

A formal description of the **repetitive nearest-neighbor algorithm** is as follows.

ALGORITHM 3: THE REPETITIVE NEAREST-NEIGHBOR ALGORITHM

- Let X be any vertex. Apply the nearest-neighbor algorithm using X as the starting vertex and calculate the total cost of the circuit obtained.
- Repeat the process using each of the other vertices of the graph as the starting vertex.
- Of the Hamilton circuits obtained, keep the best one. If there is a designated starting vertex, rewrite this circuit with that vertex as the reference point.

6.8 The Cheapest-Link Algorithm

This is the last—but not the least—of our algorithms for finding Hamilton circuits. One lesson of the repetitive nearest-neighbor algorithm is that the order in which one builds a Hamilton circuit and the order in which one actually travels the circuit do not have to be one and the same. In fact, one can build a Hamilton circuit piece by piece without requiring that the pieces be connected, so long as at the end it all comes together. People often use this strategy when putting together a large jigsaw puzzle.

The **cheapest-link algorithm** is essentially an algorithm based on this strategy. We start by grabbing the cheapest edge of the graph, wherever it may be. Once this is done, we grab the next cheapest edge of the graph, wherever it may be. We continue doing this, each time grabbing the cheapest edge available, subject to the following two restrictions:

(i) Do not allow circuits to form (other than at the very end).

(ii) Do not allow three edges to come together at a vertex.

It is clear that if we allow either of these things to happen, it would be impossible to end up with a Hamilton circuit at the end. Fortunately, these are the only two restrictions we must worry about.

EXAMPLE 6.8 A Tale of Five Cities: Part IV

To illustrate the cheapest-link algorithm, we will revisit one final time the five-city TSP in Example 6.2. Once again, we show the cost of travel between any two cities (Fig. 6-16).

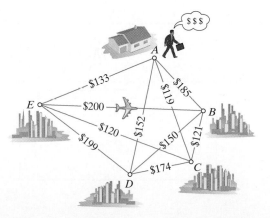

FIGURE 6-16

Our first step is to scan the graph and pick the cheapest of all possible "links," regardless of where it may be. In this case, it is the edge AC ($119). We will keep a record of the circuit-building process by marking the edges of our circuit in red. Fig. 6-17(a) shows where we are at this point. The next step is to scan the graph again, looking for the cheapest unmarked link available, which in this case is edge CE ($120). We mark it in red, as shown in Fig. 6-17(b). Once again, we scan the graph looking for the cheapest unmarked link, which in this case is edge BC ($121). But this edge can't be part of the circuit, since a circuit cannot have three edges coming together at the same vertex. This one we'll have to throw away [Fig. 6-17(c)]. After BC, the next cheapest link is given by edge AE ($133). But we have to throw this one away too—the vertices A, C, and E would be linked in a "short circuit," and a Hamilton circuit can never have a smaller circuit within it [Fig. 6-17(d)]! So we persevere, scanning the graph for the next cheapest link, which is BD ($150). This one works, so we add it to our budding circuit [Fig. 6-17(e)]. The next cheapest link available is AD ($152), and it works just fine [Fig. 6-17(f)]. At this point, we have only one way to close up the Hamilton circuit, edge BE, as shown in Fig. 6-17(g). The Hamilton circuit in red can now be described using any vertex as the reference point. Since Willy lives at A, we describe it as A, C, E, B, D, A (or its mirror image). The total cost of this circuit is $741, which is a little better than the nearest-neighbor solution but not as good as the repetitive nearest-neighbor solution.

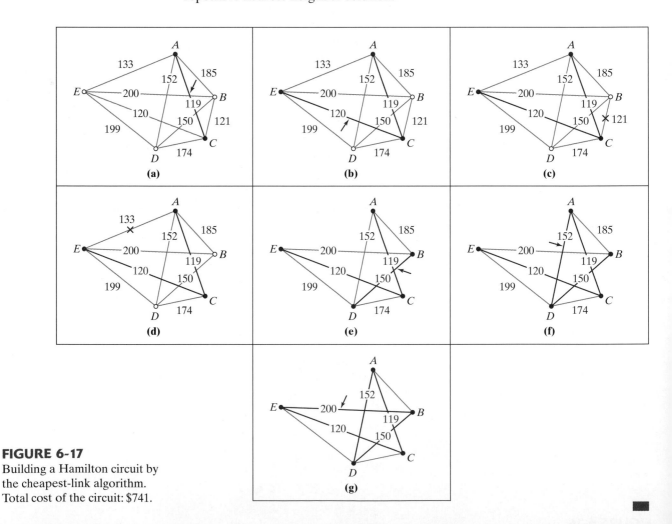

FIGURE 6-17
Building a Hamilton circuit by
the cheapest-link algorithm.
Total cost of the circuit: $741.

A formal description of the cheapest-link algorithm is as follows.

ALGORITHM 4: THE CHEAPEST-LINK ALGORITHM

- Pick the edge with the smallest weight first (in case of a tie pick one at random). Mark the edge (say in red).
- Pick the next "cheapest" edge and mark the edge in red.
- Continue picking the "cheapest" edge available and mark the edge in red except when

   (a) it closes a circuit.

   (b) it results in three edges coming out of a single vertex.

- When there are no more vertices to join, close the red circuit.

For the last example of this chapter, we will return to the problem first described in the chapter opener—that of finding an optimal route for a rover exploring Mars.

EXAMPLE 6.9

Figure 6-18(a) shows seven sites on Mars identified as those locations where some form of bacterial life is most likely to be found. Our job is to find the shortest route for a rover that will start at A, visit all the sites, and at the end, return to A. The approximate distance (in miles) between any two sites is shown in the graph in Fig. 6-18(b), a complete weighted graph with seven vertices that will serve as a model for the problem.

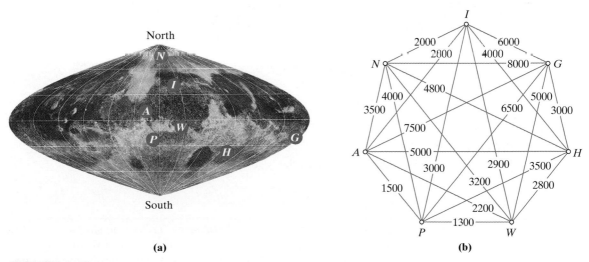

(a)   (b)

FIGURE 6-18
(a) Mars sites most likely to show evidence of life. (b) Approximate distances (in miles) between sites.

We will tackle this problem using the different approaches we have learned.

■ **The Brute-Force Approach.** This is the only method we know that is guaranteed to give us the optimal Hamilton circuit. Unfortunately, it would require us to check through 720 different Hamilton circuits (6! = 720). We will pass on that idea for now.

■ **The Cheapest-Link Approach.** This is a reasonable algorithm to use—not trivial but not too hard either. A summary of the steps is shown in Table 6-7.

TABLE 6-7

Step	Cheapest edge available	Weight	Use in circuit?
1	PW	1300	yes
2	AP	1500	yes
3	IN	2000	yes
4	AW	2200	no △
5 }	HW } tie	2800	yes
6 }	AI }	2800	yes
7	IW	2900	no ◇ & ⋎
8 }	IP } tie	3000	no △ & ⋎
9 }	GH }	3000	yes
Last	GN only way to close circuit	8000	yes

The circuit obtained using this algorithm is *A, P, W, H, G, N, I, A* with a total length of 21,400 miles (Fig. 6-19).

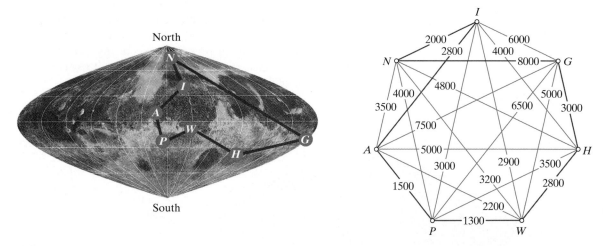

FIGURE 6-19
Hamilton circuit obtained using cheapest-link algorithm. Total length: 21,400 miles.

■ **The Nearest-Neighbor Approach.** This is the simplest of all the algorithms we learned. Starting from *A* we go to *P*, then to *W*, then to *H*, then to *G*, then to *I*, then to *N*, and finally back to *A*. The circuit obtained under this algorithm is *A, P, W, H, G, I, N, A* with a total length of 20,100 miles (Fig. 6-20). (We know that we can repeat this method with a different starting location, but we won't bother with that at this time.)

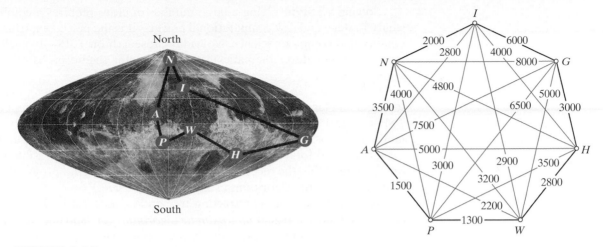

FIGURE 6-20
Hamilton circuit obtained using nearest-neighbor algorithm. Total length: 20,100 miles.

The first surprise is that the nearest-neighbor algorithm gives us a better Hamilton circuit than the cheapest-link algorithm. It happens this way about as often as it happens the other way around, so neither of the two algorithms can claim to be superior to the other one.

The second surprise is that the circuit *A, P, W, H, G, I, N, A* obtained using the nearest-neighbor algorithm turns out to be the optimal Hamilton circuit for this example. (We know this because we used a computer to find the optimal Hamilton circuit.) Essentially, this means that in this particular case, the simplest of all methods happens to produce the optimal answer—a nice turn of events. Too bad we can't count on this happening every time! Well, maybe next chapter. ◼

Conclusion

In this chapter we discussed the problem of finding *optimal Hamilton circuits* in a *complete weighted graph*. For historical reasons, this type of problem is known as the *traveling-salesman problem (TSP)* even though it may have nothing to do with traveling salesmen. In many situations, finding an optimal Hamilton circuit is reasonably easy, but a completely general algorithm that would work for every TSP has eluded mathematicians, who have been actively working on this problem for more than 50 years. This is an extremely important and at the same time notoriously difficult problem.

The *nearest-neighbor* and *cheapest-link* algorithms, which we learned in this chapter, are two fairly simple strategies for attacking TSPs. We have seen that both algorithms are *approximate algorithms*. This means that they are not likely to give us an optimal solution, although, as we saw in Example 6.9, on a lucky day, even that is possible. (By the same token, on an unlucky day, either of these two algorithms can give us very bad circuits—see Exercises 68 and 69.) In most typical problems, however, we can expect either of these algorithms to give an approximate solution that is within a reasonable margin of error. With some problems the cheapest-link algorithm gives a better solution than the nearest-neighbor algorithm; with other problems it's the other way around. Thus, while they are not the same, neither is superior to the other.

Solving a TSP involving a large number of cities requires a combination of smart strategy and raw computational power—it is the perfect marriage of mathematics and computer science. Nowadays, research on TSPs—typically done by teams that include mathematicians and computer scientists—is carried out on several fronts.

One major area of research interest is the quest for finding *optimal solutions* to larger and larger TSPs. As of the writing of this edition, the current record is a TSP with 15,112 cities, whose optimal solution was found in 2001 by a team of mathematicians and computer scientists using 110 connected computers running for a combined total of 22.6 years of computer time.[7] (It is possible that by the time you read this, the record may have been improved, but progress in this area is always slow—the algorithms used are all *inefficient*.)

A second area that is attracting tremendous interest is that of finding improved and more efficient *approximate algorithms* for "solving" very large TSPs. Here progress is much faster, and many sophisticated approximate algorithms have been developed, including algorithms based on *recombinant DNA* techniques (see Project C for details) as well as algorithms modeled after *ant colonies* (yes, that's ants, as in insects—see Project D for details). Today, the best approximate algorithms can tackle TSPs with hundreds of thousands of cities and produce solutions that are *guaranteed* to be off by no more than 1% from the optimal solution.

On a purely theoretical level, the holy grail of TSP research is to find an *optimal and efficient algorithm* for solving all TSPs, or, alternatively, to prove that such an algorithm is a mathematical impossibility. This conundrum is waiting for the next Euler to come along.

---

[7]The team consisted of David Applegate and Robert Bixby, both of Rice University, Văsek Chvátal of Rutgers, and William Cook of Georgia Tech.

A 48-city TSP with a little bite: What is the shortest circuit that visits all of the state capitals in the continental United States? The *optimal solution*, shown in red, is a Hamilton circuit approximately 12,000 miles long. Even with a fast computer and special software, it might take hundreds of computer hours to come up with this optimal answer.

For the same 48-city TSP, *approximate solutions* can be found using efficient algorithms in a matter of just minutes. This approximate solution (shown in red) was obtained by hand in less than ten minutes, using the nearest-neighbor algorithm (starting at Olympia, Washington). The total length of this trip is approximately 14,500 miles, roughly 20% longer than the optimal solution.

| PROFILE | Sir William Rowan Hamilton (1805–1865) |

Hamilton is one of the greatest and yet most tragic figures in the history of mathematics. A prodigy in languages, science, and mathematics at a very young age, he suffered from bouts of depression and fought the demon of alcohol for much of his adult life. In spite of many important mathematical discoveries, the story of his life is one of what might have been had the circumstances of his personal life turned out differently.

William Rowan Hamilton was born in Dublin, Ireland, exactly at midnight on August 3, 1805 (causing some confusion as to whether his birthday was August 3 or August 4). From a very young age he was taught by his uncle the Rev. James Hamilton, an eccentric clergyman and brilliant linguist. Under his uncle's tutelage, Hamilton learned Greek, Latin, and Hebrew by the age of 6, and by the time he was 13 he knew a dozen languages, including Persian, Arabic, and Sanskrit. At the age of 10 Hamilton became interested in mathematics, and by the age of 15 he was reading the works of Newton and Laplace. When he was just 17, Hamilton wrote his first two mathematical research papers, *On Contacts between Algebraic Curves and Surfaces*, and *Developments*. After reading the papers, John Brinkley, the Astronomer Royal of Ireland at that time, remarked, "This young man, I do not say will be, but is, the first [foremost] mathematician of his age."

In 1823, Hamilton entered Trinity College in Dublin, where he excelled in all subjects while continuing to publish groundbreaking research in mathematical physics. A year after entering college, Hamilton met and fell madly in love with a young lady from an upper class family by the name of Catherine Disney, a watershed event in his life. Although Catherine might have reciprocated his affections, there was a problem. Hamilton was just a nineteen-year-old university student and did not come from a particularly wealthy family—according to the social norms of the day, he was not a suitable match for Catherine. On the other hand, Hamilton was one of those mathematical geniuses that come around once in a generation. Not lacking for confidence, he was banking on his status and fame as a scientist to bring Catherine's parents around into eventually accepting him.

The following year Hamilton found out that Catherine's family had arranged for her marriage to another man. This was devastating news to him and he even considered suicide. Ironically, soon after Catherine was married, Hamilton's social status and financial position took a quantum leap forward. In 1827, not quite yet 22 years old and still an undergraduate, Hamilton won the appointment to the prestigious position of Astronomer Royal of Ireland as well as to a Professorship of Astronomy at Trinity College, an incredible personal triumph. Unfortunately, for him and Catherine it all happened a couple of years too late. By now Catherine was married and had a child.

Over the next twenty years, Hamilton's life was a checkered mixture of professional successes and personal setbacks. On the rebound, he soon entered into an ill-advised and unhappy marriage that bore him three children. He fought intermittent bouts of depression and started having serious problems with alcohol. At the same time, he pursued his research in mathematics and optics, where he produced many significant discoveries, including the formulas for conic refraction, which gained him much fame. In recognition of his many scientific accomplishments, Hamilton was knighted in 1835.

In 1843, Hamilton produced the mathematical discovery he is best known for, the *calculus of quaternions*. (Quaternions are essentially four-dimensional numbers that generalize the complex numbers and are ideally suited to describe mathematically many complex phenomena in physics.) Hamilton predicted that quaternions would revolutionize 19th-century physics, but he was a century ahead of his time—quaternions revolutionized *20th*-century physics, as they are a key element in quantum mechanics and Einstein's theory of relativity.

Hamilton's connection with graph theory and *Hamilton circuits* came late in his life and in a rather roundabout way. In 1857, Hamilton invented a board game he called the *Icosian game*, the purpose of which was to create a trip passing through each of 20 European cities once and only once. The cities were connected in the form of a *dodecahedral graph* (see Exercise 49 for details). Hamilton sold the rights to the game to a London game dealer for 25 pounds, and the game was actually marketed throughout Europe, a precursor to some of the "connect-the-dots" games that we know today.

KEY CONCEPTS

approximate algorithm	Hamilton path
algorithm	inefficient algorithm
brute-force algorithm	nearest-neighbor algorithm
cheapest-link algorithm	optimal
complete graph	repetitive nearest-neighbor algorithm
complete weighted graph	traveling-salesman problem (TSP)
efficient algorithm	weighted graph
factorial	weight
Hamilton circuit	

EXERCISES

WALKING

A. Hamilton Circuits and Hamilton Paths

1. For the following graph,
   (a) find three different Hamilton circuits.
   (b) find a Hamilton path that starts at A and ends at B.
   (c) find a Hamilton path that starts at D and ends at F.

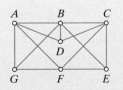

2. For the following graph,
   (a) find three different Hamilton circuits.
   (b) find a Hamilton path that starts at A and ends at B.
   (c) find a Hamilton path that starts at F and ends at I.

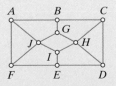

3. List all possible Hamilton circuits in the following graph.

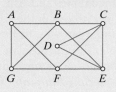

4. List all possible Hamilton circuits in the following graph.

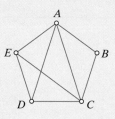

5. For the following graph,

    (a) find a Hamilton path that starts at A and ends at E.

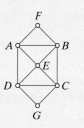

    (b) find a Hamilton circuit that starts at A and ends with the pair of vertices E, A.

    (c) find a Hamilton path that starts at A and ends at C.

    (d) find a Hamilton path that starts at F and ends at G.

6. For the following graph,

    (a) find a Hamilton path that starts at A and ends at E.

    (b) find a Hamilton circuit that starts at A and ends with the pair of vertices E, A.

    (c) find a Hamilton path that starts at A and ends at G.

    (d) find a Hamilton path that starts at F and ends at G.

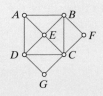

7. For the following graph,

    (a) list all Hamilton circuits that start at vertex A.

    (b) list all Hamilton circuits that start at vertex D.

    (c) explain why your answers in (a) and (b) must have the same number of circuits.

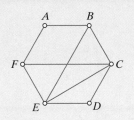

8. For the following graph,

    (a) list all Hamilton circuits that start at vertex A.

    (b) list all Hamilton circuits that start at vertex D.

    (c) explain why your answers in (a) and (b) must have the same number of circuits.

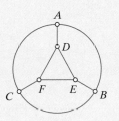

9. Explain why the following graph has neither Hamilton circuits nor Hamilton paths.

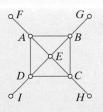

10. Explain why the following graph has no Hamilton circuit but does have a Hamilton path.

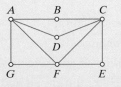

11. For the following weighted graph,

    (a) find the weight of edge *BD*.

    (b) find the weight of edge *EC*.

    (c) find a Hamilton circuit and give its weight.

    (d) find a different Hamilton circuit and give its weight.

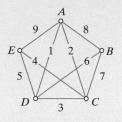

12. For the following weighted graph,

    (a) find the weight of edge *AD*.

    (b) find the weight of edge *AC*.

    (c) find a Hamilton circuit and give its weight.

    (d) find a different Hamilton circuit and give its weight.

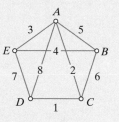

13. For the following weighted graph,

    (a) find the weight of edge *BC*.

    (b) find a Hamilton circuit and give its weight.

    (c) find a different Hamilton circuit and give its weight.

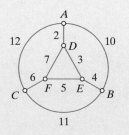

14. For the following weighted graph,

    (a) find the weight of edge *AC*.

    (b) find a Hamilton circuit and give its weight.

    (c) find a different Hamilton circuit and give its weight.

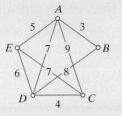

B. Factorials and Complete Graphs

15. Using a calculator, compute each of the following:

    (a) 13!

    (b) The number of distinct Hamilton circuits in K_{14}.

    (c) The number of distinct Hamilton circuits in K_{25}. (Give your answer in scientific notation.)

16. Using a calculator, compute each of the following:

    (a) 15!

    (b) The number of distinct Hamilton circuits in K_{16}.

    (c) The number of distinct Hamilton circuits in K_{30}. (Give your answer in scientific notation.)

17. Given that 9! = 362,880, compute each of the following without using a calculator:

    (a) 10!

    (b) The number of distinct Hamilton circuits in K_{10}.

    (c) The number of distinct Hamilton circuits in K_{11}.

18. Given that 20! = 2,432,902,008,176,640,000, compute each of the following without a calculator:

    (a) 19!

    (b) The number of distinct Hamilton circuits in K_{20}.

    (c) The number of distinct Hamilton circuits in K_{21}.

19. For the complete graph on 12 vertices, find

    (a) the number of edges in the graph.

    (b) the number of distinct Hamilton circuits.

20. For the complete graph on 24 vertices, find

    (a) the number of edges in the graph.

    (b) the number of distinct Hamilton circuits.

21. In each case, find the value of N.

    (a) K_N has 120 distinct Hamilton circuits.

    (b) K_N has 45 edges.

    (c) K_N has 20,100 edges.

22. In each case, find the value of N.

    (a) K_N has 720 distinct Hamilton circuits.

    (b) K_N has 66 edges.

    (c) K_N has 80,200 edges.

C. Brute-Force and Nearest-Neighbor Algorithms

23. Consider the following weighted graph.

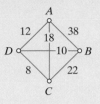

(a) Apply the brute-force algorithm to find an optimal Hamilton circuit.

(b) Apply the nearest-neighbor algorithm with starting vertex A to find a Hamilton circuit.

(c) Apply the nearest-neighbor algorithm with starting vertex B to find a Hamilton circuit.

(d) Apply the nearest-neighbor algorithm with starting vertex C to find a Hamilton circuit.

24. Consider the following weighted graph.

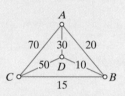

(a) Apply the brute-force algorithm to find an optimal Hamilton circuit.

(b) Apply the nearest-neighbor algorithm with starting vertex A to find a Hamilton circuit.

(c) Apply the nearest-neighbor algorithm with starting vertex C to find a Hamilton circuit.

(d) Apply the nearest-neighbor algorithm with starting vertex D to find a Hamilton circuit.

25. The following is the weighted graph of Willy's original sales-territory problem (Examples 6.2, 6.5, 6.7, 6.8).

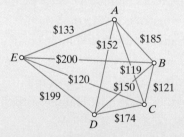

(a) Apply the nearest-neighbor algorithm with starting vertex B to find a Hamilton circuit, and verify that its weight is $722.

(b) Apply the nearest-neighbor algorithm with starting vertex C to find a Hamilton circuit, and verify that its weight is $722.

(c) Apply the nearest-neighbor algorithm with starting vertex D to find a Hamilton circuit, and verify that its weight is $722.

(d) Apply the nearest-neighbor algorithm with starting vertex E to find a Hamilton circuit, and verify that its weight is $741.

26. Sophie, a traveling salesperson, must call on customers in five different cities (*A, B, C, D,* and *E*). Sophie's trip must start and end at her home town (*A*). Each edge of the following graph shows the cost of travel between any two cities.

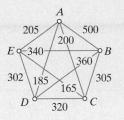

(a) Apply the nearest-neighbor algorithm with starting vertex *A* to find a Hamilton circuit in the graph. What is Sophie's cost for this Hamilton circuit?

(b) Apply the nearest-neighbor algorithm with starting vertex *B* to find a Hamilton circuit in the graph and write the circuit as it would be traveled by someone living in *A*.

(c) Apply the brute-force algorithm to find the optimal trip for Sophie. What is Sophie's cost for the optimal trip?

27. A space expedition is scheduled to visit the moons Callisto (*C*), Ganymede (*G*), Io (*I*), Mimas (*M*), and Titan (*T*) to collect rock samples at each and then return to Earth (*E*). The following graph summarizes the travel time (in years) between any two places.

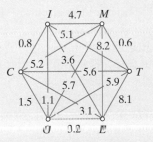

(a) Apply the nearest-neighbor algorithm with starting vertex *E* to find a Hamilton circuit in the graph. What is the travel time for this Hamilton circuit?

(b) Apply the nearest-neighbor algorithm with starting vertex *T* to find a Hamilton circuit in the graph and write the circuit as it would be traveled by an expedition started at *E*.

28. Consider the following weighted graph.

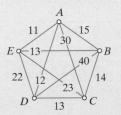

(a) Apply the nearest-neighbor algorithm with starting vertex A to find a Hamilton circuit in the graph. What is the weight of this Hamilton circuit?

(b) Apply the nearest-neighbor algorithm with starting vertex D to find a Hamilton circuit in the graph and write the circuit as it would be traveled by someone starting at vertex A.

29. Darren is a traveling salesperson whose territory consists of the six cities shown in the following mileage chart.

Mileage Chart

	Atlanta	Columbus	Kansas City	Minneapolis	Pierre	Tulsa
Atlanta	*	533	798	1068	1361	772
Columbus	533	*	656	713	1071	802
Kansas City	798	656	*	447	592	248
Minneapolis	1068	713	447	*	394	695
Pierre	1361	1071	592	394	*	760
Tulsa	772	802	248	695	760	*

Darren wants to schedule a round trip that starts and ends in his home city of Atlanta and visits each of the other cities once.

(a) Apply the nearest-neighbor algorithm with Atlanta as the starting vertex to find a Hamilton circuit in the graph. Give the total miles for this trip.

(b) Apply the nearest-neighbor algorithm with Kansas City as the starting vertex to find a Hamilton circuit in the graph and write the circuit as it would be traveled by Darren starting from his home in Atlanta. Give the total miles for this trip.

30. Jodi is a traveling salesperson whose territory consists of the seven cities shown in the following mileage chart.

Mileage Chart

	Boston	Dallas	Houston	Louisville	Nashville	Pittsburgh	St. Louis
Boston	*	1748	1804	941	1088	561	1141
Dallas	1748	*	243	819	660	1204	630
Houston	1804	243	*	928	769	1313	779
Louisville	941	819	928	*	168	388	263
Nashville	1088	660	769	168	*	553	299
Pittsburgh	561	1204	1313	388	553	*	588
St. Louis	1141	630	779	263	299	588	*

Jodi must organize a round trip that starts and ends in her home city of Nashville and visits each of the other cities once.

(a) Apply the nearest-neighbor algorithm with Nashville as the starting vertex to find a Hamilton circuit in the graph. Give the total miles for this trip.

(b) Apply the nearest-neighbor algorithm with St. Louis as the starting vertex to find a Hamilton circuit in the graph and write the circuit as it would be traveled by Jodi starting from her home in Nashville. Give the total miles for this trip.

D. Repetitive Nearest-Neighbor Algorithm

31. Apply the repetitive nearest-neighbor algorithm to find a Hamilton circuit in the following weighted graph.

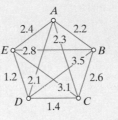

32. This exercise refers to Sophie's sales trip as discussed in Exercise 26. (The weighted graph is shown below.) Apply the repetitive nearest-neighbor algorithm to find a Hamilton circuit in the graph and write the circuit as it would be traveled by Sophie starting from her home town of *A*. Give the total cost of the trip.

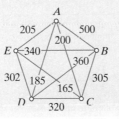

33. This exercise refers to the space expedition discussed in Example 6.3 and Exercise 27. (The weighted graph is shown below.) Apply the repetitive nearest-neighbor algorithm to find a Hamilton circuit in the graph and write the circuit as it would be traveled by an expedition starting from Earth (*E*). Give the total travel time for this trip.

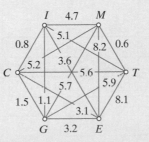

34. This exercise refers to the weighted graph discussed in Exercise 28. (The weighted graph is shown below.) Apply the repetitive nearest-neighbor algorithm to find a Hamilton circuit in the graph. Give the weight of the circuit.

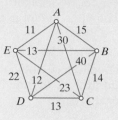

35. This exercise refers to Darren's sales trip as discussed in Exercise 29. (The mileage chart is shown below.) Apply the repetitive nearest-neighbor algorithm to find a Hamilton circuit in the graph and write the circuit as it would be traveled by Darren starting from his home in Atlanta. Give the total mileage for this circuit.

Mileage Chart

	Atlanta	Columbus	Kansas City	Minneapolis	Pierre	Tulsa
Atlanta	*	533	798	1068	1361	772
Columbus	533	*	656	713	1071	802
Kansas City	798	656	*	447	592	248
Minneapolis	1068	713	447	*	394	695
Pierre	1361	1071	592	394	*	760
Tulsa	772	802	248	695	760	*

36. This exercise refers to Jodi's sales trip as discussed in Exercise 30. (The mileage chart is shown below.) Apply the repetitive nearest-neighbor algorithm to find a Hamilton circuit in the graph and write the circuit as it would be traveled by Jodi starting from her home in Nashville. Give the total mileage for this circuit.

Mileage Chart

	Boston	Dallas	Houston	Louisville	Nashville	Pittsburgh	St. Louis
Boston	*	1748	1804	941	1088	561	1141
Dallas	1748	*	243	819	660	1204	630
Houston	1804	243	*	928	769	1313	779
Louisville	941	819	928	*	168	388	263
Nashville	1088	660	769	168	*	553	299
Pittsburgh	561	1204	1313	388	553	*	588
St. Louis	1141	630	779	263	299	588	*

E. Cheapest-Link Algorithm

37. This exercise refers to the weighted graph discussed in Exercise 31. (The weighted graph is shown below.) Apply the cheapest-link algorithm to find a Hamilton circuit in the graph.

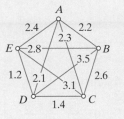

38. This exercise refers to Sophie's sales trip as discussed in Exercise 26. (The weighted graph is shown below.) Apply the cheapest-link algorithm to find a Hamilton circuit in the graph and write the circuit as it would be traveled by Sophie starting from her home town of A. Give the total cost of the trip.

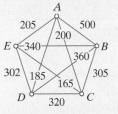

39. This exercise refers to the space expedition discussed in Example 6.3 and Exercise 27. (The weighted graph is shown below.) Apply the cheapest-link algorithm to find a Hamilton circuit in the graph and write the circuit as it would be traveled by an expedition starting from Earth (E). Give the total travel time for the trip.

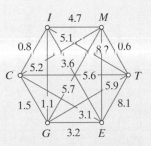

40. This exercise refers to the weighted graph discussed in Exercise 28. (The weighted graph is shown below.) Apply the cheapest-link algorithm to find a Hamilton circuit in the graph. Give the weight of the circuit.

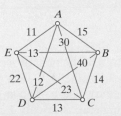

41. This exercise refers to Darren's sales trip as discussed in Exercise 29. (The mileage chart is shown below.) Apply the cheapest-link algorithm to find a Hamilton circuit in the graph and write the circuit as it would be traveled by Darren starting from his home in Atlanta. Give the total mileage for the trip.

Mileage Chart

	Atlanta	Columbus	Kansas City	Minneapolis	Pierre	Tulsa
Atlanta	*	533	798	1068	1361	772
Columbus	533	*	656	713	1071	802
Kansas City	798	656	*	447	592	248
Minneapolis	1068	713	447	*	394	695
Pierre	1361	1071	592	394	*	760
Tulsa	772	802	248	695	760	*

42. This exercise refers to Jodi's sales trip as discussed in Exercise 30. (The mileage chart is shown below.) Apply the cheapest-link algorithm to find a Hamilton circuit in the graph and write the circuit as it would be traveled by Jodi starting from her home in Nashville. Give the total mileage for the trip.

Mileage Chart

	Boston	Dallas	Houston	Louisville	Nashville	Pittsburgh	St. Louis
Boston	*	1748	1804	941	1088	561	1141
Dallas	1748	*	243	819	660	1204	630
Houston	1804	243	*	928	769	1313	779
Louisville	941	819	928	*	168	388	263
Nashville	1088	660	769	168	*	553	299
Pittsburgh	561	1204	1313	388	553	*	588
St. Louis	1141	630	779	263	299	588	*

F. Miscellaneous

43. You have a busy day ahead of you. You must run the following errands (in no particular order): go to the post office, deposit a check at the bank, pick up some French bread at the deli, visit a friend at the hospital, and get a haircut at Karl's Beauty Salon. You must start and end at home. Each block on the following map is exactly 1 mile.

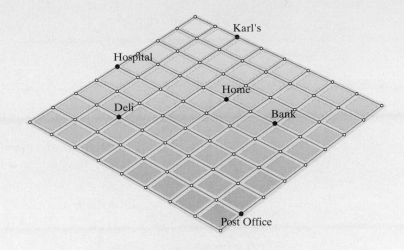

(a) Draw a weighted graph corresponding to this problem.

(b) Find the optimal (shortest) way to run all the errands. (Use any algorithm you think is appropriate.)

44. Rosa's Floral must deliver flowers to each of the five locations A, B, C, D, and E shown on the following map. The trip must start and end at the flower shop, which is located at X. Each block on the map is exactly 1 mile.

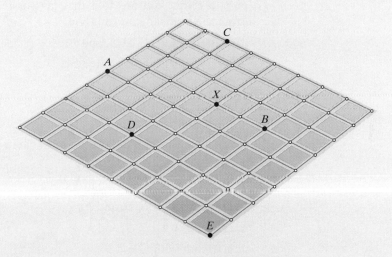

(a) Draw a weighted graph corresponding to this problem.

(b) Find the optimal (shortest) way to make all the deliveries. (Use any algorithm you think is appropriate.)

In Exercises 45 through 48 you are scheduling a dinner party for 6 people (A, B, C, D, E, and F). The guests are to be seated around a circular table, and you want to arrange the seating so that each guest is friends with the two people next to him. You can assume that all friendships are mutual (when X is a friend of Y, Y is also a friend of X).

45. Suppose that you are told that all possible friendships can be deduced from the following information.

A is friends with B and F.

B is friends with A, C, and E.

C is friends with B, D, E, and F.

E is friends with B, C, D, and F.

(a) Draw a "friendship graph" for the dinner guests.

(b) Find a possible seating arrangement for the party.

(c) Is there a possible seating arrangement in which B and E are seated next to each other? If there is, find it. If there isn't, explain why not.

46. Suppose that you are told that all possible friendships can be deduced from the following information.

A is friends with B, C, and D.

B is friends with A, C, and E.

D is friends with A, E, and F.

F is friends with C, D, and E.

(a) Draw a "friendship graph" for the dinner guests.

(b) Find a possible seating arrangement for the party.

(c) Is there a possible seating arrangement in which B and E are seated next to each other? If there is, find it. If there isn't, explain why not.

47. Suppose that you are told that all possible friendships can be deduced from the following information.

A is friends with C, D, E, and F.

B is friends with C, D, and E.

C is friends with A, B, and E.

D is friends with A, B, and E.

Explain why it is impossible to have a seating arrangement in which everybody is friends with the two people seated next to him.

48. Suppose that you are told that all possible friendships can be deduced from the following information.

A is friends with B, D, and F.

C is friends with B, D, and F.

E is friends with C and F.

Explain why it is impossible to have a seating arrangement in which everybody is friends with the two people seated next to him.

JOGGING

49. The following graph is called the *dodecahedral graph*, because it describes the relationship between the vertices and edges of a *dodecahedron*, a regular three-dimensional solid consisting of 12 faces all of which are regular pentagons. Find a Hamilton circuit in the graph. Show your answer by labeling the vertices in the order of travel.

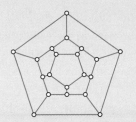

Note: This exercise has some historical importance. In 1857, Hamilton invented a game he called the Icosian game, which was essentially equivalent to this exercise, except that the purpose of the game was to schedule a grand tour that visited 20 European cities represented by the vertices of the graph. (See the historical section preceding the Exercises.)

50. The following graph is called the *icosahedral graph*, because it describes the relationship between the vertices and edges of an *icosahedron*, a regular three-dimensional solid consisting of twenty faces all of which are equilateral triangles. Find a Hamilton circuit in the graph. Show your answer by labeling the vertices in the order of travel.

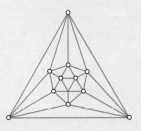

51. Find a Hamilton path in the following graph.

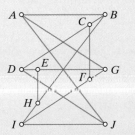

52. A 2-by-2 grid graph. The graph shown below represents a street grid that is 2 blocks by 2 blocks. (Such graph is called a *2-by-2 grid graph*.) For convenience, the vertices are labeled by type: corner vertices C_1, C_2, C_3, and C_4, boundary vertices B_1, B_2, B_3, and B_4, and the interior vertex I.

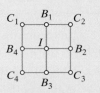

(a) Find a Hamilton path in the graph that starts at I.

(b) Find a Hamilton path in the graph that starts at one of the corner vertices and ends at a different corner vertex.

(c) Find a Hamilton path that starts at one of the corner vertices and ends at I.

(d) Find (if you can) a Hamilton path that starts at one of the corner vertices and ends at one of the boundary vertices. If this is impossible, explain why.

53. Find (if you can) a Hamilton circuit in the 2-by-2 grid graph discussed in Exercise 52. If this is impossible, explain why.

54. A 3-by-3 grid graph. The graph shown below represents a street grid that is 3 blocks by 3 blocks. The graph has four corner vertices (C_1, C_2, C_3, and C_4), eight boundary vertices (B_1 through B_8), and four interior vertices (I_1, I_2, I_3, and I_4).

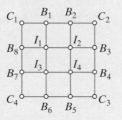

(a) Find a Hamilton circuit in the graph.

(b) Find a Hamilton path in the graph that starts at one of the corner vertices and ends at a different corner vertex.

(c) Find (if you can) a Hamilton path that starts at one of the corner vertices and ends at one of the interior vertices. If this is impossible, explain why.

(d) Given any two adjacent vertices of the graph, explain why there always is a Hamilton path that starts at one and ends at the other one.

55. A 3-by-4 grid graph. The graph that follows represents a street grid that is 3 blocks by 4 blocks.

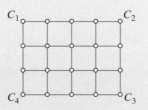

(a) Find a Hamilton circuit in the graph. (Make the circuit by labeling the vertices 1,2,3, ... right on the graph.)

(b) Find a Hamilton path in the graph that starts at C_1 and ends at C_3.

(c) Find (if you can) a Hamilton path in the graph that starts at C_1 and ends at C_2. If this is impossible, explain why.

56. Explain why the cheapest edge in any graph is always part of the Hamilton circuit obtained using the nearest-neighbor algorithm.

57. Give an example of a complete weighted graph with six vertices so that the nearest-neighbor algorithm and the cheapest-link algorithm both give the optimal Hamilton circuit. Choose the weights of the edges to be all different.

58. (a) Give an example of a graph with four vertices in which the same circuit can be both an Euler circuit and a Hamilton circuit.

    (b) Give an example of a graph with N vertices in which the same circuit can be both an Euler circuit and a Hamilton circuit. Explain why there is only one kind of graph for which this is possible.

59. Explain why the number of edges in K_N is $N(N-1)/2$.

60. Explain why 21! is more than 100 billion times bigger than 10! (i.e., show that $21! > 10^{11} \times 10!$).

Exercises 61 and 62 refer to the following situation. Nick is a traveling salesman. His territory consists of the 11 cities shown on the mileage chart below. Nick must organize a round trip that starts and ends in Dallas (that's his home) and visits each of the other 10 cities exactly once.

Mileage Chart

	Atlanta	Boston	Buffalo	Chicago	Columbus	Dallas	Denver	Houston	Kansas City	Louisville	Memphis
Atlanta	*	1037	859	674	533	795	1398	789	798	382	371
Boston	1037	*	446	963	735	1748	1949	1804	1391	941	1293
Buffalo	859	446	*	522	326	1346	1508	1460	966	532	899
Chicago	674	963	522	*	308	917	996	1067	499	292	530
Columbus	533	735	326	308	*	1028	1229	1137	656	209	576
Dallas	795	1748	1346	917	1028	*	781	243	489	819	452
Denver	1398	1949	1508	996	1229	781	*	1019	600	1120	1040
Houston	789	1804	1460	1067	1137	243	1019	*	710	928	561
Kansas City	798	1391	966	499	656	489	600	710	*	520	451
Louisville	382	941	532	292	209	819	1120	928	520	*	367
Memphis	371	1293	899	530	576	452	1040	561	451	367	*

61. Working directly from the mileage chart, implement the nearest-neighbor algorithm to find a Hamilton circuit for Nick's trip.

62. Working directly from the mileage chart, implement the cheapest-link algorithm to find a Hamilton circuit for Nick's trip.

RUNNING

63. **Complete bipartite graphs.** A complete bipartite graph is a graph with the property that the vertices can be divided into two sets A and B and each vertex in set A is adjacent to each of the vertices in set B. There are no other edges! If there are m vertices in set A and n vertices in set B, the

complete bipartite graph is written as $K_{m,n}$. The following figure gives two examples and the general case:

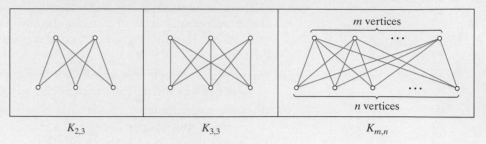

$K_{2,3}$ $K_{3,3}$ $K_{m,n}$

(a) For $n > 1$, the complete bipartite graphs of the form $K_{n,n}$ all have Hamilton circuits. Explain why.

(b) If the difference between m and n is exactly 1 (i.e., $|m - n| = 1$), the complete bipartite graph $K_{m,n}$ has a Hamilton path. Explain why.

(c) When the difference between m and n is more than 1, then the complete bipartite graph $K_{m,n}$ has no Hamilton path. Explain why.

64. *m*-by-*n* grid graphs. An m-by-n grid graph represents a rectangular street grid that is m blocks by n blocks, as indicated in the following figure. (You should try Exercises 52 through 55 before you try this one.)

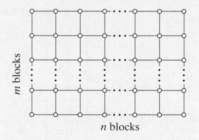

(a) If m and n are both odd, the m-by-n grid graph has a Hamilton circuit. Describe the circuit by drawing it on a generic graph.

(b) If either m or n is even and the other one is odd, then the m-by-n grid graph has a Hamilton circuit. Describe the circuit by drawing it on a generic graph.

(c) If m and n are both even, then the m-by-n grid graph does not have a Hamilton circuit. Explain why a Hamilton circuit is impossible.

65. The Petersen graph. The following graph is called the Petersen graph.

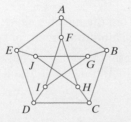

(a) Find a Hamilton path in the Petersen graph.

(b) Explain why the Petersen graph does not have a Hamilton circuit.

66. Make up an example of a complete weighted graph such that the Hamilton circuit produced by the nearest-neighbor algorithm with your choice of starting vertex has a relative percentage error of at least 100% (in other words, the weight of the Hamilton circuit produced by the nearest-neighbor algorithm is at least twice as much as the weight of the optimal Hamilton circuit).

67. Make up an example of a complete weighted graph such that the Hamilton circuit produced by the cheapest-link algorithm has a relative percentage error of at least 100% (in other words, the weight of the Hamilton circuit produced by the cheapest-link algorithm is at least twice as great as the weight of the optimal Hamilton circuit).

68. Make up an example of a complete weighted graph such that the Hamilton circuit produced by the nearest-neighbor algorithm with your choice of starting vertex gives the worst possible choice of a circuit (in other words, one whose weight is bigger than any other).

69. Make up an example of a complete weighted graph such that the Hamilton circuit produced by the cheapest-link algorithm gives the worst possible choice of a circuit.

70. **The knight's tour.** A knight is on the upper left-hand corner of a 3-by-4 "chessboard" as shown in the following figure.

(a) Draw a graph with the vertices representing the squares on the board and the edges representing the allowable chess moves of the knight (e.g., an edge joining vertices A and J means that the knight is allowed to move from square A to square J or vice versa in a single move).

(b) Find a Hamilton path starting at vertex A in the graph drawn in (a) and thus show how to move the knight so that it starts at square A and visits each square of the board exactly once.

(c) Show that the graph drawn in (a) does not have a Hamilton circuit and consequently that it is impossible for the knight to move so that it visits each square of the board exactly once and then returns to its starting point.

71. Using the ideas of Exercise 70, show that it is possible for a knight to visit each square of a 8-by-8 "chessboard" exactly once and return to its starting point, and that this is true regardless of which square is used as the knight's starting point.

PROJECTS AND PAPERS

A. The Great Kaliningrad Circus

The Great Kaliningrad Circus has been signed for an extended 21-city tour of the United States, starting and ending in Miami, Florida. The 21 cities and the travel distances between cities are shown in the mileage chart shown on p. 266. The cost of transporting an entire circus the size of the Great Kaliningrad can be estimated to be about $1000 per mile, so determining the best, or near best, route

Mileage Chart

	Atlanta	Boston	Buffalo	Chicago	Columbus	Dallas	Denver	Houston	Kansas City	Louisville	Memphis	Miami	Minneapolis	Nashville	New York	Omaha	Pierre	Pittsburgh	Raleigh	St. Louis	Tulsa
Atlanta	*	1037	859	674	533	795	1398	789	798	382	371	655	1068	242	841	986	1361	687	372	541	772
Boston	1037	*	446	963	735	1748	1949	1804	1391	941	1293	1504	1368	1088	206	1412	1726	561	685	1141	1537
Buffalo	859	446	*	522	326	1346	1508	1460	966	532	899	1409	927	700	372	971	1285	216	605	716	1112
Chicago	674	963	522	*	308	917	996	1067	499	292	530	1329	405	446	802	459	763	452	784	289	683
Columbus	533	735	326	308	*	1028	1229	1137	656	209	576	1160	713	377	542	750	1071	182	491	406	802
Dallas	795	1748	1346	917	1028	*	781	243	489	819	452	1300	936	660	1552	644	943	1204	1166	630	257
Denver	1398	1949	1508	996	1229	781	*	1019	600	1120	1040	2037	841	1156	1771	537	518	1411	1661	857	681
Houston	789	1804	1460	1067	1137	243	1019	*	710	928	561	1190	1157	769	1608	865	1186	1313	1160	779	478
Kansas City	798	1391	966	499	656	489	600	710	*	520	451	1448	447	556	1198	201	592	838	1061	257	248
Louisville	382	941	532	292	209	819	1120	928	520	*	367	1037	697	168	748	687	1055	388	541	263	659
Memphis	371	1293	899	530	576	452	1040	561	451	367	*	997	826	208	1100	652	1043	752	728	285	401
Miami	655	1504	1409	1329	1160	1300	2037	1190	1448	1037	997	*	1723	897	1308	1641	2016	1200	819	1196	1398
Minneapolis	1068	1368	927	405	713	936	841	1157	447	697	826	1723	*	826	1207	357	394	857	1189	552	695
Nashville	242	1088	700	446	377	660	1156	769	556	168	208	897	826	*	892	744	1119	553	521	299	609
New York	841	206	372	802	542	1552	1771	1608	1198	748	1100	1308	1207	892	*	1251	1565	368	489	948	1344
Omaha	986	1412	971	459	750	644	537	865	201	687	652	1641	357	744	1251	*	391	895	1214	449	387
Pierre	1361	1726	1285	763	1071	943	518	1186	592	1055	1043	2016	394	1119	1565	391	*	1215	1547	824	760
Pittsburgh	687	561	216	452	182	1204	1411	1313	838	388	752	1200	857	553	368	895	1215	*	445	588	984
Raleigh	372	685	605	784	491	1166	1661	1160	1061	541	728	819	1189	521	489	1214	1547	445	*	804	1129
St. Louis	541	1141	716	289	406	630	857	779	257	263	285	1196	552	299	948	449	824	588	804	*	396
Tulsa	772	1537	1112	683	802	257	681	478	248	659	401	1398	695	609	1344	387	760	984	1129	396	*

for the circus tour is clearly a top priority for the tour manager, which, by the way, is you!

In this project, your job is to find the best route you can for the circus tour. (You are not asked to find the optimal route, just to find the best route you can!) You should prepare a presentation to the board of directors of the circus explaining what strategies you used to come up with your route, and why you think your route is a good one.

Note: The tools at your disposal to carry out this project are everything you learned in this chapter, your ingenuity, and your time (a wall-sized map of the United States and some pins may come in handy too!). This should be a low-tech project—you should not go to the Web and try to find some software that will do this job for you!

B. The Nearest-Insertion Algorithm

The *nearest-insertion algorithm* is another approximate algorithm used for tackling TSPs. The basic idea of the algorithm is to start with a subcircuit (a circuit that includes some, but not all, of the vertices) and enlarge it, one step at a time, by adding an extra vertex—the one that is closest to some vertex in the circuit. By the time we have added all of the vertices, we have a full-fledged Hamilton circuit.

In this project, you should prepare a class presentation on the nearest-insertion algorithm. Your presentation should include a detailed description of the algorithm, at least two carefully worked out examples, and a comparison of the nearest-insertion and the nearest-neighbor algorithms.

Note: Additional information on the nearest-insertion algorithm is available at the book's Web site: *http://cwx.prenhall.com/bookbind/pubbooks/tannenbaum/ chapter6.*

C. Molecular Computing: Using DNA Molecules to Solve Math Problems

DNA is the basic molecule of life—it encodes the genetic information that characterizes all living organisms. Due to the great advances in biochemistry of the last 20 years, scientists can now snip, splice, and recombine segments of DNA almost at will. In 1994, Leonard Adleman, a professor of computer science at the University of Southern California, was able to encode a graph representing seven cities into a set of DNA segments and to use the chemical reactions of the DNA fragments to uncover the existence of a Hamilton path in the graph. Basically, he was able to use the biochemistry of DNA to solve a graph theory problem. While the actual problem solved was insignificant, the idea was revolutionary, as it opened the door for the possibility of someday using DNA computers to solve problems beyond the reach of even the most powerful of today's electronic computers.

Write a research paper telling the story of Adleman's landmark discovery. How did he encode the graph into DNA? How did he extract the mathematical solution (Hamilton path) from the chemical solution? What other kinds of problems might be solved using DNA computing? What are the implications of Adelman's discovery for the future of computing?

Notes: The following suggested readings (in this order) are an excellent start:

- Kolata, Gina, "A Novel Computing Idea: Calculating with DNA," *New York Times*, November 22, 1994.
- Bass, Thomas A., "Gene Genie," *Wired*, August 1995. Available in electronic form at *www.wired.com/wired/archive/3.08/molecular/pr.html*.
- Kolata, Gina, "A Vat of DNA May Become Fast Computer of the Future," *New York Times*, April 11, 1995.
- Adleman, Leonard, "Computing with DNA," *Scientific American*, August 1998.

D. Ant Colony Optimization Methods

An individual ant is, by most standards, a dumb little creature, but collectively, an entire ant colony can perform surprising feats of teamwork (such as lifting and carrying large leaves or branches) and self-organize to solve remarkably complex problems (finding the shortest route to a food source, optimizing foraging strategies, managing a smooth and steady traffic flow in congested ant highways). The ability of ants and other social insects to perform sophisticated group tasks goes by the name of *swarm intelligence*. Since ants don't talk to each other and don't have bosses telling them what to do, swarm intelligence is a decentralized, spontaneous type of intelligence that has many potential applications at the human scale. In recent years, computer scientists have been able to approach many difficult optimization problems (including TSPs) using *ant colony optimization* software (i.e., computer programs that use *virtual ants* to imitate the problem-solving strategies of real ants).

Write a research paper describing the concept of swarm intelligence and some of the recent developments in ant colony optimization methods and, in particular, the use of virtual ant algorithms for solving TSPs.

Notes:
- Two good sources for your paper are the article *Swarm Smarts*, by Eric Bonabeau and Guy Théraulaz, in the March 2000 issue of *Scientific American*, and the book by the same authors, *Swarm Intelligence: From Natural to Artificial Systems*, Oxford University Press, 1999.
- At the time of the writing of this edition, an Ant Colony Optimization Homepage was maintained by Marco Dorigo, of the Université Libre de Bruxelles, Belgium at *http://iridia.ulb.ac.be/ ~mdorigo/ACO/ACO.html*.

REFERENCES AND FURTHER READINGS

1. Adleman, Leonard, "Computing with DNA," *Scientific American*, 279 (August 1998), 54–61.
2. Bell, E. T., "An Irish Tragedy: Hamilton," in *Men of Mathematics*. New York: Simon and Schuster, 1986, chap. 19.
3. Bellman, R., K. L. Cooke, and J. A. Lockett, *Algorithms, Graphs and Computers*. New York: Academic Press, Inc., 1970, chap. 8.
4. Berge, C., *The Theory of Graphs and Its Applications*, New York: Wiley, 1962.

5. Bonabeau, Eric, and Théraulaz, G., "Swarm Smarts," *Scientific American*, 282 (March 2000), 73–79.

6. Chartrand, Gary, *Graphs as Mathematical Models*. Belmont, CA: Wadsworth Publishing Co., Inc., 1977, chap. 3.

7. Devlin, Keith, *Mathematics: The New Golden Age*. London: Penguin Books, 1988, chap. 11.

8. Kolata, Gina, "Analysis of Algorithms: Coping With Hard Problems," *Science*, 186 (November 1974), 520–521.

9. Lawler, E. L., J. K. Lenstra, A. H. G. Rinooy Kan, and D. B. Shmoys, *The Traveling Salesman Problem*. New York: John Wiley & Sons, Inc., 1985.

10. Lewis, H. R., and C. H. Papadimitriou, "The Efficiency of Algorithms," *Scientific American*, 238 (January 1978), 96–109.

11. Michalewicz, Z., and D. B. Fogel, *How to Solve It: Modern Heuristics*. New York: Springer-Verlag, 2000.

12. Papadimitriou, Christos H., and Kenneth Steiglitz, *Combinatorial Optimization: Algorithms and Complexity*. New York: Dover Pubns., 1998, chap. 17.

13. Peterson, Ivars, *Islands of Truth: A Mathematical Mystery Cruise*. New York: W. H. Freeman & Co., 1990, chap. 6.

14. Wilson, Robin, and John J. Watkins, *Graphs: An Introductory Approach*. New York: John Wiley & Sons, Inc., 1990.

CHAPTER 7

The Mathematics of Networks

It's All About Being Connected

As any sober person knows, the shortest distance between two points is a straight line. What, then, is the shortest distance between three points? four points? *N* points?

In 1989, a consortium of telephone companies completed the construction of a new transpacific fiber-optic trunk line called TPC-3 linking Japan, Guam, and Oahu, Hawaii. Since both the fiber-optic cable itself and the laying of it along the ocean floor are extremely expensive propositions (somewhere between $50,000 and $100,000 per mile), it was important for the telephone companies to find a way to link the three islands using the least amount of cable. This meant, of course, finding the shortest network connecting the three points. The solution is shown in the stamp on the opposite page, which was issued by the Japanese post office in commemoration of the event. We will find out later in the chapter exactly what makes this network the shortest network and how such networks can be found.

The general theme of this chapter is the problem of finding *efficient networks* connecting a set of points. In addition to telephone and Internet connections, this problem has applications in building transportation networks (roads, high-speed rail systems, canals), the construction of pipelines, and the design of computer chips.

The common thread in all these problems is twofold: (1) the need to connect all the points so that one can go from any point to any other point (when this is accomplished, we have a *network*) and (2) the desire to make the total cost of the network as small as possible. For obvious reasons, problems of this type are known as **minimum network problems**, and in this chapter we will discuss two basic variations on this theme. Once again, one of the most important tools we will use is graph theory.

cŏn · nec′ tion

(1) that which connects or unites; a tie; a bond; means of joining.

(2) a line of communication from one point to another.

Webster's New 20th Century Dictionary

7.1 Trees

EXAMPLE 7.1 An Amazonian Telephone Network

The *Amazonia Telephone Company* (ATC) is the main provider of telephone services to some of the world's most inaccessible places—small towns and villages buried deep within the Amazonian jungle. Consider, for example, the following problem. The seven villages shown in Fig. 7-1(a) are scheduled to be connected into a small regional network using state-of-the-art fiber-optic underground cable. The figure also shows the already existing network of roads connecting the seven villages. The weighted graph in Fig. 7-1(b) is a graph model describing the situation. The edges of the graph represent the existing roads and the weights of the edges represent the cost (in millions of dollars) of laying down the fiber-optic cable along each road. (Laying down cable anywhere other than along an existing road would be prohibitively expensive in the jungle.) The network that the company wants to build should link all the cities for the least money. What is the *cheapest* network?

FIGURE 7-1
(a) The seven villages in the Amazon and the roads connecting them. (b) A graph model of the possible connections. The weights of the edges are the costs (in millions) of laying fiber-optic lines.

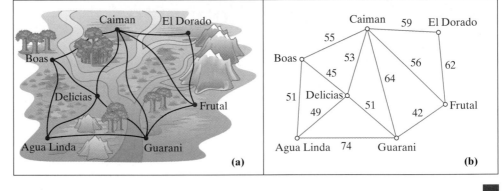

Let's start by asking a general question: What would a minimal network like the one to be built in Example 7.1 be like?

First, we observe that, as a graph, the network will have to be a **subgraph** of the original graph. In other words, building the network will require choosing some of the edges in Fig. 7-1(b), but certainly not all of them. In addition, the subgraph should include each and every vertex of the original graph; this guarantees that no village is left out of the telephone network! In the language of graph theory, a subgraph that includes *every one of the vertices of an original graph* is called a **spanning**[1] **subgraph**. To summarize, in technical terms, the network we are looking for is a *spanning* (includes all of the vertices) *subgraph* (uses only some of the edges) of the graph in Fig. 7-1(b).

But that in itself is not enough. The graph should also have the following characteristics.

- It should be *connected*. This is an obvious requirement in a network where the object is to be able to reach (place a call) from any vertex (village) to any other vertex (village).

- It should *not contain any circuits*. This reflects the requirement that this network be built in the cheapest possible way. Whenever there is a circuit, there is guaranteed to be a *redundant* connection—take away the redundant connection and the network would still work. Figure 7-2 is a closeup of a

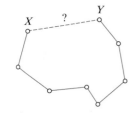

FIGURE 7-2
Is the link *XY* really necessary? Not when saving money is the object.

---

[1] **span:** to extend, reach, or stretch across. *Webster's New 20th Century Dictionary.*

hypothetical circuit. Delete any one of the edges—say, *XY*—and the telephone calls would still go through.

These last two characteristics define a very important category of graphs called trees. A **tree** is a graph that is connected and has no circuits. Because of their importance, trees have been studied extensively, and we will take a little time to get acquainted with them ourselves.

EXAMPLE 7.2

Figure 7-3 shows examples of four graphs, all of which are trees, since in each case the graph is connected and without circuits. The graph in Fig. 7-3(a) looks like a tree, but that is not a requirement. The graph in Fig. 7-3(d) represents a polyethylene molecule.

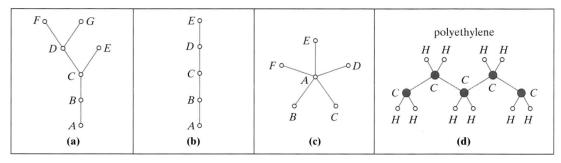

FIGURE 7-3

EXAMPLE 7.3

Figure 7-4 shows examples of graphs that are *not* trees. Figure 7-4(a) has one circuit, and Fig. 7-4(b) has several circuits. In either case, appearance notwithstanding, neither one is a tree. Figure 7-4(c) has no circuits, but is not connected, Fig. 7-4(d) fails to be a tree on both accounts—it has circuits, and it is not connected.

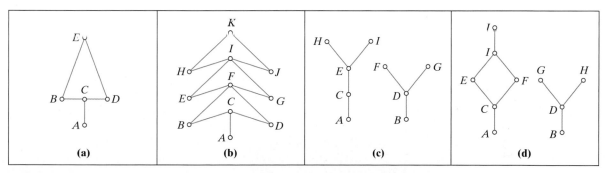

FIGURE 7-4

Properties of Trees

Trees are a very special type of graph and, as such, have some unique properties, some of which will come in quite handy in our quest for cost-efficient networks.

Let's start with the observation that in a connected graph there is always a path joining any one vertex to any other vertex, and if there are *two or more* paths joining some pair of vertices *X* and *Y* then the graph is definitely not a tree,

FIGURE 7-5
Two different paths joining X and Y make a circuit.

because the two paths joining X and Y make a circuit (Fig. 7-5). This leads to the first important property of trees.

PROPERTY 1

If a graph is a tree, there is one and only one path joining any two vertices. Conversely, if there is one and only one path joining any two vertices of a graph, the graph must be a tree.

One practical consequence of Property 1 is that a tree is connected in a very precarious way. The removal of any edge of a tree will disconnect it (see Exercise 43). We can restate this by saying that in a tree, every edge is a *bridge*.

PROPERTY 2

In a tree, every edge is a bridge. Conversely, if every edge of a connected graph is a bridge, then the graph must be a tree.

Perhaps the most important property of trees is numerical, and it relates the number of vertices to the number of edges: *The total number of edges is always one less than the number of vertices.*

PROPERTY 3

A tree with N vertices must have $N-1$ edges.

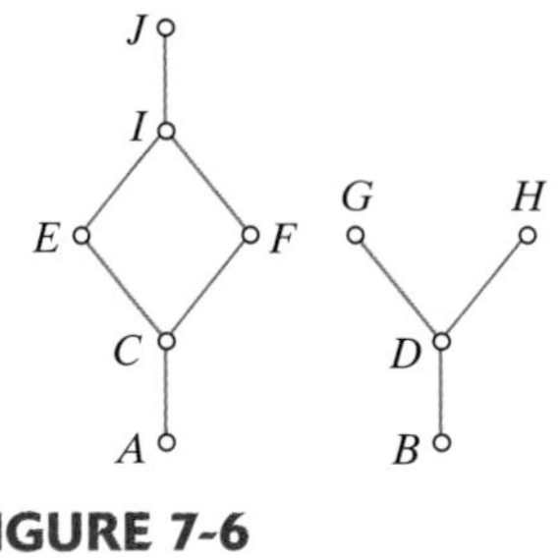

It would be nice to be able to turn Property 3 around and say that if a graph has N vertices and $N - 1$ edges, then it must be a tree. As the graph in Fig. 7-6 shows, however, this need not be the case—it has 10 vertices and 9 edges, and yet it is not a tree. What's the problem? As you may have guessed, the problem is that the graph is not connected. Fortunately, for connected graphs the converse of Property 3 is true (see Fig. 7-7).

FIGURE 7-6
The graph has 10 vertices and 9 edges, but is not a tree.

PROPERTY 4

A *connected* graph with N vertices and $N-1$ edges must be a tree.

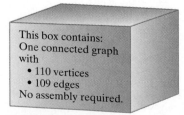

This box contains:
One connected graph with
• 110 vertices
• 109 edges
No assembly required.

FIGURE 7-7
The graph has 110 vertices and 109 edges. Since it is also connected, it is guaranteed to be a tree.

EXAMPLE 7.4

This example illustrates some of the ideas discussed so far. Let's say that we have five vertices with which to build a graph and that we start putting edges on these vertices. At first, with one, two, or three edges (Fig. 7-8), we just don't have enough edges to make the graph connected. When we get to four edges, we can, for the first time, make the graph connected. If we do so, we have a tree (Fig. 7-9). As we add even more edges, the connected graph starts picking up circuits (Fig. 7-10).

FIGURE 7-8
Graphs with five vertices and less than four edges are disconnected.

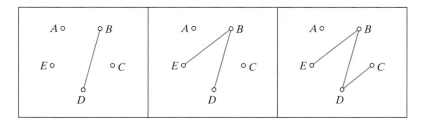

FIGURE 7-9
Graphs with five vertices and four edges—just enough to connect.

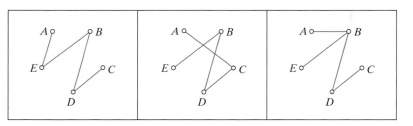

FIGURE 7-10
Graphs with five vertices and more than four edges: circuits begin to form.

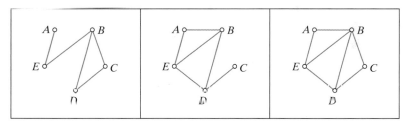

We can generalize our observations in Example 7.4 by imagining the following "connect-the-dots" game. We start with a fixed set of N vertices and without any edges. We are then allowed to start connecting the vertices by putting in edges, one at a time without duplicating edges. We will use M to denote the number of edges we have added at any point in time. The chronology of what happens goes like this (try to imagine it in your head or follow the discussion with pencil and paper): At first ($M = 1, 2, \ldots$) our graph is by necessity going to be disconnected. When we get to $M = N - 1$, we are finally able (if we were careful to avoid circuits) to make the graph connected. At this point, we have a tree—every edge is a bridge so that the graph is connected but very precariously (don't sneeze or breathe hard). From this point on, as the number of edges M increases the graph becomes more strongly connected, with an increasing number of circuits and thus, with fewer and fewer bridges (see Exercise 61). At this point, it will take more than one good sneeze to break the graph apart!

7.2 Minimum Spanning Trees

Suppose *G* is a connected graph. We can always find within *G* a tree *spanning* the vertices of *G*—such a tree is called a **spanning tree** of *G*. If *G* has *N* vertices, then any spanning tree of *G* will have the same *N* vertices (that's the *spanning* part) and exactly $N - 1$ of the edges of *G* (that's the *tree* part).

As the next example illustrates, a connected graph usually has lots of spanning trees.

EXAMPLE 7.5

The connected graph in Fig. 7-11(a) has 9 vertices and 10 edges. The graph has separate circuits *A, B, C, A* of length 3, and *D, E, F, G, H, I, D* of length 6. To find a spanning tree of the graph we need to "remove" two of the edges, but not just any two. Since a spanning tree should have no circuits, we need to remove one edge from each circuit. Figure 7-11(b) shows a spanning tree, obtained by removing the edge *AB* from the first circuit and the edge *FG* from the second. This is just one of $3 \times 6 = 18$ different possible spanning trees for this graph.

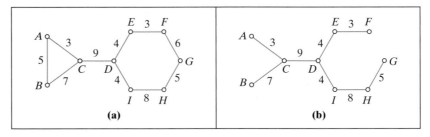

FIGURE 7-11
Busting each of the circuits in (a) gives a spanning tree (b).

Among the many possible spanning trees of a weighted graph, we are now interested in finding one with the least total weight. We call such a spanning tree a **minimum spanning tree** (or MST for short) of the graph. Given our discussion in Example 7.5, it is clear that to get the minimum spanning tree of the graph in Fig. 7-11(a) we should remove edge *BC* from the first circuit and edge *IH* from the second. Why? Because they are the most "expensive" edges in their respective circuits.

7.3 Kruskal's Algorithm

We are now ready to discuss an algorithm that finds the minimum spanning tree in any connected weighted graph. This algorithm is named after Joseph Kruskal, a mathematician at Bell Labs who proposed it in 1956.[2]

Kruskal's algorithm is a simple variation of the *cheapest-link algorithm* we discussed in Chapter 6. Much like the cheapest-link algorithm, Kruskal's algorithm works by building the solution, one edge at a time. At each step we choose the "cheapest" edge available at the time that does not close any circuits (we are building a spanning tree, so obviously we don't want any circuits!). Remarkably, this greedy and seemingly short-sighted strategy always works. Kruskal's

---

[2]Kruskal was not the first to discover this algorithm, but in one of those twists of fortune that sometimes happen with scientific discoveries, he is the one whose name is attached to the discovery.

algorithm is an *optimal* algorithm that produces, without fail, a minimum spanning tree for the graph.

EXAMPLE 7.6 The Amazonian Telephone Network: Part II

Let's use Kruskal's algorithm to find a minimum spanning tree (MST) for the graph shown in Fig. 7-12, which shows the costs (in millions of dollars) of connecting the various Amazon villages with telephone lines. The MST that we find will represent the cheapest possible network connecting the seven cities—exactly what we set out to do in Example 7.1.

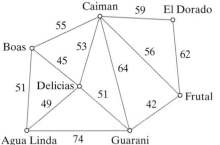

FIGURE 7-12
The graph for the Amazonian telephone-network problem.

- **Step 1.** Of all the possible connections between two villages, the cheapest one is Guarani–Frutal, at a cost of $42 million. We choose this connection first, and for the record, mark it in red.[3]

- **Step 2.** The next cheapest connection is Boas–Delicias, at a cost of $45 million. We also mark it in red, as it is also going to be part of the network.

- **Step 3.** The next cheapest connection is Agua Linda–Delicias, costing $49 million. Again, mark it in red.

- **Step 4.** The next cheapest connection is a tie between Agua Linda–Boas and Delicias–Guarani, both at $51 million. Agua Linda–Boas, however, is now a *redundant* connection, and we do not want to use it. (For bookkeeping purposes, the best thing to do is erase it.) Delicias–Guarani, on the other hand, is just fine, so we mark that connection in red.

- **Step 5.** The next cheapest connection is Caiman–Delicias, at $53 million. There are no problems here, so again, we mark it in red.

- **Step 6.** The next cheapest connection is Boas–Caiman, at $55 million, but this is a redundant connection, so we discard it. The next possible choice is Caiman–Frutal at $56 million, but this is also a redundant connection because calls between Caiman and Frutal are already possible in our budding network. The next possible choice is Caiman–El Dorado at $59 million, and this is OK, so we mark the connection Caiman–El Dorado in red.

- **Step** ... Wait a second—we are finished! We can tell we are done by just looking at the red network we have built and verifying that it is a spanning tree. Or, better yet, we can recognize that six edges—and therefore six steps—is exactly what it takes to build a tree with seven vertices.

[3]Note that these will not necessarily be the first two towns actually connected. We are putting the network together on paper, and the rules require that we follow a certain sequence, but in practice we can build the connections in any order we want.

The total cost of the telephone network we have come up with (Fig. 7–13) is $299 million, and this is, in fact, the optimal solution to the problem. There is no cheaper spanning tree!

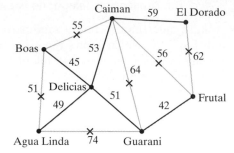

FIGURE 7-13
The MST for the Amazonian telephone-network problem (shown in red). The network has a four-way junction at Delicias.

We call the reader's attention to one other fact that will become relevant later in the chapter. The network we have built has one main junction—a four-way junction at Delicias. (In telephone networks, junction points are important because these are places where switching equipment has to be installed.)

Here is now a formal description of Kruskal's algorithm.

KRUSKAL'S ALGORITHM

- Find the "cheapest" edge in the graph. (If there is more than one, choose one at random.) Mark it in red (or any other color).

- Find the next cheapest edge in the graph. (If there is more than one, choose one at random.) Mark it in red.

- Find the next cheapest unmarked edge in the graph that does not create a red circuit. (If there is more than one, choose one at random.) Mark it in red.

- Repeat the previous step until the red edges span every vertex of the graph. The red edges form a minimum spanning tree of the graph.

The truly remarkable thing about Kruskal's algorithm is the fact that it is an *optimal algorithm*: The spanning tree that we get is guaranteed to be the cheapest possible one. In light of our experience in Chapter 6, this is a bit of a surprise. How can something so simple-minded give such great results? Kruskal's algorithm is also an *efficient algorithm*. As we increase the number of vertices and edges in the graph, the amount of work grows more or less proportionally. For the right reward (say, an "A" in your math course), it would not be unreasonable to implement Kruskal's algorithm on a graph with a couple of hundred vertices.

In short, the problem of finding minimum spanning trees represents one of those rare situations where everything falls into place. We have an important real-life problem that can be solved by means of an algorithm that is easy to understand and to carry out and that is *optimal* and *efficient*. We couldn't ask for better karma.

7.4 The Shortest Distance Between Three Points

Minimum spanning trees give us optimal networks connecting a set of locations in the case in which the connections have to be along prescribed routes. Remember, for example, that in the Amazonian telephone-network problem, the placement of the underground fiber-optic lines was restricted to be along already existing roads. But what if, in a manner of speaking, we don't have to *follow the road*? What if, we are free to design the connections any way we see fit? To clarify the distinction, let's look at a new type of telephone-network problem.

EXAMPLE 7.7 An Australian Telephone-Network Problem

This is a connection story involving three small fictional towns (Alcie Springs, Booker Creek, and Camoorea) located smack in the middle of the Australian outback, and which by sheer coincidence happen to form an equilateral triangle 500 miles on each side (Fig. 7–14). The problem, once again, is to lay telephone cable linking the three towns into a network. What is the shortest possible way to connect these towns?

While this example looks like a small-scale version of the Amazonian telephone-network problem, it is not. What makes this situation different is the nature of the terrain. The Australian outback is mostly a flat expanse of desert, and, in contrast to the Amazon situation, there is little or no advantage to laying the telephone lines along roads. (In fact, let's assume that there are no roads to speak of connecting these three towns.) Because of the flat and homogeneous nature of the terrain, we can lay the telephone cable anywhere we want, and the cost per mile is always the same.

The question now is, of all possible networks that connect the three towns, which one is the shortest? Let's start where we left off. What is a minimum spanning tree in this case? Since all three sides of the triangle are the same length, we can pick any two of them to form a minimum spanning tree, as shown in Fig. 7–15(a). The total length of the MST is 1000 miles.

It is not hard to see that the MST is not the shortest possible network connecting the three towns. Look at Fig. 7–15(b). It shows a network that is definitely shorter than 1000 miles. It has a "T"-junction at a new point we call *J*. A little high school geometry and a calculator are sufficient to verify that it is approximately 433 miles from Alcie Springs to the junction *J* (see Exercise 41), and that the network, therefore, is about 933 miles long.

Can we do even better? Why not? With a little extra thought and effort we might come up with the network shown in Fig. 7–15(c). Here, there is a

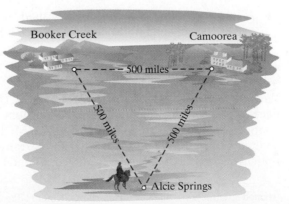

FIGURE 7-14

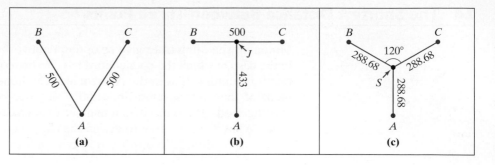

FIGURE 7-15
(a) A minimum spanning tree. (b) A shorter network with a T-junction at J. (c) The shortest network with a Y-junction at S. The three branches of the "Y" meet at equal angles.

"Y"-junction at a new point called S located at the center of the triangle. This network is approximately 866 miles long (see Exercise 42). A key feature of this network is the way the three branches come together at the junction point S, forming *equal 120° angles*. Even without a formal mathematical argument, it is not hard to be convinced that the network shown in Fig. 7–15(c) is indeed the *shortest network* connecting the three towns. After all, mathematics should not choose sides, and, because of its three-sided symmetry this is the one network that looks the same to all three towns.

Let's rephrase what we have in Fig. 7–15 in terms of the position of the *junction points* in each network. In Fig. 7–15(a), the junction point of the network is one of the original cities—in this case, the network is a *minimum spanning tree* connecting the cities. In Fig. 7–15(b) we have the junction point J, which is a *new* point (not one of the original cities). By introducing a new junction point we were able to shorten the length of the MST in Fig. 7–15(a). We can shorten the length of the network even further by placing the new junction point in a more strategic position, shown in Fig. 7–15(c). This last network is the shortest network connecting the three cities—no further improvements are possible!

In going from the MST in Fig. 7–15(a) to the shortest network shown in Fig. 7–15(c), we achieved a savings of 13.4% (the MST measures 1000 miles and the shortest network measures approximately 866 miles). This is a significant amount of savings for such a little bit of work! ■

Before we go on to our next example, let's introduce a bit of terminology. Let's assume we have a set of points (cities) that we want to connect in some kind of network.

- The network of minimum length connecting the points is called, not surprisingly, the **shortest network** for that set of points.

- A junction point in the network that is not one of the original points is called an **interior** junction point.

- A junction point in the network formed by three line segments coming together forming equal 120° angles is called a **Steiner**[4] **point** of the network.

EXAMPLE 7.8 The TPC-3 Connection

We will return to the story introduced in the chapter opener. Let's review the background for this story. In 1989, a consortium of several of the world's biggest telephone companies (among them AT&T, MCI, Sprint, and British Telephone) completed a major undertaking: the third Trans-Pacific Cable (TPC-3), a fiber-optic trunk line linking Japan to the continental United States (via Hawaii) and

[4]Named after the Swiss mathematician Jakob Steiner (1796–1863).

to Guam. For obvious reasons, the primary consideration in designing the trunk line was its cost. By and large, laying cable along the ocean floor has a fixed cost per nautical mile, so that, unlike the Amazonian telephone network problem, here we are after the shortest network. The approximate straight-line distances (in miles) between the three endpoints of TPC-3 (Chikura, Japan; Tanguisson Point, Guam; and Oahu, Hawaii) are shown in Fig. 7-16.

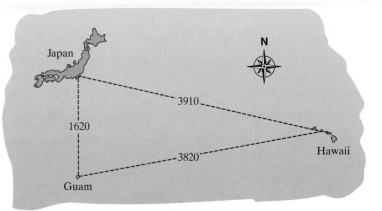

FIGURE 7-16

The distance (in miles) between the three vertices of the triangle Japan–Guam–Hawaii.

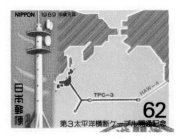

By now we have a pretty good idea that in order to find the shortest network, we are going to have to put an interior junction point inside of the Japan–Guam–Hawaii triangle. (If we don't, the best we can do is the MST, which in this example has a length of 1620 + 3820 = 5440 miles.) What might the best choice for an interior junction point be? Remarkably, it is a Steiner point. Choosing such a junction results in the shortest network, shown in Fig. 7-17 (as well as in the Japanese stamp issued to commemorate the completion of the trunk line). (For the details of *why* this is the shortest network, the reader is referred to Project B.) The total length of the network is 5180 miles.[5]
                                                                                     ∎

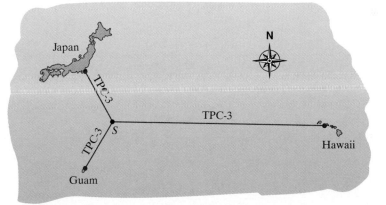

FIGURE 7-17

TPC-3: The shortest network linking Japan–Guam–Hawaii. The junction point *S* is a Steiner point. Total length of the network is 5180 miles.

From Examples 7.7 and 7.8 we are tempted to conclude that the key to finding the *shortest network* connecting three points (cities) *A*, *B*, and *C* is to locate an *interior junction* point *S* inside triangle *ABC* with the properties of a *Steiner point* (i.e., the three line segments *AS*, *BS*, and *CS* come together at *S*, forming equal 120° angles). This is true, but with one major caveat: The angles of the triangle

---

[5]The theoretical length of the network is not the same as the total amount of cable used. With cable running on the ocean floor, one has to add as much as 10% to the straight-line distance, because of the contoured nature of the ocean floor. The actual length of cable used in TPC-3 is about 5690 miles.

have to be all less than 120°. In fact, if one of the angles of the triangle is 120° or more, then talking about interior Steiner junction points is pointless—in this case there cannot possibly be such a junction point inside the triangle (see Exercise 55).

On the other hand, if all three angles of the triangle are less than 120°, Steiner points inside the triangle are indeed the secret to success. To follow up on this story, we need to go back almost 400 years.

Torricelli's Method for Finding a Steiner Point

In the early 1600s, the Italian Evangelista Torricelli tackled the problem of finding the shortest way to connect the vertices of a triangle (for more on Torricelli and the background to this problem, see the biographical profile on p. 292). Torricelli only considered triangles in which all three angles were less than 120°. For such triangles he was able to solve the problem completely by giving a remarkably simple and elegant method for locating the Steiner[6] junction point inside the triangle.

TORRICELLI'S CONSTRUCTION

Consider a triangle with vertices A, B, and C and all three angles less than 120° [Fig. 7–18(a)].

- **Step 1.** Choose any of the three sides of the triangle (say BC) and construct an equilateral triangle BCX, so that X and A are on opposite sides of BC [Fig. 7–18(b)].
- **Step 2.** Circumscribe a circle around equilateral triangle BCX [Fig. 7–18(c)].
- **Step 3.** Joint X to A with a straight line [Fig. 7–18(d)]. The point of intersection of the line segment XA with the circle is the desired Steiner point!

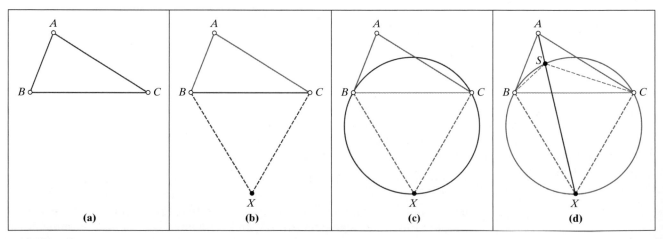

FIGURE 7-18
Finding the Steiner point S. (a) Triangle ABC. (b) Find point X opposite A such that BXC is an equilateral triangle. (c) Circumscribe triangle BXC in a circle. (d) Join X and A. The intersection of the circle and $\overline{AX}$ is S.

---

[6]Torricelli called such a point the *isogonic center* of the triangle. The point was later refered to as the *Torricelli point* of a triangle. The term *Steiner point* only came into use in the 1940s.

Torricelli's construction is a classic geometric construction that can be carried out with just a straightedge and a compass.

There are many other ways to find the Steiner point inside of a triangle besides Torricelli's method. The beauty of Torricelli's approach is that we can justify why it produces the desired Steiner point using facts from elementary geometry. Here is a brief summary of the arguments.

- Angles *BCX* and *CBX* both equal 60°. *Reason*: The triangle *BCX* is an equilateral triangle.
- Angle *BSX* equals 60°. *Reason*: *S* and *C* are both on the same arc of the circle opposite chord *BX*. It follows that angles *BSX* and *BCX* have equal measure.
- Angle *CSX* equals 60°. *Reason*: *S* and *B* are both on the same arc of the circle opposite chord *CX*. It follows that angles *CSX* and *CBX* have equal measure.

Combining all of the above we can conclude that angles *BSC*, *BSA*, and *CSA* are all 120°, and therefore *S* is the desired Steiner point. An additional bonus in Torricelli's construction is the remarkable fact that the length of the shortest network is given by the length of the line segment *XA* (see Project B).

Before we conclude our discussion on finding the shortest network connecting three points, we need to look at one more example.

EXAMPLE 7.9

Suppose that we want to build the tracks for a bullet-train network linking the cities of Los Angeles, Las Vegas, and Salt Lake City. The straight-line distances between the three cities are shown in Fig. 7-19. Once again, to minimize construction costs, we want to construct the shortest network connecting the three cities.

FIGURE 7-19

The only substantive difference between this example and Example 7.8 is that here the triangle has an angle greater than 120°, and therefore, there is no Steiner junction point inside the triangle. Without a Steiner junction point, how do we find the shortest network? The answer turns out to be surprisingly simple: *The shortest network is the minimum spanning tree*. In a triangle, this simply means that we should pick the two shortest sides. For this example, the solution is shown in Fig. 7-20.

FIGURE 7-20
The shortest network connecting the three cities is the MST.

In the opening paragraph of this chapter we raised the question, "What is the shortest distance between three points?" We finally have an answer, which comes in two parts.

THE SHORTEST DISTANCE BETWEEN THREE POINTS *A*, *B*, AND *C*

---

1. When one of the angles of the triangle *ABC* is 120° or more, the shortest network linking the three points consists of the two shortest sides of the triangle [Fig. 7-21(a)]. In this situation, *the shortest network coincides with the minimum spanning tree.*

2. When all the angles of the triangle are less than 120°, then the shortest network is obtained by finding a *Steiner point S* inside the triangle and joining *S* to each of the vertices *A*, *B*, and *C* [Fig. 7-21(b)]. The exact location of *S* can be found using Torricelli's construction.

FIGURE 7-21
The shortest network connecting *A*, *B*, and *C*. (a) The angle at *A* is greater than or equal to 120°. (b) All angles are less than 120°.

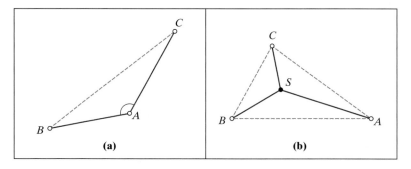

(a)   **(b)**

7.5 The Shortest Network Linking More Than Three Points

When it comes to finding shortest networks, things get really interesting when we wish to link more than three points.

EXAMPLE 7.10

Four cities (*A*, *B*, *C*, and *D*) are to be connected into a telephone network. For starters, let's imagine that the cities form the vertices of a square that is 500 miles on each side, as shown in Fig. 7-22(a). What does an optimal network connecting these cities look like? It depends on the situation.

If we *don't want to introduce any new junction points in the network* (either because we don't want to venture off the prescribed paths—as in the jungle scenario—or because the cost of creating a new junction is too high), then the answer is a *minimum spanning tree*, such as the one shown in Fig. 7-22(b). The length of the MST is 1500 miles.

On the other hand, if *interior junction points* are allowed, somewhat shorter networks are possible. One obvious improvement is the network shown in Fig. 7-22(c), having a four-way junction at *O*, the center of the square. The length of this network is approximately 1414 miles (see Exercise 57).

An even shorter network is possible by using two Steiner junction points S_1 and S_2 [Fig. 7-22(d)]. Using elementary geometry (see Exercise 58), we can find that the length of this network is approximately 1366 miles. This is it! There is no shorter network, although there is another possible network equivalent to this one [Fig. 7-22(e)].

The difference between the 1500 miles in the MST and the 1366 miles in the shortest network is 134 miles, which represents a savings of about 9% (134/1500).

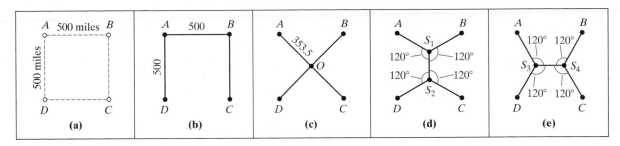

FIGURE 7-22
(a) Four cities located at the vertices of a square. (b) A minimum spanning tree network with a total length of 1500 miles. (c) A shorter network, with a total length of approximately 1414 miles, obtained by placing an interior junction point O at the center of the square. (d) A shortest network with a total length of approximately 1366 miles. The junction points S_1 and S_2 are both Steiner points. (e) A different solution, with Steiner points S_3 and S_4.

EXAMPLE 7.11

Let's repeat what we did in Example 7.10, but this time imagine that the four cities are located at the vertices of a rectangle, as shown in Fig. 7-23(a). By now, we have some experience on our side, so we can cut to the chase. We know that the minimum spanning tree network is 1000 miles long [Fig. 7-23(b)]. That's the easy part.

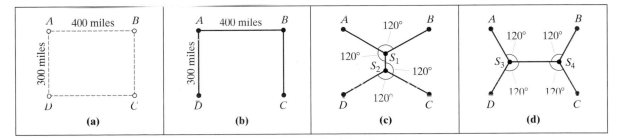

FIGURE 7-23
(a) Four cities located at the vertices of a rectangle. (b) A minimum spanning tree network (1000 miles). (c) A network with two Steiner junction points (approximately 933 miles.) (d) The shortest network also has two Steiner junction points (approximately 920 miles).

For the shortest network, let's think about what happened in the previous example, for after all, a square is just a special case of a rectangle. An obvious candidate would be a network with two interior Steiner junction points. There are two such networks, shown in Figs. 7-23(c) and 7-23(d), but this time they are not equivalent. The network shown in Fig. 7-23(c) is approximately 993 miles [see Exercise 59(a)], while the network shown in Fig. 7-23(d) is approximately 920 miles [see Exercise 59(b)]—a pretty significant difference. Obviously, Fig. 7-23(c) cannot be the shortest network, but what about Fig. 7-23(d)? If there is any justice in this mathematical world, this network fits the pattern and ought to be the shortest. In fact, it is! (But don't jump to any conclusions about justice just yet!) ∎

EXAMPLE 7.12

Let's look at four cities once more. This time, imagine that the cities are located at the vertices of a skinny trapezoid, as shown in Fig. 7-24(a). The minimum spanning tree is shown in Fig. 7-24(b), and it is 600 miles long.

FIGURE 7-24

(a) Four cities located at the vertices of a trapezoid. (b) The minimum spanning tree network (600 miles). (c) A network with an X-junction at J_1 is longer than the MST (774.6 miles). (d) A network with a couple of T-junction points at J_2 and J_3 is even worse (846.8 miles).

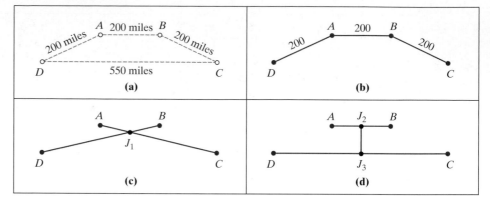

What about the shortest network? Based on our experience with Examples 7.10 and 7.11, we are fairly certain that we should be looking for a network with a couple of interior Steiner junction points. After a little trial and error, however, we realize that such a layout is impossible! The trapezoid is too skinny, or, to put it in a slightly more formal way, the angles at A and B are greater than 120°. Since no Steiner points can be placed inside the trapezoid, the shortest network, whatever it is, will have to be one without Steiner junction points.

Well, we say to ourselves, if not Steiner junction points, how about other kinds of interior junction points? How about X-junctions (Fig 7-24[c]), or T-junctions (Fig. 7-24[d]), or Y-junctions where the angles are not all 120° ? As reasonable as this idea sounds, a remarkable thing happens: *The only possible interior junction points in a shortest network are Steiner points.* For convenience, we will call this the **interior junction rule** for shortest networks. ∎

THE INTERIOR JUNCTION RULE FOR SHORTEST NETWORKS

In a shortest network, any *interior junction point* has to be a *Steiner point*.

The interior junction rule is an important and powerful piece of information in building shortest networks, and we will come back to it soon. Meanwhile, what does it tell us about the situation of Example 7.12? It tells us that the shortest network cannot have any interior junction points. Steiner junction points are impossible because of the geometry; other types of junction points do not work because of the interior junction rule. But we also know that the shortest network without interior junction points is the minimum spanning tree! Conclusion: For the four cities of Example 7.12, *the shortest network is the minimum spanning tree!* (See Fig. 7-25.)

FIGURE 7-25

The shortest network and the MST are one and the same!

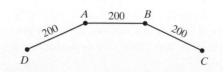

EXAMPLE 7.13

For the last time, let's look at 4 cities *A, B, C,* and *D.* This time, the cities sit as shown in Fig. 7-26(a). The minimum spanning tree is shown in Fig. 7-26(b), and its length is 1000 miles. Based on what happened in Example 7.12, we have to consider this network a serious contender for the title of shortest network. We also know that any network that is going to be shorter than this one is going to have to have some interior junction points, and if it's going to be the shortest network, then these junction points will have to be Steiner points. Because of the layout of these cities, it is geometrically impossible to build a network with two interior Steiner points (see Exercise 66). On the other hand, there are three possible networks with a single interior Steiner point [Figs. 7-26(c), (d), and (e)]. These are the only possible challengers to the minimum spanning tree. Two of them [Figs. 7-26(c) and (d)] are the same length (approximately 1325 miles), which is considerably longer than the 1000 miles in the MST. They are out! On the other hand, the network shown in Fig. 7-26(e) has a total length of approximately 982 miles. Given that there are no other possible candidates, *it must be the shortest network.*

FIGURE 7-26
(a) The distance between the four cities. (b) The minimum spanning tree (total length is 1000 miles). (c) and (d) Both of these networks have one Steiner point (S_1) connecting cities *B, C,* and *D.* Both networks have the same length (1325 miles) and are much longer than the MST. (e) A different challenger with one Steiner point (S_2) and the length of approximately 982 miles beats out (just barely) the MST. It turns out to be the shortest network connecting the four cities.

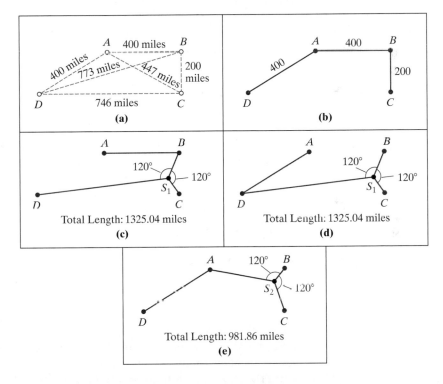

The main lesson to be learned from Examples 7.11, 7.12, and 7.13 is twofold. First, if a shortest network is not the minimum spanning tree, then it must have interior junction points and all of these junction points must be Steiner points. Any network without circuits in which every interior junction is a Steiner point is called a **Steiner tree.** The networks shown in Figs. 7-23(c) and (d), as well as Figs. 7-26(c), (d), and (e), are Steiner trees. Second, just because a network is a Steiner tree, there is no guarantee that it is the shortest network. As Example 7.13 shows, there are many possible Steiner trees, and not all of them produce shortest networks.

What happens when the number of cities gets larger? How do we look for the shortest network? Here, mathematicians face a situation analogous to the one discussed in Chapter 6—no algorithm that is both optimal and efficient is known, and there are serious doubts that such an algorithm actually exists. At this point, the best we can do is to take advantage of the following rule, which we will informally call the **shortest network rule**.

THE SHORTEST NETWORK RULE

The shortest network connecting a set of points is either
- a *minimum spanning tree* (no interior junction points) or
- a *Steiner tree* (interior junction points that are all Steiner points).

This means that we can always find the shortest network by rummaging through all possible Steiner trees, finding the shortest one among them, and comparing it with the minimum spanning tree. The shorter of these two has to be the shortest network. This sounds like a good idea, but, once again, the problem is the explosive growth of the number of possible Steiner trees. With as few as 10 cities, the possible number of Steiner trees we would have to rummage through could be in the millions; with 20 cities, it could be in the billions.

What's the alternative? Just as in Chapter 6, if we are willing to settle for an *approximate solution* (in other words, if we are willing to accept a short network that is not necessarily *the shortest*), we can tackle the problem no matter how large the number of cities. Some excellent approximate algorithms for finding short networks are presently known, and one of them happens to be Kruskal's algorithm for finding a minimum spanning tree. The reason for this is a fairly recent discovery: In 1990, mathematicians Frank Hwang of AT&T Bell Laboratories and Ding-Xhu Du of the University of Minnesota were able to prove that the percentage difference in length between the minimum spanning tree and the shortest network *is never more than 13.4%* (for details, see the article on p. 290). And this largest possible difference can only happen with three cities located at the vertices of an equilateral triangle (see Example 7.7). In fact, in most real-life applications, using the minimum spanning tree to approximate the shortest network produces a relative percentage of error that is less than 5%. Note, however, that this is still a lot of money if you are building networks whose costs can run into the billions of dollars.

Conclusion

In this chapter we discussed the problem of linking a set of points in a network in an optimal way, where "optimal" usually means least expensive or shortest distance. In practice, the points represent geographical locations (cities, telephone centrals, pumping stations, etc.), and the linkages can be rail lines, fiber-optic communication lines, pipelines, etc. Depending on the circumstances, we considered two different ways of doing this.

Version 1. In the first half of the chapter, we were required to build the network in such a way that no junction points other than the original locations were allowed, in which case the optimal network turned out to be a *minimum spanning tree*. Finding a minimum spanning tree was the great success story of the chapter—Kruskal's algorithm gives us an *optimal* and *efficient* method for doing so.

Version 2. In the second half of the chapter we considered what on the surface appeared to be a minor modification by removing the prohibition against interior junction points. In this case, the problem became one of finding the *shortest network* connecting the points. Here the situation got considerably more complicated, but we did learn a few things:

- Sometimes the minimum spanning tree and the shortest network are one and the same, but most of the time they are not. When they are not the same, the shortest network is obviously shorter than the minimum spanning tree, but by how much? It took mathematicians more than two decades to completely answer this question, but the answer is now known. The percentage difference between the minimum spanning tree and the shortest network is, in general, relatively small, and *under no circumstances can it ever be more than 13.4%.*

- In a shortest network, any interior junction points that are created must have the form of a perfect Y-junction called a *Steiner* point. A Steiner junction point is always formed by three lines joining at 120° angles. Any network without circuits connecting the original points and such that all interior junction points are Steiner points is called a *Steiner tree.* It follows that *if a shortest network has interior junction points, it must be a Steiner tree, and if it doesn't, then it must be the minimum spanning tree.*

- While, in general, the number of minimum spanning trees for a given set of points is small (usually just one) and easy to find with Kruskal's algorithm, the number of possible Steiner trees for the same set of points is usually very large and difficult to find.[7] With 7 points, for example, the possible number of Steiner trees is in the thousands, and with 10, it's in the millions.

The problem of finding the shortest network connecting a set of points has a lot of similarities to the traveling-salesman problem: No efficient optimal algorithms are known, but for most real-life problems, we can find approximate solutions that are very close (with margins of error less than 5%). For many applications, this is good enough. In other situations it isn't, and mathematicians are constantly striving to find even better algorithms. Ultimately, an embarrassingly simple question still cannot be answered: How does one find the absolutely shortest distance between N points?

---

[7]There is a fun way to find some of the Steiner trees using soap-film computers (see the Appendix "The Soap-Bubble Solution" for details), but this approach is slippery to say the least.

Solution to Old Puzzle: How Short a Shortcut?

By GINA KOLATA

Two mathematicians have solved an old problem in the design of networks that has enormous practical importance but has baffled some of the sharpest minds in the business.

Dr. Frank Hwang of A.T.&T. Bell Laboratories in Murray Hill, N.J., and Dr. Ding Xhu Du, a postdoctoral student at Princeton University, announced at a meeting of theoretical computer scientists last week that they had found a precise limit to the design of paths connecting three or more points.

Designers of things like computer circuits, long-distance telephone lines, or mail routings place great stake in finding the shortest path to connect a set of points, like the cities an airline will travel to or the switching stations for long-distance telephone lines. Dr. Hwang and Dr. Du proved, without using any calculations, that an old trick of adding extra points to a network to make the path shorter can reduce the length of the path by no more than 13.4 percent.

"This problem has been open for 22 years," said Dr. Ronald L. Graham, a research director at A.T.&T. Bell Laboratories who spent years trying in vain to solve it. "The problem is of tremendous interest at Bell Laboratories, for obvious reasons."

In 1968, two other mathematicians guessed the proper answer, but until now no one had proved or disproved their conjecture.

In the tradition of Dr. Paul Erdos, an eccentric Hungarian mathematician who offers prizes for solutions to hard problems that interest him, Dr. Graham offered $500 for the solution to this one. He said he is about to mail his check.

Add Point, Reduce Path

The problem has its roots in an observation made by mathematicians in 1640. They noticed that if you want to find a path connecting three points that form the vertices of an equilateral triangle, the best thing to do is to add an extra point in the middle of the set. By connecting these four points, you can get a shorter route around the original three than if you had simply drawn lines between those three points in the first place.

Later, mathematicians discovered many more examples of this paradoxical phenomenon. It turned out to be generally true that if you want to find the shortest path connecting a set of points, you can shorten the path by adding an extra point or two.

But it was not always clear where to add the extra points, and even more important, it was not always clear how much you would gain. In 1977, Dr. Graham and his colleagues, Dr. Michael Garey and Dr. David Johnson of Bell Laboratories, proved that there was no feasible way to find where to place the extra points, and since there are so many possibilities, trial and error is out of the question.

The Shorter Route

The length of string needed to join the three corners of a triangle is shorter if a point is added in the middle, as in the figure at right.

Source: F.K. Hwang

These researchers and others found ways to guess where an extra point or so might be beneficial, but they were always left with the nagging question of whether this was the best they could do.

The conjecture, made in 1968 by Dr. Henry Pollack, who was formerly a research director at Bell Laboratories, and Dr. Edgar Gilbert, a Bell mathematician, was that the best you could ever do by adding points was to reduce the length of the path by the square root of three divided by two, or about 13.4 percent. This was exact-

Relief for designers of computer and telephone networks.

ly the amount the path was reduced in the original problem with three vertices of an equilateral triangle.

Knowing Your Limits

In the years since, researchers proved that the conjecture was true if you wanted to connect four points, then they proved it for five.

"Everyone who contributed devised some special trick," Dr. Hwang said. "But still, if you wanted to generalize to the next number of points, you couldn't do it."

Dr. Du and Dr. Hwang's proof relied on converting the problem into one involving mathematics of continuously moving variables. The idea was to suppose that there is an arrangement for which added points make you do worse than the square root of three over two. Then, Dr. Du and Dr. Hwang showed, you can slide the points in your network around and see what happens to your attempt to make the connecting path shorter. They proved that there was no way you could slide points around to do any better than a reduction in the path length of the square root of three over two.

Dr. Graham said the proof should allow network designers to relax as they search for the shortest routes. "If you can never save more than the square route of three over two, it may not be worth all the effort it takes to look for the best solution," he said.

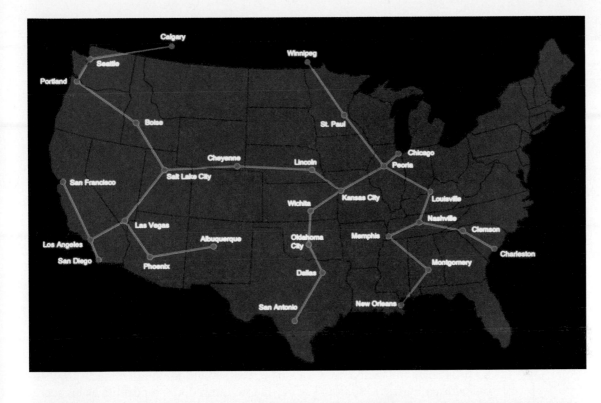

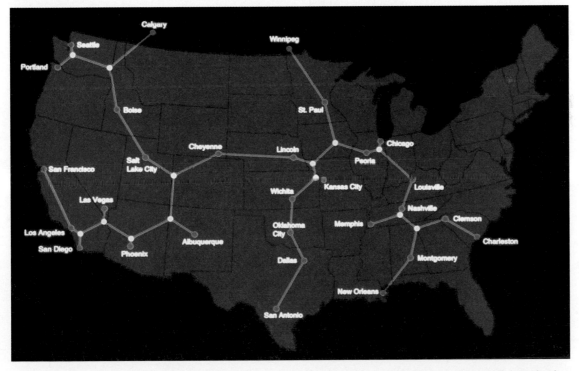

An optimal network problem involving 29 North American cities. The top plate shows the MST, found using Kruskal's algorithm in a matter of seconds. Total length: approximately 7600 miles. The bottom plate shows the shortest network, found by a computer using a sophisticated algorithm developed by researchers at the University of Victoria (Canada). Total length: approximately 7400 miles (a savings of about 3%). This network has 13 interior junction points (shown in yellow), all of which are Steiner junction points.

Evangelista Torricelli (1608–1647)

We usually first hear about Torricelli in a physics class. He was, after all, the man who invented the barometer (originally called *Torricelli's tube*), the man who discovered the laws of atmospheric pressure (called *Torricelli's laws*), and the man who gave us the equation describing the flow of a liquid discharging through an opening in the bottom of a container (known as *Torricelli's equation*). Somewhat overshadowed by his fame as a physicist is the fact that Torricelli was also one of the most gifted mathematicians of his time. In spite of a short life span, he produced many important contributions to geometry, including the elegant method for finding the shortest network connecting the vertices of a triangle that we saw earlier in the chapter (*Torricelli's construction*).

Evangelista Torricelli was born in Faenza, Italy, the son of a textile artisan of modest means. As a young man he was educated at the Jesuit school in Faenza and tutored by his uncle, a Catholic monk. At the age of 18 he was sent by his uncle to continue his studies in Rome. His tutor in Rome was Father Benedetto Castelli, a well-known scholar who had studied under Galileo. Castelli recognized that he had a brilliant student on his hands and tutored Torricelli in most of the mathematics and physics that were known at that time. Torricelli completed his studies by writing a treatise on the path of projectiles using Galileo's recently discovered laws of motion. Castelli forwarded Torricelli's completed manuscript on projectiles to Galileo, who was so impressed with the mathematical sophistication used by Torricelli in this work that he offered him a position as his secretary and personal assistant.

By this time, Galileo was in serious trouble with the Catholic church for writing a book supporting the theory that it's the Earth that moves around the Sun, and not the other way around. This was considered heresy, and Galileo was tried and placed under house arrest at his own estate near Florence. Torricelli, as much as he admired Galileo, was afraid of getting into trouble himself (guilt by association was common practice in those days), so he postponed accepting Galileo's offer for some time. For the next few years, Torricelli supported himself by working as a secretary to several professors at the University of Rome, and by grinding lenses for telescopes, a delicate craft at which he was a master. (Torricelli became known as the best telescope maker of his time and also developed an ingenious new type of microscope using tiny drops of crystal.)

In 1641 Torricelli finally decided to join Galileo and work for him, but by this time Galileo was blind due to an eye infection and in pretty poor health. Galileo died three months after Torricelli joined him. Upon Galileo's death, Torricelli was invited by the Grand Duke Ferdinand II of Tuscany to succeed Galileo as court mathematician and professor of mathematics at the University of Florence. Torricelli remained in this position and lived in the ducal palace in Florence until his death in 1647.

Having a prestigious academic post and being financially secure allowed Torricelli to concentrate on his scientific work. During this period, Torricelli conducted the experiment he is best remembered for, filling a long glass tube with mercury and turning it upside down (carefully keeping any air from getting in) into a dish also containing mercury. The level of mercury always fell a certain amount, creating a vacuum inside the tube. This single experiment secured Torricelli's name in history, as it disproved the prevailing doctrine of the time that vacuums cannot be sustained in nature. The experiment also paved the way to his discovery of the barometer.

Torricelli died in Florence at the age of 39 due to an illness of undetermined nature. A man best remembered for his practical contributions to our understanding of our physical world, he was most proud of his theoretical work in mathematics. Once, when challenged because of some inaccuracies in his calculations on the paths of cannonballs, he replied, "I write for philosophers, not bombardiers."

KEY CONCEPTS

interior junction point	spanning subgraph
interior junction rule	spanning tree
Kruskal's algorithm	Steiner point
minimum network problem	Steiner tree
minimum spanning tree (MST)	subgraph
shortest network	Torricelli's construction
shortest network rule	tree

EXERCISES

WALKING

A. Trees

1. Determine whether each of the following graphs is a tree. If it is not a tree, give a reason why.

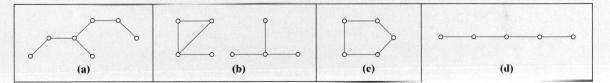

(a)　　(b)　　(c)　　(d)

2. Determine whether each of the following graphs is a tree. If it is not a tree, give a reason why.

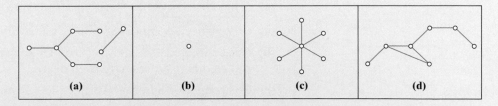

(a)　　(b)　　(c)　　(d)

In Exercises 3 through 10, you should assume that G is a graph with no loops or multiple edges, and choose from the following three options the one that best applies.

(I) The graph G being described is definitely a tree.

In this case, you should explain why the graph is a tree.

(II) The graph G being described is definitely not a tree.

In this case, you should explain why the graph cannot be a tree.

(III) The graph G being described may or may not be a tree.

In this case, you should give two examples of graphs that fit the description— one a tree and the other one not.

3. **(a)** *G* has 8 vertices and 10 edges.

   (b) *G* has 8 vertices and 5 edges.

   (c) *G* has 8 vertices and 7 edges.

   (d) *G* has 8 vertices and 7 edges, and *G* has a Hamilton circuit.

4. **(a)** *G* has 10 vertices and 8 edges.

   (b) *G* has 10 vertices and 11 edges and is a connected graph.

   (c) *G* has 10 vertices and 9 edges.

   (d) *G* has 10 vertices and 9 edges, and there is a path from any vertex to any other vertex.

5. **(a)** *G* has 8 vertices, and there is exactly one path from any vertex to any other vertex.

   (b) *G* has 8 vertices and no bridges.

   (c) *G* has 8 vertices, is connected, and every edge in *G* is a bridge.

6. **(a)** *G* has 10 vertices (*A* through *J*), and there is exactly one path from *A* to *J*.

   (b) *G* has 10 vertices (*A* through *J*), and there are two different paths from *A* to *J*.

   (c) *G* has 10 vertices and exactly 5 bridges.

7. **(a)** *G* has 8 vertices and no circuits.

   (b) *G* has 8 vertices, 7 edges, and exactly one circuit.

   (c) *G* has 8 vertices, 7 edges, and no circuits.

8. **(a)** *G* has 10 vertices, and there is a Hamilton circuit in *G*.

   (b) *G* has 10 vertices, and there is a Hamilton path in *G*.

   (c) *G* has 10 vertices, has no circuits, and there is a Hamilton path in *G*.

9. **(a)** *G* is connected and has 8 vertices, and every vertex has even degree.

   (b) *G* is connected and has 8 vertices. There are 2 vertices of odd degree, and all other vertices have even degree.

   (c) *G* is connected and has 8 vertices. The degree of every vertex is either 1 or 3.

10. **(a)** *G* is connected and has 10 vertices. Every vertex has degree 9.

    (b) *G* is connected and has 10 vertices. One of the vertices has degree 9, and all other vertices have degree less than 9.

    (c) *G* is connected and has 10 vertices, and every vertex has degree 2.

B. Spanning Trees

11. Find a spanning tree in each of the following graphs.

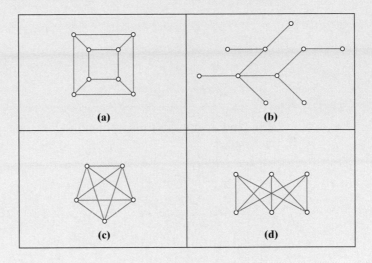

12. Find a spanning tree in each of the following graphs.

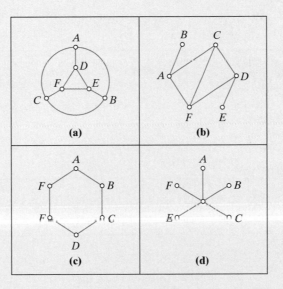

13. Find all the possible spanning trees in each of the following graphs.

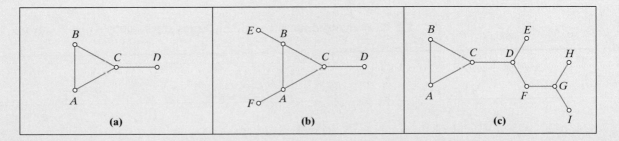

14. Find all the possible spanning trees in each of the following graphs.

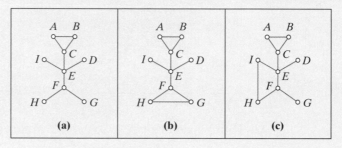

15. How many different spanning trees does each of the following graphs have?

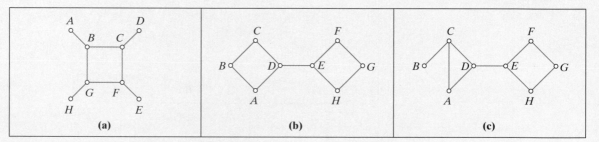

16. How many different spanning trees does each of the following graphs have?

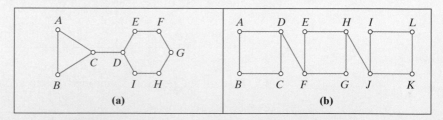

17. How many different spanning trees does each of the following graphs have?

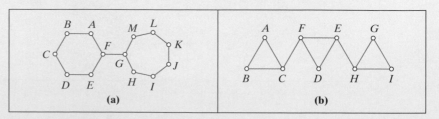

18. How many different spanning trees does each of the following graphs have?

C. Minimum Spanning Trees and Kruskal's Algorithm

19. Use Kruskal's algorithm to find a minimum spanning tree of the following weighted graph. Give the total weight of the minimum spanning tree.

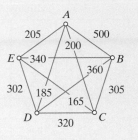

20. Use Kruskal's algorithm to find a minimum spanning tree of the following weighted graph. Give the total weight of the minimum spanning tree.

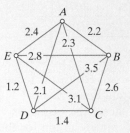

21. Use Kruskal's algorithm to find an MST of the following weighted graph. Give the total weight of the MST.

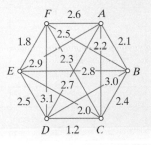

22. Use Kruskal's algorithm to find an MST of the following weighted graph. Give the total weight of the MST.

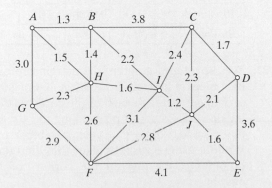

23. Find the minimum spanning tree connecting the cities of Atlanta, Georgia; Columbus, Ohio; Kansas City, Missouri; Minneapolis, Minnesota; Pierre, South Dakota; and Tulsa, Oklahoma. The following mileage chart shows the distances between cities. (*Hint:* There is no need to draw the complete graph for the problem. You should be able to implement Kruskal's algorithm directly from the chart.)

Mileage Chart

	Atlanta	Columbus	Kansas City	Minneapolis	Pierre	Tulsa
Atlanta	*	533	798	1068	1361	772
Columbus	533	*	656	713	1071	802
Kansas City	798	656	*	447	592	248
Minneapolis	1068	713	447	*	394	695
Pierre	1361	1071	592	394	*	760
Tulsa	772	802	248	695	760	*

24. Find the minimum spanning tree connecting the cities of Boston, Massachusetts; Dallas, Texas; Houston, Texas; Louisville, Kentucky; Nashville, Tennessee; Pittsburgh, Pennsylvania; and St. Louis, Missouri. The following mileage chart shows the distances between cities. (*Hint:* There is no need to draw the complete graph for the problem. You should be able to implement Kruskal's algorithm directly from the chart.)

Mileage Chart

	Boston	Dallas	Houston	Louisville	Nashville	Pittsburgh	St. Louis
Boston	*	1748	1804	941	1088	561	1141
Dallas	1748	*	243	819	660	1204	630
Houston	1804	243	*	928	769	1313	779
Louisville	941	819	928	*	168	388	263
Nashville	1088	660	769	168	*	553	299
Pittsburgh	561	1204	1313	388	553	*	588
St. Louis	1141	630	779	263	299	588	*

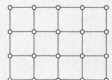

25. The 3-by-4 grid graph in the margin represents a grid of streets (3 blocks by 4 blocks) in a small subdivision. For landscaping purposes, it is necessary to get water to each of the corners (the vertices of the graph) by laying down a system of pipes along the streets. The cost of laying down the pipes is $40,000 per mile, and each block of the grid is exactly half a mile. Without using Kruskal's algorithm, find the cost of the cheapest network of pipes connecting all the corners of the subdivision. Explain your answer.

26. A weighted graph G is connected and has 121 vertices and 2565 edges. Because the graph is so large, no picture can be shown. If the weight of each edge is $10, find the weight of the minimum spanning tree for the graph.

D. Steiner Points and Shortest Networks

27. Find the length of the shortest network connecting the three cities A, B, and C shown in each of the following figures. All distances are rounded to the nearest mile. (Figures are not drawn to scale.)

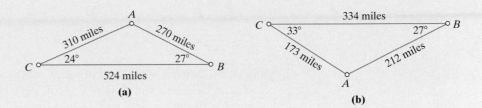

(a)   **(b)**

28. Find the length of the shortest network connecting the three cities A, B, and C shown in each of the following figures. All distances are rounded to the nearest mile. (Figures are not drawn to scale.)

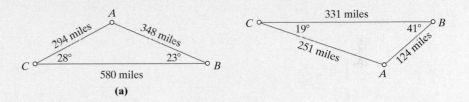

(a)

29. Find the length of the shortest network connecting the three cities A, B, and C shown in the following figure. All distances are rounded to the nearest kilometer. (Figures are not drawn to scale.)

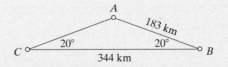

30. Find the length of the shortest network connecting the three cities A, B, and C shown in the following figure. All distances are rounded to the nearest kilometer. (Figures are not drawn to scale.)

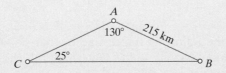

31. Find the length of the shortest network connecting the three cities A, B, and C shown in the following figure. All distances are rounded to the nearest kilometer. Explain your answer. (Figures are not drawn to scale.)

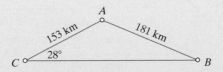

32. Find the length of the shortest network connecting the three cities A, B, and C shown in the following figure. All distances are rounded to the nearest kilometers. Explain your answer. (Figures are not drawn to scale.)

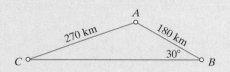

33. In the following figure, one of three points—X, Y, or Z—is a Steiner point of triangle ABC. The distances (rounded to the nearest mile) between the vertices of the triangle and each of the three points X, Y, and Z are given in the table to the right of the figure. Determine which of the three points is the Steiner point. Explain your answer. (The figure is not drawn to scale.)

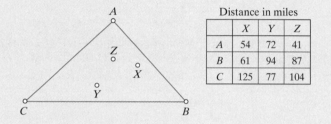

Distance in miles

	X	Y	Z
A	54	72	41
B	61	94	87
C	125	77	104

34. In the following figure, one of three points—X, Y, or Z—is a Steiner point of triangle ABC. The distances (rounded to the nearest mile) between the vertices of the triangle and each of the three points X, Y, and Z are given in the table to the right of the figure. Determine which of the three points is the Steiner point. Explain your answer. (The figure is not drawn to scale.)

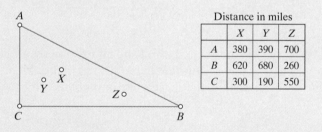

Distance in miles

	X	Y	Z
A	380	390	700
B	620	680	260
C	300	190	550

35. In the following figure, one of four points—W, X, Y, or Z—is a Steiner point of triangle ABC. Determine which one is the Steiner point. Explain your answer.

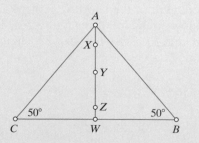

36. In the following figure, one of four points—*W, X, Y*, or *Z*—is a Steiner point of the isosceles right triangle *ABC*. Determine which one is the Steiner point. Explain your answer.

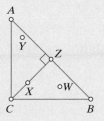

Exercises 37 and 38 refer to 5 cities (A, B, C, D, and E) located as shown in the figure in the margin. In the figure, AC = AB, DC = DB, and angle CDB = 120°.

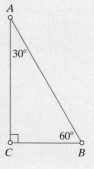

37. **(a)** Which is larger, *CD + DB* or *CE + ED + EB*? Explain.

　　(b) What is the shortest network connecting cities *C, D*, and *B*? Explain.

　　(c) What is the shortest network connecting cities *C, E*, and *B*? Explain.

38. **(a)** Which is larger, *CA + AB* or *DC + DA + DB*? Explain.

　　(b) Which is larger, *EC + EA + EB* or *DC + DA + DB*? Explain.

　　(c) What is the shortest network connecting cities *A, B*, and *C*? Explain.

E. Miscellaneous

30-60-90° Triangles. *Recall that in a 30-60-90° triangle ABC like the one in the margin, the lengths of the three sides are related by the following rules.*

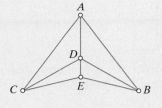

　(I)   BC = AB/ 2 (i.e., the length of the short leg is half the length of the hypotenuse).

　(II)   AC = BC√3 (i.e., the length of the long leg is the length of the short leg times √3).

　Note 1: For approximate calculations, you can use $\sqrt{3} \approx 1.732.$

　Note 2: Rule (II) is equivalent to $BC = \dfrac{AC}{\sqrt{3}} = AC\left(\dfrac{\sqrt{3}}{3}\right)$

In Exercises 39 and 40, ABC is a 30-60-90° triangle with hypotenuse AB, short leg BC, and long leg AC, as shown in the figure to the left.

39. **(a)** Given that *AB* = 20.4 cm, find *BC* and *AC* (rounded to the nearest tenth of a centimeter).

　　(b) Given that *BC* = 11.5 cm, find *AB* and *AC* (rounded to the nearest tenth of a centimeter).

　　(c) Given that *AC* = 21.0 cm, find *BC* and *AB* (rounded to the nearest tenth of a centimeter).

40. **(a)** Given that *AB* = 30.8 in., find *BC* and *AC* (rounded to the nearest tenth of an inch).

　　(b) Given that *BC* = 15.4 in., find *AB* and *AC* (rounded to the nearest tenth of an inch).

　　(c) Given that *AC* = 60.3 in., find *BC* and *AB* (rounded to the nearest tenth of an inch).

Exercises 41 and 42 refer to the Australian telephone-network problem discussed in Example 7. The towns of Booker Creek, Camoorea, and Alcie Springs form an equilateral triangle 500 miles on each side.

41. Show that the total length of the T-network shown in the following figure (rounded to the nearest mile) is 933 miles. (Use $\sqrt{3} \approx 1.732$ for your calculations.) Show the partial calculations for each branch of the network.

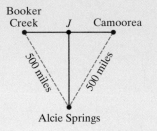

42. Show that the total length of the Y-network shown in the margin (rounded to the nearest mile) is 866 miles. (Use $\sqrt{3} \approx 1.732$ for your calculations.)

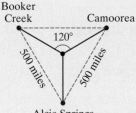

JOGGING

43. Explain why, in a tree, every edge is a bridge. (*Hint:* Use Property 1 and the fact that a single edge by itself is a path.)

44. **(a)** Can you give an example of a tree with four vertices such that the degrees of the vertices are 2, 2, 3, and 3? If yes, do so. If not, explain why not. (*Hint:* In any graph, the sum of the degrees of all the vertices is twice the number of edges.)

(b) If G is a tree with N vertices, find the sum of the degrees of all the vertices in G.

45. **(a)** Give an example of a tree with six vertices such that the degrees of the vertices are 1, 1, 2, 2, 2, and 2.

(b) Give an example of a tree with N vertices such that the degrees of the vertices are 1, 1, 2, 2, 2, ..., 2.

(c) Give an example of a tree with five vertices such that the degrees of the vertices are 1, 1, 1, 1, 4.

(d) Give an example of a tree with N vertices such that the degrees of the vertices are 1, 1, 1, ..., 1, $N - 1$.

46. **(a)** Explain why, if a single edge (but no additional vertex) is added to a tree, the resulting graph has a single circuit.

(b) Suppose that G is a connected graph with N vertices and one or more circuits. Explain why the number of bridges in G is less than $N - 1$.

(c) Explain why if G is a connected graph with N vertices and $N - 1$ bridges, G must be a tree.

47. **(a)** What is the smallest number of vertices of degree 1 that a tree with N ($N > 2$) vertices can have? Explain.

(b) What is the largest number of vertices of degree 1 that a tree with N ($N > 2$) vertices can have? Explain.

(c) Explain why, in any tree with three or more vertices, it is impossible for all the vertices to have the same degree.

48. (a) Find all possible spanning trees of the following graph.

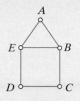

(b) How many different spanning trees does the following graph have?

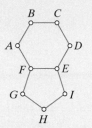

(c) How many different spanning trees does the following graph have?

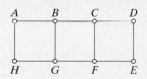

49. A theorem from graph theory (known as Cayley's theorem) states that the number of spanning trees in a complete graph with N vertices is N^{N-2}.

(a) Verify this result for the cases $N = 3$ and $N = 4$ by finding all spanning trees for complete graphs with three and four vertices.

(b) Which is larger, the number of Hamilton circuits or the number of spanning trees in a complete graph with N vertices? Explain.

50. Explain why, in a scalene triangle (all the angles are different), the minimum spanning tree consists of the two sides forming the largest angle.

51. Suppose that T is a minimum spanning tree of a weighted graph G, and suppose H is a new graph obtained by adding an additional edge (say e) of G to T. According to Exercise 46(a), the graph H has a single circuit. Explain why no edge in this circuit can have a weight larger than the weight of the new edge e.

52. A highway system connecting nine cities—$C_1, C_2, C_3, \ldots, C_9$—is to be built. Use Kruskal's algorithm to find a minimum spanning tree for this problem. The accompanying table shows the cost (in millions of dollars) of putting a highway between any two cities.

	C_1	C_2	C_3	C_4	C_5	C_6	C_7	C_8	C_9
C_1	*	1.3	3.4	6.6	2.6	3.5	5.7	1.1	3.8
C_2	1.3	*	2.4	7.9	1.7	2.3	7.0	2.4	3.9
C_3	3.4	2.4	*	9.9	3.4	1.0	9.1	4.4	6.5
C_4	6.6	7.9	9.9	*	8.2	9.7	0.9	5.5	4.9
C_5	2.6	1.7	3.4	8.2	*	4.8	7.4	3.7	3.5
C_6	3.5	2.3	1.0	9.7	4.8	*	8.9	4.4	5.8
C_7	5.7	7.0	9.1	0.9	7.4	8.9	*	4.7	3.9
C_8	1.1	2.4	4.4	5.5	3.7	4.4	4.7	*	2.8
C_9	3.8	3.9	6.5	4.9	3.5	5.8	3.9	2.8	*

53. Four cities (*A*, *B*, *C*, and *D*), shown in the following figure, must be connected into a telephone network. The cost of laying down telephone cable connecting any two of the cities is given (in millions of dollars) by the weights of the edges in the graph. In addition, in any city that serves as a junction point of the network, expensive switching equipment must be installed. The cost of installing this equipment is given (in millions of dollars) by the numbers inside the circles. Find the minimum-cost telephone network connecting these four cities.

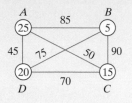

54. Five cities (*A*, *B*, *C*, *D*, and *E*) are located as shown on the following figure. The five cities need to be connected by a railroad, and the cost of building the railroad system connecting any two cities is proportional to the distance between the two cities. Find the length of the railroad network of minimum cost (assuming that no additional junction points can be added).

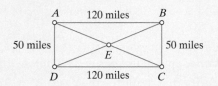

55. (a) Explain why the triangle *ABC* shown in the following figure cannot have an interior Steiner point. (*Hint:* Take *J* to be any point inside the triangle. How does the angle *BJC* compare with the angle *BAC*?)

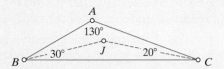

(b) Generalize your arguments in (a) to any triangle *ABC* with an angle greater than 120°.

56. Consider triangle *ABC* with equilateral triangle *EFG* inside, as shown in the following figure.

(a) Find angles *BFA*, *AEC*, and *CGB*.

(b) Explain why all the angles of triangle *ABC* are less than 120°.

(c) Explain why the Steiner point for triangle *ABC* lies inside triangle *EFG*.

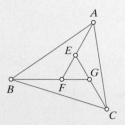

57. Use 45-45-90° triangles to show that the length of the following network is $1000\sqrt{2} \approx 1414$ miles.

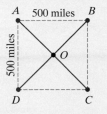

58. Use 30-60-90° triangles to show that the length of the following network is $500\sqrt{3} + 500 \approx 1366$ miles.

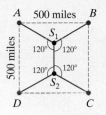

59. (a) Use 30-60-90° triangles to show that the length of the following network is $400\sqrt{3} + 300 \approx 993$ miles.

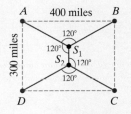

(b) Show that the length of the following network is $300\sqrt{3} + 400 \approx 919.6$ miles.

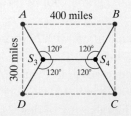

60. Suppose you are in charge of designing an optimal (i.e., shortest) fiber-optic cable network connecting the Ohio cities of Cincinnati, Toledo, and Canton. At the junction point of the network, some very expensive special equipment needs to be installed, and this equipment must be serviced by technical staff living close to the junction point. Using the map of Ohio provided on the next page, identify the city closest to the location of the Steiner point. You may use Torricelli's method or any other method you deem appropriate.

OHIO

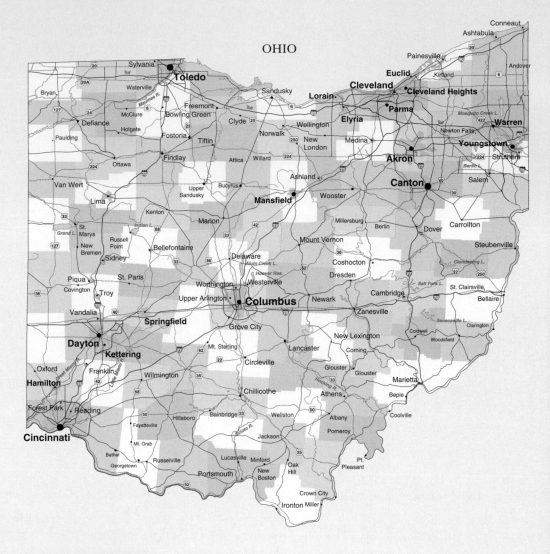

RUNNING

61. Explain why each of the following must be true.

    (a) If G is a connected graph with N vertices and N edges, then G has one circuit.

    (b) If G is a connected graph with N vertices and $N + 1$ edges, then G has at least two circuits.

    (c) If G is a connected graph with N vertices and M edges, then $M \geq N - 1$, and G has at least $M - N + 1$ circuits.

62. Show that if a tree has a vertex of degree K, then there are at least K vertices in the tree of degree 1.

63. A graph with M components, each of which is a tree, is called a **forest** (with M trees).

    (a) If G is a forest with N vertices and M trees, how many edges does G have?

    (b) What is the smallest number of vertices of degree 1 that a forest with N vertices and M trees can have? [See Exercise 47(a).]

    (c) What is the largest number of vertices of degree 1 that a forest with N vertices and M trees can have? [See Exercise 47(b).]

64. (a) Give an example of a connected graph with six vertices and six edges having no bridges.

   (b) Give an example of a connected graph with six vertices and six edges having one bridge.

   (c) Give an example of a connected graph with six vertices and six edges having two bridges.

   (d) Give an example of a connected graph with six vertices and six edges having three bridges.

   (e) Explain why a connected graph with six vertices and six edges cannot have more than three bridges.

   (f) Explain why a connected graph with N vertices and N edges ($N \geq 4$) cannot have more than $N - 3$ bridges.

65. (a) Suppose you are asked to find a minimum spanning tree for a weighted graph that must contain a given edge. Describe a modification of Kruskal's algorithm that accomplishes this.

   (b) Consider Exercise 52 again. Suppose that C_3 and C_4 are the two largest cities in the area and that the chamber of commerce insists that a section of highway directly connecting them must be built (or heads will roll). Find the minimum spanning tree that includes the section of highway between C_3 and C_4.

66. Show that the following figure (see Example 13) cannot have two interior Steiner points.

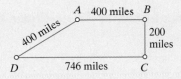

67. Consider four points (A, B, C, and D) forming the vertices of a rectangle with length b and height a as shown in the following figure, and assume that $a < b$. Determine the conditions on a and b so that the length of the minimum spanning tree is less than the length of one of the (two) Steiner trees.

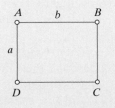

68. Find the length of the shortest network connecting three cities—A, B, and C—forming the isosceles right triangle shown in the following figure.

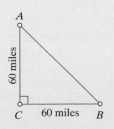

69. Eight cities (*A* through *H*) are located as shown in the following figure. (All distances are in miles.) Find the shortest network connecting the 8 cities. What is its length?

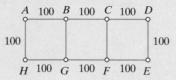

70. Find the shortest network connecting the 16 vertices of the 3-by-3 grid graph shown in the following figure. If the length of each edge of the graph is 1, what is the length of the shortest network? (*Hint:* See Exercise 58.)

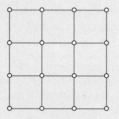

PROJECTS AND PAPERS

A. The Kruskal-Steiner Fiber-Optic Cable Network Project

Kruskal-Steiner Corporation is a new telecommunications company providing phone/Internet service to selected cities in the United States. Kruskal-Steiner plans to build a major fiber-optic cable network connecting the 21 cities shown in the mileage chart on the opposite page. The network construction costs are estimated at $100,000 per mile, so this is serious stuff.

You have been hired as a consultant for this project. Your job is to design a network connecting the 21 cities and sell your design to the board of directors. Your contract states that your consulting fee of $10,000 will be paid only if your network is better (shorter) than the *minimum spanning tree* (MST) connecting the 21 cities. In addition, you will get a bonus of $1000 for every mile that your network saves over the MST, so it's in your best interest to try to improve on the MST as much as you can.

For this project you are asked to prepare a presentation to the board. In your presentation you should (i) describe the MST and compute the total cost of its construction, (ii) describe your proposed network and show where it provides improvements over the MST, and (iii) estimate the length of your network and the savings it generates over the MST.

Notes: (1) You should start by getting two accurate, good quality maps of the United States. (2) Use Kruskal's algorithm to first find the MST, drawing it on one of the maps. (To implement Kruskal's algorithm directly on the map, you may find it convenient to first sort the distances in the mileage chart from smallest to biggest, and use this sorted list to build the MST.) (3) Now comes the fun part: Looking at the MST in the map, you should be able to locate several places where you could find shortcuts by introducing an interior Steiner junction point. (*Hint:* Anytime you see two branches of the MST coming together at an angle

Mileage Chart

	Atlanta	Boston	Buffalo	Chicago	Columbus	Dallas	Denver	Houston	Kansas City	Louisville	Memphis	Miami	Minneapolis	Nashville	New York	Omaha	Pierre	Pittsburgh	Raleigh	St. Louis	Tulsa
Atlanta	*	1037	859	674	533	795	1398	789	798	382	371	655	1068	242	841	986	1361	687	372	541	772
Boston	1037	*	446	963	735	1748	1949	1804	1391	941	1293	1504	1368	1088	206	1412	1726	561	685	1141	1537
Buffalo	859	446	*	522	326	1346	1508	1460	966	532	899	1409	927	700	372	971	1285	216	605	716	1112
Chicago	674	963	522	*	308	917	996	1067	499	292	530	1329	405	446	802	459	763	452	784	289	683
Columbus	533	735	326	308	*	1028	1229	1137	656	209	576	1160	713	377	542	750	1071	182	491	406	802
Dallas	795	1748	1346	917	1028	*	781	243	489	819	452	1300	936	660	1552	644	943	1204	1166	630	257
Denver	1398	1949	1508	996	1229	781	*	1019	600	1120	1040	2037	841	1156	1771	537	518	1411	1661	857	681
Houston	789	1804	1460	1067	1137	243	1019	*	710	928	561	1190	1157	769	1608	865	1186	1313	1160	779	478
Kansas City	798	1391	966	499	656	489	600	710	*	520	451	1448	447	556	1198	201	592	838	1061	257	248
Louisville	382	941	532	292	209	819	1120	928	520	*	367	1037	697	168	748	687	1055	388	541	263	659
Memphis	371	1293	899	530	576	452	1040	561	451	367	*	997	826	208	1100	652	1043	752	728	285	401
Miami	655	1504	1409	1329	1160	1300	2037	1190	1448	1037	997	*	1723	897	1308	1641	2016	1200	819	1196	1398
Minneapolis	1068	1368	927	405	713	936	841	1157	447	697	826	1723	*	826	1207	357	394	857	1189	552	695
Nashville	242	1088	700	446	377	660	1156	769	556	168	208	897	826	*	892	744	1119	553	521	299	609
New York	841	206	372	802	542	1552	1771	1608	1198	748	1100	1308	1207	892	*	1251	1565	368	489	948	1344
Omaha	986	1412	971	459	750	644	537	865	201	687	652	1641	357	744	1251	*	391	895	1214	449	387
Pierre	1361	1726	1285	763	1071	943	518	1186	592	1055	1043	2016	394	1119	1565	391	*	1215	1547	824	760
Pittsburgh	687	561	216	452	182	1204	1411	1313	838	388	752	1200	857	553	368	895	1215	*	445	588	984
Raleigh	372	685	605	784	491	1166	1661	1160	1061	541	728	819	1189	521	489	1214	1547	445	*	804	1129
St. Louis	541	1141	716	289	406	630	857	779	257	263	285	1196	552	299	948	449	824	588	804	*	396
Tulsa	772	1537	1112	683	802	257	681	478	248	659	401	1398	695	609	1344	387	760	984	1129	396	*

that is less than 120°, there is an opportunity for a shortcut.) (4) Estimate the length of each of the shortcuts and then the total length of your network.[8] (5) The last, and most important step: Calculate how much money you made for your efforts ($10,000 plus $1000 for every mile you saved over the MST).

B. Validating Torricelli's Construction

In the chapter we introduced Torricelli's method for finding a Steiner point inside a triangle but did not give a justification as to why the Steiner point gives the shortest network.

Torricelli proved the following three facts about a triangle ABC having all angles less than 120°:

Fact 1. There is a unique point S inside the triangle with the property that the angles ASB, ASC, and BSC are all 120° (we call it the *Steiner point* of the triangle).

Fact 2. For any point P different from S, $\overline{PA} + \overline{PB} + \overline{PC} > \overline{SA} + \overline{SB} + \overline{SC}$. (This implies that the network obtained using S as the junction point is the shortest network connecting the three vertices of the triangle.)

Fact 3. $\overline{SA} + \overline{SB} + \overline{SC} = \overline{AX}$, where X is the vertex of the equilateral triangle built on side BC and on the opposite side of A, as shown in the figure.

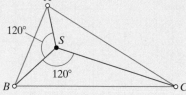

Fact 1: S is unique

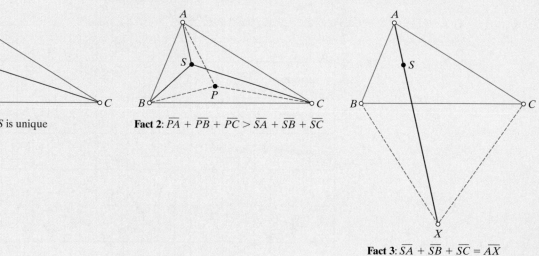

Fact 2: $\overline{PA} + \overline{PB} + \overline{PC} > \overline{SA} + \overline{SB} + \overline{SC}$

Fact 3: $\overline{SA} + \overline{SB} + \overline{SC} = \overline{AX}$

In this project you are asked to give a complete mathematical proof of each of the facts.

Notes: (1) All of the foregoing can be proved using standard facts from Euclidean geometry. (2) There are several different ways you can approach these proofs, and you can use other well-known theorems in Euclidean geometry (such as *Viviani's theorem* and *Ptolemy's theorem*) in your proofs.

C. Prim's Algorithm

Prim's algorithm is another well-known algorithm for finding the minimum

---

[8]To compute the length l of the shortest network connecting the vertices of a triangle with sides of lengths a, b, and c (when the angles of the triangle are all less than 120°), you can use the following formula:

$$l = \sqrt{\frac{a^2 + b^2 + c^2}{2} + \frac{\sqrt{3}}{2}\sqrt{2a^2b^2 + 2a^2c^2 + 2b^2c^2 - (a^4 + b^4 + c^4)}}.$$

spanning tree of a weighted graph. Like Kruskal's algorithm, Prim's algorithm is an optimal and efficient algorithm for finding MSTs.

Prepare a classroom presentation on Prim's algorithm. (i) Describe the algorithm. (ii) Illustrate how the algorithm works using the following two examples: (a) the weighted graph in Exercise 22 and (b) the mileage chart in Exercise 24. (iii) Compare Prim's algorithm to Kruskal's algorithm and discuss their differences.

Note: Prim's algorithm is covered in many books on graph theory, including references 10 and 15.

D. Minimizing with Soap Film

Occasionally, we can enlist nature's aid to help us solve reasonably interesting math problems. One of the better known example of this is the use of soap-film solutions to solve *minimal surface* and *shortest distance* problems.

For this project, prepare a classroom presentation on soap-film solutions and how they can be used to solve certain optimization problems. Discuss (i) the types of optimization problems in which soap-film solutions can be used, and (ii) the basic laws of physics that explain why and how soap-film solutions work in these problems.

Notes: You may want to read the appendix at the end of this chapter for starters. Additional reading you may find helpful are reference 1 and the following articles:

- Schechter, Bruce, "Bubbles that Bend the Mind," *Science*, March 1984.
- Isenberg, Cyril, "The Soap Film: An Analogue Computer," *American Scientist*, September–October, 1976.
- Isenberg, Cyril, "Problem Solving with Soap Films," *The Physics Teacher*, January, 1977.

A bevy of Steiner junctions: Just as a rolling ball seeks the bottom of a hill, soap films seek configurations of minimal energy. Soap bubbles trapped between glass plates always meet in sets of three, their boundaries forming 120° angles at the junction point.

APPENDIX

The Soap-Bubble Solution

Every child knows about the magic of soap bubbles. Take a wire or plastic ring, dip it into a soap-and-water solution, blow through the rings, and presto—beautiful iridescent geometric shapes magically materialize to delight and inspire our fantasy. Adults are not averse to a puff or two themselves.

What's special about soapy water that makes this happen? A very simplistic understanding of the forces of nature that create soap bubbles will help us understand how these same forces can be used to find (imagine, of all things) Steiner trees connecting a given set of points.

Take a liquid (any liquid), and put it into a container. When the liquid is at rest, there are two categories of molecules: those that are on the surface and those that are below the surface. The molecules below the surface are surrounded on all sides by other molecules like themselves and are therefore in perfect balance—the forces of attraction between molecules all cancel each other out. The molecules on the surface, however, are only partly surrounded by other like molecules and are therefore unbalanced. For this reason, an additional force called **surface tension** comes into play for these molecules. As a result of this surface tension, the surface layer of any liquid behaves exactly as if it were made of a very thin, elastic material. The amount of elasticity of this surface layer depends on the structure of the molecules in the liquid. Soap or detergent molecules are particularly well suited to create an extremely elastic surface layer. (A good soap-film solution can be obtained by adding a small amount of dishwashing liquid to water stirring gently to minimize surface bubbles, and, if necessary, adding a small amount of glycerin to make the soap film a little more stable.)

The connection between the preceding brief lesson in soapy solutions and the material in this chapter is made through one of the fundamental principles of physics: A physical system will remain in a certain configuration only if it cannot easily change to another configuration that uses less energy. Because of its extreme elasticity, the surface layer of a soapy solution has no trouble changing its shape until it feels perfectly comfortable (i.e., at a position of relatively minimal energy). When the energy is proportional to the distance, minimal energy results in minimal distance—ergo, Steiner trees.

Suppose that we have a set of points $(A_1, A_2, \ldots, A_N)$ for which we want to find a shortest network. We can find a Steiner tree that connects these points by means of an ingenious device that we will call a soap-bubble computer. To begin with, we draw the points $A_1, A_2, \ldots, A_N$ to exact scale on a piece of paper. (As much as possible, choose the scale so that points are neither too close to each other nor too far apart—somewhere from 1 to 4 inches should do just fine.) We now take two sheets of Plexiglas or Lucite and, using the paper map as a template, drill small holes on both sheets of Plexiglas at the exact locations of the points. Then we put thin metal or plastic pegs through the holes in such a way that the two sheets are held about an inch apart.

When we dip our device into a soap-and-water solution and pull it out, the soap-bubble computer goes to work. The film layer that is formed between the plates connects the various pegs. For a while it moves, seeking a configuration of minimal energy. Very shortly thereafter, it settles into a Steiner tree.

It is a bit of a disappointment that the Steiner tree we get is not necessarily the shortest network. The reasons for this are beyond the scope of our discussion. (The interested reader is referred to the excellent technical discussion of soap-film computers in Reference 1.) At the same time, we should be thankful for what nature has provided: a simple device that can compute, in seconds, what might take us hours to do with pencil and paper.

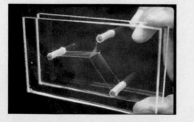 

REFERENCES AND FURTHER READINGS

1. Almgren, Fred J., Jr., and Jean E. Taylor, "The Geometry of Soap Films and Soap Bubbles," *Scientific American*, 235 (July 1976), 82–93.

2. Barabási, Albert-László, *Linked: The New Science of Networks*. Cambridge, MA: Perseus Publishing, 2002.

3. Bern, M., and R. L. Graham, "The Shortest Network Problem," *Scientific American*, 260 (January 1989), 84–89.

4. Chung, F., M. Gardner, and R. Graham, "Steiner Trees on a Checkerboard," *Mathematics Magazine*, 62 (April 1984), 83–96.

5. Cockayne, E. J., and D. E. Hewgill, "Exact Computation of Steiner Minimal Trees in the Plane," *Information Processing Letters*, 22 (1986), 151–156.

6. Du, D.-Z., and F. K. Hwang, "The Steiner Ratio Conjecture of Gilbert and Pollack is True," *Proceedings of the National Academy of Sciences*, U.S.A., 87 (December 1990), 9464–9466.

7. Gardner, Martin, "Mathematical Games: Casting a Net on a Checkerboard and Other Puzzles of the Forest," *Scientific American*, 254 (June 1986), 16–23.

8. Gilbert, E. N., and H. O. Pollack, "Steiner Minimal Trees," *SIAM Journal of Applied Mathematics*, 16 (1968), 1–29.

9. Graham, R. L., and P. Hell, "On the History of the Minimum Spanning Tree Problem," *Annals of the History of Computing*, 7 (January 1985), 43–57.

10. Gross, Jonathan, and Jay Yellen, *Graph Theory*. Boca Raton, Fla: CRC Press, 1999, Chap. 4.

11. Hwang, F. K., D. S. Richards, and P. Winter, *The Steiner Tree Problem*. Amsterdam: North Holland, 1992.

12. Melzak, Z.A., *Companion to Concrete Mathematics*. New York: John Wiley & Sons, Inc., 1973.

13. Pierce, A. R., "Bibliography on Algorithms for Shortest Path, Shortest Spanning Tree and Related Circuit Routing Problems (1956–1974)," *Networks*, 5 (1975), 129–149.

14. Robinson, P., "Evangelista Torricelli," *Mathematical Gazette*, 78 (1994), 37–47.

15. Wilson, Robin J., and John J. Watkins, *Graphs: An Introductory Approach*. New York: John Wiley & Sons, 1990, Chap 10.

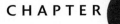

CHAPTER 8

The Mathematics of Scheduling

Directed Graphs and Critical Paths

How long does it take to build a house? Here is a deceptively simple question that defies an easy answer. Some of the factors involved are obvious: the size of the house, the type of construction, the number of workers, the kinds of tools and machinery used. Less obvious, but equally important, is another variable: the ability to organize and coordinate the timing of people, equipment, and work so that things get done in a timely way. For better or for worse, this last issue boils down to a graph-theory problem, just one example in a large family of problems that fall under the purview of what is known as the *combinatorial mathematics of scheduling*, or *combinatorial scheduling theory* for short. Discussing some of the basics of this theory will be the theme of this chapter.

L et's get back to our original question. According to the National Association of Home Builders, it takes 1092 man-hours to build the average American house. (Since we are not being picky about details, we'll just call it 1100 hours.) One way to interpret the preceding statement is this: Given just one worker (and assuming that this worker can do every single job required for building a house), it would take about 1100 hours of labor to finish this hypothetical average American house. Let's now turn the question on its head. If we had 1100 equally capable workers, could we get the same house built in one hour? Of course not! In fact, we could put tens of thousands of workers on the job and we still could not get the house built in one hour. Some inherent physical limitations to the speed with which a house can be built are outside of the builder's control. Some jobs cannot be speeded up beyond a certain point, regardless of how many workers one puts on that job. Even more significantly, certain jobs can be started only after certain other jobs have been completed. (Roofing, for example, can be started only after framing has been completed.)

Given the fact, then, that it is impossible to build the house in one hour, how fast could we build it if we had as many workers as we wanted at our disposal and we cared only about speed? (To the best of our knowledge, the record is a

> *Waste neither time nor money,*
> *but make the best use of both.*
> Ben Franklin

A simple precedence relation in construction: Framing must be finished before roofing can be started.

tad under 24 hours.) How fast could we build the house if we had 10 equally capable workers at our disposal at all times? What if we had only 3 workers? What if we needed to finish the entire project within a given time frame—say, 30 days? How many workers should we hire then? All these questions could equally well be asked if we replaced "building a house" with many other types of projects—from preparing a banquet to launching a space shuttle. These are the kinds of questions that combinatorial scheduling theory is designed to address, and while they may sound simple, they are surprisingly difficult to answer. Whenever answers are possible, the best way to find them is by means of graph models and graph algorithms, the very topics we have studied in the last three chapters.

8.1 The Basic Elements of Scheduling

We will now introduce the principal characters in every scheduling story.

The Processors

Every job requires workers. We will use the term **processors** to describe the "workers" that carry out the work. While the word *processor* may sound a little cold and impersonal, it does underscore an important point: processors need not be human beings. In scheduling, a processor could just as well be a robot, a computer, an automated teller machine, and so on. For the purposes of our discussion, we will use the notation $P_1, P_2, P_3, \ldots, P_N$ to represent the processors (where N represents the total number of processors).

The number of processors N can range from just 1 to the tens of thousands, but when $N = 1$, the whole question of scheduling is trivial and not very interesting. As far as we are concerned, real scheduling problems begin when the number of processors is 2 or more.

Processors hard at work completing their tasks. Fine restaurants deal with sophisticated scheduling problems on a daily basis.

The Tasks

In every complex project there are individual pieces of work, often called "jobs" or "tasks." We will need to be a little more precise than that, however. We will define a **task** as an indivisible unit of work that (either by nature or by choice) cannot be broken up into smaller units. Moreover, and most importantly, in our definition of the term, *a task will always be something that is, by nature, carried out by a single processor.*

To clarify the concept, let's consider a simple illustration. If a foreman assigns the wiring of a house to a single electrician, and it takes him 16 hours to do it, then we will consider wiring the house as a single 16-hour task. On the other hand, he may assign the job to 2 electricians, who together take, let's say, 7 hours. In this case, we will consider the job of wiring the house as 2 separate 7-hour tasks.

In general, we will use capital letters $A, B, C, \ldots$, to represent the tasks, although in specific situations it is convenient to use abbreviations (such as WE for "wiring the electrical system," PL for "plumbing," etc.).

At any particular moment in time throughout the project, a given task can be in one of four different states:

1. *ineligible* (the task cannot be started at this time because certain other requirements have not yet been met),
2. *ready* (the task could be started at this time),
3. *in execution* (the task is presently being carried out by one of the processors), and
4. *completed*.

The four possible states a task can be in: *ineligible* (upper left), *ready* (upper right), *in execution* (lower left), and *completed* (lower right).

The Processing Times

Associated with every task is a number called the *processing time*. The **processing time** for task X represents the amount of time, without interruption, required by one processor to carry out the task. But, one might ask, which processor? After all, how long it takes to do something often depends on who is doing it. It might take P_1 two hours to carry out task X, but it might take P_2 only one hour to carry out the same task. In general, different processors work at different rates. This surely complicates matters, so we will have to make an important concession to expediency: From now on, we will work under the assumption that *each processor can carry out each and every one of the tasks and that the processing time for a task is the same, regardless of which processor is doing it.*[1] To help things along even further, we will make a second assumption: *Once the task is started, the processor must execute it without interruption.* A processor cannot stop in the middle of a task, be it to start another task or to take a break. With these assumptions, it now makes sense to talk about *the processing time* of a task, a single nonnegative number that we will attach (in parentheses) to the right of the task's name. Thus, when we see $X(5)$, we take this to mean that it takes 5 units of time (be it minutes, hours, or whatever) to execute the task called X and that this is the case whether the task is done by P_1, P_2, or any other processor.

The Precedence Relations

These are restrictions on the order in which the tasks can be executed. A typical precedence relation is of the form *task X precedes task Y*, and it means that task Y cannot be started until task X has been completed. Such a precedence relation can be conveniently abbreviated by writing $X \rightarrow Y$, or described pictorially as in Fig. 8-1. Precedence relations arise from laws that govern the order in which things are done in the real world—we just can't put our shoes on before our socks! A single scheduling problem can have hundreds or even thousands of precedence relations, each adding another restriction on the scheduler's freedom.

At the same time, it also happens fairly often that there are no restrictions on the order of execution between two tasks in a project. When a pair of tasks X and Y have no precedence requirements between them (neither $X \rightarrow Y$ nor $Y \rightarrow X$), we say that the tasks are **independent**. When two tasks are independent, either one can be started before the other one, or they can both be started at the same time—some people put their shoes on before their shirt, others put their shirt on before their shoes, and, occasionally, some of us have been known to put our shoes and shirt on at the same time. Sometimes an entire project can be made up of all independent tasks (with no precedence relations whatsoever to worry about). We will discuss this special situation in greater detail later in the chapter.

Two final comments about precedence relations. First, precedence relations are *transitive*: if $X \rightarrow Y$ and $Y \rightarrow Z$, then it must be true that $X \rightarrow Z$. In a sense, the last precedence relation is implied by the first two, and it is really unnecessary to mention it (Fig. 8-2). Thus, we will make a distinction between two types of precedence relations: *basic* and *implicit*. Basic precedence relations are the ones that come with the problem and that we must follow in the process of creating a schedule. Once we do this, the implicit precedence relations will be taken care of automatically.

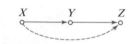

FIGURE 8-1

X precedes *Y*.

FIGURE 8-2

When $X \rightarrow Y$ and $Y \rightarrow Z$, then $X \rightarrow Z$ is implied.

---

[1]Essentially, this is the theory behind automobile repair charges. The hours charged for a given repair job (processing time) come from manuals such as *Chilton's Guide to Automobile Repairs* rather than the actual time spent on the repair.

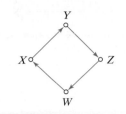

FIGURE 8-3
These tasks cannot be scheduled because of the cyclical nature of the precedence relations.

The second observation is that *we cannot have a set of precedence relations that form a cycle!* Imagine having to schedule the tasks shown in Fig. 8-3: X precedes Y, which precedes Z, which precedes W, which in turn precedes X. Clearly, this is logically impossible. From here on, we will assume that there are no cycles of precedence relations among the tasks.

Processors, tasks, processing times, and precedence relations are the basic ingredients that make up a scheduling problem. They constitute, in a manner of speaking, the hand that is dealt to us. But how do we play such a hand? To get a small inkling of what's to come, let's look at the following very simple example.

EXAMPLE 8.1 Repairing a Wreck

Imagine that you just wrecked your expensive new sports car, but thank heavens you are OK, and the insurance company will pick up the tab. You take the car to the best garage in town, operated by the Click (P_1) and Clack (P_2) brothers. The repairs on the car can be broken into four different tasks: (A) exterior body work (4 hours), (B) engine repairs (5 hours), (C) painting and exterior finish work (7 hours), and (D) repair transmission (3 hours). The only precedence relation for this set of tasks is that the painting and exterior finish work cannot be started until the exterior body work has been completed ($A \rightarrow C$). The two brothers always work together on a repair project, but each takes on a different task (so they won't argue with each other). Under these conditions, how should the different tasks be scheduled? Who should do what and when?

Even in this very simple situation, many different schedules are possible. Figure 8-4 shows several possibilities, each one illustrated by means of a timeline. Figure 8-4(a) shows a schedule that is very inefficient. All the short tasks are assigned to one mechanic (P_1) and all the long tasks to the other mechanic (P_2)—obviously not a very clever strategy. Under this schedule, the project **finishing time** (the duration of the project from the start of the first task to the completion of the last task) is 12 hours. Figure 8-4(b) shows what looks like a much better schedule, but it violates the precedence relation $A \rightarrow C$. That is, we are not allowed to start task C on the third hour. In fact, this is true not only for this schedule, but for any other schedule we might think of. No matter how clever we are (and no matter how many mechanics we have at our disposal), the painting

Scheduling the repairs for a wrecked car is easy—actually doing the repairs is not!

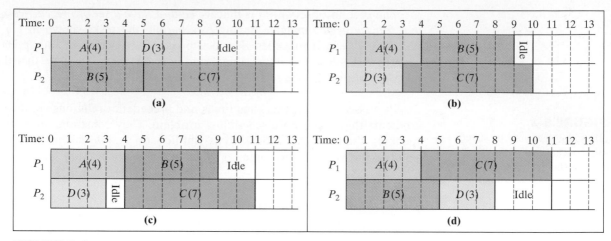

FIGURE 8-4
Some possible schedules for Example 8.1. (a) An *inefficient* schedule. Finishing time is 12 hours. (b) An *illegal* schedule. The precedence relation $A \rightarrow C$ is violated when C is started before A is completed. (c) Removing the violation in the preceding schedule is accomplished by a small adjustment: Make P_2 sit idle for one hour before starting task C. Finishing time is 11 hours. (d) A different schedule with a finishing time of 11 hours. Schedules (c) and (d) are both *optimal*, since it is impossible to finish the project in less than 11 hours.

and exterior finish work (C) *can never be started before the fourth hour.* This is because of the requirement that the exterior body work [$A(4)$] must be finished first. On the other hand, if we make P_2 sit idle for 1 hour, waiting for the green light so to speak, to start task C, we get a perfectly good schedule, shown in Fig. 8-4(c). The finishing time is 11 hours, and given that task $C(7)$ cannot be started until the fourth hour, this is as short as the schedule is going to get. *No possible schedule can complete this project in less than 11 hours.* As far as finishing time is concerned, the schedule shown in Fig. 8-4(c) is *optimal*. In general, there is likely to be more than one optimal schedule, and Fig. 8-4(d) shows a different schedule that also has a finishing time of 11 hours. ∎

As scheduling problems go, Example 8.1 was a fairly simple one. But even from this simple example, we can draw some useful lessons. First, notice that even though we had only four tasks and two processors, we were able to create several different schedules. The four we looked at were just a sampler—there are other possible schedules that we didn't bother to discuss. Imagine what would happen if we had hundreds of tasks and dozens of processors—the number of possible schedules to consider would be overwhelming. In looking for a good, or even optimal, schedule, we are going to need a systematic way to sort through the many possibilities. In other words, we are going to need some good *scheduling algorithms*.

The second useful thing we learned in Example 8.1 is that when it comes to the finishing time of a project, there is an *absolute minimum* time that no schedule can break, no matter how good an algorithm we use or how many processors we put to work. In Example 8.1, this absolute minimum was 11 hours, and, as luck would have it, we easily found a schedule [actually two—Figs. 8-4(c) and (d)] with a finishing time to match it. Every project, no matter how simple or complicated, has such an absolute minimum (called the *critical time*) that depends on the processing times and precedence relations for the tasks, and not on the number of processors used. We will return to the concept of critical time in Sec. 8.5.

To set the stage for a more formal discussion of scheduling algorithms, we will introduce the most important example of this chapter. While couched in what seems like science fiction terms, the situation it describes is not totally farfetched.

EXAMPLE 8.2 Building a Dream Home on Mars

It is the year 2050, and several human colonies have already been established on Mars. Imagine that you accept a job offer to work in one of these colonies. What will you do about housing?

Like everyone else on Mars, you will be provided with a living pod called a Martian Habitat Unit (MHU). MHUs are shipped to Mars in the form of prefabricated kits that have to be assembled on the spot—an elaborate and unpleasant job if you are going to do it yourself. A better option is to hire special workers that will do all of the assembly for you. On Mars, these workers come in the form of robots called Habitat Unit Building Robots (affectionately nicknamed *HUBRIS*), which can be rented by the hour at the local Rent-a-Robot outlet.

Here are some questions you are going to have to address: How can you get your MHU built quickly? How many *HUBRIS* should you rent? How do we create a suitable work schedule that will get the job done? (A *HUBRI* will do whatever it is told, but someone has to tell it what to do and when.)

The assembly of an MHU consists of 15 separate tasks as shown in Table 8-1, with the processing times representing *HUBRI*-hours (i.e., the number of hours it takes one *HUBRI* to execute the task). In addition, the tasks are constrained by 17 different precedence relations as shown in Table 8-2.

TABLE 8-1

Task	Symbol (Processing Time)
Assemble Pad	$AP(7)$
Assemble Flooring	$AF(5)$
Assemble Wall Units	$AW(6)$
Assemble Dome Frame	$AD(8)$
Install Floors	$IF(5)$
Install Interior Walls	$IW(7)$
Install Dome Frame	$ID(5)$
Plumbing	$PL(4)$
Install Atomic Power Plant	$IP(4)$
Install Pressurization Unit	$PU(3)$
Install Heating Units	$HU(4)$
Install Commode	$IC(1)$
Interior Finish Work	$FW(6)$
Pressurize Dome	$PD(3)$
Install Entertainment Unit (virtual reality TV, computer, music box, communication port, etc.)	$EU(2)$

TABLE 8-2

Precedence Relations
$AP \rightarrow IF$
$AF \rightarrow IF$
$IF \rightarrow IW$
$AW \rightarrow IW$
$AD \rightarrow ID$
$IW \rightarrow ID$
$IF \rightarrow PL$
$IW \rightarrow IP$
$IP \rightarrow PU$
$ID \rightarrow PU$
$IP \rightarrow HU$
$PL \rightarrow IC$
$HU \rightarrow IC$
$PU \rightarrow EU$
$HU \rightarrow EU$
$IC \rightarrow FW$
$HU \rightarrow PD$

We will return soon to the question of how best to assemble a Martian Habitat Unit, but first we will develop a few helpful concepts.

Home, sweet home (circa 2050).

8.2 Directed Graphs (Digraphs)

All of the information presented in Tables 8-1 and 8-2 can be summarized in a very convenient way, as shown in Fig. 8-5. The tasks are represented by vertices, and the precedence relations are represented by arrows pointing from one vertex to another vertex (an arrow pointing from vertex X to vertex Y indicates that X must be completed before Y can be started). This approach is consistent with what we already did in Figs. 8-1, 8-2, and 8-3.

FIGURE 8-5

A digraph describing the tasks for assembling an MHU, their completion times, and their precedence relations.

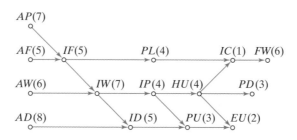

Figure 8-5 looks just like one of the graphs in the previous chapters, except that each "edge" now has a direction associated with it. A graph in which the edges have a direction associated with them is called a **directed graph**, or more commonly, a **digraph**.

Just like graphs, digraphs are used to describe relationships between objects, but in this case the nature of the relationship is such that we cannot always assume that it is reciprocal. We call such relationships *asymmetric relationships*. Being in love is a good example of an asymmetric relationship: Just because X is in love with Y, it does not necessarily follow that Y must be in love with X. Sometimes it happens, and sometimes it doesn't.

To distinguish digraphs from ordinary graphs, we use slightly different terminology. In a digraph, the edges have directions, and we call them **arcs**. Unlike an edge in an ordinary graph, an arc has a *starting vertex* and an *ending vertex*. If the starting vertex is X and the ending vertex is Y $(X \rightarrow Y)$, then we write the arc as

XY (which is now different from *YX*). A useful way to think of an arc *XY* is as a one-way road connecting *X* to *Y*.

If *XY* is an arc in the digraph, we say that vertex *X* is **incident to** vertex *Y*, or equivalently, that *Y* is **incident from** *X*. The arc *YZ* is said to be **adjacent** to the arc *XY* because the starting point of *YZ* is the ending point of *XY*. Essentially, this means one can go from *X* to *Z* by way of *Y*. In a digraph, a **path** from vertex *X* to vertex *W* consists of a sequence of arcs *XY*, *YZ*, *ZU*, ..., *VW* such that each arc is adjacent to the one before it and no arc appears more than once in the sequence—it is essentially a trip from *X* to *W* along the arcs in the digraph. The best way to describe the path is by listing the vertices in the order of travel: *X*, *Y*, *Z*, and so on.

When the path starts and ends in the same vertex, we call it a **cycle** in the digraph. Just like circuits in a regular graph, cycles in digraphs can be written in more than one way—the cycle *X*, *Y*, *Z*, *X* is the same as the cycles *Y*, *Z*, *X*, *Y* and *Z*, *X*, *Y*, *Z*.

Each vertex in a digraph has an *indegree* and an *outdegree*. The **outdegree** of *X* is the number of arcs that have *X* as their *starting point* (outgoing arcs); the **indegree** of *X* is the number of arcs that have *X* as their *ending point* (incoming arcs).

The following example illustrates some of the aforementioned concepts.

EXAMPLE 8.3

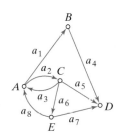

FIGURE 8-6

Consider the digraph in Fig. 8-6. This is a digraph with 5 vertices (*A*, *B*, *C*, *D*, and *E*) and 8 arcs (a_1, a_2, a_3, a_4, a_5, a_6, a_7, and a_8). In this digraph, *A* is *incident to B* and *C*, but not to *E*. By the same token, *A* is *incident from E* as well as from *C*. The indegree of vertex *A* is 2, and so is the outdegree. The indegree of vertex *C* is 1, and the outdegree is 3. We leave it to the reader to find the indegrees and outdegrees of each of the other vertices of the graph [see Exercise 1(b)].

In this digraph, there are several paths from *A* to *D*, such as *A*, *C*, *D*; *A*, *C*, *E*, *D*; *A*, *B*, *D*; and even *A*, *C*, *A*, *B*, *D*. On the other hand, *A*, *E*, *D* is not a path from *A* to *D* (because you can't travel from *A* to *E*). Are there any possible paths from *D* to *A*? Why not? As for examples of cycles in this digraph, here are two: *A*, *C*, *E*, *A* (which can also be written as *C*, *E*, *A*, *C* and *E*, *A*, *C*, *E*) and *A*, *C*, *A*. Can you find any others? Why not?

Many real-life situations can be represented by digraphs. Here are a few examples.

- **Transportation.** Here the *vertices* represent locations within a city, and the *arcs* represent one-way streets. (To represent a two-way street we use two arcs, one for each direction.)
- **The Internet.** Here the *vertices* represent sources of information, and the *arcs* represent the possible flows of information.
- **Tournaments.** Here the *vertices* represent teams (or players) and the arcs represent the games played. An arc *XY* means that *X* defeated *Y* (we assume that the games cannot end in a tie).
- **Chain of command.** In a corporation or in the military, we can use a digraph to describe the chain of command. The *vertices* are individuals, and an *arc* from *X* to *Y* indicates that *X* can give orders (is a superior) to *Y*.

EXAMPLE 8.4 Building a Dream Home on Mars: Part II

Let's return now to the problem of assembling a Martian Habitat Unit. *HUBRIS* will do the labor; we will do the thinking. We now know that the main elements of the problem can be conveniently described by the directed graph shown in Fig. 8-5, which is repeated in Fig. 8-7(a). Figure 8-7(b) shows a slight modification of Fig. 8-7(a), where we have added two fictitious tasks: START and END. These two tasks are not real—we just added them for convenience. We can now visualize the entire project as a flow that begins at START and concludes at END. By giving these fictitious tasks zero processing time, we avoid affecting the time calculations for the project. The digraph shown in Fig. 8-7(b) is called the **project digraph**.

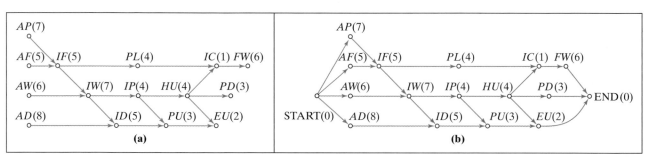

FIGURE 8-7
The project digraph for assembling a Martian Habitat Unit.

8.3 The Priority-List Model for Scheduling

The project digraph is the basic graph model used to conveniently describe all the information in a scheduling problem, but there is nothing in the project digraph itself that specifically tells us how to create a schedule. We are going to need something else, some set of instructions that indicates the order in which tasks should be executed. We can accomplish this by the simple act of prioritizing the tasks in some specified order, called a *priority list*.

A **priority list** is a list of all the tasks and a priority for their execution. If task X comes before task Y in the priority list, we should try to execute X before Y. However, if X is not yet *ready* for execution, we can skip over it and move on to the next task in the priority list. By following this simple rule, we can take any priority list and from it create a schedule. Conversely, every schedule can be traced back to at least one priority list, usually more. In other words, *all priority lists produce schedules and all schedules can be interpreted as coming from priority lists.* The trick is going to be to figure out which priority lists give us good schedules and which don't. We will come back to this topic in Sections 8.4 and 8.5.

Since each time we change the order of the tasks we get a different priority list, there are as many priority lists as there are ways to order the tasks. For 3 tasks, there are 6 possible priority lists; for 4 tasks, there are 24 priority lists; for 10 tasks, there are more than three million priority lists; and for 100 tasks, there are more priority lists than there are molecules in the universe. Clearly, a shortage of priority lists is not going to be our problem. Sound familiar? As it did in Chapter 2 and Chapter 6, the *factorial* is once again entering the scene.

For a project consisting of M tasks, the number of possible priority lists is
$$M! = 1 \times 2 \times 3 \times \cdots \times M.$$

Before we proceed, we will illustrate how the priority-list model for scheduling works with a couple of small but important examples. Even with such small examples, there is a lot to keep track of, and you are well advised to have pencil and paper in front of you as you follow the details.

EXAMPLE 8.5 Preparing for Launch

Before the launching of a satellite into space, five different system checks need to be performed by the computers on board. For simplicity, we will call the system checks $A(6)$, $B(5)$, $C(7)$, $D(2)$, and $E(5)$, with the numbers in parentheses representing the hours it takes one computer to perform that system check. In addition, there are precedence relations: D cannot be started until both A and B have been finished, and E cannot be started until C has been finished. All of the preceding information can be summarized by the project digraph shown in Fig. 8-8.

Let's suppose that two computers (P_1 and P_2) are available to carry out the system checks and that each individual system check can only be carried out by one of the computers.

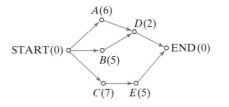

FIGURE 8-8

How does one create a schedule to get all five system checks done? To start, we will need a priority list. Let's say that we are given a priority list in which the tasks are simply listed alphabetically.

Priority List: $A(6)$, $B(5)$, $C(7)$, $D(2)$, $E(5)$

- **Time $T = 0$ hr (start of project).** $A(6)$, $B(5)$, and $C(7)$ are the only *ready* tasks. Following the priority list, we assign $A(6)$ to P_1 and $B(5)$ to P_2.

- **Time $T = 5$ hr.** P_1 is still *busy* with $A(6)$; P_2 has just *completed* $B(5)$. $C(7)$ is the only available *ready* task. We assign $C(7)$ to P_2.

- **Time $T = 6$ hr.** P_1 has just *completed* $A(6)$; P_2 is *busy* with $C(7)$. $D(2)$ has just become a *ready* task (A and B have been completed). We assign $D(2)$ to P_1.

- **Time $T = 8$ hr.** P_1 has just *completed* $D(2)$; P_2 is still *busy* with $C(7)$. There are no *ready* tasks at this time for P_1, so P_1 has to sit *idle*.

- **Time $T = 12$ hr.** P_1 is *idle*; P_2 has just *completed* $C(7)$. Both processors are *ready* for work. $E(5)$ is the only ready task, so we assign $E(5)$ to P_1, P_2 sits *idle*.

  [Note that in this situation, we could have just as well assigned $E(5)$ to P_2 and let P_1 sit idle.]

- **Time *T* = 17 hr.** P_1 has just *completed* $E(5)$. Project is completed. Finishing time is 17 hours.

The final schedule can be seen in Fig. 8–9. Is this a good schedule? All we have to do is look at all the idle time to see that it is a very bad schedule. When it comes to the finishing time, we would be hard put to come up with a worse schedule. Maybe we should try again! We can try a different strategy by simply changing the priority list.

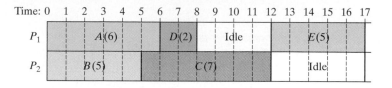

FIGURE 8-9
Final schedule for Example 8.5.

EXAMPLE 8.6 Preparing for Launch: Part II

We are going to schedule the same project with the same processors, but with a different priority list. (The project digraph from Fig. 8-8 is repeated in Fig. 8–10.)

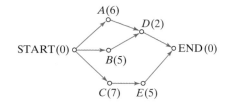

FIGURE 8-10

This time we will go in reverse alphabetical order. (Remember, we can write a priority list in any order!)

Priority List: $E(5), D(2), C(7), B(5), A(6)$

- **Time *T* = 0 hr (start of project).** $C(7)$, $B(5)$, and $A(6)$ are the only *ready* tasks. Following the priority list, we assign $C(7)$ to P_1 and $B(5)$ to P_2.

- **Time *T* = 5 hr.** P_1 is still *busy* with $C(7)$; P_2 has just *completed* $B(5)$. $A(6)$ is the only available ready task. We assign $A(6)$ to P_2.

- **Time *T* = 7 hr.** P_1 has just *completed* $C(7)$; P_2 is *busy* with $A(6)$. $E(5)$ has just become a *ready* task, and we assign it to P_1.

- **Time *T* = 11 hr.** P_2 has just *completed* $A(6)$; P_1 is *busy* with $E(5)$. $D(2)$ has just become a *ready* task, and we assign it to P_2.

- **Time *T* = 12 hr.** P_1 has just *completed* $E(5)$; P_2 is *busy* with $D(2)$. There are no tasks left, so P_1 sits idle.

- **Time *T* = 13 hr.** P_2 has just *completed* the last task, $D(2)$. Project is completed. Finishing time is 13 hours. The actual schedule is shown in Fig. 8–11.

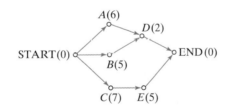

FIGURE 8-11
Final schedule for Example 8.6.

It's easy to see that this schedule is a lot better than the one in Fig. 8-9. In fact, as long as we have only two processors to do the work, this is an *optimal schedule*—the finishing time of 13 hours cannot be improved! (After all, if we add the processing times for all tasks, there is a total of 25 hours worth of work. Divide that between two processors, and we have a minimum of 12.5 hours of work for each. Since all the processing times in this project are whole numbers, the finishing time must be at least 13 hours.)

What would happen if we kept the same priority list but added a third computer to the "workforce"? One would think that this would speed things up. Let's check it out.

EXAMPLE 8.7 Preparing for Launch: Part III

This example is the same as the previous example (same project digraph and same priority list), but we will now create a schedule using three processors P_1, P_2, and P_3. For the reader's convenience the project digraph is shown again in Fig. 8-12. (It's hard to schedule without the project digraph in front of you!)

Priority List: $E(5), D(2), C(7), B(5), A(6)$

FIGURE 8-12

- **Time $T = 0$ (start of project).** $C(7)$, $B(5)$, and $A(6)$ are the *ready* tasks. We assign $C(7)$ to P_1, $B(5)$ to P_2, and $A(6)$ to P_3.
- **Time $T = 5$ hr.** P_1 is *busy* with $C(7)$; P_2 has just *completed* $B(5)$; and P_3 is *busy* with $A(6)$. There are no available ready tasks for P_2 [because $E(5)$ can't be started until $C(7)$ is done, and $D(2)$ can't be started until $A(6)$ is done], so P_2 sits idle.
- **Time $T = 6$ hr.** P_3 has just *completed* $A(6)$; P_2 is *idle*; and P_1 is still *busy* with $C(7)$. $D(2)$ has just become a *ready* task. We assign $D(2)$ to P_2 and let P_3 be idle, since there are no other ready tasks. [Note that we could have also assigned $D(2)$ to P_3 and let P_2 be idle.]
- **Time $T = 7$ hr.** P_1 has just *completed* $C(7)$ and $E(5)$ has just become a *ready* task, so we assign it to P_1. There are no other tasks to assign, so P_3 continues to sit idle.
- **Time $T = 8$ hr.** P_2 has just *completed* $D(2)$. There are no other tasks to assign, so P_2 and P_3 both sit *idle*.

- **Time T = 12 hr.** P_1 has just *completed* the last task, $E(5)$, so the project is completed. Finishing time is 12 hours.

The actual schedule is shown in Fig. 8-13. Surprisingly, adding a third processor didn't really help all that much. We'll come back to this point later in the chapter.

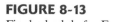

Time: 0 1 2 3 4 5 6 7 8 9 10 11 12

P_1	$C(7)$		$E(5)$
P_2	$B(5)$	Idle	$D(2)$	Idle
P_3	$A(6)$		Idle

FIGURE 8-13
Final schedule for Example 8.7.

The previous three examples give us a general sense of how to create a schedule from a project digraph *and* a priority list. We will now formalize the ground rules for the *priority-list model* for scheduling.

At any particular moment in time throughout a project, a processor can be either *busy* or *idle* and a task can be *ineligible, ready, in execution,* or *completed.* Depending on the various combinations of these, there are three different scenarios to consider:

1. *All processors are busy.* In this case, there is nothing we can do but wait.
2. *One processor is free.* In this case, we scan the priority list from left to right, looking for the first *ready* task in the priority list, which we assign to that processor. (Remember that for a task to be *ready*, all the tasks that are incident to it in the project digraph must have been completed.) If there are no ready tasks at that moment, the processor must stay idle until things change.
3. *More than one processor is free.* In this case, the first ready task on the priority list is given to one free processor, the second ready task is given to another free processor, and so on. If there are more free processors than ready tasks, some of the processors will remain idle until one or more tasks become ready. Since the processors are identical, the choice of which processor is assigned which task is totally arbitrary. (To simplify the bookkeeping, we will consistently choose the processors to go in numerical order if they have subscripts or in alphabetical order otherwise.)

It's fair to say that the basic idea behind the priority-list model is not difficult, but there is a lot of bookkeeping involved, and that becomes critical when the number of tasks is large. At each stage of the schedule we need to keep track of the status of each task—which tasks are *ready* for processing, which tasks are *in execution*, which tasks have been *completed*, which tasks are still *ineligible*. One convenient record-keeping strategy goes like this: On the priority list itself *ready* tasks are circled in red [Fig. 8-14(a)]. When a ready task is picked up by a processor and goes into *execution*, put a single red slash through the red circle [Fig. 8-14(b)]. When a task that has been in execution is completed, put a second red slash through the circle [Fig. 8-14(c)]. At this point, it is also important to check the project digraph to see if any new tasks have all of a sudden become eligible. Tasks that are *ineligible* remain unmarked [Fig. 8-14(d)].

We will now show how to implement this strategy to assemble a Martian Habitat Unit.

FIGURE 8-14
"Road" signs on a priority list.
(a) Task *X* is ready. (b) Task *X* is
in execution. (c) Task *X* is completed. (d) Task *X* is ineligible.

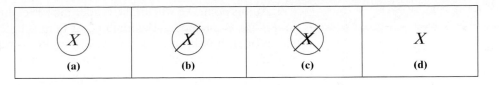

FIGURE 8-15

EXAMPLE 8.8 Assembling an MHU: Part III

We are finally ready to start the project of assembling our MHU, and, like any
good scheduler, we will first work the entire schedule out with pencil and paper.
Let's start with the assumption that maybe we can get by with just two robots (P_1
and P_2). For the reader's convenience, we show the project digraph again in
Fig. 8-15.

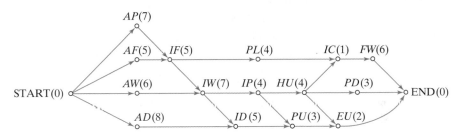

Suppose that we decide to use the following priority list.

Priority List: *AD*(8), *AW*(6), *AF*(5), *IF*(5), *AP*(7), *IW*(7), *ID*(5), *IP*(4),
PL(4), *PU*(3), *IIU*(4), *IC*(1), *PD*(3), *EU*(2), *FW*(6) (Ready tasks are circled
in red.)

- **Time: *T* = 0.**
  Status of Processors: P_1 starts *AD*; P_2 starts *AW*.
  Priority List (Updated Status):
  AD, *AW*, *AF* *IF*, *AP* *IW*, *ID*, *IP*, *PL*, *PU*, *HU*, *IC*, *PD*, *EU*, *FW*.

- **Time: *T* = 6.**
  Status of Processors: P_1 busy (executing *AD*); P_2 completed *AW* and
  starts *AF*.
  Priority List (Updated Status):
  AD, *AW*, *AF* *IF*, *AP* *IW*, *ID*, *IP*, *PL*, *PU*, *HU*, *IC*, *PD*, *EU*, *FW*.

- **Time: *T* = 8.**
  Status of Processors: P_1 completed *AD* and starts *AP*; P_2 is busy
  (executing *AF*).
  Priority List (Updated Status):
  AD, *AW*, *AF* *IF*, *AP* *IW*, *ID*, *IP*, *PL*, *PU*, *HU*, *IC*, *PD*, *EU*, *FW*.

- **Time: *T* = 11.**
  Status of Processors: P_1 busy (executing *AP*); P_2 completed *AF*, but
  since there are no ready tasks to take on, it must remain idle.
  Priority List (Updated Status):
  AD, *AW*, *AF* *IF*, *AP* *IW*, *ID*, *IP*, *PL*, *PU*, *HU*, *IC*, *PD*, *EU*, *FW*.

- **Time: *T* = 15.**
  Status of Processors: P_1 completed *AP*. Now *IF* becomes a ready
  task and is given to P_1; P_2 stays idle.
  Priority List (Updated Status):
  AD, *AW*, *AF*, *IF*, *AP* *IW*, *ID*, *IP*, *PL*, *PU*, *HU*, *IC*, *PD*, *EU*, *FW*.

At this point, we will let the reader take over and finish the schedule (see Exercise 45). Remember—the main point here is to learn how to keep track of the status of each task, and the only way to do this is with practice. (Besides, explaining the same thing over and over can get monotonous to both the explainer and the explainee.) After a fair amount of work, one obtains the final schedule shown in Fig. 8-16. The finishing time for the project is 44 hours.

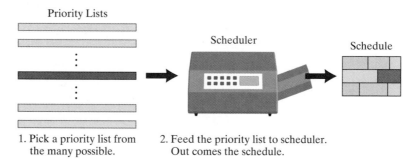

FIGURE 8-16
Final schedule for Example 8.8.

Scheduling under the priority-list model can be thought of as a two-part process: (1) Choose a priority list, and (2) use the priority list and follow the ground rules of the model to come up with a schedule. As we saw in the last example, the second part is long and tedious, but purely mechanical—it can be done by anyone (or anything) that is able to follow a set of instructions, be it a meticulous student or a properly programmed computer. We will use the term *scheduler* to describe the entity (be it student or machine) that takes a priority list as input and produces the schedule as output (see Fig. 8-17).

Ironically, it is the seemingly easiest part of this process—choosing a priority list—that is actually the most interesting. How do we know which of the many possible priority lists will give us an optimal schedule? (We will call such a priority list an *optimal priority list.*) How do we even pick a priority list that gives us a decent schedule?

In the rest of this chapter, we will try to find some answers to these questions.

FIGURE 8-17
The scheduling process.

Priority Lists

Scheduler

Schedule

1. Pick a priority list from the many possible.

2. Feed the priority list to scheduler. Out comes the schedule.

8.4 The Decreasing-Time Algorithm

Our first attempt to find a good priority list is to formalize what is a commonly used and seemingly sensible strategy: *Do the longer jobs first and leave the shorter jobs for last.* Formally, this translates into writing the priority list by listing the tasks in decreasing order of processing times, with longest first, second longest next, and so on. (When there are two or more tasks with equal processing times, choose their order at random.) We will call this the **decreasing-time list**, and we will call the process of creating a schedule using the decreasing-time list combined with the priority-list model the **decreasing-time algorithm (DTA)**.

EXAMPLE 8.9 Building an MHU Using the DTA

The following list is a decreasing-time list for the 15 tasks required to assemble an MHU.

Decreasing-Time List: $AD(8), AP(7), IW(7), AW(6), FW(6), AF(5), IF(5),$ $ID(5), IP(4), PL(4), HU(4), PU(3), PD(3), EU(2), IC(1)$

Using the decreasing-time algorithm with two processors, we get the schedule shown in Fig. 8-18, with a finishing time of 42 hours. The reader is encouraged to work out the details of this schedule independently (see Exercise 46), but in the interest of fairness, a summary of the details is given in Table 8-3 shown on the next page.

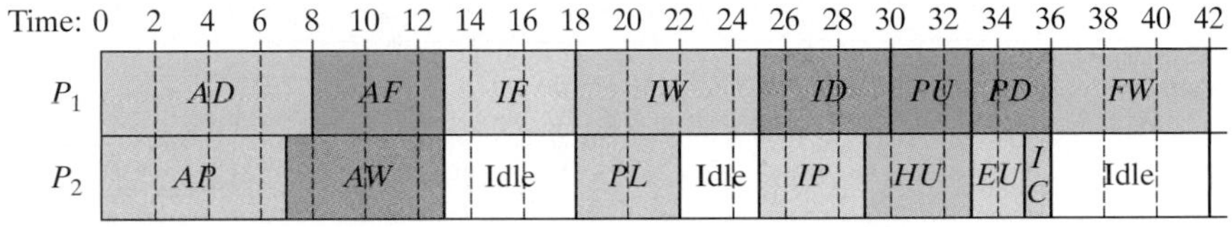

FIGURE 8-18

Schedule for assembling an MHU using the decreasing-time algorithm.

When looking at the finishing time under the DTA, one can't help but feel disappointed. This promising idea of doing the longer jobs first and the shorter jobs later turned out to be a bit of a dud—at least in this example! What went wrong? If we work our way backward from the end, we can see that we made a bad choice at $T = 33$ hours. At this point there were three ready tasks [$PD(3)$, $EU(2)$, and $IC(1)$], and both processors were available. Based on the decreasing-time priority list, we chose the two longest tasks, $PD(3)$ and $EU(2)$. Bad strategy! If we had looked at what was down the road, we would have seen that $IC(1)$ is a much more *critical* task than the other two because we can't start $FW(6)$ until we finish $IC(1)$. In short, we were shortsighted—we made our choices based on the immediate rather than the long-term benefits, and we ended up paying the price.

An even more blatant example of how the DTA can lead to bad choices in scheduling occurs at the very start of the schedule: We failed to notice that it is critical to start $AP(7)$ and $AF(5)$ as early as possible. Until we finish AP and AF, we cannot start $IF(5)$; and unless we finish IF, we cannot start $IW(7)$; and until we finish IW, we cannot start $IP(4)$ and $ID(5)$; and so on down the line. ∎

The lesson to be learned from what happened in Example 8.9 is that a task should not be prioritized by how long it takes to execute it, but rather by the sum total of all tasks that lie ahead of it. Simply put, the greater the *total amount of work lying ahead of a task*, the sooner that task should be started.

8.5 Critical Paths

To formalize some of the preceding observations, we will introduce the concepts of critical paths and critical times.

For a given vertex X of a project digraph, the **critical path for X** is the *path from X to END that has the longest total processing time*. The total processing time in the critical path for X is called the **critical time for X**. (The *total processing time* of a path is the sum of the processing times of the vertices in the path).

TABLE 8-3   The Decreasing-Time Algorithm Applied to the MHU Project (Example 8.9)

Step	Time	Priority-List Status	Schedule Status
1	$T = 0$	⊘AD(8) ⊘AP(7) IW(7) ○AW(6) FW(6) ○AF(5) IF(5) ID(5) IP(4) PL(4) HU(4) PU(3) PD(3) EU(2) IC(1)	Time: 0 2 4 6 8 10 12 14 16 18 20 22 24 26 28 30 32 34 36 38 40 42 P_1: AD P_2: AP
2	$T = 7$	⊘AD(8) ⊘AP(7) IW(7) ○AW(6) FW(6) ○AF(5) IF(5) ID(5) IP(4) PL(4) HU(4) PU(3) PD(3) EU(2) IC(1)	Time: 0 2 4 6 8 10 12 14 16 18 20 22 24 26 28 30 32 34 36 38 40 42 P_1: AD P_2: AP, AW
3	$T = 8$	⊘AD(8) ⊘AP(7) IW(7) ⊘AW(6) FW(6) ⊘AF(5) IF(5) ID(5) IP(4) PL(4) HU(4) PU(3) PD(3) EU(2) IC(1)	Time: 0 2 4 6 8 10 12 14 16 18 20 22 24 26 28 30 32 34 36 38 40 42 P_1: AD, AF P_2: AP, AW
4	$T = 13$	⊘AD(8) ⊘AP(7) IW(7) ⊘AW(6) FW(6) ⊘AF(5) ⊘IF(5) ID(5) IP(4) PL(4) HU(4) PU(3) PD(3) EU(2) IC(1)	Time: 0 2 4 6 8 10 12 14 16 18 20 22 24 26 28 30 32 34 36 38 40 42 P_1: AD, AF, IF P_2: AP, AW
5	$T = 18$	⊘AD(8) ⊘AP(7) ⊘IW(7) ⊘AW(6) FW(6) ⊘AF(5) ⊘IF(5) ID(5) IP(4) ○PL(4) HU(4) PU(3) PD(3) EU(2) IC(1)	Time: 0 2 4 6 8 10 12 14 16 18 20 22 24 26 28 30 32 34 36 38 40 42 P_1: AD, AF, IF, IW P_2: AP, AW, Idle, PL
6	$T = 22$	⊘AD(8) ⊘AP(7) ⊘IW(7) ⊘AW(6) FW(6) ⊘AF(5) ⊘IF(5) ID(5) IP(4) ⊘PL(4) HU(4) PU(3) PD(3) EU(2) IC(1)	Time: 0 2 4 6 8 10 12 14 16 18 20 22 24 26 28 30 32 34 36 38 40 42 P_1: AD, AF, IF, IW P_2: AP, AW, Idle, PL
7	$T = 25$	⊘AD(8) ⊘AP(7) ⊘IW(7) ⊘AW(6) FW(6) ⊘AF(5) ⊘IF(5) ⊘ID(5) ⊘IP(4) ⊘PL(4) HU(4) PU(3) PD(3) EU(2) IC(1)	Time: 0 2 4 6 8 10 12 14 16 18 20 22 24 26 28 30 32 34 36 38 40 42 P_1: AD, AF, IF, IW, ID P_2: AP, AW, Idle, PL, Idle, IP
8	$T = 29$	⊘AD(8) ⊘AP(7) ⊘IW(7) ⊘AW(6) FW(6) ⊘AF(5) ⊘IF(5) ⊘ID(5) ⊘IP(4) ⊘PL(4) ⊘HU(4) PU(3) PD(3) EU(2) IC(1)	Time: 0 2 4 6 8 10 12 14 16 18 20 22 24 26 28 30 32 34 36 38 40 42 P_1: AD, AF, IF, IW, ID P_2: AP, AW, Idle, PL, Idle, IP, HU
9	$T = 30$	⊘AD(8) ⊘AP(7) ⊘IW(7) ⊘AW(6) FW(6) ⊘AF(5) ⊘IF(5) ⊘ID(5) ⊘IP(4) ⊘PL(4) ⊘HU(4) ⊘PU(3) PD(3) EU(2) IC(1)	Time: 0 2 4 6 8 10 12 14 16 18 20 22 24 26 28 30 32 34 36 38 40 42 P_1: AD, AF, IF, IW, ID, PU P_2: AP, AW, Idle, PL, Idle, IP, HU
10	$T = 33$	⊘AD(8) ⊘AP(7) ⊘IW(7) ⊘AW(6) FW(6) ⊘AF(5) ⊘IF(5) ⊘ID(5) ⊘IP(4) ⊘PL(4) ⊘HU(4) ⊘PU(3) ⊘PD(3) ⊘EU(2) ○IC(1)	Time: 0 2 4 6 8 10 12 14 16 18 20 22 24 26 28 30 32 34 36 38 40 42 P_1: AD, AF, IF, IW, ID, PU, PD P_2: AP, AW, Idle, PL, Idle, IP, HU, EU
11	$T = 35$	⊘AD(8) ⊘AP(7) ⊘IW(7) ⊘AW(6) FW(6) ⊘AF(5) ⊘IF(5) ⊘ID(5) ⊘IP(4) ⊘PL(4) ⊘HU(4) ⊘PU(3) ⊘PD(3) ⊘EU(2) ⊘IC(1)	Time: 0 2 4 6 8 10 12 14 16 18 20 22 24 26 28 30 32 34 36 38 40 42 P_1: AD, AF, IF, IW, ID, PU, PD P_2: AP, AW, Idle, PL, Idle, IP, HU, EU, IC
12	$T = 36$	⊘AD(8) ⊘AP(7) ⊘IW(7) ⊘AW(6) ⊘FW(6) ⊘AF(5) ⊘IF(5) ⊘ID(5) ⊘IP(4) ⊘PL(4) ⊘HU(4) ⊘PU(3) ⊘PD(3) ⊘EU(2) ⊘IC(1)	Time: 0 2 4 6 8 10 12 14 16 18 20 22 24 26 28 30 32 34 36 38 40 42 P_1: AD, AF, IF, IW, ID, PU, PD, FW P_2: AP, AW, Idle, PL, Idle, IP, HU, EU, IC
13	$T = 42$	⊘AD(8) ⊘AP(7) ⊘IW(7) ⊘AW(6) ⊘FW(6) ⊘AF(5) ⊘IF(5) ⊘ID(5) ⊘IP(4) ⊘PL(4) ⊘HU(4) ⊘PU(3) ⊘PD(3) ⊘EU(2) ⊘IC(1)	Time: 0 2 4 6 8 10 12 14 16 18 20 22 24 26 28 30 32 34 36 38 40 42 P_1: AD, AF, IF, IW, ID, PU, PD, FW P_2: AP, AW, Idle, PL, Idle, IP, HU, EU, IC, Idle

The next three examples illustrate the concepts of critical paths and critical times using the MHU project digraph (Fig. 8-19).

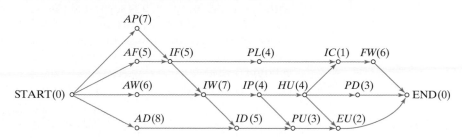

FIGURE 8-19
The MHU project digraph.

EXAMPLE 8.10

Let's try to find the critical path for vertex *HU*. There are three paths from *HU* to END. They are (i) *HU, IC, FW*, END; (ii) *HU, PD*, END; and (iii) *HU, EU*, END. The total processing time in (i) is $4 + 1 + 6 = 11$; the total processing time in (ii) is $4 + 3 = 7$; and the total processing time in (iii) is $4 + 2 = 6$. Of the three paths, path (i) has the longest processing time, so it is the *critical path for HU*. The *critical time* for *HU* is 11.

EXAMPLE 8.11

When we try to find the critical path for vertex *AD*, we notice that there is only one path from *AD* to END, namely *AD, ID, PU, EU*, END. Since this is the only path, it is the *critical path for AD*. The *critical time* for *AD* is 18.

EXAMPLE 8.12

There are quite a few paths from START to END. After looking at the project digraph for a little while, however, we can pretty much "see" that the one with the longest total processing time appears to be START, *AP, IF, IW, IP, HU, IC, FW*, END, with a total time of 34 hours. This is indeed the *critical path for the* START vertex and it is called *the critical path for the project*, or, more briefly, just *the critical path*.

The **critical path** of a project digraph is the longest path from START to END. The total processing time for the critical path is called the **critical time**. For the Martian Habitat Unit assembly project, the project's *critical path* is START, *AP, IF, IW, IP, HU, IC, FW*, END, and the *critical time* is 34 hours. We will discuss the significance of the critical time and the critical path soon, but before we do so, let's discuss how to find critical paths from any vertex of a project digraph. After all, we can hardly be expected to find critical paths in large-project digraphs the way we did in Examples 8.10, 8.11, and 8.12, where we were pretty much flying by the seat of our pants. Fortunately, there is a simple and efficient algorithm for finding critical paths and critical times. It is called the *backflow algorithm*.

The Backflow Algorithm

The basic idea behind the **backflow algorithm** is to build the critical path by working backward from the END to the START. If we know the critical times for all the vertices "in front" of a given vertex *X*, we choose among these the vertex with the *largest critical time* (call it *C*). The critical time of *X* is then obtained by adding the *processing time* of *X* to the *critical time of C* (see Fig. 8-20). To help with the record keeping, it is suggested that you write the critical time of the vertex in square brackets [ ] to distinguish it from the processing time in parentheses ( ). Once we have the critical times, the critical path for a vertex is obtained by starting at the vertex and always moving to the adjacent vertex with largest critical time.

FIGURE 8-20

Critical time for
X = processing time for
X + critical time for *C* (the
vertex incident from *X* with
largest critical time).

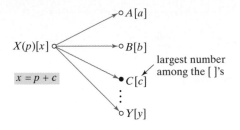

While it sounds a little complicated in words, the backflow algorithm is actually pretty easy to do, as we will show in the next example.

EXAMPLE 8.13 The Critical Path for the MHU Project

We will use the backflow algorithm to find the critical times for every task in the MHU assembly project. The project digraph is shown in Fig. 8–21.

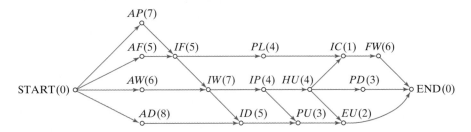

FIGURE 8-21

- **Step 1.** We start at END and assign to it a critical time of zero.
- **Step 2.** We move backward to the three vertices that are incident to END, namely, *FW*(6), *PD*(3), and *EU*(2). For each of them, the critical time is its processing time plus zero, so the critical times are *FW*[6], *PD*[3], and *EU*[2].
- **Step 3.** From *FW*[6], we move backward to *IC*(1). The only vertex incident from *IC*(1) is *FW*[6], so the critical time for *IC* is [1 + 6 = 7]. We record a [7] next to *IC* in the graph.
- **Step 4.** We move backward to *HU*(4). There are three vertices incident from it (*IC*[7], *PD*[3], and *EU*[2]), and the one with the largest critical time is *IC*[7]. It follows that the critical time for *HU* is [4 + 7 = 11]. At this stage, we can also find the critical times of *PL*(4), and *PU*(3). For *PL*(4), the only vertex incident from it is *IC*[7], so its critical time is [4 + 7 = 11]. For *PU*(3), the only vertex incident from it is *EU*[2], so its critical time is [3 + 2 = 5].
- **Step 5.** We move backward to *IP*(4). There are two vertices incident from it (*HU*[11] and *PU*[5]). The critical time for *IP* is [4 + 11 = 15]. We can also move backward to *ID*(5) and find its critical time, which is [10].
- **Step 6.** We can now move backward to *IW*(7). We leave it to the reader to verify that its critical time is [7 + 15 = 22].
- **Step 7.** We can now move backward to *IF*(5). We leave it to the reader to verify that its critical time is [5 + 22 = 27].

- **Step 8.** We now move backward to $AP(7)$, $AF(5)$, $AW(6)$, and $AD(8)$. Their respective critical times are $[7 + 27 = 34]$, $[5 + 27 = 32]$, $[6 + 22 = 28]$, and $[8 + 10 = 18]$.
- **Step 9.** Finally, we move backward to *START*. We still follow the same rule: The critical time is $[0 + 34 = 34]$. This is the critical time for the entire project!

The critical time for every vertex of the project digraph is shown (in [red]) in Fig. 8–22. To find the critical path, we just go from vertex to vertex following the path of largest critical times: START, *AP, IF, IW, IP, HU, IC, FW*, END—just as we suspected!

FIGURE 8-22
MHU project digraph: processing times in blue; critical times in red; critical path in red.

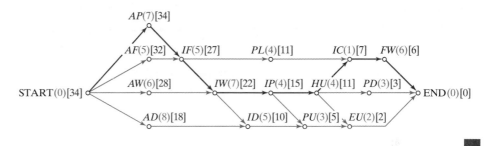

Why are the critical path and critical time of a project of special significance? There are two reasons: (1) As we discussed at the start of this chapter, in every project there is an absolute minimum time needed for the completion of the project. It is impossible to finish the project in less time, regardless of how clever the scheduler is or how many processors are used. This absolute minimum turns out to be the project's critical time. (2) For the project to be completed in the critical time, it is absolutely necessary that all the tasks in the critical path be done at the earliest possible time. Any delay in starting up one of the tasks in the critical path will necessarily delay the finishing time of the entire project. (By the way, this is why it is called the *critical path*.)

Unfortunately, it is not always possible to schedule the tasks on the critical path one after the other, bang, bang, bang without delay. For one thing, processors are not always free when we need them. (Remember that a processor cannot stop in the middle of one task to start a new task.) Another reason is the problem of uncompleted predecessor tasks. We cannot concern ourselves only with tasks along the critical path and disregard other tasks that might affect them through precedence relations. There is a whole web of interrelationships that we need to worry about. Optimal scheduling is extremely complex.

8.6   The Critical-Path Algorithm

It is possible to use the concept of critical paths to generate very good (although not necessarily optimal) schedules. The idea is the same as the one we used with the decreasing-time algorithm but at a higher level of sophistication. Instead of prioritizing the tasks in decreasing order of *processing times*, we will prioritize them in decreasing order of *critical times*. (Think of it as a mathematical version of strategic planning.) The priority list we obtain when we write the tasks in decreasing order of critical times (with ties broken randomly) is called the **critical-path list**. The process of creating a schedule using the critical-path list as the priority list is called the **critical-path algorithm (CPA)**.

EXAMPLE 8.14 Building an MHU Using the CPA

We will apply the critical-path algorithm to the MHU problem. We already calculated the critical times for each vertex, shown in red in Fig. 8–23.

AP(7)[34]
AF(5)[32]
IF(5)[27]
PL(4)[11]
IC(1)[7]
FW(6)[6]
AW(6)[28]
IW(7)[22]
IP(4)[15]
HU(4)[11]
PD(3)[3]
START(0)[34]
END(0)[0]
AD(8)[18]
ID(5)[10]
PU(3)[5]
EU(2)[2]

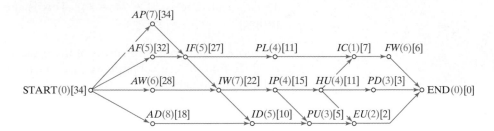

FIGURE 8-23

A critical-path list for the project is given below.

Critical-Path List: *AP*[34], *AF*[32], *AW*[28], *IF*[27], *IW*[22], *AD*[18], *IP*[15], *PL*[11], *HU*[11], *ID*[10], *IC*[7], *FW*[6], *PU*[5], *PD*[3], *EU*[2]

With two processors, the schedule that results from this priority list is shown in Fig. 8–24. This time we leave the details to the reader (see Exercise 53).

Time: 0 2 4 6 8 10 12 14 16 18 20 22 24 26 28 30 32 34 36 38 40 42
P_1
AP
IF
IW
IP
HU
I C
FW
EU
P_2
AF
AW
AD
PL
ID
PU
PD
Idle

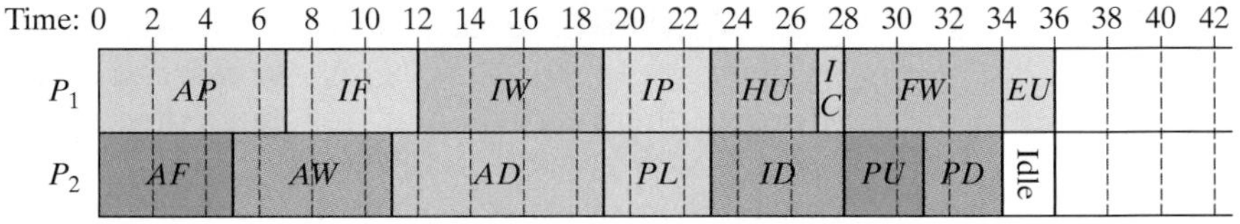

FIGURE 8-24
Schedule for the MHU project obtained using the critical-path algorithm.

In the MHU problem, the finishing time using the critical-path algorithm is 36 hours, a big improvement over the 42 hours produced by the decreasing-time algorithm. Is this an optimal solution? Figure 8–25 shows a schedule with a finishing time of 35 hours, so our 36-hour schedule, while good, is obviously not optimal.

Time: 0 2 4 6 8 10 12 14 16 18 20 22 24 26 28 30 32 34 36 38 40 42
P_1
AP
IF
IW
IP
ID
I C
FW
P_2
AF
AW
AD
PL
HU
PD
PU
EU

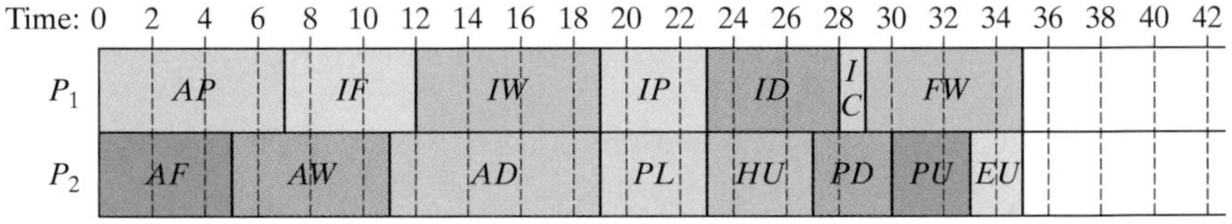

FIGURE 8-25
Optimal schedule for the MHU project using two processors.

The CPA is an excellent *approximate* algorithm for scheduling a project (in most cases far superior to the DTA), but, in general, it will not produce an optimal schedule. As it turns out, no efficient scheduling algorithm that always gives an optimal schedule is presently known. In this regard, scheduling problems are every bit as complex as traveling-salesman problems (Chapter 6) and shortest-network problems (Chapter 7): There are efficient approximate algorithms that can produce good schedules, but no efficient optimal algorithms are known. Of the standard scheduling algorithms, the critical-path algorithm is by far the most commonly used. Other, more sophisticated algorithms have been developed in the last 20 years, and under specialized circumstances they can out-perform the critical-path algorithm, but as an all-purpose algorithm for scheduling, the critical-path algorithm is hard to beat.

8.7 Scheduling with Independent Tasks

In this section, we will briefly discuss what happens to scheduling problems in the special case when there are no precedence relations to worry about. This situation arises whenever we are scheduling tasks that are all independent—for example, scheduling a group of typists in a steno pool to type a bunch of reports of various lengths.

It is tempting to think that without precedence relations hanging over one's head, scheduling becomes a simple problem, and one should be able to find optimal schedules without difficulty, but appearances are deceiving. *There are no efficient optimal algorithms known for scheduling, even when the tasks are all independent.*

While, in a theoretical sense, we are not much better able to schedule independent tasks than to schedule tasks with precedence relations, from a purely practical point of view, there are a few differences. For one thing, there is no getting around the fact that the nuts-and-bolts details of creating a schedule using a priority list become tremendously simplified when there are no precedence relations to mess with. In this case, we just assign the tasks to the processors as they become free in exactly the order given by the priority list. Second, without precedence relations, the critical-path time of a task equals its processing time. This means that the *critical-path list* and *decreasing-time list* are exactly the same list, and, therefore, the decreasing-time algorithm and the critical-path algorithm become one and the same. Before we go on, let's look at a couple of examples of scheduling with independent tasks.

EXAMPLE 8.15 Preparing for Lunch

Imagine that you and your two best friends are cooking a 9-course luncheon as part of a charity event. We will assume that each of the 9 courses is an independent task, to be done by just one of the 3 cooks (which we'll call P_1, P_2, P_3). The 9 courses are $A(70)$, $B(90)$, $C(100)$, $D(70)$, $E(80)$, $F(20)$, $G(20)$, $H(80)$, and $I(10)$, with their processing times given in minutes. Let's first use an alphabetical priority list.

Priority List: $A(70)$, $B(90)$, $C(100)$, $D(70)$, $E(80)$, $F(20)$, $G(20)$, $H(80)$, $I(10)$

Since there are no precedence relations, there are no ineligible tasks, and all tasks start out as ready tasks. As soon as a processor is free, it picks up the next

available task in the priority list. From the bookkeeping point of view, this is a piece of cake. We leave it to the reader to verify that the resulting schedule is the one in Fig. 8–26, with a finishing time of 220 minutes. It is obvious from the figure that this is not a very good schedule.

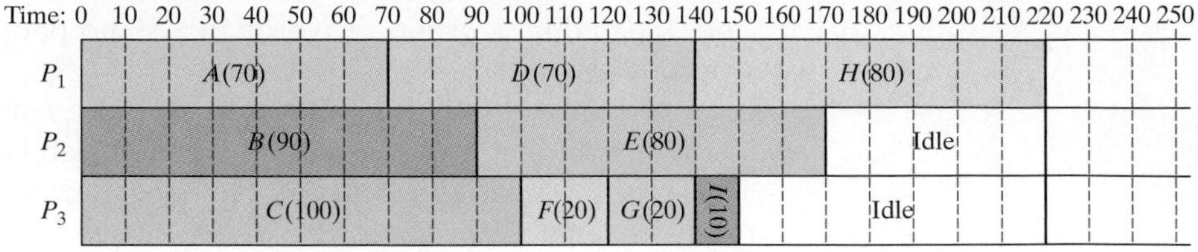

FIGURE 8-26
Schedule for Example 8.15 using an alphabetical priority list.

If we use the critical-path algorithm, the priority list is the decreasing-time list given below.

Decreasing-Time List: $C(100)$, $B(90)$, $E(80)$, $H(80)$, $A(70)$, $D(70)$, $F(20)$, $G(20)$, $I(10)$

The resulting schedule is shown in Fig. 8–27 and has a finishing time of 180 minutes. The reader is encouraged to verify the details. Clearly, this schedule is optimal, since all three processors are working for the entire time.

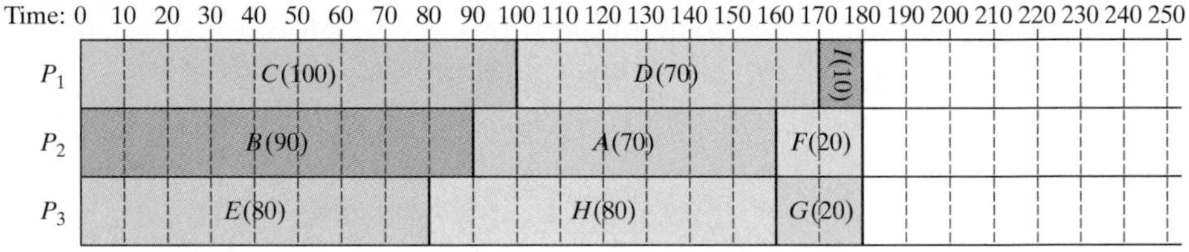

FIGURE 8-27
Schedule for Example 15 using a decreasing-time list.

In Example 8.15, the critical-path algorithm gave us the optimal schedule, but, unfortunately, this need not always be the case.

EXAMPLE 8.16 Preparing for Lunch II

After the success of your last banquet, you and your two friends are asked to prepare another banquet. This time it will be a 7-course meal. The courses are all independent tasks, and their processing times (in minutes) are $A(50)$, $B(30)$, $C(40)$, $D(30)$, $E(50)$, $F(30)$, and $G(40)$.

Using the CPA we get the priority list $A(50)$, $E(50)$, $C(40)$, $G(40)$, $B(30)$, $D(30)$, and $F(30)$. The schedule one gets is shown in Fig. 8-28, with a finishing time of 110 minutes. With a little trial and error, we can do better than this. The schedule shown in Fig. 8-29 is optimal, with a finishing time of 90 minutes.

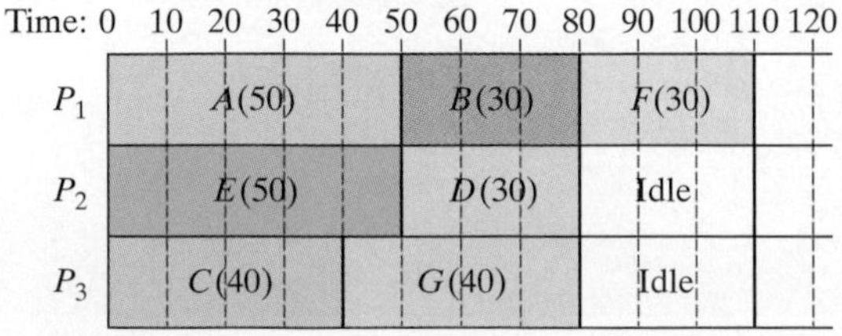

FIGURE 8-28
Schedule for Example 8.16 using a decreasing-time list.

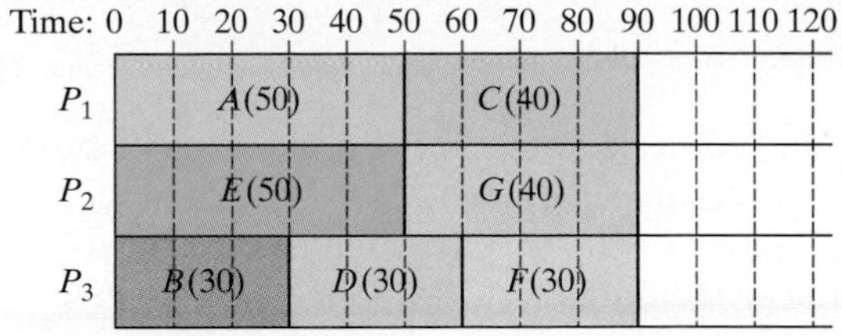

FIGURE 8-29
Optimal schedule for Example 8.16.

Once we know the optimal finishing time, we can meausre how "well" a particular scheduling algorithm does in a given situation by computing the *relative percentage of error*:

$$\text{relative percentage of error} = \frac{\text{computed finishing time}}{\text{optimal finishing time}}$$

In Example 8.16, the relative percentage of error is $(110 - 90)/90 = 20/90 \approx 0.2222 = 22.22\%$. This tells us that the CPA gave us a schedule that is 22.22% longer than the optimal schedule.

In 1969, the American mathematician Ronald L. Graham (see the biographical profile on page 341) showed that for independent tasks, the CPA will always produce schedules with finishing times that are never off by more than a fixed percentage from the optimal finishing time. Specifically, Graham proved that for independent tasks, when the number of processors is N, the relative percentage of error using the CPA is at most $(N-1)/(3N)$ (see Table 8-4). This maximum value for the relative percentage of error increases slowly as the number of processors increases but is always less than $33\frac{1}{3}\%$ (see Exercise 65). Graham's discovery essentially reassures us that when the tasks are independent, the finishing time we get from the critical-path algorithm can't be too far off from the optimal finishing time—no matter how many tasks need to be scheduled or how many processors are available to carry them out.

TABLE 8-4 Max Error (*E*) Represents the Largest Possible Percentage Error Under the CPA when the Tasks Are Independent

Number of Processors N	Max Error $E = (N-1)/3N$
2	$\frac{2-1}{3 \times 2} = \frac{1}{6} \approx 16.66\%$
3	$\frac{3-1}{3 \times 3} = \frac{2}{9} \approx 22.22\%$
4	$\frac{4-1}{3 \times 4} = \frac{3}{12} = 25\%$
5	$\frac{5-1}{3 \times 5} = \frac{4}{15} \approx 26.66\%$
$\vdots$	$\vdots$
100	$\frac{100-1}{3 \times 100} = \frac{99}{300} = 33\%$

Conclusion

In one form or another, the scheduling of human (and nonhuman) activity is a pervasive and fundamental problem of modern life. At its most informal, it is part and parcel of the way we organize our everyday living (so much so that we are often scheduling things without realizing we are doing so). In its more formal incarnation, the systematic scheduling of a set of activities for the purposes of saving either time or money is a critical issue in management science. Business, industry, government, education—wherever there is a big project, there is a schedule behind it.

By now, it should not surprise us that at their very core, scheduling problems are mathematical in nature and that the mathematics of scheduling can range from the simple to the extremely complex. By necessity, we focused on the simple side, but it is important to realize that there is a great deal more to scheduling than what we learned here (see, for example, references 2 and 3).

In this chapter we discussed scheduling problems where we are given a set of *tasks*, a set of *precedence relations* among the tasks, and a set of identical *processors*. The objective is to schedule the tasks by properly assigning tasks to processors so that the *finishing time* for all the tasks is as small as possible.

To tackle these scheduling problems systematically, we first developed a graph model of such a problem, called the *project digraph*, and a general framework by means of which we can create, compare, and analyze schedules, called the *priority-list model*. Within the priority-list model, many strategies can be followed (with each strategy leading to the creation of a specific priority list). In the chapter, we considered two basic strategies for creating schedules. The first was the *decreasing-time algorithm*, a strategy that intuitively makes a lot of sense, but which in practice often results in inefficient schedules. The second strategy, called the *critical-path algorithm*, is generally a big improvement over the decreasing-time algorithm, but it falls short of the ideal goal: an efficient optimal algorithm. The critical-path algorithm is by far the best known and most widely used algorithm for scheduling in business and industry.

When scheduling with independent tasks, the decreasing-time algorithm and the critical-path algorithm become one and the same, and the finishing times they generate are never off by much from the optimal finishing times.

Although several other, more sophisticated strategies for scheduling have been discovered by mathematicians in the last 40 years, no optimal, efficient scheduling algorithm is presently known, and the general feeling among the experts is that there is little likelihood that such an algorithm actually exists. Efficient scheduling, nonetheless, will always remain a significant human goal—another task, if you will, in the grand cosmic schedule of life.

PROFILE	Ronald L. Graham (1935–)

Ron Graham is, by all accounts, the consummate juggler (in every sense of the word). For a period of almost 40 years, as a mathematician, director, vice president and head scientist at Bell Labs, AT&T's legendary research arm, Graham was famous for his ability to keep an eclectic array of activities all going at once—doing theoretical and applied research in several areas of mathematics, administering a major research institute, lecturing and teaching all over the world, teaching himself Chinese, and *really* juggling. (In the juggling arena Graham is no slouch—he happens to be an authority on the mathematics of juggling, the inventor of many new juggling routines, and a past president of the International Jugglers Association.)

Ronald L. Graham was born in 1935 in Taft, California, a small oil town located at the southern edge of the San Joaquin Valley. Graham discovered his natural talent and love of mathematics at an early age, and by the time he was 15, without even finishing high school, he won a full scholarship to the University of Chicago. After brief stints at Chicago and the University of California, Berkeley, Graham joined the Air Force and was stationed in Fairbanks, Alaska, where he served as a communication specialist while completing his undergraduate studies at the University of Alaska. After completing his tour of duty with the Air Force, Graham returned to UC Berkeley, where he completed a Ph.D. in mathematics in 1962. While a student at Berkeley, Graham, an accomplished gymnast and trampolinist, supported himself by performing with a small circus troupe called the Bouncing Bears, which performed acrobatic stunts at schools and fairs.

Graham joined Bell Labs in 1962—where he remained for 37 years—serving in a long list of scientific and administrative roles: researcher, head of the Mathematics Division, director of the Mathematics Center, vice president of research, and head scientist. Soon after joining Bell Labs, Graham was approached by engineers working on the computer system that was being developed to manage the antiballistic missile (ABM) defense program. One of the major issues they were facing was how to schedule the computers to identify and lock on to incoming enemy missiles in an efficient way. This turned out to be a classic scheduling problem of the kind discussed in this chapter (the computers are the *processors*; identifying and destroying the incoming missiles are the *tasks*). To work on these problems, Graham developed much of the theory of scheduling as we know it today. In so doing he also created a new field in the study of algorithms known as *worst-case analysis* (see Project A on page 354).

Graham retired from Bell Labs in 1999. Not one to sit idle, he promptly accepted an endowed chair as the Irwin and Joan Jacobs Professor of Computer Science and Information Science at the University of California, San Diego. In his late 60's, Graham continues to teach, carry on with his mathematics research, lecture around the world, juggle, and in general, keep going through life at a pace that amazes everyone else. When once asked how he managed to keep so many different things going, Graham's reply was "There are 168 hours in a week."

KEY CONCEPTS		
adjacent (arcs)		critical-path algorithm
arc		critical-path list
backflow algorithm		critical time
critical path		cycle

decreasing-time algorithm   precedence relation
decreasing-time list   priority list
digraph   priority-list model
factorial   processing time
finishing time   processor
incident (to and from)   project digraph
indegree (outdegree)   task (ineligible, ready, in execution,
independent tasks   completed)
path

EXERCISES

WALKING

A. Directed Graphs

1. For each of the following digraphs, make and complete a table similar to the one shown here.

Vertex	Degree	Indegree	Outdegree	Vertex is incident to	Vertex is incident from
A					
B					
⋮					

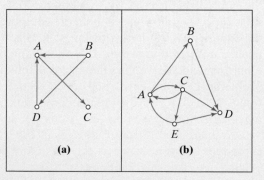

(a)   (b)

2. For each of the following digraphs, make and complete a table similar to the one shown here.

Vertex	Degree	Indegree	Outdegree	Vertex is incident to	Vertex is incident from
A					
B					
⋮					

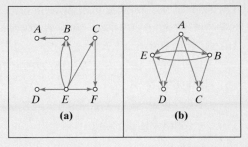

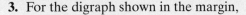

(a)   (b)

3. For the digraph shown in the margin,

  (a) list the vertices and arcs. (Use XY to represent an arc from X to Y.)

  (b) find the indegree of each vertex.

  (c) find the outdegree of each vertex.

  (d) find all vertices that are incident to E.

  (e) find all vertices that are incident from E.

  (f) find all the arcs adjacent to DE.

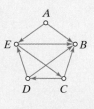

4. For the digraph shown in the margin,

  (a) list the vertices and arcs. (Use XY to represent an arc from X to Y.)

  (b) find the indegree of each vertex.

  (c) find the outdegree of each vertex.

  (d) find all vertices that are incident to B.

  (e) find all vertices that are incident from B.

  (f) find all the arcs adjacent to DB.

5. For each of the following, draw a picture of the digraph.

  (a) Vertices: A, B, C, D.

   Arcs: A is incident to B and C; D is incident from A and B.

  (b) Vertices: A, B, C, D, E.

   Arcs: A is incident to C and E; B is incident to D and E; C is incident from D and E; D is incident from C and E.

6. For each of the following, draw a picture of the digraph.

  (a) Vertices: A, B, C, D.

   Arcs: A is incident to B, C, and D; C is incident from B and D.

  (b) Vertices: V, W, X, Y, Z.

   Arcs: X is incident to V, Z, and Y; W is incident from V, Y, and Z; Z is incident to Y and incident from W and V.

7. For each of the following, draw a picture of the digraph. (*Note:* The vertex-set V is the set of vertices, and the arc-set A is the set of arcs.)

  (a) $V = \{A, B, C, D, E\}$

   $A = \{BA, BE, CE, EB, EC, ED\}$

  (b) $V = \{W, X, Y, Z\}$

   $A = \{WX, WY, WZ, XY, YX, YZ, ZW\}$

8. For each of the following, draw a picture of the digraph. (*Note*: The vertex-set $\mathcal{V}$ is the set of vertices, and the arc-set $\mathcal{A}$ is the set of arcs.)

    (a) $\mathcal{V} = \{A, B, C, D, E\}$

    $\mathcal{A} = \{AB, AE, CB, CD, DB, DE, EB, EC\}$

    (b) $\mathcal{V} = \{W, X, Y, Z\}$

    $\mathcal{A} = \{WW, WX, XX, XY, YY, YZ, ZW, ZZ\}$

9. For the following digraph,

    (a) find a path from vertex A to vertex F.

    (b) find a Hamilton path from vertex A to vertex F. (*Note:* A Hamilton path is a path that passes through every vertex of the graph once.)

    (c) find a cycle in the digraph.

    (d) explain why vertex F cannot be part of a cycle.

    (e) explain why vertex A cannot be part of a cycle.

10. For the following digraph,

    (a) find a path from vertex A to vertex D.

    (b) explain why the path you found in (a) is the only possible path from vertex A to vertex D.

    (c) find a cycle in the digraph.

    (d) explain why vertex A cannot be part of a cycle.

    (e) explain why vertex B cannot be part of a cycle.

    (f) find all the cycles in this digraph.

11. A city has several one-way streets, as well as two-way streets. The White Pine subdivision is a rectangular area 6 blocks long and 2 blocks wide. Streets alternate between one way and two way as shown in the following figure. Draw a digraph that represents the traffic flow in this neighborhood.

    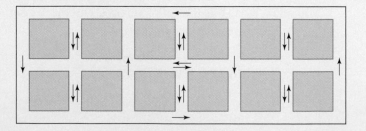

12. A mathematics textbook for liberal arts students consists of 10 chapters. While many of the chapters are independent of the others, some chapters require that previous chapters be covered first. The accompanying diagram illustrates the dependence. Draw a digraph that represents the dependence/independence relation among the chapters in the book.

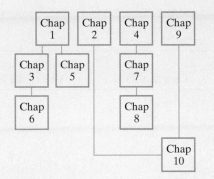

13. The digraph in the following figure is a *respect* digraph. That is, the vertices of the digraph represent members of a group, and an arc XY represents the fact that X respects Y.

(a) If you had to choose one person to be the leader of the group, whom would you pick? Explain.

(b) Who would be the worst choice to be the leader of the group? Explain.

14. The digraph in the following figure is an example of a *tournament* digraph. In this example the vertices of the digraph represent five volleyball teams in a round-robin tournament (i.e., every team plays every other team). An arc XY represents the fact that X defeated Y in the tournament. (*Note:* There are no ties in volleyball.)

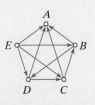

(a) Which team won the tournament? Explain.

(b) Which team came in last in the tournament? Explain.

15. Give an example of a directed graph with 4 vertices, with no loops or multiple arcs, and

(a) with each vertex having an indegree different than its outdegree.

(b) with 1 vertex of outdegree 3 and indegree 0, and with the remaining 3 vertices each having indegree 2 and outdegree 1.

16. Give an example of a directed graph with 7 vertices, no loops or multiple arcs, and 3 vertices of indegree 1 and outdegree 1, 3 vertices of indegree 2 and outdegree 2, and 1 vertex of indegree 3 and outdegree 3.

B. Project Digraphs

17. Draw a project digraph for a project consisting of the eight tasks described by the following table.

Task	Length of task	Tasks that must be completed before the task can start
A	3	
B	10	C, F, G
C	2	A
D	4	G
E	5	C
F	8	A, H
G	7	H
H	5	

18. Draw a project digraph for a project consisting of the eight tasks described by the following table.

Task	Length of task	Tasks that must be completed before the task can start
A	5	C
B	5	C, D
C	5	
D	2	G
E	15	A, B
F	6	D
G	2	
H	2	G

19. Eight computer programs need to be executed. One of the programs requires 10 minutes to complete, 2 programs require 7 minutes each to complete, 2 more require 12 minutes each to complete, and 3 of the programs require 20 minutes each to complete. Moreover, none of the 20-minute programs can be started until both of the 7-minute programs have been completed, and the 10-minute program cannot be started until both of the 12-minute programs have been completed. Draw a project digraph for this scheduling problem.

20. Ten computer programs need to be executed. Three of the programs require 4 minutes each to complete, 3 more require 7 minutes each to complete, and 4 of the programs require 15 minutes each to complete. Moreover, none of the 15-minute programs can be started until all of the 4-minute programs have been completed. Draw a project digraph for this scheduling problem.

21. Apartments Unlimited is an apartment maintenance company that refurbishes apartments before new tenants move in. The following table shows the tasks performed when refurbishing a one-bedroom apartment, the average time required for each task (measured in 15-minute increments), and the precedence relations between tasks. Draw a project digraph for refurbishing a one-bedroom apartment.

Tasks	Symbol/Time	Precedence relations
Bathrooms (clean)	$B(8)$	$L \to P$
Carpets (shampoo)	$C(4)$	$P \to K$
Filters (replace)	$F(1)$	$P \to B$
General cleaning	$G(8)$	$K \to G$
Kitchen (clean)	$K(12)$	$B \to G$
Lights (replace bulbs)	$L(1)$	$F \to G$
Paint	$P(32)$	$G \to W$
Smoke detectors (battery)	$S(1)$	$G \to S$
Windows (wash)	$W(4)$	$W \to C$
		$S \to C$

22. A ballroom is to be set up for a large wedding reception. The following table shows the tasks to be carried out, their processing times (in hours) based on one person doing that task, and the precedence relations between the tasks. Draw a project digraph for setting up the wedding reception.

Tasks	Symbol/Time	Precedence relations
Set up tables and chairs	$TC(1.5)$	$TC \to TN$
Set tablecloths and napkins	$TN(0.5)$	$TN \to PT$
Make flower arrangements	$FA(2.2)$	$CF \to PT$
Unpack crystal, china, and flatware	$CF(1.2)$	$PT \to TD$
Put place settings on table	$PT(1.8)$	$FA \to TD$
Put up table decorations (flower, balloons, etc.)	$TD(0.7)$	$TC \to SB$
Set up the sound system	$SS(1.4)$	
Set up the bar	$SB(0.8)$	

C. Schedules, Priority Lists, and the Decreasing-Time Algorithm

Exercises 23 through 26 refer to a project consisting of 11 tasks (A through K) with the following processing times (in hours): A(10), B(7), C(11), D(8), E(9), F(5), G(3), H(6), I(4), J(7), K(5).

23. (a) If a schedule with 3 processors has a completion time of 31 hours, what is the total idle time in the schedule?

(b) Explain why the completion time for a schedule with 3 processors can never be less than 25 hours.

24. (a) If a schedule with 5 processors has a completion time of 19 hours, what is the total idle time in the schedule?

(b) Explain why the completion time for a schedule with 5 processors can never be less than 15 hours.

25. Explain why the completion time for a schedule with 6 processors can never be less than 13 hours.

26. (a) Explain why the completion time for a schedule with 10 processors can never be less than 11 hours.

(b) Explain why it doesn't make sense to put more than 10 processors on this project.

Exercises 27 through 32 refer to the following project digraph.

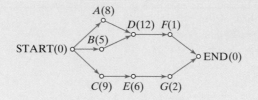

27. Using the priority list *D, C, A, E, B, G, F*, schedule the project with two processors.

28. Using the priority list *G, F, E, D, C, B, A*, schedule the project with two processors.

29. Using the priority list *D, C, A, E, B, G, F*, schedule the project with three processors.

30. Using the priority list *G, F, E, D, C, B, A*, schedule the project with three processors.

31. Using the decreasing-time algorithm, schedule the project with two processors.

32. Using the decreasing-time algorithm, schedule the project with three processors.

Exercises 33 through 38 refer to the apartment refurbishing project introduced in Exercise 21. The tasks, processing times, and precedence relations are shown in the following table.

Tasks	Symbol/Time	Precedence relations
Bathrooms (clean)	B(8)	$L \rightarrow P$
Carpets (shampoo)	C(4)	$P \rightarrow K$
Filters (replace)	F(1)	$P \rightarrow B$
General cleaning	G(8)	$K \rightarrow G$
Kitchen (clean)	K(12)	$B \rightarrow G$
Lights (replace bulbs)	L(1)	$F \rightarrow G$
Paint	P(32)	$G \rightarrow W$
Smoke detectors (battery)	S(1)	$G \rightarrow S$
Windows (wash)	W(4)	$W \rightarrow C$
		$S \rightarrow C$

33. Explain what is illegal about the following schedule for refurbishing an apartment with one worker.

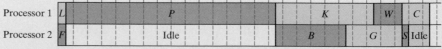

34. Explain what is illegal about the following schedule for refurbishing an apartment with two workers.

Time: 0 2 4 6 8 10 12 14 16 18 20 22 24 26 28 30 32 34 36 38 40 42 44 46 48 50 52 54

| Processor 1 | L | P | | K | W | C |
| Processor 2 | F | Idle | B | G | S Idle |

35. Using the priority list *B, C, F, G, K, L, P, S, W,*

(a) make a schedule for refurbishing an apartment with a single worker.

(b) make a schedule for refurbishing an apartment with two workers.

36. Using the priority list *W, C, G, S, K, B, L, P, F,*

(a) make a schedule for refurbishing an apartment with a single worker.

(b) make a schedule for refurbishing an apartment with two workers.

37. Using the decreasing-time algorithm, schedule the project with three workers.

38. Using the decreasing-time algorithm, schedule the project with four workers.

Exercises 39 through 44 refer to a copy center that must copy 13 court transcripts for a major trial. The times required (in hours) for the 13 jobs in increasing order are 3, 3, 4, 4, 5, 5, 5, 5, 6, 6, 7, 7, 12.

39. (a) Schedule the jobs on two copiers using the decreasing-time algorithm.

(b) Find an optimal schedule for two copiers.

40. (a) Schedule the jobs on three copiers using the decreasing-time algorithm.

(b) Find an optimal schedule for three copiers.

41. (a) Schedule the jobs on six copiers using the decreasing-time algorithm.

(b) Find an optimal schedule for six copiers.

(c) Explain why it doesn't make sense to assign more than six copiers to this project.

42. (a) Schedule the jobs on four copiers using the decreasing-time algorithm.

(b) Find an optimal schedule for four copiers.

43. (a) Schedule the jobs on eight copiers using the decreasing-time algorithm.

(b) Explain why it is impossible to schedule the jobs on eight copiers with a completion time that is less than 12 hours.

(c) Find an optimal schedule for eight copiers. (Try Exercise 41 first.)

44. (a) Schedule the jobs on five copiers using the decreasing-time algorithm.

(b) Explain why it is impossible to schedule the jobs on five copiers with a completion time that is less than 15 hours.

(c) Find an optimal schedule for 5 copiers. (*Hint:* The completion time is 15 hours.)

45. Find a schedule for building an MHU with two processors using the priority list *AD*(8), *AW*(6), *AF*(5), *IF*(5), *AP*(7), *IW*(7), *ID*(5), *IP*(4), *PL*(4), *PU*(3), *HU*(4), *IC*(1), *PD*(3), *EU*(2), *FW*(6). (See Example 8 in this chapter.)

46. Find a schedule for building an MHU with two processors using the decreasing-time algorithm. (See Example 8.9 in this chapter. Do the work on your own, and then compare with the step-by-step details shown in Table 8-3.)

D. Critical Paths and the Critical-Path Algorithm

Exercises 47 and 48 refer to the following project digraph.

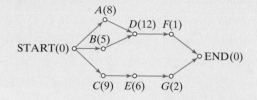

47. **(a)** Find the length of the critical path from each vertex.
    (b) Find the critical path for the project.
    (c) Use the critical-path algorithm to schedule the project using two processors.
    (d) Explain why the schedule obtained in (c) is optimal.

48. **(a)** Use the critical-path algorithm to schedule the project using three processors.
    (b) Explain why the schedule obtained in (a) is optimal.

Exercises 49 and 50 refer to the following project digraph.

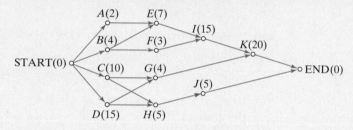

49. Use the critical-path algorithm to schedule the project using three processors.

50. **(a)** Find the length of the critical path from each vertex.
    (b) Find the critical path for the project.
    (c) Use the critical-path algorithm to schedule the project using two processors.
    (d) Explain why the schedule obtained in (c) is not optimal.

Exercises 51 and 52 refer to the project discussed in Exercise 18. The tasks, processing times and precedence relations are shown in the following table.

Task	Length of task	Tasks that must be completed before the task can start
A	5	C
B	5	C, D
C	5	
D	2	G
E	15	A, B
F	6	D
G	2	
H	2	G

51. Use the critical-path algorithm to schedule this project using three processors.

52. Use the critical-path algorithm to schedule this project using two processors.

53. Find a schedule for building an MHU with two processors using the critical-path algorithm. (See Example 14 in this chapter.)

54. Find a schedule for building an MHU with three processors using the critical-path algorithm. (See Example 14 in this chapter.)

JOGGING

55. Explain why, in any digraph, the sum of all the indegrees must equal the sum of all the outdegrees.

56. **Symmetric and totally asymmetric digraphs.** A digraph is called **symmetric** if, whenever there is an arc from vertex X to vertex Y, there is *also* an arc from vertex Y to vertex X. A digraph is called **totally asymmetric** if, whenever there is an arc from vertex X to vertex Y, there *is not* an arc from vertex Y to vertex X. For each of the following, state whether the digraph is symmetric, totally asymmetric, or neither.

(a) A digraph representing the streets of a town in which all streets are one-way streets

(b) A digraph representing the streets of a town in which all streets are two-way streets

(c) A digraph representing the streets of a town in which there are both one-way and two-way streets

(d) A digraph in which the vertices represent a bunch of men, and there is an arc from vertex X to vertex Y if X is a brother of Y

(e) A digraph in which the vertices represent a bunch of men, and there is an arc from vertex X to vertex Y if X is the father of Y

57. Determine whether each of the following is true or false. (If true, explain. If false, show with an example.)

(a) A schedule in which none of the processors is idle must be an optimal schedule.

(b) In an optimal schedule, none of the processors is idle.

58. A toy store is having a contest among its employees to find a team of two employees who can assemble a new toy on the market the quickest. The assembling of the toy involves seven tasks (A, B, C, D, E, F, and G). Two teams enter the contest: the red team (Joey and Sue) and the green team (Sharon and Jose). The rules of the contest specify that each task must be done by a single member of the team and that no team member can remain idle if there is a task to be done. The precedence relations for the tasks are shown in the following project digraph.

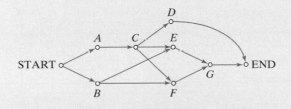

The red team practiced a lot, and both Joey and Sue are able to complete the tasks with the following times (in minutes): $A(1)$, $B(3)$, $C(1)$, $D(9)$, $E(4)$, $F(4)$, and $G(9)$.

The green team did not have as much time to practice, but both Sharon and Jose are able to complete the tasks with the following times: $A(2), B(4), C(2), D(10), E(5), F(5),$ and $G(10)$.

(a) Find an optimal schedule for the red team.

(b) Find an optimal schedule for the green team.

(c) Which team will win the contest?

(d) What would happen if the red team slowed their work a little on task C, each taking 2 minutes rather than 1 minute?

59. The following nine tasks are all independent: $A(4), B(4), C(5), D(6), E(7),$ $F(4), G(5), H(6), I(7)$. Four processors are available to carry out these tasks.

(a) Find a schedule using the critical-path algorithm.

(b) Find an optimal schedule.

60. The following seven tasks are all independent: $A(4), B(3), C(2), D(8),$ $E(5), F(3), G(5)$. Three processors are available to carry out these tasks.

(a) Find a schedule using the critical-path algorithm.

(b) Find an optimal schedule.

61. Use the critical-path algorithm to schedule independent tasks of length 1, 1, 2, 2, 5, 7, 9, 13, 14, 16, 18, and 20 using 3 processors. Is this schedule optimal? Explain.

Exercises 62 through 64 illustrate how it is sometimes possible to schedule independent tasks in such a way as to almost double the optimal completion time. The solution to these problems can be modeled after this example. The following schedule using 4 processors is optimal, having completion time of 8 hours (twice the number of processors).

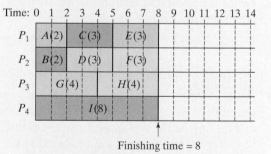

Finishing time = 8

The same independent tasks are scheduled using 4 processors again, but this time the completion time is 14 hours (2 hours less than twice the optimal completion time).

Time: 0 1 2 3 4 5 6 7 8 9 10 11 12 13 14

P_1	$A(2)$	$G(4)$			$I(8)$		
P_2	$B(2)$	$H(4)$			Idle	
P_3	$C(3)$	$D(3)$			Idle	
P_4	$E(3)$	$F(3)$			Idle	

Finishing time = 14

62. Using five processors, find two schedules for a project of independent tasks with

   (a) an optimal completion time of 10 hours (twice the number of processors).

   (b) a completion time of 18 hours (2 hours less than twice the optimal completion time).

63. For a project of independent tasks with 6 processors, find a schedule with

   (a) an optimal completion time of 12 hours (twice the number of processors).

   (b) a completion time of 22 hours (2 hours less than twice the optimal completion time).

64. For a project of independent tasks with seven processors, find a schedule with

   (a) an optimal completion time of 14 hours (twice the number of processors).

   (b) a completion time of 26 hours (2 hours less than twice the optimal completion time).

65. For independent tasks, the critical-path algorithm is never off by more than $E = (N-1)/(3N)$, where N is the number of processors.

   (a) Calculate the value of $E = (N-1)/(3N)$, for $N = 5, 6, 7, 8, 9$, and 10.

   (b) Explain why $(N-1)/(3N) < 1/3$ for any value of N.

66. In 1961, T. C. Hu of the University of California showed that in any scheduling problem in which all the tasks have equal processing times and in which the original project digraph (without the START and END vertices) is a tree, the critical-path algorithm will give an optimal schedule. Using this result, find an optimal schedule for the scheduling problem with the following project digraph, using three processors. Assume that each task takes three days. (Notice that we have omitted the START and END vertices.)

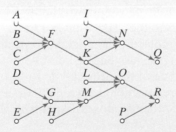

67. Let W represent the sum of all the processing times of the tasks, N be the number of processors, and F be the finishing time for a project.

   (a) Explain the meaning of the inequality $F \geq W/N$ and why it is true for any schedule.

   (b) Under what circumstances is $F = W/N$?

   (c) What does the value $NF - W$ represent?

A. Worst-Case Analysis of Scheduling Algorithms

We saw in this chapter that when creating schedules there is a wide range of possibilities, from *very bad schedules to optimal* (best possible) schedules. Our focus was on creating good (if possible optimal) schedules, but the study of bad schedules is also of considerable importance in many applications. In the late 1960s and early 1970s, Ronald Graham of AT&T Bell Labs (see the biographical profile on page 341) pioneered the study of worst-case scenarios in scheduling, a field known as *worst-case analysis*. As the name suggests, the issue in worst-case analysis is to analyze how bad a schedule can get. This research was motivated by a deadly serious question—how would the performance of the antiballistic missile defense system of the United States be affected by a failure in the computer programs that run the system?

In his research on worst-case analysis in scheduling, Graham made the critical discovery that there is a limit on how bad a schedule can be—no matter how stupidly put together, the total finishing time of a schedule can never be more than $F \cdot (2 - 1/N)$, where N is the number of processors and F is the optimal finishing time of the project.

Prepare a presentation explaining Graham's worst-case analysis result and its implications. Give examples comparing optimal schedules and the worst possible schedules for projects involving $N = 2$ and $N = 3$ processors. Explain why Graham's result tells us that given an *arbitrary* scheduling problem S, the finishing time of *any* schedule for S will always be less than twice the optimal finishing time for S.

Note: Suggested readings for this project are references 6 and/or 7.

B. Paradoxes in Scheduling

When using the priority-list model of scheduling, two situations can occur that are very counterintuitive. One of them can be informally described as the "more-is-less paradox," and it essentially says that putting additional processors to work on a project may not speed up the finishing time of the project at all—on the contrary, it may end up slowing down the project (optimal finishing time goes up). The second one is the "speed-up paradox," which says that speeding up the processors (i.e., replacing the original processing times for each task by a smaller processing time) may also end up being counterproductive and increasing the optimal finishing time of the project.

Prepare a presentation discussing both of these paradoxes. Give examples illustrating each of the paradoxes. Explain how these paradoxes are possible.

Note: Suggested readings for this project are references 6 and/or 7.

C. Dijkstra's Shortest-Path Algorithm

Just like there are weighted graphs, there are also *weighted digraphs*. A weighted digraph is a digraph in which each arc has a weight (distance, cost, time) associated with it. Weighted digraphs are used to model many important optimization problems in vehicle routing, pipeline flows, and so on. (Note that a weighted digraph is not the same as a project digraph—in the project digraph the vertices rather than the edges are the ones that have weights.) A fundamental question when working with weighted digraphs is finding among all possible paths starting at an arbitrary vertex X and ending at a different vertex Y which one is the shortest. (Of course, there may be no path at all starting at X and ending at Y, in which case the question is moot.) The classic algorithm for finding the shortest path between two vertices in a digraph is *Dijkstra's algorithm*, named after the Dutch computer scientist Edsger Dijkstra, who first discovered it in 1959.

Prepare a presentation describing Dijkstra's algorithm. Describe the algorithm carefully and illustrate how the algorithm works with a couple of examples. Discuss why Dijkstra's algorithm is important. Is it efficient? Optimal? Discuss some possible applications of the algorithm.

D. Tournaments

In the language of graph theory, a *tournament* is a digraph whose underlying graph is a complete graph. In other words, to create a tournament you can start with K_N (the complete graph on N vertices) and then change each edge into an arc by putting an arbitrary direction on it. The reason these graphs are called tournaments is because they can be used to describe the results of a tournament in which every player plays against every other player (no ties allowed).

Write a paper on the mathematics of tournaments (as defined previously). If you studied Chapter 1, you should touch on the connection between tournaments and the results of elections decided under the *method of pairwise comparisons*.

REFERENCES AND FURTHER READINGS

1. Baker, K. R., *Introduction to Sequencing and Scheduling*. New York: John Wiley & Sons, Inc., 1974.
2. Coffman, E. G., *Computer and Jobshop Scheduling Theory*. New York: John Wiley & Sons, Inc., 1976, chaps. 2 and 5.
3. Conway, R. W., W. L. Maxwell, and L. W. Miller, *Theory of Scheduling*. Reading, MA: Addison-Wesley Publishing Co., Inc., 1967.
4. Dieffenbach, R. M., "Combinatorial Scheduling," *Mathematics Teacher*, 83 (1990), 269–273.
5. Garey, M. R., R. L. Graham, and D. S. Johnson, "Performance Guarantees for Scheduling Algorithms," *Operations Research*, 26 (1978), 3–21.
6. Graham, R. L., "The Combinatorial Mathematics of Scheduling," *Scientific American*, 238 (1978), 124–132.
7. Graham, R. L., "Combinatorial Scheduling Theory," in *Mathematics Today*, ed. L. Steen. New York: Springer-Verlag, Inc., 1978, 183–211.
8. Graham, R. L., E. L. Lawler, J. K. Lenstra, and A. H. G. Rinnooy Kan, "Optimization and Approximation in Deterministic Sequencing and Scheduling: A Survey," *Annals of Discrete Mathematics*, 5 (1979), 287–326.
9. Gross, Jonathan, and Jay Yellen, *Graph Theory and Its Applications*. Boca Raton, FL: CRC Press, 1999, chap. 11.
10. Hillier, F. S., and G. J. Lieberman, *Introduction to Operations Research*, 3d ed. San Francisco: Holden-Day, Inc., 1980, chap. 6.
11. Roberts, Fred S., *Graph Theory and Its Applications to Problems of Society*, CBMS/NSF Monograph No. 29. Philadelphia: Society for Industrial and Applied Mathematics, 1978.
12. Wilson, Robin, *Introduction to Graph Theory*, 4th ed. Harlow, England: Addison-Wesley Longman Ltd., 1996.

PART THREE

Growth and Symmetry

CHAPTER 9

Spiral Growth in Nature

Fibonacci Numbers and the Golden Ratio

In nature's portfolio of architectural works, the shell of the chambered nautilus holds a place of special distinction. The spiral-shaped shell, with its revolving interior stairwell of ever-growing chambers, is more than a splendid piece of natural architecture—it is also a work of remarkable mathematical ingenuity.

Humans have imitated nature's wondrous spiral designs in their own architecture for centuries, all the while trying to understand how the magic works. What are the physical laws that govern spiral growth in nature? Why do so many different and unusual mathematical concepts come into play? How do the mathematical concepts and physical laws mesh together? What is the source of the intrinsic beauty of nature's spirals? Trying to answer these questions is the goal of this chapter.

> Come forth into the light of things, let Nature be your teacher.
>
> William Wordsworth

More than 2000 years ago, the ancient Greeks got us off to a great head start in our quest to understand nature with two great contributions: Euclidean geometry and irrational numbers. The next important mathematical connection came 800 years ago, with the serendipitous discovery by a medieval scholar named Fibonacci, of an amazing group of numbers now called Fibonacci numbers. Next came the discovery, by the French philosopher and mathematician René Descartes in 1638, that an equation he had studied for purely theoretical reasons ($r = ae^{\theta}$) is the very same equation that describes the spirals generated by seashells. Since then, other surprising connections between spiral-growing organisms and seemingly abstract mathematical concepts have been discovered, some within the last few years.

Exactly why and how these concepts play such a crucial role in the development of natural forms is not yet fully understood—not by humans, that is—a humbling reminder that nature still is the oldest and wisest of teachers.

9.1 Fibonacci Numbers

Listed in Table 9-1 is a widely known and disarmingly simple group of numbers called the **Fibonacci numbers**. They are named after the Italian Leonardo de Pisa, better known by the nickname Fibonacci (for more on Fibonacci, see the biographical profile at the end of the chapter). It doesn't take long to see the pattern these numbers follow. After the first two, which seem to stand on their own, each subsequent number is the sum of the two numbers before it: $2 = 1 + 1, 3 = 2 + 1, 5 = 3 + 2, \ldots, 144 = 89 + 55$, and so on.

TABLE 9-1 Fibonacci Numbers

$$1, 1, 2, 3, 5, 8, 13, 21, 34, 55, 89, 144, \ldots$$

Does the list of Fibonacci numbers ever end? No. The list goes on forever, with each new number in the sequence equal to the sum of the previous two. A non-ending, ordered list of numbers such as this is called an *infinite sequence* of numbers. Not surprisingly, this particular sequence is called the **Fibonacci sequence**.

As with any other sequence, there is a definite order to the Fibonacci numbers: a first Fibonacci number (1), a second (1), a third (2), ..., a seventh (13), ..., a tenth (55), an eleventh (89), and so on. Each Fibonacci number has its *place* in the Fibonacci sequence. The standard mathematical notation to describe a Fibonacci number is an F followed by a subscript indicating its place in the sequence. For example, F_8 stands for the *eighth* Fibonacci number, which is $21 (F_8 = 21)$; $F_{12} = 144$, and so on. A generic Fibonacci number can be written as F_N (with its place in the sequence being described by a generic position N). If we want to describe the Fibonacci numbers that come after F_N, we write F_{N+1}, F_{N+2}, and so on. If we want to describe the Fibonacci numbers that come before F_N, we write F_{N-1}, F_{N-2}, and so on.

With this convenient notation, the rule that *a Fibonacci number equals the sum of the two preceding Fibonacci numbers* becomes, simply,

$$\underbrace{F_N}_{\substack{\text{a generic} \\ \text{Fibonacci} \\ \text{number}}} = \underbrace{F_{N-1}}_{\substack{\text{the Fibonacci} \\ \text{number right} \\ \text{before it}}} + \underbrace{F_{N-2}}_{\substack{\text{the Fibonacci} \\ \text{number two positions} \\ \text{before it}}}.$$

Of course this rule cannot be applied to the first two Fibonacci numbers, F_1 (which has no predecessors) and F_2 (which has only one predecessor), so for a complete description, we must also give the values of the first two Fibonacci numbers: $F_1 = 1$ and $F_2 = 1$.

The three facts (i) $F_N = F_{N-1} + F_{N-2}$ (the *main rule*), (ii) $F_1 = 1$, and (iii) $F_2 = 1$ (the *seeds*) give us what is known as *a recursive definition* of the Fibonacci numbers. The definition is called **recursive**[1] because the main rule defines a number in the sequence using other (earlier) numbers in the same sequence.

---

[1] We have already encountered the idea of a recursive definition in Chapter 1, where we discussed recursive ranking methods in elections.

FIBONACCI NUMBERS (RECURSIVE DEFINITION)

- Seeds: $F_1 = 1$, $F_2 = 1$.

- Recursive rule: $F_N = F_{N-1} + F_{N-2}$, $(N \geq 3)$.

Using the above recursive definition, we can, in principle, compute any Fibonacci number, but this is easier said than done. Could we find, for example, F_{100}? You bet. How? It would be easy if we knew F_{99} and F_{98}, which we don't. In fact, there are plenty of Fibonacci numbers between $F_{12} = 144$ (the last one in Table 9-1) and F_{100} that presumably we don't know. At the same time, it is clear that if we set our minds to it, we could slowly but surely march up the Fibonacci sequence one step at a time: $F_{13} = 144 + 89 = 233$, $F_{14} = 233 + 144 = 377$, and so on. Let's cheat a little bit and say that we got to $F_{97} = 83,621,143,489,848,422,977$ and then to $F_{98} = 135,301,852,344,706,746,049$. Next comes

$$F_{99} = 135,301,852,344,706,746,049 + 83,621,143,489,848,422,977$$

$$= 218,922,995,834,555,169,026,$$

and finally

$$F_{100} = 218,922,995,834,555,169,026 + 135,301,852,344,706,746,049$$

$$= 354,224,848,179,261,915,075.$$

The moral of the preceding story is that the recursive definition gives us a blueprint as to how to calculate F_{100} (or any other Fibonacci number, for that matter), but it is an arduous climb up the hill, one step at a time. Imagine climbing up to F_{500} or F_{1000}.

The practical limitations of the recursive definition lead naturally to the question, Is there a better way? There is. A direct definition for Fibonacci numbers was discovered by Leonhard Euler (remember him from Chapter 5?) about 250 years ago. It is generally known as **Binet's formula**.[2]

FIBONACCI NUMBERS (BINET'S FORMULA)

$$F_N = \frac{\left(\frac{1 + \sqrt{5}}{2}\right)^N - \left(\frac{1 - \sqrt{5}}{2}\right)^N}{\sqrt{5}}.$$

In spite of its rather intimidating appearance, this formula has one advantage over a recursive definition: It gives us an explicit rule for calculating any Fibonacci number without having to first calculate the preceding Fibonacci numbers. For this reason, Binet's formula is called an **explicit** definition of the Fibonacci numbers.

---

[2] The formula was actually discovered first by Leonhard Euler and then rediscovered (almost 100 years later) by the Frenchman Jacques Binet, who somehow ended up getting the credit.

Three important numbers make a joint appearance in Binet's formula:

$$\sqrt{5}, \frac{1-\sqrt{5}}{2}, \text{ and } \frac{1+\sqrt{5}}{2}.$$

All three of these constants are irrational numbers, and as such, they have infinite, nonrepeating decimal expansions. The last of these numbers will be especially important to us in this chapter.

We can get to the heart of Binet's formula if we substitute the constants by letters:

$$a = \frac{1+\sqrt{5}}{2}, b = \frac{1-\sqrt{5}}{2}, \text{ and } c = \sqrt{5}$$

which gives us the much less intimidating form

$$F_N = \frac{a^N - b^N}{c}.$$

Until the advent of computers, Binet's formula was of limited practical use. Raising irrational numbers to high powers is hardly the kind of thing one would want to do by hand. With a good calculator or, better yet, with a good computer, one can use Binet's formula to directly calculate the values of fairly large Fibonacci numbers.

Fibonacci Numbers in Nature

What is so remarkable about the Fibonacci numbers? Among other things, it is the way they show up in the most unexpected places. Consistently, we find Fibonacci numbers when we count the number of petals in certain varieties of flowers: 3 (lily, iris); 5 (buttercup, columbine); 8 (cosmo, rue anemone); 13 (yellow daisy, marigold); 21 (English daisy, aster); 34 (oxeye daisy); 55 (Coral Gerber daisy), and so on.

White rue anemone   Yellow daisy   English daisy   White oxeye daisy   Coral Gerber daisy

Fibonacci numbers also appear consistently in conifers, seeds, and fruits (see Project A). The bracts in a pine cone, for example, spiral in two different directions in 8 and 13 rows; the scales in a pineapple spiral in three different directions in 8, 13, and 21 rows; the seeds in the center of a sunflower spiral in 55 and 89 rows.

Exactly why and how this unusual connection between a purely mathematical concept (Fibonacci numbers) and natural objects (daisies, sunflowers, pineapples, etc.) occurs is not yet fully understood but this connection certainly has something to do with the spiraling way in which these objects grow.

9.2 The Equation $x^2 = x + 1$ and the Golden Ratio

Before we continue our discussion of Fibonacci numbers, we will take a brief detour.

Here is a problem that looks like it came straight out of your typical algebra book: *Solve the quadratic equation $x^2 = x + 1$.*

You probably know how to do this without help, but let's go through the steps anyway. First, we move all terms to the left-hand side to get $x^2 - x - 1 = 0$. Next, we use the *quadratic formula*.[3] When the dust settles, the two solutions we get are

$$\left(\frac{1 + \sqrt{5}}{2}\right) \text{ and } \left(\frac{1 - \sqrt{5}}{2}\right)$$

We already saw these two numbers in the previous section, as they play prominent roles in Binet's formula. Both numbers are irrational and therefore have infinite, nonrepeating decimal expansions. For working purposes, we will use an approximation to three decimal places:

$$\left(\frac{1 + \sqrt{5}}{2}\right) \approx 1.618 \text{ and } \left(\frac{1 - \sqrt{5}}{2}\right) \approx -0.618.$$

It is worth noting that the first solution is positive, the second one is negative, and that their decimal expansions are identical.

The Golden Ratio

Let's focus on just the positive solution $(1 + \sqrt{5})/2$. This number is important enough to have its own symbol and name: It is called the **golden ratio** (or *golden section*) and is represented by the Greek letter Φ (phi). Thus, from here on, $\Phi = (1 + \sqrt{5})/2$.

The fact that the golden ratio Φ is a solution of the equation $x^2 = x + 1$ means that $\Phi^2 = \Phi + 1$. Using this fact repeatedly, we can calculate higher powers of Φ (which are relevant because of Binet's formula).

Multiplying both sides of the equation $\Phi^2 = \Phi + 1$ by Φ, we get $\Phi^3 = \Phi^2 + \Phi$. We now substitute $\Phi + 1$ for Φ^2 and get $\Phi^3 = (\Phi + 1) + \Phi = 2\Phi + 1$.

To find Φ^4, we multiply both sides of our last equation ($\Phi^3 = 2\Phi + 1$) by Φ and obtain $\Phi^4 = 2\Phi^2 + \Phi$. Substituting $\Phi + 1$ for Φ^2 gives us $\Phi^4 = 2(\Phi + 1) + \Phi = 3\Phi + 2$.

Continuing this way, we get

$$\Phi^5 = 3\Phi^2 + 2\Phi = 3(\Phi + 1) + 2\Phi = 5\Phi + 3,$$

$$\Phi^6 = 5\Phi^2 + 3\Phi = 5(\Phi + 1) + 3\Phi = 8\Phi + 5,$$

$$\Phi^7 = 8\Phi^2 + 5\Phi = 8(\Phi + 1) + 5\Phi = 13\Phi + 8,$$

and so on.

In each case, we can see that the coefficients of the expression on the right hand side are consecutive Fibonacci numbers. We can generalize the situation with the following useful formula.

$$\Phi^N = F_N\Phi + F_{N-1}.$$

---

[3] Just in case you forgot the quadratic formula, the solutions of $ax^2 + bx + c = 0$ are given by $x = (-b \pm \sqrt{b^2 - 4ac})/2a$. For $x^2 - x - 1 = 0$, we get $x = (1 \pm \sqrt{1 + 4})/2$. (For a brief review of the quadratic formula, see Exercises 27 through 34.)

In a sense, this formula is the reverse of Binet's. In Binet's formula we use powers of Φ to calculate Fibonacci numbers while here we use Fibonacci numbers to calculate powers of Φ.

The next connection between the Fibonacci numbers and the golden ratio is possibly the most surprising one. What do we get when we consider the ratios of consecutive Fibonacci numbers?

Table 9-2 shows the first 16 values of the ratio F_N/F_{N-1}. What's going on? It appears that, after some early fluctuation, the ratio "settles down" at a value of approximately 1.61803.

TABLE 9-2   **Ratios of Consecutive Fibonacci Numbers**

F_N	F_N/F_{N-1}	F_N	F_N/F_{N-1}	F_N	F_N/F_{N-1}
1		13	13/8 = 1.625	233	233/144 = 1.61805...
1	1/1 = 1.0	21	21/13 = 1.61538...	377	377/233 = 1.61802...
2	2/1 = 2.0	34	34/21 = 1.61904...	610	610/377 = 1.61803...
3	3/2 = 1.5	55	55/34 = 1.61764...	987	987/610 = 1.61803...
5	5/3 = 1.66666...	89	89/55 = 1.61818...	1597	1597/987 = 1.61803...
8	8/5 = 1.6	144	144/89 = 1.61797...		

If we look at the ratios of even larger Fibonacci numbers and write the decimals to many more decimal places, the pattern becomes even more apparent. Using a computer, we have calculated to 41 decimal places the ratios

$$\frac{F_{99}}{F_{98}} = \frac{218,922,995,834,555,169,026}{135,301,852,344,706,746,049} \approx 1.61803398874989484820458683436563811772033$$

and

$$\frac{F_{100}}{F_{99}} = \frac{354,224,848,179,261,915,075}{218,922,995,834,555,169,026} \approx 1.61803398874989484820458683436563811772030.$$

These two numbers, while not identical, match up everywhere except in the last digit, so that the difference between them is truly insignificant.

But there is more to it than that. The magic number that the ratios approach is $\Phi = (1 + \sqrt{5})/2$. Essentially this means that, except for the first few, *each Fibonacci number is approximately equal to ϕ times the preceding one, and the approximation gets better as the numbers get larger.*

The Golden Ratio in Art, Architecture, and Design

The golden ratio Φ is considered, right along with π (the ratio between the circumference and the diameter of a circle), among the great mathematical discoveries of antiquity. To the ancient Greeks, the golden ratio Φ was a mystical number, a gift from the gods. They called it the *divine proportion*. The famous Greek sculptor and architect Pheidias is said to have used the golden ratio in the proportions of his sculptures, as well as in his design of the Parthenon, perhaps the best-known building of ancient Greece.[4]

---

[4]In fact, the choice of Φ to represent the golden ratio comes from the fact that it's the first Greek letter in Pheidias's name.

The Parthenon, Athens, Greece. One of the architectural wonders of antiquity. Was the golden ratio used in the proportions?

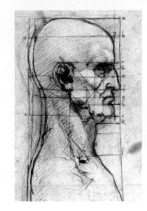

Leonardo da Vinci: *Head of an Old Man*, Accademia di Belle Arti, Venice. There is no way of telling whether Leonardo's grid governs or follows the proportions of the face.

During the Renaissance, famous artists such as Leonardo da Vinci and Botticelli knew about the golden ratio and used it in their paintings. In 1509, the friar Luca Pacioli wrote a book called, *De Divina Proportione* (*The Divine Proportion*). The book was the first mathematical treatise on the golden ratio, and its fame is due in part to its illustrations, drawn by none other than Leonardo da Vinci. Since Luca Pacioli's first book, literally hundreds of books and articles have been written about the golden ratio and its role in geometry, art, architecture, music, and nature (see Project B).

Jacopo de Barbari (1440–1515), *Fra Luca Pacioli and a Young man*, probably his pupil, Duke Guidobaldo da Montefeltro (1472-1509), (1495). Oil on wood, 99 x 120 cm. Museo Nazionale di Capodimonte, Naples, Italy. Erich Lessing/Art Resource, NY

9.3   Gnomons

The most common usage of the word *gnomon* is to describe the pin of a sundial—the part which casts the shadow which shows the time of day. The original Greek meaning of the word *gnomon* is "one who knows," so it's not surprising that the word should find its may into the vocabulary of mathematics.

In this section, we will discuss a different meaning for the word *gnomon*. Before we do so, we will take another short detour, this time revisiting a key concept from high school geometry—*similarity*.

We know from geometry that two objects are said to be **similar** if one is a scaled version of the other. When a slide projector takes the image in a slide and blows it up onto a screen, it creates a *similar* but larger image. When a photocopy machine reduces the image on a sheet of paper, it creates a *similar* but smaller image.

Here are some very basic facts about similarity that we will use in this section.

- Two triangles are similar if their sides are proportional. Alternatively, two triangles are similar if the sizes of their respective angles are the same.

- Two squares are always similar.

- Two rectangles are similar if their sides are proportional; that is, if

$$\frac{long\ side\ 1}{long\ side\ 2} = \frac{short\ side\ 1}{short\ side\ 2}.$$

- Two circles are always similar. Likewise, two circular disks are always similar.

- Two circular rings are similar if their inner and outer radii are proportional; that is, if

$$\frac{outer\ radius\ 1}{outer\ radius\ 2} = \frac{inner\ radius\ 1}{inner\ radius\ 2}.$$

We will now return to the main topic of this section—gnomons.

In geometry, a **gnomon** to a figure A is a connected figure (i.e., one without separate parts), which, when suitably *attached* to A, produces a new figure that is similar to A. By "attached," we mean that the two figures are joined together without overlapping anywhere. Informally, we will describe it this way: G is a gnomon to A if $G\&A$ is similar to A (Fig. 9.1). Here the symbol & should be taken to mean "attached in some suitable way." The study of gnomons goes back to the Greeks, presumably to Aristotle and his disciples, more than 2300 years ago.

FIGURE 9-1
(a) The original object A.
(b) The gnomon G.
(c) The two together: $G\&A$. The new object is similar to A.

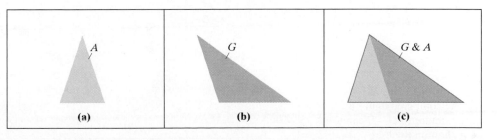

(a)   (b)   (c)

EXAMPLE 9.1

The square S in Fig. 9-2(a) has the L-shaped figure G in Fig. 9-2(b) as a gnomon, because when G is attached to S as in Fig. 9-2(c), we get a square S'.

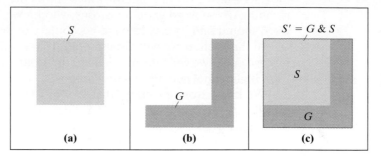

FIGURE 9-2
(a) A square S.
(b) Its gnomon G.
(c) The combined figures form a larger square.

Note that the wording is *not* reversible. The square S *is not* a gnomon to the L-shaped figure G, since there is no way to attach the two to form an L-shaped figure that is similar to G (see Exercise 54).

Gnomons to a shape are not unique, and there are other gnomons for the square S besides an L-shaped figure.

EXAMPLE 9.2

The circular disk C in Fig. 9-3(a) has as a gnomon an O-ring, like G in Fig. 9-3(b). The inner radius of G has to be r; the outer radius R can be any number greater than r. When we attach the O-ring G to C, we get a new circular disk C' [Fig. 9-3(c)] that is similar to C (all circular disks are similar).

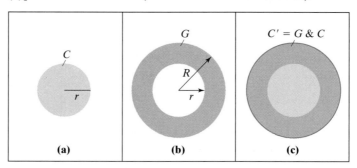

FIGURE 9-3
(a) A circular disk C.
(b) Its gnomon G.
(c) The combined figures form a larger circular disk.

EXAMPLE 9.3

Consider an O-ring O with outer radius r [Fig. 9-4(a)] and the O-ring H with inner radius r and outer radius R [Fig. 9-4(b)]. Can H be gnomon to the original O-ring O?

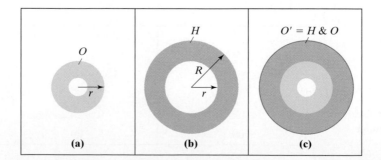

FIGURE 9-4
(a) An O-ring O.
(b) Another O-ring H.
(c) The combined figure is never similar to O—H cannot be a gnomon to O.

One is tempted to think that with the right choice of the outer radius R it might work, but it never will. No matter how we choose the outer radius of the O-ring H, when we attach the two O-rings together [Fig. 9-4(c)], O' will not be similar to O (see Exercise 56). ■

EXAMPLE 9.4

Consider a rectangle R of height h and base b as shown in Fig. 9-5(a). The L-shaped object G shown in Fig. 9-5(b) is a gnomon to the rectangle R if the rectangle R' [Fig. 9-5(c)] with height $(h + x)$ and base $(b + y)$ is similar to R. This means that the ratios $\dfrac{b}{h}$ and $\dfrac{b + y}{h + x}$ are equal, which, in turn, implies that $\dfrac{b}{h}$ and $\dfrac{y}{x}$ must also be equal (see Exercise 61).

FIGURE 9-5
(a) A rectangle R.
(b) Its gnomon G.
(c) The combined figures form a rectangle similar to R.

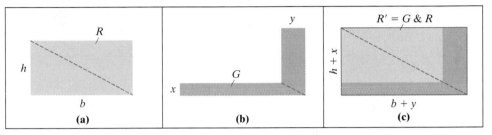

A simple geometric way to build the L-shaped gnomon G is by noticing that the line through the diagonal of the original rectangle R must also be the diagonal through the "L-corner" in G. ■

EXAMPLE 9.5

In this example, we are going to do things a little bit backward. Let's start with an isosceles triangle T, with vertices B, C, and D whose angles measure $72°$, $72°$, and $36°$, respectively, as shown in Fig. 9-6(a). On side DC, we mark a point A so that BA is congruent to BC [Fig. 9-6(b)]. (This can be done easily by centering a compass at B and drawing an arc of a circle of radius BC.) The triangle T' with vertices at C, B, and A is isosceles, with equal angles at C and A. Therefore, T' has angles of $72°$, $36°$, and $72°$. This makes T' similar to the original triangle T.

"So what?" you may ask. Where is the gnomon to triangle T? We don't have one yet! But we *do* have a gnomon to triangle T', and it is triangle G' with vertices A, B, and D [Fig. 9-6(c)]. After all, when triangle G' is attached to triangle T', we get triangle T. Note that gnomon G' is also an isosceles triangle: its angles are $36°$, $36°$, and $108°$.

FIGURE 9-6
(a) A 72-72-36° isosceles triangle T. (b) An isosceles triangle T' is constructed inside of T. (c) G' is a gnomon to T'.

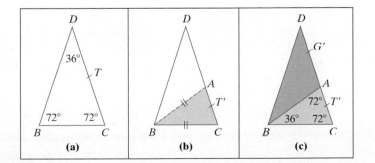

We now know how to find a gnomon not only to triangle T' but to any 72–72–36° triangle, including the original triangle T: Attach a 36–36–108° triangle to one of the longer sides [Fig. 9-7(a)]. If we repeat this process indefinitely, we get a spiraling series of ever-increasing 72–72–36° triangles [Fig. 9-7(b)]. It's not too far-fetched to use a family analogy: Triangles T and G are the *parents*, with T having the *dominant* genes; the *offspring* of their union looks just like T (but bigger). The offspring then has offspring of its own (looking exactly like grandfather T), and so on ad infinitum.

FIGURE 9-7

The process of adding a 36–36–108° gnomon G to a 72–72–36° triangle T can be repeated indefinitely, producing a spiraling chain of ever-increasing similar triangles.

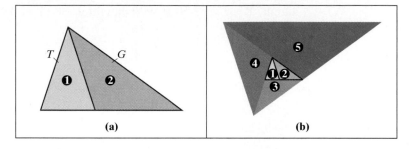

(a)   (b)

Example 9.5 is of special interest to us for two reasons. First, this is the first time we have an example where the figure and its gnomon are of the same type (isosceles triangles). Second, the isosceles triangles in this story (72–72–36° and 36–36–108°) have a property that makes them unique: In both cases, the ratio of their sides (longer side over shorter side) is the golden ratio (see Exercise 63). These are the only two isoceles triangles with this property, and for this reason they are called **golden triangles**.

EXAMPLE 9.6

In this example, we start with a rectangle R whose shorter side has a length 1 and whose longer side has an unspecified length x as shown in Fig. 9-8(a). We would like to find out if this rectangle can have a square gnomon [Fig. 9-8(b)] and, if so, what should x be? The reason we are interested in this question is that squares are one of the fundamental building blocks in nature and make for particularly nice gnomons.

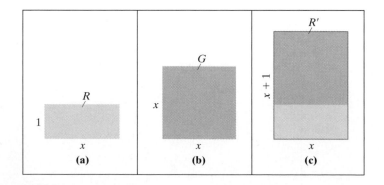

FIGURE 9-8

Can G be a gnomon to R?

For rectangle R' [Fig. 9-8(c)] to be similar to R, we must have

$$\frac{\text{long side of } R'}{\text{long side of } R} = \frac{\text{short side of } R'}{\text{short side of } R}$$

or, in other words,

$$\frac{x+1}{x} = \frac{x}{1}.$$

This equation is equivalent to the equation $x^2 = x + 1$, whose positive solution is the golden ratio Φ (see Section 9.2). Since x is the length of the side of rectangle R, it must be positive, and thus we can conclude that $x = \Phi = (1 + \sqrt{5})/2$. ■

Our choice for the dimensions of rectangle R (x and 1) was dictated by convenience (having the short side equal 1 simplifies the computations), but what is true for R is certainly true for any other rectangle similar to R. In other words, *any rectangle whose sides are in the proportion of the golden ratio has a square gnomon, and conversely, if R is a rectangle having a square gnomon, the ratio of its sides (long side/short side) equals the golden ratio Φ.*

Golden Rectangles

Rectangles whose sides are in the proportion of the golden ratio are called **golden rectangles**. In other words, a rectangle with dimensions l (long side) and s (short side) is a *golden rectangle* if $\dfrac{l}{s} = \Phi$.

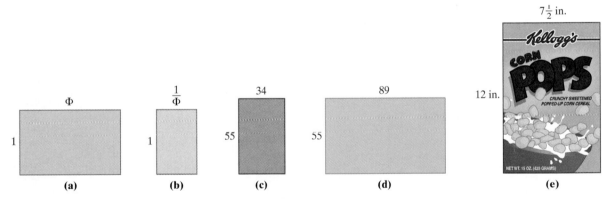

FIGURE 9-9
An assortment of golden and almost golden rectangles. (a) and (b) are exact golden rectangles, (c) and (d) are Fibonacci rectangles, and (e) has dimensions whose ratio is 1.6. To the naked eye, they all have the same proportions.

Figure 9-9 shows an assortment of rectangles. In Figs. 9-9(a) and (b), we have *exact* golden rectangles: the ratios between the longer and shorter sides are $\Phi/1 = \Phi$ and $1/(1/\Phi) = \Phi$, respectively. In Figs. 9-9(c) and (d), we have rectangles whose sides are consecutive Fibonacci numbers.[5] The ratios between their longer and shorter sides are $55/34 = 1.617647\ldots$, and $89/55 = 1.61818\ldots$. These numbers are so close to the golden ratio that we might as well call these rectangles "almost golden." The last rectangle shows the front of a box of "Corn Pops." The dimensions are 12 in. by $7\frac{1}{2}$ in., with a ratio of $12/7.5 = 1.6$, also very close to the golden ratio. It is safe to say that, at least to the naked eye, all of these rectangles look golden.

From a design perspective, golden (and almost golden) rectangles have a special appeal, and they show up in many everyday objects, from cereal boxes to

---

[5]Such rectangles are usually called **Fibonacci rectangles**.

calendars. In some sense, golden rectangles strike the perfect middle ground between being too "skinny" and being too "squarish." A prevalent theory is that human beings have an innate aesthetic bias in favor of golden rectangles, which, so the theory goes, appeal to our natural sense of beauty and proportion. This theory, known in psychology circles as the *golden ratio hypothesis*, originated with the classic experiments of the German psychologist Gustav Fechner in 1876 (readers interested in Fechner's experiment and the golden ratio hypothesis are encouraged to take a look at Project E).

Spiral Growth

In nature, where form usually follows function, the perfect balance of a golden rectangle shows up in spiral-growing organisms, often in the form of consecutive Fibonacci numbers. To see how this connection works, consider the following example, which serves as a model for certain natural growth processes.

EXAMPLE 9.7 Spiraling Fibonacci Rectangles

Start with a 1-by-1 square [marked ❶ in Fig. 9-10(a)]. Tack onto it another 1-by-1 square [marked ❷ in Fig. 9-10(b)]. Squares ❶ and ❷ together form a 2-by-1 rectangle, as shown in Fig. 9-10(b). We will call this the "second-generation" shape. For the third generation, tack on a 2-by-2 square ❸ as shown in Fig. 9-10(c). The "third-generation" shape (❶, ❷, and ❸ together) is the 2-by-3 rectangle in Fig. 9-10(c). Next, tack onto it a 3-by-3 square ❹ as shown in Fig. 9-10(d), giving a 5-by-3 rectangle. Then tack on a 5-by-5 square ❺ as shown in Fig. 9-10(e), resulting in an 8-by-5 rectangle.

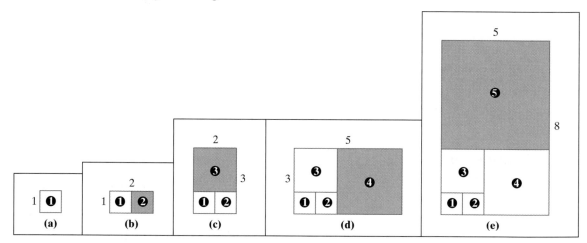

FIGURE 9-10
Fibonacci rectangles beget Fibonacci rectangles.

We can continue this process indefinitely, at each generation getting a bigger rectangle. The figures are all rectangles whose sides are consecutive Fibonacci numbers, and by the fifth generation, they are almost golden rectangles. By the time this process reaches the tenth generation, we have a 55-by-89 rectangle, with a long-side-to-short-side ratio of 1.61818—for all practical purposes, a golden rectangle. Being almost golden, the successive rectangles in Fig. 9-10 are, in practice, essentially similar rectangles (see Exercises 49 and 50). This kind of behavior—getting bigger while preserving a constant shape—is characteristic of the growth of many natural organisms. ■

The next example is a simple variation of Example 9.7.

EXAMPLE 9.8 The "Chambered" Fibonacci Rectangle

Let's repeat the process in the previous example, except that now let's add to each square an interior "chamber" in the form of a quarter-circle. We need to be a little more careful about where we attach the chambered square in each successive generation, but other than that, we can repeat the sequence of steps in Example 9.7 to get the sequence of shapes shown in Fig. 9-11. These figures depict the consecutive generations in the evolution of the *chambered Fibonacci rectangle*. The outer spiral formed by the circular arcs is often called a **Fibonacci spiral**, shown in Fig. 9-12.

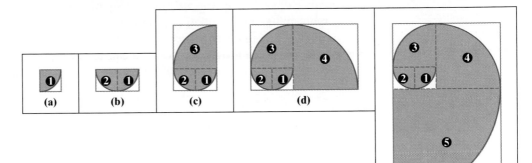

FIGURE 9-11
Revolving "chambered" Fibonacci rectangles.

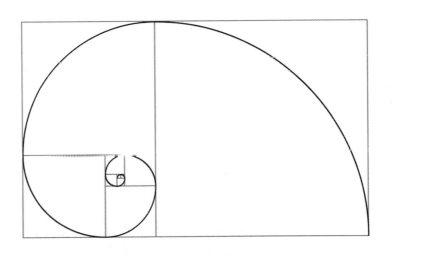

FIGURE 9-12
A Fibonacci spiral after 10 "generations."

9.4 Gnomonic Growth

As soon as humans realized that nature is a gifted builder, architect, and designer from which they could learn much about the form and function of things, understanding the laws that govern the growth and form of natural organisms became an important part of natural science. (Just for plants alone, for example, there is a discipline called *phyllotaxis*, whose primary concern is the study of the patterns of growth and distribution of *lateral organs*: leaves, petals, stalks, scales, and so on.)

Natural organisms grow in essentially two different ways. The more common type of growth (and the one we are most familiar with) is the growth exhibited by humans, animals, and many plants. This can be called *all-around growth*, in which all living parts of the organism grow simultaneously, although not necessarily at the same rate. One characteristic of this type of growth is that there is no obvious way to distinguish between the newer and the older parts of the organism. In fact, the distinction between new and old parts does not make much sense. The historical record (so to speak) of the organism's growth is lost. By the time the child becomes an adult, no identifiable traces of the child (as an organism) remain—that's why we need photographs!

Contrast this with the kind of growth exemplified by the shell of the chambered nautilus, a ram's horn, or the trunk of a redwood tree. This we may informally call *growth at one end*, or *asymmetric growth*. With this type of growth, the organism has a part added to it (either by its own or outside forces) in such a way that the old organism together with the added part form the new organism. At any stage of the growth process, we can see not only the present form of the organism, but also the organism's entire past. All the previous stages of growth are the building blocks that make up the present structure.

The second relevant fact is that most such organisms grow in a way that preserves their overall shape; in other words, they remain similar to themselves. This is where gnomons come into the picture. For the organism to retain its shape as it grows, the new growth must be a *gnomon* of the entire organism. We will call this kind of growth process **gnomonic growth**.

We have already seen abstract mathematical examples of gnomonic growth (Examples 9.7 and 9.8). Here are a pair of more realistic examples.

EXAMPLE 9.9

We know from Example 9.2 that the gnomon to a circle is an O-ring with an inner radius equal to the radius of the circle. We can thus have circular growth (Fig. 9-13). Rings added one layer at a time to a starting circular structure preserve the circular shape throughout the structure's growth. When carried to three dimensions, this is a good model for the way the trunk of a redwood tree grows. And this is why we can "read" the history of a felled redwood tree by studying its rings. ■

EXAMPLE 9.10

Figure 9-14 shows a diagram of a cross section of the chambered nautilus, the example we used to open this chapter. The chambered nautilus builds its shell in stages, each time adding another chamber to the already existing shell. At every stage of its growth, the shape of the chambered nautilus shell remains the same—the beautiful and distinctive spiral shown in the photograph. This is a classic example of gnomonic growth—each new chamber added to the shell is a gnomon of the entire shell. The gnomonic growth of the shell proceeds, in essence, as follows:

FIGURE 9-13

The growth rings in a redwood tree—an example of circular gnomonic growth.

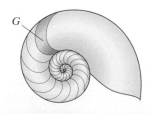

FIGURE 9-14

Gnomonic growth in the chambered nautilus. The spiral curve it generates is a logarithmic spiral.

Starting with its initial shell (which is a tiny spiral similar in all respects to the adult spiral shape), the animal builds a chamber (by producing a special secretion around its body that calcifies and hardens). The resulting, slightly enlarged spiral shell is similar to the original one. The process then repeats itself over many stages, each one a season in the growth of the animal. Each new chamber is a gnomon to the shell, creating an enlarged shell that is larger, but in all other respects similar to the younger shell. This process is a real-life variation of the mathematical spiral-building process discussed in Example 9.8. The curve generated by the outer edge of a nautilus shell's cross section is called a *logarithmic spiral* (see Project C for more on logarithmic spirals). ■

More complex examples of gnomonic growth occur in sunflowers, daisies, pineapples, pine cones, and so on. Here, the rules that govern growth are somewhat more involved, but Fibonacci numbers and the golden ratio once again play a prominent role. The interested reader is encouraged to follow up on this subject by way of Project A.

Conclusion

Some of the most beautiful shapes in nature arise from a basic principle of design: *Form follows function*. The beauty of natural shapes is a result of their inherent elegance and efficiency, and imitating nature's designs has helped humans design and build beautiful and efficient structures of their own.

In this chapter, we examined a special type of growth—gnomonic growth—where an organism grows by the addition of gnomons, thereby preserving its basic shape even as it grows. Many beautiful spiral-shaped organisms, from sea shells to flowers, exhibit this type of growth.

To us, understanding the basic principles behind spiral growth was relevant because it introduced us to some important mathematical concepts that have been known and studied in their own right for centuries: Fibonacci numbers, the golden ratio, gnomons, golden rectangles, and Fibonacci spirals.

To humans, these abstract mathematical concepts have been, by and large, intellectual curiosities. To nature—the consummate artist and builder—they are the building tools for some of its most beautiful creations. Whatever lesson one draws from this, it should include something about the inherent value of good mathematics.

Left: Sunflowers, by Vincent Van Gogh. 1888. Neue Pinakothek, Munich, Germany. *Center:* Sunflower. *Right:* Rome Sports Palace dome design (Pier Paolo Nervi, architect).

| P R O F I L E | Leonardo Fibonacci (circa 1175–1250) |

Leonardo Fibonacci was born in Pisa, Italy sometime around 1175 (there are no reliable records of Leonardo's exact date of birth). His actual family name was Bonaccio, so the name Fibonacci, which he preferred and by which he is now known, is a sort of nickname derived from the Latin for *filius Bonacci*, meaning "son of Bonaccio." Pisa, best known nowadays as the home of the famous leaning tower, was at the time an independent and prosperous city-state with important trade routes throughout the Mediterranean. Leonardo's father, Guglielmo Bonaccio, was a civil servant/diplomat who served as the agent and representative for Pisan business interests in North Africa. When Leonardo was a young boy he moved with his father to North Africa, where he spent most of his youth. Thus, rather than receiving a traditional European education, Leonardo was educated in the Moorish tradition, a happenstance that might very well have affected the entire course of Western civilization.

As his father wanted him to be a merchant, a significant part of Leonardo's education was the study of mathematics and accounting. Under the Moors, Leonardo learned arithmetic and algebra using the Hindu-Arabic decimal system of numeration—the very same system we use today and a system leaps and bounds superior to the Roman system that was then used in Europe. Leonardo was not the first European to learn about Hindu-Arabic numerals and the algorithms to manipulate them, but he was the first one to truly grasp their power.

Sometime around the year 1200, when he was about 25, Leonardo returned to Pisa. For the next 25 years, he dedicated himself to bringing the Hindu-Arabic methods of mathematics to Europe. During this period he published at least four books (these are the ones for which manuscripts still exist—it is believed that there were two other books for which the manuscripts are lost). His first book, *Liber Abaci (The Book of the Abacus)*, was published in 1202.[6]

One can argue that the publication of *Liber Abaci* is one of those little-known events that changed the course of history. In *Liber Abaci*, Fibonacci described the workings of the Hindu-Arabic number system and richly illustrated it with marvelous problems and examples. The advantages of this place-value based decimal system for accounting and bookkeeping were so obvious that in a very short time it became the de facto standard for conducting business, the Roman system was abandoned, and the rest is history.

A certain man put a pair of rabbits in a place surrounded on all sides by a wall. How many pairs of rabbits can be produced from that pair in a year if it is supposed that every month each pair begets a new pair which from the second month on becomes productive?

One of the many problems that Fibonacci introduced in *Liber Abaci* was the famous rabbit-breeding problem whose solution is the sequence 1, 1, 2, 3, 5, 8, 13, 21, 34, ..., which we now call the *Fibonacci sequence* (for details see Example 10.1 in Chapter 10). To Fibonacci, the rabbit problem was just another problem among the many discussed in his book—little did he know that the numbers generated by the solution to the problem would be his ticket to immortality. It is a great irony that the man who brought the Western world its modern system for doing mathematics is best remembered now for the reproductive pattern of a bunch of fictitious rabbits.

By the time Fibonacci published his last book, *Liber Quadratorum*, he was already a famous man, so much so that in 1225 Frederick II, then the Holy Roman Emperor, asked to meet him during a visit to Pisa. Sadly, not much is known about the latter part of Fibonacci's life or about the date and circumstances of his death, believed to be between 1240 and 1250.

---

[6]An English translation of *Liber Abaci*, by Laurence Sigler (reference 18), has recently been published.

Binet's formula
explicit definition (of a sequence)
Fibonacci number
Fibonacci rectangle
Fibonacci sequence
Fibonacci spiral
gnomon

gnomonic growth
golden ratio
golden rectangle
golden triangle
recursive definition (of a sequence)
similarity

WALKING

A. Fibonacci Numbers

1. Compute the value of each of the following.
   (a) F_{10}
   (b) $F_{10} + 2$
   (c) F_{10+2}
   (d) $F_{10} - 8$
   (e) F_{10-8}
   (f) $3F_4$
   (g) $F_{3 \times 4}$

2. Compute the value of each of the following.
   (a) F_{12}
   (b) $F_{12} - 1$
   (c) F_{12-1}
   (d) $F_{12}/2$
   (e) $F_{12/2}$

3. Describe in words what each of the expressions represents.
   (a) $3F_N + 1$
   (b) $3F_{N+1}$
   (c) $F_{3N} + 1$
   (d) F_{3N+1}

4. Describe in words what each of the expressions represents.
   (a) $F_{2N} - 3$
   (b) F_{2N-3}
   (c) $2F_N - 3$
   (d) $2F_{N-3}$

5. Given that $F_{36} = 14,930,352$ and $F_{37} = 24,157,817$,

   (a) find F_{38}.

   (b) find F_{35}.

6. Given that $F_{31} = 1,346,269$ and $F_{33} = 3,524,578$,

   (a) find F_{32}.

   (b) find F_{34}.

7. Determine which of the following two rules (I or II) is an equivalent formulation of the recursive rule for Fibonacci numbers given in the chapter $(F_N = F_{N-1} + F_{N-2}, N \geq 3)$. Explain.

   (I) $F_{N+2} = F_{N+1} + F_N, N > 0$

   (II) $F_N = F_{N+1} + F_{N+2}, N > 0$

8. Determine which of the following two rules (I or II) is an equivalent formulation of the recursive rule for Fibonacci numbers given in the chapter $(F_N = F_{N-1} + F_{N-2}, N \geq 3)$. Explain.

   (I) $F_{N-1} - F_N = F_{N+1}, N > 1$

   (II) $F_{N+1} - F_N = F_{N-1}, N > 1$

9. Write each of the following integers as the sum of *distinct* Fibonacci numbers.

   (a) 47

   (b) 48

   (c) 207

   (d) 210

10. Write each of the following integers as the sum of *distinct* Fibonacci numbers.

    (a) 52

    (b) 53

    (c) 107

    (d) 112

Exercises 11 through 16 refer to various known relationships among the Fibonacci numbers.

11. Consider the following sequence of equations involving Fibonacci numbers.

    $1 + 2 = 3$
    $1 + 2 + 5 = 8$
    $1 + 2 + 5 + 13 = 21$
    $1 + 2 + 5 + 13 + 34 = 55$
    $\vdots$

    (a) Write down a reasonable choice for the fifth equation in this sequence.

    (b) Find the subscript that will make the following equation true.

    $$F_1 + F_3 + F_5 + \ldots + F_{21} = F_?$$

12. Consider the following sequence of equations involving Fibonacci numbers.

$$2(2) - 3 = 1$$
$$2(3) - 5 = 1$$
$$2(5) - 8 = 2$$
$$2(8) - 13 = 3$$
$$\vdots$$

(a) Write down a reasonable choice for the fifth equation in this sequence.

(b) Find the subscript that will make the following equation true.

$$2(F_?) - F_{15} = F_{12}$$

(c) Find the subscript that will make the following equation true.

$$2(F_{N+2}) - F_{N+3} = F_?$$

13. Consider the following sequence of equations involving Fibonacci numbers.

$$(1 + 1) + 1 = 3$$
$$(1 + 1 + 2) + 1 = 5$$
$$(1 + 1 + 2 + 3) + 1 = 8$$
$$(1 + 1 + 2 + 3 + 5) + 1 = 13$$
$$\vdots$$

(a) Write down reasonable choices for the fifth and sixth equations in this sequence.

(b) Find the subscript that will make the following equation true.

$$(F_1 + F_2 + F_3 + \ldots + F_7 + F_8) + 1 = F_?$$

(c) Find the subscript that will make the following equation true.

$$(F_1 + F_2 + F_3 + \ldots + F_N) + 1 = F_?$$

14. Fact: *If we make a list of any 10 consecutive Fibonacci numbers, the sum of all these numbers divided by 11 is always equal to the seventh number on the list.*

(a) Verify this fact is true of the first 10 Fibonacci numbers.

(b) Verify this fact for the set of 10 consecutive Fibonacci numbers that start with F_5.

(c) Using F_N as the first Fibonacci number on the list, write this fact as a mathematical equation.

15. Fact: *If we make a list of any four consecutive Fibonacci numbers, twice the third one minus the fourth one is always equal to the first one.*

(a) Verify this fact is true of the first four Fibonacci numbers.

(b) Verify this fact for the set of four consecutive Fibonacci numbers that start with F_4.

(c) Using F_N as the first Fibonacci number on the list, write this fact as a mathematical equation.

16. Fact: *If we make a list of any four consecutive Fibonacci numbers, the first one times the fourth one is always equal to the third one squared minus the second one squared.*

(a) Verify this fact is true of the first four Fibonacci numbers.

(b) Verify this fact for the set of four consecutive Fibonacci numbers that start with F_8.

(c) Using F_N as the first Fibonacci number on the list, write this fact as a mathematical equation.

B. The Golden Ratio

Exercises 17 through 20 require the use of a calculator with an exponent key. (On most calculators, the exponent key looks something like $\boxed{y^x}$. To calculate an exponent, say, $(2.3)^7$, first enter 2.3, then enter $\boxed{y^x}$ and finally enter 7 followed by $\boxed{=}$. On other calculators, the exponent key looks like $\boxed{\wedge}$. With such a calculator to calculate the exponent $(2.3)^7$, first enter 2.3. Then enter $\boxed{\wedge}$, and finally enter 7 followed by $\boxed{enter}$.)

17. Calculate each of the following to five decimal places.

    (a) $\left(\dfrac{1+\sqrt{5}}{2}\right)^8$

    (b) $21\left(\dfrac{1+\sqrt{5}}{2}\right) + 13$

    (c) $\dfrac{\left(\dfrac{1+\sqrt{5}}{2}\right)^8 - \left(\dfrac{1-\sqrt{5}}{2}\right)^8}{\sqrt{5}}$

18. Calculate each of the following to five decimal places.

    (a) $\left(\dfrac{1+\sqrt{5}}{2}\right)^{10}$

    (b) $55\left(\dfrac{1+\sqrt{5}}{2}\right) + 34$

    (c) $\dfrac{\left(\dfrac{1+\sqrt{5}}{2}\right)^{10} - \left(\dfrac{1-\sqrt{5}}{2}\right)^{10}}{\sqrt{5}}$

19. Using $\Phi = (1+\sqrt{5})/2$, calculate each of the following, rounded to the nearest integer.

    (a) $\Phi^8/\sqrt{5}$

    (b) $\Phi^9/\sqrt{5}$

    (c) Without using a calculator, first try to guess the value of $\Phi^7/\sqrt{5}$, rounded to the nearest integer. Verify your guess with a calculator.

20. Using $\Phi = (1+\sqrt{5})/2$, calculate each of the following, rounded to the nearest integer.

    (a) $\Phi^{10}/\sqrt{5}$

    (b) $\Phi^{11}/\sqrt{5}$

    (c) Without using a calculator, first try to guess the value of $\Phi^{12}/\sqrt{5}$, rounded to the nearest integer. Verify your guess with a calculator.

In Exercises 21 and 22, use the formula $\Phi^N = F_N\Phi + F_{N-1}$ to express the given powers of Φ. Do not use a calculator.

21. Find each of the following. (Your answer should be given in terms of integers and $\sqrt{5}$.)

(a) Φ^9

(b) Φ^{12}

22. Find each of the following. (Your answer should be given in terms of integers and $\sqrt{5}$.)

(a) Φ^6

(b) Φ^{15}

23. Recall that each Fibonacci number is approximately Φ times the preceding Fibonacci number. An approximate value of F_{499} in scientific notation is 8.6168×10^{103}. Give a corresponding value for F_{500} in scientific notation.

24. Recall that each Fibonacci number is approximately Φ times the preceding Fibonacci number. An approximate value of F_{1002} in scientific notation is 1.138×10^{209}. Give a corresponding value for F_{1000} in scientific notation.

25. The *Fibonacci sequence of order 2* is the sequence of numbers 1, 2, 5, 12, 29, 70, ... Each term in this sequence (from the third term on) equals two times the term before it plus the term two places before it; in other words, $A_N = 2A_{N-1} + A_{N-2}$ $(N \geq 3)$.

(a) Compute A_7.

(b) Use your calculator to compute to five decimal places the ratio A_N/A_{N-1} for $N = 7$.

(c) Use your calculator to compute to five decimal places the ratio A_N/A_{N-1} for $N = 11$.

(d) Guess the value (to five decimal places) of the ratio A_N/A_{N-1} when $N > 11$.

26. The *Fibonacci sequence of order 3* is the sequence of numbers 1, 3, 10, 33, 109, ... Each term in this sequence (from the third term on) equals three times the term before it plus the term two places before it, in other words, $A_N = 3A_{N-1} + A_{N-2}$ $(N > 3)$.

(a) Compute A_6.

(b) Use your calculator to compute to five decimal places the ratio A_N/A_{N-1} for $N = 6$.

(c) Guess the value (to five decimal places) of the ratio A_N/A_{N-1} when $N > 6$.

C. Fibonacci Numbers and Quadratic Equations

*Exercises 27 through 34 involve quadratic equations with coefficients that are Fibonacci numbers. Recall that any quadratic equation can be solved by first putting it in the form $ax^2 + bx + c = 0$ and then using the **quadratic formula** $x = (-b \pm \sqrt{b^2 - 4ac})/2a$.*

27. (a) Use the quadratic formula to find the two solutions of $x^2 = 2x + 1$. Use a calculator to approximate the solutions to five decimal places.

(b) The positive solution of $x^2 = 2x + 1$ was the basis of a system of proportions used by the Romans during the first and second centuries. Compare this value to that found in Exercise 25(d).

28. (a) Use the quadratic formula to find the two solutions of $x^2 = 3x + 1$. Use a calculator to approximate the solutions to five decimal places.

(b) Compare the positive solution of $x^2 = 3x + 1$ to the value found in Exercise 26(c).

29. Use the quadratic formula to find the two solutions of $3x^2 = 5x + 8$. Use a calculator to approximate the solutions to five decimal places.

30. Use the quadratic formula to find the two solutions of $8x^2 = 5x + 3$. Use a calculator to approximate the solutions to five decimal places.

31. Consider the quadratic equation $55x^2 = 34x + 21$.

(a) Without using the quadratic formula, find one of the solutions to this equation. (*Hint:* One of the solutions is a small integer.)

(b) Without using the quadratic formula, find the other solution to the equation. (*Hint:* The sum of the two solutions of the quadratic equation $ax^2 + bx + c = 0$ equals $-b/a$.)

32. Consider the quadratic equation $21x^2 = 34x + 55$.

(a) Without using the quadratic formula, find one of the solutions to this equation. (*Hint:* One of the solutions is a negative integer.)

(b) Without using the quadratic formula, find the other solution to the equation. (*Hint:* The sum of the two solutions of the quadratic equation $ax^2 + bx + c = 0$ equals $-b/a$.)

33. Consider the quadratic equation $F_N x^2 = F_{N-1}x + F_{N-2}$.

(a) Explain why $x = 1$ is one of the solutions to the equation.

(b) Explain why the other solution is given by $x = (F_{N-1}/F_N) - 1$. [*Hint:* See the hint for Exercise 31(b).]

34. Consider the quadratic equation $F_N x^2 = F_{N+1}x + F_{N+2}$.

(a) Explain why $x = -1$ is one of the solutions to the equation.

(b) Explain why the other solution is given by $x = (F_{N+1}/F_N) + 1$. [*Hint:* See the hint for Exercise 32(b).]

D. Gnomons and Similarity

35. T and T' are similar triangles. (The triangles are not drawn to scale.)

(a) If the perimeter of T is 13 in., what is the perimeter of T' (in meters)?

(b) If the area of T is 20 sq. in., what is the area of T' (in square meters)?

36. P and P' are similar polygons.

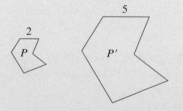

(a) If the perimeter of P is 10, what is the perimeter of P'?

(b) If the area of P is 30, what is the area of P'?

37. Find the length c of the shaded rectangle so that it is a gnomon to the white rectangle with sides 3 and 9.

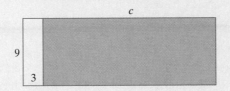

38. Find the value of x so that the shaded figure is a gnomon to the white rectangle.

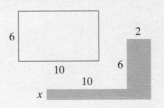

39. Rectangle A is 10 by 20. Rectangle B is a gnomon to rectangle A. What are the dimensions of rectangle B?

40. Find the value of x so that the shaded figure is a gnomon to the white rectangle.

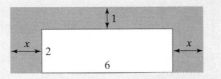

41. Find the value of x so that the shaded "rectangular ring" is a gnomon to the white rectangle.

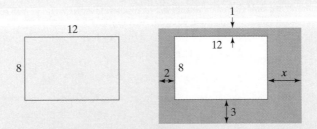

42. Find the value of x so that the shaded "rectangular ring" is a gnomon to the white rectangle.

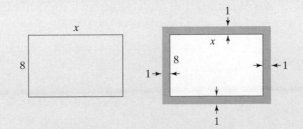

43. Find the values of x and y so that the shaded figure is a gnomon to the white triangle.

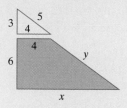

44. Find the values of x and y so that the shaded triangle is a gnomon to the white triangle ABC.

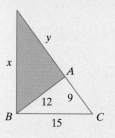

45. A rectangle has a 10-by-10 square gnomon. What are the dimensions of the rectangle?

46. What are the dimensions of a rectangle that is a gnomon to itself? (*Hint:* Label the sides of the rectangle x and 1, as shown in the figure.)

47. What are the dimensions of rectangle $ABCD$ having the shaded double square as its gnomon? (*Hint:* Label the sides of the rectangle x and 1, as shown in the figure.)

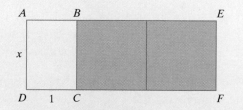

48. What are the dimensions of a right triangle that is a gnomon to itself? (*Hint:* Label the legs of the triangle x and 1, as shown in the figure.)

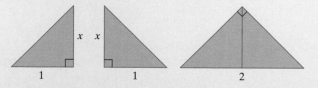

49. For each of the following rectangles, determine whether the rectangle is (I) a golden rectangle, (II) not quite a golden rectangle but very close to one, or (III) neither (I) nor (II).

   (a) A Fibonacci rectangle of dimensions 1 by 2.

   (b) A Fibonacci rectangle of dimensions 34 by 55.

   (c) A rectangle with dimensions 1 and Φ.

   (d) A rectangle with dimensions Φ and Φ^2.

50. For each of the following rectangles, determine if the rectangle is (I) a golden rectangle, (II) not quite a golden rectangle but very close to it, or (III) neither (I) nor (II).

   (a) A Fibonacci rectangle of dimensions 89 by 55.

   (b) A rectangle with dimensions 1 and $1/\Phi$.

   (c) A rectangle similar to a 2-by-5 rectangle.

   (d) A rectangle with dimensions $\Phi + 1$ and $\Phi^2 + \Phi$.

JOGGING

51. Consider the following sequence of numbers: $5, 5, 10, 15, 25, 40, 65, \ldots$. If A_N is the Nth term of this sequence, write A_N in terms of F_N.

52. This problem investigates the relationship between the *Lucas sequence* given by the numbers

   $1, 3, 4, 7, 11, 18, 29, \ldots$

   and the Fibonacci sequence

   $1, 1, 2, 3, 5, 8, 13, 21, \ldots$.

   If L_N represents the Nth term of the Lucas sequence,

   (a) write L_N in terms of the Fibonacci numbers F_{N-1} and F_{N+1}.

   (b) write L_N in terms of the Fibonacci numbers F_N and F_{N+1}.

53. Suppose that $T_N = aF_{N+1} + bF_N$, where a and b are fixed integers.

   (a) What is T_1?

   (b) What is T_2?

   (c) Show that $T_N = T_{N-1} + T_{N-2}$.

54. Explain why a square (regardless of size) cannot be a gnomon to the L-shaped figure.

55. Find the values of *x, y*, and *z* so that the shaded figure has an area eight times the area of the white triangle and at the same time is a gnomon to the white triangle.

(*Hint:* The ratio of the area of similar triangles is the square of the ratio of the sides.)

56. (a) Which of the following O-rings is similar to I? Explain your answer.

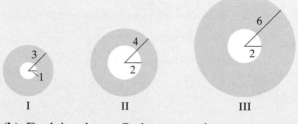

I II III

(b) Explain why an O-ring cannot have a gnomon.

57. Find the values of *x* and *y* so that the shaded figure has an area of 75 and at the same time is a gnomon to the white rectangle.

(*Hint:* The ratio of the area of similar rectangles is the square of the ratio of the sides.)

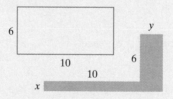

58. Find the values of *x* and *y* so that the shaded triangle *ABD* is a gnomon to the white triangle *ABC*.

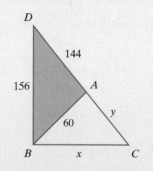

59. (a) Use the fact that $\Phi^2 = \Phi + 1$ to compute the exact value of

$$\sqrt{1 + \sqrt{1 + \sqrt{1 + \Phi}}}.$$

(b) Use the fact that $\Phi^2 = \Phi + 1$ to compute the value of

$$1 + \cfrac{1}{1 + \cfrac{1}{1 + \cfrac{1}{\Phi}}}.$$

60. Let *ABCD* be an arbitrary rectangle as shown in the following figure. Let *AE* be perpendicular to the diagonal *BD* and *EF* perpendicular to *AB* as shown. Show that the rectangle *BCEF* is a gnomon to the rectangle *ADEF*. (*Hint:* Show that the rectangle *ADEF* is similar to the rectangle *ABCD*.)

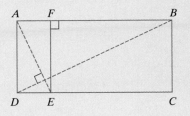

61. Show that the L-shaped object in the figure is a gnomon to rectangle *A* if and only if $\dfrac{b}{h} = \dfrac{y}{x}$.

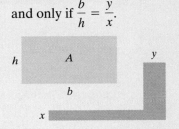

62. A rectangle has a square gnomon. The new rectangle obtained by attaching the square gnomon to the original rectangle has longer leg equal to 20. What are the dimensions of the original rectangle?

63. In the figure, triangle *BCD* is a 72–72–36° triangle with base of length 1 and longer side of length *x*. (Using this choice of values, the ratio of the longer side to the shorter side is $x/1 = x$.)

(a) Show that $x = \Phi = (1 + \sqrt{5})/2$. (*Hint:* Use the fact that triangle *ACB* is similar to triangle *BCD*.)

(b) What are the interior angles of triangle *DAB*?

(c) Show that in the isosceles triangle *DAB*, the ratio of the longer to the shorter side is also Φ.

64. The regular pentagon in the following figure has sides of length 1. Show that the length of any one of its diagonals is Φ.

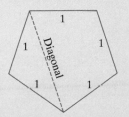

65. **(a)** A regular decagon (10 sides) is inscribed in a circle of radius 1. Find the perimeter in terms of Φ.

   (b) Repeat (a) with radius r. Find the perimeter in terms of Φ and r.

RUNNING

66. In the following figure, $ABCD$ is a square and the three triangles I, II, and III have equal areas. Show that x/y is the golden ratio.

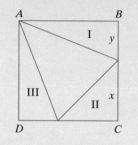

67. Show that $F_1 + F_2 + F_3 + \cdots + F_N = F_{N+2} - 1$.

68. Show that $F_1 + F_3 + F_5 + \cdots + F_N = F_{N+1}$. (Note that on the left side of the equation we are adding the Fibonacci numbers with odd subscripts up to N.)

69. Show that every positive integer greater than 2 can be written as the sum of distinct Fibonacci numbers.

70. Consider the following equation relating various terms of the Fibonacci sequence.

   $$F_{N+2}^2 - F_{N+1}^2 = F_N \cdot F_{N+3}$$

   Using the algebraic identitiy $A^2 - B^2 = (A - B)(A + B)$, show that the equation is true for every positive integer N.

71. Show that the sum of any 10 consecutive Fibonacci numbers is a multiple of 11.

72. Suppose that T is a Fibonacci-type sequence; that is, $T_N = T_{N-1} + T_{N-2}$, but $T_1 = c$ and $T_2 = d$.

   (a) Show that there are constants a and b such that $T_N = aF_{N+1} + bF_N$ (where F is the Fibonacci sequence).

   (b) Show that $T_{N+1}/T_N \approx \Phi$ when N is large.

73. **(a)** Show that the ratio of alternate Fibonacci numbers F_{N+2}/F_N approximates the value Φ^2 as N gets large. (*Hint:* The ratio F_{N+1}/F_N approximates Φ as N gets large.)

   (b) What number do the ratios F_{N+3}/F_N approximate as N gets large?

   (c) What number do you think the ratios F_{N+k}/F_N approximate as N gets large? Explain your answer.

74. Find the values of x, y, and z so that the shaded "triangular ring" is a gnomon to the white triangle.

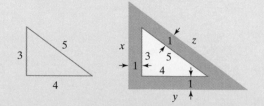

75. During the time of the Greeks the star pentagram was a symbol of the Brotherhood of Pythagoras. A typical diagonal of the large outside regular pentagon is broken up into three segments of lengths x, y, and z, as shown in the following figure.

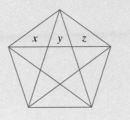

(a) Show that $\dfrac{x}{y} = \Phi$, $\dfrac{x + y}{z} = \Phi$ and $\dfrac{x + y + z}{x + y} = \Phi$.

(b) Show that if $y = 1$, then $x = \Phi$, $x + y = \Phi^2$, and $x + y + z = \Phi^3$.

PROJECTS AND PAPERS

A. Fibonacci Numbers, the Golden Ratio, and Phyllotaxis

In this chapter we touched on the fact that Fibonacci numbers and the golden ratio often show up in natural organisms, both in the plant and animal worlds. In this project, you are asked to expand on this topic.

1. Give several detailed examples of the appearance of Fibonacci numbers or the golden ratio (or both) in the plant world. Include examples of branch formation in plants, leaf arrangements around stems, and seed arrangements on circular seedheads (such as sunflower heads).

2. Briefly discuss *phyllotaxis*. What is it, and what are some of the mathematical theories behind it?

Notes: (1) The literature on Fibonacci numbers, the golden ratio, and phyllotaxis is extensive. A search on the Web should provide plenty of information. Two excellent Web sites on this subject are Ron Knott's site at the University of Surrey, England (*http://www.mcs.surrey.ac.uk/Personal/R.Knott/Fibonacci*), and the site *Phyllotaxis: An Interactive Site for the Mathematical Study of Plant Pattern Formation* (*http://www.math.smith.edu/~phyllo/*) based at Smith College. (2) Suggested readings on the mathematics of phyllotaxis are references 4, 11, and 15.

B. The Golden Ratio in Art, Architecture, and Music

It is often claimed that from the time of the ancient Greeks through the Renaissance to modern times, artists, architects, and musicians have been fascinated by the golden ratio. Choose one of the three fields (art, architecture, or music) and write a paper discussing the history of the golden ratio in that field. Describe famous works of art, architecture, or music in which the golden ratio is alleged to have been used. How? Who were the artists, architects, composers?

Be forewarned that there are plenty of conjectures, unsubstantiated historical facts, controversies, claims, and counterclaims, surrounding some of the alleged uses of the golden ratio and Fibonacci numbers. Whenever appropriate, you should present both sides to a story.

C. Logarithmic Spirals

In this project, you are to investigate logarithmic spirals. You should explain what a logarithmic spiral is and how it is related to Fibonacci numbers and the golden ratio, and you should reference several examples of where they occur in the natural world.

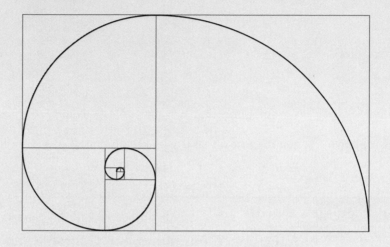

D. Figurate Numbers

The *triangular* numbers are numbers in the sequence $1, 3, 6, 10, 15, \ldots$. (The *N*th **triangular** number T_N is given by the sum $1 + 2 + 3 + \ldots + N$.) In a similar way, *square, pentagonal*, and *hexagonal* numbers can be defined. In this project, you are to investigate these types of numbers (called *figurate* numbers), give some of their more interesting properties, and discuss the relationship between these numbers and *gnomons*.

E. The "Golden Ratio Hypothesis"

A long-held belief among those who study how humans perceive the outside world (mostly psychologists and psychobiologists) is that the *golden ratio* plays a special and prominent role in the human interpretation of "beauty." Shapes and objects whose proportions are close to the golden ratio are believed to be more pleasing to human sensibilities than those that are not. This theory, generally known as *the golden ratio hypothesis*, originated with the experiments of the famous psychologist Gustav Fechner in the late 1800s. In a classic experiment, Fechner showed rectangles of various proportions to hundreds of subjects and asked them to choose. His results showed that the rectangles that were close to the proportions of the golden ratio were overwhelmingly preferred over the rest. Since Fechner's original experiment, there has been a lot of controversy about the golden ratio hypothesis, and many modern experiments have cast serious doubts about its validity.

Write a paper describing the history of the golden ratio hypothesis. Start with a description of Fechner's original experiment. Follow up with other experiments duplicating Fechner's results and some of the more recent experiments that seem to disprove the golden ratio hypothesis. Conclude with your own analysis.

REFERENCES AND FURTHER READINGS

1. Conway, J. H., and R. K. Guy, *The Book of Numbers*. New York: Springer-Verlag, 1996.
2. Coxeter, H. S. M., "The Golden Section, Phyllotaxis, and Wythoff's Game," *Scripta Mathematica*, 19 (1953), 135–143.
3. Coxeter, H. S. M., *Introduction to Geometry*. New York: John Wiley & Sons, Inc., 1961, chap. 11.
4. Douady, S., and Y. Couder, "Phyllotaxis as a Self-Organized Growth Process," in *Growth Patterns in Physical Sciences and Biology*, eds. J. M. Garcia-Ruiz et al. New York: Plenum Press, 1983.
5. Erickson, R. O., "The Geometry of Phyllotaxis," in *The Growth and Functioning of Leaves*, eds. J. E. Dale and F. L. Milthrope. New York: Cambridge University Press, 1983.
6. Gardner, Martin, "About Phi, an Irrational Number That Has Some Remarkable Geometrical Expressions," *Scientific American*, 201 (August 1959), 128–134.
7. Gardner, Martin, "The Multiple Fascinations of the Fibonacci Sequence," *Scientific American*, 220 (March 1969), 116–120.
8. Gazalé, M. J., *Gnomon: From Pharaohs to Fractals*. Princeton, N.J.: Princeton University Press, 1999.
9. Gullberg, Jan, *Mathematics: From the Birth of Numbers*. New York: W. W. Norton, 1997.
10. Herz-Fischler, Roger, *A Mathematical History of the Golden Number*. New York: Dover, 1998.
11. Jean, R. V., *Mathematical Approach to Pattern and Form in Plant Growth*. New York: John Wiley & Sons, Inc., 1984.
12. Livio, Mario, *The Golden Ratio: The Story of Phi, the World's Most Astonishing Number*. New York: Random House, 2002.
13. May, Mike, "Did Mozart Use the Golden Section?," *American Scientist*, 84 (March-April 1996), 118.
14. Markovsky, George, "Misconceptions About the Golden Ratio," *College Mathematics Journal*, 23 (1992), 2–19.
15. Neill, William, and Pat Murphy, *By Nature's Design*. San Francisco, CA: Chronicle Books, 1993.
16. Prusinkiewicz, P., and A. Lindenmayer, *The Algorithmic Beauty of Plants*. New York: Springer-Verlag, 1990, chap. 4.
17. Putz, John F., "The Golden Section and the Piano Sonatas of Mozart," *Mathematics Magazine*, 68 (1995), 275–282.
18. Sigler, Laurence, *Fibonacci's Liber Abaci*. New York: Springer-Verlag, 2002.
19. Stewart, Ian, "Daisy, Daisy, Give Me Your Answer, Do," *Scientific American* (January 1995), 96–99.
20. Thompson, D'Arcy, *On Growth and Form*. New York: Dover, 1992, chaps. 11, 13, and 14.

"... you, be ye fruitful and multiply."

Genesis 9:7

The Mathematics of Population Growth

There Is Strength in Numbers

In Chapter 9 we discussed the concept of *growth* as applied to a single organism. In this chapter we will discuss the concept of *growth* as it applies to entire populations.

The connection between the study of populations and mathematics goes back to the very beginnings of civilization. One of the reasons that humans invented the first numbering systems was their need to handle the rudiments of counting populations—how many sheep in the flock, how many people in the tribe, and so on. By biblical times, simple models of population growth were being used to measure crop production and even to estimate the yields of future crops.

Today, mathematical models of population growth are a fundamental tool in our efforts to understand the rise and fall of endangered wildlife populations, fishery stocks, agricultural pests (such as locusts, cicadas, and boll weevils), infectious diseases, radioactive waste, ordinary trash, and so on. Entire modern disciplines, such as *mathematical ecology, population biology*, and *biostatistics* are built around the mathematics of population growth.

As the role of mathematics in the study of population growth has expanded, so has its complexity, and the mathematical tools in use today to study populations can be quite sophisticated. The overall mathematical principles involved, however, are reasonably simple. (The devil is always in the details!) This chapter deals with the basic principles behind the mathematics of population growth and presents some of the simpler models that can be used in studying its *dynamics*.

Before we proceed, let us clarify some terminology. In its modern usage, the term *population growth* has become very broad, owing primarily to the broad meanings given nowadays to both *population* and *growth*. The Latin root of *population* is *populus* (which means "people"), so that in its original interpretation the word refers to human populations. Over time, this scope has been expanded to include any collection of objects (animate or inanimate) about which

we want to make a numerical or quantitative statement. Thus, we can also use the word *population* when speaking of penguins, used tires, and dollars.

Second, we normally think of the word *growth* as being applied to things that get bigger, but in this chapter we will ascribe a slightly more technical meaning to it. *Growth* can mean *negative growth* or *decay* (i.e., a population getting smaller), as well as *positive growth* (i.e., a population getting bigger). This is convenient, because often we don't know ahead of time how a population is going to change: Is it going to increase or decrease? By allowing *growth* to mean either, we need not concern ourselves with making the distinction.

10.1 The Dynamics of Population Growth

The growth of a population is a *dynamical process*, meaning that it represents a situation that changes over time. Mathematicians distinguish between two kinds of situations: continuous growth and discrete growth. In **continuous growth** the dynamics of change are in effect all the time—every hour, every minute, every second, there is change. The classic example of this kind of growth is represented by money left in an account that is drawing interest on a continuous basis (see Project A). We will not study continuous growth in this chapter because the mathematics involved (calculus) is beyond the scope of this book.

The second type of growth, **discrete growth**, is the most common and natural way by which populations change. We can think of it as a *stop-and-go* type of situation. For a while nothing happens; then, there is a sudden change in the population. We call such a change a **transition**. Then, for a while nothing happens again; then another transition takes place, and so on. Of course, the period between transitions can be 100 years, an hour, a second, or a nanosecond. To us, the length of time between transitions will not make a difference. The human population of our planet is an example of what we mean. Nothing happens until someone is born or someone dies, at which point there is a change ($+1$ or -1); then, there is no change until the next birth or death. However, since someone is born every fraction of a second and someone dies slightly less often, it is somewhat tempting to think that the world's human population is, for all practical purposes, changing in a continuous way. But the laws of growth affecting the world's population are only quantitatively different from the laws affecting the population of Loving County, Texas (population 140[1]), where a change in the population may not come about for months or even years.

The basic problem of population growth is to figure out what happens to a given population over time. Sometimes, we talk about a specific period. ("The

---

[1]*http://www.naco.org/counties*

[United States] population is projected to grow to 394 million by 2050—nearly double its 1995 size."[2]) Other times, we may talk about the long-term behavior of the population. ("The black rhino population is heading for extinction."[3]) In either case, the most basic way to deal with the question of growth of a particular population is to find the rules that govern the transitions. We will call these the **transition rules**. After all, if we have a way to figure out how the population changes each time there is a transition, then (with a little help from mathematics) we can usually figure out how the population changes after many transitions.

The ebb and flow of a particular population over time can be conveniently thought of as a list of numbers called the **population sequence**. Every population sequence starts with an initial population P_0 (the "0th generation") and continues with P_1, P_2, etc., where P_N is the size of the population in the Nth generation. Note that by choosing to start the sequence with P_0 the subscripts match the generations: P_N describes the population that is N transitions removed from the initial population. Figure 10-1 is a schematic illustration of how a population sequence is generated.

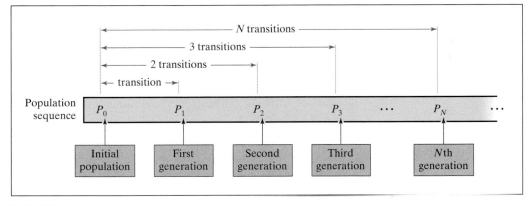

FIGURE 10-1
A generic population sequence. P_N is the population size in the Nth generation.

Just to get our feet wet, we will start with a classic example of a problem in population growth, first raised by Fibonacci in 1202 (see the biographical profile of Fibonacci on p. 376).

EXAMPLE 10.1 Fibonacci's Rabbits

In his famous book *Liber Abaci*,[4] Fibonacci posed the following question, which we quote directly from the book:

A certain man put a pair of rabbits in a place surrounded on all sides by a wall. How many pairs of rabbits can be produced from that pair in a year if it is supposed that every month each pair begets a new pair which from the second month on becomes productive?

---

[2]U.S. Department of Commerce, Bureau of the Census, *Current Population Reports. Population Projections of the United States by Age, Sex, Race, and Hispanic Origin: 1995 to 2050.* Feb. 1996, p. 5.
[3]*New York Times*, May 7, 1991, B5.
[4]Leonardo Pisano Fibonacci, *Liber Abaci*, ca. 1202. An English translation of *Liber Abaci* has recently been published (see reference 12).

For the sake of convenience, we will count Fibonacci's rabbits in male–female pairs, and, following the notation we have just adopted, we will let P_0 represent the initial population and P_1, P_2, P_3, ... represent the number of pairs in the first, second, third, etc., generations. Figure 10-2 illustrates the pattern of growth for the first six months. The blue arrows represent the fact that each mature pair begets a baby pair; the red arrows represent the fact that every pair is a mature pair in the next generation.

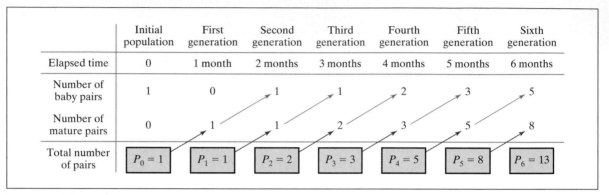

FIGURE 10-2

Fibonacci's rabbits: P_N is the number of male–female pairs in the Nth generation.

We can see from Fig. 10-2 that $P_0 = 1$, $P_1 = 1$, $P_2 = P_1 + P_0$, $P_3 = P_2 + P_1$, and so on. These P's are Fibonacci numbers (see Chapter 9), offset by 1 because we started with P_0. Thus,

$$P_N = F_{N+1}.$$

It follows that after one year (12 generations) we will have $P_{12} = F_{13} = 233$ pairs of rabbits—a grand total of 466 rabbits! ∎

As long as all the rabbits stay alive and continue breeding according to our rules, we can describe the generic transition rule for passing from one generation to the next by the following recursive rule:

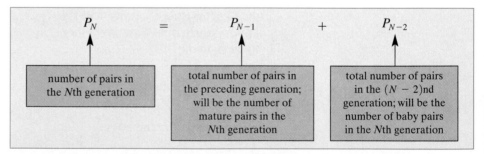

Of course, real-life rabbits are not as accommodating—they live and breed by considerably more complicated rules, which we could never hope to capture in a simple equation, and in general, this is true about most other types of equations that attempt to model the growth of natural populations.

Is there any use, then, for simplistic mathematical models of how populations grow? The answer is yes! We can make excellent predictions about the growth of a population over time, even when we don't have a completely realistic set of transition rules. The secret is to capture the variables that are really influential in determining how the population grows, put them into a few transition rules that describe how the variables interact, and forget about the small details. This, of course, is easier said than done. In essence, it is what population biologists and mathematical ecologists do for a living, and it is as much an art as it is a science.

In the rest of this chapter we will discuss three of the most basic models of population growth: the *linear growth model*, the *exponential growth model*, and the *logistic growth model*.

10.2   The Linear Growth Model

The linear growth model is the simplest of all models of population growth. In this model, in each generation the *population increases (or decreases) by a fixed amount.* The easiest way to see how the model works is with an example.

EXAMPLE 10.2

The city of Cleansburg is considering a new law that would restrict the monthly amount of garbage allowed to be dumped in the local landfill to a maximum of 120 tons a month. There is concern among local officials that, unless this restriction on dumping is imposed, the landfill will reach its maximum capacity of 20,000 tons in a few years. Currently, there are 8000 tons of garbage already in the landfill. Assuming that the law is passed and the landfill collects exactly 120 tons of garbage each month, how much garbage will there be in the landfill 5 years from now? How long before the landfill reaches its 20,000-ton capacity?

While the circumstances are fictitious, the questions raised are realistic and important. The *population* in this example is the garbage in the landfill, and since we only care about monthly totals, we define the transitions as happening once a month. The key assumption about this population-growth problem is that the monthly garbage at the landfill grows by a constant of 120 tons a month.

Our starting population (P_0) is 8000 tons. Thus, we have the following population sequence:

$$P_0 = 8000; \; P_1 = 8120; \; P_2 = 8240; \; P_3 = 8360; \; \ldots.$$

In 5 years we will have had 60 transitions, each representing an increase of 120 tons. The population after 5 years is given by P_{60}, which is obtained by adding 60 transitions of 120 tons each to the existing 8000 tons in the landfill. In other words,

$$P_{60} = 8000 + 60(120) = 15{,}200.$$

To determine the number of transitions it would take for the landfill to reach its 20,000-ton maximum, we set up the equation

$$8000 + 120N = 20{,}000,$$

which has as a solution $N = 100$. This means that it will take 100 monthly transitions (8 years and 4 months) for the landfill to reach its maximum capacity. Based on this information, local officials should start making plans soon for a new landfill. ■

Example 10.2 is a typical example of the general linear growth model, whose basic characteristic is that in each transition a constant amount—call it d—is added to the previous population. Mathematically, the general linear growth model can be described as follows.

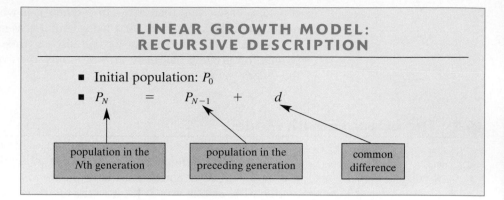

A population that grows according to a linear growth model produces a type of sequence commonly known as an **arithmetic sequence**. Technically speaking, the arithmetic sequence is just the numerical description of a population that is growing according to a linear growth model—informally, linear growth and arithmetic sequences can be considered synonymous. The number d is called the **common difference** for the arithmetic sequence, because any two consecutive values of the arithmetic sequence will always differ by the amount d.

The equation $P_N = P_{N-1} + d$ gives a **recursive description** of a population sequence, because values of the sequence are expressed in terms of preceding values of that sequence. While recursive descriptions tend to be nice and tidy, they have one major drawback: To calculate one value in a population sequence, we essentially have to first calculate all the earlier values. As we learned in Chapter 9 with the Fibonacci numbers, this can be quite an inconvenience.

Fortunately, in the present case there is a very convenient way to describe a linearly growing population sequence that does not require a recursive description.

LINEAR GROWTH MODEL: EXPLICIT DESCRIPTION

$$P_N = P_0 + N \times d.$$

The equation follows from the fact that, to get to P_N, we need to go through N transitions, each of which consists of adding d. (See Fig. 10-3.)

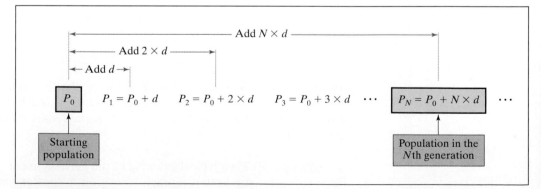

FIGURE 10-3
Population growth based on a linear growth model.

The equation $P_N = P_0 + N \times d$ gives an **explicit description** of the population sequence, because it allows one to calculate any value of the sequence explicitly, without having to know any of the preceding values except for the starting population P_0.

EXAMPLE 10.3

A population grows according to a linear growth model. The starting population is given by $P_0 = 37$, and the common difference is given by $d = 6$. What is the population size in the 15th generation? How many generations will it take for the population to exceed 200?

Using the explicit description of linear growth, we immediately find that

$$P_{15} = 37 + 15 \times 6 = 127,$$

which means that the size of the population in the 15th generation is 127.

To determine how many generations it will take for the population to exceed 200, we must find the smallest integer N such that $P_N > 200$. We write

$$P_N = 37 + 6N$$

and solve

$$37 + 6N > 200.$$

This gives

$$N > 163/6 \approx 27.167.$$

Thus, it will take 28 generations for the population to exceed 200. ▬

EXAMPLE 10.4

A population grows according to a linear growth model. Unfortunately, all population records have been lost in a fire except for two values of the population sequence: $P_9 = 1324$ and $P_{25} = 2684$.

We would like to find an explicit description for P_N. In order to do so, we need to find P_0 and d. Fortunately, we can set up a nice little system of two equations and two unknowns:

$$\begin{cases} P_0 + 25d = 2684, \\ P_0 + 9d = 1324. \end{cases}$$

If we subtract the second equation from the first, we get

$$16d = 1360,$$

and thus

$$d = 85.$$

Replacing d by 85 in either of the two equations gives $P_0 = 559$. Once we have d and P_0 we are in business—our population sequence is given by the explicit formula

$$P_N = 559 + 85N.$$

▬

Plotting Population Growth

A very convenient way to describe population growth is by means of a *plot* or *graph*. The horizontal axis usually represents time (with the tick marks generally corresponding to the transitions), and the vertical axis usually represents the size of the population. A population plot can consist of just marks (such as dots) indicating the population size at each generation or of dots joined by lines, which sometimes helps the visual effect. The former is called a **scatter plot**, the latter a **line graph**. Figure 10-4 shows a scatter plot and a line graph for the same population sequence.

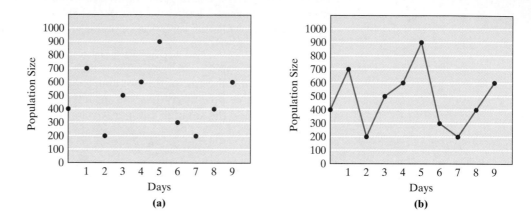

FIGURE 10-4
(a) Scatter plot,
(b) line graph.

(a)   (b)

Because we have complete freedom in choosing both the horizontal and vertical scales, plots can be misleading. Consider Figs. 10-5(a) and (b). Which population sequence is growing faster? Actually, both plots represent the growth of the same population: the garbage problem in Example 10.2. These plots illustrate why linear growth is called linear growth—no matter how we plot, it, the values of the population line up in a straight line.

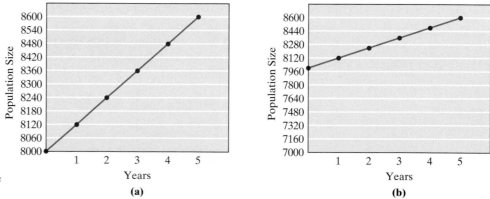

FIGURE 10-5
Two line graphs describing the same population sequence.

(a)   (b)

Adding Terms of an Arithmetic Sequence

EXAMPLE 10.5

Jane Doe is a company that manufactures tractors. The company has decided to start production on a new model and, in preparation for its national release, will need to store tractors at several of its warehouses. Each week, for the next 72 weeks, three new tractors will have to be moved into storage. If the storage costs are $10 per tractor per week, what is the total storage cost to the company over the 72-week period?

In this problem the weekly storage bill grows linearly, with a common difference d equal to $30 (3 new tractors at $10). Thus, the weekly storage costs to the company for weeks 1 through 72 are given by the arithmetic sequence $30, 60, 90,$ $\ldots, 2160$. Note that the last term in the sequence is the storage bill for week 72 (30×72).

The total cost to the company over the 72-week period is the sum of all the terms in the sequence. Calling this total cost S, the challenge is to compute this long sum: $S = 30 + 60 + 90 + \ldots + 2160$. We could, of course, add these 72 numbers, with or without a calculator. Either way, it's a horrible thought! Luckily, there is a much easier way to do this.

Let's write the sum we want to compute twice, once forward and once backward.

$$S = \quad 30 \ + \ 60 \ + \ 90 \ + \ \ldots \ + \ 2160 \quad \text{(forward)}$$

$$S = 2160 + 2130 + 2100 + \ \ldots \ + \ \ 30 \quad \text{(backward)}$$

Notice that lined up this way, each of the 72 columns on the right-hand side of the equal signs adds up to 2190. The grand total when adding *all* the numbers is 72×2190. Since this is twice the sum S we are trying to compute, we get

$$S = \frac{72 \times 2190}{2} = 78{,}840.$$

■

EXAMPLE 10.6

Consider the sum

$$S = \underbrace{5 + 12 + 19 + 26 + 33 + \cdots}_{132 \text{ terms}}$$

Here we are adding 132 consecutive terms of an arithmetic sequence. The first term is $A_0 = 5$; the common difference is $d = 7$. We need to find the last term, which is A_{131}. (Remember that we start with a first term labeled A_0.) We already know how to do this: $A_{131} = 5 + 131 \times 7 = 922$. We now apply the same approach we used in Example 10.5.

$$S = \quad 5 \ + \ 12 \ + \ 19 \ + \ \cdots \ + \ 922 \quad \text{(forward)}$$

$$S = 922 + 915 + 908 + \ \cdots \ + \ \ 5 \quad \text{(backward)}$$

Note how each column adds up to 927, the sum of the first and last terms in S. Adding all 132 columns gives a grand total of $2 \times S = 132 \times 927$, or

$$S = \frac{132 \times 927}{2} = 61{,}182.$$

■

The approach used in Examples 10.5 and 10.6 works with *any* arithmetic sequence $A_0, A_1, A_2, \ldots$. (Note the change in the choice of letters. We use A's rather than P's to emphasize the fact that we are working with an arithmetic sequence.) We can add up any number of consecutive terms easily with the following formula, which we will informally refer to as the *arithmetic sum formula*.

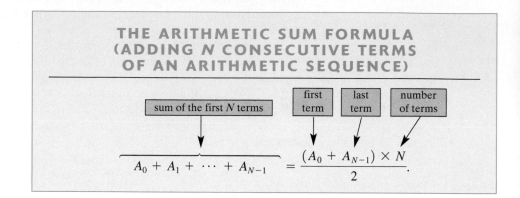

THE ARITHMETIC SUM FORMULA (ADDING N CONSECUTIVE TERMS OF AN ARITHMETIC SEQUENCE)

sum of the first N terms

first term

last term

number of terms

$$A_0 + A_1 + \cdots + A_{N-1} = \frac{(A_0 + A_{N-1}) \times N}{2}.$$

Before we go on, a couple of comments about the arithmetic sum formula. First, it is an extremely useful and important formula. It's not known who first discovered it, but it goes back a long way, and the first known reference to it can be found in Book IX of Euclid's *Elements*, written around 300 B.C. At a minimum the formula has been known for 2300 years, possibly much longer. Second, why did we write the last term of the sum as A_{N-1} rather than A_N? (Surely, A_N seems a lot more user friendly). The reason is that the sum starts with A_0, which makes the number of terms and the subscript for the last term offset by 1. In short, the Nth term is A_{N-1}, and if you want A_N you need to go to the $(N + 1)$st term.

Sometimes the following informal version of the formula gets us by without the hassle of worrying about subscripts:

$$Sum = \frac{(first\ term\ +\ last\ term) \times (number\ of\ terms)}{2}.$$

A note of warning: Don't get carried away! This formula applies *only* to a sum of consecutive terms in an *arithmetic sequence*.

EXAMPLE 10.7

Compute the sum

$$4 + 13 + 22 + 31 + 40 + \cdots + 922.$$

Here we are adding the terms of an arithmetic sequence with $A_0 = 4$ and common difference $d = 9$. Before we can apply our recently discovered formula, we need to first find the number of terms N. To find N we set up an equation: $A_{N-1} = 922 = 4 + 9(N - 1)$. From the equation, we get $9(N - 1) = 918$, which gives $N - 1 = 102$ and $N = 103$. It follows that

$$4 + 13 + 22 + 31 + 40 + \cdots + 922 = \frac{(4 + 922) \times 103}{2} = 47,689.$$

10.3 The Exponential Growth Model

The exponential growth model is another classic model of population growth. The main characteristic of this model is that in each transition, the population changes by a *fixed proportion*.

Before we start our discussion of exponential growth in earnest, let's develop some background. The next two examples have to do with the use of percentages to calculate increases and decreases.

EXAMPLE 10.8

A firm manufactures an item at a cost of C dollars. The item is marked up 10% and sold to a distributor. The distributor then marks the item up 20% (based on the price he or she paid) and sells the item to a retailer. The retailer marks that price up 50% and sells the item to the public. By what percent has the item been marked up over its original cost?

- Original cost of item: C.
- Cost to distributor after 10% markup:
  $$D = 110\% \text{ of } C = (1.1)C.$$
- Cost to retailer after 20% markup:
  $$R = 120\% \text{ of } D = (1.2)D = (1.2)(1.1)C = (1.32)C.$$
- Price (P) to the public after 50% markup:
  $$P = 150\% \text{ of } R = (1.5)R = (1.5)(1.32)C = (1.98)C.$$

Therefore, the markup over the original cost is 98%.

EXAMPLE 10.9

A retailer buys an item for C dollars and marks it up 80%. He then puts the item on sale for 40% off the marked price. What is the net percentage markup on this item?

- Original cost of item: C.
- Price after 80% markup: $P = 180\%$ of $C = (1.8)C$.
- Sale price after 40% discount: $S = 60\%$ of $P = (0.6)P = (0.6)(1.8)C = (1.08)C$.

The net markup is 8%.

The main point of Examples 10.8 and 10.9 is the following: Increasing a number C by $x\%$ is equivalent to multiplying C by the quantity $(1 + x/100)$. The $x/100$ represents $x\%$ in decimal form; adding the 1 represents the fact that we are *increasing* the original number C.

Let's return now to the exponential growth model.

EXAMPLE 10.10

The sum of $1000 is deposited in a retirement account that pays 10% *annual* interest. (That is, interest is paid once a year at the end of the year.) How much money is there in the account after 25 years if the interest is left in the account? How much money is there in the account after N years?

Table 10-1 will help us get started.

TABLE 10-1

	Account balance at beginning of year	Interest earned for the year	Account balance at end of year
Year 1	$1000	$100	$1100
Year 2	1100	110	1210
Year 3	1210	121	1331
⋮	⋮	⋮	⋮
Year 24	?	?	?
Year 25	?	?	?

The critical observation is that the account balance at the end of the year 1 is obtained by adding the *principal* ($1000) and the interest earned for the year (10% of $1000), which is the same as taking 110% of $1000—in other words, $1000 × 1.1. Repeating the argument for year 2, we find that the account balance at the end of the second year is

$$\underbrace{\$1000 \times (1.1)}_{\text{Account balance at beginning of year 2}} \times (1.1) = \$1000 \times (1.1)^2.$$

Similarly, the account balance at the end of the third year is

$$1000 \times (1.1)^2 \times (1.1) = 1000 \times (1.1)^3.$$

Following this pattern, we can see that the account balance after 25 years (in other words, at the start of year 26) is

$$\$1000 \times (1.1)^{25} \approx \$10{,}834.71.$$

It isn't hard to see what's happening: Each transition (which occurs at the end of a year) corresponds to taking 110% of the balance at the start of that year, which is the same as multiplying the balance at the start of the year by 1.1.

We can now give a general rule describing the balance in the account after any number of years. Starting with a principal $P_0 = \$1000$ compounded at an annual interest rate of 10% a year, at the end of the Nth year the balance in the account is

$$P_N = \$1000 \times (1.1)^N.$$

Figure 10-6(a) plots the growth of the money in the account for the first 8 years. Figure 10-6(b) plots the growth of the money in the account for the first 30 years.

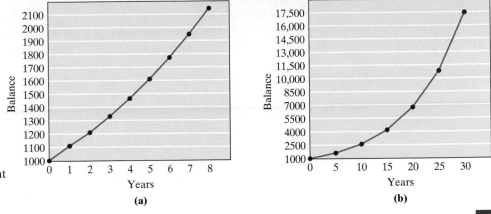

FIGURE 10-6
Cumulative growth of $1000 at 10% interest compounded annually.

Example 10.10 is a classic example of exponential growth. The money draws interest, then the money plus the interest draw interest, and so on. While the most familiar examples of exponential growth have to do with the growth of money, the exponential growth model is useful in the study of biological populations as well. The essence of exponential growth is *repeated multiplication*: each transition consists of multiplying the size of the population by a constant factor. In Example 10.10, the constant factor is 1.1.

A sequence defined by repeated multiplication—every term in the sequence after the first is obtained by multiplying the preceding term by a fixed amount r—is called a **geometric sequence**. The constant factor r is called the **common ratio** of the geometric sequence   it is the ratio of two successive terms in the sequence. (To insure that the population sequence does not have negative numbers, we will restrict the values of the common ratio r to positive numbers, although no such restriction is necessary when dealing with geometric sequences in general.)

The general exponential growth model can be described recursively by the following formula.

EXPONENTIAL GROWTH MODEL: RECURSIVE DESCRIPTION

- Initial population: P_0
- $P_N = P_{N-1} \times r \quad (r > 0)$

As in the case of arithmetic sequences, we can also define the terms of the geometric sequence explicitly by the following formula.

EXPONENTIAL GROWTH MODEL: EXPLICIT DESCRIPTION

$$P_N = P_0 \times r^N$$

A common misconception is that exponential growth implies that the population always increases. This need not be the case.

EXAMPLE 10.11

A population grows according to an exponential growth model with common ratio $r = 0.3$, starting population $P_0 = 1,000,000$. If the transitions take place once a year, what is the size of the population at the end of 6 years?

In this case, we have $P_6 = 1,000,000 \times (0.3)^6 = 729$. Figure 10-7 plots the "growth" of this population for the first 6 years, and we can clearly see that it is heading toward extinction.

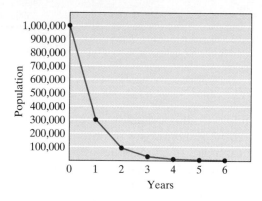

FIGURE 10-7

Exponential growth with $r = 0.3$.

It is often convenient to distinguish between exponential growth situations in which populations increase (as in Example 10.10) and those in which populations decrease (as in Example 10.11). The latter situation is commonly referred to as **exponential decay**. The difference between growth and decay is in the value of the common difference r. When $r < 1$ the populations decrease (decay); for $r > 1$ the populations increase (true growth); and for $r = 1$ we have a constant population.

Putting Your Money Where Your Math Is

Let's discuss now a general version of Example 10.10: A certain sum of money P_0 (called the *principal*) is deposited in an account that draws interest at an *annual* interest rate i (i.e., the interest is paid once a year at the end of the year). If the principal and interest are left in the account to accumulate, how much money is in the account at the end of N years?

We know now that we are dealing with a geometric population sequence whose terms are given explicitly by the formula

$$P_N = P_0 \times r^N.$$

How do we find the common ratio r? When the annual interest was 10% (Example 10.10), we got the common ratio 1.1 (110%). If the annual interest had been 12%, the common ratio would have been 1.12 (112%), and if the annual interest had been $6\frac{3}{4}\%$, the common ratio would have been 1.0675 (106.75%). In general, if we write the annual interest rate as a decimal i (rather than as a percent), then the common ratio r will be $(1 + i)$. Replacing r with $(1 + i)$ in the exponential growth model formula gives the general rule for interest that is compounded annually:

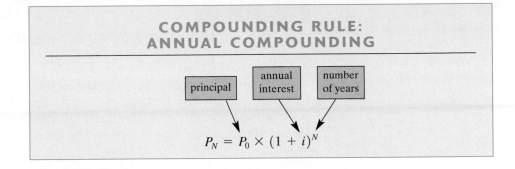

COMPOUNDING RULE: ANNUAL COMPOUNDING

$$P_N = P_0 \times (1 + i)^N$$

EXAMPLE 10.12

Suppose you deposit $367.51 in a savings account that pays an annual interest rate of $9\frac{1}{2}\%$ a year, and you leave both the principal and the interest in the account for a full 7 years. How much money will there be in the account at the end of the 7 years? Here $P_0 = 367.51$, $i = 0.095(r = 1.095)$, and $N = 7$. The answer is

$$P_7 = \$367.51 \times (1.095)^7 \approx \$693.69.$$

EXAMPLE 10.13

Let's now consider a variation of Example 10.10. Suppose we find a bank that pays 10% *annual interest with the interest compounded monthly*. If we deposit $1000 (and leave the interest in the account), how much money will there be in the account at the end of 5 years?

This problem is still one in exponential growth. The big difference now is that the period between transitions is a month (instead of a year). At the end of 5 years, we will have gone through 60 transitions, so in this example we want to find P_{60}. Since the population sequence is a geometric sequence, we have

$$P_{60} = \$1000 \times r^{60}.$$

Just as before, we need to find the value of r. Since the interest rate of 10% is *annual*, but the transitions occur *monthly*, we must divide the annual 10% interest rate by 12. This gives the **periodic interest rate**

$$p = \frac{0.10}{12} = 0.0083333\ldots.$$

The common ratio is then $r = 1.0083333\ldots$, and therefore,

$$P_{60} = \$1000 \times (1.0083333\ldots)^{60} \approx \$1645.31.$$

What would happen if we left the money in the account for 25 years? Everything is the same as in our last computation, except that the number of transitions is now $25 \times 12 = 300$. It follows that after 25 years the amount of money in the account would be

$$P_{300} = \$1000 \times (1.0083333\ldots)^{300} \approx \$12,056.95.$$

This is a significant improvement over the $10,834.71 that we would get on the same investment under annual compounding (see Example 10.10).

EXAMPLE 10.14

Now suppose we find a bank that pays 10% annual interest compounded *daily*. If we deposit $1000 for 5 years (just as in Example 10.13), how much will we have at the end of the 5 years?

In this case, the period between transitions is one day. The total number of transitions in 5 years is $365 \times 5 = 1825$, and the periodic interest rate is $p = 0.10/365 \approx 0.00027397$ (obtained by dividing the annual interest of 0.10 by the 365 compounding days in a year). The value of r for this exponential growth problem is

$$r = \left(1 + \frac{0.10}{365}\right) \approx 1.00027397,$$

and the final answer is

$$\$1000\left(1 + \frac{0.10}{365}\right)^{365 \times 5} \approx \$1000(1.00027397)^{1825} \approx \$1648.61.$$

Examples 10.13 and 10.14 illustrate the general rule for growth under compound interest, which is given by the following formula.

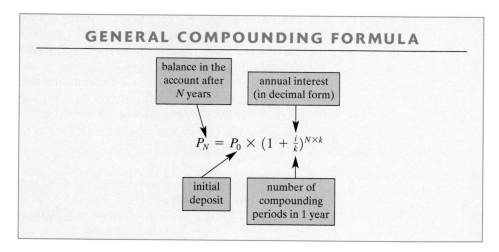

GENERAL COMPOUNDING FORMULA

balance in the account after N years

annual interest (in decimal form)

$$P_N = P_0 \times \left(1 + \frac{i}{k}\right)^{N \times k}$$

initial deposit

number of compounding periods in 1 year

EXAMPLE 10.15 Shopping for a Bank

You have an undisclosed amount of money to invest. Bank A offers savings accounts that pay 10% annual interest *compounded yearly*. Bank B offers accounts that pay 9.75% annual interest *compounded monthly*. Bank C offers accounts that pay 9.5% annual interest *compounded daily*. Which bank offers the best deal?

Note that the problem does not indicate the amount of money we invest or the length of time we plan to leave the money in the account. The answer to the problem depends only on the annual interest and the number of compounding periods in a year. The way to compare these different accounts is to use a common yardstick—for example, how much does $1 grow in 1 year?

At bank A, offering 10% interest compounded once a year, in 1 year $1 becomes $1.10.

At bank B, offering 9.75% annual interest compounded monthly, in one year $1 becomes

$$\$\left(1 + \frac{0.0975}{12}\right)^{12} \approx \$1.102.$$

And with bank C, offering 9.5% annual interest compounded daily, in one year $1 becomes

$$\$\left(1 + \frac{0.095}{365}\right)^{365} \approx \$1.0996.$$

We can now see that bank B offers the best deal. The differences between the three banks may appear insignificant, but they are significant when we leave the money for an extended period (see Exercise 27).   ■

These same calculations are described by banks in a slightly different form called the *annual yield*. The **annual yield** is the percentage increase that the investment will produce in one year. In Example 10.15, the annual yield for bank A is 10%, for bank B is 10.2%, and for bank C is 9.96%. These numbers can be read directly from the preceding calculations.

Adding Terms in a Geometric Sequence

We learned in Example 10.5 that a straightforward formula allows us to add up the consecutive terms of any arithmetic sequence. There is an equally useful formula that conveniently allows us to add consecutive terms in a geometric sequence. To simplify the notation, we use a for P_0 and r for the common ratio. The basic fact that we need is given by the following formula, which we will informally call this *geometric sum formula*.

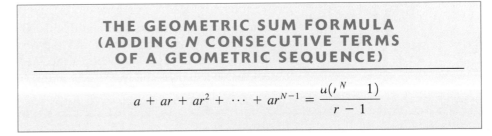

**THE GEOMETRIC SUM FORMULA
(ADDING N CONSECUTIVE TERMS
OF A GEOMETRIC SEQUENCE)**

$$a + ar + ar^2 + \cdots + ar^{N-1} = \frac{a(r^N - 1)}{r - 1}$$

The geometric sum formula works for all values of the common ratio r except $r = 1$. For $r = 1$ the formula breaks down because we get a zero denominator on the right-hand side. Fortunately, the case $r = 1$ is trivial, since then every term in the sum is a, and the sum equals $N \cdot a$.

The geometric sum formula can be verified using high school algebra—all one has to do is carry out the polynomial multiplication $(r - 1)(a + ar + ar^2 + \cdots + ar^{N-1})$ and confirm that this product equals $a(r^N - 1)$ (see Exercise 71).

An even better trick than memorizing the formula is to notice that multiplying r by each term in S gives us $rS = ar + ar^2 + ar^3 + \cdots + ar^{N-1} + ar^N$, which matches S in all of its terms except the last. It follows that when we subtract S from rS we get $rS - S = ar^N - a$ (notice all the beautiful cancellations). Thus, $(r - 1)S = a(r^N - 1)$, which gives us $S = \frac{a(r^N - 1)}{r - 1}$, the good old geometric sum formula once again.

Like its arithmetic counterpart, the geometric sum formula is an important tool in many real-life applications. We will illustrate a couple of these applications in the next two examples.

EXAMPLE 10.16

At the emerging stages, the spread of many infectious diseases—such as HIV and the West Nile virus—often follows an exponential growth model. Let's consider the case of an imaginary infectious disease called the *X-virus*. The first appearance of the deadly *X-virus* occurred in 2000 (year 0), when a total of 5000 cases of the disease were recorded in the United States. Epidemiologists estimate that until a vaccine is developed, the virus will spread at a 40% annual rate of growth, and it is expected that it will take at least 10 years to develop a vaccine. Under these assumptions, how many estimated cases of the *X-virus* will occur in the United States over the 10-year period 2000–2009?

Tracking the spread of the virus, we have

- 2000 (Year 0): 5000 cases
- 2001 (Year 1): $5000 \times 1.4 = 7000$ cases
- 2002 (Year 2): $5000 \times (1.4)^2 = 9800$ cases
  $\vdots$
- 2009 (Year 9): $5000 \times (1.4)^9$ cases

To compute the total number of cases over the 10-year period we add the terms $5000 + 5000 \times 1.4 + 5000 \times (1.4)^2 + \cdots + 5000 \times (1.4)^9$. Using the geometric sum formula with $a = 5000$, $r = 1.4$, and $N = 10$, the total number of cases then is

$$\frac{5000(1.4^{10} - 1)}{(1.4 - 1)} \approx 349{,}068.$$

Our computation shows that about 350,000 people will contract the *X-virus* over the next 10 years. And what would happen if, due to budgetary or technical problems, it takes 15 years to develop a vaccine?

All we have to do is change N to 15 in the geometric sum formula:

$$\frac{5000(1.4^{15} - 1)}{(1.4 - 1)} \approx 1{,}932{,}101.$$

These are sobering numbers—unless some kind of treatment is developed, our model predicts that between 2009 and 2014, the number of *X-virus* cases would grow from 350,000 to almost 2 million! ■

EXAMPLE 10.17

A mother decides to set up a college trust fund for her newborn child. The plan is to deposit $100 a month for the next 18 years (i.e., 216 deposits) in a savings account that pays 6% annual interest compounded monthly. How much money will there be in the account at the end of 18 years?

This is a problem of exponential growth with a twist: Each $100 deposit grows at the same monthly rate [$r = 1 + (0.06/12) = 1.005$], but the number of periods it compounds is different for each deposit. Let's make a list.

- First deposit of $100 draws interest compounded for 216 months, producing $\$100(1.005)^{216}$
- Second deposit of $100 draws interest compounded for 215 months, producing $\$100(1.005)^{215}$
- Third deposit of $100 draws interest compounded for 214 months, producing $\$100(1.005)^{214}$
  $\vdots$
- Two-hundred-sixteenth deposit of $100 draws interest for 1 month, producing $\$100(1.005)$

The total amount in the account at the end of 18 years will be

$$100(1.005)^{216} + 100(1.005)^{215} + \cdots + 100(1.005).$$

The mother, who knows a fair amount of mathematics, computes this total using the geometric sum formula (see Exercise 68):

$$\$\frac{100(1.005)(1.005^{216} - 1)}{0.005} \approx \$38,929.$$

Act II. At this point, mom realizes that by the time Junior goes off to college this sum will cover maybe a year of college expenses. She wonders if she should double the monthly deposits and make it $200 a month. How much money would then be in the trust fund at the end of 18 years?

We can answer this question easily by just thinking about what impact doubling the monthly deposits would have on the geometric sum formula. Just looking at the preceding formula, we can see that the *only* factor that would change would be the $100, which would now be $200. The net effect of this change, then, is simply to double the total—there would be $77,858 in the trust fund under this plan.

The plan is looking better, but mom still wonders if it is going to be enough (what a mom!).

Act III. The last scheme she considers is to make monthly deposits of $400, but to only do it for 9 years (in other words, double the amount again but cut the number of payments in half.) We are now talking of $400 payments for 108 months. After making the necessary adjustments (see Exercise 68), the geometric sum formula gives

$$\$\frac{400(1.005)[(1.005)^{108} - 1]}{0.005} \approx \$57,381.40.$$

This is the balance in the trust fund at the end of 9 years. At this point mom stops making additional deposits, but the money remains in the trust fund for another 9 years and continues to accrue interest at a 6% annual rate compounded monthly. Using the general compounding formula given on page 408 (with $P_0 = \$57,381.4$, $i = 0.06$, $k = 12$, and $N = 9$) gives

$$\$57381.4\left(1 + \frac{0.06}{12}\right)^{108} \approx \$98,334.50.$$

Now that makes for a healthy-looking trust fund!

10.4 The Logistic Growth Model

When dealing with animal populations, the two models we have studied so far are mostly inadequate. As we now know, *linear growth* models situations where there is a fixed amount of growth during each period between transitions. This model might work for populations of inanimate objects (garbage, production goods, sales figures, and so on) but fails completely when some form of breeding must be taken into account. *Exponential growth*, on the other hand, represents the case in which there is unrestrained breeding (e.g., money left to compound in a bank account, and sometimes in the early stages of an actual animal population). In population biology, however, it is generally the case that the rate of growth of an animal population is not always the same. Instead, it depends on the relative sizes of other interacting populations (predators, prey, and so on—see Project F on page 431) and, even more importantly, on the relative size of the population itself. When the relative size of the population is small and there is plenty of room to grow, the rate of growth is high. As the population gets more crowded, there is less room to grow and the growth rate starts to taper off. Sometimes the population gets too large for its own good, leading to decay and possibly to extinction.

A series of classic studies on the effects of population density on behavior were conducted in the 1950s by John B. Calhoun. Calhoun found that when rats were placed in a closed environment, their behavior and growth rates were normal as long as the rats were not too crowded. When their environment became too crowded, the rats started to exhibit abnormal behaviors, such as infertility and cannibalism, which effectively put a brake on the population growth rate. In extreme cases, the entire rat population became extinct.

Calhoun's experiments with rats are but one classic illustration of a general principle in population biology: *A population's growth rate is negatively impacted by the population's density.* This principle is particularly important in cases where the population is confined to a limited environment.

Population biologists call such an environment the **habitat**. The habitat might be a cage (as in Calhoun's rat experiments), a lake (as for a population of fish), a garden (as for a population of snails), and, of course, the planet itself, which is everyone's habitat.

Of the many mathematical models that attempt to deal with a variable growth rate in a fixed habitat, the simplest is a model first proposed in 1838 by the Belgian mathematician Pierre François Verhulst. Verhulst called his model the **logistic growth model**. To put it very informally, the key idea in the logistic growth model is that the rate of growth of the population is directly proportional to the amount of "elbow room" available in the population's habitat. Thus, lots of elbow room means a high growth rate. Little elbow room means a low growth rate (possibly less than 1, which, as we know, means that the population is actually decreasing). And finally, if the habitat is ever completely saturated, the population will die out.

There are two equivalent ways we can describe the situation mathematically. Suppose C is some constant that describes the total saturation point of the habitat. (Population biologists call C the **carrying capacity** of the habitat.) Then for a population of size P_N, we can say that the amount of elbow room is the difference between the carrying capacity and the population size, namely, $C - P_N$. When the growth rate is proportional to the amount of elbow room, we have

$$\text{growth rate for period } N = R(C - P_N)$$

where R is a constant of proportionality that depends only on the particular population we are studying. Using the fact that (population for period $N + 1$) = (population at period N) × (growth rate at period N), we get the following transition rule for the logistic growth model.

$$P_{N+1} = R(C - P_N)P_N.$$

There are two constants in the foregoing transition rule: R, which depends on the population we are studying, and C, which depends on the habitat.

A slightly more convenient way to describe the same thing is to put everything in relative terms. The maximum of the population is 1 (i.e., 100% of the habitat is taken up by the population). The minimum is 0 (i.e., the population is extinct). Every other possible population size P_N is identified with the ratio P_N/C, which we will denote by p_N (to distinguish it from P_N). The relative amount of elbow room is then $(1 - p_N)$, and the transition rules for the logistic model can be rewritten in the form of the following equation, called the **logistic equation**.

LOGISTIC EQUATION

$$p_{N+1} = r(1 - p_N)p_N$$

In this equation, the value p_N represents the fraction of the habitat's carrying capacity taken up by the actual population P_N, and the constant r depends on both the original growth rate R and the habitat's carrying capacity C. We will call r the **growth parameter**.

Because it measures the population growth using a single common yardstick (the fraction of its habitat's carrying capacity taken up by the population), the logistic equation is particularly convenient when making growth comparisons between populations and is preferred by ecologists and population biologists. We will stick to it ourselves. In the examples that follow, we will look at the growth pattern of an imaginary population under the logistic growth model. In each case, all we need to get started is the original population p_0 (given as a fraction of the habitat's carrying capacity) and the value of the growth parameter r. (Note that p_0 should always be between 0 and 1, and, for mathematical reasons, we will restrict r to be between 0 and 4). The logistic equation and a good calculator (or better yet, a spreadsheet) will do the rest. Be forewarned, however, that the calculations shown in the examples that follow were done with a computer and carried to 16 decimal places before being rounded off to 3 or 4 decimal places; thus, they may not match exactly with the same calculations done with a hand calculator.

EXAMPLE 10.18

Suppose we are planning to go into the business of fish farming. We have a pond in which we plan to raise a special and expensive variety of trout. Let's say that the growth parameter for this type of trout is $r = 2.5$.

We decide to start the business by stocking the pond with 20% of its carrying capacity. In the language of the logistic growth model, this is the same as saying $p_0 = 0.2$. Now let's see what the logistic growth model predicts for our future business.

After the first breeding season,[5] we have

$$p_1 = 2.5 \times (1 - 0.2) \times (0.2) = 0.4.$$

The population of the pond has doubled, and things are looking good! Since the fish are small, we decide to continue with the program. After the second breeding season, we have

$$p_2 = 2.5 \times (1 - 0.4) \times (0.4) = 0.6. \text{ (Not too bad!)}$$

After the third breeding season, we have

$$p_3 = 2.5 \times (1 - 0.6) \times (0.6) = 0.6. \text{ (A surprise!)}$$

Stubbornly, we try one more breeding season:

$$p_4 = 2.5 \times (1 - 0.6) \times (0.6) = 0.6.$$

It is quite clear that by the third generation the trout population has stabilized at 60% of the pond's carrying capacity, and unless some external change is made, it will remain at the same level for all future generations. Figure 10-8 shows a line graph of the population sequence. We can see that this is a very steady situation. ∎

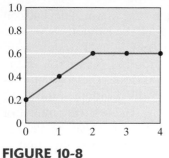

FIGURE 10-8
$r = 2.5, p_0 = 0.2.$

EXAMPLE 10.19

Suppose that we have the same pond and the same variety of trout as in Example 10.18 (in other words, we still have $r = 2.5$), but this time we stock the pond differently—let's say we start with $p_0 = 0.3$. We now have

$$p_1 = 2.5 \times (1 - 0.3) \times (0.3) = 0.525,$$

$$p_2 = 2.5 \times (1 - 0.525) \times (0.525) \approx 0.6234,$$

$$p_3 = 2.5 \times (1 - 0.6234) \times (0.6234) \approx 0.5869,$$

$$p_4 = 2.5 \times (1 - 0.5869) \times (0.5869) \approx 0.6061,$$

$$p_5 = 2.5 \times (1 - 0.6061) \times (0.6061) \approx 0.5968,$$

$$p_6 = 2.5 \times (1 - 0.5968) \times (0.5968) \approx 0.6016.$$

Something different is happening now, or is it? After the second breeding season, the population of the pond starts fluctuating—up, down, up again, back down—but in a rather special way. We leave it to the reader to verify that as one continues with the population sequence, the p-values inch closer and closer to 0.6 in an oscillating (up, down, up, down. . . .) manner. Figure 10-9 shows a line graph for the first six generations of this population. ∎

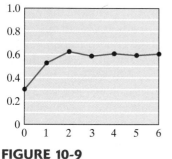

FIGURE 10-9
$r = 2.5, p_0 = 0.3.$

EXAMPLE 10.20

What happens in Example 10.19 if $p_0 = 0.7$? After the first generation, the population behaves identically with that in Example 19. This follows from the fact that in both cases we get the same value for p_1:

$$p_1 = 2.5 \times 0.3 \times 0.7 = 2.5 \times 0.7 \times 0.3 = 0.525.$$

[5]In animal populations, the transitions usually correspond to breeding seasons.

A useful general rule about logistic growth can be spotted here. If we replace p_0 with its complement $(1 - p_0)$, then after the first generation the populations will behave identically. ▪

EXAMPLE 10.21

Let's say that based on what we learned from the previous examples we decide to try to raise a different population of fish—a special variety of catfish for which the growth parameter is $r = 3.1$.

What happens if we start with $p_0 = 0.2$? For the sake of brevity, we will write the values of the population in sequence form and leave the calculations to the reader.

$$p_0 = 0.2, \quad p_1 = 0.496, \quad p_2 \approx 0.775, \quad p_3 \approx 0.541,$$

$$p_4 \approx 0.770, \quad p_5 \approx 0.549, \quad p_6 \approx 0.767, \quad p_7 \approx 0.553,$$

$$p_8 \approx 0.766, \quad p_9 \approx 0.555, \quad p_{10} \approx 0.766, \quad p_{11} \approx 0.556,$$

$$p_{12} \approx 0.765, \quad p_{13} \approx 0.557, \quad p_{14} \approx 0.765, \quad p_{15} \approx 0.557, \quad \ldots$$

An interesting pattern emerges here. After a few breeding seasons, the population settles into a two-period cycle, alternating between a high-population period at 0.765 and a low-population period at 0.557. Figure 10-10 convincingly illustrates the oscillating nature of the population sequence.

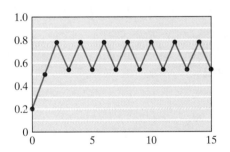

FIGURE 10-10
$r = 3.1$, $p_0 = 0.2$.

▪

There are many animal populations whose behavior parallels that of the fish population in Example 10.21—a lean season followed by a boom season followed by a lean season, and so on.

EXAMPLE 10.22

We are now out of the fish-farming business and have acquired an interest in entomology—the study of insects. We are going to study the behavior of a type of flour beetle with a growth parameter given by $r = 3.5$.

Let's suppose that the starting population is given by $p_0 = 0.56$ and we use the logistic equation to predict the growth of this population. We leave it to the reader to verify these numbers and fill in the missing details. (A calculator is all that is needed.) We have

$$p_0 = 0.560, \quad p_1 \approx 0.862, \quad p_2 \approx 0.415, \quad p_3 \approx 0.850,$$

$$p_4 \approx 0.446, \quad p_5 \approx 0.865, \quad \ldots \quad p_{20} \approx 0.497,$$

$$p_{21} \approx 0.875, \quad p_{22} \approx 0.383, \quad p_{23} \approx 0.827, \quad p_{24} \approx 0.501,$$

$$p_{25} \approx 0.875, \quad \ldots\ldots$$

It took a while, but we can now see a pattern: Since $p_{25} = p_{21}$, the population will repeat itself in a four-period cycle ($p_{26} = p_{22}, p_{27} = p_{23}, p_{28} = p_{24}, p_{29} = p_{25} = p_{21}$, etc.), an interesting and surprising turn of events. Figure 10-11 shows the first 25 generations in this population sequence.

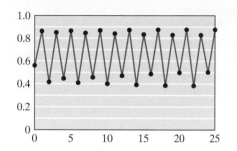

FIGURE 10-11
$r = 3.5, p_0 = 0.56$.

The cyclical behavior exhibited in Example 10.22 is not unusual, and many insect populations follow cyclical patterns of various lengths—7-year cycles (locusts), 17-year cycles (cicadas), and so on.

EXAMPLE 10.23

Our last and most remarkable example is a population sequence determined by the logistic growth model with a growth parameter of $r = 4$. Let's start with $p_0 = 0.2$. The first 20 values of the population sequence are given by

$p_0 = 0.2000,$ $\quad p_1 = 0.640,$ $\quad p_2 \approx 0.9216,$ $\quad p_3 \approx 0.2890,$

$p_4 \approx 0.8219,$ $\quad p_5 \approx 0.5854,$ $\quad p_6 \approx 0.9708,$ $\quad p_7 \approx 0.1133,$

$p_8 \approx 0.4020,$ $\quad p_9 \approx 0.9616,$ $\quad p_{10} \approx 0.1478,$ $\quad p_{11} \approx 0.5039,$

$p_{12} \approx 0.9999,$ $\quad p_{13} \approx 0.0002,$ $\quad p_{14} \approx 0.0010,$ $\quad p_{15} \approx 0.0039,$

$p_{16} \approx 0.0157,$ $\quad p_{17} \approx 0.0617,$ $\quad p_{18} \approx 0.2317,$ $\quad p_{19} \approx 0.7121.$

Figure 10-12 plots the behavior of the population for the first 19 generations. The reader is encouraged to chart this population for a few additional generations. The surprise here is the absence of any predictable pattern. Even though the population sequence is governed by a very precise rule (the logistic equation), to an outside observer the pattern of growth appears to be quite erratic and seemingly random.

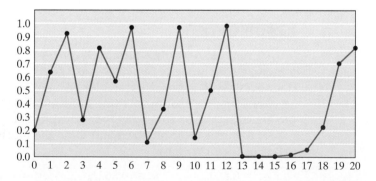

FIGURE 10-12
$r = 4.0, p_0 = 0.2$.

The behavior of the logistic growth model exhibits many interesting surprises. In addition to trying Exercises 39 through 46 at the end of the chapter, the reader is encouraged to experiment on his or her own in a manner similar to the work we did in the preceding examples. (Choose a p_1 between 0 and 1, choose an r between 3 and 4, and fire up both your calculator and your imagination!) An excellent nontechnical account of the surprising patterns produced by the logistic growth model can be found in references 1 and 2. More technical accounts of the logistic equation can be found in references 3, 8, and 9.

Conclusion

In this chapter we studied three simple models that describe the way that populations grow.

In the *linear model* of population growth, the population is described by an arithmetic sequence, and at each transition period, the population grows by a constant amount called the *common difference*. Linear growth is most common with populations consisting of inanimate objects—commodities, resources, garbage, etc.

In the *exponential model* of population growth, the population is described by a geometric sequence. In each transition period, the population is multiplied by a constant amount called the *common ratio*. Exponential growth is typical of situations in which there is unrestricted breeding. Money drawing interest in a bank account is one such example; the spread of emerging infectious diseases is another.

The *logistic model* of population growth represents situations in which the rate of growth of the population varies from one season to the next, depending on the amount of space available in the population's habitat. When confined to a single-species habitat, many animal populations are governed by the logistic model or simple variations of it.

Most serious studies of population growth involve models with much more complicated mathematical descriptions, but to us, that is neither here nor there. Ultimately, the details are not as important as the overall picture: a realization that mathematics can be useful even in its most simplistic forms to describe and predict the rise and fall of populations in many fields—from the human realms of industry, finance, and public health to the natural world of population biology and ecology.

PROFILE   Sir Robert May (1936–   )

If there is one consistent thread that runs through my life from when I was young, it's that I enjoy playing games, whether it's Monopoly, Risk, chess, or contract bridge. And my professional interest is in working out the rules of the games that nature plays.[6]

Robert May was born in Sydney, Australia, in 1936. A man of many talents and interests, May could very well be the most versatile scientist of his generation. After starting his undergraduate studies in chemical engineering, May switched to physics for his doctoral work. He received a Ph.D. in theoretical physics from the University of Sydney in 1959, and in 1962, at the age of 26, he became a professor of physics there. Soon after, he changed direction again, dropping a successful career in physics research to start a new career in population ecology. An ordinary Joe changing careers too many times is called a "flake," but May was no ordinary Joe. By the mid-1970s May had become the world's foremost authority in mathematical ecology, and in 1973 he was appointed Professor of Biology at Princeton University, where he remained until 1988.

While at Princeton, May produced his pioneering work in mathematical models of population growth. The conventional wisdom at the time among ecologists was that to model a complex ecosystem one needs a complicated mathematical model, and the more complexity shown in the system the more sophisticated the equations and variables of the model need to be. The flip side of this was the notion that a simple mathematical model could not possibly reproduce complicated patterns of behavior. In a groundbreaking paper entitled "Simple Mathematical Models with Very Complicated Dynamics" published in the scientific journal *Nature* in 1976 (reference 9), May showed that the extremely simple (by mathematical ecology standards) *logistic growth model* could exhibit surprisingly exotic patterns, such as periodicity, bifurcation, and seemingly random fluctuation. May's 1976 paper paved the way for the development of a new and revolutionary branch of mathematics called *chaos theory* (for details, see reference 2).

After Princeton, May's career took another turn, as he became an important public figure and world advocate for science, the environment, and public policy. Between 1988 and 1995, May served as a Royal Society Research Professor of Zoology at the University of Oxford, England. In 1995 he was appointed Chief Scientific Advisor to the Government of the United Kingdom, and in 2000 he was appointed president of The Royal Society of London, one of the most distinguished scientific posts in the world. May is the recipient of many important prizes and awards, including the prestigious MacArthur "genius" award (1984), the Linnean Medal in Zoology (1991), and the Crafoord Prize in Biosciences (1996). He was awarded the honorary titles of Knight of the United Kingdom (1996), Companion of the Order of Australia (1998), and in 2001 was made a Lord of the United Kingdom for his many contributions to science and society.

---

[6]Ferry, Georgina, "Sir Robert May, Complexity and Real World Problems." *Sante Fe Institute Bulletin*, Vol. 16, no. 1, 2001, 1–6.

KEY CONCEPTS

annual yield
arithmetic sequence
arithmetic sum formula
carrying capacity
common difference
common ratio
continuous growth
discrete growth
explicit description (of a sequence)
exponential decay
exponential growth model
geometric sequence
geometric sum formula

growth parameter
habitat
linear growth model
line graph
logistic equation
logistic growth model
periodic interest rate
population sequence
recursive description (of a sequence)
scatter plot
transition
transition rule

EXERCISES

WALKING

A. Linear Growth and Arithmetic Sequences

1. Consider a population that grows according to the recursive rule $P_N = P_{N-1} + 125$, with initial population $P_0 = 80$.
   (a) Find P_1, P_2, and P_3.
   (b) Find P_{100}.
   (c) Give an explicit description of the population sequence.

2. Consider a population that grows according to the recursive rule $P_N = P_{N-1} + 23$, with initial population $P_0 = 57$.
   (a) Find P_1, P_2, and P_3.
   (b) Find P_{200}.
   (c) Give an explicit description of the population sequence.

3. Consider a population that grows according to a linear growth model. The initial population is $P_0 = 75$, and the common difference is $d = 5$.
   (a) Find P_{30}.
   (b) How many generations will it take for the population to reach 1000?
   (c) How many generations will it take for the population to reach 1002?

4. Consider a population that decays according to a linear model. The initial population is $P_0 = 520$, and the common difference is $d = -20$.
   (a) Find P_{24}.
   (b) How many generations will it take for the population to reach 10?
   (c) How many generations will it take for the population to become extinct?

5. Consider a population that grows according to a linear growth model. The initial population is $P_0 = 8$, and the population in the 10th generation is $P_{10} = 38$.
   (a) Find the common difference d.

   (b) Find P_{50}.

   (c) Give an explicit description of the population sequence.

6. Consider a population that grows according to a linear growth model. The population in the fifth generation is $P_5 = 37$, and the population in the seventh generation is $P_7 = 47$.

   (a) Find the common difference d.

   (b) Find the initial population P_0.

   (c) Give an explicit description of the population sequence.

7. An arithmetic sequence has $A_1 = 11$ and $A_2 = -4$.

   (a) Find A_3.

   (b) Find A_0.

   (c) How many terms in the sequence are bigger than 30? Explain.

8. An arithmetic sequence has $A_5 = 20$ and $A_6 = -5$.

   (a) Find A_7.

   (b) Find A_4.

   (c) How many terms in the sequence are bigger than 50? Explain.

9. Mr. G.Q. is a snappy dresser and has an incredible collection of neckties. Each month, he buys himself 5 new neckties. Let P_0 represent the number of neckties he starts out with and P_N be the number of neckties in his collection at the end of the Nth month. Assume that he started out with just 3 neckties and that he never throws neckties away.

   (a) Give a recursive description for P_N.

   (b) Give an explicit description for P_N.

   (c) Find P_{300}.

10. A nuclear power plant produces 12 lb of radioactive waste every month. The radioactive waste must be stored in a special storage tank. On Jan. 1, 2000, there were 25 lb of radioactive waste in the tank. Let P_N represent the amount of radioactive waste (in pounds) in the storage tank after N months.

    (a) Give a recursive description for P_N.

    (b) Give an explicit description for P_N.

    (c) If the maximum capacity of the storage tank is 500 lb, when will the tank reach its maximum capacity?

11. Find $\underbrace{2 + 7 + 12 + \cdots}_{100 \text{ terms}}$.

12. Find $\underbrace{21 + 28 + 35 + \cdots}_{57 \text{ terms}}$.

13. Find $12 + 15 + 18 + \cdots + 309$.

14. Find $1 + 10 + 19 + \cdots + 2701$.

15. Consider a population that grows according to a linear growth model. The initial population is $P_0 = 23$, and the common difference is $d = 7$.

    (a) Find $P_0 + P_1 + P_2 + \cdots + P_{999}$.

    (b) Find $P_{100} + P_{101} + \cdots + P_{999}$.

16. Consider a population that grows according to a linear growth model. The initial population is $P_0 = 7$, and the population in the first generation is $P_1 = 11$.

(a) Find $P_0 + P_1 + P_2 + \cdots + P_{500}$.

(b) Find $P_{100} + P_{101} + \cdots + P_{500}$.

17. The city of Lightsville currently has 137 street lights. As part of an urban renewal program, the city council has decided to install and have operational 2 additional street lights at the end of each week for the next 52 weeks. Each street light costs $1 to operate for 1 week.

    (a) How many street lights will the city have at the end of 38 weeks?

    (b) How many street lights will the city have at the end of N weeks? ($N \leq 52$.)

    (c) What is the cost of operating the original 137 lights for 52 weeks?

    (d) What is the additional cost for operating the newly installed lights for the 52-week period during which they are being installed?

18. A manufacturer currently has on hand 387 widgets. During the next 2 years, the manufacturer will be increasing his inventory by 37 widgets per week. (Assume that there are exactly 52 weeks in one year.) Each widget costs 10 cents a week to store.

    (a) How many widgets will the manufacturer have on hand after 20 weeks?

    (b) How many widgets will the manufacturer have on hand after N weeks? (Assume $N \leq 104$.)

    (c) What is the cost of storing the original 387 widgets for 2 years (104 weeks)?

    (d) What is the additional cost of storing the increased inventory of widgets for the next 2 years?

B. Exponential Growth and Geometric Sequences

19. Suppose you deposit $3250 in a savings account that pays 9% annual interest, with interest credited to the account at the end of each year. Assuming that no withdrawals are made, how much money will be in the account after 4 years?

20. Suppose you deposit $1237.50 in a savings account that pays 8.25% annual interest, with interest credited to the account at the end of each year. Assuming that no withdrawals are made, how much money will be in the account after 3 years?

21. Suppose you deposited $3420 on Jan. 1, 1997, in a savings account paying $6\frac{5}{8}\%$ annual interest, with interest credited to the account on December 31 of each year. On Jan. 1, 1999, you withdrew $1500, and on Jan. 1, 2000, you withdrew $1000. If you make no other withdrawals, what will be the balance in your account on Jan. 1, 2003?

22. Suppose you deposited $2500 on Jan. 1, 1996, in a savings account paying $5\frac{3}{8}\%$ annual interest, with interest credited to the account on December 31 of each year. On Jan. 1, 1999, you withdrew $850. If you make no other withdrawals, what will be the balance in your account on Jan. 1, 2004?

23. (a) The amount of $5000 is deposited in a savings account that pays 12% annual interest compounded monthly. Assuming that no withdrawals are made, how much money will be in the account after 5 years?

    (b) What is the annual yield on this account?

24. (a) The amount of $874.83 is deposited in a savings account that pays $7\frac{3}{4}\%$ annual interest compounded daily. Assuming that no withdrawals are made, how much money will be in the account after 2 years?

    (b) What is the annual yield on this account?

25. You have some money to invest. The Great Bulldog Bank offers accounts that pay 6% annual interest compounded yearly. The First Northern Bank offers accounts that pay 5.75% annual interest compounded monthly. The Bank of Wonderland offers 5.5% annual interest compounded daily. What is the annual yield for each bank?

26. Complete the following table.

Annual interest rate	Compounded	Annual yield
12%	Yearly	12%
12%	Semiannually	?
12%	Quarterly	?
12%	Monthly	?
12%	Daily	?
12%	Hourly	?

27. Consider the three banks discussed in Example 10.15. A savings account at Bank A has an annual yield of 10%, a savings account at Bank B has an annual yield of 10.2%, and a savings account at Bank C has an annual yield of 9.96%.

   (a) If you deposit $1000 in Bank A and make no withdrawals, how much money will there be in the account at the end of 25 years?

   (b) If you deposit $1000 in Bank B and make no withdrawals, how much money will there be in the account at the end of 25 years?

   (c) If you deposit $1000 in Bank C and make no withdrawals, how much money will there be in the account at the end of 25 years?

28. Your bank is offering a special promotion for its preferred customers. If you buy a $500 Certificate of Deposit (CD), at the end of the year you can cash the CD for $555. What is the annual yield of this investment?

29. A population grows according to an exponential growth model. The initial population is $P_0 = 11$, and the common ratio is $r = 1.25$.

   (a) Find P_1.

   (b) Find P_9.

   (c) Give an explicit description for the population sequence.

30. A population grows according to an exponential growth model. The initial population is $P_0 = 8$, and the common ratio is $r = 1.5$.

   (a) Find P_1.

   (b) Find P_9.

   (c) Give an explicit description for the population sequence.

31. Consider the geometric sequence with first term $P_0 = 3$ and common ratio $r = 2$.

   (a) Find P_{100}.

   (b) Find P_N.

   (c) Find $P_0 + P_1 + \cdots + P_{100}$.

   (d) Find $P_{50} + P_{51} + \cdots + P_{100}$.

32. Consider the geometric sequence with first four terms 1, 3, 9, and 27.

   (a) Find P_{100}.

   (b) Find P_N.

(c) Find $P_0 + P_1 + \cdots + P_{100}$.

(d) Find $P_{50} + P_{51} + \cdots + P_{100}$.

33. You decide to open a Christmas Club account at a bank that pays 6% annual interest compounded monthly. You deposit $100 on the first of January and on the first of each succeeding month through November. How much will you have in your account on the first of December?

34. You decide to save money to buy a car by opening a special account at a bank that pays 8% annual interest compounded monthly. You deposit $300 on the first of each month for 36 months. How much will you have in your account at the end of the 36th month?

35. You are interested in buying a car 5 years from now, and you estimate that the future cost of the car will be $10,000. You decide to deposit money today in an account that pays interest, so that 5 years hence you will have the $10,000. How much money do you need to deposit if the account you deposit your money in

(a) has an interest rate of 10% compounded annually?

(b) has an interest rate of 10% compounded quarterly?

(c) has an interest rate of 10% compounded monthly?

36. You have $1000 to invest. Suppose you find an investment that guarantees an $8\frac{1}{2}$% annual yield, with the interest paid once a year at the end of the year. How many years will it take for you to at least double your original investment?

C. Logistic Growth Model

Exercises 37 through 52 refer to the logistic growth model and the logistic equation $p_{N+1} = r(1 - p_N)p_N$. *For most of these exercises, either a calculator with a memory register or a spreadsheet is suggested.*

37. A population grows according to the logistic growth model, with growth parameter $r = 0.8$. Starting with an initial population given by $p_0 = 0.3$,

(a) find p_1.

(b) find p_2.

(c) determine what percent of the habitat's carrying capacity is taken up by the third generation.

38. A population grows according to the logistic growth model, with growth parameter $r = 0.6$. Starting with an initial population given by $p_0 = 0.7$,

(a) find p_1.

(b) find p_2.

(c) determine what percent of the habitat's carrying capacity is taken up by the third generation.

39. For the population discussed in Exercise 37 ($r = 0.8$, $p_0 = 0.3$),

(a) find the values of p_1 through p_{10}.

(b) what does the logistic growth model predict in the long term for this population?

40. For the population discussed in Exercise 38 ($r = 0.6$, $p_0 = 0.7$),

(a) find the values of p_1 through p_{10}.

    (b) what does the logistic growth model predict in the long term for this population?

41. A population grows according to the logistic growth model, with growth parameter $r = 1.8$. Starting with an initial population given by $p_0 = 0.4$,

    (a) find the values of p_1 through p_{10}.

    (b) what does the logistic growth model predict in the long term for this population?

42. A population grows according to the logistic growth model, with growth parameter $r = 1.5$. Starting with an initial population given by $p_0 = 0.8$,

    (a) find the values of p_1 through p_{10}.

    (b) what does the logistic growth model predict in the long term for this population?

43. A population grows according to the logistic growth model, with growth parameter $r = 2.8$. Starting with an initial population given by $p_0 = 0.15$,

    (a) find the values of p_1 through p_{10}.

    (b) what does the logistic growth model predict in the long term for this population?

44. A population grows according to the logistic growth model, with growth parameter $r = 2.5$. Starting with an initial population given by $p_0 = 0.2$,

    (a) find the values of p_1 through p_{10}.

    (b) what does the logistic growth model predict in the long term for this population?

45. A population grows according to the logistic growth model, with growth parameter $r = 3.25$. Starting with an initial population given by $p_0 = 0.2$,

    (a) find the values of p_1 through p_{10}.

    (b) what does the logistic growth model predict in the long term for this population?

46. A population grows according to the logistic growth model, with growth parameter $r = 3.51$. Starting with an initial population given by $p_0 = 0.4$,

    (a) find the values of p_1 through p_{10}.

    (b) what does the logistic growth model predict in the long term for this population?

D. Miscellaneous

47. Each of the following sequences follows either a linear, exponential, or a logistic growth model. For each sequence, determine which model applies (if more than one applies, then indicate all the ones that apply).

    (a) 2, 4, 8, 16, 32, ...

    (b) 2, 4, 6, 8, 10, ...

    (c) 0.8, 0.4, 0.6, 0.6, 0.6, ...

    (d) 0.81, 0.27, 0.09, 0.03, 0.01, ...

    (e) 0.49512, 0.81242, 0.49528, 0.81243, 0.49528, ...

    (f) 0.9, 0.75, 0.6, 0.45, 0.3, ...

    (g) 0.7, 0.7, 0.7, 0.7, 0.7, ...

48. Each of the following graphs describes a population that grows according to a linear, exponential, or logistic model. For each line graph, determine which model applies.

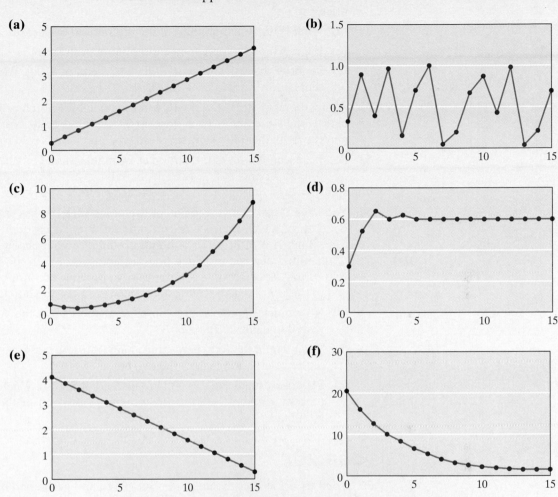

(a)

(b)

(c)

(d)

(e)

(f)

49. A population of laboratory rats grows according to the following transition rule: $P_N = P_{N-1} + 2P_{N-2}$. The initial population is $P_0 = 6$, and the population in the first generation is $P_1 = 10$.

(a) Find P_2.

(b) Find P_3.

(c) Explain why there is always an even number of rats.

50. A population of guinea pigs grows according to the following transition rule: $P_N = 2P_{N-1} - P_{N-2}$. The initial population is $P_0 = 3$, and the population in the first generation is $P_1 = 5$.

(a) Find P_2.

(b) Find P_3.

(c) Explain why there is always an odd number of guinea pigs.

(d) Give an explicit description of the population sequence.

51. You have a coupon worth 15% off any item (including sale items) in a store. The particular item you want is on sale at 30% off the marked price of $100. The store policy allows you to use your coupon before the 30%

discount or after the 30% discount. (That is, you can take 15% off the marked price first and then take 30% off the resulting price, or you can take 30% off the marked price first and then take 15% off the resulting price.)

(a) What is the dollar amount of the discount in each case?

(b) What is the total percentage discount in each case?

(c) Suppose the article costs P dollars (instead of $100). What is the percentage discount in each case?

52. A membership store gives a 10% discount on all purchases to its members. If the store marks each item up 50% (based on its cost), what is the markup actually realized by the store when an item is sold to a member?

53. For 3 consecutive years, the tuition at Tasmania State University increased by 10%, 15%, and 10%, respectively. What was the total percentage increase overall during the 3-year period? (*Hint:* The answer is not 35%!)

54. You have $1000 to invest in one of two competing banks (bank A or bank B), both of which are paying 10% annual interest on deposits left for 1 year. Bank A is offering a 5% bonus credited to your account at the time of the initial deposit, provided that the funds are left in the account for a year. Bank B is offering a 5% bonus paid on your account balance at the end of the year after the interest has been credited to your account.

(a) How much money would you have at the end of the year if you invested in bank A? In bank B?

(b) What would the total percentage gain (interest plus bonus) at the end of the year be for each of the two banks?

(c) Suppose you invested P dollars (instead of $1000). What will the total percentage gain (interest plus bonus) at the end of the year be for each bank?

JOGGING

55. You buy a $500 certificate of deposit (CD), and at the end of 2 years you cash it for $561.80. What is the annual yield of this investment?

56. **(a)** The *Happyville Gazette* wants to sign a one-year contract with a Web service provider to have each edition (including back issues) of its newspaper available online. The contract specifies the cost of storage will be 2 cents per edition per day. Determine the cost of this one-year contract.

(b) Suppose that the *Gazette* also needs to hire a Web designer to format the online newspaper. The Web designer charges $3000 for the first month, and his fee increases by 2% every month thereafter. Determine the yearly cost of hiring the Web designer.

57. Find those right triangles for which the sides are in an arithmetic sequence. [*Hint:* Find the value(s) of a in the following right triangle.]

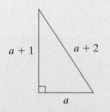

58. Find those right triangles for which the sides are in a geometric sequence. [*Hint:* Find the value(s) of k in the following right triangle.]

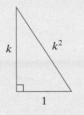

59. How much should a retailer mark up her goods so that when she has a 25%-off sale, the resulting prices will still reflect a 50% markup (on her cost)?

60. What annual interest rate compounded semiannually gives an annual yield of 21%?

61. Before Annie set off for college, Daddy Warbucks offered her a choice between the following two incentive programs:

- **Option 1.** A $100 reward for every A she gets in a college course
- **Option 2.** One cent for her first A, 2 cents for the second A, 4 cents for the third A, 8 cents for the fourth A, and so on

Annie chose Option 1. After getting a total of 30 A's in her college career, Annie is happy with her reward of $100 \times 30 = 3000. Unfortunately, Annie did not get an A in math. Help her figure out how much she would have made had she chosen option 2.

62. Suppose that $P_0, P_1, P_2, \ldots, P_N$ are the terms of a geometric sequence. Suppose, moreover, that the sequence satisfies the recursive rule $P_N = P_{N-1} + P_{N-2}$, for $N \geq 2$. Find the common ratio r.

63. Give an example of a geometric sequence in which P_0, P_1, P_2, and P_3 are integers, and all the terms from P_4 on are fractions.

64. Consider a population that grows according to the logistic growth model with initial population given by $p_0 = 0.7$. What growth parameter r would keep the population constant?

65. Suppose that you are in charge of stocking a lake with a certain type of alligator with a growth parameter $r = 0.8$. Assuming that the population of alligators grows according to the logistic growth model, is it possible for you to stock the lake so that the alligator population is constant? Explain.

66. Consider a population that grows according to the logistic growth model with growth parameter r ($r > 1$). Find p_0 in terms of r so that the population is constant.

67. Suppose the habitat of a population of snails has a carrying capacity of $C = 20,000$ and the current population is 5,000. Suppose also that the growth parameter for this particular type of snail is $r = 3.0$. What does the logistic growth model predict for this population after four transition periods?

68. The purpose of this exercise is to fill in all the missing details of Example 10.17.

(a) In Example 10.17 it is claimed that

$$100(1.005)^{216} + 100(1.005)^{215} + \cdots + 100(1.005) = \frac{100(1.005)(1.005^{216} - 1)}{0.005}.$$

Use the geometric sum formula to explain why the preceding statement is true. (What is a? What is r?)

(b) If $a(1.005)^{216} + a(1.005)^{215} + \cdots + a(1.005) = S$, then express the sum $b(1.005)^{216} + b(1.005)^{215} + \cdots + b(1.005)$ in terms of a, b, and S.

(c) If you deposit \$400 each month in a savings account that pays 6% annual interest compounded monthly and leave all the money in the account, at the end of 9 years you will have a total of $\dfrac{400(1.005)(1.005^{108} - 1)}{0.005}$. Fill in the details.

69. Compute the sum

$$1 + 5 + 3 + 8 + 5 + 11 + 7 + 14 + \cdots + 99 + 152.$$

(*Hint:* Break the sum up into two separate sums.)

70. Compute the sum

$$1 + 1 + 2 + \frac{1}{2} + 4 + \frac{1}{4} + 8 + \frac{1}{8} + \cdots + 4096 + \frac{1}{4096}.$$

71. Prove the geometric sum formula

$$a + ar + ar^2 + \cdots + ar^{N-1} = \frac{a(r^N - 1)}{r - 1}.$$

(*Hint:* Multiply the left-hand side by $r - 1$ and do the algebra.)

72. Show that the sum of the first N terms of an arithmetic sequence with first term c and common difference d is

$$\frac{N}{2}[2c + (N - 1)d].$$

RUNNING

73. You are purchasing a home for \$120,000 and are shopping for a loan. You have a total of \$31,000 to put down, including the closing costs of \$1000 and any loan fee that might be charged. Bank A offers a 10%-annual-interest loan amortized over 30 years with 360 equal monthly payments. There is no loan fee. Bank B offers a 9.5%-annual-interest loan amortized over 30 years with 360 equal monthly payments. There is a 3% loan fee (i.e., a one-time up-front charge of 3% of the loan). Which loan is better?

74. A friend of yours sells his car to a college student and takes a personal note (cosigned by the student's rich uncle) for \$1200 with no interest, payable at \$100 per month for 12 months. Your friend immediately approaches you and offers to sell you this note. How much should you pay for the note if you want an annual yield of 12% on your investment?

75. The purpose of this exercise is to understand why we assume that, under the logistic growth model, the growth parameter r is between 0 and 4.

(a) What does the logistic equation give for p_{N+1} if $p_N = 0.5$ and $r > 4$. Is this a problem?

(b) What does the logistic equation predict for future generations if $p_N = 0.5$ and $r = 4$?

(c) If $0 \le p \le 1$, what is the largest possible value of $(1 - p)p$?

(d) Explain why, if $0 < p_0 < 1$ and $0 < r < 4$, then $0 < P_N < 1$, for every positive integer N.

76. Suppose $r > 3$. Using the logistic growth model, find a population p_0 such that $p_0 = p_2 = p_4 \cdots$, but $p_0 \neq p_1$.

77. Show that if $P_0, P_1, P_2, \ldots$ is an arithmetic sequence, then $2^{P_0}, 2^{P_1}, 2^{P_2}, \ldots$ must be a geometric sequence.

A. The Many Faces of e

Just as the golden ratio Φ is tied to growth in nature (see Chapter 9), the irrational number e, named after Euler (see page 202), is closely related to many population growth models. For example, if $1 is deposited in an account which earns 100% interest and is compounded *continuously* (yes, that is one busy banker!), the $1 will grow to e (approximately $2.72) after one year.

Much like Φ and π, e is a ubiquitous irrational number with many remarkable mathematical properties. In this research project you are asked to present and discuss five of the more interesting mathematical properties of e. For each property, give a historical background (if possible), a simple mathematical explanation (avoid technical details if you can), and a real-life application (if possible).

Note: A good starting point for this project is the book *e: The Story of a Number*, by Eli Maor (Princeton University Press, 1998).

B. Annuities

In this project you are to explain, with the aid of a concrete example, the concept of an *annuity*. In doing so, you will also want to discuss the concepts of *future value* and *present value*. Pay particular attention to how annuities relate to the geometric sum formula discussed in this chapter.

C. The Malthusian Doctrine

In 1798, Thomas Malthus wrote his famous *Essay on the Principle of Population*. In this essay, Malthus put forth the principle *that population grows according to an exponential growth model, whereas food and resources grow according to a linear growth model.* Based on this doctrine, Malthus predicted that at some point in the future, the demand for food would be much greater that the supply.

Write an analysis paper detailing some of the consequences of Malthus's doctrine. Does the doctrine apply in a modern technological world? Can the doctrine be the explanation for the famines in sub-Saharan Africa? Discuss the many possible criticisms that can be leveled against Malthus's doctrine. To what extent do you agree with Malthus's doctrine?

D. The Logistic Equation and the United States Population

The logistic growth model, first discovered by Verhulst, was rediscovered in 1920 by the American population ecologists Raymond Pearl and Lowell Reed. Pearl and Reed compared the population data for the United States between 1790 and 1920 with what would be predicted using a logistic equation and found that the numbers produced by the equation and the real data matched quite well.

In this project, you are to discuss and analyze Pearl and Reed's 1920s paper (reference 11). Here are some suggested questions you may want to discuss: Is the logistic model a good model to use with human populations? What might be a reasonable estimate for the carrying capacity of the United States? What happens with Pearl and Reed's model when you expand the census population data all the way to 2000?

E. World Population Growth

How do demographers model world population? Is this different from how they model, say, the population of the United States? How does this process compare with that used by biologists in determining the size of a future salmon spawn?

In this project, you will compare and contrast the process demographers use to model human population growth with that which biologists use to model animal populations.

F. The Lotka-Volterra Two-Species Population Model

The population growth models discussed in this chapter involve only a single species. Most of the time, several species share a habitat, and in this case predator-prey interactions play an important role in population growth models. One of the first models to incorporate interactions between predators and prey was proposed in 1925 by the American biophysicist Alfred Lotka and independently by the Italian mathematician Vito Volterra in 1926. It is now known as the *Lotka-Volterra model*.

In this project you should describe the basic ideas behind the Lotka-Volterra model. Present the information in the simplest possible terms and avoid the technical details.

REFERENCES AND FURTHER READINGS

1. Cipra, Barry, "Beetlemania: Chaos in Ecology," in *What's Happening in the Mathematical Sciences 1998–1999*. Providence, RI: American Mathematical Society, 1999.

2. Gleick, James, *Chaos: Making a New Science*. New York: Viking Penguin, Inc., 1987, chap. 3.

3. Gordon, W. B., "Period Three Trajectories of the Logistic Map," *Mathematics Magazine*, 69 (1996), 118–120.

4. Hoppensteadt, Frank, *Mathematical Methods of Population Biology*. Cambridge: Cambridge University Press, 1982.

5. Hoppensteadt, Frank, *Mathematical Theories of Populations: Demographics, Genetics and Epidemics*. Philadelphia: Society for Industrial and Applied Mathematics, 1975.

6. Hoppensteadt, Frank, and Charles Peskin, *Mathematics in Medicine and the Life Sciences*. New York: Springer-Verlag, 1992.

7. Kingsland, Sharon E., *Modeling Nature: Episodes in the History of Population Ecology*. Chicago: University of Chicago Press, 1985.

8. May, Robert M., "Biological Populations with Nonoverlapping Generations: Stable Points, Stable Cycles and Chaos," *Science*, 186 (1974), 645–647.

9. May, Robert M., "Simple Mathematical Models with Very Complicated Dynamics," *Nature*, 261 (1976), 459–467.

10. May, Robert M., and George F. Oster, "Bifurcations and Dynamic Complexity in Simple Ecological Models," *American Naturalist*, 110 (1976), 573–599.

11. Pearl, Raymond, and Lowell J. Reed, "On the Rate of Growth of the Population of the United States since 1790 and Its Mathematical Representation," Proceedings of the National Academy of Sciences USA, 1920.

12. Sigler, Laurence, *Fibonacci's Liber Abaci*. New York: Springer-Verlag, 2002.

13. Smith, J. Maynard, *Mathematical Ideas in Biology*. Cambridge: Cambridge University Press, 1968.

14. Swerdlow, Joel, "Population," *National Geographic* (October, 1998), 2–35.

Symmetry

Mirror, Mirror, Off The Wall . . .

It is said that Eskimos have dozens of different words for ice. Ice is, after all, a universal theme in the Eskimo's world. Along the same lines, we would expect science and mathematics to have dozens of different words to describe the notion of symmetry, since symmetry is a recurrent theme in the world around us. Surprisingly, just the opposite is the case. We use a single word—*symmetry*—to cover an incredibly diverse set of situations and ideas.

E xactly what is symmetry? The answer depends very much on the context of the question. In everyday language, *symmetry* is most often taken to mean *mirror symmetry* (also called *bilateral* or *left-right symmetry*) such as the almost but not quite perfect left-right symmetry exhibited externally[1] by the human body (famously illustrated in Leonardo da Vinci's *Vitruvian Man*). In everyday language, symmetry is also used to describe an aesthetic value—people think of something *symmetric* as well balanced, pleasing to the eye, well proportioned. This is often how the word is used in art and architecture. Along the same lines, *symmetry* is used in musical composition to describe special melodic effects. The music of Bach, for example, is often described and analyzed in terms of its symmetry.[2] Even in poetry and literature, symmetry is used as an important element of literary form.[3]

> *Symmetry, as wide or as narrow as you may define its meaning, is one idea by which man through the ages has tried to comprehend and create order, beauty, and perfection.*
>
> Hermann Weyl

In this chapter we will take a geometric perspective of *symmetry*, focusing on its application to both real-world physical objects and abstract geometric shapes. Simply put, this chapter is about how to *see, read*, and *make sense of* the symmetry of the world around us.

---

[1]Internally, the human body is not even close to having left-right symmetry. The heart and stomach, for example, are essentially on the left side; the liver on the right.
[2]See, for example, reference 8.
[3]A trip to a medium-sized university library produced approximately 150 books with the word *Symmetry* somewhere in the title. Of these, roughly 10% were in poetry, literature, or music; another 10% in art and/or architecture; approximately 50% in chemistry, physics, or engineering; and 30% in pure or applied mathematics.

11.1 Geometric Symmetry

When applied to physical objects and geometric shapes, symmetry is often referred to as *geometric symmetry*, and this is the context in which we will use the term "symmetry" in this chapter. So, in this context, what do we mean by symmetry? The famous Russian crystallographer E. S. Fedorov defined it as "the property of figures to repeat their parts, or more precisely, their property of coinciding with their original position when in different positions."[4] Sounds confusing. A check in Webster's[5] dictionary shows several entries under **sym' me · try**; the one most appropriate to our discussion is "the correspondence of parts or relations; similarity of arrangement." Not much help there, either. Geometric symmetry is one of those concepts that is almost harder to define than to understand.

A simple example should help us get started.

EXAMPLE 11.1

Figure 11–1 shows three triangles: triangle 1 is equilateral; triangle 2 is isosceles; triangle 3 is scalene (all three sides are different). Imagine a very tiny, almost microscopic observer standing at one of the vertices of triangle 1 and looking inward. The observer would see exactly the same thing whether standing at vertex *A*, *B*, or *C*. In fact, if the vertices were not labeled and there were no other frames of reference, the observer would be unable to distinguish one position from the other. In triangle 2, the observer would see the same thing when standing at *B* or *C* but not when standing at *A*. In triangle 3, the observer would see a different thing at each vertex. Informally, we might say that triangle 1 has more symmetry than triangle 2, which in turn has more symmetry than triangle 3.

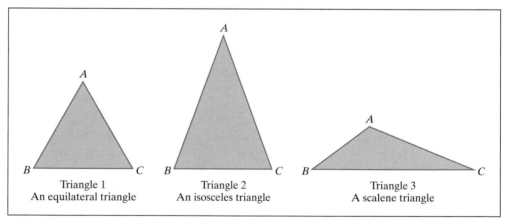

Triangle 1
An equilateral triangle

Triangle 2
An isosceles triangle

Triangle 3
A scalene triangle

FIGURE 11-1

Well, this is still a little vague, but a beginning nonetheless. We will start by saying that an object has **symmetry** if it looks exactly the same when seen from two or more different vantage points. What if rather than moving the observer, we move the object itself? For example, saying that triangle 2 looks the same to an observer whether he stands at vertex *B* or vertex *C* is equivalent to saying that

[4]Quoted in I. Hargittai and M. Hargittai, *Symmetry through the Eyes of a Chemist*, 2d ed. (New York: Plenum Press, 1995).
[5]*Webster's New Twentieth Century Dictionary*, 2d ed. (New York: Simon and Schuster, 1979).

we can move triangle 2, so that vertices *B* and *C* swap locations and the triangle as a whole looks exactly as before.

Informally, an object's symmetry is somehow related to the fact that we can move the object in such a way that when all the moving is done, the object looks exactly as it did before. Thus, to fully understand symmetry, we need to understand the different ways in which we can "move" an object.

11.2 Rigid Motions

The act of taking an object and moving it from some starting position to some ending position *without altering its shape or size* is called a **rigid motion** (and sometimes an *isometry*). When, in the process of moving the object, we stretch it, tear it, or generally alter its shape or size, that's *not* a rigid motion (see Fig. 11-2). Since in a rigid motion the size and shape of an object are not altered, distances between points are preserved: *The distance between any two points X and Y in the starting position is the same as the distance between the same two points in the ending position.*

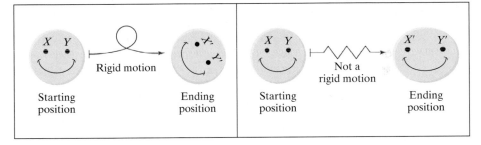

FIGURE 11-2
A rigid motion always preserves distances between points.

In studying rigid motions, *the only things that we will care about are the starting and ending positions; we will not care about what happens in between.* To illustrate this point, consider the adventures of a humble quarter sitting on top of a dresser. In the morning we might pick it up, put it in a pocket, drive around town with it, take it out of the pocket, flip it in the air, put it back in a different pocket, go home, take it out of the pocket, and finally put it back on top of the dresser again. While the actual trip taken by the quarter was long and eventful, the end result certainly wasn't: The quarter started somewhere on top of the dresser and ended somewhere else on the dresser. From the quarter's perspective, we could have accomplished the whole thing in a much simpler way—possibly a little slide along the top of the dresser, and maybe a single flip (if opposite sides of the quarter were facing up in the starting and final positions).

When two rigid motions accomplish the same net effect, they are said to be **equivalent rigid motions**. Thus, when a quarter "moves" from point A to point B on top of a dresser, there is just one rigid motion involved, regardless of what the actual trip may have been like. It is a remarkable fact that every rigid motion, no matter how complicated, is always equivalent to something very basic.

To keep things simple, in this chapter we will concentrate on *two-dimensional* objects and shapes—the world of the page, if you will. The study of symmetry for three-dimensional objects is similar, albeit somewhat more complicated (see Project B).

For two-dimensional objects in a plane, every rigid motion is equivalent to a rigid motion of *one of only four possible kinds*: it's either a *reflection*, a *rotation*, a *translation*, or a *glide reflection*. We will call these four types of rigid motions the **basic rigid motions of the plane**.

A rigid motion (let's call it M) in the plane moves each point in the plane from its starting position P to an ending position P', also in the plane. We will call the point P' the **image** of the point P under the rigid motion M and describe this informally by saying that M *moves* P to P'. (Throughout the chapter, we will stick to the convention that the image point has the same label as the original point but with a prime symbol added.) It is possible for a point P to end up back where it started under the rigid motion ($P' = P$), in which case, we call P a **fixed point** of the rigid motion.

It would appear that to "completely know" a rigid motion, one would need to know how every point of the plane moves. But fortunately, as we will see, we need to know how only a few points (three at the most) move. Because the motion is rigid, the behavior of those few points forces all the rest of the points to follow their lead. (We can think of this as a sort of "Pied Piper effect.")

We will now discuss each of the basic rigid motions in the plane in a little more detail.

11.3 Reflections

A **reflection** in the plane is a rigid motion that moves an object into a new position that is a mirror image of the starting position. In two dimensions, the "mirror" is just a line, called the **axis** of the reflection.

A reflection is completely described by its axis. (In other words, if we know the axis of the reflection, we know everything we need to know about the reflection.) Figures 11-3 and 11-4 show examples of reflections.

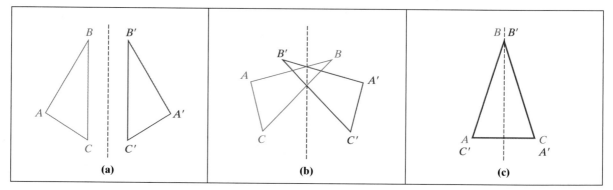

FIGURE 11-3
Original figure in blue; reflected figure in red; axis of reflection is dashed line.

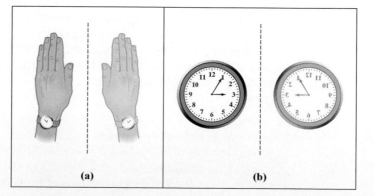

FIGURE 11-4
Reflections are improper—the image of a left hand is a right hand, and the image of a clock is a "counterclock."

An important characteristic of a reflection is that it reverses all the traditional frames of reference one uses for orientation. As illustrated in Fig. 11-4, in a reflection, left is interchanged with right, and clockwise with counterclockwise. We will say that reflection is an **improper** rigid motion to indicate the fact that it reverses the left-right and clockwise-counterclockwise orientations.

From a purely geometric point of view, a reflection can best be described by showing how it moves a generic point P. Given the axis of the reflection, the image of a point P is found by drawing a line through P perpendicular to the axis and finding the point P' that is on this line on the opposite side of the axis and at the same distance as P from the axis (Fig. 11-5). If P is on the axis itself, it is a fixed point of the reflection. Conversely, if we know any point P and its image P' under the reflection, we can find the axis—it is the perpendicular bisector of the segment joining the two points (Fig. 11-5).[6] A useful consequence of this is that given a distinct pair of points P and P', we can determine the axis and thus completely specify the reflection.

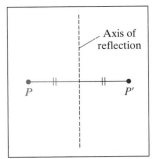

FIGURE 11-5

The axis of a reflection is the perpendicular bisector of the segment joining a point P to its image P'.

Another important fact about reflections is that if we apply the same reflection twice, every point ends up exactly where it started. In other words, *the net effect of applying the same reflection twice is the same as not having moved the object at all*. This leads us to an interesting semantic question. Should not moving an object at all be considered in itself a rigid motion? On the one hand, it seems rather absurd to say yes. If we are talking about motion, then there should be some kind of movement, however small. On the other hand, we are equally compelled to argue that the result of combining two (or more) consecutive rigid motions should itself be a rigid motion regardless of what the net effect is. If this is the case, then combining two consecutive reflections with the same axis (which produces the same result as no motion at all) should be a rigid motion. We will opt for the latter alternative, because it is the mathematically correct way to look

---

[6]If $P = P'$ is a fixed point, then the axis passes through P and an additional fixed point is needed.

at things. We will formally agree that *not moving an object at all* is itself a very special kind of rigid motion of the object, which we will call the **identity motion.**

11.4 Rotations

The second category of rigid motions we will discuss are *rotations*. For two-dimensional figures, a **rotation** is described by specifying a point called the **center** of the rotation (or **rotocenter** for short), and an **angle of rotation** indicating the *amount* of the rotation. Figures 11-6 and 11-7 show examples of rotations.

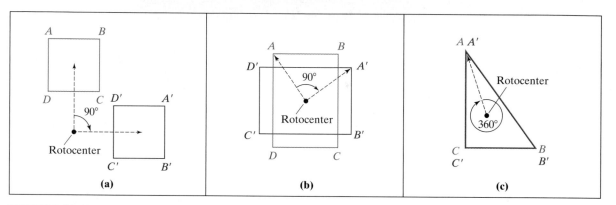

FIGURE 11-6

(a) A 90° clockwise rotation with rotocenter outside the figure, (b) a 90° clockwise rotation with rotocenter inside the figure, and (c) a 360° clockwise rotation is the indentity motion regardless of where the rotocenter is. Original figures in blue; rotated figures in red.

A few comments about the rotations shown in Fig. 11-6. In each example, we have specified the angle of rotation in degrees. This is strictly a matter of personal choice—some people prefer degrees, others radians.[7]

Our second observation starts with the well-known fact that a rotation by 360° leaves the figure unchanged—it is the identity motion. This has several useful consequences. First, any rotation by an angle that is more than 360° is equivalent to another rotation with the same center by an angle that is between 0° and 360°—all we have to do is divide the angle by 360 and take the remainder. For example, as a rigid motion, a clockwise rotation by 759° is the same as a clockwise rotation by an angle of 39°, because 759 divided by 360 gives a quotient of 2 and a remainder of 39. Second, any rotation that is specified in a clockwise orientation can just as well be specified in a counterclockwise orientation. In the special

FIGURE 11-7

(a) 180° rotation of the hour hand only. (b) 180° rotation of the clock itself.

---

[7]Throughout the chapter we will stick with degrees, but one can always change degrees to radians by using the equation *radians* $= (\pi/180) \times degrees$.

case of a *half-turn* (a rotation by an angle of 180°), clockwise and counterclockwise give the same results.

A common misconception is to confuse a 180° rotation with a reflection, but the two are very different. A rotation, unlike a reflection, is always a **proper** rigid motion—in other words, it preserves left-right and clockwise-counterclockwise orientations within the rotated object.

Given two points P and P', there are infinitely many possible rotations that move P to P'—all we have to do is to choose a rotocenter on the perpendicular bisector of the segment PP' [Fig. 11-8(a)]. This implies that we need another pair of points Q and Q' to nail down a specific rotation. The rotocenter will be the unique point where the perpendicular bisectors of PP' and QQ' meet [Fig. 11-8(b)]. In the special case where PP' and QQ' are parallel, the center of rotation is the intersection of PQ and $P'Q'$ [Fig. 11-8(c)].

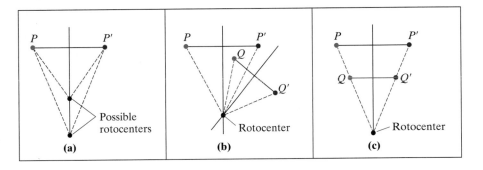

FIGURE 11-8
Finding the rotocenter requires at least two pairs of points P, P' and Q, Q'.

11.5 Translations

A **translation** is essentially a slide of an object in the plane. It is completely specified by the direction and amount of the slide. These two pieces of information are combined in the form of a **vector**. A vector can be represented by an arrow giving its direction and length. As long as the arrow points in the proper direction and has the right length, its actual placement is immaterial, as shown in Fig. 11-9.

Translations, like rotations, are *proper* rigid motions of the plane because they do not change the left-right or clockwise-counterclockwise orientations. On the other hand, translations are like reflections in the sense that they can be completely described by giving a point P and its image P'. The arrow joining P to P' gives us the vector of the translation. Once we know the vector of translation, we know where the translation moves any other point.

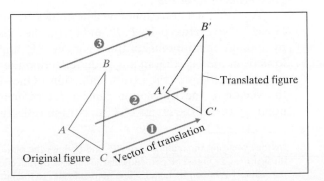

FIGURE 11-9
Any one of the arrows can be used to indicate the vector of translation.

11.6 Glide Reflections

A **glide reflection**, as the name suggests, is a rigid motion consisting of a translation (the glide part) followed by a reflection. The axis of the reflection *must* be parallel to the direction of the translation. The wording "translation followed by a reflection" is somewhat misleading. We can just as well do the reflection first and the translation second, and the end result will be the same (Fig. 11-10).

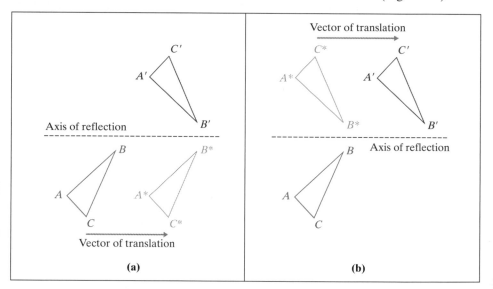

FIGURE 11-10
(a) A glide reflection consists of a translation followed by a reflection on a parallel axis, or (b) a reflection first followed by the translation. In either case, the results are the same. (original figure in blue; image in red)

A glide reflection is an *improper* rigid motion—it changes left-right and clockwise-counterclockwise orientations. We can thank the reflection part of the glide reflection for that.

A glide reflection cannot be determined by just one point P and its image P'. However, given two pairs P, P' and Q, Q', the axis of the reflection can be found by joining the midpoints of the segments PP' and QQ' [Fig. 11-11(a)]. (This follows from the fact that in any glide reflection the midpoint between a point and its image belongs to the axis of reflection.) Once the axis of reflection is known, the vector of the translation can be determined by locating the intermediate point $P*$ that is the image of P' under the reflection [Fig. 11-11(b)].[8]

---

[8]In the event that the midpoints of PP' and QQ' are the same point M, then the line passing through P and Q must be perpendicular to the axis of reflection [Fig. 11-11(c)]. Here the axis of reflection is obtained by taking a line perpendicular to the line PQ and passing through the common midpoint M.

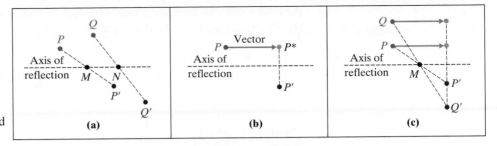

FIGURE 11-11
A glide reflection is determined by two pairs P, P' and Q, Q'.

Of the four basic rigid motions in the plane, the glide reflection is unique in that it is defined as a combination of two other rigid motions—a translation and a reflection. For this particular combination, there is no simpler way to describe the resulting rigid motion. Surprisingly, any other combination of motions, no matter how complex, is guaranteed to be equivalent to one (that's it—just one!) of the four basic rigid motions.

And what about the identity motion? Where does it fit in this picture? There are two possible answers: The identity can be thought of as a *translation* by a "zero vector" or, alternatively, a *rotation* by an angle of 360° (or 720°, or 0°, etc.). Of the two choices, the latter seems to be most natural and is the one we will use throughout this chapter.

A summary of the key facts about the four basic rigid motions in the plane is given in Table 11-1.

TABLE 11-1 **The Four Basic Rigid Motions**

Rigid motion	Specified by	Proper or improper	Fixed points	Point/Image pairs that determine it
reflection	axis of reflection	improper	infinitely many	one pair
rotation[*]	rotocenter and angle	proper	one	two pairs
translation	vector of translation	proper	none	one pair
glide reflection	vector of translation and axis of reflection	improper	none	two pairs
[*] identity motion (360° rotation)	N/A	proper	all	N/A

11.7 Symmetry Revisited

With an understanding of rigid motions and their classification, we will be able to consider the concept of geometric symmetry in a much more precise way. Here, finally, is a good definition of geometric symmetry, one that probably would not have made much sense at the start of this chapter. *A symmetry of an object or shape is a rigid motion that moves the object back onto itself.* In other words, in a symmetry one cannot tell, at the end of the motion, that the object has been moved. It is important to note that this does not necessarily force the rigid motion to be the identity motion. Individual points may be moved to different positions, even though the whole object is moved back into itself. And of course, the identity motion is itself a symmetry, one possessed by every object and that from now on we will call simply the **identity**.

For two-dimensional objects there are only four possible types of symmetry corresponding to the four possible types of rigid motions: *reflection symmetry, rotation symmetry, translation symmetry*, and *glide reflection symmetry*.

EXAMPLE 11.2 The Symmetries of a Square

What are the possible rigid motions that move the square in Fig. 11-12(a) back onto itself? First, there are *reflection symmetries*. For example, if we use the line l_1 in Fig. 11-12(b) as the axis of reflection, the square falls back into itself with points A and B interchanging places and C and D interchanging places. It is not hard to think of three other reflection symmetries, with axes l_2, l_3, and l_4 shown in Fig. 11-12(b). Are there any other symmetries? Yes. There are *rotation symmetries* with rotocenter O, the center of the square. The angles of rotation are 90°, 180°, 270°, and 360°—this last one being none other than the identity.

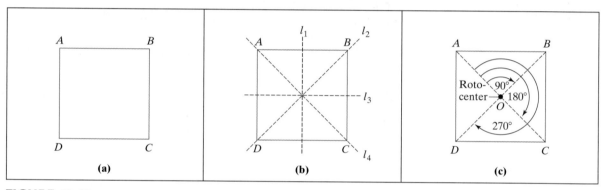

FIGURE 11-12
(a) The original square, (b) its four reflection symmetries (axes are l_1, l_2, l_3, and l_4), (c) its four rotation symmetries with rotocenter O (90°, 180°, 270°, and the identity).

All in all, we have easily found 8 symmetries for the square in Fig. 11-12(a): four of them are reflections, and four are rotations. Could there be more? What if we combined one of the reflections together with one of the rotations? A symmetry combined with another symmetry, after all, has to be itself a symmetry. It turns out that the eight symmetries we listed are all there are—no matter how we combine them, we always end up with one of the eight (see Exercise 69) ∎.

EXAMPLE 11.3 The Symmetries of a Propeller

Consider the four-bladed propeller shown in Fig. 11-13(a). What can we say about its symmetries? It's not hard to see that, once again, there are four reflection symmetries [Fig. 11-13(b)], as well as four rotations: 90°, 180°, and 270°, and the identity. And there are no other possible symmetries. ∎

An important lesson lurks behind Examples 11.2 and 11.3: *Two different-looking objects can have exactly the same set of symmetries.* A good way to think about this is that the square and the propeller, while certainly different objects, are blood relatives—both members of the same "symmetry family."

FIGURE 11-13
(a) The propeller, (b) its four reflection symmetries, and (c) its four rotation symmetries (including the identity).

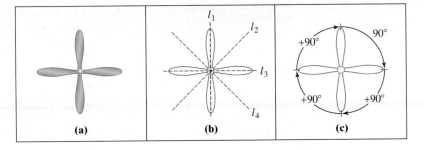

Formally, we will say that two objects or shapes are of the same **symmetry type** if they have exactly the same set of symmetries. The symmetry type for the square (as well as the propeller) is called D_4—which is shorthand for four reflections and four rotations. Figure 11-14 shows several objects with symmetry type D_4.

FIGURE 11-14
Objects with symmetry type D_4: (a) a propeller, (b) a plus sign, and (c) a mosaic.

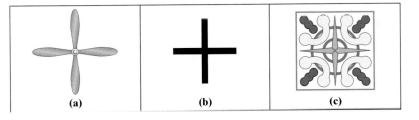

EXAMPLE 11.4 The Symmetries of a Propeller: Part II

Let's consider now the object shown in Fig. 11-15(a), a slightly different propeller from the one in Example 11.3. The difference is subtle, but from the symmetry point of view significant. Does this figure have four reflection symmetries? Certainly not! A vertical reflection, for example, would not give us an identical propeller [Fig. 11-15(b)], and for that matter, neither would a horizontal or any other kind of reflection. This propeller has no reflection symmetries at all! On the other hand, it still has four rotations (identity, 90°, 180°, and 270°). This propeller has *only* the four rotation symmetries (see Exercise 65) and belongs therefore to a new symmetry family called Z_4 (which is shorthand for the symmetry type of objects having four rotations only).

FIGURE 11-15
This propeller has four rotation symmetries only. Reflections don't work. (Symmetry type: Z_4.)

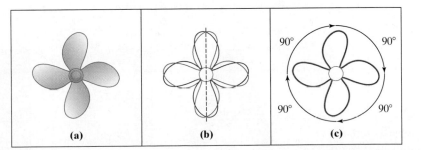

EXAMPLE 11.5 The Symmetries of a Propeller: Part III

Here is one last propeller example. Every once in a while a propeller looks like the one in Fig. 11-16(a), which is kind of a cross between Figs. 11-15(a) and 11-14(a): Only opposite blades are the same. This figure has no reflection symmetries (try it!), and a 90° rotation won't work either [Fig. 11-16(b)]. Only the identity and a 180° rotation are possible as symmetries of this propeller. Any object having only these symmetries is of symmetry type Z_2. Figure 11-17 shows several additional examples of shapes of symmetry type Z_2.

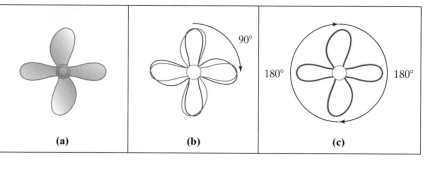

FIGURE 11-16
A propeller with only two rotation symmetries and no reflection symmetries. (Symmetry type: Z_2.)

(a) (b) (c)

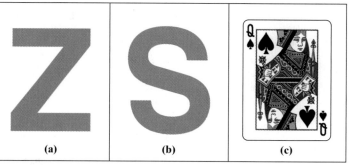

FIGURE 11-17
Objects with symmetry type Z_2. (a) The letter Z, (b) the letter S (in some fonts but not in others), and (c) the Queen of Spades (and many other cards in the deck).

(a) (b) (c)

EXAMPLE 11.6 The Symmetries of a Butterfly, etc.

One of the most common symmetry types occurring in nature is that of objects having only 1 reflection symmetry and 1 rotation symmetry (the identity). This symmetry type is called D_1. Figure 11-18 shows several examples of shapes and objects having symmetry type D_1. Notice that it doesn't matter if the axis of reflection is vertical, horizontal, or anywhere in between: If the figure has just one reflection symmetry, it is guaranteed to be of symmetry type D_1. Note that we don't have to worry about the one rotation symmetry—it is the identity—and every object has this symmetry.

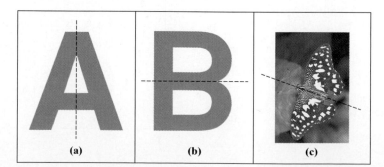

FIGURE 11-18
Shapes and objects with only 1 reflection symmetry (symmetry type D_1). The axis of the reflection is shown in black.

(a) (b) (c)

EXAMPLE 11.7 The Symmetries of "Shapes with no Symmetry"

Many objects and shapes are informally considered to have no symmetry at all, but this is a little misleading, since *every object has at least the identity symmetry*. Objects whose only symmetry is the identity are said to have symmetry type Z_1. Figure 11-19 shows a few examples of objects of symmetry type Z_1—there are plenty of such objects around.

FIGURE 11-19
Shapes and objects with only the identity symmetry (Symmetry type Z_1). Why doesn't the six of clubs have a 180° rotation symmetry? (The answer is in the two middle clubs.)

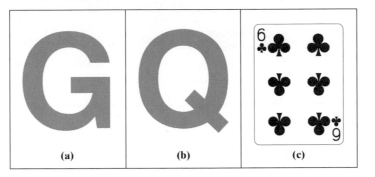

(a)   (b)   (c)

EXAMPLE 11.8 Shapes with Many Rotations and Many Reflections

In everyday language, certain objects and shapes are said to be "highly symmetric" when they have lots of rotation and reflection symmetries. Figure 11-20(a) shows a snowflake, with six reflection symmetries. (Can you find all six axes of symmetry?) It also has six rotation symmetries. (The rotocenter is the center of the snowflake and the angles are 60°, 120°, 180°, 240°, 300°, and 360°.) The snowflake, like all other snowflakes, has symmetry type D_6 (short for 6 reflections and 6 rotations).[9] Figure 11-20(b) shows a ceramic plate. It has nine reflections and nine rotation symmetries, and its symmetry type (not surprisingly) is called D_9. Finally, in Fig. 11-20(c) we have a picture of the dome of the Sports Palace in Rome, Italy, with 36 reflection and 36 rotation symmetries (symmetry type D_{36}).

FIGURE 11-20
(a) Snowflake (symmetry type D_6); (b) ceramic plate (symmetry type D_9); (c) dome design (symmetry type D_{36}).

[9]This symmetry type occurs often in nature; it is commonly known as *hexagonal symmetry* because it is the symmetry type of the regular hexagon.

In each case illustrated in Example 11.8, the number of reflections matches the number of rotations. This was also true in Examples 11.2, 11.3, and 11.6. Coincidence? Not at all. When a finite object or shape has *both* reflection and rotation symmetries, the number of rotation symmetries (which includes the identity) has to match the number of reflection symmetries! Any finite object or shape with exactly N reflection symmetries and N rotation symmetries is said to have symmetry type D_N. The standard example for a shape with symmetry type D_N is the regular polygon with N sides, commonly known as the *regular N-gon*.

EXAMPLE 11.9 Shapes with Infinitely Many Rotations and Reflections

If we are looking for a two-dimensional shape that has *as much symmetry as possible*, we don't have to look past the wheels of a car. The wheel works so wonderfully well as a means of locomotion because of the infinitely many rotation and reflection symmetries of the circle. In a circle, a rotation with center at the center of the circle and by any angle whatsoever is a symmetry, and any line passing through the center of the circle can be used as an axis of reflection symmetry. We call the symmetry type of the circle D_{infinity}. ■

EXAMPLE 11.10 Shapes with Rotations, but No Reflections

We now know that if a finite two-dimensional shape has rotations *and* reflections, then it must have exactly the same number of each. In this case, the shape belongs to the D family of symmetries, specifically, it has symmetry type D_N.[10] However, we also saw in Examples 11.4, 11.5, and 11.7 shapes that have rotations, *but no* reflections. In this case, we used the letter Z to describe the symmetry type, with a subscript indicating the actual number of rotations. Figure 11–21 shows examples of shapes having symmetry types of the *Z-something* variety.

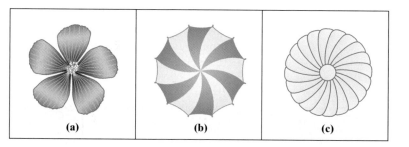

FIGURE 11-21
(a) Hibiscus (symmetry type Z_5), (b) top view of a parasol (symmetry type Z_6), and (c) turbine (symmetry type Z_{20}).

(a)   (b)   (c)

■

SYMMETRY TYPES OF FINITE SHAPES

- D_N ($N = 1, 2, 3, \ldots$): This is the symmetry type of an object with N reflection symmetries and N rotation symmetries. With the N reflections, the rotations automatically follow—an object can't have reflection symmetries without an equal number of rotation symmetries.

- Z_N ($N = 1, 2, 3, \ldots$): This is the symmetry type of an object with N rotation symmetries but no reflection symmetries. Rotations without reflections are possible.

- D_∞: This is the symmetry type of a circle.

---

[10]The formal mathematical name for this class of symmetry types is *dihedral symmetry*.

11.8   Patterns

Well, we've come a long way, but we have yet to see examples of shapes having translation and/or glide reflection symmetry. In fact, if we think of objects and shapes as being finite, then translation symmetry is impossible. (There is no way that a finite object can be slid a certain distance and still be exactly where it was before!) On the other hand, if we broaden our interpretation of a "shape" and include infinitely repeating patterns in our definition of a "shape," these translation symmetries are indeed possible.

We will formally define a **pattern** as an infinite "shape" consisting of an infinitely repeating basic design (sometimes called the *motif*). The reason we have "shape" in quotation marks is that a pattern is really an abstraction—in the real world there are no infinite objects as such, although the idea of an infinitely repeating motif is familiar to us from wallpaper, textiles, carpets, ribbons, and so on.

Just like finite shapes, patterns fall into specific symmetry types. The classification of patterns according to their symmetry type is of fundamental importance in the study of molecular and crystal organization in chemistry, so it is not surprising that some of the first people to seriously investigate the symmetry types of patterns were crystallographers. Archeologists and anthropologists have also found that the symmetry types characteristic of a particular culture (in their textile and pottery) can be used as a means to gain a better understanding of that culture (see reference 13).

We will briefly discuss the symmetry types of one-dimensional and two-dimensional patterns. A comprehensive study of patterns is beyond the scope of this book, so we will not go into as much detail as we did with finite shapes.

Border (One-Dimensional) Patterns

In a one-dimensional pattern we have a basic design (the *motif*) that keeps repeating indefinitely along a single direction. One-dimensional patterns are commonly known as **border patterns** (and sometimes as *frieze patterns*), and they are found in ribbons, architectural friezes, baskets, pottery, and so on.

The most common direction in a border pattern (what we will call the *direction of the pattern*) is horizontal, but in general a border pattern can be in any direction. (For typesetting in a book, it is more efficient to display a border pattern horizontally, and we will do so from now on. Thus, when we say "horizontal direction," we really mean *the direction of the pattern*, and it follows that when we say "vertical direction," we really mean *the direction perpendicular to the direction of the pattern*.)

In a border pattern, a theme repeats itself in one direction, be it a straight line or around a circle. (*Center:* Enamelled brick frieze from the Palace of Artaxerxes II, Susa, 5th c. B.C. Louvre, Paris, France).

We will now discuss the possible symmetries of a border pattern. At first, one might think that there are more possibilities for symmetry in a border pattern than in a finite shape, but in fact the opposite is true. The possibilities for symmetry in a border pattern are fairly limited.

- **Translations.** A border pattern always has translation symmetries—a *basic* translation (the smallest slide that moves the pattern back onto itself) and multiples of the basic translation obtained by applying the basic translation more than once (Fig. 11-22).

FIGURE 11-22
The basic translation symmetry is shown in red. Any multiples of the red translation are themselves translation symmetries.

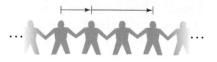

- **Reflections.** A border pattern can have (i) no reflection symmetry [Fig. 11-23(a)], (ii) only *horizontal* (i.e., in the direction of the pattern) reflection symmetry [Fig. 11-23(b)], (iii) only *vertical* reflection symmetries [Fig. 11-23(c)], or (iv) both *horizontal* and *vertical* reflection symmetries [Fig. 11-23(d)]. No other reflection symmetries are possible (see Exercise 61).

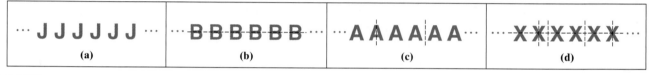

FIGURE 11-23
Border patterns with (a) no reflection symmetry, (b) horizontal symmetry only, (c) vertical symmetries only (many axes are possible) and, (d) both.

- **Rotations.** A border pattern can be rotated back onto itself only through a half-turn (180° rotation), or a full turn (the identity). Thus, a border pattern has either (i) one rotation symmetry—the identity [Fig. 11-24(a)] or (ii) two rotation symmetries—the identity and 180° rotation [Fig. 11-24(b)]. No other rotation symmetries are possible (see Exercise 62).

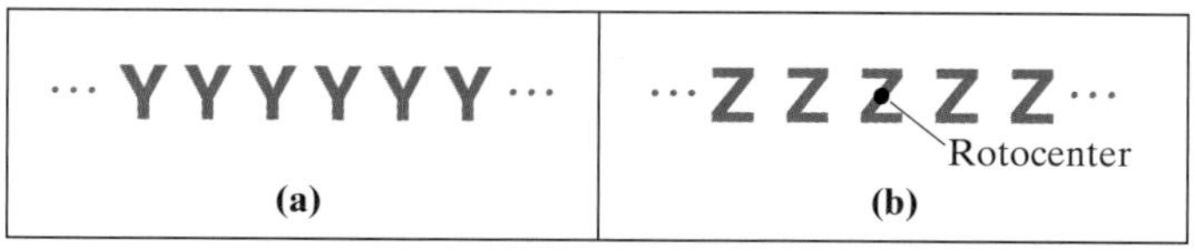

FIGURE 11-24
Border patterns with (a) one rotation symmetry (the identity) and (b) two rotation symmetries (the identity and 180°).

... Y Y Y Y Y Y ...
(a)

... Z Z Z Z Z ...
￼Rotocenter
(b)

- **Glide reflections.** A border pattern can have (i) no glide reflection symmetry [Fig. 11-25(a)] or (ii) a basic glide reflection symmetry. The latter can happen only under fairly restrictive conditions: the axis of reflection *has* to be a line along the center of the pattern, and the reflection in the glide reflection cannot itself be a symmetry of the pattern. Thus, a pattern such as the one in Fig. 11-25(a), which has both translation and horizontal reflection symmetries separately, is *not* considered to have glide reflection symmetry.

FIGURE 11-25
(a) This pattern has separate reflection and translation symmetries—it does *not* have glide reflection symmetry. (b) This pattern has no reflection symmetry, but when a reflection is combined with a glide, the pattern falls back onto itself—it has glide reflection symmetry.

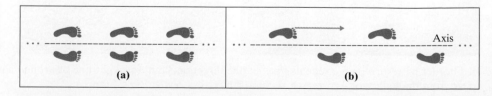

We can summarize the preceding observations by listing the possible symmetries of a border pattern:

- the *identity* (think of it as a 360° rotation)
- *translations* (always in the direction of the pattern)
- *horizontal reflections* (the axis of reflection runs through the pattern)
- *vertical reflections* (the axis of reflection is perpendicular to the pattern)
- *half-turns* (180° rotations)
- *glide reflections* (a glide along the direction of the pattern and a reflection with axis running through the pattern)

The number of combinations of these symmetries into symmetry types is surprisingly small: *There are only seven different symmetry types for border patterns.* Table 11-2 shows each of the seven symmetry types, with the checkmarks in each column indicating the set of symmetries of that symmetry type.[11]

TABLE 11-2   The Seven Symmetry Types for Border Patterns

Symmetries		*m*1	*mm*	*mg*	11	12	1*m*	1*g*
	Translations	✓	✓	✓	✓	✓	✓	✓
First symbol	*m*: vertical reflections	✓	✓	✓	No	No	No	No
Second symbol	*m*: horizontal reflection	No	✓	No	No	No	✓	No
	g: glide reflections	No	No	✓	No	No	No	✓
	2: 180° rotations	No	✓	✓	No	✓	No	No

---

[11]Like most people, you may wonder about the oddball two-symbol codes used to name the seven symmetry types. These codes are based on the *standard crystallographic notation*—a classification scheme originally developed by crystallographers. In this notation, the first symbol is either an *m* (which indicates that the pattern has vertical reflections) or a 1 (indicating no vertical reflections). The second symbol specifies the existence of a horizontal reflection (*m*), a glide reflection (*g*), a half-turn (2), or none of these (1).

Wallpaper (Two-Dimensional) Patterns

Wallpaper patterns are patterns that fill the plane by repeating a *motif* indefinitely along several (two or more) nonparallel directions. Typical examples of such patterns can be found in wallpaper (of course), carpets, textiles, and so on.

With wallpaper patterns things get a bit more complicated, so we will skip the details. The possible symmetries of a wallpaper pattern are as follows:

- **Translations.** Every wallpaper pattern has translation symmetry in at least two different (nonparallel) directions (Fig. 11-26).

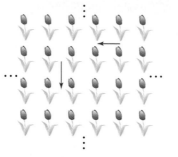

FIGURE 11-26

- **Reflections.** A wallpaper pattern can have (i) no reflections, (ii) reflections in only one direction, (iii) reflections in two nonparallel directions, (iv) reflections in three nonparallel directions, (v) reflections in four nonparallel directions, (vi) reflections in six nonparallel directions. There are no other possibilities. (Examples are shown in the chapter appendix.) Note that particularly conspicuous in its absence is the case of reflections in exactly five different directions.

- **Rotations.** In terms of rotation symmetries, a wallpaper pattern can have (i) the identity only, (ii) two rotations (identity and 180°), (iii) three rotations (identity, 120°, and 240°), (iv) four rotations (identity, 90°, 180°, and 270°), and (v) six rotations (identity, 60°, 120°, 180°, 240°, and 300°). There are no other possibilities. (Examples are shown in the chapter appendix.) Once again, note that a wallpaper pattern cannot have exactly five different rotations.

- **Glide reflections.** A wallpaper pattern can have (i) no glide reflections, (ii) glide reflections in only one direction, (iii) glide reflections in two nonparallel directions, (iv) glide reflections in three nonparallel directions, (v) glide reflections in four nonparallel directions, and (vi) glide reflections in six nonparallel directions. There are no other possibilities. (Examples are shown in the chapter appendix.)

It is a truly remarkable fact that in spite of all these possibilities, the symmetries of a wallpaper pattern can be combined into only 17 *distinct symmetry types*. The hundreds and thousands of wallpapers one can find at a decorating store all fall into just 17 different symmetry families. They are listed and illustrated in the chapter appendix.

Conclusion

Real-life tangible physical objects as well as abstract shapes from geometry, art, and ornamental design are often judged and measured by a yardstick that can be both mathematical and aesthetic: *How much symmetry and what kinds of symmetry does it have?*

The possibilities, while limitless, fall into a small and well-defined set of categories. For two-dimensional objects and shapes that are finite, there are really only two possible scenarios: The object has rotation symmetries only (a *Z-something* kind of shape), or it has both rotation and reflection symmetries in equal amounts (a *D-something* kind of shape). It is quite remarkable that there are no other possibilities. Nowhere in the universe of finite two-dimensional shapes does there exist, for example, a shape with three reflection symmetries and five rotation symmetries—it just can't happen.

Patterns—that is, shapes with an infinitely repeating theme—are even more surprising in their symmetry pedigrees. One-dimensional patterns, commonly known as *border patterns*, fall into just *seven* different symmetry types, whereas two-dimensional patterns, such as those found in wallpapers and textiles, fall into just *seventeen* different symmetry types. It wasn't until 1924 that a rigorous mathematical proof of the latter fact was given by the Hungarian mathematician George Polya.

In this chapter we learned that there is a lot more to symmetry than a reflection in a mirror, and that the key to unlocking its mysteries can be found in mathematics. We conclude with a brief quote from the great mathematician Hermann Weyl:

Symmetry is a vast subject, significant in art and nature. Mathematics lies at its root, and it would be hard to find a better one on which to demonstrate the working of the mathematical intellect.

Sir Roger Penrose (1931–)

Much like the tiling of a bathroom floor, a mathematical *tiling* of the plane is an arrangement of tiles (of one or several shapes) that fully cover the plane without overlaps or gaps. A tiling of the plane is an abstract concept—unlike a bathroom floor, which has a finite area, a plane goes on forever. And yet, there is an easy method for tiling a plane: Find a basic design (a single tile or a combination of several tiles) that can be fitted snugly with other copies of itself, and just repeat the same design, mindlessly marching off to infinity in all directions. This kind of tiling, called a *periodic tiling*, is essentially equivalent to a wallpaper pattern. With the right *motif* (the basic design that repeats), incredibly beautiful and exotic periodic tilings are possible (witness the amazing tilings found in some of M. C. Escher's graphic designs). Because tiling a plane is an infinite task, it stands to reason that the only way it can be done is by means of a periodic tiling, and this was the conventional wisdom for centuries. Artists, designers, chemists, and physicists all banked on this assumption. Enter Roger Penrose. In 1974, Penrose, a mathematical physicist already famous for his contributions to cosmology (he proved mathematically that black holes can exist) found an *aperiodic* tiling of the plane—that is, a method for tiling the plane not based on the periodic repetition of one motif. Using only two basic tile shapes, Penrose was able to show that it is possible to tile a plane with infinitely changing patterns (see Project C for more details).

Roger Penrose was born in Colchester, England. His father, a famous medical geneticist, moved the family to Canada during World War II, and Roger spent his early school years in London, Ontario. After the war, the family returned to the other London, where Penrose began his university studies, completing his B.Sc. degree in mathematics from University College, and a Ph.D. in mathematical physics from Cambridge University.

Sir Roger (he was knighted in 1994) has made significant contributions to pure mathematics, mathematical physics, and cosmology, and in 1988 he shared the prestigious Wolf Prize in Physics with Stephen Hawking. Penrose has held appointments at many academic institutions, including Cambridge University, Princeton University, and the University of Texas. He is currently the Rouse Ball Professor of Mathematics at the University of Oxford, England.

In addition to his serious professional research in mathematical physics, Penrose is famous for his contributions to recreational mathematics. In fact, his original interest in tilings of the plane was strictly recreational (he just loves challenging puzzles). Ironically, his discovery of aperiodic tilings turned out to have amazing and unexpected practical implications. In 1982, following Penrose's lead, chemists working at the National Institute of Standards and Technology discovered the existence of *quasicrystals*, crystal structures with nonrepetitive patterns analogous to Penrose's aperiodic tilings. The discovery of quasicrystals turned the world of crystallography and materials science on its head. The aperiodic structure of quasicrystals defies all the traditional rules of crystal formation in nature and has led to the discovery of heretofore unknown properties of solid matter. As new alloys based on quasicrystalline structure are being developed (the latest is a nonscratch coating for frying pans), it is hard to imagine that we owe it all to one man's singular love of patterns and puzzles.

KEY CONCEPTS

angle of rotation
axis (of reflection, of symmetry)
basic rigid motions of the plane
bilateral symmetry
border pattern (one-dimensional
  pattern)
equivalent rigid motion
fixed point
glide reflection (rigid motion)
glide reflection symmetry
identity (rigid motion)
image
improper (rigid motion)
pattern

proper (rigid motion)
reflection (rigid motion)
reflection symmetry
rigid motion (isometry)
rotation (rigid motion)
rotation symmetry
rotocenter
symmetry (geometric symmetry)
symmetry type
translation (rigid motion)
translation symmetry
vector (of translation)
wallpaper pattern (two-dimensional
  pattern)

EXERCISES

WALKING

A. Reflections

1. Which point in the figure is the image of P under
   (a) the reflection with axis l_1?
   (b) the reflection with axis l_2?
   (c) the reflection with axis l_3?
   (d) the reflection with axis l_4?

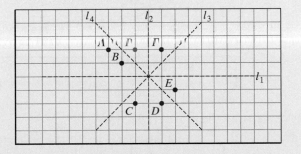

2. Which point in the figure is the image of P under
   (a) the reflection with axis l_1?
   (b) the reflection with axis l_2?
   (c) the reflection with axis l_3?
   (d) the reflection with axis l_4?

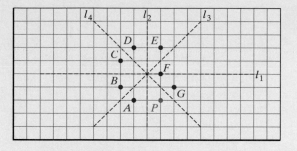

3. Given a reflection with the axis of reflection as shown in the figure, find
   (a) S' (the image of S) under the reflection.
   (b) the image of quadrilateral $PQRS$ under the reflection.

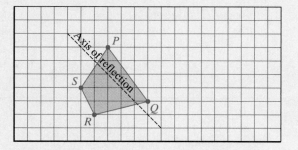

4. Given a reflection with the axis of reflection as shown in the figure, find
   (a) P' (the image of P) under the reflection.
   (b) the image of triangle PQR under the reflection.

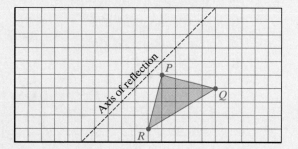

5. Given a reflection that sends the point P to the point P' as shown in the figure, find
   (a) the axis of reflection.
   (b) Q' (the image of Q) under the reflection.
   (c) the image of triangle PQR under the reflection.

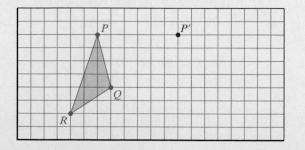

6. Given a reflection that sends the point P to the point P' as shown in the figure, find

    (a) the axis of reflection.

    (b) S' (the image of S) under the reflection.

    (c) the image of quadrilateral $PQRS$ under the reflection.

    (d) a point on the quadrilateral $PQRS$ that is a fixed point of the reflection.

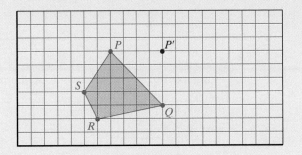

7. Given a reflection that sends the point P to the point P' as shown in the figure, find

    (a) the axis of reflection.

    (b) the image of triangle PQR under the reflection.

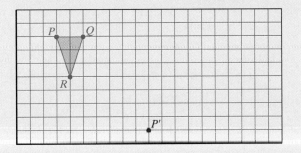

8. Given a reflection that sends the point R to the point R' as shown in the figure, find

    (a) the axis of reflection.

    (b) the image of quadrilateral $PQRS$ under the reflection.

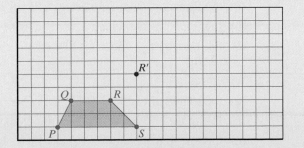

9. Consider a reflection for which A and B in the figure are fixed points. Find the image of the shaded region under the reflection.

10. Consider a reflection for which A and B in the figure are fixed points. Find the image of the shaded region under the reflection.

B. Rotations

Exercises 11 and 12 refer to the following figure.

11. Which point in the figure is
    (a) the image of B under a 90° clockwise rotation with rotocenter A?
    (b) the image of B under a 180° rotation with rotocenter A?
    (c) the image of A under a 90° clockwise rotation with rotocenter B?
    (d) the image of D under a 60° clockwise rotation with rotocenter B?
    (e) the image of D under a 120° clockwise rotation with rotocenter B?
    (f) the image of D under a 120° counterclockwise rotation with rotocenter B?

12. Which point in the figure is
    (a) the image of C under a 90° clockwise rotation with rotocenter B?
    (b) the image of C under a 90° counterclockwise rotation with rotocenter B?
    (c) the image of H under a 90° clockwise rotation with rotocenter B?

   (d) the image of F under a 60° clockwise rotation with rotocenter A?

   (e) the image of F under a 120° clockwise rotation with rotocenter B?

   (f) the image of I under a 90° clockwise rotation with rotocenter H?

13. In each of the following give an answer between 0° and 360°.

   (a) A clockwise rotation by an angle of 250° is equivalent to a counter-clockwise rotation by an angle of _____.

   (b) A clockwise rotation by an angle of 710° is equivalent to a clockwise rotation by an angle of _____.

   (c) A counterclockwise rotation by an angle of 710° is equivalent to a clockwise rotation by an angle of _____.

   (d) A clockwise rotation by an angle of 3681° is equivalent to a clockwise rotation by an angle of _____.

14. In each of the following give an answer between 0° and 360°.

   (a) A clockwise rotation by an angle of 500° is equivalent to a clockwise rotation by an angle of _____.

   (b) A clockwise rotation by an angle of 500° is equivalent to a counter-clockwise rotation by an angle of _____.

   (c) A clockwise rotation by an angle of 5000° is equivalent to a clockwise rotation by an angle of _____.

   (d) A clockwise rotation by an angle of 50,000° is equivalent to a clock-wise rotation by an angle of _____.

15. Given a rotation that moves the point B to the point B' and the point C to the point C' as shown in the figure, find

   (a) the rotocenter.

   (b) the image of triangle ABC under the rotation.

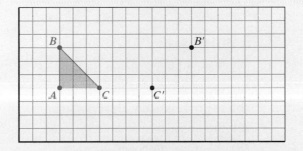

16. Given a rotation that moves the point A to the point A' and the point B to the point B' as shown in the figure, find

   (a) the rotocenter.

   (b) the image of triangle ABC under the rotation.

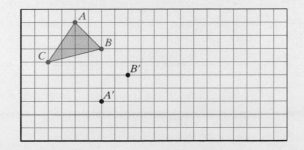

17. Given a 90° clockwise rotation that moves the point B to the point B' as shown in the figure, find

    (a) the rotocenter.

    (b) the image of triangle ABC under the rotation.

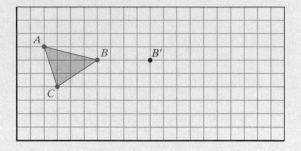

18. Given a half-turn (180° rotation) that moves the point A to the point A' as shown in the figure, find

    (a) the rotocenter.

    (b) the image of the shaded region under the rotation.

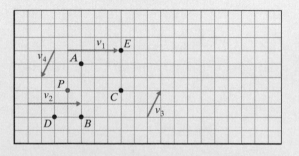

C. Translations

19. Which point in the figure is the image of P under

    (a) the translation with vector v_1?

    (b) the translation with vector v_2?

    (c) the translation with vector v_3?

    (d) the translation with vector v_4?

20. Which point in the figure is the image of P under

    (a) the translation with vector v_1?

(b) the translation with vector v_2?

(c) the translation with vector v_3?

(d) the translation with vector v_4?

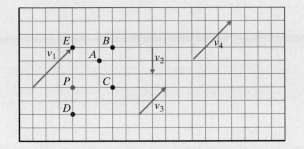

21. Given a translation that sends the point E to the point E' as shown in the figure, find

    (a) the image of A under the translation.

    (b) the image of figure $ABCDE$ under the translation.

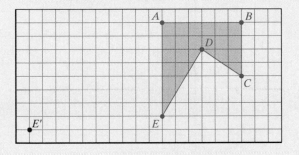

22. Given a translation that sends the point Q to the point Q' as shown in the figure, find

    (a) the image of P under the translation.

    (b) the image of figure $PQRS$ under the translation.

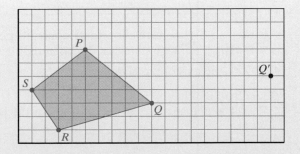

D. Glide Reflections

23. Given a glide reflection with vector v and axis l as shown in the figure, find the image of the triangle ABC under the glide reflection.

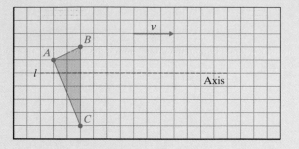

24. Given a glide reflection with vector v and axis l as shown in the figure, find the image of the quadrilateral $ABCD$ under the glide reflection.

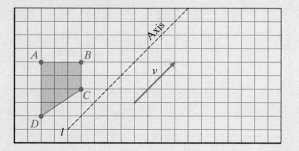

25. Given a glide reflection that sends the point B to the point B' and the point D to the point D' as shown in the figure, find

  (a) the axis of the glide reflection.

  (b) A' (the image of A) under the glide reflection.

  (c) the image of figure $ABCDE$ under the glide reflection.

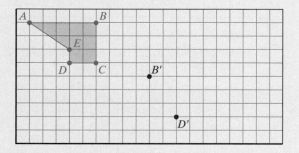

26. Given a glide reflection that sends the point A to the point A' and the point C to the point C' as shown in the figure, find

  (a) the axis of the glide reflection.

  (b) B' (the image of B) under the glide reflection.

  (c) the image of figure $ABCD$ under the glide reflection.

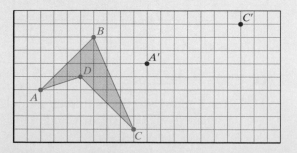

27. Given a glide reflection that sends the point B to the point B' and the point C to the point C' as shown in the figure, find

  (a) the axis of the glide reflection.

  (b) the image of the figure $ABCDE$ under the glide reflection.

28. Given a glide reflection that sends the point P to the point P' and the point Q to the point Q' as shown in the figure, find

  (a) the axis of the glide reflection.

  (b) the image of the figure $PQRS$ under the glide reflection.

E. Symmetries of Finite Shapes

In Exercises 29 through 32, list all the symmetries of each figure. Describe each symmetry by giving specifics—the axes of reflection, the centers and angles of rotation, etc.

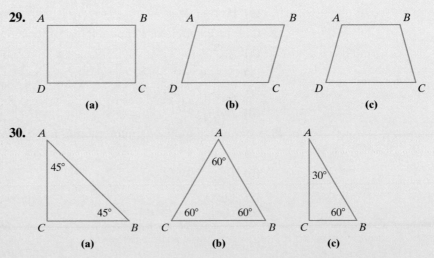

29.

(a)   (b)   (c)

30.

(a)   (b)   (c)

31.

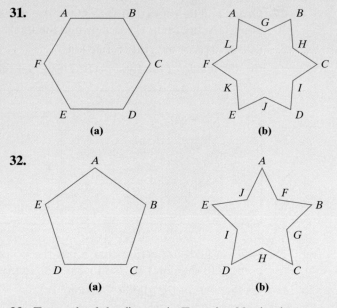

(a)   (b)

32.

(a)   (b)

33. For each of the figures in Exercise 29, give its symmetry type.

34. For each of the figures in Exercise 30, give its symmetry type.

35. For each of the figures in Exercise 31, give its symmetry type.

36. For each of the figures in Exercise 32, give its symmetry type.

37. Find the symmetry type for each of the following letters.

(a) A

(b) D

(c) L

(d) Z

(e) Q

38. Find the symmetry type for each of the following letters.

(a) T

(b) E

(c) N

(d) R

(e) H

39. Give an example of a capital letter of the alphabet that has symmetry type

(a) Z_1.

(b) D_1.

(c) Z_2.

(d) D_2.

40. Give an example of a numeral that has symmetry type

(a) Z_1.

(b) D_1.

(c) Z_2.

(d) D_2.

41. (a) Give an example of a natural object (plant, animal, mineral) that has symmetry type D_5. Explain your answer.

   (b) Give an example of a human-made object (logo, gadget, consumer product, etc.) that has symmetry type D_5. Explain your answer.

42. (a) Give an example of a natural object (plant, animal, mineral) that has symmetry type D_6. Explain your answer.

   (b) Give an example of a human-made object (logo, gadget, consumer product, etc.) that has symmetry type D_6. Explain your answer.

43. (a) Give an example of a natural object (plant, animal, mineral) that has symmetry type Z_1. Explain your answer.

   (b) Give an example of a human-made object (logo, gadget, consumer product, etc.) that has symmetry type Z_1. Explain your answer.

44. (a) Give an example of a natural object (plant, animal, mineral) that has symmetry type Z_2. Explain your answer.

   (b) Give an example of a human-made object (logo, gadget, consumer product, etc.) that has symmetry type Z_2. Explain your answer.

F. Symmetries of Border Patterns

45. For each of the following border patterns, give its symmetry type using the standard crystallography notation ($mm, mg, m1, 1m, 1g, 12, 11$).

   (a) … A A A A A …
   (b) … D D D D D …
   (c) … Z Z Z Z Z …
   (d) … L L L L L …

46. For each of the following border patterns, give its symmetry type using the standard crystallography notation ($mm, mg, m1, 1m, 1g, 12, 11$).

   (a) … J J J J J …
   (b) … T T T T T …
   (c) … C C C C C …
   (d) … N N N N N …

47. For each of the following border patterns, give its symmetry type using the standard crystallography notation ($mm, mg, m1, 1m, 1g, 12, 11$).

   (a) … WMWMWM …
   (b) … pdpdpdpd …
   (c) … pbpbpbpb …
   (d) … pqbdpqbd …

48. For each of the following border patterns, give its symmetry type using the standard crystallography notation ($mm, mg, m1, 1m, 1g, 12, 11$).

   (a) … qbqbqbqb …
   (b) … qdqdqdqd …
   (c) … dbdbdbdb …
   (d) … qpdbqpdb …

G. Miscellaneous

49. Explain why any proper rigid motion that has a fixed point must be equivalent to a rotation.

50. Explain why any rigid motion other than the identity that has two or more fixed points must be equivalent to a reflection.

Exercises 51 through 60 refer to the product of two rigid motions. Given two rigid motions M and N, we can combine the two rigid motions by first applying M and then applying N to the result. The rigid motion defined by combining M and N in that order is called the product of M and N. Note that the product of N and M (where we apply N first and M second) is not the same as the product of M and N.

51. Find the image of P under the product of

    (a) the reflection with axis l_1 and the reflection with axis l_2.

    (b) the reflection with axis l_2 and the reflection with axis l_1.

    (c) the reflection with axis l_2 and the reflection with axis l_3.

    (d) the reflection with axis l_3 and the reflection with axis l_2.

    (e) the reflection with axis l_1 and the reflection with axis l_4.

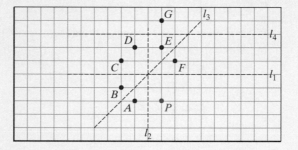

52. Find the image of P under the product of

    (a) the reflection with axis l and the 90° clockwise rotation with rotocenter A.

    (b) the 90° clockwise rotation with rotocenter A and the reflection with axis l.

    (c) the reflection with axis l and the 180° rotation with rotocenter A.

    (d) the 180° rotation with rotocenter A followed by the reflection with axis l.

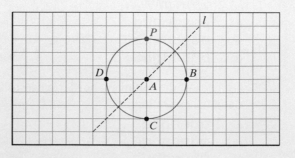

53. In each case, state whether the rigid motion M is proper or improper.

    (a) M is the product of a proper rigid motion and an improper rigid motion.

    (b) $\mathcal{M}$ is the product of an improper rigid motion and an improper rigid motion.

    (c) $\mathcal{M}$ is the product of a reflection and a rotation.

    (d) $\mathcal{M}$ is the product of a reflection and a reflection.

54. In each case, state whether the rigid motion $\mathcal{M}$ has (i) no fixed points, (ii) exactly one fixed point, or (iii) infinitely many fixed points.

    (a) $\mathcal{M}$ is the product of a reflection with axis l_1 and a reflection with axis l_2. Assume the lines l_1 and l_2 intersect at a point C.

    (b) $\mathcal{M}$ is the product of a reflection with axis l_1 and a reflection with axis l_3. Assume the lines l_1 and l_3 are parallel.

55. Suppose that a rigid motion $\mathcal{M}$ is the product of a reflection with axis l_1 and a reflection with axis l_2, where l_1 and l_2 intersect at a point C. Explain why $\mathcal{M}$ must be a rotation with center C. [*Hint*: See Exercises 53 (d) and 54 (a).]

56. Suppose that the rigid motion $\mathcal{M}$ is the product of the reflection with axis l_1 and the reflection with axis l_3, where l_1 and l_3 are parallel. Explain why $\mathcal{M}$ must be a translation. [*Hint*: See Exercises 53 (d) and 54 (b).]

Jogging

57. Suppose that lines l_1 and l_2 intersect at C and that the angle between them as shown in the figure is α.

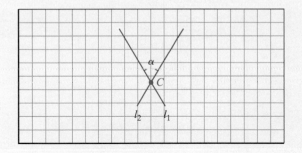

    (a) Give the rotocenter, angle, and direction of rotation of the product of the reflection with axis l_1 and the reflection with axis l_2. (*Hint:* See Exercises 51 and 55.)

    (b) Give the rotocenter, angle, and direction of rotation of the product of the reflection with axis l_2 and the reflection with axis l_1.

58. Suppose that lines l_1 and l_3 are parallel and that the distance between them as shown in the figure is d.

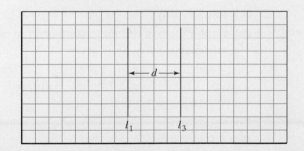

(a) Give the length and direction of the vector of the product of the reflection with axis l_1 and the reflection with axis l_3. [*Hint:* See Exercises 51(e) and 56.]

(b) Give the length and direction of the vector of the product of the reflection with axis l_3 and the reflection with axis l_1.

59. Translation 1 moves point P to point P'; translation 2 moves point Q to point Q', as shown in the figure.

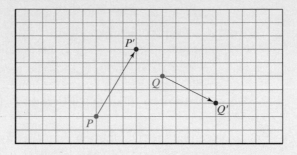

(a) Find the images of P and Q under the product of translation 1 and translation 2.

(b) Show that the product of translation 1 and translation 2 is a translation. Give a geometric description of the vector of the translation.

60. (a) Given a glide reflection with axis l and vector v as shown, find the image of the triangle ABC under the product of the glide reflection with itself.

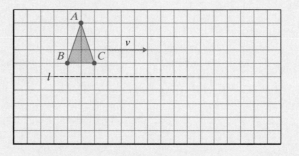

(b) Show that the product of a glide reflection with itself is a translation. Describe the direction and amount of the translation in terms of the direction and amount of the original glide.

61. (a) Explain why a border pattern cannot have a reflection symmetry along an axis forming 45° with the direction of the pattern.

(b) Explain why a border pattern can have only horizontal and/or vertical reflection symmetry.

62. (a) Explain why a border pattern cannot have a rotation symmetry of 90°.

(b) Explain why a border pattern can have only the identity or a 180° rotation symmetry.

63. Using only the letter "F," the border pattern … F⊣F⊣F⊣ … of symmetry type 12 can be constructed. Using only repetitions of the letter "F," construct a border pattern of type

(a) *mg*.

(b) *m*1.

(c) 1g.

(d) 11.

64. Using only the letter "**E**," a border pattern ... **E∃E∃E∃** ... of symmetry type *mm* can be constructed. Using only repetitions of the letter "**E**," construct a border pattern of type

(a) *mg*.

(b) *m*1.

(c) 1*m*.

(d) 11.

65. Explain why the propeller shown in the picture doesn't have any reflection symmetries.

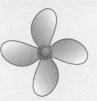

66. A *palindrome* is a word that is the same when read forward or backward. MOM is a palindrome and so is ANNA. (For simplicity, we will assume all letters are capitals.)

(a) Explain why if a word has vertical reflection symmetry, then it must be a palindrome.

(b) Give an example of a palindrome (other than ANNA) that doesn't have vertical reflection symmetry.

(c) If a palindrome has vertical reflection symmetry, what can you say about the symmetries of the individual letters in the word?

(d) Find a palindrome with 180° rotational symmetry.

67. A rigid motion $\mathcal{M}$ moves the triangle PQR into the triangle $P'Q'R'$ as shown in the figure. Explain why the rigid motion $\mathcal{M}$ must be a glide reflection.

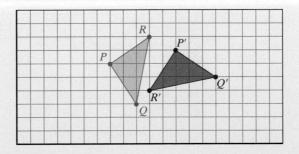

Running

68. Consider the equilateral triangle ABC shown in the margin and its six symmetries:

r_1: reflection with axis l_1 (passing through A and the midpoint of BC),

r_2: reflection with axis l_2 (passing through B and the midpoint of AC),

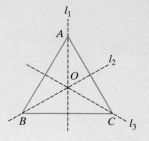

r_3: reflection with axis l_3 (passing through C and the midpoint of AB),

R_1: 120° clockwise rotation with rotocenter O,

R_2: 240° clockwise rotation with rotocenter O,

I: the identity symmetry.

For each row and column of the table that follows, enter the symmetry which results when applying the symmetry in that row followed by the symmetry in that column. (For example the entry in row r_1 column r_2 is R_1 because the reflection r_1 followed by the reflection r_2 equals the rotation R_1.)

	r_1	r_2	r_3	R_1	R_2	I
r_1		R_1				
r_2						
r_3						
R_1						
R_2						
I						

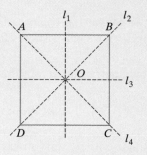

69. Consider the square $ABCD$ shown in the margin and its eight symmetries:

r_1: reflection with axis l_1,

r_2: reflection with axis l_2,

r_3: reflection with axis l_3,

r_4: reflection with axis l_4,

R_1: 90° clockwise rotation with rotocenter O,

R_2: 180° clockwise rotation with rotocenter O,

R_3: 270° clockwise rotation with rotocenter O,

I: the identity symmetry.

For each row and column of the table that follows, enter the symmetry which results when applying the symmetry in that row followed by the symmetry in that column.

	r_1	r_2	r_3	r_4	R_1	R_2	R_3	I
r_1								
r_2								
r_3								
r_4								
R_1								
R_2								
R_3								
I								

70. Find the symmetry type of the wallpaper pattern. (*Hint:* Use the flow chart on p. 474.)

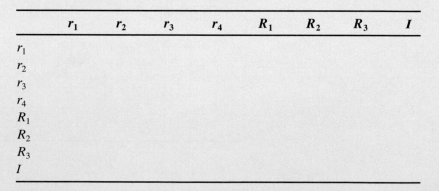

71. Find the symmetry type of the wallpaper pattern. (*Hint:* Use the flow chart on p. 474.)

72. Find the symmetry type of the wallpaper pattern. (*Hint:* Use the flow chart on p. 474.)

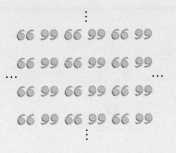

73. Find the symmetry type of the wallpaper pattern. (*Hint:* Use the flow chart on p. 474.)

74. Find the symmetry type of the wallpaper pattern. (*Hint:* Use the flow chart on p. 474.)

75. Find the symmetry type of the wallpaper pattern. (*Hint:* Use the flow chart on p. 474.)

PROJECTS AND PAPERS

A. Patterns Everywhere

Border patterns can be found in many objects from the real world—ribbons, wallpaper borders, and architectural friezes. Even ceramic pots and woven baskets exhibit border patterns (when the pattern goes around in a circle it can be unraveled as if it were going on a straight line). Likewise, wallpaper patterns can be found in wallpapers, textiles, neckties, rugs, and gift-wrapping paper. Patterns are truly everywhere.

This project consists of two separate subprojects.

Part 1. Find examples from the real world of each of the seven possible border-pattern symmetry types. Do not use photographs from a book or designs you have just downloaded from some Web site. This part is not too hard, and it is a warm-up for Part 2, the real challenge.

Part 2. Find examples from the real world of each of the 17 wallpaper-pattern symmetry types. The same rules apply as for Part 1. Use the flow chart on p. 474 to classify the wallpaper patterns.

Notes: (1) Your best bet is to look at wallpaper patterns and borders at a wallpaper store and/or gift-wrapping paper and ribbons at a paper store. You will have to do some digging—a few of the wallpaper-pattern symmetry types are hard to find. (2) For ideas, you may want to visit Steve Edwards' excellent Web site *Tiling Plane and Fancy* at *http://www2.spsu.edu/math/tile/index.htm*.

B. Three-Dimensional Rigid Motions

For two-dimensional objects, we have seen that every rigid motion is of one of four basic types. For three-dimensional objects moving in three-dimensional space there are *six* possible types of rigid motion. Specifically, every rigid motion in three-dimensional space is equivalent to either a **reflection**, a **rotation**, a **translation**, a **glide reflection**, a **rotary reflection**, or a **screw displacement**.

Prepare a presentation on the six possible types of rigid motions in three-dimensional space. For each of the rigid motions give a precise definition, describe some of its most important features and give illustrations (hand drawings are OK) as well as real-world examples.

C. Penrose Tilings of the Plane

In the mid-1970s, the British mathematician Roger Penrose (see the biographical profile on p. 452) discovered several different ways in which a plane could be tiled using a nonperiodic pattern. One of the simplest and most surprising aperiodic tilings discovered by Penrose uses just two shapes—the two rhombi shown.

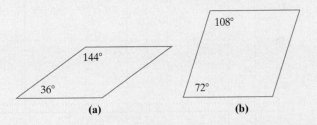

(a)   (b)

Prepare a presentation discussing and describing Penrose's tilings based on figures (a) and (b). Include in your presentation the connection between figures (a) and (b) and the golden ratio, as well as the connection between Penrose tilings and quasicrystals.

Suggested readings:

- Von Baeyer, Hans, "Impossible Crystals," *Discover*, February, 1990.

D. Symmetry and Beauty

Write a paper discussing the connections between symmetry and physical beauty.

Suggested readings:

- Cowley, Geoffrey, "The Biology of Beauty," *Newsweek*, June 3, 1996.
- Lemley, Brad, "Isn't She Lovely?", *Discover*, February, 2000.
- Etcoff, Nancy, *Survival of the Prettiest: The Science of Beauty*. Anchor Books (2000).

APPENDIX

The Seventeen Wallpaper Symmetry Types

APPENDIX The Seventeen Wallpaper Symmetry Types

Symmetry Type	Translation (2 or more directions)	Rotations (given by the smallest angle)					Reflections (Number of directions)					Glide Reflections (Number of directions)					Example
		identity 0	2-fold 180	4-fold 90	3-fold 120	6-fold 60	1	2	4	3	6	1	2	4	3	6	
p1	✓	✓															
pm	✓	✓					✓										
pg	✓	✓										✓					
cm	✓	✓					✓					✓					
p2	✓	✓	✓														
pmg	✓	✓	✓				✓					✓					
pmm	✓	✓	✓					✓									
pgg	✓	✓	✓										✓				
cmm	✓	✓	✓					✓					✓				

Symmetry Type	Translation (2 or more directions)	Rotations (smallest angle) — identity 0°	2-fold 180°	4-fold 90°	3-fold 120°	6-fold 60°	Reflections (directions) 1	2	4	3	6	Glide Reflections (directions) 1	2	4	3	6	Example
$p4$	✓	✓	✓^{a}	✓^{b}													
$p4m$	✓	✓	✓^{a}	✓^{b}					✓				✓				
$p4g$	✓	✓	✓^{a}	✓^{b}				✓						✓			
$p3$	✓	✓			✓												
$p3m1$	✓	✓			✓^{c}					✓					✓		
$p31m$	✓	✓			✓^{d}					✓					✓		
$p6$	✓	✓	✓		✓	✓											
$p6m$	✓	✓	✓		✓	✓					✓					✓	

a, b : Different rotocenters.

c : All rotocenters on axes of reflection.

d : Not all rotocenters on axes of reflection.

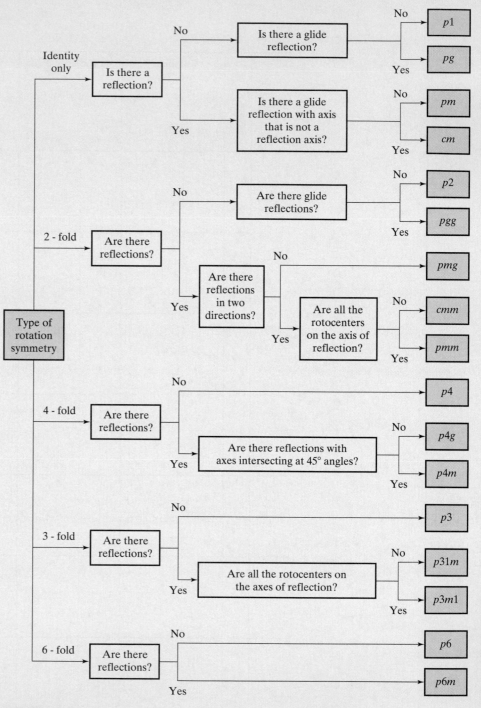

FIGURE A-1
Flow chart for classifying a wallpaper pattern.

REFERENCES AND FURTHER READINGS

1. Bunch, Bryan, *Reality's Mirror: Exploring the Mathematics of Symmetry*. New York: John Wiley & Sons, Inc., 1989.

2. Coxeter, H. S. M., *Introduction to Geometry*, 2d ed. New York: John Wiley & Sons, Inc., 1969.

3. Crowe, Donald W., "Symmetry, Rigid Motions, and Patterns," *UMAP Journal*, 8 (1987), 206–236.

4. Field, M., and M. Golubitsky, *Symmetry in Chaos*. New York: Oxford University Press, 1992.

5. Gardner, Martin, *The New Ambidextrous Universe: Symmetry and Asymmetry from Mirror Reflections to Superstrings*, 3d ed. New York: W. H. Freeman & Co., 1990.

6. Grünbaum, Branko, and G. C. Shephard, *Tilings and Patterns: An Introduction*. New York: W. H. Freeman & Co., 1989.

7. Hargittai, I., and M. Hargittai, *Symmetry: A Unifying Concept*. Bolinas, CA: Shelter Pubns., 1994.

8. Hofstadter, Douglas R., *Gödel, Escher, Bach: An Eternal Golden Braid*. New York: Vintage Books, 1980.

9. Martin, George E., *Transformation Geometry: An Introduction to Symmetry*. New York: Springer-Verlag, 1994.

10. Rose, Bruce, and Robert D. Stafford, "An Elementary Course in Mathematical Symmetry," *American Mathematical Monthly*, 88 (1981), 59–64.

11. Schattsneider, Doris, *Visions of Symmetry: Notebooks, Periodic Drawings, and Related Work of M. C. Escher*. New York: W. H. Freeman & Co., 1990.

12. Shubnikov, A. V., and V. A. Koptsik, *Symmetry in Science and Art*. New York: Plenum Publishing Corp., 1974.

13. Washburn, Dorothy K., and Donald W. Crowe, *Symmetries of Culture*. Seattle, WA: University of Washington Press, 1988.

14. Weyl, Hermann, *Symmetry*. Princeton, NJ: Princeton University Press, 1952.

15. Wigner, Eugene, *Symmetries and Reflections*. Bloomington, IN: Indiana University Press, 1967.

Fractal Geometry

Fractally Speaking

There is something unique and distinctive about many of nature's most beautiful creations. Mountains always look like mountains, even as they differ in their details. There is an undefinable, but unmistakable, "mountain look." And we can always tell a fake mountain from a real mountain, can't we? It is practically impossible to capture in an image, outside of a photograph, the subtle feel that makes a mountain look real—isn't it? And what's true for

> Nature is a mutable cloud which is always and never the same.
> Ralph Waldo Emerson

mountains is true for trees, rivers, clouds, and so on. So, which of the images on the facing page are real photographs and which are fakes? (For the answer, check the photo credits on p. 727.)

Over the last 20 years, an entirely new type of geometry has allowed humans to understand and reconstruct many of nature's most complex images, ranging from the everyday world of mountains and heads of cauliflower to the microscopic world of the body's vascular system to the otherworldly look of galaxies and nebulae.

In this chapter we will introduce the basic ideas behind this new geometry of natural shapes, called **fractal geometry**. The conceptual building blocks of fractal geometry are the notions of *recursive replacement rules* and *self-similarity,* and we will illustrate both of these concepts by means of several important examples.

Left: Head of cauliflower. *Center:* Veins and arteries. *Right:* Spiral galaxy.

12.1 The Koch Snowflake

The Koch snowflake is a remarkable geometric shape first studied by the Swedish mathematician Helge von Koch in the early 1900s. The construction of the Koch snowflake proceeds as follows.

- **Start.** Start with a solid *equilateral* triangle of arbitrary size [Fig. 12-1(a)]. (For simplicity we will assume that the sides of the triangle are of length 1.)
- **Step 1.** *Procedure KS: Attach in the middle of each side an equilateral triangle, with sides of length one third of the previous side* [Fig. 12-1(b)]. When we are done, the result is a "star of David" with 12 sides, each of length 1/3 [Fig. 12-1(c)].

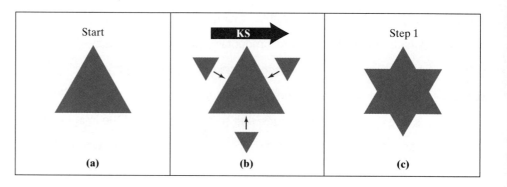

FIGURE 12-1
(a) A solid blue equilateral triangle. (b) Smaller copies of the original are added on each side. (c) A solid blue star.

- **Step 2.** For each of the 12 sides of the star of David in Step 1, repeat *procedure KS*: In the middle of each side attach an equilateral triangle (with dimensions one-third of the dimensions of the side). The resulting shape has 48 sides, each of length 1/9 [Fig. 12-2(a)].
- **Steps 3, 4, etc.** Continue repeating *procedure KS* ad infinitum [Fig. 12-2(b), (c), etc.].

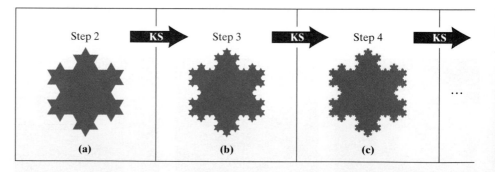

FIGURE 12-2
Successive steps in the recursive process leading toward the Koch snowflake.

At each step of this construction, the figure changes a little, but after a while, the changes are less and less noticeable. By the seventh or eighth step, the process has become *visually stable*. (That is, we really can't see the difference between the seventh and eighth steps with the naked eye.) For all practical purposes, what we are seeing is the ultimate shape we want: the **Koch snowflake** (Fig. 12-3).

FIGURE 12-3
A rendering of the Koch snowflake.

It is clear that, because the process of building the Koch snowflake is infinite, a perfect picture of it is impossible. However, this should not deter us from rendering good versions of it (as in Fig. 12-3) or from using such renderings to study its mathematical properties. (This is very similar to the situation in elementary geometry where we learned a lot about squares, triangles, and circles, even when our drawings of them were far from perfect.)

Recursive Replacement Rules

The construction of the Koch snowflake is an example of a *recursive process,* a process in which the same set of rules is applied over and over, with the end product at each step becoming the starting point for the next step. The concept of a recursive process is not new. It appeared in Chapter 1 (recursive ranking methods), Chapter 9 (Fibonacci numbers), and Chapter 10 (logistic model for population growth). In the case of the Koch snowflake, the objects of the recursive process are shapes rather than numbers; other than that, the basic principles are quite similar.

One main advantage of recursive processes is that they allow for very simple and efficient descriptions of objects, even when the objects themselves are quite complicated. The Koch snowflake, for example, is a fairly complicated geometric shape, but we could describe it in two lines using a form of shorthand we will call a **recursive replacement rule**—a rule that specifies how to substitute one piece for another.

THE KOCH SNOWFLAKE: RECURSIVE REPLACEMENT RULE

- Start with a solid equilateral triangle ▲ .
- Whenever you see a boundary line segment ——outside/inside—— replace it with ——▲inside——.

If we look at the Koch snowflake from the perspective of traditional geometry, we find that it has some very unusual properties. Let's start by discussing two typical questions that always come up in traditional geometry: *perimeter* and *area*.

Perimeter of the Koch Snowflake

The most interesting part of the Koch snowflake is its boundary. If we forget about the solid interior and look at just the boundary, we get an extremely jagged curve (Fig. 12-4) commonly known as the **Koch curve** (or *snowflake curve*).

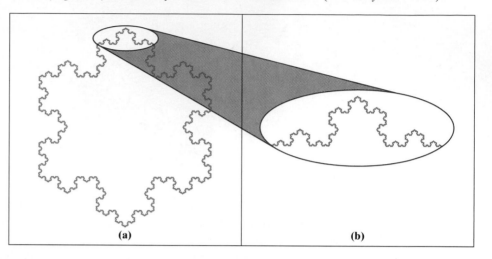

FIGURE 12-4

(a) The Koch curve. (b) A portion of the curve in detail (magnified by 3).

(a)   (b)

How long is the Koch curve? Figure 12-5 shows the perimeter for the first few generations of the Koch curve. In each step, the perimeter grows by a factor of $\frac{4}{3}$ (in other words, at each step the curve is $33\frac{1}{3}\%$ longer than in the previous step). After infinitely many such steps, the length of the curve is infinite. This is our first important fact.

The boundary of the Koch snowflake has infinite length.

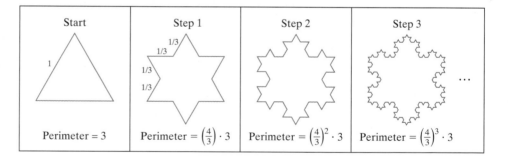

FIGURE 12-5

At each step of the recursive process, the length of the curve is multiplied by 4/3.

Start — Perimeter = 3

Step 1 — Perimeter = $\left(\frac{4}{3}\right) \cdot 3$

Step 2 — Perimeter = $\left(\frac{4}{3}\right)^2 \cdot 3$

Step 3 — Perimeter = $\left(\frac{4}{3}\right)^3 \cdot 3$

Area of the Koch Snowflake

Here we will start with the facts.

The area of the Koch snowflake is 1.6 times the area of the starting equilateral triangle.

This fact is, at first glance, very surprising. The Koch snowflake represents a shape with a finite area enclosed within an infinite boundary—something that seems contrary to our geometric intuition, but that is characteristic of many important shapes in nature. The vascular system of veins and arteries in the human body, for

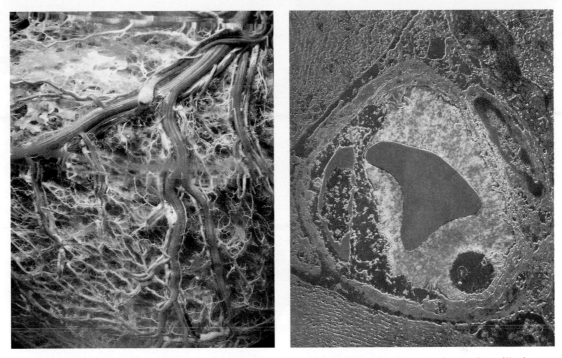

Left: The vascular network of the human body. Forty-thousand miles of veins, arteries, and capillaries packed inside small quarters. *Right:* Cross section of a blood capillary, with a single red blood cell in the center (Magnification: 7070 times).

example, occupies a small fraction of the body and has a relatively small volume, yet its length is enormous: Laid end to end, the veins, arteries, and capillaries of a single human being would reach over 40,000 miles.

In what follows, we will give an outline of the argument showing that the area of the Koch snowflake is 1.6 times the area of the original equilateral triangle, leaving the technical details as exercises for the reader. In fact, the reader who wishes to do so may skip the forthcoming explanation without prejudice.

The key to calculating the area of the Koch snowflake can be found by studying Fig. 12-6 (on the next page) carefully. At each step, we can compute how many new triangles are being added and the area of each one. From Fig. 12-6, we can see that in the Nth step we are adding a total of $3(4^{N-1})$ new triangles, each having an area of $(1/9)^N \cdot A$, which gives an added area of $(4/9)^{N-1}(1/3)A$. The total area at the Nth step is the sum of the original equilateral triangle's area and the areas added at each step:

$$A + \left(\frac{1}{3}\right)A + \left(\frac{4}{9}\right)\left(\frac{1}{3}\right)A + \left(\frac{4}{9}\right)^2\left(\frac{1}{3}\right)A + \cdots + \left(\frac{4}{9}\right)^{N-1}\left(\frac{1}{3}\right)A.$$

Except for the first term, we are looking at the sum of terms of a geometric sequence with common ratio $r = 4/9$. Using the formula given in Chapter 10 for adding the consecutive terms of a geometric sequence, we can simplify the preceding expression to

$$A + \left(\frac{3}{5}\right)A\left[1 - \left(\frac{4}{9}\right)^N\right].$$

We leave the technical details to the reader (see Exercise 50).

We are now ready to wrap this up. We need only figure out what happens to the expression $(4/9)^N$ as N becomes increasingly larger. In fact, what happens to

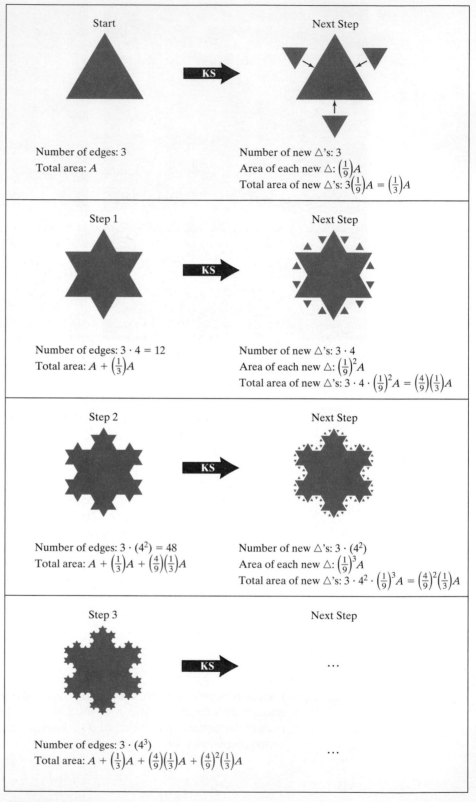

FIGURE 12-6
Calculating the area of the Koch snowflake.

any positive number less than 1 when we raise it to higher and higher powers? If you know the answer, then you are finished. If you don't, take a calculator, enter a number between 0 and 1, and multiply it by itself repeatedly. You will readily convince yourself that the result gets closer and closer to 0. The bottom line is that as N gets increasingly larger, the expression inside the square brackets approaches 1, and therefore, the area approaches $A + (3/5)A = (1.6)A$.

Self-Similarity

What does the fine detail of the Koch curve look like? Pick a small section along the Koch curve. If we magnify it [Fig. 12-7(a)], we get the image shown in Fig. 12-7(b). Further magnification is not much help—Fig. 12-7(c) shows a detail of the Koch curve after magnifying it by a factor of almost 100.

We can see from Fig. 12-7 that something very surprising is happening: Anywhere we look, the fine detail of the Koch curve looks exactly the same as the rough detail! This remarkable characteristic of the Koch curve is called *self-similarity* or *symmetry of scale*. As the name suggests, it is a symmetry that carries itself across different scales—a symmetry between the large-scale structure and the small-scale structure of an object.

We will say that a shape has **self-similarity** (or *symmetry of scale*) if parts of the shape appear at infinitely many different scales. In the case of the Koch curve, there is a specific pattern[1] that shows up everywhere and at every scale.

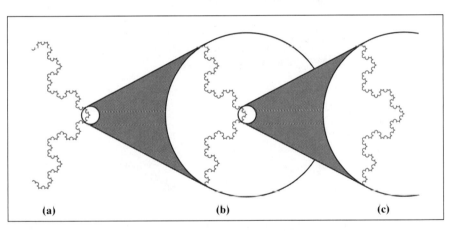

FIGURE 12-7
(a) A section of the Koch curve. (b) Detail of a small section of the Koch curve (magnified by 9). (c) Detail of a tiny section of the Koch curve (magnified by 81).

12.2 The Sierpinski Gasket

This is another interesting shape exhibiting self-similarity. It was first studied (in a slightly modified form) by the Polish mathematician Waclaw Sierpinski around 1915.

The construction starts with an arbitrary triangle ABC [see Fig. 12-8(a)] but this time, instead of *adding* smaller copies of the original triangle, we will *remove* smaller copies of the original triangle according to the following procedure:

- **Start.** Start with an arbitrary solid triangle ABC.

---

[1]The recurring pattern is ⌐⋀⌐ .
    inside

- **Step 1.** ***Procedure SG:*** *Remove the triangle whose vertices are the midpoints of the sides of the triangle.* (We'll call this triangle the *middle* triangle.) This leaves a white triangular hole in the original solid triangle, and three solid triangles, each of which is a half-scale version of the original [Fig. 12-8(b)].

- **Step 2.** For each of the solid triangles in the previous step, repeat *procedure SG*. (That is, remove its middle triangle.) This leaves us with 9 solid triangles (all similar to the original triangle *ABC*) and 4 triangular white holes [Fig. 12-8(c)].

- **Steps 3, 4, etc.** Continue repeating *procedure SG* on every solid triangle, ad infinitum.

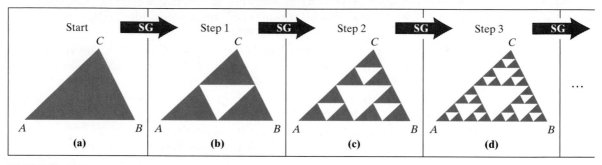

FIGURE 12-8
First three steps in the construction of the Sierpinski gasket.

After seven or eight steps the figure becomes visually stable. Shown in Fig. 12-9, the resulting shape resembles an exotic gasket from which it gets its name: the **Sierpinski gasket.**

FIGURE 12-9
A rendering of the Sierpinski gasket.

The Sierpinski gasket can be described in a very convenient way by a recursive replacement rule.

SIERPINSKI GASKET: RECURSIVE REPLACEMENT RULE

- Start with an arbitrary solid triangle ▲ .
- Whenever you see a ▲ , replace it with a ⧊ .

We leave the following two facts as exercises to be verified by the reader:

- The Sierpinski gasket has zero area. (See Exercise 41.)

- The Sierpinski gasket has an infinitely long boundary. (See Exercise 42.)

Looking at Fig. 12-9, it appears that the Sierpinski gasket is made of a huge number of tiny solid triangles, but this is the result of poor eyesight and the inadequacies of printing. The Sierpinski gasket has no solid triangles! If we were to magnify any one of those small solid specks, we would see another Sierpinski gasket (Fig. 12-10). This, of course, is another example of self-similarity.

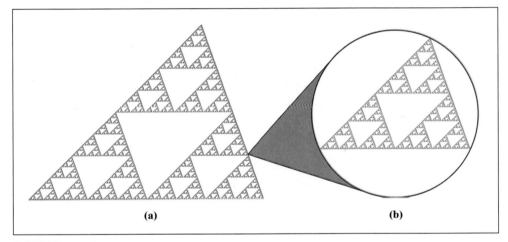

(a)   **(b)**

FIGURE 12-10
Detail of a small section of the Sierpinski gasket (magnified by 256).

An important modern application of the Sierpinski gasket can be found in the area of wireless communication: Because of their self-similarity, Sierpinski gaskets make extremely versatile and efficient internal antennas for cell phones, wireless modems, and GPS receivers (see Project D).

12.3 The Chaos Game

This example involves the laws of chance. We start with an arbitrary triangle with vertices *A*, *B*, and *C* and an honest die [Fig. 12-11(a)]. To each of the vertices of the triangle we assign two of the six possible outcomes of rolling the die. For example, *A* is the "winner" if we roll a 1 or a 2; *B* is the "winner" if we roll a 3 or a 4; and *C* is the "winner" if we roll a 5 or a 6. We are now ready to play the game.

- **Start.** Roll the die. Start at the "winning" vertex. Say we roll a 5. We then start at vertex *C* [Fig. 12-11(b)].

- **Step 1.** Roll the die again. Say we roll a 2, so the winner is vertex *A*. *We now move straight from our previous position toward the winning vertex, but stop halfway.* Mark the new position (*M*₁) [Fig. 12-11(c)].

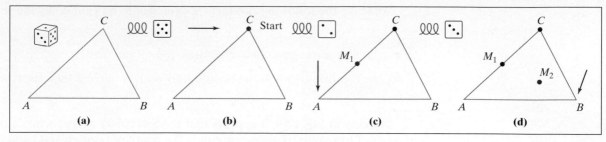

FIGURE 12-11
The Chaos Game: Always move from the previous position towards the chosen vertex, and stop halfway.

- **Step 2.** *Roll the die again, and move straight from the last position toward the winning vertex, but stop halfway.* (If the roll is 3, for example, stop at M_2 halfway between M_1 and B [Fig. 12-11(d)].) Mark your new position.

- **Steps 3, 4, etc.** Continue rolling the die, each time moving to a point halfway from the last position to the winning vertex.

Figure 12-12(a) shows the trail of points after 50 rolls of the die—just a bunch of scattered dots. Figure 12-12(b) shows the trail of points after 500 rolls. Figure 12-12(c) shows the trail of points after 5000 rolls. The pattern is unmistakable: a Sierpinski gasket! After 10,000 rolls, it would be impossible to tell the difference between the trail of points and the Sierpinski gasket (see Fig. 12-9). This is a truly surprising turn of events. After all, the pattern of points created by the chaos game is ruled by the laws of chance, and one would not expect that any predictable pattern would appear. Instead, we get an approximation to the Sierpinski gasket, and the longer we play the chaos game, the better the approximation gets.

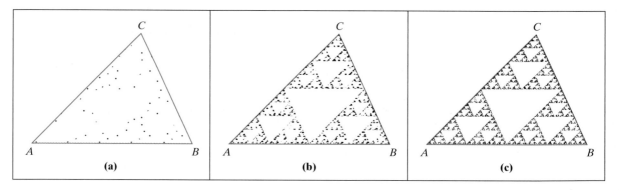

FIGURE 12-12
The "footprint" of the chaos game after (a) 50 rolls of the die, (b) 500 rolls of the die, and (c) 5000 rolls of the die.

12.4 The Twisted Sierpinski Gasket

Our next example is a simple variation of the original Sierpinski gasket. For lack of a better name, we will call it the **twisted Sierpinski gasket**.

The construction starts out exactly like the one for the regular Sierpinski gasket, with a solid triangle ABC [Fig. 12-13(a)] from which we remove the middle triangle, whose vertices we will call M, N, and L [Fig. 12-13(b)]. The next move (which we will call the "twist") is new. Each of the points M, N, and L is moved a small amount in a random direction (as if an earthquake had randomly displaced them to new positions M', N', and L'). One possible resulting shape is shown in Fig. 12-13(c).

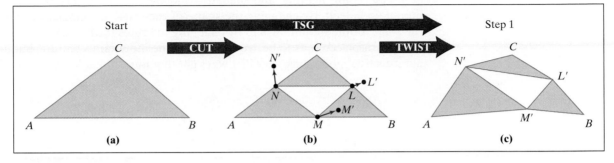

FIGURE 12-13
The two moves in *procedure TSG*: The *cut* and the *twist*.

When the process of cutting and twisting is repeated ad infinitum, we get the twisted Sierpinski gasket.

The following is a formal description of the process for building a twisted Sierpinski gasket:

- **Start.** Start with an arbitrary solid triangle ABC.

- **Step 1.** Apply procedure TSG to the starting solid triangle.
    - *Procedure TSG:*
        (a) Cut. *Remove the middle triangle from a solid triangle* [Fig. 12-13(b)].
        (b) Twist. *Move each of the midpoints of the triangle in an arbitrary direction and by a random amount that is small[2] in relation to the length of the corresponding side* [Fig. 12-13(c)].

After step 1 is complete, we end up with 3 twisted solid triangles and 1 twisted hole in the middle, as shown in Fig. 12-14(b).

- **Step 2.** For each of the solid triangles in the previous step, repeat *procedure TSG*. This leaves us with nine twisted solid triangles and four twisted white holes [Fig. 12-14(c)].

- **Steps 3, 4, etc.** Continue repeating *procedure TSG* on each solid triangle.

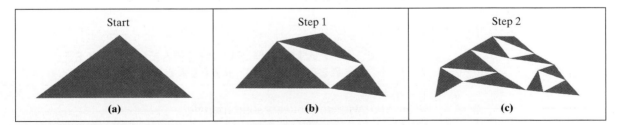

FIGURE 12-14
The first two steps in generating a twisted Sierpinski gasket.

---

[2]In Fig. 12-13, we did not allow the displacement to be more than 25% of the length of the corresponding side.

Figure 12-15 shows an example of a twisted Sierpinski gasket after eight steps. Even without touch up we can see that the twisted Sierpinski gasket has the unmistakable look of a mountain. Add a few of the standard tools of computer graphics—color, lighting, and shading—and we can get a very realistic-looking mountain indeed. By changing the shape of the starting triangle, we can change the shape of the mountain, and by changing the rules for how large we allow the random displacements to be, we can change the mountain's texture, but just as in nature's true mountains, we will always get that unmistakable "mountain look."

FIGURE 12-15
A twisted Sierpinski gasket after eight steps.

Carolina by Ken Musgrave and Benoit Mandelbrot. This computer-generated mountain scene was created using random displacement methods similar to those used in the twisted Sierpinski gasket.

The most remarkable thing of all is that these complicated-looking geometric shapes can be described in just a few lines by means of a simple recursive replacement rule.

TWISTED SIERPINSKI GASKET: RECURSIVE REPLACEMENT RULE

- Start with an arbitrary solid triangle.
- Wherever you see a solid triangle, apply procedure TSG to it.

What about self-similarity? Does the twisted Sierpinski gasket have it? Well, not exactly. Whenever we magnify a part of it, we don't see exactly the same things, but we do see random variations of a common theme. That special mountain look is going to show up at every scale.

Approximate self-similarity in nature. *Left*: Closeup of a head of cauliflower. *Right*: Further closeup on one of the florets, or wait … is it the other way around?

When an object or shape has the kind of symmetry of scale where approximate (but not identical) versions of a common theme appear at every scale, we will say it has **approximate self-similarity** (which from now on will stand in contrast to **exact self-similarity**, such as the one exhibited by the Koch snowflake and the ordinary Sierpinski gasket).

Approximate self-similarity is a common property of many natural objects and shapes: mountains, trees, plants, clouds, lightning, the human vascular system, and so on. In fact, it is what gives these things their distinctive natural look, and only by understanding the mathematical details of this type of symmetry can we hope to realistically imitate and understand the objects themselves.

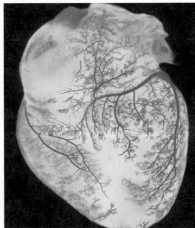

The same pattern of approximate self-similarity sometimes shows up at different scales in nature. *Left*: Branching in an oak tree. *Right*: Vascular branching inside a human heart.

12.5   Self-Similarity in Art and Literature

The notion of self-similarity is not unique. It pops up in various forms in art, poetry, and literature.

A frequently used construct in art is that of an *infinite regress*, an arrangement where an infinite sequence of repeatedly smaller scale versions of the same image converge toward a vanishing point called an *attractor*.

Figure 12-16 shows two examples of "infinite" regress in art. (As in any work of the human hand, the infinite regress in these pictures is only illusory; at some point, the detail has to stop.) Figure 12-16(a) is a woodcut by the famous Dutch artist M. C. Escher. The attractor for this incredibly sophisticated infinite regress is at the exact center of the picture. Explaining the details of how he created the image, Escher wrote:

> *In this woodcut I have consistently and almost maniacally continued the reduction down to the limit of practical execution. I was dependent on four factors: the quality of my wood material, the sharpness of my tools, the steadiness of my hand, and especially my keen-sightedness. . . .* [3]

Figure 12-16(b) shows a more simple-minded example of an infinite regress. In this cover of *TV Week*, a man is sitting in an armchair holding a remote control in his left hand and an issue of the very same *TV Week* in his right hand, with himself on it holding another *TV Week*, and so on ad infinitum, or at least that's the idea. In reality, the regress only goes four steps deep. Interestingly enough, this piece of art can be described by means of the following fairly simple recursive replacement rule:

- **Start.** Start with the big picture of the man with the TV, the remote, the armchair, and so on (everything except the issue of *TV Week* in his right hand).
- **Procedure TVG.** Reduce the picture down to 35% of its original size; (b) translate and rotate the reduced picture so that it "slips" into the man's right hand in the old picture; (c) bring the man's fingers into the foreground.

FIGURE 12-16
Left: Smaller and Smaller by M.C. Escher. In this infinite regress the attractor is in the center of the picture. (© 1956 M.C. Escher/Cordon Art, Baarn, Holland). *Right: Couch Potato.* The attractor of this infinite regress is the red point on the lower right. (Reproduced by permission of the *Fresno Bee*.)

---

[3]M. C. Escher, *Escher on Escher* (New York: H. N. Abrams, 1989).

Procedure TVG could be repeated indefinitely, creating an infinite regress that converges toward the point marked by the red dot. If we kept zooming in at that point, we would continue seeing little men with remote controls and *TV Week*s in their hands.

In literature, infinite regresses are implied: A plot has a subplot that parallels the original plot. The subplot, in turn, has its own parallel subplot. The rest is usually left to the imagination of the reader. Variations on the literary form of infinite regress have been used by many modern writers, including E. E. Cummings, Aldous Huxley, and Norman Mailer.

We conclude this section with a short poem by Jonathan Swift—about fleas, of all things. The infinite regress is in the mind's eye.

> *So Nat'ralists observe, A Flea*
> *Hath smaller Fleas that on him prey*
> *and these have smaller Fleas to bite 'em*
> *And so proceed, ad infinitum.*

12.6 The Mandelbrot Set

We now return to a much more mathematical example. In fact, the mathematics in this example goes a bit beyond the level of this book, so we will describe the overall idea in general terms. The actual purpose here is not to get bogged down in the details, but rather to illustrate one of the most interesting and beautiful geometric objects ever created by the human "hand." The object is called the *Mandelbrot set* (and sometimes simply the *M-set*) after the Polish-born mathematician Benoit Mandelbrot. Mandelbrot was the first person to extensively study and fully appreciate the importance of this beautiful and complex mathematical object. (For more on Mandelbrot, see the biographical profile on p. 498.)

Before we discuss the mathematics behind the Mandelbrot set, let's take a brief visual tour. The Mandelbrot set is the black object (blob?) seen in Fig. 12-17(a). When color is added to the region near the boundary, as in Fig. 12-17(b), the M-set comes to life.

What does the M-set look like? In the minds of many people, it resembles some sort of bug—a flea from some exotic planet, maybe? The overall structure of this "flea" can be described as consisting of a body, a head, and an antenna coming out of the middle of the head. Both the head and the body are full of smaller fleas. We can even see in Fig. 12-17 that these smaller fleas have fleas of their own. Maybe Jonathan Swift was onto something.

FIGURE 12-17
(a) The Mandelbrot set (M-set) shown on a stark white background. (b) The M-set lights up like a neon sign when color is added to region near the boundary.

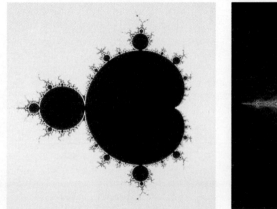
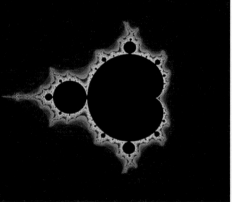

(a)   **(b)**

Notice that one thing is different in the main flea that distinguishes it from the smaller fleas on it: It has "buttocks." There are hundreds of fleas of various sizes—some pretty big; others so small that we can hardly see them—surrounding the big flea. When we magnify the view around two of these (Fig. 12-18), we can see the common theme (fleas of many sizes crawling all over the place), but we can also see surrounding the boundary, as in an otherworldly coral reef, new and surprising formations—exotic "starfish" in Fig. 12-18(a); "urchins" and "seahorse tails" in Fig. 12-18(b). If we look carefully, we can also see, floating in the midst of this bizarre but beautiful seascape, tiny copies of the M-set—wayward children carrying the mother set's genes.

(a)

FIGURE 12-18
(a) Finely detailed closeup of a small region on the boundary of the M-set. Like a wayward child, a tiny copy of the original M-set can be seen at the top center of the picture.
(*Magnification*: × 10.)
(b) A closeup of a different region on the boundary of the M-set. Barely visible in the center of the picture is another tiny copy of the original.
(*Magnification*: × 10.)

(b)

Figure 12-19(a) shows a closeup of one of the seahorse tails from Fig. 12-18(b). A further closeup of a section of Fig. 12-19(a) is shown in Fig. 12-19(b), and an even further magnification in Fig. 12-19(c), revealing a small M-set surrounded by a beautiful arrangement of swirls, spirals, and seahorse tails. The

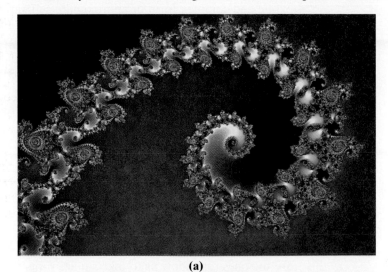

(a)

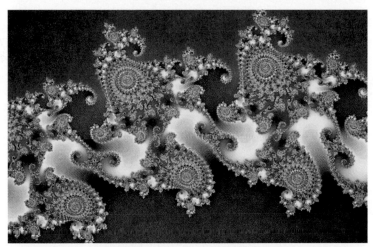

(b)

FIGURE 12-19
(a) A closeup of one of the "seahorse tails" in Fig. 12-18(b). (*Approximate magnification*: × 100.) (b) A further closeup of a section of 12-19(a) shows the refined detail and ... more seahorse tails. (*Approximate magnification*: × 200.) (c) At every level of magnification, recurring themes continue to show up. (*Approximate magnification*: × 10,000.)

(c)

M-set in this picture is about 50,000 times smaller than the mother set. Anywhere we choose to look in this picture (or any of the others), we will find (if we magnify enough) copies of the original Mandelbrot set, always surrounded by an infinitely changing, but always stunning, background. The Mandelbrot set has a very exotic and complex form of approximate self-similarity—infinite repetition and infinite variety mingle together at every scale in a landscape as diverse as nature itself.

Constructing the Mandelbrot Set: Mandelbrot Sequences

The Mandelbrot set has been rightfully described as "the most complex object ever devised by man," even though it wasn't until the advent of powerful computers in the last 20 years that images such as those shown on Figs. 12-18 and 12-19 could be generated.

How does this magnificent mix of beauty and complexity called the Mandelbrot set come about? Incredibly, the Mandelbrot set itself can be described mathematically by a very simple process involving just numbers. The only rub is that the numbers are *complex numbers*.

Complex Numbers   You may recall having seen such numbers before. These are the ones that allow us to take square roots of negative numbers, solve quadratic equations of any kind, and so on. The basic building block for complex numbers is the number $\sqrt{-1} = i$. Using i, we can build all other complex numbers, such as $(3 + 2i)$, $(\frac{5}{3} - \frac{4}{3}i)$, and the generic complex number $(a + bi)$.

For our purposes, the most important fact about complex numbers is that each of them can be identified with a unique point in a coordinate plane. The basic idea is to identify the complex number $(a + bi)$ with the point (a, b) in a Cartesian coordinate system (Fig. 12-20). Once we realize this, we can talk about complex numbers and points in the plane as being one and the same. (For a review of the basic facts about complex numbers, see a good algebra text.)

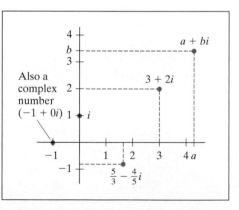

FIGURE 12-20
Every complex number is a point in the plane.

Mandelbrot Sequences   Our basic construction will be to start with a complex number (point in the plane) and from it create an infinite sequence of numbers (points) that depend on the starting number. This sequence of numbers we will call a **Mandelbrot sequence**, and the starting point we will call the **seed** of the sequence. The basic recursive rule for a Mandelbrot sequence is that *each number in the sequence equals the preceding number in the sequence squared plus the seed*. The general description of a Mandelbrot sequence with seed s is shown in Fig. 12-21.

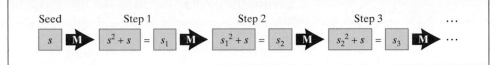

FIGURE 12-21
A Mandelbrot sequence with seed s.

A Mandelbrot sequence can also be easily described by means of a recursive rule (just as we did with the Koch snowflake and Sierpinski gasket).

> ## MANDELBROT SEQUENCE: RECURSIVE RULE
>
> **Start:** Set $s_0 = s$ (the *seed*).
>
> **Procedure M:** $s_{N+1} = s_N^2 + s$. (*Each term in the sequence equals the preceding term squared plus the seed.*)

The choice of the name "seed" for the starting value of each sequence gives us a convenient metaphor: Each seed, when planted, produces a different sequence of numbers (the tree). The recursive replacement rule is like a rule telling the tree how to grow from one season to the next.

Let's look at some examples of Mandelbrot sequences.

EXAMPLE 12.1

Figure 12–22 shows the first four steps in the Mandelbrot sequence with seed $s = 1$. (Please note that integers and decimals are also complex numbers, so they make perfectly acceptable seeds.)

The pattern that emerges is clear: The numbers are becoming increasingly larger. The corresponding points are getting further and further away from the origin. We will call this kind of sequence an **escaping** Mandelbrot sequence. For such a sequence, the point in the plane identified with the seed will be a *nonblack point*. (It's a funny way to put it, but we are noncommittal—the only thing certain is that it will *not* be a black point!)

FIGURE 12-22
Mandelbrot sequence with seed $s = 1$ (*escaping*).

EXAMPLE 12.2

Figure 12-23 shows the first three steps in the Mandelbrot sequence with seed $s = -1$.

The pattern that emerges here is also clear: the numbers alternate between 0 and -1. We say in this case that the Mandelbrot sequence is **periodic**. For such a sequence, the point in the plane identified with the seed *will always be a black point*.

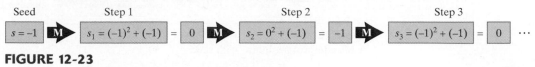

FIGURE 12-23
Mandelbrot sequence with seed $s = -1$ (*periodic*).

EXAMPLE 12.3

Figure 12-24 shows the first four steps in the Mandelbrot sequence with seed $s = -0.75$. Here a calculator will probably come in handy.

In this case, the pattern is not obvious, and additional terms of the sequence are needed. As an exercise (Exercise 51), the reader should carry this example out for another 20 steps and verify that the values of the Mandelbrot sequence approach -0.5. In this case, we will say that -0.5 is an **attractor** for the Mandelbrot sequence, or, equivalently, that the sequence is *attracted* to -0.5.

When a Mandelbrot sequence is attracted to a number the point in the plane identified with the seed *will always be a black point* (just as for periodic sequences).

FIGURE 12-24
Mandelbrot sequence with seed $s = -0.75$ (*attracted*).

EXAMPLE 12.4

Figure 12-25 shows the first four steps in the Mandelbrot sequence with seed $s = i$. Here for the first time we are dealing with an imaginary number. (Reminder: $i^2 = -1$.)

Just as in Example 12.2, the Mandelbrot sequence here is periodic; thus, in the Cartesian plane, the seed will be a black point.

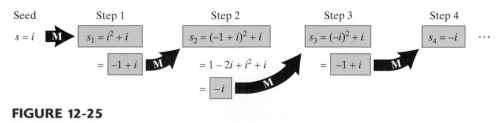

FIGURE 12-25
Mandelbrot sequence with seed $s = i$ (*periodic*).

The Mandelbrot Set (Definition)

We are finally ready to explain how the Mandelbrot set comes about. At this point, the definition will sound simple: The **Mandelbrot set** consists of all the points in the plane (complex numbers) that are black seeds of Mandelbrot sequences. Thus, our entire discussion can be summarized by the following logical sequence:

- Each point in the Cartesian plane is a complex number and can be used as a seed for a Mandelbrot sequence.

- If the Mandelbrot sequence is *periodic* or *attracted*, the point is part of the Mandelbrot set. If the sequence is *escaping*, the point is *not* in the Mandelbrot set. In the latter case, the point can be given different colors, depending on the speed of escape (for example, "hot" colors, such as red, yellow, and orange, if it escapes slowly, or cool colors, such as blue, and purple, if it escapes quickly). The coloring of the escaping points is what livens up the amazing pictures that we saw in Figs. 12–18 and 12–19.

Because the Mandelbrot set provides a bounty of aesthetic returns for a relatively small mathematical investment, it has become one of the most popular mathematical playthings of our time. There are now literally hundreds of software programs available (many of them shareware) that allow one to explore the beautiful landscapes surrounding the Mandelbrot set.

Conclusion

Fractals

The word **fractal** (from the Latin *fractus*, meaning "broken up or fragmented") was coined by Benoit Mandelbrot (see his biographical profile on page 499) in the mid-1970s to describe objects as diverse as the Koch curve, the Sierpinski gasket, the twisted Sierpinski gasket, and the Mandelbrot set, as well as many shapes in nature, such as clouds, trees, mountains, lightning, and the vascular system in the human body.

These objects share one key characteristic. They all have some form of self-similarity. (This is not the only defining characteristic of a fractal—others, such as *fractional dimension*, would take us beyond the scope of this chapter.)

The discovery and study of fractals and their geometric structure has become one of the hottest mathematical topics of the last 20 years. It is a part of mathematics that combines complex and interesting theories, beautiful graphics, and extreme relevance to the real world.

There is a striking visual difference between the kinds of shapes we discussed in this chapter and the shapes of traditional geometry. It is difficult to mistake one for the other. The shapes of traditional geometry (squares, circles, cones, etc.) and the objects we build based on them (bridges, machines, buildings, etc.) have a distinct human-made look. Many of the shapes of nature (mountains, trees, clouds, etc.) have a completely different kind of look, one that humans have always had difficulty re-creating. Mandelbrot, who is the father of *fractal geometry*, was the first to realize that the foundation of this natural look is some form of self-similarity and that geometric objects built on the principles of self-similarity can be used to model many shapes and patterns in nature. Today, the principles of fractal geometry are used to study the patterns of clouds and how they affect the weather, to diagnose the pattern of contractions of a human heart, to analyze the behavior of the stock market, to design more efficient antennas, and to create the truly incredible computer graphics that animate many of the latest science fiction movies.

Geometry as we have known it in the past was developed by the Greeks about 2000 years ago and passed on to us essentially unchanged. It was (and still is) a great triumph of the human mind, and it has allowed us to develop much of our technology, engineering, architecture, and so on. As a tool and a language for modeling and representing nature, however, Greek geometry has by and large been a failure. The discovery of fractal geometry seems to have given science the right mathematical language to overcome this failure, and thus it promises to be one of the great achievements of 20th-century mathematics.

Left: The geometry of humans—Euclidean and smooth. *Right*: The geometry of nature—fractal and infinitely textured.

Benoit Mandelbrot (1924–   )

With the publication in 1983 of his classic book *The Fractal Geometry of Nature*, Benoit Mandelbrot became somewhat of a scientific celebrity. Later, with images of the Mandelbrot set and other exotic fractals becoming part of our popular culture (through screensavers, tee-shirt designs, and television ads) the Mandelbrot name became an icon of the computer age.

In *The Fractal Geometry of Nature* as well as in many of his other writings, Mandelbrot popularized a radically different way to look at the natural world—using a new geometry built around the paradigms of self-similarity and fractional dimension. In his own words,

> *Why is [traditional] geometry often described as "cold" and "dry"? One reason lies in its inability to describe the shape of a cloud, a mountain, a coastline, or a tree. Clouds are not spheres, mountains are not cones, coastlines are not circles, and bark is not smooth nor does lightning travel in a straight line. . . .*

Born in Warsaw, Poland in 1924, Mandelbrot's family moved to France when he was 11 years old. These were difficult times—the beginning of World War II—and Mandelbrot received little formal education during his formative years. One of his uncles was a mathematics professor and Mandelbrot learned some mathematics from him, but by and large, the mathematics that Mandelbrot learned in his youth was self-taught. To Mandelbrot, this turned out to be an asset rather than a liability, as he credits much of his refined mathematical intuition and his ability to think "outside the box" to the unstructured nature of his early education.

After the war, Mandelbrot was able to receive a first rate university education: an undergraduate degree in mathematics from the École Polytechnique in Paris in 1947, an M.S. degree in Aeronautics from the California Institute of Technology in 1948, and a Ph.D. in Mathematics from the University of Paris in 1952. After a brief stint doing research and teaching in Europe, Mandelbrot accepted an appointment in the research division of IBM and moved to the United States in 1958. For the next thirty years Mandlebrot worked at IBM's T. J. Watson Research Center in New York State, first as a Research Fellow and eventually as an IBM Fellow, the most prestigious research position within the company. It was during his years at IBM that Mandelbrot developed and refined his theories on fractal geometry.

Upon his retirement from IBM in 1987, Mandelbrot accepted an endowed chair at Yale University, where he is currently the Abraham Robinson Professor of Mathematical Sciences.

approximate self-similarity	Mandelbrot set
attractor	Mandelbrot sequence
chaos game	periodic (sequence)
escaping (sequence)	recursive replacement rule
exact self-similarity	seed (of a Mandelbrot sequence)
fractal	Sierpinski gasket
fractal geometry	self-similarity (symmetry of scale)
Koch curve (snowflake curve)	twisted Sierpinski gasket
Koch snowflake	

WALKING

A. The Koch Snowflake and Variations

*Exercises 1 through 4 refer to the **square snowflake**, a variation of the Koch snowflake defined by the following recursive procedure:*

- **Start.** *Start with a solid square.*
- **Step 1. Procedure SS:** *Divide each side of the square into three equal segments. Attach to the middle segment of each side of the figure a solid square with dimensions equal to one-third that side.*
- **Steps 2, 3, etc.** *Repeat procedure SS on the sides of the figure obtained in the previous step.*

1. Carefully draw the figures at steps 1, 2, and 3 of the construction of the square snowflake. (*Hint:* Use graph paper and start with a 9-by-9 square.)

2. Determine the number of sides of
   - **(a)** the figures at steps 2, 3, and 4 of the construction of the square snowflake.
   - **(b)** the figure at step N of the construction of the square snowflake.

3. If the starting square has sides of length a, find the perimeter of
   - **(a)** the figures at steps 1, 2, and 3 of the construction of the square snowflake.
   - **(b)** the figure at step N of the construction of the square snowflake.
   - **(c)** the square snowflake.

4. If the starting square has area X, find the area of the figure at
   - **(a)** step 1 of the construction of the square snowflake.
   - **(b)** step 2 of the construction of the square snowflake.
   - **(c)** step 3 of the construction of the square snowflake.
   - **(d)** step 4 of the construction of the square snowflake.

*Exercises 5 through 8 refer to the **Koch antisnowflake**, a shape obtained by essentially reversing the process for constructing the Koch snowflake. At each stage, instead of adding a solid equilateral triangle to the middle of each side of the figure, we cut out the equilateral triangle. The Koch antisnowflake can be described by the following recursive replacement rule.*

- Start with a solid equilateral triangle: △ .
- Whenever you see a boundary line segment $\overline{}^{\text{outside}}$, replace it with $\underset{\vee}{\overset{\text{outside}}{}}$.

5. **(a)** Using graph paper, carefully draw the figures at steps 1 and 2 of the construction of the Koch antisnowflake.
   - **(b)** Determine the number of sides of the figures at steps 1, 2, 3, and 4 of the construction of the Koch antisnowflake.
   - **(c)** Determine the number of sides of the figure at step N of the construction of the Koch antisnowflake.

6. If the starting equilateral triangle has sides of length 1, find the perimeter of
   (a) the figures at steps 1, 2, 3, and 4 of the construction of the Koch antisnowflake.
   (b) the figure at step N of the construction of the Koch antisnowflake.
   (c) the Koch antisnowflake.

7. Find the number of new triangles that are removed from
   (a) the figures at steps 1, 2, and 3 of the construction of the Koch antisnowflake.
   (b) the figure at step N of the construction of the Koch antisnowflake. (*Hint:* The number of triangles removed at step N is the same as the number of sides at step $N - 1$.)

8. If the area of the starting triangle is A, find the area of the figure at
   (a) step 1 of the construction of the Koch antisnowflake.
   (b) step 2 of the construction of the Koch antisnowflake.
   (c) step 3 of the construction of the Koch antisnowflake.

   (*Hint:* Remember that the construction of the Koch antisnowflake parallels the construction of the Koch snowflake with triangles being removed instead of added.)

*Exercises 9 through 12 refer to the **quadratic Koch island**, a variation of the Koch curve defined by the following recursive replacement rule:*

- Start with a square: ⬚

- Whenever you see a horizontal segment _____, replace it with a ⌐L⌐ ,

  and whenever you see a vertical segment ⎸ , replace it with ⌐⌐ .

9. Carefully draw the figures at steps 1 and 2 of the construction of the quadratic Koch island. (*Hint:* Use graph paper and start with an 8-by-8 or 16-by-16 square.)

10. Determine the number of sides of
    (a) the figures at steps 1, 2, and 3 of the construction of the quadratic Koch island.
    (b) the figure at step N of the construction of the quadratic Koch island.

11. Suppose that the starting square has perimeter P. Find the perimeter (expressed in terms of P) of
    (a) each of the figures at steps 1, 2, and 3 of the construction of the quadratic Koch island.
    (b) the figure at step N of the construction of the quadratic Koch island.
    (c) the quadratic Koch island.

12. If the starting square encloses an area X, find the area enclosed by
    (a) the figure at step 1 of the construction of the quadratic Koch island.
    (b) the figure at step 2 of the construction of the quadratic Koch island.
    (c) the figure at step N of the construction of the quadratic Koch island.
    (d) the quadratic Koch island.

B. The Sierpinski Gasket and Variations

Exercises 13 and 14 refer to the construction of a Sierpinski gasket with a starting triangle of sides 3, 4, and 5. The starting triangle and the figure obtained in step 1 of the construction are shown in the following figure.

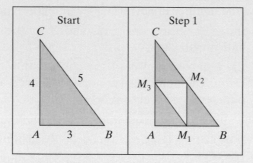

13. Find the *area* of
    (a) triangle ABC. (*Hint: ABC* is a special triangle.)
    (b) triangle $M_1 AM_3$. Explain the significance of the answer.
    (c) triangle $M_1 M_2 M_3$.
    (d) the shaded region in step 1.

14. Find the *perimeter* of
    (a) triangle $M_1 AM_3$.
    (b) triangle $M_1 M_2 M_3$.
    (c) the shaded region in step 1.

*Exercises 15 and 16 refer to the construction of an **equilateral Sierpinski gasket**— that is, a Sierpinski gasket constructed from a starting triangle that is equilateral.*

15. Carefully draw the figures at steps 1, 2, and 3 of the construction of an equilateral Sierpinski gasket.

16. If the area of the starting equilateral triangle is A,
    (a) find the areas of the figures at steps 1, 2, and 3 of the construction of the equilateral Sierpinski gasket.
    (b) find the area of the figure at step N of the construction of the equilateral Sierpinski gasket.
    (c) Explain why the area of the equilateral Sierpinski gasket is 0.

*Exercises 17 through 20 refer to the **Sierpinski carpet** (or Sierpinski square), a variation of the Sierpinski gasket. The Sierpinski carpet is defined by the following recursive procedure.*

- **Start.** *Start with a solid square.*
- **Step 1. Procedure SC:** *Subdivide the square into nine equal subsquares and remove the central subsquare.*
- **Steps 2, 3, etc.** *On every remaining solid square, repeat procedure SC ad infinitum.*

The starting square and the figure obtained at Step 1 of the construction of the Sierpinski carpet are shown in the following figure.

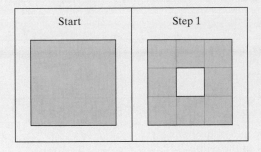

17. Using graph paper, carefully draw the figures at steps 2 and 3 of the construction of the Sierpinski carpet.

18. Complete the following table.

	Start	Step 1	Step 2	Step 3	...	Step N
Number of square holes	0	1				

19. Suppose the starting square has sides of length 1.

    (a) Complete the following table. (Enter your answers as fractions.)

	Start	Step 1	Step 2	Step 3	Step 4
Length of the boundary of the figure	4	16/3			

    (b) If the length of the boundary of the figure at step N is given by L, give the length of the boundary of the figure at step $N + 1$ in terms of L.

20. Suppose that the area of the starting square is 1.

    (a) Complete the following table.

	Start	Step 1	Step 2	Step 3	Step 4
Area of the figure	1	8/9			

    (b) If the area of the figure at step N is given by A, give the area of the figure at step $N + 1$ in terms of A.

*Exercises 21 through 24 refer to a variation of the Sierpinski gasket that we will call the **triplet gasket**. The triplet gasket is defined by the following recursive replacement rule.*

    ■ *Start with a solid equilateral triangle* ▲ .

    ■ *Whenever you see a* ▲ *, replace it with a* ▲▲ .

21. Complete the following table.

	Start	Step 1	Step 2	Step 3	...	Step N
Number of solid triangles	1	6				

22. Suppose the perimeter of the starting triangle is P. Complete the following table.

	Start	Step 1	Step 2	Step 3
Length of the boundary of the figure	P			

23. Suppose the area of the starting triangle is A. Complete the following table.

	Start	Step 1	Step 2	Step 3
Area of the figure	A	$(2/3)A$		

24. Suppose that the starting equilateral triangle has sides of length 1.

(a) Give the length of the boundary of the figure at step N in terms of N. (*Hint:* Do Exercise 22 first.)

(b) Give the area of the figure at step N in terms of N. (*Hint:* Do Exercise 23 first, and use the fact that the area of the starting equilateral triangle is $\dfrac{\sqrt{3}}{4}$.)

(c) Using a calculator with an exponent key, find the areas of the figures at steps 10, 15, 20, and 25 of the construction of the triplet gasket rounded to six decimal places. What is your conclusion?

C. The Chaos Game and Variations

Exercises 25 through 28 refer to the chaos game as described in the chapter. Start with an isosceles right triangle ABC with AB = AC = 32, as shown in the figure on p. 504. You should use graph paper with 10 squares per inch or make a copy of the figure and work directly on it. Assume that vertex A corresponds to numbers 1 and 2, vertex B to numbers 3 and 4, and vertex C to numbers 5 and 6.

25. Suppose that an honest die is rolled 6 times and that the outcomes are 3, 1, 6, 4, 5, and 5. Carefully draw the points P_1 through P_6 corresponding to these outcomes. (*Note:* Each of the points P_1 through P_6 falls on a grid point of the graph. You should be able to identify the location of each point without using a ruler.)

26. Suppose that an honest die is rolled 6 times and that the outcomes are 2, 6, 1, 4, 3, and 6. Carefully draw the points P_1 through P_6 corresponding to these outcomes. (*Note:* Each of the points P_1 through P_6 falls on a grid point of the graph. You should be able to identify the location of each point without using a ruler.)

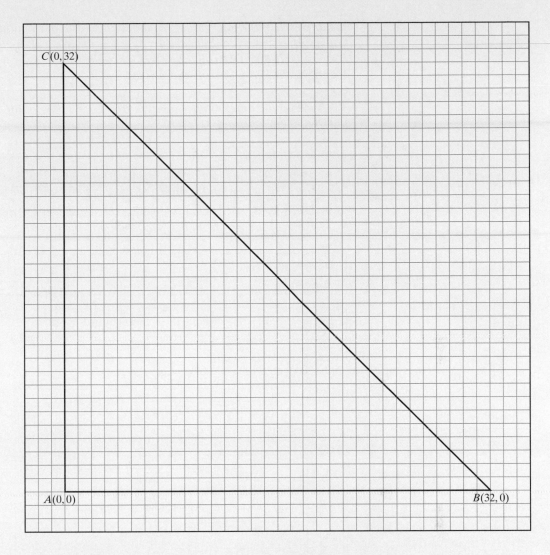

27. Using a rectangular coordinate system with A at $(0,0)$, B at $(32,0)$, and C at $(0,32)$, complete the following table.

Number rolled	3	1	2	3	5	5
Point	P_1	P_2	P_3	P_4	P_5	P_6
Coordinates	(32,0)	(16,0)				

28. Using a rectangular coordinate system with A at $(0,0)$, B at $(32,0)$, and C at $(0,32)$, complete the following table.

Number rolled	2	6	5	1	3	6
Point	P_1	P_2	P_3	P_4	P_5	P_6
Coordinates	(0,0)	(0,16)				

Exercises 29 through 32 refer to a game that is a variation of the chaos game discussed in the chapter. When played a large number of times, the set of points generated by this game approximates a Sierpinski carpet. (See Exercises 17 through 20.) Here we start with a square ABCD, such as the one shown in the figure. We need to identify each of the four vertices of the square with four equally likely random outcomes. An easy way to do this is to roll a fair die. We will say that A is the "winner" if we roll a 1, B is if we roll a 2, C is if we roll a 3, and D is if we roll a 4. If we roll a 5 or 6, we disregard the roll and roll again.

Each roll of the die generates a point inside or on the boundary of the square according to the following rules.

- **Start.** *Roll the die. Mark the "winning" vertex and call it P_1.*
- **Step 1.** *Roll the die again. From P_1 move two-thirds of the way straight toward the next winning vertex. Mark this point and call it P_2.*
- **Steps 2, 3, etc.** *Continue rolling the die, each time moving to a point two-thirds of the way from the last position to the winning vertex.*

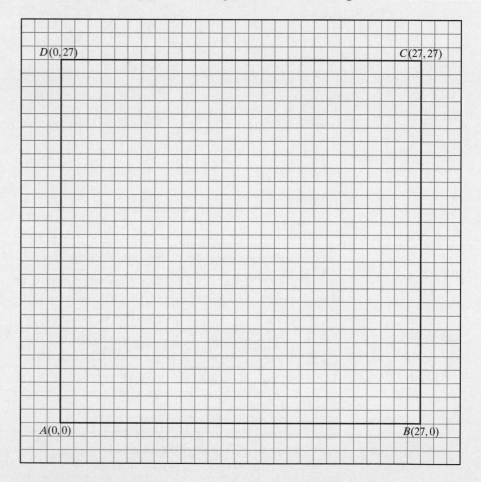

29. Using graph paper, carefully locate the points P_1, P_2, P_3, and P_4 corresponding to
    (a) the sequence of rolls $4, 2, 1, 2$.
    (b) the sequence of rolls $3, 2, 1, 2$.
    (c) the sequence of rolls $3, 3, 1, 1$.

30. Using graph paper, carefully locate the points P_1, P_2, P_3, and P_4 corresponding to
    (a) the sequence of rolls $2, 2, 4, 4$.

(b) the sequence of rolls 2, 3, 4, 1.

(c) the sequence of rolls 1, 3, 4, 1.

31. Using a rectangular coordinate system with A at $(0,0)$, B at $(27,0)$, C at $(27,27)$, and D at $(0,27)$, find the coordinates of the points P_1, P_2, P_3, and P_4 corresponding to

(a) the sequence of rolls 4, 2, 1, 2.

(b) the sequence of rolls 3, 1, 1, 3.

(c) the sequence of rolls 1, 3, 4, 2.

32. Using a rectangular coordinate system with A at $(0,0)$, B at $(27,0)$, C at $(27,27)$, and D at $(0,27)$, find the coordinates of the points P_1, P_2, P_3, and P_4 corresponding to

(a) the sequence of rolls 2, 3, 4, 1.

(b) the sequence of rolls 4, 2, 2, 4.

(c) the sequence of rolls 3, 1, 2, 4.

D. Mandelbrot Sequences

Exercises 33 through 40 refer to Mandelbrot sequences as discussed in the chapter.

33. Consider the Mandelbrot sequence with seed $s = -2$.

(a) Find s_1, s_2, s_3, and s_4.

(b) Find s_{100}.

(c) Is this Mandelbrot sequence *escaping, periodic*, or *attracted*? Explain.

34. Consider the Mandelbrot sequence with seed $s = 2$.

(a) Find s_1, s_2, s_3, and s_4.

(b) Is this Mandelbrot sequence *escaping, periodic*, or *attracted*? Explain.

35. Consider the Mandelbrot sequence with seed $s = -0.5$.

(a) Using a calculator, find s_1 through s_5, rounded to four decimal places.

(b) Suppose you are given $s_N = -0.366$. Using a calculator, find s_{N+1}, rounded to four decimal places.

(c) Is this Mandelbrot sequence *escaping, periodic*, or *attracted*? Explain.

36. Consider the Mandelbrot sequence with seed $s = -0.25$.

(a) Using a calculator, find s_1 through s_{10}, rounded to six decimal places.

(b) Suppose you are given $s_N = -0.207107$. Using a calculator, find s_{N+1}, rounded to six decimal places.

(c) Is this Mandelbrot sequence *escaping, periodic*, or *attracted*? Explain.

37. Consider the Mandelbrot sequence with seed $s = 1/2$.

(a) Find s_1, s_2, and s_3 without using a calculator. Give the answers in fractional form.

(b) Suppose that $s_N > 1$. Explain why this implies that $s_{N+1} > s_N$.

(c) Is this Mandelbrot sequence *escaping, periodic*, or *attracted*? Explain.

38. Consider the Mandelbrot sequence with seed $s = -1.75$.

(a) Using a calculator, find s_1 through s_{12}, rounded to five decimal places.

(b) Is this Mandelbrot sequence *escaping, periodic*, or *attracted*? Explain.

39. Suppose $s_N = 6$, and $s_{N+1} = 38$ are two consecutive terms of a Mandelbrot sequence.

    (a) Find the seed s.

    (b) If $s_N = 6$, what is the value of N? [*Hint:* Use the seed you found in (a).]

40. Suppose $s_N = -15/16$, and $s_{N+1} = -159/256$ are two consecutive terms of a Mandelbrot sequence.

    (a) Find the seed s.

    (b) If $s_N = -15/16$, what is the value of N? [*Hint:* Use the seed you found in (a).]

JOGGING

Exercises 41 through 44 refer to the construction of the Sierpinski gasket discussed in the chapter, as well as in Exercises 13 through 16.

41. Suppose that the area of the starting triangle ABC is 1.

    (a) Give the area of the figure at step N in terms of N.

    (b) Use your answer in (a) and a calculator with an exponent key to find the area, rounded to four decimal places, of the figures at steps $10, 20$, and 40 of the construction of the Sierpinski gasket.

    (c) Explain why the Sierpinski gasket has zero area.

42. Suppose that the perimeter of the starting triangle ABC is P.

    (a) Find the lengths of the boundaries of the figures at steps $1, 2$, and 3 of the construction of the Sierpinski gasket.

    (b) Given that the length of the boundary of the figure at step N is L, express the length of the boundary of the figure at step $N + 1$ in terms of L.

    (c) Give the length of the boundary of the figure at step N in terms of N.

    (d) Explain why the Sierpinski gasket has an infinitely long boundary.

43. Explain why the construction of the Sierpinski gasket does not end up in an all-white triangle.

44. The total number of white triangles at step N of the construction is given by $(3^N - 1)/2$. Explain how this formula can be derived. (*Hint:* You need to use the formula for the sum of the terms in a geometric sequence given in Chapter 10.)

Exercises 45 and 46 refer to the Menger sponge, *a three-dimensional cousin of the Sierpinski carpet of Exercises 17–20. The* **Menger sponge** *is defined by the following recursive procedure.*

- **Start.** *Start with a solid cube.*
- **Step 1. Procedure MS:** *Subdivide the cube into 27 equal subcubes and remove the cube in the center and the 6 cubes in the centers of each face.*
- **Steps 2, 3, etc.** *On every remaining solid cube, repeat procedure MS ad infinitum.*

The starting cube and the figures obtained at steps 1, 2, and 3 of the construction of the Menger sponge are shown in the following figure.

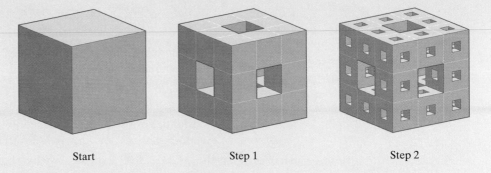

Start   Step 1   Step 2

45. Complete the following table.

	Start	Step 1	Step 2	Step 3	...	Step N
Number of cubes removed	0	7				

46. Suppose the starting cube has sides of length 1.

(a) Complete the following table.

	Start	Step 1	Step 2	Step 3	...	Step N
Volume of the figure	1	20/27				

(b) Explain why the Menger sponge has zero volume.

(c) Calculate the surface area at the start and at step 1 in the construction of the Menger sponge.

(d) Explain why the surface area at each step of the construction of the Menger sponge increases.

Exercises 47 through 49 refer to reflection and rotation symmetries as discussed in Chapter 11, and thus require a good understanding of the material in that chapter.

47. Consider a Sierpinski gasket constructed on an equilateral triangle *ABC*.

(a) Describe all the reflection symmetries of the gasket.

(b) Describe all the rotation symmetries of the gasket.

(c) What is the symmetry type of the Sierpinski gasket?

48. Consider the Koch snowflake as discussed in the chapter.

(a) Describe all the reflection symmetries of the snowflake.

(b) Describe all the rotation symmetries of the snowflake.

(c) What is the symmetry type of the Koch snowflake?

49. Consider the Sierpinski carpet discussed in Exercises 17 through 20. What is its symmetry type?

50. Use the formula for adding consecutive terms of a geometric sequence (see Chapter 10) to show that

(a) $1 + \left(\dfrac{4}{9}\right) + \left(\dfrac{4}{9}\right)^2 + \cdots + \left(\dfrac{4}{9}\right)^{N-1} = \dfrac{9}{5}\left[1 - \left(\dfrac{4}{9}\right)^N\right].$

(b) $\left(\dfrac{1}{3}\right)A + \left(\dfrac{4}{9}\right)\left(\dfrac{1}{3}\right)A + \left(\dfrac{4}{9}\right)^2\left(\dfrac{1}{3}\right)A + \cdots + \left(\dfrac{4}{9}\right)^{N-1}\left(\dfrac{1}{3}\right)A$

$= \dfrac{3}{5}A\left[1 - \left(\dfrac{4}{9}\right)^N\right].$

Exercises 51 through 56 are about Mandelbrot sequences. (You will need a calculator.)

51. Consider the Mandelbrot sequence with seed $s = -0.75$. Toward what number is this Mandelbrot sequence attracted?

52. Consider the Mandelbrot sequence with seed $s = 0.2$. Is this Mandelbrot sequence *escaping, periodic,* or *attracted?* If attracted, to what number?

53. Consider the Mandelbrot sequence with seed $s = 0.25$. Is this Mandelbrot sequence *escaping, periodic,* or *attracted?* If attracted, to what number?

54. Consider the Mandelbrot sequence with seed $s = -1.25$. Is this Mandelbrot sequence *escaping, periodic,* or *attracted?* If attracted, to what number?

55. Consider the Mandelbrot sequence with seed $s = \sqrt{2}$. Is this Mandelbrot sequence *escaping, periodic,* or *attracted?* If attracted, to what number?

56. Consider the Mandelbrot sequence with seed $s = -\sqrt{2}$. Is this Mandelbrot sequence *escaping, periodic,* or *attracted?* If attracted, to what number?

RUNNING

57. This exercise refers to the Koch antisnowflake discussed in Exercises 5 through 8. Suppose the area of the starting triangle is A. Find the area of the Koch antisnowflake.

58. This exercise refers to the square snowflake discussed in Exercises 1 through 4. Suppose the area of the starting square is A. Find the area of the square snowflake.

59. Suppose that we play the chaos game using triangle ABC and that M_1, M_2, and M_3 are the midpoints of the three sides of the triangle. Explain why it is impossible at any time during the game to land inside triangle $M_1M_2M_3$.

60. Find the first few terms (as many as necessary) of the Mandelbrot sequence for the seed

  (a) $s = 1 + i$. Is s in the Mandelbrot set or not?

  (b) $s = 1 - i$. Is s in the Mandelbrot set or not?

61. Find the first few terms (as many as necessary) of the Mandelbrot sequence for the seed

  (a) $s = -0.25 + 0.25i$. Is s in the Mandelbrot set or not?

  (b) $s = -0.25 - 0.25i$. Is s in the Mandelbrot set or not?

62. Show that the Mandelbrot set has a reflection symmetry. (*Hint:* See Exercises 60 and 61.)

63. Consider the Menger sponge as discussed in Exercises 45 and 46. Suppose the starting cube has sides of length 1 (giving a surface area of 6).

  (a) Compute the surface area at step 1 of the construction of the Menger sponge. (*Hint:* There are two different types of subcubes: 8 corner cubes and 12 side cubes.)

  (b) At step 2 of the construction, $20 \cdot 6$ cubes having surface area are removed. Each such cube changes from contributing at most $(1/9)^2$ to the surface area, to four times that much. The net effect is that each of the $20 \cdot 6$ removed cubes contributes at least $3 \cdot (1/9)^2$ to the surface area. Use this to determine a lower bound for the surface area of the object in step 2.

(c) At step 3 of the construction, $20^2 \cdot 6$ cubes having surface area are removed. Each such cube changes from contributing at most $(1/27)^2$ to the surface area, to four times that much. Use (b) to determine a lower bound for the surface area at step 3.

(d) Argue that at step $N(N \geq 2)$ of the construction, at least $\dfrac{2 \cdot 20^{N-1}}{9^{N-1}} \geq 4$ is being added to the surface area. Hence, the surface area of the Menger sponge is infinite.

PROJECTS AND PAPERS

A. The Dimension of a Fractal

How does one measure the dimension of an object? For example, the dimensions of the line segment, square, and cube in (a), (b), and (c) are, as we all learned in school, 1, 2, and 3, respectively. But what is the dimension of the Sierpinski gasket in (d)?

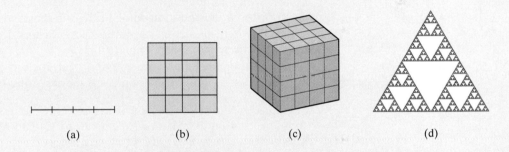

(a)   (b)   (c)   (d)

Looking at the dimensions of the objects in (a), (b), and (c) produces a pattern: $4 = 4^1$ (a segment of size 4 consists of 4^1 segments of size 1); $16 = 4^2$ (a square of size 4 consists of 4^2 squares of size 1); and $64 = 4^3$ (a cube of size 4 consists of 4^3 cubes of size 1). Thus, we can see that dimension is an exponent, specifically the exponent D in the equation $N = S^D$, where N is the number of self-similar pieces and S is the scale in which they compare. In the case of the Sierpinski gasket in (d), $N = 3$ and $S = 2$. Solving for D in the equation $3 = 2^D$ (you have to use *logarithms* to do so) gives the value $D = \dfrac{\log 3}{\log 2} \approx 1.585$. Thus, the Sierpinski gasket has a dimension that is an irrational number, approximately 1.585.

In this project, you prepare a presentation discussing the concept of dimension as it applies to fractal shapes. You should explain in greater detail how to calculate the dimension of a fractal shape and illustrate the calculations with several examples, including the *Koch snowflake*, the *Sierpinski carpet* (see Exercises 17–20), and the *Menger sponge* (see Exercises 45 and 46).

B. Fractals and Music

The hallmark of a fractal shape is the property of *self-similarity*—the fact that there are themes that repeat themselves (either exactly or approximately) at many different scales. This type of repetition also works in music, and the application of fractal concepts to musical composition has produced many intriguing results.

Write a paper discussing the connections between fractals and music.

Suggested readings:
- "J. S. Bach + Fractals = New Music," by Malcolm W. Browne, *New York Times*, April 16, 1991. (You can find a copy of this article at this book's Web site, *http://cwx.prenhall.com/bookbind/pubbooks/tannenbaum/chapter12.*)
- "Bach to Chaos," by Ivars Peterson, *Science News*, Dec. 24–31, 1994, pp. 428–429.
- "Art and Biology Give Songs from the Heart," by Malcolm W. Browne, *New York Times*, Feb. 12, 1996. (A copy of this article is available at this book's Web site, *http://cwx.prenhall.com/bookbind/pubbooks/tannenbaum/chapter12.*)

C. Book Review: *The Fractal Murders*

If you enjoy mystery novels, this project is for you.

The Fractal Murders, by Mark Cohen (Muddy Gap Press, 2002) is a *who-done-it* with a mathematical backdrop. In addition to the standard elements of a classic murder mystery (including brilliant but eccentric detective) this novel has a fractal twist: The victims are all mathematicians doing research in the field of fractal geometry.

Read the novel and write a review of it. Include in your review a critique of both the literary and the mathematical merits of the book.

Suggested reading: To get some ideas as to how to write a good book review, you should check out the *New York Times Book Review* section, which appears every Sunday in the *New York Times* (*www.nytimes.com*).

D. Fractal Antennas

One of the truly innovative practical uses of fractals is in the design of small but powerful antennas that go inside wireless communication devices such as cell phones, wireless modems, and GPS receivers. The application of fractal geometry to antenna design follows from the discovery in 1999 by radio astronomers Nathan Cohen and Robert Hohlfeld of Boston University that an antenna that has a self-similar shape has the ability to work equally well at many different frequencies of the radio spectrum.

Write a paper discussing the application of the concepts of fractal geometry to the design of antennas.

REFERENCES AND FURTHER READINGS

1. Berkowitz, Jeff, *Fractal Cosmos: The Art of Mathematical Design*. Oakland, CA: Amber Lotus, 1998.
2. Briggs, John, *Fractals: The Patterns of Chaos*. New York: Touchstone Books, 1992.
3. Cohen, Mark, *The Fractal Murders*. Boulder, CO: Muddy Gap Press, 2002.

4. Dewdney, A. K., "Computer Recreations: A computer microscope zooms in for a look at the most complex object in mathematics," *Scientific American*, 253 (August 1985), 16–24.

5. Dewdney, A. K., "Computer Recreations: A tour of the Mandelbrot set aboard the Mandelbus," *Scientific American*, 260 (February 1989), 108–111.

6. Dewdney, A. K., "Computer Recreations: Beauty and profundity. The Mandelbrot set and a flock of its cousins called Julia," *Scientific American*, 257 (November 1987), 140–145.

7. Flake, Gary W., *The Computational Beauty of Nature: Computer Explorations of Fractals, Chaos, Complex Systems, and Adaptation*. Boston, MA: MIT Press, 2000.

8. Gleick, James, *Chaos: Making a New Science*. New York: Viking Penguin, Inc., 1987, Chap. 4.

9. Hastings, Harold, and G. Sugihara, *Fractals: A User's Guide for the Natural Sciences*. New York: Oxford University Press, 1995.

10. Jurgens, H., H. O. Peitgen, and D. Saupe, "The Language of Fractals," *Scientific American*, 263 (August 1990), 60–67.

11. Mandelbrot, Benoit, *The Fractal Geometry of Nature*. New York: W. H. Freeman & Co., 1983.

12. Musser, George, "Practical Fractals," *Scientific American*, 281 (July 1999), 38.

13. Peitgen, H. O., H. Jurgens, and D. Saupe, *Chaos and Fractals: New Frontiers of Science*. New York: Springer-Verlag, Inc., 1992.

14. Peitgen, H. O., H. Jurgens, and D. Saupe, *Fractals for the Classroom*. New York: Springer-Verlag, Inc., 1992.

15. Peitgen, H. O., and P. H. Richter, *The Beauty of Fractals*. New York: Springer-Verlag, Inc., 1986.

16. Peterson, Ivars, *The Mathematical Tourist*. New York: W. H. Freeman & Co., 1988, Chap. 5.

17. Schechter, Bruce, "A New Geometry of Nature," *Discover*, 3 (June 1982), 66–68.

18. Schroeder, Manfred, *Fractals, Chaos, Power Laws: Minutes from an Infinite Paradise*. New York: W. H. Freeman & Co., 1991.

19. Wahl, Bernt, *Exploring Fractals on the Macintosh*. Boston, MA: Addison-Wesley, 1994.

PART FOUR
Statistics

www.census.gov

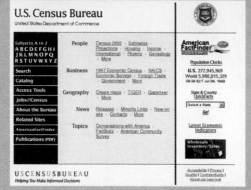

Fast, Free, and Easy Access for All Information Users!

Whether you're a business person *looking* for new markets... a community leader *developing* after-school youth programs... an urban planner or housing developer *tracking* population trends... an economist *projecting* inflation... a journalist *gathering* facts for a story... a student *researching* a paper... or a census survey respondent *learning* how the data collected are used, there's valuable information at our web site — for everyone with access to the Internet.

And, there's no password required!

Critical Information Organized the Way You Want It

Locating the information you need from the wealth of social, demographic, and economic statistics available from the U.S. Census Bureau has never been simpler!

The newly streamlined site features:

- clear organization
- consistent navigation
- mapping capability
- faster page loading
- links to other statistical agencies and other related sites

The newly redesigned U.S. Census Bureau web site is *arranged by topic* so you can go straight to the information you want. Under each category includes topics such as:

People — Census 2000, Population, Income, Housing, and Poverty Statistics

Business — Indicators, Classifications, and Foreign Trade Numbers

Geography — Maps, TIGER Products, Gazetteer, and Census Geographic Concepts

News — Press releases on Census 2000, latest Tipsheets, *Census Briefs*, radio broadcasts, and upcoming events

Topics — New items of interest

Thousands of Statistics & Hundreds of Topics From Which to Choose!

Once logged on to the U.S. Census Bureau's web site, novice and experienced data miners alike will find:

- immediate access to an "A to Z" subject index
- timely U.S. and world population information (updated every 5 minutes)
- latest economic indicators
- state and county profiles
- online product catalog
- information about the Census Bureau

...and **Access Tools** such as:

American FactFinder — Our NEW interactive database engine for data from the latest Economic Census... the American Community Survey... the 1990 Census... Census 2000 Dress Rehearsal... and eventually, Census 2000 as information is released!

CHAPTER 13

Collecting Statistical Data

Censuses, Surveys, and Clinical Studies

Some 60 years ago, the great novelist and science fiction writer H. G. Wells predicted: "Statistical reasoning will one day be as necessary for efficient citizenship as the ability to read and write." That day has arrived with a vengeance. Information has become the primary currency of the 21st century, and, by and large, wherever there is information, statistics are not far behind. Open today's paper and look at the sports page—plenty of stats there, no question about that. Don't like sports? Check the health section, or the business section, or the weather news. All of them are spiked with statistics.

S tatistically speaking, today's world is a jungle. Our goal in this part of the book is to introduce and discuss the basic tools of statistical literacy needed to safely and efficiently move through this jungle—what H. G. Wells called "statistical reasoning ... for efficient citizenship."

What is *statistics*? Statistics is, in a sense the blending of two fundamental skills we all learn separately in school: handling information and manipulating numbers. When we put our numerical abilities at the service of our communication needs, we are doing some form of statistical work. Here is a slightly more formal description. When information is packaged in numerical form, it is called **data**, and to put it in a nutshell, *statistics* is the science of dealing with data. This includes gathering data, organizing data, interpreting data, and understanding data. We will discuss all of these things in the next four chapters.

> "Data! Data! Data!" he cried impatiently. "I can't make bricks without clay."
>
> Sherlock Holmes

Behind every statistical statement there is a story, and like any story, it has a beginning, a middle, an end, and a moral. In this chapter we will discuss the beginning of the story, which, in statistics, typically means the process of gathering or collecting data. Data are the raw material of which statistical information is made, and in order to get good statistical information one needs good data.

Collecting data seems deceptively simple. However, doing so in an efficient and timely manner is often the most difficult part of the statistical story. This chapter illustrates the do's and don'ts of gathering data. We will do this in two ways: simple examples and in-depth case studies.

13.1 The Population

Every statistical statement refers, directly or indirectly, to some group of individuals or objects. In statistical terminology, this collection of individuals or objects is called the **population**. The first question we should ask ourselves when trying to make sense of a statistical statement is, "What is the population to which the statement applies?" If we are lucky, the population is clearly defined and we are off to a good start. Most often, this happens with statistical statements that are very specific and direct.

EXAMPLE 13.1

GRAY-WHALE NUMBERS SHRINK BY THOUSANDS

By Eric Sorensen

The population of Eastern North Pacific gray whales has dropped in the past four years from an estimated high of more than 26,000 to less than 18,000. ... The 1997–1998 estimate of 26,700 is often quoted as the current population but that number has decreased since then.

The figure is calculated by counting whales as they migrate southward from Alaska to Mexico past Central California between December and February. But the counts have to be adjusted to account for whales passing at night, times when they can't be easily seen from shore, their distances from shore and other factors. ...

Seattle Times, May 18, 2002

This is a story about a very clearly defined population—the population of Eastern North Pacific gray whales. Because of their enormous size and predictable migratory patterns, whales are easier to count that most other animal populations and reasonably accurate population figures are possible. By the same token, the Pacific is a big ocean, and a complete accounting of every whale is out of the question. This explains the need for *adjustments* to the counts. ∎

The *N*-Value

If one were able to make an accurate head count of every member of a population, one would get a whole number *N*, sometimes informally called the *N-value* for the population. It's important to remember that the *N*-value is just one specific measurement of a population, not to be confused with the population itself.

As we learned in Chapter 10, populations change, and thus *N*-values must often be discussed within the context of time. Example 13.1 illustrates this point well. The *N*-value of the Eastern North Pacific gray whale population in 1997 was not the same as the *N*-value of the same population in 2002.

The next example illustrates the fact that in some situations, there can be different plausible interpretations of what the population (and thus its *N*-value) is.

EXAMPLE 13.2

A child has a coin jar full of quarters. He is hoping that there is enough money to buy a new baseball glove. Dad says to go count them, and if there isn't enough, he will make up the difference. The child comes back with a count of 116 quarters. So what is the *N*-value here? The answer depends on how we define our population.

To Dad, who will probably end up stuck with all the quarters, the total number of coins might be the most relevant issue. In this case, we might say that the N-value is $N = 116$. To the child, the most relevant issue is the total amount of money in the jar. From his point of view, it is the $29 in the jar that matters the most, and we might say then that $N = 29$.   ■

Knowing the size of a population is often an important part of drawing reliable statistical conclusions, but it is not always as easy as it was in Example 13.2 to determine the N-value of a population. In Example 13.1 we saw that even in the relatively simple case of counting whales (which are not only hard to miss because of their size but also cooperate by migrating along established routes), only an estimate of the N-value was possible.

The most notoriously difficult N-value question around is: What is the N-value of the national population of the United States? This is exactly the question the United States Census is meant to answer every ten years. Unfortunately, in modern American society, an accurate head count of the entire nation is an essentially impossible task, in spite of the tremendous expense and effort the government puts into it. In fact, the U.S. Census is the largest and most expensive peace-time effort taken on by the federal government (the 2000 Census employed over 850,000 people and cost about 6.5 billion dollars) and yet, as we will see in the next example, it missed between 3 and 4 million people.

EXAMPLE 13.3

POLITICAL FIGHT BREWS OVER CENSUS CORRECTION

By Haya El Nasser

The 2000 Census did a better job counting people than the last Census, especially minorities and children, but it still missed about 2.7 million to 4 million people. ...

Preliminary estimates show that the net number of people missed falls between 0.96% and 1.4%. In 1990, the undercount was 1.6%, or 4 million people. There was a significant drop in the undercount of blacks, Hispanics, American Indians and children, population groups that were disproportionately missed in 1990. ...

The estimates are bound to heat up political infighting. The Census Bureau must decide whether the numbers should be adjusted to compensate for the undercount. ...

Census numbers are used to redraw political districts. An adjusted count would include more minorities, which could reshape key political districts. Republicans worry an adjusted count would help Democrats. ...

The Census Bureau estimates the number of people missed through the same method that would be used to adjust the numbers. It surveys 314,000 sample households and checks to see whether those households filled out Census forms.

New York Times, February 15, 2001

Given the critical importance of the U.S. Census and given the tremendous resources put behind the effort by the federal government, why is the head count so far off? How can the best intentions and tremendous resources of our government fail so miserably in an activity that on a smaller scale can be carried out by a single child (see Example 13.2)? Our first case study gives a brief overview of the ins and outs of the modern U.S. Census and of the tremendous complexity of accurately counting our population.   ■

CASE STUDY 1 THE U.S. CENSUS

Article 1, Section 2, of the Constitution of the United States mandates that a national census be conducted every 10 years. The original intent of the census was to "count heads" for a twofold purpose: taxes and political representation. Like everything else in the Constitution, Article 1, Section 2, was a compromise of many competing interests: The count was to exclude "Indians not taxed" and to count slaves as "three-fifths of a free Person." Since then, the scope and purpose of the U.S. Census has been modified and expanded by the 14th Amendment and the courts in many ways:

- Besides counting heads, the U.S. Census Bureau now collects additional information about the population: sex, age, race, ethnicity, marital status, housing, income, and employment data. Some of this information is updated on a regular basis, not just every 10 years.

- Census data are now used for many important purposes beyond its original ones of *taxation* and *representation*: the allocation of billions of federal dollars to states, counties, cities, and municipalities; the collection of other important government statistics such as the Consumer Price Index and the Current Population Survey; the redrawing of legislative districts within each state; and the strategic planning of production and services by business and industry.

- For the purposes of the census, the United States population is defined as consisting of "all persons *physically present* and *permanently residing* in the United States." Citizens, legal resident aliens, and even illegal aliens are meant to be included.

Nowadays, the notion that if we put enough money and effort into it, all individuals living in the United States can be counted like coins in a jar is unrealistic. In 1790, when the first U.S. Census was carried out, the population was smaller and relatively homogeneous, as people tended to stay in one place, and, by and large, they felt comfortable in their dealings with the government. Under these conditions it might have been possible for census takers to accurately count heads. Today's conditions, are completely different. People are constantly on the move. Many distrust the government. In large urban areas many people are homeless or don't want to be counted. And then there is the apathy of many people who think of a census form as another piece of junk mail.

If the Census undercount was consistent among all segments of the population, the undercount problem could be easily solved. Unfortunately, the modern U.S. Census is plagued by what is known as a *differential undercount*. Ethnic minorities, migrant workers, and the urban poor populations have significantly larger undercount rates than the population at large, and the undercount rates vary significantly within these groups.

Using modern statistical techniques, it is possible to make adjustments to the raw Census figures that correct some of the inaccuracy caused by the differential undercount. However, in 1999, the Supreme Court ruled that only the raw numbers (and not statistically adjusted numbers) can be used for the purposes of apportionment of Congressional seats among the states,[1] insuring that the futile attempt to conduct a full head count will continue in the future.

---

[1]Department of Commerce et al. *v.* United States House of Representatives et al., 1999.

13.2 Surveys

A much more economical alternative to collecting data from each and every member of a population is to collect data only from a selected subgroup and then to use these data to draw conclusions and make statistical inferences about the entire population. Statisticians call this approach a **survey**. A survey starts with a **sampling frame**, from which a subgroup called a **sample** is chosen and data are collected. Ideally, the sampling frame is the same as the target population. In reality, however, there are often differences. For example, to collect data on the population of San Francisco, the city phone directory could be used as the sampling frame. The sample might consist of all phone numbers ending with the digits "75."

The basic idea behind a survey is simple and well understood. If we have a sample that is "representative" of the entire population, then whatever we want to know about a population can be found out by getting the information from the sample.

Implementing a survey is far from simple, however. The critical issues are (a) finding a sample that is representative of the population, and (b) determining how big the sample should be. These two issues go hand in hand, and we will discuss them next.

Sometimes, a very small sample can be used to get reliable information about a population, no matter how large the population is. This is the case when the population is highly homogeneous. For example, with the exception of identical twins, every person's DNA is different. Yet, the DNA sampled from just one human cell is sufficient to characterize and identify all the DNA of that individual. Similarly, a person's blood is essentially the same everywhere in the body, which explains why a small blood sample drawn from an arm can be used to draw conclusions about about all of the patient's blood.

The more heterogeneous a population gets, the more difficult it is to find a representative sample. The perils and difficulties of surveys of large, heterogeneous populations can be illustrated by examples from the history of *public opinion polls*.

Public Opinion Polls

We are all familiar with public opinion polls, such as the Gallup poll, the Harris poll, and many others. A public opinion poll is a survey in which the members of the sample provide information by answering specific questions from an "interviewer." The question–answer exchange can be done through a questionnaire, a personal telephone interview, or a direct face-to-face interview.

Nowadays, public opinion polls are used regularly to measure "the pulse of the nation." They give us statistical information ranging from voters' preferences before an election to opinions on issues such as the environment, abortion, and the economy.

Given their widespread use and the influence they exert, it is important to ask how much we can trust the information that we get from public opinion polls. This is a complex question that goes to the very heart of mathematical statistics. We'll start our exploration of it with some historical examples.

CASE STUDY 2 THE 1936 *LITERARY DIGEST* POLL

The U.S. presidential election of 1936 pitted Alfred Landon, the Republican governor of Kansas, against the incumbent Democratic President, Franklin D. Roosevelt. At the time of the election, the nation had not yet emerged from the Great Depression, and economic issues such as unemployment and government spending were the dominant themes of the campaign.

The *Literary Digest*, one of the most respected magazines of the time, conducted a poll a couple weeks before the election. The magazine had been polling the electorate since 1916, always accurately predicting the results of the election. The 1936 poll, based on a huge sample of approximately 2.4 million people, was one of the largest and most expensive polls over conducted. The sample for the *Literary Digest* poll was chosen by putting together, in one enormous list, the names of every person listed in a telephone directory anywhere in the United States, as well as the names of people on magazine subscription lists and rosters of clubs and professional associations. Altogether, a mailing list of about 10 million names was created. Every name on this list was mailed a mock ballot and asked to mark it and return it to the magazine.

One cannot help but be impressed by the sheer ambition of such a project, and it is not surprising that the magazine's confidence in the forthcoming results was in direct proportion to the magnitude of the effort. In its issue of August 22, 1936, the *Literary Digest* crowed:

> *Once again, [we are] asking more than ten million voters—one out of four, representing every county in the United States—to settle November's election in October.*
>
> *Next week, the first answers from these ten million will begin the incoming tide of marked ballots, to be triple-checked, verified, five-times cross-classified and totaled. When the last figure has been totted and checked, if past experience is a criterion, the country will know to within a fraction of 1 percent the actual popular vote of forty million [voters].*

Based on the feedback, the *Literary Digest* predicted that Landon would get 57% of the vote against Roosevelt's 43%. The actual results of the election were 62% for Roosevelt against 38% for Landon. The difference between the poll's prediction and the actual election results was a whopping 19%, the largest error ever in a major public opinion poll. The results damaged the credibility of the magazine so much so that soon after the election its sales dried up and it went out of business—the victim of a statistical *faux pas*.

For the same election, a young pollster named George Gallup (for more on Gallup, see the biographical profile on p. 534) was able to predict accurately a victory for Roosevelt based on a much smaller sample of approximately 50,000 people. In fact, he *also* publicly predicted, to within 1%, the incorrect results that the *Literary Digest* would get by sampling 3,000 people from the same lists the magazine was using. Why was it that Gallup's results were so superior to those of the *Literary Digest*?

The first thing seriously wrong with the *Literary Digest* poll was in the selection process for the names on the mailing list. Names were taken from telephone directories, rosters of club members, lists of magazine subscribers, etc. Such a list was inherently slanted toward members of the middle and upper classes. Telephones in 1936 were something of a luxury. So, too, were club memberships and

The cover of the *Literary Digest* the week after the election.

magazine subscriptions, at a time when 9 million people were unemployed. At least with regard to economic status, the *Literary Digest* mailing list was far from being a representative cross section of the population. This was a critical problem, because voters often vote on economic issues, and given the economic conditions of the time, this was especially true in 1936.

When the choice of the sample has a built-in tendency (whether intentional or not) to exclude a particular group or characteristic within the population, we say that a survey suffers from **selection bias**. It is obvious that selection bias must be avoided, but it is not always easy to detect it ahead of time. Even the most scrupulous attempts to eliminate selection bias can fall short (as will become apparent in our next case study).

The second serious problem with the *Literary Digest* poll was that out of the 10 million people whose names were on the original mailing list, only about 2.4 million responded to the survey. Thus, the number of respondents was about one-fourth of the size of the original sample.[2] When the proportion of respondents to the total number of people in the sample, called the **response rate**, is low, a survey is said to suffer from **nonresponse bias**. For the *Literary Digest* poll the response rate was 24%, which is extremely low.

It is well known that people who respond to surveys are different from people who don't, not only in the obvious way (their attitude toward the usefulness of surveys), but also in more subtle ways. They tend to be better educated and in higher economic brackets, and are, in fact, more likely to vote Republican. Thus, nonresponse bias is a special type of selection bias—it excludes from the sample reluctant and disinterested people. But don't we want them represented?

---

[2]Recall that by its own admission the *Literary Digest* was expecting about 10 million respondents.

Eliminating nonresponse bias from a survey is difficult. In a free country we cannot force people to participate, and paying them is hardly ever a solution, since it can introduce other forms of bias. Some ways of minimizing nonresponse bias are known, however. The *Literary Digest* survey was conducted by mail. This approach is the most likely to magnify nonresponse bias, because people often consider a mailed questionnaire just another form of junk mail. Of course, considering the size of the sample, the *Literary Digest* really had no other choice but to use mailed questionnaires. Here we see how a big sample size can be more of a liability than an asset.

The *Literary Digest* story has two morals: (1) *You'll do better with a well-chosen small sample than with a badly chosen large one*, and (2) *watch out for selection bias and nonresponse bias*.

Nowadays, almost all legitimate public opinion polls are conducted either by telephone or by personal interviews. Telephone polling is subject to slightly more nonresponse bias than personal interviews, but it is considerably cheaper. In some special situations, however, telephone polls can be so biased as to be useless.[3]

Our next case study illustrates how difficult it can be, even with the very best intentions, to get rid of selection bias.

CASE STUDY 3 THE 1948 PRESIDENTIAL ELECTION

Despite the fiasco of 1936, and possibly because of the lessons learned from it, by 1948 the use of public opinion polls to measure the American electorate was thriving. Three major polls competed for the prize of correctly predicting the outcome of the national elections: the Gallup poll, the Roper poll, and the Crossley poll.

In 1948, these three polls were using a much more scientific method for choosing their samples: **quota sampling**. George Gallup had introduced quota sampling as early as 1935 and had successfully used it to predict the winner of the 1936, 1940, and 1944 presidential elections. Quota sampling is a systematic effort to force the sample to fit a certain national profile by using quotas. The sample should have so many women, so many men, so many blacks, so many whites, so many people living in urban areas, so many people living in rural areas, and so on. The proportions in each category in the sample should be the same as those in the electorate at large.

If we assume that every important characteristic of the population is taken into account when the quotas are set up, it is reasonable to expect that quota sampling will produce a good cross section of the population and therefore lead to accurate predictions.

For the 1948 election between Thomas Dewey and Harry Truman, Gallup conducted a poll with a sample size of approximately 3250 people. Each individual in the sample was interviewed in person by a professional interviewer to

---

[3]A blatant example of selection bias occurs when the sample is self-selected—you are in the sample because you volunteer to be in it. The worst instances of this are Area Code 900 telephone polls, where a person actually has to pay (sometimes as much as $2) to be part of the sample. People who are willing to pay to express their opinions are hardly representative of the general public, and information collected from such polls should be considered totally unreliable.

minimize nonresponse bias, and each interviewer was given a very detailed set of quotas to meet—for example, 7 white males under 40 living in a rural area, 5 black males over 40 living in a rural area, 6 white females under 40 living in a rural area, and so on. By the time all the interviewers met their quotas, the entire sample was expected to accurately represent the entire population in every respect: gender, race, age, and so on.

Based on his sample, Gallup predicted a victory for Dewey, the Republican candidate. The predicted breakdown of the vote was 49.5% for Dewey, 44.5% for Truman, and 6% for third-party candidates Strom Thurmond and Henry Wallace. The other two polls made similar predictions. In fact, after an early September poll showed Truman trailing Dewey by 13 percentage points, Roper announced that he would discontinue polling since the outcome was already so obvious. However, much would change from the time each organization conducted its final poll. The actual results of the election turned out to be almost the exact reverse of Gallup's prediction: 49.9% for Truman, 44.5% for Dewey, and 5% for the third-party candidates.

"Ain't the way I heard it." Truman gloats while holding an early edition of the *Chicago Daily Tribune* in which the headline erroneously claimed a Dewey victory based on the predictions of all the polls.

Truman's victory was a great surprise to the nation as a whole. So convinced was the *Chicago Daily Tribune* of Dewey's victory that it went to press on its early edition for November 4, 1948, with the headline "Dewey defeats Truman"—a blunder that led to Truman's famous post-election retort, "Ain't the way I heard it." The picture of Truman holding aloft a copy of the *Tribune* (see photo) has become part of our national folklore. To pollsters and statisticians, the erroneous predictions of the 1948 election had two lessons: (1) *Poll until election day*, and (2) *quota sampling is intrinsically flawed*.

What's wrong with quota sampling? After all, the basic idea behind it appears to be a good one: force the sample to be a representative cross section of the population by having each important characteristic of the population proportionally represented in the sample. Since income is an important factor in determining how people vote, the sample should have all income groups represented in the same proportion as the population at large. The same should be true for sex, race, age, and so on. Right away, we can see a potential problem: Where do we stop? No matter how careful we might be, we might miss some criterion that would affect the way people vote, and the sample could be deficient in this regard.

An even more serious flaw in quota sampling is that, other than meeting the quotas, the interviewers are free to choose whom they interview. This opens the door to selection bias.

Looking back over the history of quota sampling, one can see a clear tendency to overestimate the Republican vote. In 1936, using quota sampling, Gallup predicted that the Republican candidate would get 44% of the vote, but the actual number was 38%. In 1940, the prediction was 48%, and the actual vote was 45%; in 1944, the prediction was 48%, and the actual vote was 46%. Nonetheless, Gallup was able to predict the winner correctly in each of these elections, mostly because the spread between the candidates was large enough to cover the error. In 1948, Gallup (and all the other pollsters) simply ran out of luck. It was time to ditch quota sampling.

The failure of quota sampling as a method for getting representative samples has a simple moral: *Even with the most carefully laid plans, human intervention in choosing the sample can result in selection bias.*

13.3 Random Sampling

If human intervention in choosing the sample is always subject to bias, what are the alternatives? The answer is to let the laws of chance decide who is in the sample—to draw the names, as it were, out of a hat. Some people find this method hard to believe in. Isn't it possible, they wonder, to get by sheer chance a sample that is very biased (say, for example, all female)? In theory, such an outcome is possible, but in practice, when the sample is large enough, the odds of it happening are so low that we can pretty much rule it out. Most present-day methods of quality control in industry, corporate audits in business, and public opinion polling are based on **random sampling** methods—that is, methods for choosing the sample in which chance intervenes in one form or another. The reliability of data collected by random sampling methods is supported by both practical experience and mathematical theory. We will discuss some of the details of this theory in Chapter 16.

Simple Random Sampling

The most basic form of random sampling is called **simple random sampling**. It is based on the same principle a lottery is. Any set of numbers of a given size has an equal chance of being chosen as any other set of numbers of that size. Thus, if a lottery ticket consists of 6 winning numbers, a fair lottery is one in which any combination of 6 numbers has the same chance of winning as any other combination of 6 numbers. In sampling, this means that any group of members of the population should have the same chance of being the sample as any other group of the same size.

In theory, simple random sampling is easy to implement. We put the name of each individual in the population in "a hat," mix the names well, and then draw as many names as we need for our sample. Of course "a hat" is just a metaphor. If our population is 100 million voters and we want to choose a simple random sample of 2000, we will not be putting all the names in a real hat and then drawing 2000 names one by one. The modern way to do any serious simple random sampling is by computer. Make a list of members of the population, enter it into the computer, and then let the computer randomly select the names.

While simple random sampling works well in many cases, for national surveys and public opinion polls it presents some serious practical difficulties. First, it requires us to have a list of all the members of the population. The population, however, may not be clearly defined, or, even if it is, a complete list of the members may not be available. Second, implementing simple random sampling in national public opinion polls raises problems of expediency and cost. Interviewing several thousand people chosen by simple random sampling means chasing people all over the country. This requires an inordinate amount of time and money. For most public opinion polls—especially those done on a regular basis—the time and money needed to do this are simply not available.

Our next case study describes the sampling method currently used in most public opinion polls.

CASE STUDY **4**   **MODERN PUBLIC OPINION POLLS**

Present-day methods for conducting public opinion polls need to take into account two sets of considerations: (1) minimizing sample bias and (2) choosing a sample that is accessible in a cost-efficient, timely manner. A random sampling method that deals in a satisfactory way with both these issues is **stratified sampling**. The basic idea of stratified sampling is to break the sampling frame into categories, called **strata**, and then (unlike quota sampling) *randomly* choose a sample from these strata. The chosen strata are then further divided into categories, called *substrata*, and a random sample is taken from these substrata. The selected substrata are further subdivided, a random sample is taken from them, and so on. The process goes on for a predetermined number of layers.

In public opinion polls, the strata and substrata are usually defined by criteria that involve a combination of geographic and demographic elements. For example, at the first level, the nation is divided into "size of community" strata (big cities, medium cities, small cities, villages, rural areas, etc.). Each of these strata is then subdivided by geographical region (New England, Middle Atlantic, East Central, etc.). Within each geographical region and within each size of community stratum, some communities are selected by simple random sampling. The selected communities (called *sampling locations*) are the only places where interviews will be conducted. To further randomize things, each of the selected sampling locations is subdivided into geographical units, called *wards*, and within each sampling location some of its wards are once again selected by simple random sampling. The selected wards are then divided into smaller units, called *precincts*, and within each ward some of its precincts are selected by simple random sampling. At the last stage, *households* are selected for interviewing by simple random sampling within each precinct. The interviewers are then given specific instructions as to which households in their assigned area they must conduct interviews in and the order that they must follow.

The efficiency of stratified sampling compared to simple random sampling in terms of cost and time is clear. The members of the sample are clustered in well-defined and easily manageable areas, significantly reducing the cost of conducting interviews, as well as the response time needed to collect the data. For a large, heterogeneous nation like the United States, stratified sampling has generally proved to be a reliable way to collect national data, and most modern public opinion polls are based on stratified samples.

THE GALLUP POLL
Design of the Sample

The design of the sample used by the Gallup Poll for its standard surveys of public opinion is that of a replicated area probability *stratified sample* down to the block level in the case of urban areas and to segments of townships in the case of rural areas.

After stratifying the nation geographically and by size of community in order to insure conformity of the sample with the 1990 Census distribution of the population, over 360 different sampling locations or areas are selected on a mathematically random basis from within cities, towns, and countries which have in turn, been selected on a mathematically random basis. The interviewers have no choice whatsoever concerning the part of the city, town, or country in which they conduct their interviews.

Approximately five interviews are conducted in each randomly selected sampling point. Interviewers are given maps of the area to which they are assigned and are required to follow a specific travel pattern on contacting households. At each occupied dwelling unit, interviewers are instructed to select respondents by following a prescribed systematic method. This procedure is followed until the assigned number of interviews with male and female adults have been completed. . . .

Since this sampling procedure is designed to produce a sample which approximates the adult civilian population (18 and older) living in private households (that is, excluding those in prisons and hospitals, hotels, religious and educational institutions, and on military bases) the survey results can be applied to this population for the purpose of projecting percentages into numbers of people. The manner in which the sample is drawn also produces a sample which approximates the population of private households in the United States. Therefore, survey results also can be projected in terms of numbers of households.

Sampling Error

In interpreting survey results, it should be borne in mind that all sample surveys are subject to *sampling error*, that is, *the extent to which the results may differ from those that would be obtained if the whole population surveyed had been interviewed.*

Source: *The Gallup Report.* Princeton, NJ: American Institute of Public Opinion, 1991.

What about the size of the sample? Surprisingly, it does not have to be very large. Typically, a Gallup poll is based on samples consisting of approximately 1500 individuals, and roughly the same size sample can be used to poll the population of a small city as the population of the United States. *The size of the sample does not have to be proportional to the size of the population.* How can this be? George Gallup, explained it beautifully in this often quoted soup analogy:

> *Whether you poll the United States or New York State or Baton Rouge (Louisiana) ... you need ... the same number of interviews or samples. It's no mystery really—if a cook has two pots of soup on the stove, one far larger than the other, and thoroughly stirs them both, he doesn't have to take more spoonfuls from one than the other to sample the taste accurately.*[4]

Before we continue with our examples and case studies, let's review some of the key concepts in sampling and introduce some new terminology.

[4]As quoted in "The Man Who Knows How We Think," *Modern Maturity*, 17, no. 2 (April–May 1974).

13.4 Sampling: Terminology and Key Concepts

As we now know, except for a *census*, the common way to collect statistical information about a population is by means of a *survey*. In a survey, we use a subset of the population, called a *sample*, as the source of our information, and from this sample, we try to generalize and draw conclusions about the entire population. Statisticians use the term **statistic** to describe any kind of numerical information drawn from a sample. A statistic is always an estimate for some unknown measure, called a **parameter**, of the population. Let's put it this way: A *parameter* is the numerical information we would like to have—the pot of gold at the end of the statistical rainbow, so to speak. Calculating a parameter is difficult and often impossible, since the only way to get the exact value for a parameter is to use a census. If we use a sample, then we can get only an estimate for the parameter, and this estimate is called a *statistic*.

We will use the term **sampling error** to describe the difference between a parameter and a statistic used to estimate that parameter. In the case of a public opinion poll, the sampling error is "the extent to which the results [of the poll] differ from those that would be obtained if the whole population surveyed had been interviewed" (see the last paragraph of the Gallup poll box). Since the very point of sampling is to avoid interviewing the whole population, the exact amount of the sampling error is usually not known. The best we can do when it comes to sampling errors is to estimate their maximum size. This is usually paraphrased in stock phrases of the form "The margin of sampling error for this poll was plus or minus 3 percent." We will discuss in greater detail the exact meaning of such statements in Chapter 16.

Sampling error can be attributed to two factors: *chance error* and *sampling bias*.

- **Chance error** is the result of the basic fact that a statistic cannot give exact information about the population, because it is, by definition, based on partial information (the sample). In surveys, chance error is the result of **sampling variability**: the fact that two different samples are likely to give two different statistics, even when the samples are chosen using the same sampling method. While sampling variability, and thus chance error, are unavoidable, with careful selection of the sample and the right choice of sample size they can be kept to a minimum.

- **Sample bias** is the result of having a poorly chosen sample. Even with the best intentions, getting a sample that is representative of the entire population can be very difficult and can be affected by many subtle factors. Sample bias is the result. As opposed to chance error, sample bias can be eliminated by using proper methods of sample selection.

Lastly, we shall make a few comments about the size of the sample, usually denoted by the letter n to contrast with N, which is the size of the population. The ratio n/N is called the **sampling rate**. A sampling rate of x% tells us that the sample is x% of the population. We now know that in sampling, it is not the sampling rate that matters, but rather choosing a good sample of a reasonable size. In public opinion polls, for example, whether $N = 100{,}000$ or $N = 250$ million, a well-chosen sample of $n = 1500$ is sufficient to get reliable statistics.

The Capture–Recapture Method: Estimating the *N*-value of a Population by Sampling

We have already observed that finding the exact N-value of a large and elusive population can be extremely difficult and sometimes impossible. In many cases, a good estimate is all we really need, and such estimates are possible through sampling methods. The simplest sampling method for estimating the N-value of a population is called the **capture–recapture method**. The method consists of two steps, which we will describe in the jargon of the field in which it is most frequently used: population biology.

- **Step 1. The Capture:** Capture (choose) a sample of size n_1, *tag* (mark, identify) the animals (objects, people), and release them back into the general population.
- **Step 2. The Recapture:** After a certain period of time, capture a new sample of size n_2, and take an exact head count of the *tagged* individuals (i.e., those that were also in the first sample). Let's call this number k.

If we can assume that the recaptured sample is representative of the entire population, then the proportion of tagged individuals in it is approximately equal to the proportion of the tagged individuals in the population. In other words, the ratio k/n_2 is approximately equal to the ratio n_1/N. From this we can solve for N and get $N \approx n_1 n_2 / k$.

EXAMPLE 13.4

A large pond is stocked with catfish. You decide to use the capture-recapture method to estimate how many catfish there are in the pond.

- **Step 1.** Capture a sample of n_1 catfish, say, for example, $n_1 = 200$. The fish are tagged and released unharmed back in the pond.
- **Step 2.** After giving enough time for the released fish to mingle and disperse throughout the pond, capture a second sample of n_2 catfish. While n_2 does not have to equal n_1, it is a good idea for the two samples to be of approximately the same order of magnitude. Let's say that $n_2 = 250$. Of the 250 catfish in the second sample, 35 have tags.

If the second sample is representative of the catfish population in the pond, we can assume that the ratio of tagged fish in the second sample (35/250) and the ratio of tagged fish in the pond (200/N) are approximately the same. Solving

$$35/250 \approx 200/N$$

for N gives us

$$N \approx 200 \times 250/35 \approx 1428.57.$$

Obviously, the foregoing value of N cannot be taken literally, since N must be a whole number. Besides, even in the best of cases, the computation is only an estimate. A sensible conclusion is that there are approximately $N = 1400$ catfish in the pond. ■

13.5 Clinical Studies

A survey typically deals with issues and questions that can be answered in a direct and objective manner. *If the election were held today, would you vote for candidate X or candidate Y? How many people live in your household? What types of music do you listen to?*

A different type of data collection attempts to answer questions for which there is no clear, immediate answer. *Does smoking increase your chances of lung disease? Does continued exposure to second-hand smoke significantly increase your risk for developing lung cancer? Does a high daily intake of caffeine reduce your chances of developing type 2 diabetes? Do the benefits of hormone replacement therapy for women over 50 outweigh the risk?* These kinds of questions have two things in common: (1) They involve a cause and an effect, and (2) the answers require observation over an extended period of time.

When one wants to know if a certain cause X produces a certain effect Y, one sets up a *study* in which cause X is produced and its effects are observed. If the effect Y is observed, then it is possible that X was indeed the cause of Y. We have established an *association* between the cause X and the effect Y. The problem, however, is the nagging possibility that some other cause Z different from X produced the effect Y and that X had nothing to do with it. Just because we established an association, we have not established a cause–effect relation between the variables. Statisticians like to explain this by a simple saying: *Association is not causation.*

Let's illustrate with a fictitious example. Suppose we want to find out if eating lots of chocolate increases one's chance of developing diabetes. Here the cause X is eating chocolate, and the effect Y is diabetes. We set up an experiment in which 100 laboratory rats are fed 8 ounces of chocolate a day for a period of six months. At the end of the six-month period, 15 of the 100 rats have diabetes. Since in the general rat population only 3% are diabetic, we are tempted to conclude that the diabetes in the rats is indeed caused by the excessive chocolate in the diet. The problem is that there is no absolute certainty that the chocolate diet was the cause. Could there be another unknown reason for the observed effect? Even if there isn't, can we necessarily conclude that just because chocolate produces diabetes in rats, it will do the same for humans?

If you think that our fictititous chocolate story is too far-fetched to be realistic, consider the next case study.

CASE STUDY 5 THE ALAR SCARE

Alar is a chemical used by apple growers to regulate the rate at which apples ripen. Until 1989, practically all apples sold in grocery stores were sprayed with Alar. But in 1989 Alar became bad news, denounced in newspapers and on TV as a potent cancer-causing agent and a primary cause of cancer in children. As a result of these reports, people stopped buying apples, schools all over the country removed apple juice from their lunch menus, and the Washington state apple industry lost an estimated $375 million.

The case against Alar was based on a single 1973 study in which laboratory mice were exposed to the active chemicals in Alar. The dosage used in the study was eight times greater than the maximum tolerated dosage—a concentration at which even harmless substances can produce tissue damage. In fact, a child would have to eat about 200,000 apples a day to be exposed to equivalent dosage of the chemical. Subsequent studies conducted by the National Cancer Institute and the Environmental Protection Agency failed to show any cause-and-effect relationship between Alar and cancer in children.

While it is generally accepted now that Alar does not cause cancer, because of potential legal liability, it is no longer used. The Alar scare turned out to be a false alarm based on a poor understanding of the statistical evidence.

Unfortunately, it left in its wake a long list of casualties, among them the apple industry, the product's manufacturer, the media, and the public's confidence in the system.

For most cause-and-effect situations, especially those complicated by the involvement of human beings, a single effect can have many possible and actual causes. What causes cancer? Unfortunately, there is no single cause—diet, lifestyle, the environment, stress, and heredity are all known to be contributory causes. The extent to which each of these causes contributes individually and the extent to which they interact with each other are extremely difficult questions that can be answered only by means of carefully designed statistical studies.

For the remainder of this chapter we will illustrate an important type of study called a **clinical study** or **clinical trial**. Generally, clinical studies are concerned with determining whether a single variable or treatment (usually a vaccine, a drug, therapy, etc.) can cause a certain effect (a disease, a symptom, a cure, etc.). The importance of such clinical studies is self-evident: Every new vaccine, drug, or treatment must prove itself by means of clinical study before it is officially approved for public use. Likewise, almost everything that is bad for us (cigarettes, caffeine, cholesterol, etc.) gets its official certification of badness by means of a clinical study.

Properly designing a clinical study can be both difficult and controversial, and as a result, we are often bombarded with conflicting information produced by different studies examining the same cause-and-effect question. The basic principles guiding a clinical study, however, are pretty much established by statistical practice.

The first and most important issue in any clinical study is to isolate the cause (treatment, drug, vaccine, therapy, etc.) that is under investigation from all other possible contributing causes (called **confounding variables**) that could produce the same effect. Generally, this is best done by *controlling* the study.

In a **controlled study**, the subjects are divided into two different groups: the *treatment group* and the *control group*. The **treatment group** consists of those subjects receiving the actual treatment, the **control group** consists of subjects that are not receiving any treatment—they are there for comparison purposes only (that's why the control group is sometimes also called the *comparison* group). If a real cause-and-effect relationship exists between the treatment and the effect being studied, then the treatment group should show the effects of the treatment and the control group should not.

To eliminate the many potential confounding variables that can bias its results, a well-designed controlled study should have control and treatment groups that are similar in every characteristic other than the fact that one group is being treated and the other one is not. (It would be a very bad idea, for example, to have a treatment group that is all female and a control group that is all male.) The most reliable way to get equally representative treatment and control groups is to use a *randomized controlled study*. In a **randomized controlled study**, the subjects are assigned to the treatment group or the control group randomly (typically by a computer program).

When the randomization part of a randomized controlled study is properly done, treatment and control groups can be assumed to be statistically similar. But there is still one major difference between the two groups that can significantly affect the validity of the study—a critical confounding variable known as the *placebo effect*. The **placebo effect** follows from the generally accepted principle that *just the idea that one is getting a treatment, even if it is a phantom treatment,* *can produce positive results.*[5] Thus, when subjects in a study are getting a pill or a vaccine or some other kind of treatment, how can the researchers separate positive results that are consequences of the treatment itself from those that might be caused by the placebo effect? When possible, the standard way to handle this problem is to give the control group a *placebo*. A **placebo** is a *make-believe* form of treatment—a harmless pill, an injection of saline solution, or any other fake type of treatment intended to mimic the real treatment. A controlled study in which the subjects in the control group are given a placebo is called a **controlled placebo study**.

By giving all subjects a seemingly equal treatment (the treatment group gets the real treatment and the control group gets a placebo which mimics the real treatment), we do not eliminate the placebo effect but rather control it—whatever its effect might be, it impacts all subjects equally. It goes without saying that the use of placebos is pointless if the subject knows he or she is getting a placebo. Thus, a second key element of a good controlled placebo study is that all subjects be kept in the dark as to whether they are being treated with a real treatment or a placebo. A study in which neither the members of the treatment group nor the members of the control group know to which of the two groups they belong is called a **blind** study.

Blindness is a key requirement of a controlled placebo study, but not the only one. To keep the interpretation of the results (which can often be ambiguous) totally objective, it is important that the scientists conducting the study and collecting the data also be in the dark when it comes to who got the treatment and who got the placebo. A controlled placebo study in which neither the subjects nor the scientists conducting the experiment know which subjects are in the treatment group and which are in the control group is called a **double-blind study**.

Our next case study illustrates one of the most famous and important double-blind studies in the annals of clinical research.

CASE STUDY 6 THE 1954 SALK POLIO VACCINE FIELD TRIALS

Polio (infantile paralysis) has been practically eradicated in the Western world. In the first half of the twentieth century, however, it was a major public health problem. Over one-half million cases of polio were reported between 1930 and 1950, and the actual number may have been considerably higher.

Because polio attacks mostly children and because its effects can be so serious (paralysis or death), eradication of the disease became a top public health priority in the United States. By the late 1940s, it was known that polio is a virus and, as such, can best be treated by a vaccine which is itself made up of a virus. The vaccine virus can be a closely related virus that does not have the same harmful effects, or it can be the actual virus that produces the disease but which has been killed by a special treatment. The former is known as a *live-virus vaccine*, the latter as a *killed-virus vaccine*. In response to either vaccine, the body is known to produce *antibodies* that remain in the system and give the individual immunity against an attack by the real virus.

Both the live-virus and the killed-virus approaches have their advantages and disadvantages. The live-virus approach produces a stronger reaction and better

[5]Some researchers estimate that as many as 30% of patients in a study can be impacted by the placebo effect. At the other end of the spectrum are researchers that question whether the placebo effect really exists. For a more in-depth look at this issue, see Project D.

immunity, but at the same time, it is also more likely to cause a harmful reaction and, in some cases, even to produce the very disease it is supposed to prevent. The killed-virus approach is safer in terms of the likelihood of producing a harmful reaction, but it is also less effective in providing the desired level of immunity.

These facts are important because they help us understand the extraordinary amount of caution that went into the design of the study that tested the effectiveness of the polio vaccine. By 1953, several potential vaccines had been developed, one of the more promising of which was a killed-virus vaccine developed by Jonas Salk at the University of Pittsburgh. The killed-virus approach was chosen because there was a great potential risk in testing a live-virus vaccine in a large-scale study. (A large-scale study was needed to collect enough information on polio, which, in the 1950s, had a rate of incidence among children of about 1 in 2000.)

The testing of any new vaccine or drug creates many ethical dilemmas that have to be taken into account in the design of the study. With a killed-virus vaccine the risk of harmful consequences produced by the vaccine itself is small. So one possible approach would have been to distribute the vaccine widely among the population and then follow up on whether there was a decline in the national incidence of polio in subsequent years. This approach, which was not possible at the time because supplies were limited, is called the *vital statistics* approach and is the simplest way to test a vaccine. This is essentially the way the smallpox vaccine was determined to be effective. The problem with such an approach for polio is that polio is an epidemic type of disease, which means that there is a great variation in the incidence of the disease from one year to the next. In 1952, there were close to 60,000 reported cases of polio in the United States, but in 1953, the number of reported cases had dropped to almost half that (about 35,000). Since no vaccine or treatment was used, the cause of the drop was the natural variability that is typical of epidemic diseases. But, if an ineffective polio vaccine had been tested in 1952 without a control group, the observed effect of a large drop in the incidence of polio in 1953 could have been incorrectly interpreted as statistical evidence that the vaccine worked.

The final decision on how best to test the effectiveness of the Salk vaccine was left to an advisory committee of doctors, public officials, and statisticians convened by the National Foundation for Infantile Paralysis and the Public Health Service. It was a highly controversial decision, but at the end, a large-scale, randomized, double-blind, controlled placebo study was chosen. Approximately 750,000 children were randomly selected to participate in the study. Of these, about 340,000 declined to participate, and another 8500 dropped out in the middle of the experiment. The remaining children were randomly divided into two groups—a treatment group and a control group—with approximately 200,000 children in each group. Neither the families of the children nor the researchers collecting the data knew if a particular child was getting the actual vaccine or a shot of harmless solution. The latter was critical because polio is not an easy disease to diagnose—it comes in many different forms and degrees. Sometimes, it can be a borderline call, and if the doctor collecting the data had prior knowledge of whether the subject had received the real vaccine or the placebo, the diagnosis could have been subjectively tipped one way or the other.

A summary of the results of the Salk vaccine field trials is shown in Table 13-1. These data were taken as conclusive evidence that the Salk vaccine was an effective treatment for polio and on the basis of this study, a massive inoculation campaign was put into effect. Today, all children are routinely inoculated against polio, and the disease has essentially been eradicated in the United States. Statistics played a key role in this important public health breakthrough.

TABLE 13-1	Results of the Salk Vaccine Field Trials			
	Number of Children	Number of Reported Cases of Polio	Number of Paralytic Cases of Polio	Number of Fatal Cases of Polio
Treatment group	200,745	82	33	0
Control group	201,229	162	115	4
Declined to participate in the study	338,778	182*	121*	0*
Dropped out in the middle	8,484	2*	1*	0*
Total	749,236	428	270	4

*These figures are not a reliable indicator of the actual number of cases—they are only self-reported cases. (Adapted from Thomas Francis, Jr., et al., "An Evaluation of the 1954 Poliomyelitis Vaccine Trials—Summary Report." *American Journal of Public Health*, 45 (1955) 25.)

Conclusion

In this chapter we have discussed different methods for collecting data. In principle, the most accurate method is a *census*, a method that relies on collecting data from each member of the population. In most cases, because of considerations of cost and time a census is an unrealistic strategy. When data are collected from only a subset of the population (called a *sample*), the data collection method is called a *survey*. The most important rule in designing good surveys is to eliminate or minimize *sample bias*. Today, almost all strategies for collecting data are based on surveys in which the laws of chance are used to determine how the sample is selected, and these methods for collecting data are called *random sampling* methods. Random sampling is the best way known to minimize or eliminate sample bias. Two of the most common random sampling methods are *simple random sampling* and *stratified sampling*. In some special situations, other, more complicated types of random sampling can be used.

Sometimes, identifying the sample is not enough. In cases in which cause-and-effect questions are involved, the data may come to the surface only after an extensive study has been carried out. In these cases, isolating the cause variable under consideration from other possible causes (called *confounding variables*) is an essential prerequisite for getting reliable data. The standard strategy for doing this is a *controlled study* in which the sample is broken up into a *treatment group* and a *control group*. Controlled studies are now used (and sometimes abused) to settle issues affecting every aspect of our lives. We can thank this area of statistics for many breakthroughs in social science, medicine, and public health, as well as for the constant and dire warnings about our health, our diet, and practically anything that is fun.

George Gallup (1901–1984)

George Gallup was a man of many talents—journalist, sociologist, political analyst, businessman, and statistician. It was the combination of all these talents, together with a strong entrepreneurial spirit and an unlimited amount of self-confidence, that made him a unique fixture of 20th-century American life—the man trusted by generations to best measure the nation's pulse. Gallup did not invent the modern day public opinion poll, but he certainly set the standard for how to do it right. Under his tenure, the poll that bears his name became the most widely read and credible public opinion poll in the world.

George Horace Gallup Jr. was born in Jefferson, Iowa, a small farm town in America's heartland. He studied journalism at the University of Iowa, paying his way through school partly by working as the editor of the student newspaper. He earned a bachelor's degree in journalism in 1923, a master's degree in psychology in 1925, and a Ph.D. in journalism in 1928. In his doctoral thesis, entitled *About Unbiased Methodology of Exploring Readers' Interest in the Content of Newspapers*, Gallup showed that public opinion could be scientifically collected from a very small sample. His doctoral work was a precursor of the sampling methods he would later develop to conduct public opinion polls.

After working as a professor of journalism at Drake University and at Northwestern University, Gallup was recruited by New York advertising agency Young and Rubicam to become the head of their newly developed market-research department. But Gallup was a journalist at heart, and by 1935 he had enough of the advertising business, notwithstanding the fact that he had made millions in it. He moved to Princeton, New Jersey, where he founded the American Institute of Public Opinion and started a weekly syndicated newspaper column immodestly entitled "America Speaks," which featured the results of public opinion polls he designed and conducted. From these humble beginnings, the Gallup poll was born. The organization he founded (now called the Gallup Organization and run by his two sons, George III and Alec) has grown to become the largest and most respected private polling organization in the world.

Today, public opinion polling is used extensively throughout the world, but the increased use of polls in modern life has generated many criticisms about their impact and influence on the political process. As the father of polling, Gallup spent his later years trying to defend polls as an important and useful instrument of democracy. "When a president, or any other leader, pays attention to poll results, he is, in effect, paying attention to the views of the people," he once said.

George Gallup died in 1984 in Switzerland, where he had spent a good part of his last years, partly for its beauty and partly because, as he put it, the country was "virtually run by polls." Once, as a young editor of the University of Iowa student newspaper, Gallup wrote, "Doubt everything, Question everything. Be a radical!" Even as he became an American institution, George Gallup remained always true to this credo.

blind study	control group
capture–recapture method	controlled study
census	controlled placebo study
chance error	data
clinical study (clinical trial)	double-blind study
confounding variable	nonresponse bias

parameter
placebo
placebo effect
population
quota sampling
randomized controlled study
random sampling
response rate
sample
sample bias
sampling error

sampling frame
sampling rate
sampling variability
selection bias
simple random sampling
statistic
strata
stratified sampling
survey
treatment group

EXERCISES

WALKING

A. Surveys and Public Opinion Polls

Exercises 1 through 4 refer to the following situation. As part of a sixth-grade statistics project, the teacher brings to class a candy jar full of gumballs of two different colors: red and green. The students are told that there are 200 gumballs in the jar, and their job is to estimate the number that are red. To do this, the jar is shaken well, and one of the students draws 25 gumballs from the jar. Of these, 8 are red and 17 are green.

1. What is the population of this survey?
2. **(a)** Describe the sample for this survey.
   (b) What is the sampling rate for this survey?
   (c) Name the sampling method used for this survey.
3. Estimate the number of red gumballs in the candy jar.
4. **(a)** Given that the actual number of red gumballs in the jar is 50, find the sampling error, expressed as a percent.
   (b) Is this sampling error a result of sampling variability or sampling bias? Explain.

Exercises 5 through 8 refer to the following situation. The city of Cleansburg has 8325 registered voters. There is an election for mayor of Cleansburg, and there are three candidates for the position: Smith, Jones, and Brown. The day before the election, a telephone poll of 680 randomly chosen registered voters produced the following results: 306 people surveyed indicated that they would vote for Smith, 272 indicated that they would vote for Jones, and 102 indicated that they would vote for Brown.

5. **(a)** Describe the sample for this survey.
   (b) What is the sampling rate for this survey?
6. **(a)** What is the population in this example?
   (b) What is the N-value?
7. Given that in the actual election candidate Smith received 42% of the vote, candidate Jones 43% of the vote, and candidate Brown 15% of the vote, find the sampling error expressed as a percent.

8. Do you think that the sampling error in this example is due primarily to chance error or to sample bias? Explain your answer.

Exercises 9 through 12 refer to the following survey. In 1988, "Dear Abby" asked her readers to let her know whether they had cheated on their spouses or not. The readers' responses are summarized in the accompanying table.

Status	Women	Men
Faithful	127,318	44,807
Unfaithful	22,468	15,743
Total	149,786	60,550

Based on the results of this survey, "Dear Abby" concluded that the amount of cheating among married couples is much less than people believe. (In her words, "The results were astonishing. There are far more faithfully wed couples than I had surmised.")

9. **(a)** Describe as specifically as you can the sampling frame for this survey.
   (b) Compare and contrast the target population and the sampling frame for this survey.
   (c) How was the sample chosen?
   (d) Eighty-five percent of the women who responded to this survey claimed to be faithful. Is 85% a parameter? A statistic? Neither? Explain your answer.
10. **(a)** Explain why this survey was subject to selection bias.
    (b) Explain why this survey was subject to nonresponse bias.
11. **(a)** Based on the "Dear Abby" data, estimate the percentage of married men who are faithful to their spouses.
    (b) Based on the "Dear Abby" data, estimate the percentage of married people who are faithful to their spouses.
    (c) How accurate do you think these estimates are? Explain.
12. If money were no object, could you devise a survey that might give more reliable results than the "Dear Abby" survey? Describe briefly what you would do.

Exercises 13 through 16 refer to the following hypothetical situation. The Cleansburg Planning Department is trying to determine what percent of the people in the city want to spend public funds to revitalize the downtown mall. To do so, they decide to conduct a survey. Five professional interviewers (A, B, C, D, and E) are hired, and each is asked to pick a street corner of their choice within the city limits. Everyday between 4:00 and 6:00 P.M., the interviewers are to ask each passerby if he or she wishes to respond to a survey sponsored by Cleansburg City Hall and to make a record of their response. If the response is yes, the person is asked if he or she is in favor of spending public funds to revitalize the downtown mall. The interviewers are asked to return to the same street corner as many days as are necessary until each one has conducted a total of 100 interviews. The data collected are seen in Table 13-2.

TABLE 13-2

Interviewer	Yes[a]	No[b]	Nonrespondents[c]
A	35	65	321
B	21	79	208
C	58	42	103
D	78	22	87
E[d]	12	63	594

[a]In favor of spending public funds to revitalize the downtown mall.
[b]Opposed to spending public funds to revitalize the downtown mall.
[c]Declined to be interviewed.
[d]Got frustrated and quit.

13. **(a)** Describe as specifically as you can the target population for this survey.

    (b) Compare and contrast the target population and the sampling frame for this survey.

14. **(a)** What is the size of the sample?

    (b) Calculate the response rate in this survey. Was this survey was subject to nonresponse bias?

15. **(a)** Can you explain the big difference in the data from interviewer to interviewer?

    (b) One of the interviewers conducted the interviews at a street corner downtown. Which interviewer? Explain.

    (c) Do you think the survey was subject to selection bias? Explain.

    (d) Was the sampling method used in this survey the same as quota sampling? Explain.

16. **(a)** Do you think this was a good survey? If you were a consultant to the Cleansburg Planning Department, could you suggest some improvements? Be specific.

Exercises 17 through 20 refer to the following survey. The dean of students at Tasmania State University wants to determine the percent of undergraduates who tried but could not enroll in Math 101 this semester because of insufficient space. There are 15,000 undergraduates at TSU, so it is decided that the cost of checking with each and every one would be prohibitive. The following method (called systematic sampling) is proposed to choose a representative sample of undergraduates to interview. Start with the registrar's alphabetical listing containing the names of all undergraduates. Randomly pick a number between 1 and 100, and count that far down the list. Take that name and every 100th name after it. (For example, if the random number chosen is 73, then pick the 73rd, 173rd, 273rd, etc., names on the list.) Assume that the survey has a response rate of 0.90.

17. **(a)** Compare and contrast the sampling frame and the target population for this survey.

    (b) Give the exact N-value of the population.

18. **(a)** Find the size n of the sample.

    (b) Find the sampling rate.

19. **(a)** Explain why the method used for choosing the sample is not simple random sampling.

    (b) If all those responding claimed they could not enroll in Math 101, is it more likely the result of sampling variability or sampling bias? Explain.

20. **(a)** If 8 of the students who responded said that they were unable to enroll in Math 101, give a reasonable estimate for the total number of students at the university that were unable to enroll in Math 101.

    (b) Do you think the results of this survey will be reliable? Explain.

Exercises 21 and 22 refer to the following survey. An orange grower wishes to compute the average yield from his orchard. The orchard contains three varieties of trees—50% of his trees are of variety A, 25% of variety B, and 25% of variety C.

21. **(a)** Suppose the grower samples randomly from 300 trees of variety A, 150 trees of variety B, and 150 trees of variety C. What type of sampling is being used?

    (b) Suppose the grower is unable to tell which trees are of which variety. For his sample, he chooses 300 trees that look like they are of variety A, 150 trees that look like variety B, and 150 that appear to be of variety C. What type of sampling is being used?

22. **(a)** Suppose that in his survey, the grower found that each tree of variety A averages 100 oranges, each tree of variety B averages 50 oranges, and each tree of variety C averages 70 oranges. Estimate the average yield per tree of his orchard.

    (b) Is the yield you found in (a) a parameter or a statistic? Explain.

B. The Capture-Recapture Method

23. You want to estimate how many fish there are in a small pond. Let's suppose that you capture $n_1 = 500$ fish, tag them, and throw them back in the pond. After a couple of days, you go back to the pond and capture $n_2 = 120$ fish, of which $k = 30$ are tagged. Give an estimate of the N-value of the fish population in the pond.

24. The following example is based on data given in D. G. Chapman and A. M. Johnson, "Estimation of Fur Seal Pup Populations by Randomized Sampling," *Transactions of the American Fisheries Society*, 97 (July 1968), 264–270. To estimate the population in a rookery, 4965 fur seal pups were captured and tagged in early August. In late August, 900 fur seal pups were captured. Of these, 218 had been tagged. Based on these figures, estimate the population of fur seal pups in the rookery to the nearest hundred.

Exercises 25 through 28 refer to the following situation. You have a very large coin jar full of nickels, dimes, and quarters. You want to have an approximate idea of how much money you have, but you don't want to go through the trouble of counting them all, so you decide to use the capture-recapture method.

For the first sample, you shake the jar well and randomly draw 50 coins. You get 12 quarters, 15 nickels, and 23 dimes. Using a black marker, you mark the 50 coins with a black dot and put them back in the jar.

For the second sample, you shake the jar well and randomly draw another set of 100 coins. You get 28 quarters, 4 of which have black dots; 29 nickels, 5 of which have black dots; and 43 dimes, 8 of which have black dots.

25. Estimate the total number of quarters in the jar.
26. Estimate the total number of nickels in the jar.
27. Estimate the total number of dimes in the jar.
28. Do you think the capture-recapture method is a reliable way to estimate the number of coins in the jar? Explain your answer. Discuss some of the potential pitfalls and issues one should be concerned about.

C. Clinical Studies

Exercises 29 through 32 refer to the following hypothetical study. The manufacturer of a new vitamin (vitamin X) decides to sponsor a study to determine its effectiveness in curing the common cold. Five hundred college students in the San Diego area who are suffering from colds are paid to participate as subjects in this study. They are all given two tablets of vitamin X a day. Based on information provided by the subjects themselves, 457 out of the 500 subjects are cured of their colds within 3 days. The average number of days a cold lasts is 4.87 days. As a result of this study, the manufacturer launches an advertising campaign, claiming that "vitamin X is more than 90% effective in curing the common cold."

29. **(a)** Describe as specifically as you can the target population for this study.
    (b) Compare and contrast the target population and the sampling frame for this study.
    (c) Is selection bias present in the sample?
    (d) Was this health study a controlled experiment?
30. **(a)** Do you think the placebo effect could have played a role in this study?
    (b) List three possible causes other than the effectiveness of vitamin X itself that could have confounded the results of this study.
31. List four different problems with this study that indicate poor design.
32. Make some suggestions for improving the study.

Exercises 33 through 36 refer to the following. A study by a team of Harvard University scientists [Science News, 138, no. 20 (November 17, 1990), 308] found that regular doses of beta carotene (a nutrient common in carrots, papayas, and apricots) may help prevent the buildup of plaque-produced arteriosclerosis (clogging of the arteries), which is the primary cause of heart attacks. The subjects in the study were 333 volunteer male doctors, all of whom had shown some early signs of coronary artery disease. The subjects were randomly divided into two groups. One group was given a 50-milligram beta carotene pill every other day for six years, and the other group was given a similar-looking placebo pill. The study found that the men taking the beta carotene pills suffered 50% fewer heart attacks and strokes than the men taking the placebo pills.

33. Describe as specifically as you can the target population for this study.
34. **(a)** Describe the sample.
    (b) What was the size n of the sample?
    (c) Was the sample chosen by random sampling? Explain.
35. **(a)** Explain why this study can be described as a controlled placebo experiment.
    (b) Describe the treatment group in this study.

    (c) Explain why this study can be described as a randomized controlled experiment.

36. (a) Mention two possible confounding variables in this study.

    (b) Carefully state what a legitimate conclusion from this study might be.

Exercises 37–40 refer to the following. A team of researchers and surgeons at the Houston VA Medical Center randomly divided 180 potential knee surgery patients into three groups. (There were 324 participants who met inclusion criteria for the study, but 144 declined to participate.) The first group received arthroscopic debridement. A second group received arthroscopic lavage. Patients in the third group received skin incisions and underwent a simulated procedure ("sham" surgery) without actual insertion of the arthroscope. The patients in the study [New England Journal of Medicine, 347, No. 2 (July 11, 2002), 81–88] did not know which group they were being divided into and therefore did not know if they were receiving the real or simulated surgery. All the patients who participated in the study were evaluated for two years after the procedure. In the two-year follow-up, all three groups said they had slightly less pain and better knee movement. However, the sham-surgery group often reported the best results.

37. Describe as specifically as you can the target population for this study.

38. (a) Describe the sample.

    (b) What was the size n of the sample?

    (c) Was the sample chosen by random sampling? Explain.

39. (a) Was this study a controlled placebo experiment? Explain.

    (b) Describe the treatment group(s) in this study.

    (c) Could this study be considered a randomized controlled experiment? Explain.

    (d) Was this experiment blind, double blind, or neither?

40. (a) Discuss any ethical dilemmas in this type of study involving sham surgery.

    (b) Carefully state what a legitimate conclusion from this study might be.

Exercises 41 through 44 refer to the following hypothetical situation. A college professor has a theory that a dose of about 500 milligrams of caffeine a day can actually improve students' performance in their college courses. To test his theory, he chooses the 13 students in his Psychology 101 class who failed the first midterm and asks them to come to his office three times a week for individual tutoring. When the students come to his office, he engages them in friendly conversation, while at the same time pouring them several cups of strong coffee (a total of 500 milligrams of caffeine per student). After a month of doing this, he observes that of the 13 students, 8 show significant improvement in their second midterm scores, 3 show some improvement, and 2 show no improvement at all. Based on this, he concludes that his theory about caffeine is correct.

41. Which of the following terms best describes the professor's study: (i) randomized controlled experiment, (ii) double-blind experiment, (iii) controlled placebo experiment, or (iv) clinical study? Explain your choice and why you ruled out the other choices.

42. (a) Describe the target population and the sample of this study.

    (b) What was the value of n?

(c) Which of the following percentages best describes the sampling rate for this study: (i) 10%, (ii) 1%, (iii) 0.1%, (iv) 0.01%, or (v) less than 0.01%? Explain.

43. (a) Was the study blind, double blind, or neither? Explain.

(b) List at least three possible causes other than caffeine that could have confounded the results of this study.

44. Make some suggestions to the poor professor as to how he might improve the study.

JOGGING

45. **Informal surveys.** In everyday life, we are constantly involved in activities that can be described as *informal surveys*, often without even realizing it. Here are some examples.

(i) Al gets up in the morning and wants to know what kind of day it is going to be, so he peeks out the window. He doesn't see any dark clouds, so he figures it's not going to rain.

(ii) Betty takes a sip from a cup of coffee and burns her lips. She concludes the coffee is too hot and decides to add a tad of cold water to it.

(iii) Carla got her first Math 101 exam back with a C grade on it. The students sitting on each side of her also received C grades. She concludes that the entire Math 101 class received a C on the first exam.

For each of the preceding examples,

(a) describe the population.

(b) discuss whether the sample is random or not.

(c) discuss the validity of the conclusions drawn. (There is no right or wrong answer to this question, but you should be able to make a reasonable case for your position.)

46. Read the examples of informal surveys given in Exercise 45. Give three more examples of your own. Make them as different as possible from the ones given in Exercise 45 [for example, changing coffee to tea or soup in (ii) is not acceptable].

47. **Leading-question bias.** The way the questions in many surveys are phrased can itself be a source of bias. When a question is worded in such a way as to predispose the respondent to provide a particular response, the results of the survey are tainted by a special type of bias called *leading-question bias*. The following is an extreme hypothetical situation intended to drive the point home.

In an effort to find out how the American taxpayer feels about a tax increase, the institute conducts a "scientific" one-question poll.

Are you in favor of paying higher taxes to bail the federal government out of its disastrous economic policies and its mismanagement of the federal budget? Yes _____. No _____.

Ninety-five percent of the respondents answered no.

(a) Explain why the results of this survey might be invalid.

(b) Rephrase the question in a neutral way. Pay particular attention to highly charged words.

(c) Make up your own (more subtle) example of leading-question bias. Analyze the critical words that are the cause of bias.

48. Consider the following hypothetical survey designed to find out what percentage of people cheat on their income taxes. Fifteen hundred taxpayers are randomly selected from the Internal Revenue Service (IRS) rolls. These individuals are then interviewed in person by representatives of the IRS and read the following statement.

> *This survey is for information purposes only. Your answer will be held in strict confidence. Have you ever cheated on your income taxes? Yes _____. No _____.*

Twelve percent of the respondents answered yes.

(a) Explain why the above figure might be unreliable.

(b) Can you think of ways in which a survey of this type might be designed so that more reliable information could be obtained? In particular, discuss who should be sponsoring the survey and how the interviews should be carried out.

49. Listing bias. Today, most consumer marketing surveys are conducted by telephone. In selecting a sample of households that are representative of all the households in a given geographical area, the two basic techniques used are (i) randomly selecting telephone numbers to call from the local telephone directory or directories, and (ii) using a computer to randomly generate 7-digit numbers to try that are compatible with the local phone numbers.

(a) Briefly discuss the advantages and disadvantages of each technique. In your opinion, which of the two will produce the more reliable data? Explain.

(b) Suppose that you are trying to market burglar alarms in New York City. Which of the two techniques for selecting the sample would you use? Explain your reasons.

50. The following two surveys were conducted in January 1991 in order to assess how the American public viewed media coverage of the Persian Gulf war.

Survey 1 was an Area Code 900 telephone poll survey conducted by "ABC News." Viewers were asked to call a certain 900 number if they felt that the media was doing a good job of covering the war and a different 900 number if they felt that the media was not doing a good job in covering the war. Each call cost 50 cents. Of the 60,000 respondents, 83% felt that the media was not doing a good job.

Survey 2 was a telephone poll of 1500 randomly selected households across the United States conducted by the *Times-Mirror* survey organization. In this poll, 80% of the respondents indicated that they approved of the press coverage of the war.

(a) Briefly discuss survey 1, indicating any possible types of bias.

(b) Briefly discuss survey 2, indicating any possible types of bias.

(c) Can you explain the discrepancy between the results of the two surveys?

(d) In your opinion, which of the two surveys gives the more reliable data?

51. (a) For the capture–recapture method to give a reasonable estimate of N, what assumptions about the two samples must be true?

(b) Give reasons why in many situations, the assumptions in (a) may not hold true.

52. An article in the *Providence Journal* about automobile accident fatalities includes the following observation: "Forty-two percent of all fatalities occurred on Friday, Saturday, and Sunday, apparently because of increased drinking on the weekends."

(a) Give a possible argument as to why the conclusion drawn may not be justified by the data.

(b) Give a different possible argument as to why the conclusion drawn may be justified by the data after all.

PROJECTS AND PAPERS

A. What's the Latest?

In this project you are to do an in-depth report on a recent study. Find a recent article from a newspaper or news magazine reporting the results of a major study and write an analysis of the study. Discuss the extent to which the article gives the reader enough information to assess the validity of the study's conclusions. If in your opinion there is information missing in the article, generate a list of questions that would help you further assess the validity of the study's conclusions. Pay particular attention to those ideas discussed in this chapter including the target population, sample size, sampling bias, randomness, controls, and so on.

Note: The best reporting on clinical studies can be found in major newspapers such as the *New York Times, Washington Post*, and *Los Angeles Times* or weekly news magazines such as *Time* or *Newsweek*. All of these have Web sites in which their most recent articles can be downloaded for free.

B. The Governing "Body"

One of the biggest political upsets in recent years was the election of former wrestler Jesse "The Body" Ventura to governor of Minnesota in 1998. Ventura, the Reform Party candidate, won the three-way race against Republican Norm Coleman and Democrat Hubert "Skip" Humphrey III by a vote of 37 to 34 to 29%, respectively. But the day before the election, the *Star-Tribune*/KMSP-TV Minnesota poll showed Coleman leading at 36% and Ventura tied with Humphrey at 29%. A few days earlier, the *St. Paul Pioneer Press* showed 34% support for Humphrey, 33% for Coleman, and only 23% for Ventura. In fact, not one major poll before the election showed that Ventura had much of a chance.

Write an analysis paper on the 1998 Minnesota gubernatorial election. Discuss the methodology used by pollsters leading up to this election and hypothesize about why these polls failed to predict a Ventura victory. Then develop a list of lessons learned by polling organizations from this election.

C. The U.S. Census: A Continuing Political Battle

In January 1999, the United States Supreme Court ruled 5 to 4 (*Department of Commerce et al. v. U.S. House of Representatives et al.*) that the state population figures used for the apportionment of seats in the House of Representatives (see Chapter 4) cannot be determined by means of statistical sampling methods and that an actual census of the population is required by the Census Act. The ruling was based on arguments presented in the earlier case *Glavin v. Clinton*. In an effort to gain a seat in the House of Representatives, the state of Utah has recently unsuccessfully raised similar arguments with the Court ([*Utah et al. v. Evans, Secretary of Commerce, No. 01-283, 2002*] and [*Utah et al. v. Evans, Secretary of Commerce, No. 01-714, 2002*]).

In this project, you are to use the Court decisions cited as a guide in writing a summary of arguments both for and against the Census Bureau's current methodologies. Such arguments may mirror those made by Republicans and Democrats in this continuing political battle.

D. The Placebo Effect: Myth or Reality?

There is no consensus among researchers conducting clinical studies as to the true impact of the placebo effect. According to some researchers, as many as 30% of patients in a clinical trial can be impacted by it; according to others the placebo effect is a myth.

Write a paper on the *placebo effect*. Discuss the history of this idea, the most recent controversies regarding whether it truly exists or not and why is it important to determine its true impact.

E. The Pepsi Challenge

In 1975, Pepsi introduced a marketing campaign called the Pepsi Challenge. During blind taste tests, participants sipped both Coca-Cola and Pepsi-Cola. Even when Coca-Cola tried running their own such blind tests, a majority of participants chose Pepsi over Coke. Despite market research that indicated a change in the taste of Coca-Cola would hurt sales, these taste tests and the associated campaign led Coca-Cola to introduce "New Coke" in 1985. Of course, "New Coke" turned out to be one of the biggest marketing disasters in corporate history.

In this project, you are to discuss the scientific accuracy of the cola taste tests, reasons why the taste test results may have been skewed, and reasons why "New Coke" failed. Then assume you are a statistical consultant to Coca-Cola and develop a strategy using surveys and clinical studies designed to improve the company's market share.

F. Ethical Issues in Clinical Studies

Until the world was exposed to the atrocities committed by Nazi physicians during World War II, there was little consensus regarding the ethics of medical experiments involving human subjects. Following the trial of the Nazi physicians in 1946, the Nuremberg Code of Ethics was developed to deal with the ethics of clinical studies on human subjects. Sadly, unethical experimentation continued. In the Tuskegee syphilis study starting in 1932, 600 low-income African-American males (400 infected with syphilis) were monitored. Even though a proven cure (penicillin) became available in the 1950s, the study continued until 1972 with participants denied treatment. As many as 100 subjects died from the disease.

In this project you are to write a paper summarizing current ethical standards for the treatment of humans in experimental studies. Sources for your summary may include the Belmont Report (1979), the Declaration of Helsinki (1964), and the Nuremberg Code (1947).

REFERENCES AND FURTHER READINGS

1. Anderson, M. J., and S. E. Fienberg, *Who Counts? The Politics of Census-Taking in Contemporary America*. New York: The Russell Sage Foundation, 1999.
2. Day, Simon, *Dictionary for Clinical Trials*. New York: John Wiley and Sons, Inc., 1999.
3. Francis, Thomas, Jr., et al., "An Evaluation of the 1954 Poliomyelitis Vaccine Trials—Summary Report, *American Journal of Public Health*, 45 (1955), 1–63.
4. Freedman, D., R. Pisani, R. Purves, and A. Adhikari, *Statistics*, 2d ed. New York: W.W. Norton, Inc., 1991, chaps. 19 and 20.
5. Gallup, George, *The Sophisticated Poll Watcher's Guide*. Princeton, N.J.: Princeton Public Opinion Press, 1972.
6. Gleick, James, "The Census: Why We Can't Count," *New York Times Magazine* (July 15, 1990), 22–26, 54.
7. Matthews, J. N. S., *An Introduction to Randomized Controlled Clinical Trials*. London, England: Edward Arnold, 2000.
8. Meier, Paul, "The Biggest Public Health Experiment Ever: The 1954 Field Trial of the Salk Poliomyelitis Vaccine," in *Statistics: A Guide to the Unknown*, 3d ed., ed. Judith M. Tanur et al. Belmont, CA: Wadsworth, Inc., 1989, 3–14.
9. Mosteller, F., et al., *The Pre-election Polls of 1948*. New York: Social Science Research Council, 1949
10. Paul, John, *A History of Poliomyelitis*. New Haven, CT: Yale University Press, 1971.
11. Scheaffer, R. L., W. Mendenhall, and L. Ott, *Elementary Survey Sampling*. Boston: PWS-Kent, 1990.
12. Utts, Jessica M., *Seeing Through Statistics*. Belmont, CA: Wadsworth, Inc., 1996.
13. Warwick, D. P., and C. A. Lininger, *The Sample Survey: Theory and Practice*. New York: McGraw-Hill Book Co., 1975.
14. Yates, Frank, *Sampling Methods for Censuses and Surveys*. New York: Macmillan Publishing Co., Inc., 1981.
15. Zivin, Justin A., "Understanding Clinical Trials," *Scientific American*, 282 (April, 2000), 69–75.

Descriptive Statistics

Graphing and Summarizing Data

Data[1] are the building blocks in the language of statistics. The primary purpose of collecting data is to give meaning to a statistical story, to uncover some new fact about our world, and—last but certainly not least—to make a point, no matter how outlandish. But how is this done?

Imagine yourself following the ups and downs of one company's stock in the stock market—a perfectly reasonable thing to imagine. There are data to track, but the data come in small doses, and we are able to make sense out of it. Now try to imagine doing the same thing for an entire stock market, with hundreds or thousands of individual stocks to follow. The amount of data to deal with becomes overwhelming—a huge babble of numbers.

> In God we trust;
> All others must bring data!
> W. Edwards Deming

There comes a point when a list of numbers is too long for the human mind to digest or comprehend, and that point comes surprisingly early (at five or, at the most, six numbers, psychologists believe). What do we do when we have too much data?

One important purpose of statistics is to describe large amounts of data in a way that is understandable, useful, and, if need be, convincing. This area, called *descriptive statistics*, is the subject of this chapter.

There are two strategies for describing data. One is to present the data in the form of pictures or graphs; the other is to use numerical summaries that serve as "snapshots" of the data. Sometimes we even combine the two strategies, using pictures and numerical summaries together.

We will start our discussion of descriptive statistics with an introduction to some of the basic methods for displaying data graphically.

---

[1]In current usage, the word *data* can represent both the singular and plural forms of the word, though usually the plural form is used in print.

14.1 Graphical Descriptions of Data

Data Sets

A **data set** is a collection of data values. Statisticians often refer to the individual data values in a data set as **data points**. For the sake of simplicity, we will work with data sets in which each data point consists of a single number, but in more complicated settings, a single data point can consist of many numbers.

As usual, we will use the letter N to represent the size of the data set. In real-life applications, data sets can range in size from reasonably small (a dozen or so data points) to very large (hundreds of millions of data points), and the larger the data set is, the more we need a good way to describe and summarize it.

To illustrate many of the ideas of this chapter we will need a reasonable data set—big enough to be realistic, but not so big that it will bog us down. Example 14.1, which we will revisit several times in the chapter, provides such a data set. This is a fictitious data set from a hypothetical statistics class, but except for the details, it describes a situation that is familiar to every college student.

EXAMPLE 14.1 The Stat 101 Midterm Scores

As usual, the day after the midterm exam in his Stat 101 class, Professor Blackbeard has posted the results in the hallway outside his office (Table 14-1). The data set consists of $N = 75$ data points. Each data point (listed in the second column) is a raw score on the midterm between 0 and 25 (Professor Blackbeard gives no partial credit). Note that the student IDs in Table 14-1 are numbers but not data—they are used as a substitute for names to protect the students' rights of privacy.

TABLE 14-1 Stat 101 Midterm Exam Scores (25 Points Possible); $N = 75$.

ID	Score	ID	Score	ID	Score	ID	Score	ID	Score
1257	12	2651	10	4355	8	6336	11	8007	13
1297	16	2658	11	4396	7	6510	13	8041	9
1348	11	2794	9	4445	11	6622	11	8129	11
1379	24	2795	13	4787	11	6754	8	8366	13
1450	9	2833	10	4855	14	6798	9	8493	8
1506	10	2905	10	4944	6	6873	9	8522	8
1731	14	3269	13	5298	11	6931	12	8664	10
1753	8	3284	15	5434	13	7041	13	8767	7
1818	12	3310	11	5604	10	7196	13	9128	10
2030	12	3596	9	5644	9	7292	12	9380	9
2058	11	3906	14	5689	11	7362	10	9424	10
2462	10	4042	10	5736	10	7503	10	9541	8
2489	11	4124	12	5852	9	7616	14	9928	15
2542	10	4204	12	5877	9	7629	14	9953	11
2619	1	4224	10	5906	12	7961	12	9973	10

Like students everywhere, the students in the Stat 101 class have one question foremost on their mind when they look at Table 14-1: How did I do? Each student can answer this question directly from the table. It's the next question that is statistically much more interesting. How did the class as a whole do?

To answer this last question, we will have to find a way to package the information in Table 14-1 into a compact, organized, and intelligible whole. There are indeed many ways to do this—from simple to fancy, from straightforward to misleading, from functional to artistic.

In the next few sections, we will discuss some of the basic ways in which to describe data using visual images. (For a more in-depth treatment of the art and science of graphically displaying data, the reader is encouraged to look at references 2, 8, 9, 10, and 14.)

Bar Graphs and Variations Thereof

The first important step in summarizing the information in Table 14-1 is to put the scores into a **frequency table**, as shown in Table 14-2.

TABLE 14-2	**Frequency Table for the Stat 101 Data Set**												
Exam score	1	6	7	8	9	10	11	12	13	14	15	16	24
Frequency	1	1	2	6	10	16	13	9	8	5	2	1	1

The number below each score represents the **frequency** of the score—that is, the number of students getting that particular score. We can readily see from Table 14-2 how many students got what score. There was one student with a score of 1, one with a score of 6, two with a score of 7, six with a score of 8, and so on. Note that the scores with a frequency of zero are not listed in the table.

While Table 14-2 is a considerable improvement over Table 14-1, we can do even better. Figure 14-1 shows the same information in a much more visual way called a **bar graph**, with the test scores listed in increasing order on a horizontal axis and the frequency of each test score displayed by the *height* of the column above that test score. Notice that in the bar graph, even the test scores with a frequency of zero show up—there simply is no column above these scores.

FIGURE 14-1
Bar graph for the Stat 101 data set.

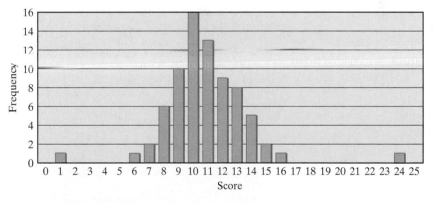

Bar graphs are easy to read, and they are a nice way to present a good general picture of the data. With a bar graph, for example, it is easy to detect **outliers**—extreme data points that do not fit into the overall pattern of the data. In this example there are two obvious outliers—the score of 24 (head and shoulders above the rest of the class) and the score of 1 (lagging way behind the pack).

If the frequencies are large numbers, it is customary to describe the bar graph in term of *relative frequencies*—that is, the frequencies expressed as percentages of the total population. Figure 14-2 shows a *relative frequency bar graph* for the Stat 101 data set. Note that we indicated on the graph that we are dealing with percentages rather than total counts and that the size of the data set is $N = 75$. Letting the viewer know the size of the data set in a relative frequency bar graph is important, because it allows anyone who wishes to do so to compute the actual frequencies (*actual frequency = percentage × N/*100). The change from actual frequencies to percentages does not change the shape of the graph—it is basically a change of scale.

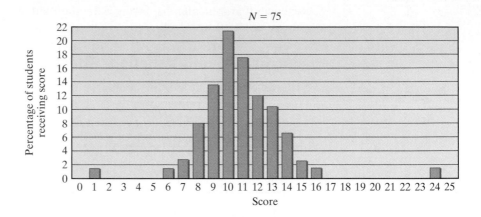

FIGURE 14-2

Relative frequency bar graph for the Stat 101 data set.

While the term *bar graph* is most commonly used for graphs like the ones in Figs. 14-1 and 14-2, devices other than bars can be used to add a little extra flair or to subtly influence the content of the information given by the raw data. Professor Blackbeard, for example, could have chosen to display the midterm data using a graph like the one shown in Fig. 14-3, which conveys all the information of the original bar graph (Fig. 14-1) and also includes a subtle individual message to each student.

The general point here is that a bar graph is often used not only to inform, but also to impress and persuade, and, in such cases, a well-chosen design for the frequency columns can be more effective than just a bar. Graphs such as those in Figs. 14-3 and 14-4, which use icons instead of bars to show the frequencies, are commonly referred to as **pictograms**.

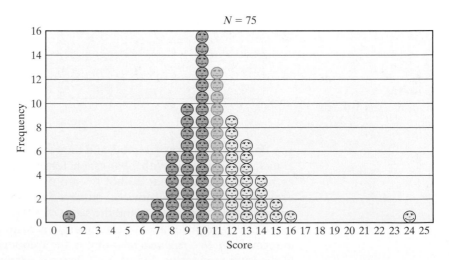

FIGURE 14-3

Frequency chart for the Stat 101 data set.

EXAMPLE 14.2

Figure 14-4 is a pictogram showing the growth in yearly sales of the XYZ Corporation over the period from 1997 through 2002. It looks very impressive, but the picture is actually quite misleading. Figure 14-5 shows a pictogram for exactly the same data with a much more accurate and sobering picture of how well XYZ corporation had been doing.

The difference between the two pictograms can be attributed to a couple of standard tricks of the trade: (1) stretching the scale of the vertical axis and (2) "cheating" on the choice of starting value on the vertical axis. As an educated consumer, you should always be on the lookout for these tricks. In graphical descriptions of data, a fine line separates objectivity from propaganda. (For more on this topic, see Project A.)

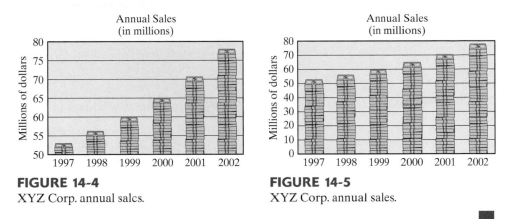

FIGURE 14-4
XYZ Corp. annual sales.

FIGURE 14-5
XYZ Corp. annual sales.

14.2   Variables: Quantitative and Qualitative; Continuous and Discrete

Before we continue with our discussion of graphs, we need to discuss briefly the concept of a **variable**. In statistical usage, a variable is any characteristic that varies with the members of a population. The students in Professor Blackbeard's Stat 101 course (the population) did not all perform equally on the exam. Thus, the *test score* is a variable, which, in this particular case is a whole number between 0 and 25. In some instances, such as when the instructor gives partial credit or when there is subjective grading, a test score may take on a fractional value, such as 18.5 or 18.25. Even in these cases, however, the possible increments for the values of the variable are given by some minimum amount: a quarter-point, a half-point, whatever. In contrast to this situation, consider a different variable: the *amount of time* each student studied for the exam. In this case the variable can take on values that differ by any amount: an hour, a minute, a second, a tenth of a second, and so on.

A variable that represents a measurable quantity is called a **numerical** (or *quantitative*) variable. When the difference between the values of a numerical variable can be arbitrarily small, we call the variable **continuous**; when possible values of the numerical variable change by minimum increments, the variable is called **discrete**. Examples of *discrete* variables are a person's IQ, an SAT score, a person's shoe size, and the number of points scored in a basketball game.

Examples of *continuous* variables are a person's height, weight, foot size (as opposed to shoe size), and the time it takes them to run a mile.

Sometimes, in the real world, the distinction between continuous and discrete variables is blurred. Height, weight, and age are all continuous variables in theory, but in practice they are frequently rounded off to the nearest inch, ounce, and year (or month in the case of babies), respectively, at which point they become discrete variables. On the other hand, money, which is in theory a discrete variable (because the difference between two values cannot be less than a penny), is almost always thought of as continuous, because in most real-life situations a penny can be thought of as an infinitesimally small amount of money.

Variables can also describe characteristics that cannot be measured numerically: nationality, gender, hair color, and so on. Variables of this type are called **categorical** (or *qualitative*) variables.

In some ways, categorical variables must be treated differently from numerical variables—they cannot, for example, be added, multiplied, or averaged. In other ways, categorical variables can be treated much like discrete numerical variables, particularly when it comes to graphical descriptions, such as bar graphs and pictograms.

EXAMPLE 14.3 Enrollment (by School) at Tasmania State University

Table 14-3 shows undergraduate enrollments in each of the 5 schools at Tasmania State University. A sixth category ("Other") includes undeclared students, interdisciplinary majors, and so on.

TABLE 14-3
Undergraduate Enrollments at TSU

School	Enrollment
Agriculture	2400
Business	1250
Education	2840
Humanities	3350
Science	4870
Other	290

Figure 14-6 displays the information in Table 14-3 in the form of a bar graph. The bar graph is very similar to the one in Fig. 14-1 except that the variable along the horizontal axis is categorical.

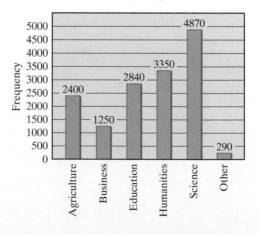

FIGURE 14-6
Undergraduate enrollments at TSU (by school).

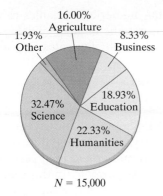

N = 15,000

FIGURE 14-7
Pie chart showing undergraduate enrollments at TSU (by school).

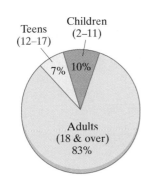

FIGURE 14-8
Audience composition for prime-time TV viewership by age group.
(***Source:*** Nielsen Media Research.)

When the number of categories is small, as it is in Example 14.3, another commonly used way to describe relative frequencies of a population by categories is the **pie chart**. The "pie" represents the entire population (100%), and the "slices" represent the categories or classes, with the size (area) of each slice being proportional to the relative frequency of the corresponding category. Some relative frequencies, such as 50% and 25%, are very easy to sketch; but how do we accurately draw the slice corresponding to a more complicated frequency, say, 32.47%? Here, a little elementary geometry comes in handy. Since 100% equals $360°$, 1% corresponds to an angle of $360°/100 = 3.6°$. It follows that the frequency 32.47% is given by $32.47 \times 3.6° = 117°$ (rounded to the nearest degree, which is generally good enough for most practical purposes). Figure 14-7 shows the school-enrollment data in Example 14.3 described by a pie chart.

Bar graphs and pie charts are excellent ways to graphically display categorical data, but, as always, we should be wary of jumping to hasty conclusions based on what we see on a graph. Our next example illustrates this point.

EXAMPLE 14.4 Who's Watching the Boob Tube Tonight?

According to Nielsen Media Research data, the percentages of the TV audience watching TV during prime time (8 P.M. to 11 P.M.), broken up by age group, are as follows: adults (18 years and over), 83%; teenagers (12–17 years), 7%; children (2–11 years), 10%.[2] The pie chart in Fig. 14-8 shows this breakdown of audience composition by age group.

When looking at this pie chart, one is tempted to conclude that, at least during prime time, children and teenagers do not watch much TV. Could all the reports we read about how much TV young people watch be wrong?

The problem with this pie chart is that, while accurate, it is also very misleading. Children (2–11 years) make up only 15% of the population at large, teens (12–17) make up only 8%, and adults make up the rest. Given that there are more than 5 times as many adults as there are children, is it any wonder that there are more prime-time TV-viewing adults than there are prime-time TV-viewing children? Likewise, in absolute terms, there are more TV-viewing children than TV-viewing teenagers, but then, there are more children than there are teenagers. In relative terms, a higher percentage of teenagers (taken out of the total teenage population) watch prime-time TV than children. (This is not all that surprising, given that most children's bedtimes are around 8 P.M.). ∎

The moral of Example 14.4 is that using absolute percentages, as we did in Fig. 14-8, can be quite misleading. When comparing characteristics of a population that is broken up into categories, it is essential to take into account the relative sizes of the various categories.

Class Intervals

While the distinction between qualitative and quantitative data is important in many aspects of statistics, when it comes to deciding how best to display graphically the frequencies of a population, a critical issue is the number of categories into which the data can fall. When the number of categories is too big (say, in the dozens), a bar graph or pictogram can become muddled and ineffective. This

---

[2]These figures are rough approximations based on information taken from *The World Almanac* and averaged over several years. The exact figures vary from year to year.

happens more often than not with quantitative data—numerical variables can take on infinitely many values, and even when they don't, the number of values can be too large for any reasonable graph. Our next example illustrates how to deal with this situation.

EXAMPLE 14.5 SAT Scores

In 2001, 1,276,320 college-bound seniors took the SAT. Suppose that we want to look at the math SAT scores for these students. In theory, this situation is no different from the one illustrated in Example 14.1 (Stat 101 midterm scores). Just as in Example 14.1, our data represent a *discrete quantitative variable* (in this case, math SAT scores). In practice, because of the extremely large number of possible SAT scores (which are given in 10-point increments and range between 200 and 800) we must deal with such data differently. The standard way to display bar graphs in this situation is to break up the range of scores into **class intervals**.

The decision as to how the class intervals are defined and how many there are will depend on the statistical story you are trying to tell, but as a general rule of thumb, the number of class intervals should be somewhere between 5 and 20.

In this example, a sensible thing to do might be to break up the SAT scores into 12 class intervals of essentially the same size (the only exception being the class interval 750–800 which has one more test score than the others). In this case a bar graph for the SAT scores would look something like Fig. 14-9.

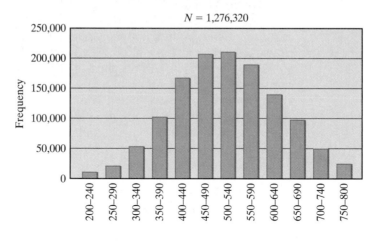

FIGURE 14-9
2001 SAT math scores.
Source: The College Board.

Class intervals need not be of equal size and sometimes, as our next example shows, it makes more sense to choose class intervals of different sizes.

EXAMPLE 14.6 Midterm Grades

Imagine now that Professor Blackbeard wants to convert the test scores in the Stat 101 data set into letter grades. In our terminology, this means converting a *numerical* variable (test score) into a *categorical* one (letter grade) by defining *class intervals* associated with each grade category (A, B, C, D, and F). In this case there is a good reason not to use class intervals of equal length. Professor Blackbeard decides to define the grade intervals for this particular exam according to the breakdown shown in Table 14-4.

TABLE 14-4		TABLE 14-5		Stat 101 Grade Distribution
Grade	**Exam score**	**Grade**	**Frequency**	**Percentage**
A	18–25	F	10	13.33%
B	14–17	D	26	34.67%
C	11–13	C	30	40%
D	9–10	B	8	10.67%
F	0–8	A	1	1.33%

If we combine the Stat 101 exam scores (see Table 14-2) with the grade intervals as defined in Table 14-4, we get a new frequency table (Table 14-5) and a corresponding bar graph for the grade distribution in the exam (Fig. 14-10).

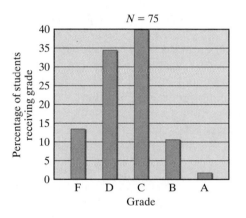

FIGURE 14-10
Bar graph for the Stat 101 test grades based on Table 14-5.

Histograms

When a numerical variable is continuous, its possible values can vary by infinitesimally small increments. As a consequence, there are no gaps between the class intervals, and our old way of doing things (using separated columns or stacks) will no longer work. In this case we use a variation of a bar graph called a **histogram**. We illustrate the concept of a histogram in the next example.

EXAMPLE 14.7 Starting Salaries of TSU Graduates

Suppose we want to use a graph to display the distribution of starting salaries for last year's graduating class at Tasmania State University.

The starting salaries of the $N = 3258$ graduates range from a low of $40,350 to a high of $74,800. Based on this range and the amount of detail we want to show, we must decide on the length of the class intervals. A reasonable choice would be to use class intervals defined in increments of $5000. Table 14-6 is a frequency table for the data based on these class intervals. We chose a starting value of $40,000 for convenience. (The third column in the table shows the data as a percentage of the population.)

TABLE 14-6	Starting Salaries of First-Year TSU Graduates	
Salary	**Number of students**	**Percentage**
40,000⁺–45,000	228	7%
45,000⁺–50,000	456	14%
50,000⁺–55,000	1043	32%
55,000⁺–60,000	912	28%
60,000⁺–65,000	391	12%
65,000⁺–70,000	163	5%
70,000⁺–75,000	65	2%
Total	3258	100%

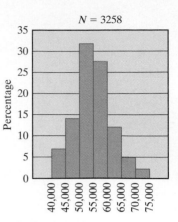

FIGURE 14-11
Histogram for starting salaries of first-year graduates of TSU (with class intervals of $5000).

The histogram showing the relative frequency of each class interval is shown in Fig. 14-11. As we can see, a histogram is very similar to a bar graph. Several important distinctions must be made, however. To begin with, because a histogram is used for continuous variables, there can be no gaps between the class intervals, and it follows, therefore, that the columns of a histogram must touch each other. Among other things, this forces us to make an arbitrary decision as to what happens to a value that falls exactly on the boundary between two class intervals. Should it always belong to the class interval to the left or to the one to the right? This is called the *endpoint convention*. The superscript "plus" marks in Table 14-6 indicate how we chose to deal with the endpoint convention in Fig. 14-11. A starting salary of exactly $50,000, for example, would be listed under the 45,000⁺–50,000 class interval, rather than the 50,000⁺–55,000 class interval. ■

When creating histograms, we should try, as much as possible, to define class intervals of equal length. When the class intervals are of unequal length, the rules for creating a histogram are considerably more complicated, since it is no longer appropriate to use the heights of the columns to indicate the frequencies of the class intervals. We will not discuss the details of this situation here, but we refer the interested reader to Exercises 75 and 76 at the end of the chapter.

14.3 Numerical Summaries of Data

As we have seen, a picture can be an excellent tool for summarizing large data sets. Unfortunately, circumstances do not always lend themselves equally well to the use of pictures, and bar graphs and pie charts cannot be readily used in everyday conversation. A different and very important approach is to use a few well-chosen numbers to summarize an entire data set.

In this section we will discuss several of the most commonly used *numerical summaries* of a data set. We will discuss two types of numerical summaries of data: numbers that tell us something about the actual values of the data set, and numbers that tell us something about the spread of the values in the data set. The former numbers are called **measures of location**, the latter **measures of spread**. The most important measures of location are the *average* (or *mean*), the *median*, and the *quartiles*. The most important measures of spread are the *range*, the *interquartile range*, and the *standard deviation*. We will discuss each of these in order.

The Average

The best known of all numerical summaries of data is the *average*, also called the *mean*. (As much as possible, we will stay with the word "average" but in some settings the word "mean" is more common, and in such cases we will follow the custom.) The **average** of a set of N numbers is obtained by adding the numbers and dividing by N. When the set of numbers is small, one can often calculate the average in one's head; for larger data sets, pencil and paper or a calculator can be helpful. In either case, the idea is very straightforward.

EXAMPLE 14.8 Average Points per Game

In 10 playoff games, NBA star "Hoops" Tallman scores 8, 5, 11, 7, 15, 0, 7, 4, 11, and 14 points. His total is 82 points in 10 games. His playoff average is 8.2 points per game. Note that it is actually impossible to score 8.2 points. As it often happens, an average can be an impossible data value. ∎

EXAMPLE 14.9 The Average Test Score in the Stat 101 Test

Table 14-7 is the same as Table 14-2, shown again for the reader's convenience. We want to calculate the average test score.

TABLE 14-7   Frequency Table for the Stat 101 Data Set

Exam score	1	6	7	8	9	10	11	12	13	14	15	16	24
Frequency	1	1	2	6	10	16	13	9	8	5	2	1	1

The 75 data values can be totaled by taking each score, multiplying it by its corresponding frequency, and adding. In this case, we get

$$(1 \times 1) + (6 \times 1) + (7 \times 2) + (8 \times 6) + \ldots + (16 \times 1) + (24 \times 1) = 814$$

The average score on the midterm exam (rounded to two decimal places) is

$$814 \div 75 \approx 10.85 \text{ points.}$$

Intuitively, we think of this average as representing a typical student's score. If all test scores had been about the same, then, given the same total, each score would have been about 10.85 points. ∎

Table 14-8 shows a generic frequency table. To find the average of the data we do the following:

TABLE 14-8

Data	Frequency
s_1	f_1
s_2	f_2
$\vdots$	$\vdots$
s_k	f_k

- **Step 1.** Calculate the total of the data.
  Total $= s_1 \cdot f_1 + s_2 \cdot f_2 + \cdots + s_k \cdot f_k$
- **Step 2.** Calculate N.
  $N = f_1 + f_2 + \cdots + f_k$
- **Step 3.** Calculate the average.
  Average $=$ total $\div N$

Sometimes, averages can be deceiving, as illustrated in our next example.

EXAMPLE 14.10 Starting Salaries of Philosophy Majors

The average annual starting salary for the 75 philosophy majors who recently graduated from Tasmania State University is $76,400. This is an impressive figure, but before we all rush out to change majors, consider the fact that one of these graduates is professional basketball star "Hoops" Tallman, whose starting salary is a whopping $3.5 million a year.

If we were to disregard this one outlier, the average annual starting salary for the remaining 74 philosophy majors could be computed as follows

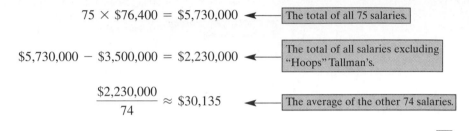

$$75 \times \$76,400 = \$5,730,000 \quad \longleftarrow \boxed{\text{The total of all 75 salaries.}}$$

$$\$5,730,000 - \$3,500,000 = \$2,230,000 \quad \longleftarrow \boxed{\begin{array}{l}\text{The total of all salaries excluding} \\ \text{"Hoops" Tallman's.}\end{array}}$$

$$\frac{\$2,230,000}{74} \approx \$30,135 \quad \longleftarrow \boxed{\text{The average of the other 74 salaries.}}$$

Example 14.10 underscores the point that even a single outlier can have a big impact on the average. We must always be alert to the possibility that an average may have been distorted by one or more outliers. On the other hand, if we know that the data set does not have outliers, we can rely on the average as a useful numerical summary.

So far, all our examples have involved data values that are positive, but negative data values are also possible, and when both negative and positive data values are averaged, the results can be misleading.

EXAMPLE 14.11

The monthly savings (monthly income minus monthly spending) of a college student over a one-year period is shown in Table 14-9. A negative amount indicates that, rather than saving money that month, the student spent more than his monthly income.

TABLE 14-9

Month	Savings (in $)	Comment
Jan.	−732	← Christmas bills
Feb.	−158	
Mar.	−71	
Apr.	−238	
May	1839	← $2000 lottery winnings
Jun.	−103	
Jul.	−148	
Aug.	−162	
Sep.	−85	
Oct.	−147	
Nov.	−183	
Dec.	500	← Christmas present from mom

The average monthly "savings" of this college student over the one-year period is given by

$$\frac{-732 - 158 - 71 - 238 + 1839 - 103 - 148 - 162 - 85 - 147 - 183 + 500}{12} = \$26$$

which is an accurate but deceptive figure. The true picture is that of a student living well beyond his means and bailed out by a lucky break and a generous mom. ■

Percentiles

While a single numerical summary—such as the average—can be useful, it is rarely sufficient to give a meaningful description of a data set. A better picture of the data set can be presented by using a well-organized cadre of numerical summaries. The most common way to do this is by means of *percentiles*.

The *p*th **percentile** of a data set is a value such that *p* percent of the numbers fall *at or below* this value and the rest fall *at or above* it. It essentially splits a data set into two parts: the lower *p*% of the data values and the upper $(100 - p)$% of the data values.

Most college students are familiar with percentiles, if for no other reason than the way they pop up in SAT reports. In all SAT reports, a given score—say a score of 610 in the math section—is identified with a percentile, say the 81st percentile. This can be interpreted to mean that 81% of those taking the test scored 610 or less, or, looking up instead of down, that 19% of those taking the test scored 610 or more.

There are several different ways to compute percentiles that will satisfy the definition, and different statistics books describe different methods. We will illustrate one such method below.

The first step in finding the *p*th *percentile* of a data set of *N* numbers is to *sort the numbers by size*. In other words, we *must* rewrite the numbers in the data set in *increasing order from smallest to largest*.

To simplify the explanation, we will now introduce a bit of notation. Let's denote the sorted data set by $\{d_1, d_2, d_3, \ldots, d_N\}$. In this notation, d_1 represents the first number in the sorted data set (the smallest number), d_2 the second number, d_{10} the tenth number, and so on. Sometimes we will also need to talk about the average of two consecutive numbers in our list, so we will use more exotic subscripts such as $d_{3.5}$ to represent the average of the data values d_3 and d_4; $d_{7.5}$ to represent the average of the data values d_7 and d_8, and so on.

The next, and most important step, is to identify which *d* represents the *p*th percentile of the data set. To do this, we compute the *p*th *percent of N*, which we will call the **locator** and denote by the letter *L*. [In other words, $L = (p/100) \cdot N$.] If *L* happens to be a whole number, then the *p*th *percentile* will be $d_{L.5}$ (the average of d_L and d_{L+1}). If *L* is not a whole number, then the *p*th *percentile* will be d_{L^+}, where L^+ represents the value of *L rounded up*.

The entire procedure for finding the *p*th *percentile* of a data set is summarized as follows:

FINDING THE *P*TH PERCENTILE OF *N* NUMBERS

- **Step 1.** Sort the original data set.
    (Let $\{d_1, d_2, d_3, \ldots, d_N\}$ represent the sorted data set.)
- **Step 2.** Compute the value of the *locator*: $L = (p/100) \cdot N$.
- **Step 3.** The *p*th *percentile* is given by
    - **(a)** $d_{L.5}$ if L is a whole number,
    - **(b)** d_{L^+} if L is not a whole number.

[*Notation*: L^+ is L rounded up; $d_{L.5}$ is the average of d_L and d_{L+1}.]

The following example illustrates the procedure for finding percentiles of a data set.

EXAMPLE 14.12

Consider a set of 15 students all of whom are members of the debate team. Their respective GPAs are

3.4, 3.9, 3.3, 3.6, 3.5, 3.4, 4.0, 3.7, 3.3, 3.8, 3.6, 3.9, 3.7, 3.4, 3.6.

This is our original data set, and $N = 15$.

Suppose the university has committed to giving a $5000 scholarship to all students with GPAs at or above the 80th percentile within this group. Which students get a $5000 scholarship?

To compute the 80th percentile, we first sort the GPAs from smallest to largest. The sorted list of GPAs is

$\{3.3, 3.3, 3.4, 3.4, 3.4, 3.5, 3.6, 3.6, 3.6, 3.7, 3.7, 3.8, 3.9, 3.9, 4.0\}$.

The locator L for the 80th percentile is $L = (0.8) \cdot 15 = 12$. Here L is a whole number, so the 80th percentile will be $d_{12.5}$, which denotes the average between $d_{12} = 3.8$ and $d_{13} = 3.9$. Thus, the 80th percentile of this set of GPAs is 3.85. There are exactly three GPAs above 3.85, and those students get the $5000 scholarship.

Now suppose that for some reason we also need to compute the 55th percentile of the above set of GPAs. The locator for the 55th percentile is $L = (0.55) \cdot 15 = 8.25$, which is not a whole number. In this case we round the locator *up* to 9, and the 55th percentile is $d_9 = 3.6$. ∎

The Median and the Quartiles

Undoubtedly, the most commonly used percentile is the 50th percentile. The 50th percentile of a data set is known as the **median** and denoted by M. The median splits a data set into two halves—half of the data is at or below the median and half of the data is at or above the median. The locator for the median is given by $L = (50/100) \cdot N = N/2$. Thus, when N is even, the locator $N/2$ is a whole number and the median is the average of $d_{N/2}$ and $d_{(N/2)+1}$. On the other hand, when N is odd, $N/2$ is not a whole number; when we round it up we get $(N+1)/2$, and the median is $d_{(N+1)/2}$.

FINDING THE MEDIAN OF *N* NUMBERS

- Sort the original data set.
  (Let $\{d_1, d_2, d_3, \ldots, d_N\}$ represent the sorted data set.)
- The *median M* is given by
  (a) the average between $d_{N/2}$ and $d_{(N/2)+1}$ if N is even,
  (b) $d_{(N+1)/2}$ if N is odd.

EXAMPLE 14.13   The Median Test Score for the Stat 101 Test

We will now find the median score for the Stat 101 data set given in Table 14-10.

TABLE 14-10   Frequency Table for the Stat 101 Data Set

Exam score	1	6	7	8	9	10	11	12	13	14	15	16	24
Frequency	1	1	2	6	10	16	13	9	8	5	2	1	1

Having the frequency table available eliminates the need for sorting the scores—the frequency table has, in fact, done this for us. The total number of scores is $N = 75$, which means that the median can be found in position $(N + 1)/2 = 38$ from the left in the frequency table. To find the 38th number in Table 14-10, we tally frequencies as we move from left to right: $1 + 1 = 2$; $1 + 1 + 2 = 4$; $1 + 1 + 2 + 6 = 10$; $1 + 1 + 2 + 6 + 10 = 20$; $1 + 1 + 2 + 6 + 10 + 16 = 36$. At this point, we know that the 36th test score on the list is a 10 (the last of the 10s) and the next 13 scores are all 11s. We can conclude that the 38th test score is 11. Thus, $M = 11$. ∎

After the median, the next most commonly used set of percentiles are the *first* and *third quartiles*. The **first quartile** (denoted by Q_1) is the 25th percentile, and the **third quartile** (denoted by Q_3) is the 75th percentile. Thus, 25% of the data in a data set is less than or equal to Q_1, and 25% of the data is greater than or equal to Q_3.

EXAMPLE 14.14

During the last year, 11 homes sold in the Green Hills subdivision. The selling prices, in chronological order, were $167,000, $152,000, $128,000, $134,000, $192,000, $163,000, $121,000, $145,000, $170,000, $138,000, and $155,000. We would like to find the *median* and the *quartiles* of the $N = 11$ home prices.

If we sort the prices from smallest to largest and drop the 000s, we get the data set

121, 128, 134, 138, 145, 152, 155, 163, 167, 170, 192.

To compute the median home price, we first find the value of the locator. In this case $L = N/2 = 11/2 = 5.5$. Since this number is not a whole number, we round up to 6. Thus, the median is $d_6 = 152$ ($\$152,000$).

To compute the first quartile of the home prices, we find the corresponding locator. In this case $L = 0.25(11) = 2.75$. Since once again this number is not a whole number, we round it up to 3. The first quartile is $d_3 = 134$, so $Q_1 = \$134,000$.

Finally, to find the third quartile we compute its locator $L = 0.75(11) = 8.25$ and round it up to 9. This gives us the third quartile at $d_9 = 167$ (i.e., $Q_3 = \$167,000$). ∎

EXAMPLE 14.14 (continued)

Oops! We just heard that this morning, another home sold in Green Hills for $\$164,000$. We need to recalculate the median and quartiles for what are now $N = 12$ home prices.

We can use the sorted data set that we already had—all we have to do is insert the new home price (164) in the right spot (remember, we drop the 000s!). This gives

$$121, 128, 134, 138, 145, 152, 155, 163, \mathbf{164}, 167, 170, 192.$$

For this data set of 12 numbers, the locator for the median is $L = 0.5(12) = 6$. Since this is a whole number, the median is the average of $d_6 = 152$ and $d_7 = 155$ (i.e., $M = 153.5$).

Note that in this case the median $M = 153.5$ is not an actual number in the data set, but it still cuts the data set exactly in half—half of the homes sold for less than the median and half of the homes sold for more.

The locator for the first quartile is $L = 0.25(12) = 3$. Here again, the locator is a whole number so the first quartile is the average of $d_3 = 134$ and $d_4 = 138$, which means that $Q_1 = 136$.

Similarly, the third quartile is $Q_3 = 165.5$—the average of $d_9 = 164$ and $d_{10} = 167$. ∎

A note of warning: Medians, quartiles, and general percentiles are often computed using statistical calculators or statistical software packages, which is all well and fine, since the whole process can be a bit tedious. The problem is that there is no universally agreed procedure for computing percentiles, so different types of calculators and different statistical packages may give different answers from each other and from those given in this book for quartiles and other percentiles (everyone agrees on the median). *Keep this in mind when doing the exercises—the answer given by your calculator may be slightly different from the one you would get from the procedure we use in the book.*

EXAMPLE 14.15

Let's revisit the Stat 101 midterm data set. Recall that in Example 14.13 we found the median ($M = 11$) working directly from the frequency table. Let's do it again, and now find the first and third quartiles. Once again, it is helpful to have the frequency table in front of you and work directly out of the frequency table.

TABLE 14-11 Frequency Table for the Stat 101 Data Set

Exam score	1	6	7	8	9	10	11	12	13	14	15	16	24
Frequency	1	1	2	6	10	16	13	9	8	5	2	1	1

Since $N = 75$, the locator for the first quartile is $L = 0.25(75) = 18.75$. Thus, $Q_1 = d_{19}$. To find d_{19} in the frequency table (Table 14-11), we tally frequencies from left to right: $1 + 1 = 2$; $1 + 1 + 2 = 4$; $1 + 1 + 2 + 6 = 10$; $1 + 1 + 2 + 6 + 10 = 20$. At this point we realize that $d_{10} = 8$ (the last of the 8s) and that d_{11} through d_{20} all equal 9. Hence, the first quartile of the Stat 101 midterm scores is $d_{19} = 9$.

We leave it to the reader to verify that the third quartile of the Stat 101 scores is d_{57}, and that $d_{57} = 12$. ∎

EXAMPLE 14.16   SAT Scores

College-bound high school seniors have to jump through many hoops, one of the more memorable of which is taking the SAT exam. As most of us know, the SAT exam consists of a verbal section and a math section. In each section, the scores range from a minimum of 200 to a maximum of 800 and go up in increments of 10 points. Because the number of students taking the SAT is very large and heterogeneous, every possible score between 200 and 800 does show up more than once, from the ridiculous 200 (pranksters, one would hope) to the sublime 800.

In 2001, a total of $N = 1,276,320$ high school seniors took the SAT. For the math section, the third quartile score was 590, the median score was 510, and the first quartile score was 440. So what does this mean?

Informally, it means that *about 25% of the students scored 440 or less, about 50% of the students scored 510 or less, and about 75% of the students scored 590 or less*. This is a correct, but somewhat informal way to interpret the data. We are now in a position to be a bit more precise.

Since $N = 1,276,320$ is an even number, the median score is the average of $d_{638,160}$ and $d_{638,161}$, where the d's represent the SAT scores as they would appear if they were sorted in a list (see Exercise 40). This tells us that there were *at least* 638,160 students who scored 510 or less. Why did we use "at least" in the preceding sentence? Could there have been more than that number who scored 510 or less? Yes, almost surely. Since the number of students who scored 510 is in the thousands, it is very unlikely that the 638,160th score is the last of the 510s.

In a similar vein, we can say that the first quartile of 440 points is the average of $d_{319,080}$ and $d_{319,081}$ [see Exercise 40(b)]. This tells us that there were at least 319,080 students who scored 440 points or less. Finally, the fact that the third quartile was 590 points means that there were at least 957,240 students who scored 590 points or less. Of course, this also means that there were 319,080 students who scored 590 points *or more*. ∎

The Five-Number Summary

A good profile of a large data set can be provided by giving the lowest value of the data (called the *Min*), the first quartile (Q_1), the median (M), the third quartile (Q_3), and the largest value of the data (called the *Max*). These five numbers constitute the **five-number summary** of the data set.

EXAMPLE 14.17

The five-number summary of the Stat 101 data set is

$$Min = 1, Q_1 = 9, M = 11, Q_3 = 12, Max = 24.$$

(We found that $M = 11$ in Example 14.13 and that $Q_1 = 9, Q_3 = 12$ in Example 14.15. Just by looking at the frequency table we can see that $Min = 1$ and $Max = 24$.)

Note that without the two quartiles we would get a distorted picture of the Stat 101 data set, since both $Min = 1$ and $Max = 24$ are outliers. As we know, the scores were not evenly spread out in the range between 1 and 24. With the quartiles, we get a much better idea of what happened—the middle half of the Stat 101 test scores were concentrated in a very narrow range (between 9 and 12 points); the remaining half of the test scores fell between 12 and 25 or between 1 and 9. ∎

Box Plots

Invented in 1977 by statistician John Tukey, a *box plot* (also known as a *box-and-whisker* plot) is a picture of the five-number summary of a data set. The **box plot** consists of a rectangular box that sits above a scale and extends from the first quartile Q_1 to the third quartile Q_3 on that scale. A vertical line crosses the box, indicating the position of the median M. On both sides of the box are "whiskers" extending to the smallest value, Min, and largest value, Max, of the data.

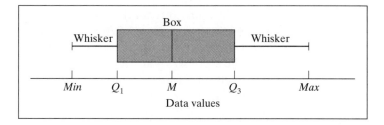

FIGURE 14-12

A generic box plot.

Figure 14-12 shows a generic box plot for a data set. Figure 14-13(a) shows a box plot for the Stat 101 data set. The long whiskers in this box plot are largely due to the outliers 1 and 24. Figure 14-13(b) shows a variation of the same box plot, but with the two outliers, marked with 2 crosses, separated from the rest of the data. This last box plot is a much more accurate picture of the data set.

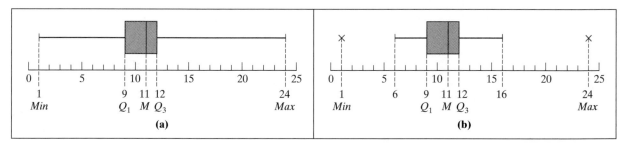

FIGURE 14-13

(a) Box plot for the Stat 101 data set. (b) Same box plot with the outliers separated from the rest of the data.

Box plots are particularly useful when comparing similar data for two or more populations. This is illustrated in the next example.

EXAMPLE 14.18

Figure 14-14 shows box plots for the starting salaries of two different populations: first-year agriculture and engineering graduates of Tasmania State University. Superimposing the two box plots on the same scale allows us to make some useful comparisons. It is clear, for instance, that engineering graduates are doing better overall than agriculture graduates, even though at the very top levels, agriculture graduates are better paid. Another interesting point is that the median salary of agriculture graduates is less than the first quartile of the salaries of engineering graduates. The very short whisker on the left side of the agriculture box plot tells us that the bottom 25% of agriculture salaries are concentrated in a very narrow salary range. We can also see that agriculture salaries are much more spread out than engineering salaries, even though most of the spread occurs at the higher end of the salary scale.

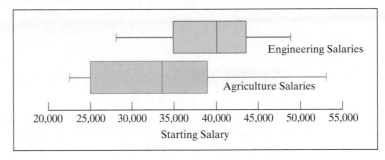

FIGURE 14-14
Comparison of starting salaries of first-year graduates in agriculture and engineering.

We can see that the old chestnut "a picture is worth a thousand words" applies well to statistics. We can learn a lot even from a simple picture like the one in Fig. 14-14—if we know how to read it!

14.4   Measures of Spread

An important aspect of summarizing numerical data is to give an idea of how *spread out* the data values are.

EXAMPLE 14.19

Consider the following two data sets:

$$\text{data set } 1 = \{45, 46, 47, 48, 49, 51, 52, 53, 54, 55\};$$

$$\text{data set } 2 = \{1, 12, 20, 31, 41, 59, 70, 78, 89, 99\}.$$

We leave it to the reader to verify that for both data sets the average is 50 and the median is 50. If we just used the average (or the median) to summarize these sets, we would convey no significant difference between them, which is clearly not the case. It is obvious to the naked eye that the two data sets differ in their spread: the numbers in data set 2 are much more spread out than those in data set 1.

There are several different ways to describe the spread of a data set; in this section we will describe the three most commonly used ones.

The Range

An obvious approach to describing the spread of a data set is to take the difference between the highest and lowest values of the data ($R = Max - Min$). This difference is called the **range**. For data set 1 in Example 14.19, the range is $55 - 45 = 10$, and for data set 2, the range is $99 - 1 = 98$.

As a measure of spread the range is useful only if there are no outliers, since outliers can significantly affect the range. For example, for the Stat 101 data set, the range of the exam scores is $24 - 1 = 23$ points, but without the two outliers, the range would be $16 - 6 = 10$. [See Fig. 14-13(b).]

The Interquartile Range

To eliminate the possible distortion caused by outliers, a common practice when measuring the spread of a data set is to use the **interquartile range** (IQR). The interquartile range is the difference between the third quartile and the first quartile ($IQR = Q_3 - Q_1$), and it tells us how spread out the middle 50% of the data values are. For many types of real-world data, the interquartile range is a useful measure of spread. When the five-number summary is used, both the range and the interquartile range are easy to find.

EXAMPLE 14.20

For the 2001 SAT math scores discussed in Example 14.16, we have a range of 600 ($800 - 200$), because, as we discussed earlier, there were students who scored the minimum 200 and students who scored the maximum 800. Given that the first quartile was $Q_1 = 440$ and the third quartile was $Q_3 = 590$, the interquartile range is $Q_3 - Q_1 = 150$ points. ■

The Standard Deviation

The most important and most commonly used measure of spread for a data set is the *standard deviation*. The key concept for understanding the standard deviation is the concept of *deviation from the mean*. The idea is to measure spread by looking at how far each data point is from a fixed reference point. If we pick a good reference point, the "distances" between it and each data point could give a good description of the spread of the data. The reference point we will use is the mean (average) of the data set. Imagine that we plant a flag there and that we measure how far each data point is from the flag by taking the difference (*data value* − *mean value*). These numbers are called the **deviations from the mean**.

EXAMPLE 14.21 Deviations from the Mean in the Stat 101 Test

Once again we return to the Stat 101 data set. We calculated in Example 14.9 the average (mean) test score to be 10.85. For each possible test score, we can now calculate how far that score is from the average score of 10.85 (as shown in the middle column of Table 14-12).

TABLE 14-12　　Stat 101 Data: Deviations from the Mean

Test score (x)	Deviation from the mean ($x - 10.85$)	Frequency
1	−9.85	1
6	−4.85	1
7	−3.85	2
8	−2.85	6
9	−1.85	10
10	−0.85	16
11	0.15	13
12	1.15	9
13	2.15	8
14	3.15	5
15	4.15	2
16	5.15	1
24	13.15	1

The deviations from the mean are themselves a data set, which we would like to summarize. One way would be to average them, but if we do that, the negative deviations and the positive deviations will always cancel each other out, so that we end up with an average of 0 (see Exercise 81). This, of course, makes the average useless in this case. The cancellation of positive and negative deviations can be avoided by squaring each of the deviations. The squared deviations are never negative, and if we average them out, we get an important measure of spread called the **variance**.[3] If we take the square root of the variance, we get the **standard deviation**, usually denoted by the Greek letter σ. The process is complicated, but not necessarily difficult, if we take it one step at a time.

EXAMPLE 14.22

Let's find the standard deviation of the 10 numbers 45, 46, 47, 48, 49, 51, 52, 53, 54, 55. The first step is to find the mean (average) of the data set, which we will call A. Here, $A = 50$. The second step is to calculate the *deviations from the mean*. These are shown in the second column of Table 14-13. The third step is to square each of the preceding deviations. These are shown in the third column of Table 14-13. Next, we average these numbers. This gives the variance $V = 11$. Finally, we take the square root of the variance to get the standard deviation $\sigma = \sqrt{11} \approx 3.317$.

TABLE 14-13

Data	Deviation from $A = 50$	Squared deviations
45	-5	25
46	-4	16
47	-3	9
48	-2	4
49	-1	1
51	1	1
52	2	4
53	3	9
54	4	16
55	5	25
$A = 50$	average $= 0$	$V = 11$

Standard deviations of large data sets are not fun to calculate by hand, and they are rarely found that way. The standard procedure for calculating standard deviations is to use a computer or a good scientific or business calculator, which often are pre-programmed to do all the steps automatically. Be that as it may, it is still good to know what steps are involved in calculating a standard deviation, even when the actual grunt work is going to be performed by a machine.

[3]In many statistics books and statistical computer programs, the variance is defined by dividing the sum of the squared deviations by $N - 1$ (instead of by N, as one would in an ordinary average). There are reasons that this definition is appropriate in some circumstances, but a full explanation would take us beyond the purpose and scope of this chapter. In any case, except for small values of N, the difference between the two definitions tends to be very small.

FINDING THE STANDARD DEVIATION OF A DATA SET

- **Step 1.** Find the *average* (mean) of the data set. Call it *A*.
- **Step 2.** For each number *x* in the data set, find *x* − *A*, the *deviation from the mean*.
- **Step 3.** Square each of the deviations found in step 2. These are the *squared deviations*.
- **Step 4.** Find the average of the squared deviations. This number is called the *variance V*.
- **Step 5.** Take the square root of the variance. This is the *standard deviation σ*.

The standard deviation is arguably the most important and frequently used measure of data spread. Yet, it is not a particularly intuitive concept. If the standard deviation of a data set is 15 and the standard deviation of a different data set is 150, what conclusions, if any, can we draw? Here are a few basic guidelines that might help make some sense out of standard deviations.

- The standard deviation of a data set is measured in the same units as the original data. For example, if the data are points on a test, then the standard deviation is also given in points. Conversely, if the standard deviation is given in dollars, we can conclude that the original data must have been money—home prices, salaries, or something like that. For sure, the data couldn't have been test scores on an exam.

- It is pointless to compare standard deviations of data sets that are given in different units. Even for data sets that are given in the same units, say, for example, test scores, the underlying scale should be the same. We should not try to compare standard deviations for SAT scores that are given on a scale of 200–800 points with standard deviations of a Stat 101 quiz given on a scale of 0–25 points.

- For data sets that are based on the same underlying scale, a comparison of standard deviations can tell us something about the spread of the data. In the extreme case, there is the utterly boring data set in which all data points are equal. In this case, the standard deviation is 0. Conversely, if the standard deviation is 0, the data points must all be equal—there is no spread! (See Exercise 74.) If the standard deviation is small, we can conclude that the data points are all bunched together—there is very little spread. As the standard deviation increases, we can conclude that the data points are beginning to spread out. The more spread out they are, the larger the standard deviation becomes.

As a measure of spread, the standard deviation is particularly useful for analyzing real-life data. We will come to appreciate its importance in this context in Chapter 16.

Conclusion

Whether we like to or not, as we navigate through life in the information age, we are awash in a sea of data. Today, data is the common currency of scientific, social, and economic discourse. Powerful satellites constantly scan our planet, collecting prodigious amounts of weather, geological, and geographical data. Government agencies, such as the Bureau of the Census and the Bureau of Labor Statistics, collect millions of numbers a year about our living, working, spending, and dying habits. Even in our less serious pursuits, such as sports, we are flooded with data, not all of it great.

Faced with the common problem of data overload, statisticians and scientists have devised many ingenious ways to organize, display, and summarize large amounts of data. In this chapter we discussed some of the basic concepts in this area of statistics.

Graphical summaries of data can be produced by bar graphs, pictograms, pie charts, histograms, and so on. (There are many other types of graphical descriptions that we did not discuss in the chapter.) The kind of graph that is the most appropriate for a situation depends on many factors, and creating a good "picture" of a data set is as much an art as a science.

Numerical summaries of data, when properly used, help us understand the overall pattern of a data set without getting bogged down in the details. They fall into two categories: (1) measures of location, such as the *average*, the *median*, and the *quartiles*, and (2) measures of spread, such as the *range*, the *interquartile range*, and the *standard deviation*. Sometimes, we even combine numerical summaries and graphical displays, as in the case of the *box plot*. We touched upon all of these in this chapter, but the subject is a big one, and by necessity, we only scratched the surface.

In this day and age, we are all consumers of data, and at one time or another, we are likely to be providers of data as well. Thus, understanding the basics of how data are organized and summarized has become an essential requirement for personal success and good citizenship.

W. Edwards Deming (1900–1993)

W. E. Deming was a pioneer in the application of statistics to industry. He is best known for his theories of quality control in manufacturing, which have been widely adopted, first by Japanese and later by American companies such as Xerox and Ford. Like all great ideas, Deming's ideas about quality control were built on a simple observation: all industrial processes are subject to some level of statistical variation, and this variation negatively impacts quality. From this Deming developed the principle that to improve manufacturing quality one has to reduce the causes of statistical variation.

William Edwards Deming was born in Sioux City, Iowa in 1900, into a family of very modest means. As a young man, much of his life revolved around study and work. As an undergraduate at the University of Wyoming, he supported himself by doing all sorts of odd jobs, from janitor's aide to cleaning boilers at an oil refinery. He earned a B.S. in electrical engineering from Wyoming in 1921, a master's degree in mathematics and physics from the University of Colorado in 1925, and a Ph.D. in mathematical physics from Yale University in 1928. Deming became famous as a statistician and management guru, but he was trained in classical mathematics and physics and was an accomplished musician who played several instruments and composed religious music.

With a doctorate from Yale in hand and a family to support, Deming joined the Department of Agriculture in Washington, D.C., where he performed laboratory research on fertilizers and developed statistical methods to boost productivity. In 1939, Deming joined the Census Bureau where, as head mathematician, he helped develop many currently used statistical sampling methods. In 1946 he retired from the Census Bureau and became a statistical consultant and Professor at New York University.

After World War II, the methods of *statistical process control* (SPC) developed by Deming were quickly implemented by Japanese industry. Deming's methods focused on reducing variation in assembly line production, thereby minimizing defects and increasing productivity. Though a folk hero in Japan, Deming was virtually ignored by the American business community until 1980 when NBC aired the documentary "If Japan can, why can't we?" detailing the superior quality and growing popularity of Japanese products. Though already 80 years old at the time, many American industrial giants belatedly adopted Deming's techniques.

Deming remained active up until his death at the age of 93, and his lectures and management seminars were attended by the thousands. He was the author of many successful books and over 170 papers and was the recipient of many honors and awards. The *Deming Prize*, established in his honor by the Japanese Union of Scientists and Engineers, is today the highest award a company can achieve for its excellence in management.

KEY CONCEPTS

average (mean)	deviations from the mean
bar graph	discrete variable
box plot (box and whisker plot)	five-number summary
categorical (qualitative) variable	frequency
category (class)	frequency table
class interval	histogram
continuous variable	interquartile range
data set	locator
data values (data points)	measures of location

measures of spread	pie chart
median	quartiles
numerical (quantitative) variable	range
outlier	standard deviation
percentile	variable
pictogram	variance

EXERCISES

WALKING

A. Frequency Tables, Bar Graphs, and Pie Charts

Exercises 1 through 4 refer to the scores in a Chem 103 final exam consisting of 10 questions worth 10 points each. The scores on the exam are given in the following table.

Chem 103 Final Exam Scores

Student ID	Score	Student ID	Score	Student ID	Score	Student ID	Score
1362	50	2877	80	4315	70	6921	50
1486	70	2964	60	4719	70	8317	70
1721	80	3217	70	4951	60	8854	100
1932	60	3588	80	5321	60	8964	80
2489	70	3780	80	5872	100	9158	60
2766	10	3921	60	6433	50	9347	60

1. Make a frequency table for the Chem 103 final exam scores.
2. Make a bar graph showing the actual frequencies of the scores on the exam.
3. Using the scale A: 80–100. B: 70–79, C: 60–69, D: 50–59, and F: 0–49,
   (a) find the grade distribution for the Chem 103 final exam.
   (b) make a bar graph showing the grade distribution for the Chem 103 final exam.
4. Make a pie chart showing the grade distribution for the Chem 103 final exam.

Exercises 5 and 6 refer to the following situation. Every year, the first-grade students at Cleansburg Elementary are given a musical aptitude test. Based on the results of the test, the children are scored from 0 (no musical aptitude) to 5 (extremely talented). This year's results are as follows:

Aptitude score	0	1	2	3	4	5
Frequency	24	16	20	12	5	3

5. **(a)** How many children were given the aptitude test?
   (b) What percent of the students tested showed no musical aptitude?
   (c) Make a relative frequency bar graph showing the results of the musical aptitude test.
6. Make a pie chart showing the results of the musical aptitude test.

Exercises 7 through 10 refer to the following table, which gives the distance from home to school (measured to the closest half-mile) for each kindergarten student at Cleansburg Elementary School.

Distance from Home to School for Cleansburg Elementary School Kindergarten Students

Student ID	Distance to school (miles)	Student ID	Distance to school (miles)	Student ID	Distance to school (miles)	Student ID	Distance to school (miles)
1362	1.5	2877	1.0	4355	1.0	6573	0.5
1486	2.0	2964	0.5	4454	1.5	8436	3.0
1587	1.0	3491	0.0	4561	1.5	8592	0.0
1877	0.0	3588	0.5	5482	2.5	8854	0.0
1932	1.5	3711	1.5	5533	1.0	8964	2.0
1946	0.0	3780	2.0	5717	8.5		
2103	2.5	3921	5.0	6307	1.5		

7. Make a frequency table for the data set.

8. Make a bar graph for the data set.

9. Suppose that class intervals for the distances from home to school for the kindergarteners at Cleansburg Elementary School are defined as follows.

   Very close: Less than 1 mile

   Close: 1 mile up to and including 1.5 miles

   Nearby: 2 miles up to and including 2.5 miles

   Not too far: 3 miles up to and including 4.5 miles

   Far: 5 miles or more

   (a) Make a frequency table for the class intervals.

   (b) Draw a pie chart for the percentage of students in each class interval.

10. Make a bar graph for the class intervals defined in Exercise 9.

Exercises 11 and 12 refer to the scores on a math quiz, the results of which are shown in the following bar graph.

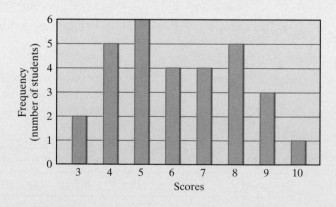

11. **(a)** How many students took the math quiz?

    (b) What percentage of the students scored 2 points?

    (c) If a grade of 6 or more was needed to pass the quiz, what percentage of the students passed?

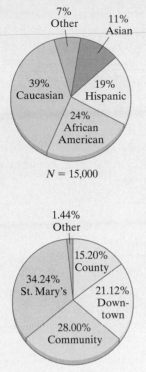

7%
Other

11%
Asian

39%
Caucasian

19%
Hispanic

24%
African
American

$N = 15{,}000$

1.44%
Other

15.20%
County

34.24%
St. Mary's

21.12%
Down-
town

28.00%
Community

$N = 625$

12. Make a pie chart showing the results of the quiz.

Exercises 13 and 14 refer to the pie chart in the margin, which shows the breakdown of the student body at Tasmania State University by ethnicity.

13. Calculate the size of the angle (to the nearest degree) for each of the slices shown in the pie chart.

14. **(a)** Give a frequency table showing the actual frequencies for each category.

(b) Draw the bar graph corresponding to the frequency table in (a).

Exercises 15 and 16 refer to the pie chart in the margin, which shows the percentage of babies born at each of the four hospitals in the city of Cleansburg during the last year.

15. **(a)** How many babies were born at Downtown Hospital?

(b) How many babies were born outside one of the four hospitals (e.g., at home, on the way to the hospital, and so on)?

16. Calculate the size of the angle (in degrees) for each of the slices shown in the pie chart used in Exercise 15.

17. The percentage of the U.S. population enrolled in HMOs for the years 1991–2000 is given in the following table (***Source:*** *The World Almanac and Book of Facts 2002*, p. 889).

Percentage of the U.S. Population in HMOs			
Year	**Percent in HMOs**	**Year**	**Percent in HMOs**
1991	13.6	1996	22.3
1992	14.3	1997	25.2
1993	15.1	1998	28.6
1994	17.3	1999	30.1
1995	19.4	2000	30.0

Using the ideas of Example 14.2 make two different-looking pictograms showing the growth in the percentage of the population enrolled in HMOs from 1991–2000. In the first pictogram, you are trying to convince your audience that HMOs are growing very fast. The second pictogram should give a more conservative picture.

18. The percentage sales of recorded music on CDs from 1991 to 2000 is given in the following table (***Source:*** *http://www.riaa.com*).

Percentage of CD Sales (out of total recorded music sales)			
Year	**Percent CD Sales**	**Year**	**Percent CD Sales**
1991	38.9	1996	68.4
1992	46.5	1997	70.2
1993	51.2	1998	74.8
1994	58.4	1999	83.2
1995	65.0	2000	89.3

Using the ideas of Example 14.2, make two different-looking pictograms showing the growth in the percentage of recorded music sold on compact discs from 1991–2000. In the first pictogram, you are trying to convince your audience that CDs are not taking over the market and are not a threat to other music media. The second pictogram should give a more realistic picture.

B. Histograms

Exercises 19 through 22 refer to the data in the following table, which shows the weights (in ounces) of the 625 babies born in the city of Cleansburg in the last year.

Weights of Babies Born in Cleansburg Last Year

Weight (in ounces)		
More than	Less than or equal to	Frequencies
48	60	15
60	72	24
72	84	41
84	96	67
96	108	119
108	120	184
120	132	142
132	144	26
144	156	5
156	168	2

19. **(a)** Give the length of each class interval (in ounces).

    (b) Suppose a baby weighs exactly 5 pounds 4 ounces. What class interval does she belong to? Describe the endpoint convention.

20. Write a new table for these data values using class intervals of length equal to 24 ounces.

21. Draw the histogram corresponding to these data values using the class intervals as shown in the original table.

22. Draw the histogram corresponding to the same data when class intervals of 24 ounces are used.

C. Averages and Medians

23. Consider the data set $\{3, -5, 7, 4, 8, 2, 8, -3, -6\}$.
    (a) Find the average.
    (b) Find the median.
    (c) Consider the data set $\{3, -5, 7, 4, 8, 2, 8, -3, -6, 2\}$ obtained by adding one more data point to the original data set. Find the average and median of this data set.

24. Consider the data set $\{-3.8, -7.3, -4.5, 8.3, 8.3, -9.1, -3.8, 13.2\}$.
    (a) Find the average.
    (b) Find the median.
    (c) Consider the data set $\{-3.8, -7.3, -4.5, 8.3, 8.3, -9.1, -3.8\}$ having one less data point than the original set. Find the average and the median of this data set.

25. For each data set, find the average and the median.

    (a) {0, 1, 2, 3, 4, 5, 6, 7, 8, 9}

    (b) {1, 2, 3, 4, 5, 6, 7, 8, 9}

    (c) {1, 2, 3, 4, 5, 6, 7, 8, 9, 10}

26. For each data set, find the average and the median.

    (a) {1, 2, 1, 2, 1, 2, 1, 2, 1, 2}

    (b) {1, 2, 3, 4, 1, 2, 3, 4, 1, 2, 3, 4, 1, 2, 3, 4}

    (c) {1, 2, 3, 4, 5, 5, 4, 3, 2, 1}

27. For the data set {1, 2, 3, 4, 5, ..., 98, 99}, find

    (a) the average. [*Hint:* Recall that

$$1 + 2 + 3 + \cdots + N = N \times (N+1)/2.]$$

    (b) the median.

28. For the data set {1, 2, 3, 4, 5, ..., 997, 998, 999, 1000}, find

    (a) the average. [*Hint:* Recall that

$$1 + 2 + 3 + \cdots + N = N \times (N+1)/2.]$$

    (b) the median.

29. This exercise refers to the musical aptitude test discussed in Exercises 5 and 6. The results of the test are given by the following frequency table.

Aptitude score	0	1	2	3	4	5
Frequency	24	16	20	12	5	3

    (a) Find the average aptitude score.

    (b) Find the median aptitude score.

30. The ages of the firemen in the City of Cleansburg Fire Department are given in the following frequency table.

Age	25	27	28	29	30	31	32	33	37	39
Frequency	2	7	6	9	15	12	9	9	6	4

    (a) Find the average age rounded to 2 decimal places.

    (b) Find the median age.

31. The results of a 10-point philosophy quiz are shown in the following pie chart.

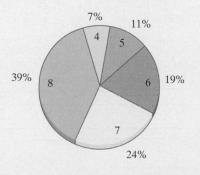

    (a) Find the average quiz score.

    (b) Find the median quiz score.

32. The results of a 10-point math quiz are shown in the following bar graph.

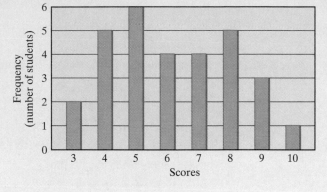

(a) Find the average quiz score.

(b) Find the median quiz score.

D. Percentiles and Quartiles

33. Consider the data set $\{3, -5, 7, 4, 8, 2, 8, -3, -6\}$.

(a) Find the first quartile.

(b) Find the third quartile.

(c) Consider the data set $\{3, -5, 7, 4, 8, 2, 8, -3, -6, 2\}$ obtained by adding one more data point to the original data set. Find the first and third quartiles of this data set.

34. Consider the data set $\{-3.8, -7.3, -4.5, 8.3, 8.3, -9.1, -3.8, 13.2\}$.

(a) Find the first quartile.

(b) Find the third quartile.

(c) Consider the data set $\{-3.8, -7.3, -4.5, 8.3, 8.3, -9.1, -3.8\}$ obtained by deleting one data point from the original data set. Find the first and third quartiles of this data set.

35. For each data set, find the 75th and the 90th percentiles.

(a) $\{1, 2, 3, 4, \ldots, 98, 99, 100\}$

(b) $\{0, 1, 2, 3, 4, \ldots, 98, 99, 100\}$

(c) $\{1, 2, 3, 4, \ldots, 98, 99\}$

(d) $\{1, 2, 3, 4, \ldots, 98\}$

36. For each data set, find the 10th and the 25th percentiles.

(a) $\{1, 2, 3, \ldots, 49, 50, 50, 49, \ldots, 3, 2, 1\}$

(b) $\{1, 2, 3, \ldots, 49, 50, 49, \ldots, 3, 2, 1\}$

(c) $\{1, 2, 3, \ldots, 49, 49, \ldots, 3, 2, 1\}$

37. This exercise refers to the data given in Exercise 30.

Age	25	27	28	29	30	31	32	33	37	39
Frequency	2	7	6	9	15	12	9	9	6	4

(a) Find the first quartile.

(b) Find the third quartile.

(c) Find the 90th percentile.

38. This exercise refers to the math quiz discussed in Exercise 32. The results of the quiz are shown in the following bar graph.

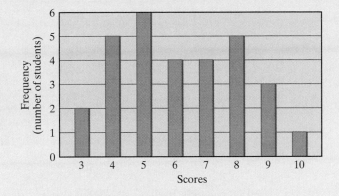

(a) Find the first quartile.

(b) Find the third quartile.

(c) Find the 70th percentile.

39. In 1998, a total of $N = 1,172,779$ college-bound seniors took the SAT test. Assume that the test scores are sorted from lowest to highest and that the sorted data set is $\{d_1, d_2, \ldots, d_{1,172,779}\}$.

(a) Determine the position of the median.

(b) Determine the position of the first quartile.

(c) Determine the position of the third quartile.

40. In 2001, a total of $N = 1,276,320$ college-bound seniors took the SAT test. Assume that the test scores are sorted from lowest to highest and that the sorted data set is $\{d_1, d_2, \ldots, d_{1,276,320}\}$.

(a) Determine the position of the median.

(b) Determine the position of the first quartile.

(c) Determine the position of the third quartile.

E. Box Plots and Five-Number Summaries

41. For the data set $\{3, -5, 7, 4, 8, 2, 8, -3, -6\}$,

(a) find the five-number summary. [Use the results of Exercises 23(b) and 33.]

(b) draw a box plot.

42. For the data set $\{-3.8, -7.3, -4.5, 8.3, 8.3, -9.1, -3.8, 13.2\}$,

(a) find the five-number summary. [See Exercises 24(b) and 34.]

(b) draw a box plot.

43. This exercise refers to the ages of the firemen discussed in Exercises 30 and 37.

(a) Find the five-number summary.

(b) Draw a box plot.

44. This exercise refers to the math quiz discussed in Exercises 32 and 38. The results of the quiz are shown in the following bar graph.

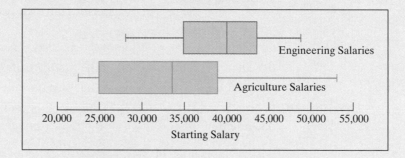

(a) Find the five-number summary for the quiz scores.

(b) Draw a box plot for the quiz scores.

Exercises 45 and 46 refer to the following figure describing the starting salaries for Tasmania State University first-year graduates in agriculture and engineering. (These are the two box plots discussed in Example 14.18.)

45. (a) Approximately how much is the median salary for agriculture majors?

(b) Approximately how much is the median salary for engineering majors?

(c) Explain how we can tell that the median salary for engineering majors is more than the third quartile of the salaries for agriculture majors.

46. (a) There were 612 engineering majors. How many of them had a starting salary of $35,000 or more?

(b) If 240 agriculture majors made less than $25,000, what was the total number of agriculture majors?

F. Ranges and Interquartile Ranges

47. For the data set $\{3, -5, 7, 4, 8, 2, 8, -3, -6\}$, find

(a) the range.

(b) the interquartile range (see Exercise 33).

48. For the data set $\{-3.8, -7.3, -4.5, 8.3, 8.3, -9.1, -3.8, 13.2\}$, find

(a) the range.

(b) the interquartile range (see Exercise 34).

49. A realty company has sold $N = 341$ homes in the last year. The five-number summary for the sale prices is $Min = \$97,000$, $Q_1 = \$115,000$, $M = \$143,000$, $Q_3 = \$156,000$, and $Max = \$249,000$.

   (a) Find the interquartile range of the home sale prices.

   (b) How many homes sold for a price between $115,000 and $156,000 (inclusive)? (*Note*: If you don't believe you have enough information to give the exact answer, you should give the answer in the form of "at least ___" or "at most ___.")

50. For the starting salaries of Tasmania State University first-year graduates in agriculture and engineering discussed in Exercises 45 and 46,

   (a) Estimate the interquartile range for the starting salaries of agriculture majors.

   (b) Estimate the interquartile range for the starting salaries of engineering majors.

   (c) There were 612 engineering majors. Determine how many starting salaries of engineering majors were between $Q_1 = \$35,000$ and $Q_3 = \$43,500$ (inclusive). (*Note*: If you don't believe you have enough information to give the exact answer, you should give the answer in the form of "at least ___" or "at most ___.")

Exercises 51 through 54 refer to the use of the IQR to identify outliers. While there is no uniform agreement among statisticians as to how to precisely identify outliers, many statisticians use the following criterion: An outlier is any data value that is larger than the third quartile by more than 1.5 times the IQR, or below the first quartile by more than 1.5 times the IQR.[4]

51. Using the preceding definition, determine the value(s) of any outliers for the Stat 101 data set discussed throughout the chapter (see Example 14.17). If no outliers exist, explain why not.

52. Using the preceding definition, determine the value(s) of any outliers for the ages of the firemen in the City of Cleansburg Fire Department discussed in Exercises 30 and 37. (*Hint:* Do Exercise 37 first.)

53. Using the preceding definition, determine the value(s) of any outliers for the math quiz scores discussed in Exercises 32 and 38. (*Hint:* Do Exercise 38 first.)

54. Using the preceding definition, determine the value(s) of any outliers for the starting salaries for Tasmania State University first-year graduates in agriculture and engineering discussed in Example 14.18 as well as Exercises 45, 46, and 50.

G. Standard Deviations

55. Find the standard deviation of each of the following data sets.

   (a) $\{5, 5, 5, 5\}$

   (b) $\{0, 5, 5, 10\}$

   (c) $\{-5, 0, 0, 25\}$

56. Find the standard deviation of each of the following data sets.

   (a) $\{10, 10, 10, 10\}$

---

[4]What we have termed outliers are sometimes called **mild outliers**. **Extreme outliers** are then defined as data values that are either above Q_3 or below Q_1 by three IQRs.

      (b) $\{1, 6, 13, 20\}$

      (c) $\{1, 1, 18, 20\}$

57. Find the standard deviation of each of the following data sets.

      (a) $\{0, 1, 2, 3, 4, 5, 6, 7, 8, 9\}$

      (b) $\{1, 2, 3, 4, 5, 6, 7, 8, 9, 10\}$

      (c) $\{6, 7, 8, 9, 10, 11, 12, 13, 14, 15\}$

58. Find the standard deviation of the Stat 101 test scores. (See Example 14.21.)

H. Miscellaneous

The Mode. *The mode of a data set is the data point that occurs with the highest frequency. In a frequency table, we look for the largest number in the frequency row; the corresponding data point is the mode. In a bar graph, we look for the tallest bar; the corresponding data point (or category) is the mode. In a pie chart, we look for the largest slice; the corresponding category is the mode.*

When there are several data points (or categories) tied for the most frequent, each of them is a mode, but if all data points have the same frequency, rather than say that every data point is a mode, it is customary to say that there is no mode.

In Exercises 59 through 66, you should find the mode or modes of the given data sets. If there is no mode, your answer should indicate that.

59. The Stat 101 data set given by the following frequency table.

Frequency Table for the Stat 101 Data Set

Exam score	1	6	7	8	9	10	11	12	13	14	15	16	24
Frequency	1	1	2	6	10	16	13	9	8	5	2	1	1

60. The history midterm exam scores given by the following table.

History Midterm Exam Scores

Student ID	Score	Student ID	Score	Student ID	Score	Student ID	Score	Student ID	Score
1075	74%	1998	75%	3491	57%	4713	83%	6234	77%
1367	83%	2103	59%	3711	70%	4822	55%	6573	55%
1587	70%	2169	92%	3827	52%	5102	78%	7109	51%
1877	55%	2381	56%	4355	74%	5381	13%	7986	70%
1946	76%	2741	50%	4531	77%	5717	74%	8436	57%

61. The Math 100 quiz scores given by the following bar graph.

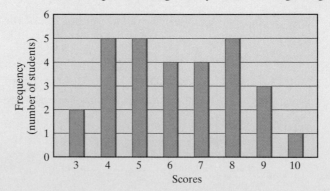

62. The data set $\{3, -5, 7, 4, 8, 2, 11, -3, -6, 9\}$.

63. The data set given by the following pie chart.

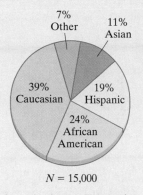

$N = 15,000$

Student body at Tasmania
State University by ethnicity.

64. The data set given by the following bar graph.

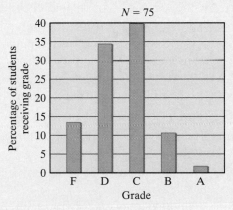

Bar graph for the Stat 101 test grades.

65. The data set given by the following bar graph.

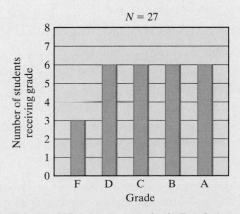

Bar graph for the final grades in Psych 4.

66. The data set given by the following pie chart.

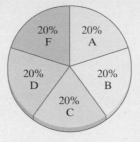

Pie chart for the final grades in Speech 1.

JOGGING

67. Mike's average on the first five exams in Econ 1A is 88. What must he earn on the next exam in order to raise his overall average to 90?

68. Sarah's overall average in Physics 101 was 93%. Her average was based on four exams each worth 100 points and a final worth 200 points. What is the lowest possible score she could have made on the first exam?

69. Josh and Ramon each have an 80% average on the five exams given in Psychology 4. Ramon, however, did better than Josh on all of the exams except one. Give an example that illustrates this situation.

70. Kelly and Karen each have an average of 75 on the six exams given in Botany 1. Kelly's scores have a small standard deviation, and Karen's scores have a large standard deviation. Give an example that illustrates this situation.

71. **(a)** Give an example of 10 numbers with an average less than the median.

(b) Give an example of 10 numbers with a median less than the average.

(c) Give an example of 10 numbers with an average less than the first quartile.

(d) Give an example of 10 numbers with an average more than the third quartile.

72. Suppose that the average of 10 numbers is 7.5 and that the smallest of them is $Min = 3$.

(a) What is the smallest possible value of Max?

(b) What is the largest possible value of Max?

73. What happens to the five-number summary of the Stat 101 data set (see Example 14.17) if

(a) two points are added to each score?

(b) ten percent is added to each score?

74. A data set is called **constant** if every value in the data set is the same. A constant data set can be described by $\{a, a, a, \dots, a\}$.

(a) Show that the standard deviation of a constant data set is 0.

(b) Show that if the standard deviation of a data set is 0, it must be a constant data set.

Exercises 75 and 76 refer to histograms with unequal class intervals. When sketching such histograms, the columns must be drawn so that the frequencies or percentages are proportional to the area of the column. The accompanying figure illustrates this. If the column over class interval 1 represents 10% of the population, then the column over class interval 2, also representing 10% of the population, must be one-third as high, because the class interval is three times as large.

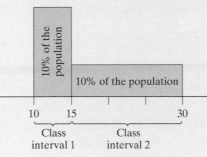

75. If the height of the column over the class interval 20–30 is 1 unit and the column represents 25% of the population, then

   (a) how high should the column over the interval 30–35 be if 50% of the population falls into this class interval?

   (b) how high should the column over the interval 35–45 be if 10% of the population falls into this class interval?

   (c) how high should the column over the interval 45–60 be if 15% of the population falls into this class interval?

76. Two hundred senior citizens are tested for fitness and rated on their times on a 1-mile walk. These ratings and associated frequencies are given in the following table.

Time	Rating	Frequency
6^+–10 minutes	Fast	10
10^+–16 minutes	Fit	90
16^+–24 minutes	Average	80
24^+–40 minutes	Slow	20

Draw a histogram for these data based on the categories given by the ratings in the table.

77. News media have accused Tasmania State University of discriminating against women in the admission policies in its schools of architecture and engineering. The *Tasmania Gazette* states that, "68% of all male applicants to the schools of architecture or engineering are admitted, while only 51% of the female applicants to these same schools are admitted." The actual data are given in the following table.

	School of Architecture		School of Engineering	
	Applied	Admitted	Applied	Admitted
Male	200	20	1000	800
Female	500	100	400	360

   (a) What percent of the male applicants to the School of Architecture were admitted? What percent of the female applicants to this same school were admitted?

   (b) What percent of the male applicants to the School of Engineering were admitted? What percent of the female applicants to this same school were admitted?

(c) How did the *Tasmania Gazette* come up with its figures?

(d) Explain how it is possible for the results in (a) and (b) and the *Tasmania Gazette* statement all to be true.

78. Given that the numbers $x_1, x_2, x_3, \ldots, x_N$ have average a and median M, explain why the numbers $x_1 + c, x_2 + c, x_3 + c, \ldots, x_N + c$ have

(a) average $A + c$

(b) median $M + c$

79. The data set for this exercise consists of the number of exercises per chapter in this book. Thus, $N = 16$.

(a) Find the average number of exercises per chapter in this book.

(b) Find the five-number summary for this data set.

(c) Find the mode for this data set. (See the definition of mode preceding Exercise 59.)

(d) Find the standard deviation for this data set.

RUNNING

80. Show that the standard deviation of any set of numbers is always less than or equal to the range of the set of numbers.

81. (a) Calculate the average of the deviations from the mean for the Stat 101 data set (see Example 14.21).

(b) Show that if $\{x_1, x_2, x_3, \ldots, x_N\}$ is a data set with average A, then the average of $x_1 - A, x_2 - A, x_3 - A, \ldots, x_N - A$ is 0.

82. (a) Show that if $\{x_1, x_2, x_3, \ldots, x_N\}$ is a data set with average A and standard deviation s, then $s\sqrt{N} \geq |x_i - A|$ for every data value x_i.

(b) Use (a) to show that

$$A - s\sqrt{N} \leq x_i \leq A + s\sqrt{N}$$

for every data value x_i.

83. (a) Find two numbers (expressed in terms of A and s) whose average is A and whose standard deviation is s.

(b) Find three equally spaced numbers whose average is A and whose standard deviation is s.

(c) Generalize the preceding by finding N equally spaced numbers whose average is A and whose standard deviation is s. (*Hint:* Consider N even and N odd separately.)

84. Show that the median and the average of the numbers $1, 2, 3, \ldots, N$ are always the same.

85. Suppose that the average of the numbers $x_1, x_2, x_3, \ldots, x_N$ is A and that the variance of these same numbers is V. Suppose also that the average of the numbers $x_1^2, x_2^2, x_3^2, \ldots, x_N^2$ is B. Show that $V = B - A^2$. (In other words, for any data set, if we take the average of the squared data values and subtract the square of the average of the data values, we get the variance.)

86. Given that the numbers $x_1, x_2, x_3, \ldots, x_N$ have standard deviation s, explain why the numbers $x_1 + c, x_2 + c, x_3 + c, \ldots, x_N + c$ also have standard deviation s.

87. Given that the numbers $x_1, x_2, x_3, \ldots, x_N$ have standard deviation s, explain why the numbers $a \cdot x_1, a \cdot x_2, a \cdot x_3, \ldots, a \cdot x_N$ (where a is a positive number) have standard deviation $a \cdot s$.

88. Using the formula $1^2 + 2^2 + 3^2 + \cdots + N^2 = N(N + 1)(2N + 1)/6$,

    (a) find the standard deviation of the data set $\{1, 2, 3, \ldots, 98, 99\}$. (*Hint:* Use Exercise 85.)

    (b) find the standard deviation of the data set $\{1, 2, 3, \ldots, N\}$.

89. (a) Find the average and standard deviation of the data set $\{315, 316, \ldots, 412, 413\}$. [*Hint:* Use Exercises 27(a), 78(a), 86, and 88.]

    (b) Find the average and standard deviation of the data set $\{k + 1, k + 2, \ldots, k + N\}$.

90. Chebyshev's Theorem. The Russian mathematician P. L. Chebyshev (1821–1894) showed that for any data set and any constant k greater than 1, at least $1 - \dfrac{1}{k^2}$ of the data must lie within k standard deviations on either side of the average A. For example, when $k = 2$, this says that $1 - \dfrac{1}{4} = \dfrac{3}{4}$ (i.e., 75%) of the data must lie within two standard deviations of A (i.e., in the interval $[A - 2 \cdot s, A + 2 \cdot s]$).

    (a) Using Chebyshev's theorem, what percentage of a data set must lie within three standard deviations of the mean?

    (b) How many standard deviations on each side of the mean must we take to be assured of including 99% of the data?

    (c) For the Stat 101 data set of Example 14.1, determine the value of k such that 100% of the data lies within k standard deviations of the average A. [*Hint:* Do Exercise 58 first and use that result.]

    (d) Suppose the average of a data set is A. Explain why there is no number k of standard deviations s for which we can be certain that 100% of the data lies within k standard deviations on either side of the average A.

PROJECTS AND PAPERS

A. Lies, Damn Lies, and Statistics

There are three kinds of lies: lies, damned lies, and statistics.

Mark Twain*

Statistics are often used to exaggerate, distort, and misinform, and this is most commonly done by the misuse of graphs and charts. The following two bar graphs, for example, illustrate the same set of data—the average faculty salaries at a certain university (that will remain nameless) between 1997 and 2002. The

*In his autobiography, Mark Twain attributed this witticism to British Prime Minister Benjamin Disraeli. Thus, the saying is often credited to Disraeli although there is no record that Disraeli ever said it.

left bar graph, produced by the faculty union, makes the point that salaries have gone up very slowly. The second bar graph, produced by the university administration, makes the point that salaries have grown dramatically.

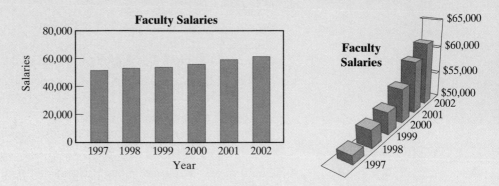

In this project you are to discuss the different graphical "tricks" that can be used to mislead or slant the information presented in a picture. Attempt to include items from recent newspapers, magazines, and other media.

Note: A good starting point for this project can be found in references 2, 9, 13, and 14.

B. Skewness

Measures of location and *measures of spread* are not the only types of numerical summaries of a data set. It is also possible to measure the amount of asymmetry in a data set by means of a statistical measure called the **skewness** of the data set. Consider the following three data sets. The first has a positive skew (skews right), the second has no skew, and the third has a negative skew (skews left).

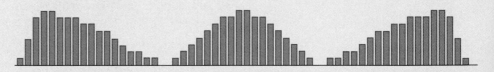

In this project, you are to discuss the concept of *skewness* in a data set. You should include a precise mathematical definition of skewness, and give examples of five different data sets (each containing at least 10 data points) and having skewness of −1, −0.5, 0, 0.5, and 1, respectively.

Note: You may want to use a spreadsheet program such as *Excel* to experiment with the data sets, compute their skewness, and generate graphs. You can also do this with a graphing calculator.

C. Data in Your Daily Life

Which month is the best one to invest in the stock market? During which day of the week are you most likely to get in an automobile accident? Which airline is the safest to travel with? Who is the best placekicker in the National Football League?

In this project, you are to formulate a question from everyday life (similar to one of the aforementioned questions) that is amenable to a statistical analysis. Then you will need to find relevant data that attempt to answer this question and summarize that data using the methods discussed in this chapter. Present your final conclusions and defend them using appropriate charts and graphs.

D. Book Review: *Curve Ball*

Baseball is the ultimate statistical sport, and statistics are as much a part of baseball as peanuts and Cracker Jack. The book *Curve Ball: Baseball, Statistics, and the Role of Chance in the Game*, written by statisticians and baseball lovers Jim Albert and Jay Bennett, is a compilation of all sorts of fascinating statistical baseball issues, from how to better measure a player's offensive performance to what is the true value of home field advantage.

In this project you are to pick one of the many topics discussed in *Curve Ball* and write an analysis paper on the topic.

REFERENCES AND FURTHER READINGS

1. Albert, Jim and Jay Bennett, *Curve Ball: Baseball, Statistics, and the Role of Chance in the Game*. New York: Springer-Verlag, 2001, chap. 2.
2. Cleveland, W. S., *The Elements of Graphing Data*, rev. ed. New York: Van Nostrand Reinhold Co., 1994.
3. Cleveland, W. S., *Visualizing Data*. Summit, NJ: Hobart Press, 1993.
4. Gabor, Andrea, *The Man Who Discovered Quality*. New York: Penguin Books, 1992.
5. Harris, Robert, *Information Graphics: A Comprehensive Illustrated Reference*. New York: Oxford University Press, 2000.
6. Mosteller, F., and W. Kruskal, et al., *Statistics by Example: Exploring Data*. Reading, MA: Addison-Wesley, 1973.
7. Tanner, Martin, *Investigations for a Course in Statistics*. New York: Macmillan Publishing Co., Inc., 1990.
8. Tufte, Edward, *Envisioning Information*. Cheshire, CT: Graphics Press, 1990.
9. Tufte, Edward, *The Visual Display of Quantitative Information*. Cheshire, CT: Graphics Press, 1983.
10. Tufte, Edward, *Visual Explanations: Images and Quantities, Evidence and Narrative*. Cheshire, CT: Graphics Press, 1997.
11. Tukey, John W., *Exploratory Data Analysis*. Reading, MA: Addison-Wesley, 1977.
12. Utts, Jessica, *Seeing Through Statistics*. Belmont, CA: Wadsworth Publishing Co., 1996.
13. Wainer, H., "How to Display Data Badly," *The American Statistician*, 38 (1984), 137–147.
14. Wainer, H., *Visual Revelations*. New York: Springer-Verlag, 1997.
15. Wildbur, Peter, *Information Graphics*. New York: Van Nostrand Reinhold Co., 1989.

Dow-Jones

11,000

10,000

9,000

8,000

7,000

01 2002

O N D J F M A M J J A S O N D

Chances, Probabilities, and Odds

Measuring Uncertainty

With the possible exception of death and taxes, pretty much everything else in our

lives is shrouded with uncertainty. That's why we spend some of our

hard-earned dollars on car insurance (we just might be involved in

an automobile accident), don't invest all of our money in the stock

market (we might lose our shirts), and breathlessly watch the flight of

the ball on that last second, game-winning free throw (it might not go in).

> *Nothing in life is certain except death and taxes.*
> *Benjamin Franklin*

While we are all familiar with the idea of uncertainty, we don't always have a good

grasp on how to measure it. Sometimes there is very little uncertainty ("I'm pretty sure

I'm going to get an A on my statistics exam"), and sometimes there is a lot of uncer-

tainty ("I have no clue as to how I did in that statistics exam"), and sometimes the un-

certainty is somewhere in between. To quantify the amount of uncertainty in the

many uncertain events that affect our lives, we use the concept of *probability*.

*C*hance, *probability, odds*—we use these words carelessly in casual conversation, and most of the time we can get away with it. Technically speaking, however, each of these words represents a slightly different way to
express the measure of uncertainty of an event. When we speak of the *chance* of
some event happening, we state the answer in terms of percentages, such as when
the weatherperson reports "the chance of rain tomorrow is 40%." When we
speak of the *probability* of some event happening, we should state the answer in
terms of a fraction (such as 2/5) or a decimal (0.4). Finally, when we speak of the
odds in favor of an event, we use a pair of numbers, as in "the odds for the Giants
to win the World Series are 2 to 3."

In this chapter we will learn how to interpret and work with probabilities,
chances, and odds in a formal mathematical context. This will be our very brief introduction to the mathematical theory of probability, a relatively young branch of
mathematics that has become critically important to many aspects of modern life.
Insurance, public health, science, sports, gambling, the stock market—wherever
there is uncertainty to be tamed—the mathematical theory of probability plays a
significant role.

Our discussion in this chapter is broken up into two parts. In the first half we lay down the basic theoretical framework needed for a meaningful discussion of probabilities; in the second half we define and calculate probabilities in a practical setting.

15.1 Random Experiments and Sample Spaces

In broad terms, probability is the *quantification of uncertainty*. To understand what that means, we may start by formalizing the notion of uncertainty.

We will use the term **random experiment** to describe an activity or process *whose outcome cannot be predicted ahead of time*. Typical examples of random experiments are tossing a coin, rolling a pair of dice, pulling the arm of a slot machine, and predicting the direction of the stock market.

Associated with every random experiment is the *set* of all of its possible outcomes, called the **sample space** of the experiment. For the sake of simplicity, we will concentrate on experiments for which there is only a finite set of outcomes, although experiments with infinitely many outcomes are both possible and important.

We illustrate the importance of the sample space by means of several examples. Since the sample space of any experiment is a set of outcomes, we will use set notation to describe it. We will consistently use the letter S to denote a sample space, and N to denote its size (i.e., the number of outcomes in S).

EXAMPLE 15.1 A Coin Toss

One simple random experiment is to *toss a quarter* and *observe whether it lands heads or tails*. The sample space can be described by $S = \{H, T\}$, where H stands for heads and T for tails. Here $N = 2$.

A couple of comments about coins are in order here. First, the fact that the coin in Example 15.1 was a quarter is essentially irrelevant. Practically all coins have an obvious "heads" side (and thus a "tails" side), and even when they don't—as in a "buffalo nickel"—we can agree ahead of time which side is which. Second, there are fake coins out there on which both sides are "heads." Tossing such a coin does not fit our definition of a random experiment, so from now on, we will assume that all coins used in our experiments have two different sides, which we will call H and T.

EXAMPLE 15.2 A Double Coin Toss

(a) Suppose we *toss a coin twice and observe on each toss whether it lands heads or tails*. The sample space now is $S = \{HH, HT, TH, TT\}$, where HT means that the first toss came up H and the second toss came up T, which is a different outcome from TH (first toss T and second toss H). For this sample space, $N = 4$.

(b) Suppose again that we toss a coin twice, but now we are only interested in the *number of heads that come up*. This random experiment is different from the one in (a), and its sample space is $\{0, 1, 2\}$.

(c) Suppose now we *toss two coins, say, a nickel and a quarter, at the same time*. This random experiment appears to be different from the one in (a), but the sample space is $S = \{HH, HT, TH, TT\}$. (Here we must agree what the order of the symbols is—for example, the first symbol describes the quarter and the second the nickel.) Because they have the same sample space, we will consider the random experiments described in (a) and (c) as the same. ■

The important point made in Example 15.2 is that a random experiment is defined by two things: the action (such as tossing coins) and the observation that follows that action (what it is that we are really looking for).

EXAMPLE 15.3 Shooting Free Throws (in the abstract)

Suppose *a basketball player shoots two consecutive free throws*.

(a) We might consider this a random experiment with sample space $S = \{ss, sf, fs, ff\}$, where *s* means success and *f* means failure. Here, as in Example 15.2(a), $N = 4$.

(b) We could also consider a random experiment in which we observe *how many free throws the player made*. In this case, the sample space would be $S = \{0, 1, 2\}$ and $N = 3$.

■

We will now discuss a few examples of random experiments involving dice. A die[1] is a cube, usually made of plastic, whose six faces are marked with dots (from 1 through 6) called "pips." Random experiments using dice have a long-standing tradition in our culture and are a part of both gambling and recreational games (such as Monopoly, Yahtzee, etc.).

EXAMPLE 15.4 Rolling a Die

Suppose we *roll a single die*. Typically, we are interested in the "number" rolled, and thus, the sample space for this experiment is $S = \{\boxdot, \boxdot, \boxdot, \boxdot, \boxdot, \boxdot\}$. Here $N = 6$. ■

---

[1] Singular, die; plural, dice.

EXAMPLE 15.5 Rolling a Pair of Dice: Part I

Suppose we *roll a pair of dice* and observe the number that comes up on each die. The sample space is

Here, $N = 36$. Notice that, as we did with the coins, we are treating the dice as *distinguishable* objects (as if one were white and the other red), so that ⚀⚄ and ⚄⚀ are considered different outcomes. ∎

EXAMPLE 15.6 Rolling a Pair of Dice: Part II

In many games (such as Monopoly, craps, etc.), we roll a pair of dice and what really matters is the *total*, rather than the actual numbers, rolled. In this situation the sample space is $S = \{2, 3, 4, 5, 6, 7, 8, 9, 10, 11, 12\}$ and $N = 11$. ∎

EXAMPLE 15.7 Ranking the Top 3 Candidates in an Election

Five candidates (A, B, C, D, and E) are running in an election. The top 3 finishers are chosen President, Vice President, and Secretary, in that order. The election itself can be considered a random experiment with sample space $S = \{ABC, ACB, BAC, BCA, CAB, CBA, ABD, ADB, \ldots\}$, where the outcome ABC signifies that candidate A is elected President, B is elected Vice President, and C is elected Secretary. The "$\ldots$" at the end of the sample space is another way of saying "and so on." ∎

What happened in Example 15.7 is commonplace. Once we realized that the sample space S is big, we decided against writing each and every outcome down. The critical task is to find the actual size N of the sample space without having to list each individual outcome, and this can be done by using a few basic rules of "counting." We will learn how to do this next.

15.2 Counting: The Multiplication Rule

EXAMPLE 15.8 Triple Coin Toss

Suppose we *toss a coin three times* and *observe on each toss whether it lands heads or tails*. Here the sample is $S = \{HHH, HHT, HTH, HTT, THH, THT, TTH, TTT\}$, with $N = 8$. ∎

Example 15.8 sets the stage for the next example.

EXAMPLE 15.9 Multiple Coin Toss

Suppose we *toss a coin 8 times* and *observe on each toss whether it lands heads or tails*. This sample space is too big to write down in full. Nonetheless, we can find its size in a relatively painless way.

The first thing we should ask is, What does a random outcome look like? Taking our cue from Example 15.8, we can say that a random outcome can be described by a string of 8 consecutive letters, where the letters can be either Hs or

*T*s. For example, the string *THHTHTHH* represents a *single* outcome in our sample space—the one in which the first toss came up *T*, the second toss came up *H*, the third toss came up *H*, and so on. To count *all* the outcomes, we will argue as follows: (1) The number of possibilities for the first letter is 2 (*H* or *T*); (2) the number of possibilities for the second letter is also 2, ...; (8) the number of possibilities for the last letter is 2. The total number of outcomes *N* is given by *multiplying* all of these numbers: $N = 2 \times 2 \times 2 \times 2 \times 2 \times 2 \times 2 \times 2 = 256$.

The basic rule we used in Example 15.9 is called (for obvious reasons) the **multiplication rule**. Informally stated, the multiplication rule says that *when something takes place in several stages, to find the total number of ways it can occur, multiply the number of ways each individual stage can occur.*

The easiest way to understand the multiplication rule is through examples.

EXAMPLE 15.10 Buying Ice Cream

Imagine that you want to buy a *single* scoop of ice cream. There are two types of cones available (sugar and regular) and three flavors to choose from (strawberry, chocolate chip, and chocolate). Figure 15-1 shows all the possible combinations.

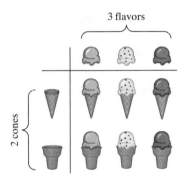

3 flavors

2 cones

FIGURE 15-1
2 cones and 3 flavors make $2 \times 3 = 6$ combinations.

The multiplication rule is undoubtedly the most important tool used in solving *counting problems* (that is, problems that ask in how many different ways can one thing or another happen). These are exactly the kinds of questions one needs to answer to carry out the basic probability calculations that we will want to do later in this chapter. Before we get to that, we will take a brief detour to explore a few of the subtleties of counting. This is an important and rich subject, full of interesting twists and turns, but our detour, by necessity, will be brief.

Our detour starts with a straightforward application of the multiplication rule. As we move on, the level of sophistication will gradually increase, with each successive example showing some variation of the original theme.

EXAMPLE 15.11 The Making of a Wardrobe: Part I

Dolores is a young saleswoman planning her next business trip. She is thinking about packing 3 different pairs of shoes, 4 skirts, 6 blouses, and 2 jackets. If all the items are color coordinated, how many different *outfits* will she be able to make out of these items?

To answer this question, we must first define what we mean by an "outfit." Let's assume that an outfit consists of a pair of shoes, a skirt, a blouse, and a jacket. Here we can use the multiplication rule directly. The total number of possible outfits Dolores can make is $3 \times 4 \times 6 \times 2 = 144$. (Color coordination obviously pays—Dolores can be on the road for over 4 months and never have to wear the same outfit twice!)

The next example is a more subtle variation of Example 15.11.

EXAMPLE 15.12 The Making of a Wardrobe: Part II

Once again, Dolores is packing for a business trip. This time, she packs 3 pairs of shoes, 4 skirts, 3 pairs of slacks, 6 blouses, 3 turtlenecks, and 2 jackets. As before, we can assume that she coordinates the colors so that everything goes with everything else. This time, we will define an outfit as consisting of a pair of shoes, a choice of "lower wear" (either a skirt *or* a pair of slacks), a choice of "upper wear" (it could be a blouse *or* a turtleneck *or both*), and, finally, she may or may not choose to wear a jacket.

Once again, we want to count how many different such outfits are possible. Our strategy will be to think of an outfit as being put together in stages and to draw a box for each of the stages. We then separately count the number of choices at each stage and enter that number in the corresponding box. (Some of these calculations can themselves be mini-counting problems.) The last step is to multiply the numbers in each box. The details are illustrated in Fig. 15-2. The final count for the number of different outfits is $N = 3 \times 7 \times 27 \times 3 = 1701$.

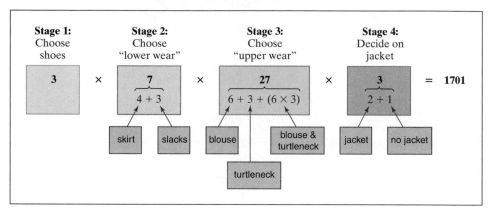

FIGURE 15-2
Counting all possible outfits using a box model.

The method of drawing boxes representing the successive stages in a process and putting the number of choices for each stage inside the box is a convenient device that often helps clarify one's thinking. Silly as it may seem, we strongly recommend it. For ease of reference, we will call it the *box model* for counting.

EXAMPLE 15.13 Ranking the Top 3 Candidates in a 5-Person Election

We are back to the question raised in Example 7. Five candidates are running in an election, with the top 3 getting elected (in order) as President, Vice President, and Secretary. We want to know how big the sample space is. Using a box model, we see that this becomes a reasonably easy counting problem, as illustrated in Fig. 15-3.

FIGURE 15-3
Ranking 3 out of 5 candidates using a box model.

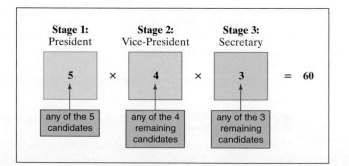

15.3 **Permutations and Combinations**

In many counting problems, the multiplication rule and the box model are by themselves not enough, and we need to add some new tools to our toolbox. Take, for example, the question of ordering ice cream at Baskin-Robbins.

EXAMPLE 15.14 "True Doubles" at Baskin-Robbins

Baskin-Robbins offers 31 different flavors of ice cream. A "true double" is the name some kids use for two scoops of ice cream of two *different* flavors. How many different bowls of *true doubles* are possible?

It would appear at first glance that this is a simple variation of Example 15.13 and that the total number of possible true doubles is $31 \times 30 = 930$, as shown in Fig. 15-4. But if we give the matter a little careful thought, we will realize that we have double counted. Double counting true doubles? Why? Most people would agree that the order in which the scoops of ice cream are put in a bowl is irrelevant and that picking strawberry first and chocolate second is no different from picking chocolate first and strawberry second. But in a box model, *there is always a definite order to things*, and strawberry–chocolate is counted separately from chocolate–strawberry, so our count of 930 is wrong. Fortunately, we also know exactly how and why the count of 930 is wrong—it is double what it should be! Thus, dividing 930 by 2 gives us the correct count: A total of 465 true doubles are possible at Baskin-Robbins.

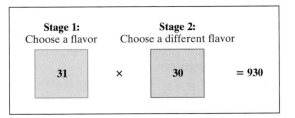

FIGURE 15-4
True doubles at Baskin-Robins are only half of this count.

Example 15.14 is an important one. It warns us that we have to be careful about how we use the multiplication rule and box models, especially in problems where changing the order in which we choose the parts does not change the whole.

EXAMPLE 15.15 "True Triples" at Baskin-Robbins

Let's carry the ideas of Example 15.14 one step further. Let's say that a *true triple* consists of 3 scoops of ice cream, each of the 3 scoops being a different flavor. How many different bowls of *true triples* can be ordered at Baskin-Robbins?

Starting with a box model, we have 31 choices for the "first" flavor, 30 choices for the "second" flavor, and 29 choices for the "third" flavor, for an apparent grand total of $31 \times 30 \times 29 = 26{,}970$ combinations. But once again, this is not the correct answer. In fact, the correct answer is the above number divided by 6 (see Fig. 15-5), giving a total count of 4495 true triples.

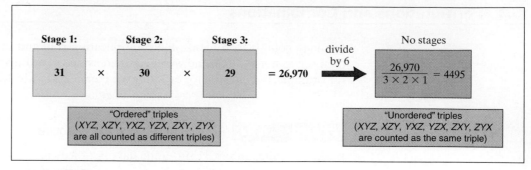

FIGURE 15-5
Counting true triples at Baskin-Robbins.

The key question is, Why did we divide by 6? For any combination of three flavors (call them *X*, *Y*, and *Z*), there are 6 orders in which these flavors can be listed (*XYZ*, *XZY*, *YXZ*, *YZX*, *ZXY*, and *ZYX*). We can call these "ordered triples." Each one of these is counted as a different triple under the multiplication rule, but when we consider them as unordered triples, they are all the same. *Changing from ordered triples to unordered triples is accomplished by dividing by 6.* ▪

It is helpful to think of the final answer to Example 15.15 (4495) in terms of how it comes about: $4495 = (31 \times 30 \times 29)/(3 \times 2 \times 1)$. The numerator $(31 \times 30 \times 29)$ comes from counting ordered triples using the multiplication rule; the denominator $(3 \times 2 \times 1)$ comes from the fact that there are that many ways to shuffle around 3 things (in this case, the 3 flavors in a triple). The denominator $3 \times 2 \times 1$ is already familiar to us—it is the *factorial* of 3. (The factorial of a positive integer *N* is denoted by *N*! and is the number $N \times (N - 1) \times (N - 2) \times \ldots \times 2 \times 1$. It represents, among other things, the number of ways in which *N* objects can be ordered, or, if you will, shuffled. We discussed the factorial in Chapters 2 and 6, so we won't dwell on it here.)

Our next example will deal with poker, a popular card game and the source of many interesting counting questions. Poker is played with a standard deck of 52 cards. (The cards are divided into 4 *suits*, and there are 13 *values* in each suit.) Many variations of poker are played (5-card poker, 7-card poker, draw poker, stud poker, etc.). We will dispense with most of the details; what will be most important to us is the distinction between *down* cards (seen only by the player who gets them) and *up* cards (which are dealt face up and can be seen by all the players).

EXAMPLE 15.16 Five-Card Poker Hands

(a) Let's start by counting the number of possible 5-card *stud poker hands*. In stud poker, the first card is dealt down, and the remaining 4 cards are dealt up, one at a time. In between successive cards, there is a round of betting. In this situation, the order in which the cards are dealt is extremely important. If you get "good cards" as your second and third cards, then the other players know that you have a good hand, so they are not likely to stick around and lose their money to you. If your best card is the first card, which is a down card, that is much better, as your opponents won't know that you have a strong hand. In any case, since the order of the 5 cards matters, we can make direct use of the multiplication rule, as shown in Fig. 15-6(a). The total number of 5-card stud poker hands is an enormous number: $52 \times 51 \times 50 \times 49 \times 48 = 311,875,200$.

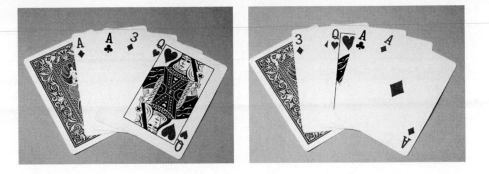

In stud poker hands, the order in which the cards are dealt matters. *Left*: Early "good" cards might scare other players away. *Right*: Late "good" cards might keep other players around.

(b) Now let's consider the number of possible 5-card hands in *draw poker*. In draw poker, all cards are down cards. This means that the order in which the cards are dealt is irrelevant—the only one that sees the cards is the player who gets them. Once the cards are in the player's hand, the player can shuffle them around any way he or she sees fit. In fact, we can be more specific: 5 cards can be shuffled in $5! = 120$ ways. Using the idea of Example 15.15, if we divide the total number of ordered hands by 5!, we get the total number of unordered (draw poker) hands: $(52 \times 51 \times 50 \times 49 \times 48)/5! = 2,598,960$ [see Fig. 15-6(b)].

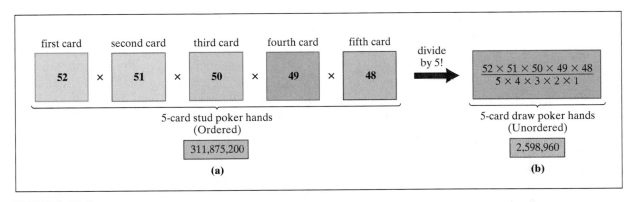

FIGURE 15-6
(a) Number of 5-card stud poker hands and (b) number of 5-card draw poker hands.

We should now be able to generalize the ideas we learned in Examples 15.14, 15.15, and, especially, 15.16. Imagine that we have a set of n distinct objects and that we want to select r different objects from this set. The number of ways that this can be done depends on whether the selections are ordered or unordered. Ordered selections are the generalization of stud poker hands—change the order of selection of the same objects and you get something different. Unordered selections are the generalization of draw poker hands—change the order of selection of the same objects and you get nothing new. This distinction is of fundamental importance in counting, and mathematicians have a name for each scenario. Ordered selections are called **permutations**, whereas unordered selections are called **combinations**. For a given number of objects n and a given selection size r, we can talk about the "number of permutations of n objects taken r at a time" and the "number of combinations of n objects taken r at a time," and these two extremely important families of numbers are denoted $_nP_r$ and $_nC_r$, respectively. Essentially, the numbers $_nP_r$ can be computed directly using the multiplication rule, as in Example 15.13. The numbers $_nC_r$ can be computed by dividing the corresponding $_nP_r$ by $r!$, as in Examples 15.14, 15.15, and 15.16.[2] A summary of the essential facts about these numbers is given in Table 15-1.

TABLE 15-1   Selecting r Different Objects out of n Objects $(r \leq n)$

	Ordered selections	Unordered selections
Name	Permutations	Combinations
Symbol	$_nP_r$	$_nC_r \left[\text{also} \begin{pmatrix} n \\ r \end{pmatrix} \right]$
Formula (with factorials)	$\dfrac{n!}{(n-r)!}$	$\dfrac{n!}{(n-r)!r!}$
Formula (without factorials)	$n \times (n-1) \times \cdots \times (n-r+1)$	$\dfrac{n \times (n-1) \times \cdots \times (n-r+1)}{r \times (r-1) \times \cdots \times 1}$
Applications	Rankings; stud poker hands; committees (when each member has a different job).	Subsets; draw poker hands; lottery tickets; coalitions.

EXAMPLE 15.17 The Florida Lotto

To play the Florida Lotto, a person has to pick 6 out of 53 numbers (after paying $1 for the privilege). If the person picks exactly the same 6 numbers as the ones drawn by the lottery, he or she can win mountains of money (usually a few million but it can be as much as 100 million). How many different Florida Lotto tickets are possible?

The key question is, Are the lottery tickets *ordered* or *unordered* selections of 6 out of 53 numbers? In the Florida Lotto, the order in which the numbers come up is irrelevant, so lottery draws are unordered. The rest is easy. The number of Florida Lotto ticket combinations is $_{53}C_6 = 22,957,480$. ■

We now leave the wonderful world of counting problems and return to the main theme of the chapter—calculating probabilities.

15.4   What Is a Probability?

Suppose that we toss a coin in the air. What is the probability that it will land heads up? That is not a deep mathematical question, and almost everybody agrees on the answer, although not necessarily for the same reason. The standard answer given is 1/2 (or 50%, or 1 out of 2). But why is the answer 1/2, and what does such an answer mean?

One common explanation given for the probability of 1/2 is that when we toss a coin, there are two possible outcomes (H and T), and since H represents one of the two possibilities, the probability of an H outcome must be 1 out of 2 or 1/2. This logic, while correct in the case of an honest coin, has a lot of holes in it. Consider how the same argument would sound in a different scenario.

---

[2]Most business and scientific calculators have built-in $_nP_r$ and $_nC_r$ keys. The sequence of keystrokes used to compute $_nP_r$ (or $_nC_r$) varies from brand to brand, but in most cases you enter the value of n first, push the $_nP_r$ (or $_nC_r$) key next, then enter the value of r, and, finally, press the equals key.

EXAMPLE 15.18 Shooting Free Throws (in the real world)

Imagine Mark Price, the most accurate free-throw shooter in NBA history, at the free-throw line. Once again, there are two possible outcomes to the free throw (success or failure), but it would be absurd to conclude in this case that the probability of success is 1/2. The two outcomes, while both possible, are not both equally likely, and thus their probabilities should be different. What then is the probability that Price will make the free throw?

Checking the NBA Web page (*www.nba.com*), we found that over a 13-year professional career in the NBA, Mark Price shot 2362 free throws and made 2135. The ratio 2135/2362 = 0.904 gives us a number that can be interpreted as an approximate value of the probability based on empirical data—a large number of repetitions of the same random experiment. Thus, we can say that for Mark Price, the approximate probability of making a free throw is 0.904. ∎

Example 15.18 leads us to an empirical interpretation of probabilities and to make a different argument about the coin toss: When tossing an honest coin the probability of heads is 1/2, because, if we toss the coin over and over—hundreds, possibly thousands of times—in the long run about half of the tosses will turn out to be heads. We don't actually have to do it—we instinctively believe that this is true, and that's why we give the answer of 1/2.

15.5   Probability Spaces

The argument as to exactly how to interpret the statement "the probability of *X* is such and such" goes back to the late 1600s, and it wasn't until the 1930s that a formal mathematical theory of probabilities was developed by the Russian mathematician A. N. Kolmogorov. This theory has made probability one of the most useful and important concepts in modern mathematics. In the remainder of this chapter we will discuss some of the basic concepts of probability theory.

We will start by drawing a distinction between an *individual outcome* and a *set of possible outcomes*. When drawing a card from a deck of cards we may be interested in the probability of drawing the ace of spades (an individual outcome), or we may be interested in the probability of drawing a spade (many possible outcomes). To deal with these latter types of probabilities we first introduce the concept of an *event*.

Events

An **event** is any subset of the sample space. That is, an event is a set of individual outcomes.

EXAMPLE 15.19

In Example 15.8 we considered the random experiment of tossing a coin 3 times and saw that the sample space was $S = \{HHH, HHT, HTH, HTT, THH, THT, TTH, TTT\}$. There are many possible events for this sample space. Table 15-2 shows just a few of them.

TABLE 15-2	Some of the Many Possible Events in a Sample Space	
Event	**Set of Outcomes**	**Size of Event**
1. Toss 2 or more heads	$\{HHT, HTH, THH, HHH\}$	4
2. Toss more than 2 heads	$\{HHH\}$	1
3. Toss 2 heads or less	$\{TTT, TTH, THT, HTT, THH, HTH, HHT\}$	7
4. Toss no tails	$\{HHH\}$	1
5. Toss exactly 1 tail	$\{HHT, HTH, THH\}$	3
6. Toss exactly 1 head	$\{HTT, THT, TTH\}$	3
7. First toss is heads	$\{HHH, HHT, HTH, HTT\}$	4
8. Toss same number of heads as tails	$\{\ \}$	0
9. Toss at most 3 heads	S	8
10. First toss is heads and at least 2 tails are tossed	$\{HTT\}$	1

There are many ways of combining outcomes in a sample space to make an event, and the same event can be described (in English) in more than one way (e.g., events 2 and 4 in Table 15-2). The actual number of individual outcomes in an event can be as low as 0 and as high as N (the size of the sample space). In the case in which the number of outcomes is 0 (as in event 8 in Table 15-2), the event is called the **impossible event**. In the case in which the event is the whole sample space S (as in event 9 in Table 15-2), it is called the **certain event**.

Probability Assignments

Let's return to free-throw shooting, as it is a useful metaphor for many probability questions.

EXAMPLE 15.20 More on Shooting Free Throws

A person shoots a free throw. We know nothing about his or her abilities. (For all we know, the person could be Mark Price or Joe Schmoe.) What is the probability that he or she will make the free throw?

It seems that there is no way to answer this question, since we know nothing about the shooter. One could argue that the probability could be just about any number, as long as it is not a negative number and it is not greater than 1 (a perfect 100% free-throw shooter). We learned how to handle situations like this in algebra. We just make our unknown probability a variable, say p.

What can we say about the probability that our shooter misses the free throw? A lot. Since there are only two possible outcomes in the sample space $S = \{s, f\}$, the probability of success (s) and the probability of failure (f) must complement each other—in other words, must add up to 1. This means that the probability of missing the free throw must be $1 - p$.

Table 15-3 is a summary of the line on a generic free-throw shooter.

TABLE 15-3   A Generic Free-Throw Shooter

Event	Probability
Neither success nor failure: { }	0
Success: $\{s\}$	p
Failure: $\{f\}$	$1 - p$
Success or failure: $\{s, f\}$	1

Table 15-3, humble as it may seem, gives a complete model of free-throw shooting. It works when the free-throw shooter is Mark Price (make $p = 0.904$), Joe Schmoe (make $p = 0.45$), or any other Tom, Dick, or Shaq in between. Each one of the choices results in a different assignment of numbers to the outcomes in the sample space.

Example 15.20 illustrates the concept of a *probability assignment*. A **probability assignment** is a function that assigns to each event E a number between 0 and 1, which represents the probability of the event E and which we denote by $\Pr(E)$.[3] A probability assignment always assigns probability 0 to the *impossible event* [$\Pr(\{ \}) = 0$], and probability 1 to the *certain event* [$\Pr(S) = 1$].

With finite sample spaces of the type we are considering in this chapter, a valid probability assignment can be made by simply assigning probabilities to each of the individual outcomes in the sample space—the only requirement is that these numbers are between 0 and 1 and add up to 1. Once we do this, we can find the probability of any event by simply *adding the probabilities of the individual outcomes that make up that event*. The next example gives a basic illustration of how it all works.

EXAMPLE 15.21 Tennis Anyone?

Two men (*Andre* and *Pete*) and three women (*Ana, Venus,* and *Serena*) reach the finals of a tennis tournament. We are interested in computing various probabilities for winning the tournament. The sample space consists of the five finalists: $S = \{Andre, Pete, Ana, Venus, Serena\}$. Typically, each of the players is assigned a probability for winning the tournament, which we denote by $\Pr(player's\ name)$. While these probabilities are subjective, they must satisfy the rules for a valid probability assignment, and from them we can reconstruct the probability of any event.

For example, suppose that we are given $\Pr(Andre) = 0.25$, $\Pr(Pete) = 0.28$, $\Pr(Venus) = 0.18$, and $\Pr(Serena) = 0.19$. Notice that the value of $\Pr(Ana)$ is missing, but we can determine that $\Pr(Ana) = 0.10$ from the rule that the sum of the probabilities of all the outcomes must be 1.

---

[3]For events consisting of just a single outcome, say for example $\{o\}$, we will cheat a little and for the sake of simplicity speak of "the probability of the outcome o," when we really should say "the probability of the event consisting of the single outcome o." Accordingly, we will write $\Pr(o)$ as a substitute for $\Pr(\{o\})$.

Once we have the probabilities of each of the outcomes in the sample space, we can compute the probability of any event. For example, the probability that a player whose name starts with A wins the tournament is given by $\Pr(\{Andre, Ana\}) = 0.25 + 0.10 = 0.35$. Likewise, the probability that a woman wins the tournament is $\Pr(\{Venus, Serena, Ana\}) = 0.47$ (just add them up!). ■

When we combine a specific sample space with a specific assignment of probabilities to the individual outcomes of that sample space we have a **probability space**. We will now summarize the various elements that constitute a probability space.

THE ELEMENTS OF A FINITE PROBABILITY SPACE

1. A finite *sample space* $S = \{o_1, o_2, \ldots, o_N\}$. (The o's are the individual outcomes).

2. *Events.* Any subset of S is an event. Two special events are $\{\ \}$ (called the *impossible* event) and S itself (called the *certain* event).

3. A *probability assignment* for S. To each individual outcome o_i, we assign a number $\Pr(o_i)$. The two rules for a probability assignment are $0 \leq \Pr(o_i) \leq 1$ and $\Pr(o_1) + \Pr(o_2) + \cdots + \Pr(o_N) = 1$.

4. *Probabilities of events.* The probability of an event is obtained by adding the probabilities of the individual outcomes that make up the event. In particular, $\Pr(\{\ \}) = 0$ and $\Pr(S) = 1$.

15.6 Probability Spaces with Equally Likely Outcomes

An important special case of a probability space is the one in which every individual outcome has an equal probability assigned to it. This is the case when we toss an honest coin, roll an honest die, or draw a card from a well-shuffled deck of cards.

When the probability of each individual outcome in the sample space is the same, then calculating probabilities becomes simply a matter of counting. For a sample space of size N, the probability of each individual outcome must be $1/N$ (because these probabilities must add up to 1), and the probability of an event with several outcomes is the number of outcomes in the event divided by N.

COMPUTING PROBABILITIES WHEN ALL OUTCOMES ARE EQUALLY LIKELY

- Size of sample space $= N$.

- If E is an event, $\Pr(E) = \dfrac{\text{number of outcomes in } E}{N}$.

EXAMPLE 15.22 The Probability of Drawing an Ace

The top card is drawn from a well-shuffled standard deck of 52 cards. What is the probability of drawing an ace?

Here, $N = 52$, and the event "top card is an ace" is given by the set

$$\left\{ \boxed{\heartsuit}, \boxed{\clubsuit}, \boxed{\spadesuit}, \boxed{\diamondsuit} \right\}.$$

Since each of the outcomes has probability 1/52, it follows that

$$\Pr(\text{"top card is an ace"}) = \frac{4}{52} = \frac{1}{13} \approx 0.077.$$

Suppose now that we want to know the probability that the tenth card in the deck is an ace. Is it different than for the top card? When the deck is well shuffled, the aces can be anywhere in the deck—there is nothing about the top position that makes it special or different from the tenth position or, for that matter, from any of the other positions in the deck. The probability of the tenth card in the deck being an ace is still 1/13. ▬

EXAMPLE 15.23

Suppose that we roll a pair of honest dice.

Here, $N = 36$ (see Example 15.5), and since the dice are honest, each of the 36 outcomes has probability 1/36.

What is the probability of rolling a total of 11? There are 2 ways of rolling a total of 11 ("roll 11" = $\left\{ \boxed{\cdot\cdot}\boxed{::}, \boxed{::}\boxed{\cdot\cdot} \right\}$). Thus,

$$\Pr(\text{"roll 11"}) = \frac{2}{36} = \frac{1}{18} \approx 0.056.$$

What is the probability of rolling a total of 7? There are 6 ways of rolling a total of 7 ("roll 7" = $\left\{ \boxed{\cdot}\boxed{::}, \boxed{\cdot}\boxed{:\cdot}, \boxed{\cdot\cdot}\boxed{::}, \boxed{::}\boxed{\cdot\cdot}, \boxed{:\cdot}\boxed{\cdot\cdot}, \boxed{::}\boxed{\cdot} \right\}$). Thus,

$$\Pr(\text{"roll 7"}) = \frac{6}{36} = \frac{1}{6} \approx 0.167.$$

What is the probability of rolling a total of 7 or 11? The event "roll 7 or 11" has 8 possible outcomes (the 6 in "roll 7" and the 2 in "roll 11"), so

$$\Pr(\text{"roll 7 or 11"}) = \frac{8}{36} = \frac{2}{9} \approx 0.222.$$

▬

EXAMPLE 15.24

If we roll a pair of honest dice, what is the probability that at least one of them is a $\boxed{\cdot}$?

We know that each individual outcome in the sample space has probability of 1/36. We will show three different ways to solve this problem.

Solution 1 (The brute-force approach). If we just write down the event E, which is "we will roll at least one $\boxed{\cdot}$," we have

$$E = \left\{ \boxed{\cdot}\boxed{\cdot}, \boxed{\cdot}\boxed{\cdot\cdot}, \boxed{\cdot}\boxed{:\cdot}, \boxed{\cdot}\boxed{::}, \boxed{\cdot}\boxed{:\cdot}, \boxed{\cdot}\boxed{::}, \boxed{\cdot\cdot}\boxed{\cdot}, \boxed{:\cdot}\boxed{\cdot}, \boxed{::}\boxed{\cdot}, \boxed{:\cdot}\boxed{\cdot}, \boxed{::}\boxed{\cdot} \right\}.$$

Thus, it follows that $\Pr(E) = 11/36 \approx 0.306$.

Solution 2 (The roundabout approach). Let's say, for the sake of argument, that we will win if at least one of the two dice comes up a ⚀ and that we will lose otherwise. This means that we will lose if both dice come up with a number other than ⚀ Let's calculate first the probability that we will lose. (This is called the roundabout way of doing things.) Using the multiplication principle, we can calculate the number of individual outcomes in the event "we lose."

- Number of ways first die can come up not a ⚀ = 5.
- Number of ways the second die can come up not a ⚀ = 5.
- Total number of ways neither die comes up ⚀ = 5 × 5 = 25.

Probability that we will lose: $\Pr(\text{lose}) = \dfrac{25}{36} \approx 0.694$.

Probability that we will win: $\Pr(\text{win}) = 1 - \dfrac{25}{36} = \dfrac{11}{36} \approx 0.306$.

Solution 3 (Independent events). In this solution, we consider each die separately. In fact, we will find it slightly more convenient to think of rolling a single honest die twice (which, mathematically, is exactly the same thing as rolling a pair of honest dice once).

Let's start with the first roll. The probability that we won't roll a ⚀ is 5/6. (There are 6 possible outcomes, 5 of which are not a ⚀.) For the same reason, the probability that the second roll will not be a ⚀ is also 5/6.

Now comes a critical observation. The probability that neither of the first two rolls will be a ⚀ is $(5/6) \times (5/6) = 25/36$. The reason that we can multiply the probabilities of the two events ("first roll is not a ⚀") and ("second roll is not a ⚀") is that these two events are **independent**. That is, the outcome of the first roll does not in any way affect the outcome of the second roll.

We finish the problem exactly as in solution 2:

$$\Pr(\text{lose}) = \frac{25}{36} \text{ and, therefore, } \Pr(\text{win}) = 1 - \frac{25}{36} = \frac{11}{36} \approx 0.306.$$

Of the three solutions to Example 15.24, solution 3 appears to be the most complicated, but, in fact, it shows us the most useful approach. It is based on what we will call the *multiplication principle for independent events*.

Independent events. Two events are said to be independent if the outcome of one event does not affect the outcome of the other.

The multiplication principle for independent events. When a complex event E can be broken down into a combination of two simpler events that are *independent* (call them F and G), then we can calculate the probability of E by multiplying the probabilities of F and G.

The multiplication principle for independent events in an important and useful rule, but it works only when the parts are independent. The next two examples illustrate the usefulness of this principle.

EXAMPLE 15.25

If we roll an honest die 4 times, what is the probability that we will roll a ⊡ at least once?

Let's try the same approach we used in solution 3 of Example 15.24. We will win if we roll a ⊡ at least once, and we will lose if none of the four rolls comes up as a ⊡. Thus, we know that

$$\text{Pr}(1^{\text{st}} \text{ roll is not a } ⊡) = 5/6,$$

$$\text{Pr}(2^{\text{nd}} \text{ roll is not a } ⊡) = 5/6,$$

$$\text{Pr}(3^{\text{rd}} \text{ roll is not a } ⊡) = 5/6,$$

$$\text{Pr}(4^{\text{th}} \text{ roll is not a } ⊡) = 5/6.$$

Because each roll is independent of the preceding ones, we can use the multiplication principle for independent events.

$$\text{Pr}(\text{lose}) = \text{Pr}(\text{not rolling any } ⊡\text{'s in four rolls}) = (5/6)^4 \approx 0.482.$$

Thus, it follows that

$$\text{Pr}(\text{win}) = \text{Pr}(\text{rolling at least one } ⊡ \text{ in four rolls}) \approx 0.518.$$

We now leave the dice and move on to two examples that illustrate the variety of problems we are now able to solve.

EXAMPLE 15.26 The Probability of Four Aces

What is the probability of getting four aces in a 5-card draw poker hand? The sample space for this question is the set of all possible 5-card draw poker hands. We saw in Example 15.16 that for this sample space, $N = 2{,}598{,}960$. The event $E = $ "draw four aces" has 48 different outcomes (as four of the cards are aces, and the fifth card can be any one of the 48 other cards). Thus,

$$\text{Pr}(E) = \frac{48}{2{,}598{,}960} = \frac{1}{54{,}145} \approx 0.0000185.$$

Thus, the probability of drawing 4 aces in a 5-card draw poker hand is roughly 2 in 100,000.

EXAMPLE 15.27

If we toss an honest coin 10 times, what is the probability of getting 5 Hs and 5 Ts? (This is an important question, and you might find the answer surprising. Before you read on, you are encouraged to make a rough guess.)

The size of the sample space when tossing 10 coins is $N = 2^{10} = 1024$. How many of the 1024 equally likely strings of 10 Hs and Ts have exactly 5 Hs and 5 Ts? To count these we count the possible ways in which we can choose the 5 "slots" for the Hs. These are unordered selections, and thus the answer is $_{10}C_5 = 252$. It follows that the probability of tossing 5 Hs and 5 Ts is $252/1024 \approx 0.246$.

15.7 Odds

Dealing with probabilities as numbers that are always between 0 and 1 is the mathematician's way of having a consistent terminology. To the everyday user, consistency is not that much of a concern, and we know that people talk about *chances* (probabilities expressed as percentages) and *odds*, which are most frequently used to describe probabilities associated with gambling events. In this section, we will briefly discuss how to interpret and calculate odds. To simplify our discussion we will consider only the situation in which all outcomes are equally likely.

> **Odds in favor of an event.** The odds in favor of event E are given by the ratio of the number of ways event E can occur to the number of ways in which event E cannot occur.

EXAMPLE 15.28

If we roll a pair of honest dice, what are the odds *in favor* of rolling a total of 7?

We saw in Example 15.23(b) that of the 36 different outcomes that are possible when rolling a pair of dice, 6 are *favorable* (i.e., result in a total of 7) and the other 30 are *unfavorable* (i.e., result in a total that is not 7). Thus, it follows that the odds in favor of rolling a 7 are 6 to 30, or, equivalently, 1 to 5. ∎

EXAMPLE 15.29

If we roll a pair of honest dice, what are the odds *against* rolling a total of 7?

This question is essentially the reverse of the one asked in Example 15.28. In this case, of the 36 possible outcomes, 30 are favorable (i.e., result in a total that is not 7) and 6 are unfavorable (i.e., result in a total of 7). Thus, the odds against rolling a 7 are 30 to 6, or 5 to 1, which are the same numbers as in Example 15.28 but reversed. ∎

> **Odds against an event.** If the odds in favor of event E are m to n, then the odds against event E are n to m.

Sometimes we want to calculate the odds in favor of an event, but all we have to go on is the probability of that event.

EXAMPLE 15.30

When Mark Price shoots a free throw, the empirical probability that he will make it is 0.904 (see Example 15.18). We can interpret this to mean that on the average, out of every 1000 free throws he attempts, he will make 904 and miss 96, so it follows that the odds in favor of his making a free throw are 904 to 96, or, reduced to simplest form, 113 to 12. Typically, this kind of accuracy is rarely called for, and it is customary to round off the odds to more manageable numbers, in this case, 9 to 1. ∎

The general rule for converting probabilities to odds is as follows.

> If $\Pr(E) = a/b$, the *odds in favor* of E are a to $b - a$ and the *odds against* E are $b - a$ to a.

The general rule for converting odds to probabilities is as follows.

> If the *odds* in *favor* of an event E are m to n, then $\Pr(E) = \dfrac{m}{m + n}$.

A word of caution: There is a difference between odds as discussed in this section and the *payoff odds* posted by casinos or bookmakers in sports gambling situations. Suppose we read in the newspaper, for example, that the Las Vegas sports books have established that "the odds that the Los Angeles Lakers will win the NBA championship are 5 to 2." What this means is that if you want to bet in favor of the Lakers, for every \$2 that you bet, you can win \$5 if the Lakers win. This ratio may be taken as some indication of the actual odds in favor of the Lakers winning, but several other factors affect payoff odds, and the connection between payoff odds and actual odds is tenuous at best.

Conclusion

While the average citizen thinks of probabilities, chances, and odds as vague, informal concepts that are useful primarily when discussing the weather or playing the lottery, scientists and mathematicians think of probability as a formal framework within which the laws that govern chance events can be understood. The basic elements of this framework are a *sample space* (which represents a precise mathematical description of all the possible outcomes of a *random experiment*), *events* (collections of these outcomes), and a *probability assignment* (which associates a numerical value to each of these events).

Of the many ways in which probabilities can be assigned, a particularly important case is the one in which all individual outcomes have the same probability. When this happens, the critical steps in calculating probabilities revolve around two basic (but not necessarily easy) questions: (1) Given a sample space, what is its size? and (2) Given an event, what is its size? To answer these kinds of questions, knowing how to "count" large sets is critical.

When one stops to think how much of life is ruled by fate and chance, the importance of probability theory in almost every walk of life is hardly surprising. As the great French poet and philosopher Voltaire put it, "His Sacred Majesty, Chance, decides everything."

P R O F I L E   Persi Diaconis (1945–   )

Persi Diaconis picks up an ordinary deck of cards, fresh from the box, and writes a word in Magic Marker on one side: RANDOM. He shuffles the deck once. The letters have re-formed themselves into six bizarre runes that still look vaguely like the letters R, A, and so on. Diaconis shuffles again, and the markings on the side become indecipherable. After two more shuffles, you can't even tell that there used to be six letters. The side of the pack looks just like the static on a television set. It didn't look random before, but it sure looks random now.

Keep watching. After three more shuffles, the word RANDOM miraculously reappears on the side of the deck–only it is written twice, in letters half the original size. After one more shuffle, the original letters materialize at the original size. Diaconis turns the cards over and spreads them out with a magician's flourish, and there they are in their exact original sequence, from the ace of spades to the king of diamonds.

Diaconis has just performed eight perfect shuffles in a row. There's no hocus-pocus, just skill perfected in his youth: Diaconis ran away from home at 14 to become a magician's assistant and later became a professional magician and blackjack player. Even now, at 57, he is one of a couple of dozen people on the planet who can do eight perfect shuffles in less than a minute.

Diaconis's work these days involves much more than nimbleness of hand. He is a professor of mathematics and statistics at Stanford University. But he is also the world's leading expert on shuffling. He knows that what seems to be random often isn't, and he has devoted much of his career to exploring the difference. His work has applications to filing systems for computers and the reshuffling of the genome during evolution. And it has led him back to Las Vegas, where instead of trying to beat the casinos, he now works for them.

A card counter in blackjack memorizes the cards that have already been played to get better odds by making bets based on his knowledge of what has yet to turn up. If the deck has a lot of face cards and 10s left in it, for instance, and he needs a 10 for a good hand, he will bet more because he's more likely to get it. A good card counter, Diaconis estimates, has a 1 to 2 percent advantage over the casino. On a bad day, a good card counter can still lose $10,000 in a hurry. And on a good day, he may get a tap on the shoulder by a large person who will say, "You can call it a day now." By his mid-twenties, Diaconis had figured out that doing mathematics was an easier way to make a living.

Two years ago, Diaconis himself got a tap on the shoulder. A letter arrived from a manufacturer of casino equipment, asking him to figure out whether its card-shuffling machines produced random shuffles. To Diaconis's surprise, the company gave him and his Stanford colleague, Susan Holmes, carte blanche to study the inner workings of the machine. It was like taking a Russian spy on a tour of the CIA and asking him to find the leaks.

When shuffling machines first came out, Diaconis says, they were transparent, so gamblers could actually see the cutting and riffling inside. But gamblers stopped caring after a while, and the shuffling machines turned into closed boxes. They also stopped shuffling cards the way humans do. In the machine that Diaconis and Holmes looked at, each card gets randomly directed, one at a time, to one of 10 shelves. The shuffling machine can put each new card either on the top of the cards already on the shelf or on the bottom, but not between them.

"Already I could see there was something wrong," says Holmes. If you start out with all the red cards at the top of the deck and all the black cards at the bottom, after one pass through the shuffling machine you will find that each shelf contains a red-black sandwich. The red cards, which got placed on the shelves first, form the middle of each sandwich. The black cards, which came later, form the outside. Since there are only 10 shelves, there are at most 20 places where a red card is followed by a black one or vice versa—fewer than the average number of color changes (26) that one would expect from a random shuffle.

The nonrandomness can be seen more vividly if the cards are numbered 1 to 52. After they have passed through the shuffling machine, the numbers on the cards form a zigzag pattern. The top card on the top shelf is usually a high number. Then the numbers decrease until they hit the middle of the first red-black sandwich; then they increase and decrease again, and so on, at most 10 times.

Diaconis and Holmes figured out the exact probability that any given card would end up in any given location after one pass through the machine. But that didn't indicate whether a gambler could use this information to beat the house.

So Holmes worked out a demonstration. It was based on a simple game. You take cards from a deck one by one and each time try to predict what you've selected before you look at it. If you keep track of all the cards, you'll always get the last one right. You'll guess the second-to-last card right half the time, the third-to-last a third of the time, and so on. On average, you will guess about 4.5 cards correctly out of 52.

By exploiting the zigzag pattern in the cards that pass through the shuffling machine, Holmes found a way to double the success rate. She started by predicting that the highest possible card (52) would be on top. If it turned out to be 49, then she predicted 48—the next highest number—for the second card. She kept going this way until her prediction was too low—predicting, say, 15 when the card was actually 18. That meant the shuffling machine had reached the bottom of a zigzag and the numbers would start climbing again. So she would predict 19 for the next card. Over the long run, Holmes (or, more precisely, her computer) could guess nine out of every 52 cards correctly.

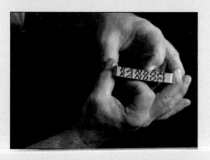

To a gambler, the implications are staggering. Imagine playing blackjack and knowing one-sixth of the cards before they are turned over! In reality, a blackjack player would not have such a big advantage, because some cards are hidden and six full decks are used. Still, Diaconis says, "I'm sure it would double or triple the advantage of the ordinary card counter."

Diaconis and Holmes offered the equipment manufacturer some advice: Feed the cards through the machine twice. The alternative would be more expensive: Build a 52-shelf machine.

A small victory for shuffling theory, one might say. But randomization applies to more than just cards. Evolution randomizes the order of genes on a chromosome in several ways. One of the most common mutations is called a "chromosome inversion," in which the arm of a chromosome gets cut in two random places, flipped over end-to-end, and reattached, with the genes in reverse order. In fruit flies, inversions happen at a rate of roughly one per every million years. This is very similar to a shuffling method called transposition that Diaconis studied 20 years ago. Using his methods, mathematical biologists have estimated how many inversions it takes to get from one species of fruit fly to another, or to a completely random genome. That, Diaconis suggests, is the real magic he ran away from home to find. "I find it amazing," he says, "that mathematics developed for purely aesthetic reasons would mesh perfectly with what engineers or chromosomes do when they want to make a mess."

Source: Dana Mackenzie, "The Mathematics of Shuffling," *Discover* (October 2002), pp. 22-23; Reprinted by permission.

certain event
combinations
event
impossible event
independent events
multiplication principle for
  independent events

multiplication rule
odds
permutations
probability assignment
probability space
random experiment
sample space

WALKING

A. Random Experiments and Sample Spaces

1. Write out the sample space for each of the following random experiments.

  (a) A coin is tossed three times in a row and we observe on each toss whether it lands heads or tails.

  (b) A coin is tossed three times in a row and we observe the number of times it lands tails.

  (c) A person shoots ten consecutive free throws and we observe the number of missed free throws.

2. Write out the sample space for each of the following random experiments.

  (a) A coin is tossed four times in a row and we observe on each toss whether it lands heads or tails.

  (b) A student takes a four-question true–false quiz and we observe the student's answers. (Assume that the student answers all the questions, but do not make any assumptions about the student's knowledge or lack thereof.)

3. Four names (*A, B, C,* and *D*) are written each on a separate slip of paper, put in a hat, and mixed well. The slips are randomly taken out of the hat, one at a time, and the names recorded.

  (a) Write out the sample space for this random experiment.

  (b) Find *N*.

4. A gumball machine has gumballs of 4 different flavors: cherry (*C*), grape (*G*), lemon (*L*), and sour apple (*S*). When a quarter is put into the machine, 2 random gumballs come out.

  (a) Write out the sample space for this random experiment.

  (b) Find *N*.

In Exercises 5 through 10, the sample spaces are too big to write down in full. In these exercises, you should describe the sample space either by describing a generic outcome or by listing some outcomes and then using the ... notation. In the latter case, you should write down enough outcomes to make the description reasonably clear.

5. A coin is tossed 10 times in a row and we observe on each toss whether it lands heads or tails.

(a) Describe the sample space.

(b) Find N.

6. A student takes a ten-question true–false quiz and we observe the student's answers. (Assume that the student answers all the questions, but do not make any assumptions about the student's knowledge or lack thereof.)

   (a) Describe the sample space.

   (b) Find N.

7. A die is rolled 4 times in a row and we observe the number that comes up on each roll.

   (a) Describe the sample space.

   (b) Find N.

8. In the game of Yahtzee, a set of five dice is rolled at once, and the number on each die is observed.

   (a) Describe the sample space.

   (b) Find N.

9. Ten names ($A, B, C, D, E, F, G, H, I$, and J) are written each on a separate slip of paper, put in a hat, and mixed well. Four slips are randomly taken out of the hat, one at a time, and the names recorded. Assume that the order in which the names are drawn matters.

   (a) Describe the sample space.

   (b) Find N.

10. A gumball machine has gumballs of 4 different flavors: cherry (C), grape (G), lemon (L), and sour apple (S). When a fifty-cent piece is put into the machine, 5 random gumballs come out. Describe the sample space.

B. The Multiplication Rule

11. A California license plate starts with a digit other than 0, followed by three capital letters followed by three more digits (0 through 9).

    (a) How many different California license plates are possible?

    (b) How many different California license plates start with a 5 and end with a 9?

    (c) How many different California license plates have no repeated symbols (all the digits are different and all the letters are different)?

12. A computer password consists of four letters (A through Z) followed by a single digit (0 through 9). Assume that the passwords are not case sensitive (i.e., that an uppercase letter is the same as a lowercase letter).

    (a) How many different passwords are possible?

    (b) How many different passwords end in 1?

    (c) How many different passwords do not start with Z?

    (d) How many different passwords have no Zs in them?

13. A computer password consists of four letters (A through Z) followed by a single digit (0 through 9). Assume that the passwords are case sensitive (i.e., uppercase letters are considered different from lowercase letters).

    (a) How many different passwords are possible?

(b) How many different passwords start with Z?

(c) How many different passwords do not start with either z or Z?

(d) How many different passwords have no Zs in them (uppercase or lowercase)?

14. A French restaurant offers a menu consisting of 3 different appetizers, 2 different soups, 4 different salads, 9 different main courses, and 5 different desserts.

(a) A fixed-price lunch meal consists of a choice of appetizer, salad, and main course. How many different lunch fixed-price meals are possible?

(b) A fixed-price dinner meal consists of a choice of appetizer, a choice of soup or salad, a main course, and a dessert. How many different dinner fixed-price meals are possible?

(c) A dinner special consists of a choice of soup, or salad, or both, and a main course. How many dinner specials are possible?

15. A set of reference books consists of 8 volumes numbered 1 through 8.

(a) In how many ways can the 8 books be arranged on a shelf?

(b) In how many ways can the 8 books be arranged on a shelf so that at least 1 book is out of order?

16. Four men and 4 women line up at a checkout stand in a grocery store.

(a) In how many ways can they line up?

(b) In how many ways can they line up if the first person in line must be a woman?

(c) In how many ways can they line up if they must alternate woman, man, woman, man, and so on and if a woman must always be first in line?

17. The ski club at Tasmania State University has 35 members (15 females and 20 males). A committee of 3 members—a President, a Vice President, and a Treasurer—must be chosen.

(a) How many different 3-member committees can be chosen?

(b) How many different 3-member committees can be chosen if the president must be a female?

(c) How many different 3-member committees can be chosen if the committee cannot have all females or all males?

18. The ski club at Tasmania State University has 35 members (15 females and 20 males). A committee of 4 members—a President, a Vice President, a Treasurer, and a Secretary—must be chosen.

(a) How many different 4-member committees can be chosen?

(b) How many different 4-member committees can be chosen if the president and treasurer must be females?

(c) How many different 4-member committees can be chosen if the president and treasurer must both be female and the vice president and secretary must both be male?

(d) How many different 4-member committees can be chosen if the committee must have two females and two males?

19. How many 7-digit numbers (i.e., numbers between 1,000,000 and 9,999,999)

    (a) are even?

    (b) are divisible by 5?

    (c) are divisible by 25?

20. How many 10-digit numbers (i.e., numbers between 1,000,000,000 and 9,999,999,999)

    (a) have no repeated digits?

    (b) are palindromes? (A palindrome is a number such as 37473 that reads the same whether read from left to right or from right to left.)

C. Permutations and Combinations

In Exercises 21 through 24, give your answer in symbolic form, using the notation $_nP_r$ or $_nC_r$.

21. The board of directors of the XYZ corporation has 15 members. In how many ways can one choose

    (a) a committee of 4 members (President, Vice President, Treasurer, and Secretary)?

    (b) a delegation of 4 members where all members have equal standing?

22. There are 10 horses entered in a race. In how many ways can one pick

    (a) the top three finishers regardless of order?

    (b) the first-, second-, and third-place finishers in the race?

23. There are 20 singers auditioning for a musical. In how many different ways can the director choose

    (a) a duet?

    (b) a lead singer and a backup?

    (c) a quintet?

24. There are 117 Division I-A college football teams.

    (a) How many Top 25 rankings are possible?

    (b) How many ways are there to choose 8 teams for a playoff?

In Exercises 25 through 32, compute each of the following without using a calculator.

25. **(a)** $_{10}P_2$

    (b) $_{10}C_2$

    (c) $_{10}P_3$

    (d) $_{10}C_3$

26. **(a)** $_{11}P_2$

    (b) $_{11}C_2$

    (c) $_{20}P_2$

    (d) $_{20}C_2$

27. **(a)** $_{10}C_9$

    (b) $_{10}C_8$

(c) $_{100}C_{99}$

(d) $_{100}C_{98}$

28. (a) $_{12}P_2$

(b) $_{12}P_3$

(c) $_{12}P_4$

(d) $_{12}P_5$

29. (a) $_{20}C_2$

(b) $_{20}C_{18}$

(c) $_{20}C_3$

(d) $_{20}C_{17}$

30. (a) $_{10}C_3 + {}_{10}C_4$

(b) $_{11}C_4$

(c) $_9C_6 + {}_9C_7$

(d) $_{10}C_7$

31. (a) $_3C_0 + {}_3C_1 + {}_3C_2 + {}_3C_3$

(b) $_4C_0 + {}_4C_1 + {}_4C_2 + {}_4C_3 + {}_4C_4$

(c) $_5C_0 + {}_5C_1 + {}_5C_2 + {}_5C_3 + {}_5C_4 + {}_5C_5$

(d) $_{10}C_0 + {}_{10}C_1 + {}_{10}C_2 + \cdots + {}_{10}C_{10}$

[*Hint*: Look for the pattern in (a) through(c).]

32. (a) $_9C_4 - {}_9C_5$

(b) $_{10}C_3 - {}_{10}C_7$

(c) $_{100}C_{10} - {}_{100}C_{90}$

(d) $_{1000}C_{498} - {}_{1000}C_{592}$

For Exercises 33 and 34, you should use a calculator with built-in $_nP_r$ and $_nC_r$ keys. Most scientific and business calculators have such keys. The sequence of key strokes to compute $_nC_r$ (or $_nP_r$) varies from calculator to calculator, but the most common sequence is to enter the value of n first, then press the $_nC_r$ (or $_nP_r$) key, then enter the value of r, and finally press the equals key. (If this sequence doesn't work, you should consult the instruction booklet that came with your calculator.)

33. Using a calculator, compute each of the following. If you cannot get an answer, explain why not.

(a) $_{20}P_{10}$

(b) $_{52}C_{20}$

(c) $_{52}C_{32}$

(d) $_{100}P_{25}$

(e) $_{3650}C_{1000}$

34. Using a calculator, compute each of the following. If you cannot get an answer, explain why not.

(a) $_{18}P_{10}$

(b) $_{51}C_{21}$

(c) $_{51}C_{30}$

(d) $_{80}P_{25}$

(e) $_{1999}C_{300}$

D. General Probability Spaces

35. Consider the sample space $S = \{o_1, o_2, o_3, o_4, o_5\}$. Suppose you are given $\Pr(o_1) = 0.22$ and $\Pr(o_2) = 0.24$.

   (a) If o_3, o_4, and o_5 all have the same probability, find $\Pr(o_3)$.

   (b) If o_3 has the same probability as o_4 and o_5 combined, find $\Pr(o_3)$.

   (c) If o_3 has the same probability as o_4 and o_5 combined and if $\Pr(o_5) = 0.1$, give the probability assignment for this probability space.

36. Consider the sample space $S = \{o_1, o_2, o_3, o_4\}$. Suppose you are given $\Pr(o_1) + \Pr(o_2) = \Pr(o_3) + \Pr(o_4)$.

   (a) If $\Pr(o_1) = 0.15$, find $\Pr(o_2)$.

   (b) If $\Pr(o_1) = 0.15$, and $\Pr(o_3) = 0.22$, give the probability assignment for this probability space.

37. There are 7 players (call them $P_1, P_2, \ldots, P_7$) entered in a tennis tournament. According to one expert, P_1 is twice as likely to win as any of the other players, and $P_2, P_3, \ldots, P_7$ all have an equal chance of winning. Write down the sample space, and find the probability assignment for the probability space defined by this expert's opinion.

38. There are 8 players (call them $P_1, P_2, \ldots, P_8$) entered in a chess tournament. According to an expert, P_1 has a 25% chance of winning the tournament, P_2 a 15% chance, P_3 a 5% chance, and all the other players an equal chance. Write down the sample space, and find the probability assignment for the probability space defined by this expert's opinion.

39. A teacher's circular spinner has the 5 regions shown in the figure in the margin. The red region corresponds to a 108° angle, the blue and white regions correspond to 72° angles, and the green and yellow regions correspond to 54° angles. A game is played by spinning the needle. Depending on the color the arrow points to, different colored crayons are awarded to the class. When the needle falls exactly on the line between regions, the needle is spun again.

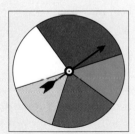

   (a) Find the probability that the needle points to the red region.

   (b) Describe a sample space for this game.

   (c) Give the probability assignment for this probability space.

40. A game is played by spinning the needle on the spinner shown in the margin and then moving a piece on a game board the indicated number of squares. When the needle falls exactly on the line between regions, the needle is spun again. The region numbered 5 corresponds to a 36° angle, the region numbered 3 corresponds to a 54° angle, the region numbered 2 corresponds to a 126° angle, and the region numbered 1 corresponds to a 144° angle.

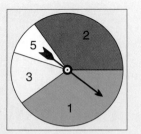

   (a) Find the probability that the player gets to move 5 squares.

   (b) Describe a sample space for this game.

   (c) Give the probability assignment for this probability space.

E. Events

41. Consider the random experiment of tossing a coin 3 times in a row. [See Exercise 1(a).] Write out the event described by each of the following statements as a set.

(a) E_1: "toss exactly 2 heads."

(b) E_2: "all tosses come out the same."

(c) E_3: "half of the tosses are heads, and half are tails."

(d) E_4: "first two tosses are tails."

42. Consider the random experiment from Exercise 2(b) where a student takes a four-question true–false quiz. Write out the event described by each of the following statements as a set.

(a) E_1: "exactly 2 of the answers given are Ts." (*Note:* T = True, F = False.)

(b) E_2: "at least 2 of the answers given are Ts."

(c) E_3: "at most 2 of the answers given are Ts."

(d) E_4: "the first 2 answers given are Ts."

43. Consider the random experiment from Example 15.5 of rolling a pair of dice. Write out the event described by each of the following statements as a set.

(a) E_1: "roll two of a kind." (That is, both numbers are equal.)

(b) E_2: "roll a total of 3 or less."

(c) E_3: "don't roll a total of 7 or less."

44. Consider the random experiment of drawing 1 card out of an ordinary deck of 52 cards. Write out the event described by each of the following statements as a set.

(a) E_1: "the card drawn is a queen."

(b) E_2: "the card drawn is a heart."

(c) E_3: "the card drawn is the queen of hearts."

(d) E_4: "the card drawn is a face card." (A face card is a jack, queen, or king.)

45. Consider the random experiment from Exercise 5 of tossing a coin 10 times in a row. Write out the event described by each of the following statements as a set.

(a) E_1: "toss no tails."

(b) E_2: "toss exactly 1 tail."

(c) E_3: "toss exactly twice as many heads as tails."

46. Consider the random experiment of drawing 2 cards out of an ordinary deck of 52 cards. (Here the order of the cards does not matter.) Write out the event described by each of the following statements as a set.

(a) E_1: "draw a pair of queens."

(b) E_2: "draw a pair." (A pair is two cards of the same value—two 7s, two Jacks, etc.)

47. Consider the sample space $S = \{A, B, C\}$. Make a list of all the possible events for this sample space. (Remember that an event is any subset of S including $\{\ \}$ and S itself.)

48. Consider the sample space $S = \{A, B, C, D\}$. Make a list of all the possible events for this sample space. (Remember that an event is any subset of S including $\{\ \}$ and S itself.)

F. Probability Spaces with Equally Likely Outcomes

49. Consider the random experiment of tossing an honest coin 3 times in a row. Find the probability of each of the following events. (*Hint*: See Exercises 1 and 41.)

   (a) E_1: "toss exactly 2 heads."

   (b) E_2: "all tosses come out the same."

   (c) E_3: "half of the tosses are heads, and half are tails."

   (d) E_4: "first two tosses are tails."

50. Consider the random experiment where a student takes a four-question true–false quiz. Assume now that the student randomly guesses the answer for each question. Find the probability of each of the following events. (*Hint*: See Exercises 2 and 42.).

   (a) E_1: "exactly 2 of the answers given are Ts." (*Note*: T = True, F = False.)

   (b) E_2: "at least 2 of the answers given are Ts."

   (c) E_3: "at most 2 of the answers given are Ts."

   (d) E_4: "the first 2 answers given are Ts."

51. Consider the random experiment of rolling a pair of honest dice. Find the probability of each of the following events. (*Hint*: See Exercise 43.)

   (a) E_1: "roll two of a kind" (i.e., both numbers are equal).

   (b) E_2: "roll a total of 3 or less."

   (c) E_3: "don't roll a total of 7 or less."

52. Consider the random experiment of drawing 1 card out of a well-shuffled, honest deck of 52 cards. Find the probability of each of the following events. (*Hint*: See Exercise 44.)

   (a) E_1: "the card drawn is a queen."

   (b) E_2: "the card drawn is a heart."

   (c) E_3: "the card drawn is the queen of hearts."

   (d) E_4: "the card drawn is a face card." (A face card is a Jack, queen, or king.)

53. Consider the random experiment of tossing an honest coin 10 times in a row. Find the probability of each of the following events. (*Hint*: See Exercises 5 and 45.)

   (a) E_1: "toss no tails."

   (b) E_2: "toss exactly 1 tail."

   (c) E_3: "toss exactly twice as many heads as tails."

54. Consider the random experiment of drawing 2 cards out of a well-shuffled, honest deck of 52 cards. (Here the order of the cards does not matter.) Find the probability of each of the following events. (*Hint*: See Exercise 46.)

   (a) E_1: "draw a pair of queens."

   (b) E_2: "draw a pair." (A pair is two cards of the same value—two 7s, two Jacks, etc.)

55. If a pair of honest dice are rolled once, find the probability of

(a) rolling a total of 8.

(b) not rolling a total of 8.

(c) rolling a total of 8 or 9.

(d) rolling a total of 8 or more.

56. Suppose a student takes a 10-question true–false quiz and the student randomly guesses the answer for each question. (That is, the probability that the students gets the right answer is equal to the probability that she gets the wrong answer.) Assume that each correct answer is worth 1 point. Find the probability that the student

(a) gets 10 points.

(b) gets 0 points.

(c) gets 9 points.

(d) gets 9 or more points.

(e) gets 7 or more points.

57. Ten names (A, B, C, D, E, F, G, H, I, J) are written, each on a separate slip of paper, put in a hat, and mixed well. Four names are randomly taken out of the hat, one at a time. Assume that the order in which the names are drawn matters. (See Exercise 9.) Find the probability that

(a) A is the first name chosen.

(b) A is one of the four names chosen.

(c) A is not one of the four names chosen.

(d) The four names chosen are A, B, C, and D in that order.

58. A gumball machine has gumballs of 4 different flavors: cherry (C), grape (G), lemon (L), and sour apple (S). When a fifty-cent piece is put into the machine, 5 random gumballs come out. Find the probability that

(a) each gumball is a different flavor.

(b) at least 2 of the gumballs are the same flavor.

59. A club has 15 members. A delegation of 4 members must be chosen to represent the club at a convention. All delegates are equal, so the order in which they are chosen doesn't matter. Assume that the delegation is chosen randomly by drawing the names out of a hat. Find the probability that

(a) Alice (one of the members of the club) is selected.

(b) Alice is not selected.

(c) club members Alice, Bert, Cathy, and Dale are selected.

60. Suppose that the probability of giving birth to a boy and the probability of giving birth to a girl are both 0.5. In a family of 4 children, what is the probability that

(a) all 4 children are girls.

(b) there are 2 girls and 2 boys.

(c) the youngest child is a girl.

G. Odds

61. Find the odds in favor of each of the following events.

(a) An event E with $\Pr(E) = 4/7$.

(b) An event E with $\Pr(E) = 0.6$.

62. Find the odds in favor of each of the following events.

    (a) An event E with $\Pr(E) = 3/11$

    (b) An event E with $\Pr(E) = 0.7$.

63. In each case, find the probability of an event E having the given odds.

    (a) The odds in favor of E are 3 to 5.

    (b) The odds against E are 8 to 15.

    (c) The odds in favor of E are 1 to 1.

64. In each case, find the probability of an event E having the given odds.

    (a) The odds in favor of E are 4 to 3.

    (b) The odds against E are 12 to 5.

    (c) The odds in favor of E are the same as the odds against E.

JOGGING

65. Two teams (call them X and Y) play against each other in the World Series. The World Series is a best-of-7 series. This means that the first team to win 4 games wins the series. (Games cannot end in a tie.) We can describe an outcome for the World Series by writing a string of letters that indicate (in order) the winner of each game. For example, the string $XYXXYX$ represents the outcome: X wins game 1, Y wins game 2, X wins game 3, and so on.

    (a) Describe the event "X wins in 6 games."

    (b) Describe the event "the series lasts 6 games."

66. A pizza parlor offers 6 toppings—pepperoni, Canadian bacon, sausage, mushroom, anchovies, and olives—that can be put on their basic cheese pizza. How many different pizzas can be made? (A pizza can have anywhere from no toppings to all 6 toppings.)

67. **(a)** In how many different ways can 10 people form a line?

    (b) In how many different ways can 10 people be seated around a circular table? [*Hint:* The answer to (b) is much less than the answer to (a). There are many different ways in which the same circle of 10 people can be broken up to form a line. How many?]

    (c) In how many different ways can 5 boys and 5 girls get in line so that boys and girls alternate (boy, girl, boy, ..., or girl, boy, girl, ...)?

    (d) In how many different ways can 5 boys and 5 girls sit around a circular table so that boys and girls alternate?

68. Eight points are taken on a circle.

    (a) How many chords can be drawn by joining all possible pairs of the points?

    (b) How many triangles can be made using these points as vertices?

69. Dolores wants to walk from point A to point B (a total of 6 blocks), which are shown on the street map. Assuming that she always walks toward B (i.e., up or to the right), how many different ways can she take this walk?

70. A study group of 15 students is to be split into 3 groups of 5 students each. In how many ways can this be done?

71. If we toss an honest coin 20 times, what is the probability of

(a) getting 10 *H*s and 10 *T*s?

(b) getting 3 *H*s and 17 *T*s?

(c) getting 3 or more *H*s?

Exercises 72 through 76 refer to 5-card draw poker hands. [See Example 15.16(b).]

72. What is the probability of getting "4 of a kind" (4 cards of the same value)?

73. What is the probability of getting all 5 cards of the same color?

74. What is the probability of getting a "flush" (all 5 cards of the same suit)?

75. What is the probability of getting an "ace-high straight" (10, J, Q, K, A of any suit but not all of the same suit)?

76. What is the probability of getting a "full house" (3 cards of equal value and 2 other cards of equal value)?

77. Consider the following game. We roll a pair of honest dice 5 times. If we roll a total of 7 at least once, we win; otherwise, we lose. What is the probability that we will win?

78. Which makes the better bet, *Bet 1* or *Bet 2*? *Bet 1*: In 4 consecutive rolls of an honest die, a 6 will come up at least once. *Bet 2*: In 24 consecutive rolls of a pair of dice, two 6s (called *boxcars*) will come up at least once.

Historical note: This question was raised by the French nobleman and gambler Chevalier de Méré (1607–1684), who asked his friend, the famous mathematician Blaise Pascal, to come up with the answer. Pascal's solution of this question is considered by many historians to have laid the groundwork for the formal mathematical theory of probability. (See also Project B.)

79. A factory assembles car stereos. From random testing at the factory, it is known that, on the average, 1 out of every 50 car stereos will be defective (which means that the probability that a car stereo randomly chosen from the assembly line will be defective is 0.02). After manufacture, car stereos are packaged in boxes of 12 for delivery to the stores.

(a) What is the probability that in a box of 12, there are no defective car stereos? What assumptions are you making?

(b) What is the probability that in a box of 12, there is at most 1 defective car stereo?

RUNNING

80. If an honest coin is tossed *N* times, what is the probability of getting the same number of *H*s as *T*s? (*Hint*: Consider two cases: *N* even and *N* odd.)

81. How many different "words" (they don't have to mean anything) can be formed using all the letters in

(a) the word PARSLEY. (*Note:* This one is easy!)

(b) the word PEPPER. [*Note:* This one is much harder! Think about the difference between (a) and (b).]

82. In the game of craps, the player's first roll of a pair of dice is very important. If the first roll is 7 or 11, the player wins. If the first roll is 2, 3, or 12, the player loses. If the first roll is any other number (4, 5, 6, 8, 9, 10), this number is called the player's "point." The player then continues to roll until either the point reappears, in which case the player wins, or until a 7 shows up before the point, in which case the player loses. What is the probability that the player will win? (Assume that the dice are honest.)

83. The birthday problem. There are 30 people in a room. What is the probability that at least 2 of these people have the same birthday—that is, have their birthdays on the same day and month?

84. Yahtzee is a dice game in which 5 standard dice are rolled at one time.

    (a) What is the probability of scoring "Yahtzee" with one roll of the dice? (You score Yahtzee when all 5 dice match.)

    (b) What is the probability of a *four of a kind* with one roll of the dice? (A *four of a kind* is rolled when four of the five dice match.)

    (c) What is the probability of rolling a *large straight* in one roll of the dice? (A *large straight* consists of 5 numbers in succession on the dice.)

    (d) What is the probability of rolling *trips* with one roll of the dice? (*Trips* are rolled when three of the five dice match, and the other two dice do not match the trips or each other.)

85. Heartless poker. Complete the following table for five-card draw poker hands from a deck in which all the hearts have been removed.

Hand	Probability
Four of a kind	
Flush (five cards of same suit)	
Full house (3 cards of equal value and two others of equal value)	
Three of a kind	
Two pair	

PROJECTS AND PAPERS

A. A History of Gambling

Games of chance, be they for religious purposes or for profit, can be traced all the way back to some of the earliest civilizations—the Babylonians, Assyrians, and ancient Egyptians were all known to play primitive dice games as well as board games of various kinds.

    In this project you are to write a paper discussing the history of gambling and the historical role of games of chance in the development of probability theory.

Suggested readings: There are many books available on the history of gambling. Two very readable ones are

- Bennett, Deborah, *Randomness*. Harvard University Press, 1998.
- Bernstein, Peter, *Against the Gods: The Remarkable Story of Risk*. Wiley, 1996.

B. The Letters Between Pascal and Fermat

In the summer of 1654, Pierre de Fermat and Blaise Pascal—two of the most famous mathematicians of their time—exchanged a series of famous letters discussing certain mathematical problems related to gambling. The problems were originally proposed to Pascal by his friend, a certain Chevalier de Méré—a nobleman and notorious gambler with a flair for raising interesting mathematical questions. The exchange of letters between Pascal and Fermat is of particular interest because it laid the foundation for the future development of the mathematical theory of probability. While not all of the correspondence between Fermat and Pascal has survived, enough of the letters exist to see how both mathematicians approached, each in his own way, de Méré's questions.

In this project, you are to write a paper discussing the famous exchange of letters between Pascal and Fermat. Describe the questions that were discussed and the approach that each mathematician took to solve them.

Suggested readings: Two good references for this project are
- Rényi, Alfréd, *Letters on Probability*. Wayne State University Press, 1972.
- David, Florence N., *Games, Gods, and Gambling*. Dover Publications, 1998.

C. The Largest Number Game

Suppose you are offered a chance to play the following game: A box contains 100 tickets. Each ticket has a different number written on it. The numbers can be any 100 numbers, large or small, integer or rational—anything goes as long as they are all different. Among all the tickets in the box, the one with the largest number is the *winning ticket*—if you can turn it in you win $100. For any other ticket, you get nothing. The ground rules for this game are as follows: You can draw a ticket out of the box. If you don't like it, you can draw another ticket from the box, but first you must tear up the other ticket. You can continue drawing from the box until you decide to stop or there are no more tickets in the box, but you must always tear up the previous ticket before you can draw again. In other words, the only ticket you can use is the one you drew last.

In this project you are to discuss and explain a strategy for playing this game that will give you a 25% or better chance of winning the $100. You should also discuss what is a reasonable sum that you are willing to pay for the right to play the game.

Note: This game and variations of it are sometimes discussed in the literature under the name *The Secretary Problem*.

D. Book Review: *Conned Again, Watson! Cautionary Tales of Logic, Math, and Probability*

As its title suggests, *Conned Again, Watson! Cautionary Tales of Logic, Math, and Probability*, by Colin Bruce (Perseus Publishing, 2002), is a collection of mathematical tales set in the form of Sherlock Holmes adventures. Each of the stories has a mathematical connection and several touch on issues related to the theme of this chapter (particularly relevant are "The Case of the Gambling Nobleman" and "The Case of the Surprise Heir," but several others deal with probability topics as well).

Choose one of the stories in the book with a probabilistic connection and write a review of the story. Pay special attention to the mathematical issues raised in the story.

REFERENCES AND FURTHER READINGS

1. Bennett, Deborah, *Randomness*. Cambridge, MA: Harvard University Press, 1998.
2. Bernstein, Peter, *Against the Gods: The Remarkable Story of Risk*. New York: John Wiley & Sons, 1996.
3. Bruce, Colin, *Conned Again, Watson! Cautionary Tales of Logic, Math, and Probability*. Cambridge, MA: Perseus Publishing, 2002.
4. David, Florence N., *Games, Gods, and Gambling*. New York: Dover Publications, 1998.
5. de Finetti, B., *Theory of Probability*. New York: John Wiley & Sons, 1970.
6. Everitt, Brian S., *Chance Rules: An Informal Guide to Probability, Risk, and Statistics*. New York: Springer-Verlag, 1999.
7. Gigerenzer, Gerd, *Calculated Risks*. New York: Simon & Schuster, 2002.
8. Gnedenko, B. V., and A. Y. Khinchin, *An Elementary Introduction to the Theory of Probability*. New York: Dover Publications, 1962.
9. Haigh, John, *Taking Chances*. New York: Oxford University Press, 1999.
10. Keynes, John M., *A Treatise on Probability*. New York: Harper and Row, 1962.
11. Krantz, Les, *What the Odds Are*. New York: HarperCollins Publishers, Inc., 1992.
12. Levinson, Horace, *Chance, Luck and Statistics: The Science of Chance*. New York: Dover Publications, 1963.
13. McGervey, John D., *Probabilities in Everyday Life*. New York: Ivy Books, 1992.
14. Mosteller, Frederick, *Fifty Challenging Problems in Probability with Solutions*. New York: Dover Publications, 1965.
15. Packel, Edward, *The Mathematics of Games and Gambling*. Washington, DC: Mathematical Association of America, 1981.
16. Weaver, Warren, *Lady Luck: The Theory of Probability*. New York: Dover Publications, Inc., 1963.

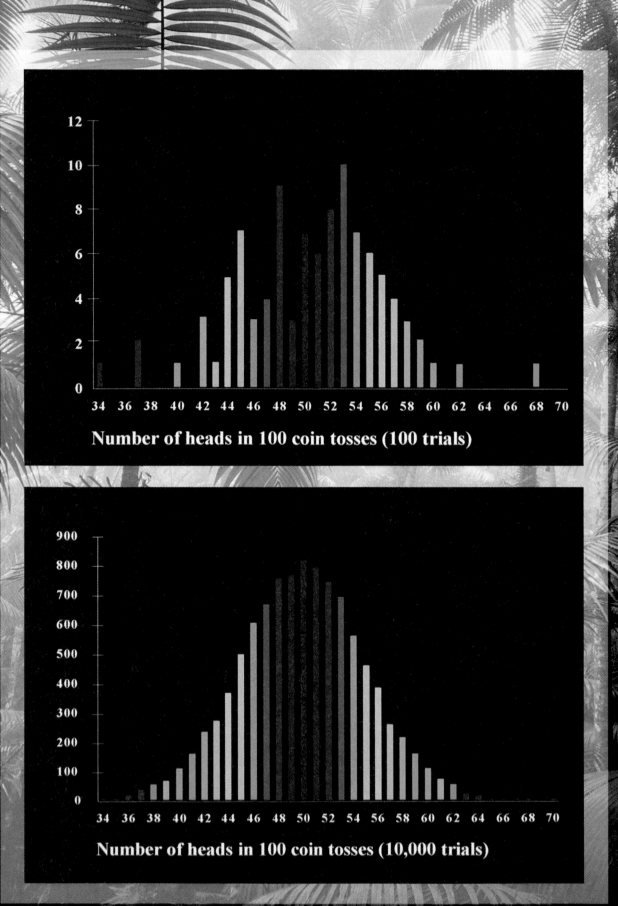

Number of heads in 100 coin tosses (100 trials)

Number of heads in 100 coin tosses (10,000 trials)

CHAPTER 16

Normal Distributions

Everything Is Back to Normal (Almost)

What does a scientist do when he has nothing but time on his hands? Some 60 years ago, the South African mathematician John Kerrich spent five years as a German prisoner of war. To pass the time, Kerrich decided to try a coin-tossing experiment. He tossed a coin 100 times and tallied the number of heads. He tallied 44 heads. He decided to do it again. The second time he tallied 54 heads. Undaunted, he repeated his coin-tossing experiment (tossing 100 times and tallying the number of heads) again and again. By the time he was done, he had tossed the coin 10,000 times and had meticulous records of the number of heads in every 100 tosses.

> *The normal is what you find but rarely. The normal is an ideal.*
>
> W. Somerset Maugham

The top bar graph on the opposite page shows the actual results of Kerrich's 10,000 coin tosses, broken up in groups of 100s.[1] Admittedly the picture is not particularly interesting, so why do we care about Kerrich's data? Essentially, Kerrich quit his experiment too soon. What would have happened if Kerrich had continued the coin-tossing experiment for a few more months? The remarkable picture we see at the bottom of the opposite page shows the results of 1 million coin tosses (where the number of heads in sets of 100 tosses is plotted). These results were obtained by a computer simulation rather than an actual coin toss marathon, but are just as valid as if they were done for real. In fact, had Kerrich tossed a coin 1 million times, his data would have produced a bell-shaped bar graph looking just like the bottom graph on the opposite page. We can say this with full confidence even though it never happened. Like many other phenomena, the long-term behavior of a tossed coin is guaranteed to produce a bell-shaped distribution—it is a fundamental law of probability known as the *central limit theorem*.

More than any other type of regular pattern, bell-shaped patterns rule the statistical world. The purpose of this chapter is to gain an understanding of these patterns and how they can be used to draw inferences about the way things are and the way things ought to be.

---

[1]*Source:* John Kerrich, *An Experimental Introduction to the Theory of Probability*, 1964.

16.1 Approximately Normal Distributions of Data

We start with a pair of examples.

EXAMPLE 16.1 Heights of NBA Players

Table 16-1 shows a frequency table for the heights of all 415 National Basketball Association players listed on team rosters at the start of the 2002–2003 NBA season.

TABLE 16-1 Heights of $N = 415$ NBA Players (2002–2003 season)

Height	Frequency	Height	Frequency	Height	Frequency	Height	Frequency
5′5″	1	6′3″	22	6′8″	36	7′2″	4
5′10″	2	6′4″	23	6′9″	53	7′3″	1
5′11″	6	6′5″	27	6′10″	47		
6′0″	6	6′6″	29	6′11″	43	7′6″	1
6′1″	21	6′7″	33	7′0″	27		
6′2″	22			7′1″	11		

Source: National Basketball Association (*http://www.nba.com*)

A bar graph for the height distribution is shown in Fig. 16-1. We can say that the bar graph is roughly shaped like a bell (with the emphasis on roughly). A mathematical idealization of what a perfect bell shape would look like is superimposed on the bar graph (in red). If it wasn't for all those extra 6′9″ to 7′ players, the bell-shaped nature of the height distribution would be more apparent.[2]

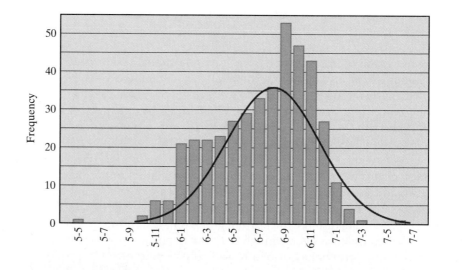

FIGURE 16-1
Height distribution of NBA players (2002–2003 season).

EXAMPLE 16.2 2001 SAT Scores (Verbal)

Table 16-2 is a relative-frequency table for scores on the 2001 SAT examination (verbal). The scores range from 200 to 800 and are grouped in class intervals of 50 points. The population for this data consists of college-bound seniors, and the size of the data set is $N = 1,276,320$. A bar graph for the data is shown in

---

[2]The excessive number of players in the 6′9″ -to-7′ range is a recent phenomenon—a consequence of modern NBA playing styles and not a quirk of nature.

Fig. 16-2. A smooth bell-shaped curve showing a mathematical idealization of the data is superimposed on the bar graph, and in this example, the bar graph fits the curve quite well.

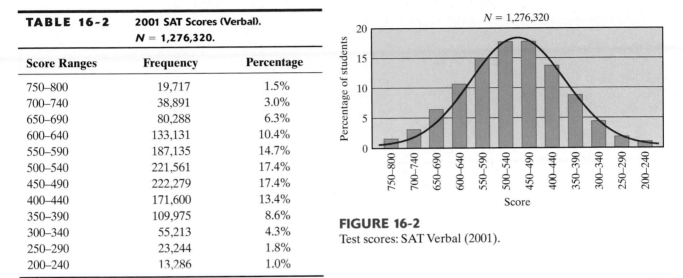

TABLE 16-2	2001 SAT Scores (Verbal). N = 1,276,320.	
Score Ranges	**Frequency**	**Percentage**
750–800	19,717	1.5%
700–740	38,891	3.0%
650–690	80,288	6.3%
600–640	133,131	10.4%
550–590	187,135	14.7%
500–540	221,561	17.4%
450–490	222,279	17.4%
400–440	171,600	13.4%
350–390	109,975	8.6%
300–340	55,213	4.3%
250–290	23,244	1.8%
200–240	13,286	1.0%

FIGURE 16-2
Test scores: SAT Verbal (2001).

Source: The College Board National Report, 2001

The two very different situations illustrated in Examples 16.1 and 16.2 have one thing in common: Both data sets can be described as fitting an approximately bell-shaped pattern. In Example 16.1, the fit is very crude; in Example 16.2, it is very good. In either case, we say that the data set has an **approximately normal distribution**. The word "normal" in this context is to be interpreted as synonymous with "bell shaped." A distribution of data that has a perfect bell shape is called a **normal distribution**. Real-world bell-shaped data are always approximately normal with some, as we have seen, more approximate than others.

When we have a bar graph for data with a normal distribution, we can connect the tops of the bars into a smooth bell-shaped curve. Perfect bell curves are called **normal curves**. The study of normal curves can be traced back to the work of the great German mathematician Carl Friedrich Gauss (for more on Gauss, see the biographical profile on p. 641) and for this reason, normal curves are sometimes known as *Gaussian curves*.

When the data set has an approximately normal distribution, an appropriate normal curve (such as the curves shown in red in Figs. 16-1 and 16-2) represents an idealization of the data (what things would look like in a perfect world). This is not wishful thinking—it is mathematical modeling, a powerful tool for understanding and describing the data. Thus, to fully understand real-world data sets with an approximately normal distribution, we first need to learn some of the mathematical properties of normal curves.

16.2 Normal Curves and Normal Distributions

Normal curves all share the same basic shape—that of a bell. Other than that, they can differ widely in their appearance. Some bells are tall and skinny, others are short and squat, and others fall somewhere in between. Mathematically speaking, however, they all share the same genes. In fact, whether a normal curve is skinny and tall or short and squat or somewhere in between (Fig. 16-3) depends on the way we scale the units on the axes. With the proper choice of scale, any two normal curves can be made to look the same.

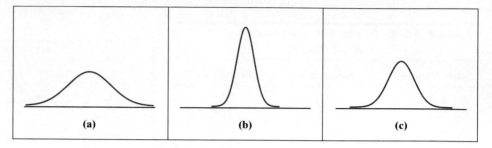

FIGURE 16-3
Three normal curves. (a) Short and squat. (b) Tall and skinny. (c) In between.

What follows is a summary of some of the essential facts about normal curves and their associated normal distributions. These facts are going to help us greatly later on in the chapter.

■ **Symmetry.** Every normal curve is symmetric about a vertical axis (see Fig. 16-4). The axis of symmetry splits the bell-shaped region outlined by the curve into two identical halves. This is the only line of symmetry of a normal curve, so we can refer to it without ambiguity as *the line of symmetry*.

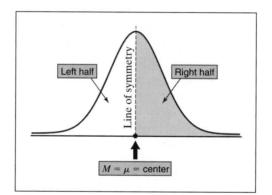

FIGURE 16-4
A normal curve has one axis of reflection symmetry. The line of symmetry crosses the horizontal axis at the center, which is both the median and the average of the data.

■ **Median = Mean = Center.** An important data value for a normal distribution can be found at the point where the line of symmetry crosses the horizontal axis. This point is called the **center** of the distribution, and it corresponds to both the median and the mean (average) of the data. We use the Greek letter μ (mu) to denote this value. Thus, in a normally distributed data set, the median is indeed the same as the mean. The fact that the median equals the mean implies that 50% of the data are less than or equal to the average and 50% of the data are greater than or equal to the average. (Beware: There is a common misconception that this is always true—for a data set that does not have a normal distribution it can be very wrong to assume this!)

> In a *normal distribution*,
> Center = Median(M) = Mean(μ)

For a real-life data set with an approximately normal distribution, the median and the mean may not be exactly the same number, but they will be close.

In an *approximately normal distribution,*
Median ≈ Mean

- **Standard Deviation.** We discussed the standard deviation—traditionally denoted by the Greek letter σ (sigma)—in Chapter 14. The standard deviation is an important measure of spread in general, but it is particularly important when dealing with normal (or approximately normal) distributions, as we will see shortly.

The easiest way to describe how to find the standard deviation of a normal distribution is in geometric terms. Pretend that you want to bend a piece of wire into the bell shape of a normal curve. At the very top, you must bend the wire downward [see Fig. 16-5(a)], and at the bottom, you must bend the wire upward [see Fig. 16-5(b)]. As we move our hands shaping the wire, the curvature gradually changes, and there is one point on each side of the curve where the transition from being bent downward to being bent upward takes place. Such a point [P in Fig. 16-5(c)] is called a **point of inflection** of the curve. Every normal curve has two points of inflection (P and P' in Fig. 16-6), and the horizontal distance between the axis of symmetry of the curve and either of these points is the standard deviation.

In a *normal distribution,* the standard deviation σ equals the distance between a point of inflection and the axis of symmetry.

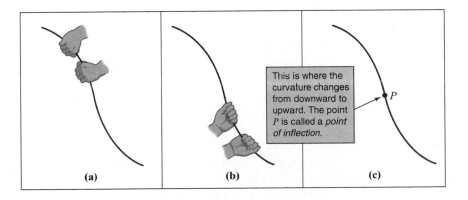

This is where the curvature changes from downward to upward. The point P is called a *point of inflection.*

(a)   (b)   (c)

FIGURE 16-5
(a) At the top, the wire has "downward" curvature. (b) At the bottom, the wire has "upward" curvature. (c) At P, the transition takes place.

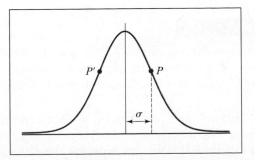

FIGURE 16-6
The horizontal distance between a point of inflection and the axis of symmetry equals the standard deviation (σ).

- **Quartiles.** We learned in Chapter 14 how to find the quartiles of a data set. For a normally distributed data set, we can find an approximate value of the first and third quartiles easily once we know the mean μ and the standard deviation σ. The secret is to memorize a single number: 0.675. Multiplying the standard deviation σ by 0.675 tells us how far to go to the right or left of the mean to locate the quartiles. In other words,

$$Q_3 \approx \mu + (0.675)\sigma$$
$$Q_1 \approx \mu - (0.675)\sigma.$$

EXAMPLE 16.3

Let's suppose that we have to analyze a data set having a normal distribution with mean $\mu = 506$ and standard deviation $\sigma = 111$. Here are all the things we can now say about the data:

- The median is 506. Thus, we know that half of the data are less than or equal to 506 and that half are greater than or equal to 506.
- The first quartile is $Q_1 \approx 506 - 0.675 \times 111 \approx 431$. This means that one-fourth of the data are less than or equal to 431, and another one-fourth fall between 431 and 506.
- The third quartile is $Q_3 \approx 506 + 0.675 \times 111 \approx 581$. This means that one-fourth of the data fall between 506 and 581, and one-fourth of the data are greater than or equal to 581. ■

16.3 Standardizing Normal Data Sets

We have seen that normal curves don't all look alike, but this is only a matter of perception. In fact, all normal distributions tell the same underlying story, but use slightly different dialects to do it. One way to understand the story of any given normal distribution is to rephrase it in a simple common language—a language that uses the mean (μ) and the standard deviation (σ) as its only vocabulary. The process is called **standardizing** the data set, and it essentially consists of measuring, in standard deviations, how far a data value has strayed from the mean. The best way to illustrate how to standardize data is by means of a couple of examples.

EXAMPLE 16.4

We will consider a normal distribution with mean $\mu = 45$ and standard deviation $\sigma = 10$. Let's imagine that the data corresponds to some measurement given in feet, and look at several different measurements in this data set.

- A measurement given by $x_1 = 55$ ft represents a data point that is 10 ft above the mean $\mu = 45$ ft. Coincidentally, 10 ft equals 1 standard deviation in this data set. We can rephrase the fact that $x_1 = 55$ ft is a data value 1 standard deviation above the mean (see Fig. 16-7) by saying that $x_1 = 55$ has a *standardized value* of 1.

- A measurement given by $x_2 = 35$ ft represents a data point that is 10 ft, or 1 standard deviation below the mean (see Fig. 16-7). A data value 1 standard deviation below the mean is said to have a *standardized value* of -1.

- A measurement given by $x_3 = 50$ ft represents a data point that is 5 ft, or half a standard deviation above the mean (see Fig 16-7), and this corresponds to a standardized value of 0.5. (We could have said that the standardized value is 1/2, but it is customary to use decimals to describe standardized values.)

- What about a more complicated data point, such as $x_4 = 21.58$ ft? How do we standardize it? First, we can measure how far the data point is from the mean by subtracting the mean from the data value. In this case we get 21.58 ft $-$ 45 ft $= -23.42$ ft. (Notice that for values below the mean this difference will be negative.) If we divide this difference by 10 ft, we get the standardized value of -2.342, which tells us exactly how far the data value is from the mean, measured in standard deviations (see Fig. 16-7).

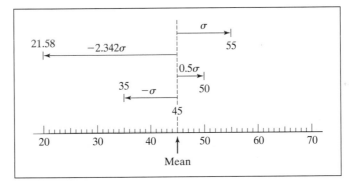

FIGURE 16-7
Standardized values are given by number of standard deviations above (positive) or below (negative) the mean.

In Example 16.4 we were somewhat fortunate in that the standard deviation was $\sigma = 10$, an especially easy number to work with. It helped us get our feet wet. What do we do in more realistic situations, where the mean and standard deviation may not be such nice round numbers? Other than the fact that we may not be able to do the arithmetic in our heads, the procedure we used in Example 16.4 remains the same: *For a normal distribution with mean μ and standard deviation σ, a data value x has a standardized value obtained by subtracting the mean μ from x and dividing the result by the standard deviation σ.*

Original data value	Standardized data value
x	$z = (x - \mu)/\sigma$

EXAMPLE 16.5

Consider a normal distribution with mean $\mu = 63.18$ and standard deviation $\sigma = 13.27$. What is the standardized value of the data point $x = 91.54$?

This looks nasty, but with a calculator, it's a piece of cake:

$$(91.54 - 63.18)/13.27 = 28.36/13.27 = 2.13715\ldots$$

Rounding off to two decimal places gives us a standardized value of $z \approx 2.14$.

Conversely, we can reverse this process and from a standardized value z compute the corresponding data value x. All we have to do is take the relation $z = (x - \mu)/\sigma$ and solve for x in terms of z. This gives $x = \mu + \sigma \cdot z$.

Standardized data value	Original data value
z	$x = \mu + \sigma \cdot z$

EXAMPLE 16.6

Consider a normal distribution with mean $\mu = 235.7$ m and standard deviation $\sigma = 41.58$ m. What data point has a standardized value of -3.45?

To compute -3.45 standard deviations, we multiply -3.45×41.58 m $= -143.451$ m. The negative value indicates that we are to the left of the mean. The data point we are looking for is given by 235.7 m $- 143.451$ m $= 92.249$ m. ■

16.4 The 68–95–99.7 Rule

When we look at any normal distribution, we can see that most of the data are concentrated in the neighborhood of the center. As we move away from the center, the heights of the columns drop rather fast, and if we go far enough away from the center, there are essentially no data to be found. These are all rather informal observations, but there is a more formal way to phrase this called the **68–95–99.7 rule**. This useful rule is obtained by using 1, 2, and 3 standard deviations above and below the mean as special landmarks, and in effect, it is three separate rules in one.

1. In every normal distribution, 68% of all the data values fall within 1 standard deviation above and below the mean. In other words, 68% of all the data have standardized values between -1 and 1. The remaining 32% of the data have standardized values greater than or equal to 1 or less than or equal to -1. By symmetry, there is an equal amount of each. [See Fig. 16-8(a).]

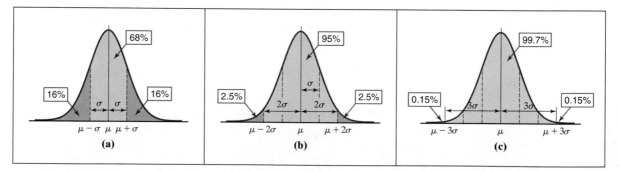

FIGURE 16-8
The 68–95–99.7 rule.

2. In every normal distribution, 95% of all the data values fall within 2 standard deviations above and below the mean. In other words, 95% of all the data have standardized values between -2 and 2. The remaining 5% of the data are divided equally between data with standardized values less than or equal to -2 and data with standardized values greater than or equal to 2. [See Fig. 16-8(b).]

3. In every normal distribution, 99.7% (which is practically 100%) of all the data values fall within 3 standard deviations above and below the mean. In other words, 99.7% of all the data have standardized values between −3 and 3. There is a minuscule amount of data with standardized values outside this range. [See Fig. 16-8(c).]

For approximately normal distributions, it is often convenient to round off the 99.7% to 100% and work under the assumption that all of the data fall within three standard deviations above and below the mean. This means that there are approximately 6 standard deviations separating the smallest (*Min*) and the largest (*Max*) values of the data. In Chapter 14 we defined the range R of a data set ($R = Max - Min$), and, in the case of an approximately normal distribution, we can conclude that the range is about 6 standard deviations.

> In an *approximately normal distribution*,
> $$R \approx 6\sigma.$$

16.5 Normal Curves as Models of Real-Life Data Sets

The reason we like to idealize a real-life, approximately normal data set by means of a normal distribution is that we can use many of the properties we just learned about normal distributions to draw useful conclusions about our data. For example, the 68–95–99.7 rule for normal curves can be reinterpreted in the context of an approximately normal data set as follows:

1. About 68% of the data values fall within (plus or minus) 1 standard deviation of the mean.

2. About 95% of the data values fall within (plus or minus) 2 standard deviations of the mean.

3. About 99.7%, or practically 100%, of the data values fall within (plus or minus) 3 standard deviations of the mean.

EXAMPLE 16.7 Analyzing SAT Scores

In 2001, a total of 1,276,320 college-bound high school seniors took the SAT. We saw in Example 16.2 that the scores in the verbal part fit very nicely an approximately normal distribution. The mean and standard deviation of the test scores were $\mu = 506$ and $\sigma = 111$, respectively.[3]

Without ever looking at the data (which are given in Table 16-2, Example 16.2), we can estimate the median and the quartiles using some of the facts we learned in this chapter.

- Since the median should be approximately equal to the mean $\mu = 506$, we can estimate the median score at about 506 points. The actual median score was 500.
- The first quartile is approximately $(0.675)\sigma = 0.675 \times 111 = 74.925$ points below the mean of 506. Rounding 74.925 to 75 gives us an estimate of 431 points for the first quartile. Since SAT scores come in multiples of 10, we would estimate the first quartile to be 430 points. This turns out to be right on the money—the first quartile was 430 points! This tells us that about 320,000

[3]*Source:* The College Board, National Report, 2001.

students (25% of the 1,276,320 students taking the test) scored 430 points or less on the verbal part of the 2001 SAT.

■ The third quartile should be approximately $506 + 75 = 581$ points. Once again, since SAT scores come in multiples of 10 we estimate the third quartile at 580, and this was indeed the third quartile score. This tells us that about 960,000 students scored 580 or less on the verbal part of the 2001 SAT.

Additional information about the distribution of scores can be obtained using the 68–95–99.7 rule.

■ The percentage of students scoring within 1 standard deviation of the mean should be about 68%. In this case, that means scores between $506 - 111 = 395$ and $506 + 111 = 617$ points. Since SAT scores can only come in multiples of 10, we can estimate that about two-thirds of students had scores between 400 and 610. The remaining third are equally divided between those scoring 610 or more and those scoring 400 or less.

■ The percentage of students scoring within 2 standard deviations of the mean should be about 95%. In this case, that means scores between $506 - 222 = 284$ and $506 + 222 = 728$ points. Since SAT scores can come only in multiples of 10, this really means scores between 290 and 720.

■ The 99.7 part of the 68–95–99.7 rule is not much help in this example. Essentially, it says that practically 100% of the students had test scores between 173 and 839, which does not tell us anything useful, since everybody's score has to fall between 200 and 800. ■■

16.6 Normal Distributions of Random Events

We are now ready to take up another important aspect of normal curves—their connection with random events and, through that, their critical role in margins of error of public opinion polls. Our starting point is the following important example.

EXAMPLE 16.8 A Coin-Tossing Experiment

In the opening of this chapter, we discussed the coin-tossing experiment performed by John Kerrich while he was a prisoner of war during World War II. Kerrich tossed a coin 10,000 times and kept records of the number of heads in groups of 100 tosses.

With modern technology, one can repeat Kerrich's experiment and take it much further. Practically any computer can imitate the tossing of a coin by means of a random-number generator. If we use this technique we can "toss coins" in mind-boggling numbers—millions of times if we so choose.

We will start modestly. We will toss our make-believe coin 100 times and count the number of heads, which we will denote by X. Before we do that, let's say a few words about X. Since we cannot predict ahead of time its exact value—we are tempted to think that it should be 50, but, in principle, it could be anything from 0 to 100—we call X a **random variable**. The possible values of the random variable X are governed by the laws of probability: Some values of X are extremely unlikely ($X = 0$, $X = 100$); others are much more likely ($X = 50$), although the likelihood of $X = 50$ is not as great as one would think. It also seems reasonable that (assuming that the coin is fair and heads and tails are equally likely) the likelihood of $X = 49$ should be the same as the likelihood $X = 51$, the likelihood of $X = 48$, should be the same as the likelihood of $X = 52$, and so on.

While all of the preceding statements are true, we still don't have a clue as to what is going to happen when we toss the coin 100 times. One way to get a sense of the probabilities of the different values of X is to repeat the experiment many times and check the frequencies of the various outcomes. Finally, we are ready to do some experimenting!

Our first trial results in 46 heads out of 100 tosses ($X = 46$). The first 10 trials give, in order, $X = 46, 49, 51, 53, 49, 52, 47, 46, 53, 49$. Figure 16-9(a) shows a bar graph for these data.

Continuing this way, we collect data for the values of X in 100, 500, 1000, 5000, and 10,000 trials. The bar graphs are shown in Figs. 16-9(b)–(f), respectively.

Figure 16-9 paints a pretty clear picture of what happens: As the number of trials increases, the distribution of the data becomes more and more bell shaped.

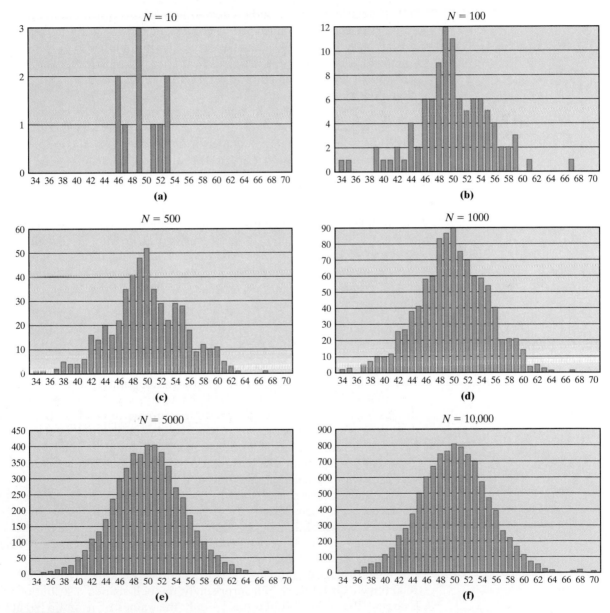

FIGURE 16-9
Distribution of random variable X (number of heads in 100 coin tosses) (a) 10 times, (b) 100 times, (c) 500 times, (d) 1000 times, (e) 5000 times, and (f) 10,000 times.

At the end, we have data from 10,000 trials, and the bar graph gives an almost perfect normal distribution!

What would happen if someone else decided to repeat what we did—toss an honest coin (be it by hand or by computer) 100 times, count the number of heads, and repeat this experiment a few times? The first 10 trials are likely to produce results very different from ours, but as the number of trials increases, their results and our results will begin to look more and more alike. After 10,000 trials, their bar graph will be almost identical to the bar graph shown in Fig. 16-9(f). In a sense, this says that doing the experiments a second time is a total waste of time— in fact, it was even a waste the first time! *Everything that happened at the end could have been predicted without ever tossing a coin*!

Knowing that the random variable X has an approximately normal distribution is, as we have seen, quite useful. The clincher would be to find out the values of the mean μ and the standard deviation σ of this distribution. Looking at Fig. 16-9(f), we can pretty much see where the mean is—right at 50. This is not surprising, since the axis of symmetry of the distribution has to pass through 50 as a simple consequence of the fact that the coin is honest. The value of the standard deviation is less obvious. For now, let's accept the fact that it is $\sigma = 5$. We will explain how we got this value shortly.

Let's summarize what we now know. An honest coin is tossed 100 times. The number of heads in the 100 tosses is a random variable, which we call X. If we repeat this experiment a large number of times (call it N), the random variable X will have an approximately normal distribution with mean $\mu = 50$ and standard deviation $\sigma = 5$, and the larger the value of N is, the better this approximation will be.

The real significance of these facts is that they are true not because we took the trouble to toss a coin a million times. Even if we did not toss a coin at all, all of these statements would still be true. *For a sufficiently large number of repetitions of the experiment of tossing an honest coin 100 times, the number of heads X is a random variable that has an approximately normal distribution with center* $\mu = 50$ *heads and standard deviation* $\sigma = 5$ *heads*. This is a mathematical, rather than an experimental, fact. ∎

16.7 Statistical Inference

Next, we are going to take our first tentative leap into statistical inference. Suppose that we have an honest coin and plan to toss it 100 times. We are going to do this just once, and we will call the resulting number of heads X. Been there, done that! What's new now is that we a have a solid understanding of the statistical behavior of this random variable—it has an approximately normal distribution with mean $\mu = 50$ and standard deviation $\sigma = 5$—and this allows us to make some very reasonable predictions about what is to happen.

For starters, we can predict the chance that the number of heads will fall somewhere between 45 and 55 (1 standard deviation below and above the mean)—it is 68%. Likewise, we know that the chance that the number of heads will fall somewhere between 40 and 60 is 95%, and between 35 and 65 is a whopping 99.7%.

What if, instead of tossing the coin 100 times, we were to toss it 500 times? Or 1000 times? Or n times? Not surprisingly, the bell-shaped distribution we saw in Example 16.8 would still be there—only the values of μ and σ would change. Specifically, the number of heads X would be a random variable with an approximately normal distribution with mean $\mu = n/2$ heads and standard deviation

$\sigma = \sqrt{n}/2$ heads. This is an important fact for which we have coined the name the **honest-coin principle**.

THE HONEST-COIN PRINCIPLE

Suppose an honest coin is tossed n times and that X denotes the number of heads that come up. The random variable X has an *approximately normal* distribution with *mean* $\mu = n/2$ and *standard deviation* $\sigma = \sqrt{n}/2$.

When we apply the honest-coin principle to $n = 100$ tosses, we get the mean number of heads tossed to be $\mu = 100/2 = 50$ and the standard deviation to be $\sigma = \sqrt{100}/2 = 10/2 = 5$ heads.

EXAMPLE 16.9 Betting on the Outcome of 256 Coin Tosses

An honest coin is going to be tossed 256 times. Before this is done, we have the opportunity to make some bets. Let's say that we can make a bet (with even odds) that if the number of heads tossed falls somewhere between 120 and 136, we will win; otherwise we will lose. Should we make such a bet?

Let X denote the number of heads in 256 tosses of an honest coin. By the honest-coin principle, X is a random variable having a distribution that is approximately normal with mean $\mu = 256/2 = 128$ heads and standard deviation $\sigma = \sqrt{256}/2 = 8$ heads. The values 120 to 136 are exactly 1 standard deviation below and above the mean of 128, which means that there is a 68% chance that the number of heads will fall somewhere between 120 and 136. We should indeed make this bet! A similar calculation tells us that there is a 95% chance that the number of heads will fall somewhere between 112 and 144, and the chance that the number of heads will fall somewhere between 104 and 152 is 99.7%. ∎

What happens when the coin being tossed is not an honest coin? Surprisingly, the distribution of the number of heads X in n tosses of such a coin is still approximately normal, as long as the number n is not too small.[4] All we need now is a **dishonest-coin principle** to tell us how to find the mean and the standard deviation.

THE DISHONEST-COIN PRINCIPLE

Suppose an arbitrary coin is tossed n times ($n \geq 30$), and that X denotes the number of heads that come up. Suppose also that p is the probability of the coin landing heads, and $(1 - p)$ is the probability of the coin landing tails. Then, the random variable X has an approximately normal distribution with mean $\mu = n \cdot p$ and standard deviation $\sigma = \sqrt{n \cdot p \cdot (1 - p)}$.

---

[4]An accepted rule of thumb in statistics is that n should be at least 30.

EXAMPLE 16.10 Tossing a Dishonest Coin

A coin is rigged so that it comes up heads only 20% of the time (i.e., $p = 0.20$). The coin is tossed 100 times ($n = 100$), and X is the number of heads in the 100 tosses. What can we say about X?

According to the dishonest-coin principle, the distribution of the X is approximately normal with mean $\mu = 100 \times 0.20 = 20$ heads and standard deviation $\sigma = \sqrt{100 \times 0.20 \times 0.80} = 4$ heads.

Based on these facts, we can now make the following assertions.

- There is a 68% chance that the number of heads will fall somewhere between 16 and 24, which represents one standard deviation below and above the mean.

- There is a 95% chance that the number of heads will fall somewhere between 12 and 28, which represents standardized values between -2 and 2.

- The number of heads is almost guaranteed (a 99.7% chance) to fall somewhere between 8 and 32.

Note that in this example, heads and tails are no longer interchangeable concepts—heads is an outcome with probability $p = 0.2$, while tails is an outcome with much higher probability (0.8). We can, however, apply the principle equally well to describe the distribution of the number of tails in 100 coin tosses of the same dishonest coin: The distribution for the number of tails is approximately normal with mean $\mu = 100 \times 0.80 = 80$ and standard deviation $\sigma = \sqrt{100 \times 0.80 \times 0.20} = 4$. Note that σ is still the same. ■

The dishonest-coin principle can be applied to any coin, even one that is fair ($p = 1/2$). In the case $p = 1/2$, the honest- and dishonest-coin principles say the same thing (see Exercise 71).

The dishonest-coin principle is a down-to-earth version of one of the most important laws in statistics, a law generally known as the *central limit theorem*. We will now briefly illustrate why the importance of the dishonest-coin principle goes beyond the tossing of coins.

EXAMPLE 16.11 Sampling for Defective Light Bulbs

An assembly line produces 100,000 light bulbs a day, 20% of which generally turn out to be defective. Suppose we draw a random sample of $n = 100$ light bulbs. Let X represent the *number of defective light bulbs* in the sample. What can we say about X?

A moment's reflection will show that, statistically, this example is almost identical to Example 16.10—the approximate probability that each light bulb chosen is defective is 0.20 (just as the probability that the coin will come up heads in Example 16.10). We can use the dishonest-coin principle to infer that the number of defective light bulbs in the sample is a random variable having an approximately normal distribution with a mean of 20 light bulbs and standard deviation of 4 light bulbs. Thus,

- There is a 68% chance that the number of defective light bulbs in the sample will fall somewhere between 16 and 24.

- There is a 95% chance that the number of defective light bulbs in the sample will fall somewhere between 12 and 28.

- The number of defective light bulbs in the sample is practically guaranteed (a 99.7% chance) to fall somewhere between 8 and 32.

Probably the most important point here is that each of the preceding facts can be rephrased in terms of sampling errors, a concept we first discussed in Chapter 13. For example, say we had 24 defective light bulbs in the sample; in other words, 24% of the sample (24 out of 100) are defective light bulbs. If we use this statistic to estimate the percent of defective light bulbs overall, then the sampling error would be 4% (because the estimate is 24% and the value of the parameter is 20%). By the same token, if we had 16 defective light bulbs in the sample, the sampling error would be −4%. Coincidentally, the standard deviation is $\sigma = 4$ light bulbs, or 4% of the sample. (We computed it in Example 16.10). Thus, we can rephrase our previous assertions about sampling errors as follows:

- When estimating the proportion of defective light bulbs coming out of the assembly line by using a sample of 100 light bulbs, there is a 68% chance that the sampling error will fall somewhere between −4 and 4%.

- When estimating the proportion of defective light bulbs coming out of the assembly line by using a sample of 100 light bulbs, there is a 95% chance that the sampling error will fall somewhere between −8 and 8%.

- When estimating the proportion of defective light bulbs coming out of the assembly line by using a sample of 100 light bulbs, there is a 99.7% chance that the sampling error will fall somewhere between −12 and 12%. ■

EXAMPLE 16.12 Sampling with Larger Samples

Suppose we have the same assembly line as in Example 16.11, but, this time, we are going to take a really big sample of $n = 1600$ light bulbs. Before we even count the number of defective light bulbs in the sample, let's see how much mileage we can get out of the dishonest-coin principle. The standard deviation for the distribution of defective light bulbs in the sample is $\sqrt{1600 \times 0.2 \times 0.8} = 16$, which just happens to be exactly 1% of the sample ($16/1600 = 1\%$). This means that when we estimate the proportion of defective light bulbs coming out of the assembly line using this sample, we can have some sort of a handle on the sampling error.

- We can say with some confidence (68%) that the sampling error will fall somewhere between −1 and 1%.

- We can say with a lot of confidence (95%) that the sampling error will fall somewhere between −2 and 2%.

- We can say with tremendous confidence (99.7%) that the sampling error will fall somewhere between −3 and 3%. ■

The next and last example shows how the dishonest-coin principle can be used to estimate the margin of error in a public opinion poll, an issue of considerable importance in modern statistics.

EXAMPLE 16.13 Measuring the Margin of Error of a Poll

In California, school bond measures require a 66.67% vote for approval. Suppose that an important school bond measure is on the ballot in the upcoming election. In the most recent poll of 1200 randomly chosen voters, 744 of the 1200 voters sampled, or 62%, indicated that they would vote for the school bond measure. Let's assume that the poll was properly conducted and that the 1200 voters sampled represent an unbiased sample of the entire population. What are the chances that the 62% statistic is the result of sampling variability and that the actual vote for the bond measure will be 66.67% or more?

Here, we will use a variation of the dishonest-coin principle, with each voter being likened to a coin toss. Voting for the bond measure is like the coin coming up heads; against is tails. The probability (p) of "heads" for this "coin" will turn out to be the proportion of voters in the population that support the bond measure: If p turns out to be 0.6667 or more, the bond measure will pass. Our problem is that we don't know p, so how can we use the dishonest-coin principle to estimate the mean and standard deviation of the sampling distribution?

The idea here is to use the 62% (0.62) statistic from the sample as an estimate for the actual value of p in the formula for the standard deviation given by the dishonest-coin principle. (Even though we know that this is only a rough estimate for p, this generally turns out to give us a good estimate for the standard deviation.) In our example, the approximate standard deviation for the number of "heads" in the sample turns out to be $\sqrt{1200 \times 0.62 \times 0.38} \approx 16.8$ voters. When we convert this number to percentages, we get a standard deviation that is approximately 1.4% of the sample ($16.8/1200 = 0.014$).

The standard deviation for the sampling distribution expressed as a percentage of the entire sample is called the **standard error**. (For our example, we have found that the standard error is approximately 1.4%.) In sampling and public opinion polls, it is customary to express the information about the population in terms of **confidence intervals**, which are themselves based on standard errors: A 95% confidence interval is given by 2 standard errors below and above the statistic obtained from the sample; a 99.7% confidence interval is given by going 3 standard errors below and above the sample statistic.

In our example, we have a 95% confidence interval of 62% plus or minus 2.8%, which means that we can say with 95% confidence (we would be right 95 out of 100 times) that the actual vote for the bond measure will fall somewhere between 59.2% ($62 - 2.8$) and 64.8% ($62 + 2.8$), and thus, that the bond measure will lose. Want even more certainty? Take a 99.7% confidence interval of 62% plus or minus 4.2%—it is almost certain that the actual vote will turn out somewhere in that range. Even in the most optimistic scenario, the vote will not reach the 66.67% needed to pass the bond measure. ■

Conclusion

From coin tosses to test scores, many real-life data sets follow the call of the bell. In this chapter we studied bell-shaped (normal) curves, some of their mathematical properties, and how these properties can be used to analyze real-life bell-shaped data sets. It was a brief introduction to what is undoubtedly one of the most widely used and sophisticated tools of modern mathematical statistics.

In this chapter we also got a brief glimpse of the concept of statistical inference. The process of drawing conclusions and inferences based on limited data is an essential part of statistics. It gives us a way not only to analyze what has already taken place, but also to make reasonably accurate large-scale predictions of what will happen in certain random situations. Casinos know, without any shadow of a doubt, that in the long run, they will make a profit—it is a mathematical law! A similar law gives us the confidence to trust the results of surveys and public opinion polls (up to a point!), the quality of the products we buy, and even the statistical data our government uses to make many of its decisions. In all of these cases, bell-shaped distributions of data and the laws of probability come together to give us insight into what was, is, and most likely will be.

Carl Friedrich Gauss (1777–1855)

In spite of the obvious connection between mathematics and money, mathematicians don't usually make the cut when it comes to getting their picture featured in a bank note. The notable exception is Carl Friedrich Gauss, the man often referred to as "the prince of mathematicians." The German ten mark note shown on the left celebrates both Gauss and one of his most famous "discoveries"—the normal curve (shown above the 10).

Carl Friedrich Gauss was born in the Duchy of Brunswick (now Germany) into a poor, working-class family of little education. From a very early age, young Gauss showed a prodigious talent for numbers and abstract mathematical thinking. One of the most famous stories about Gauss was his solution to the addition problem $1 + 2 + 3 + \cdots + 99 + 100 = ?$ The problem was assigned one day in class by a teacher hoping to keep the children busy for a while. No sooner had the teacher finished assigning the problem, Gauss had the solution. His insight: The sum consists of 50 pairs $(1 + 100, 2 + 99, \cdots, 50 + 51)$, each of which adds up to 101 and thus equals $50 \times 101 = 5050$. He was seven years old at the time.

Recognizing Gauss's mathematical genius, the Duke of Brunswick agreed to be the young man's sponsor and financially support his education. Gauss first attended the Collegium Carolinum in Brunswick, but eventually transferred to the more prestigious Göttingen University. By the time he was 21, Gauss had produced several major mathematical discoveries including the complete characterization of all regular polygons that can be constructed with straightedge and compass, a classic problem that could be traced all the way back to the ancient Greeks. In 1799 Gauss received his doctorate in mathematics from the University of Helmstedt for his work on what is now known as the *fundamental theorem of algebra*. Two years later he published his first and greatest book, *Disquisitiones Arithmeticae*, a pioneering treatise on the theory of numbers. With the publication of *Disquisitiones Arithmeticae* Gauss came to be recognized as one of the greatest mathematician of his age. He was 24 years old at the time.

Gauss's scientific genius went well beyond theoretical mathematics and he made major contributions to empirical sciences such as astronomy and surveying, where precise observation and measurement are critical. He greatly added to his growing scientific reputation when he was able to predict the exact orbit of the asteroid Ceres, which had been discovered in 1801 by the Italian astronomer Giuseppe Piazzi. Piazzi had observed and tracked what he thought was a new planet for only a few weeks before it disappeared behind the Sun. Using Piazzi's limited data and a method of his own invention, Gauss predicted where and when Ceres would be seen again. Just about a year after its disappearance Ceres reappeared, almost in the exact position Gauss had predicted. On the strength of this achievement, Gauss was appointed Director of the Göttingen Observatory and Professor of Astronomy at Göttingen University.

As part of his astronomical calculations, Gauss became interested in the study of measurement errors and their probability distributions. This led him to the discovery of the bell-shaped normal distribution, which is also called the *Gaussian distribution* in his honor. Always a practical man, towards the later part of his life Gauss became increasingly interested in the applications of mathematics to physics and made important discoveries in the theory of magnetism, developed Kirchhoff's laws in electricity, and constructed an early version of the telegraph. He also became interested in the analysis and prediction of financial markets, and made a tremendous amount of money speculating in stocks.

By the time he died in 1855 at the age of 78, Gauss had achieved a remarkable measure of fame and fortune—he was the preeminent mathematician of his time as well as an extremely wealthy man.

KEY CONCEPTS

68–95–99.7 rule	normal distribution
approximately normal distribution	point of inflection
center (mean and median)	random variable
confidence interval	standard deviation
dishonest-coin principle	standard error
honest-coin principle	standardized value
normal curve	

EXERCISES

WALKING

A. Normal Curves

1. For the normal distribution described by the curve shown in the figure, find
   (a) the mean.
   (b) the median.
   (c) the standard deviation. (*Note: P* is a point of inflection of the curve.)

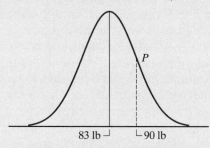

83 lb ⌐   ⌐90 lb

2. For the normal distribution described by the curve shown in the figure, find
   (a) the mean.
   (b) the median.
   (c) the standard deviation. (*Note: P* is a point of inflection of the curve.)

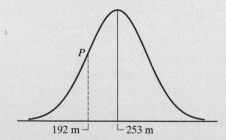

192 m ⌐   ⌐253 m

3. In the normal curve in the figure, P and P' are the inflection points. For the normal distribution described by the curve, find
   (a) the center.
   (b) the standard deviation.
   (c) the first and third quartiles (rounded to the nearest inch).

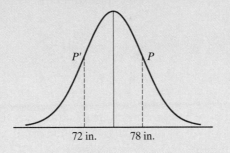

72 in.   78 in.

4. In the normal curve in the figure, P and P' are the inflection points. For the normal distribution described by the curve, find
   (a) the center.
   (b) the standard deviation.
   (c) the first and third quartiles (rounded to the nearest point).

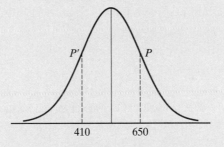

410   650

5. For a normal distribution with mean $\mu = 81.2$ lb and standard deviation $\sigma = 12.4$ lb, find
   (a) the first quartile (rounded to the nearest tenth of a pound).
   (b) the third quartile (rounded to the nearest tenth of a pound).

6. For a normal distribution with mean $\mu = 2354$ points and standard deviation $\sigma = 468$ points, find
   (a) the first quartile (rounded to the nearest point).
   (b) the third quartile (rounded to the nearest point).

7. Find the standard deviation of a normal distribution with mean $\mu = 81.2$ in. and third quartile $Q_3 = 94.7$ in.

8. Find the standard deviation of a normal distribution with mean $\mu = \$18{,}565$ and first quartile $Q_1 = \$15{,}514$.

In Exercises 9 through 12, you are given some information about a distribution. Explain why the distribution is not normal. (Note: M denotes the median.)

9. A distribution with $M = 82$, $\mu = 71$, and $\sigma = 11$.
10. A distribution with $M = 210$, $\mu = 195$, and $\sigma = 15$.

11. A distribution with $M = 453$, $\mu = 453$, $Q_1 = 343$, and $Q_3 = 553$.

12. A distribution with $M = 47$, $\mu = 47$, $Q_1 = 35$, and $Q_3 = 61$.

B. Standardizing Data

13. Suppose that a normal distribution has mean $\mu = 30$ and standard deviation $\sigma = 15$. Find the standardized value of each of the following numbers.

    (a) 45

    (b) 54

    (c) 0

    (d) 3

14. Suppose that a normal distribution has mean $\mu = 110$ and standard deviation $\sigma = 12$. Find the standardized value of each of the following numbers.

    (a) 128

    (b) 100

    (c) 110

    (d) 71

15. Suppose that a normal distribution has mean $\mu = 253.45$ ft and third quartile $Q_3 = 278.58$ ft. Find the standardized value of each of the following numbers (rounded to the nearest hundredth).

    (a) 261.71 ft

    (b) 185.79 ft

    (c) 253.45 ft

16. Suppose that a normal distribution has mean $\mu = 47.3$ lb and first quartile $Q_1 = 44.1$ lb. Find the standardized value of each of the following numbers (rounded to the nearest tenth).

    (a) 56.9 lb

    (b) 36.9 lb

    (c) 59.1 lb

    (d) 31.6 lb

17. Suppose that a normal distribution has mean $\mu = 183.5$ ft and standard deviation $\sigma = 31.2$ ft. Find the data value having a standardized value of

    (a) -1.

    (b) 0.5.

    (c) -2.3.

    (d) 0.

18. Suppose that a normal distribution has mean $\mu = 83.2$ gal and standard deviation $\sigma = 4.6$ gal. Find the data value having a standardized value of

    (a) 2.

    (b) -1.5.

    (c) -0.43.

    (d) 0.

19. Suppose that a normal distribution has mean $\mu = 50$ and that a data value of 84 has a standardized value of 2. Find the standard deviation.

20. Suppose that a normal distribution has mean $\mu = 30$ and that a data value of -60 has a standardized value of -3. Find the standard deviation.

21. Suppose that a normal distribution has standard deviation $\sigma = 15$ and that a data value of 50 has a standardized value of 3. Find the mean.

22. Suppose that a normal distribution has standard deviation $\sigma = 20$ and that a data value of 10 has a standardized value of -2. Find the mean.

23. Suppose that in a normal distribution a data value of 20 has a standardized value of -2, and a data value of 100 has a standardized value of 3. Find the mean μ and the standard deviation σ.

24. Suppose that in a normal distribution a data value of -10 has a standardized value of 0, and a data value of 50 has a standardized value of 2. Find the mean μ and the standard deviation σ.

C. The 68-95-99.7 Rule

25. Find the mean μ and standard deviation σ for the normal distribution described by the curve shown in the figure.

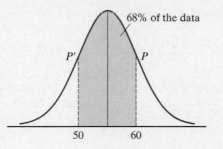

26. Find the mean μ and standard deviation σ for the normal distribution described by the curve shown in the figure.

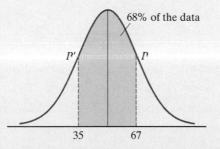

27. Find the mean μ and standard deviation σ for the normal distribution described by the curve shown in the figure.

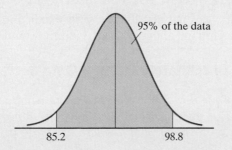

28. Find the mean μ and standard deviation σ for the normal distribution described by the curve shown in the figure.

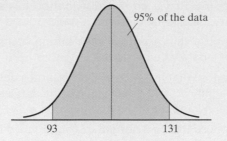

95% of the data

93   131

29. Find the mean μ and standard deviation σ for the normal distribution described by the curve shown in the figure.

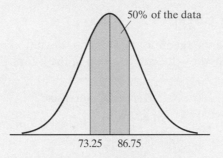

50% of the data

73.25   86.75

30. Find the mean μ and standard deviation σ for the normal distribution described by the curve shown in the figure.

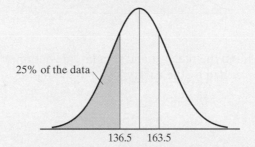

25% of the data

136.5   163.5

31. In a normal distribution, what percent of the data fall

  (a) below the point two standard deviations above the mean?

  (b) between one and two standard deviations above the mean?

32. In a normal distribution, what percent of the data fall

  (a) above the point three standard deviations below the mean?

  (b) between two and three standard deviations below the mean?

33. For a normal distribution with mean $\mu = \$84$, suppose that 97.5% of the data fall below \$114. Find the standard deviation σ.

34. For a normal distribution with mean $\mu = 56.3$ cm, suppose that 84% of the data fall above 50.2 cm. Find the standard deviation σ.

D. Approximately Normal Data Sets

Exercises 35 through 38 refer to the following: 2500 students take a college entrance exam. The scores on the exam have an approximately normal distribution with mean $\mu = 52$ points and standard deviation $\sigma = 11$ points.

35. (a) Estimate the average score on the exam.

(b) Estimate what percent of the students scored 52 points or more.

(c) Estimate what percent of the students scored between 41 and 63 points.

(d) Estimate what percent of the students scored 63 points or more.

36. (a) Estimate how many students scored between 30 and 74 points.

(b) Estimate how many students scored 30 points or less.

(c) Estimate how many students scored 19 points or less.

37. (a) Estimate the first-quartile score for this exam.

(b) Estimate the third-quartile score for this exam.

(c) Estimate the interquartile range for this exam.

38. For each of the following scores, estimate in what percentile of the students taking the exam the score would place you.

(a) 52

(b) 63

(c) 60

(d) 85

Exercises 39 through 42 refer to the following: As part of a research project, the blood pressures of 2000 patients in a hospital are recorded. The systolic blood pressures (given in millimeters) have an approximately normal distribution with mean $\mu = 125$ and standard deviation $\sigma = 13$.

39. (a) Estimate the number of patients whose blood pressure was between 99 and 151 millimeters.

(b) Estimate the number of patients whose blood pressure was between 112 and 151 millimeters.

40. (a) Estimate the number of patients whose blood pressure was 99 millimeters or higher.

(b) Estimate the number of patients whose blood pressure was between 99 and 138 millimeters.

41. For each of the following blood pressures, estimate the percentile of the patient population to which they correspond.

(a) 100 millimeters

(b) 112 millimeters

(c) 115 millimeters

(d) 138 millimeters

(e) 164 millimeters

42. (a) Estimate the value of the lowest (Min) and the highest (Max) blood pressures. (Assume that there were no outliers, and use the 68–95–99.7 rule.)

(b) Assuming that there were no outliers, give an estimate of the five-number summary (Min, Q_1, μ, Q_3, Max) for the distribution of blood pressures.

Exercises 43 through 46 refer to the following: Packaged foods sold at supermarkets are not always the weight indicated on the package. Variability always crops up in the manufacturing and packaging process. Suppose that the exact *weight of a "12-ounce" bag of potato chips is a random variable that has an approximately normal distribution with mean $\mu = 12$ ounces and standard deviation $\sigma = 0.5$ ounce.*

43. If a "12-ounce" bag of potato chips is chosen at random, what are the chances that

   (a) it weighs somewhere between 11 and 13 ounces?

   (b) it weighs somewhere between 12 and 13 ounces?

   (c) it weighs more than 11 ounces?

44. If a "12-ounce" bag of potato chips is chosen at random, what are the chances that

   (a) it weighs somewhere between 11.5 and 12.5 ounces?

   (b) it weighs somewhere between 12 and 12.5 ounces?

   (c) it weighs more than 12.5 ounces?

45. Suppose that 500 "12-ounce" bags of potato chips are chosen at random. Estimate the number of bags with weight

   (a) 11 ounces or less.

   (b) 11.5 ounces or less.

   (c) 12 ounces or less.

   (d) 12.5 ounces or less.

   (e) 13 ounces or less.

   (f) 13.5 ounces or less.

46. Suppose that 1500 "12-ounce" bags of potato chips are chosen at random. Estimate the number of bags of potato chips with weight

   (a) between 11 and 11.5 ounces.

   (b) between 11.5 and 12 ounces.

   (c) between 12 and 12.5 ounces.

   (d) between 12.5 and 13 ounces.

   (e) between 13 and 13.5 ounces.

Exercises 47 through 50 refer to the following: The distribution of weights for children of a given age and sex is approximately normal. This fact allows a doctor or nurse to find from a child's weight the weight percentile of the population (all children of the same age and sex) to which the child belongs. Typically, this is done using special charts provided to the doctor or nurse, but these percentiles can also be computed using facts about approximately normal distributions, such as the ones we learned in this chapter. (Note: The numbers in these examples are 1977 figures taken from charts produced by the National Center for Health Statistics, U.S. Department of Health and Human Services.)

47. The distribution of weights for 6-month-old baby boys is approximately normal with mean $\mu = 17.25$ pounds and standard deviation $\sigma = 2$ pounds.

   (a) Suppose that a 6-month-old boy weighs 15.25 pounds. Approximately what weight percentile is he in?

   (b) Suppose that a 6-month-old boy weighs 21.25 pounds. Approximately what weight percentile is he in?

(c) Suppose that a 6-month-old boy is in the 75th percentile in weight. Estimate his weight.

48. The distribution of weights for 12-month-old baby girls is approximately normal with mean $\mu = 21$ pounds and standard deviation $\sigma = 2.2$ pounds.

(a) Suppose that a 12-month-old girl weighs 16.6 pounds. Approximately what weight percentile is she in?

(b) Suppose that a 12-month-old girl weighs 18.8 pounds. Approximately what weight percentile is she in?

(c) Suppose that a 12-month-old girl is in the 75th percentile in weight. Estimate her weight.

49. The distribution of weights for 1-month-old baby girls is approximately normal with mean $\mu = 8.75$ pounds and standard deviation $\sigma = 1.1$ pounds.

(a) Suppose that a 1-month-old girl weighs 11 pounds. Approximately what weight percentile is she in?

(b) Suppose that a 1-month-old girl weighs 12 pounds. Approximately what weight percentile is she in?

(c) Suppose that a 1-month-old girl is in the 25th percentile in weight. Estimate her weight.

50. The distribution of weights for 12-month-old baby boys is approximately normal with mean $\mu = 22.5$ pounds and standard deviation $\sigma = 2.2$ pounds.

(a) Suppose that a 12-month-old boy weighs 24 pounds. Approximately what weight percentile is he in?

(b) Suppose that a 12-month-old boy weighs 21 pounds. Approximately what weight percentile is he in?

(c) Suppose that a 12-month-old boy is in the 84th percentile in weight. Estimate his weight.

51. In this chapter we discussed the fact that the range of an approximately normal distribution is approximately 6 standard deviations. Estimate how many standard deviations describe the interquartile range of an approximately normal distribution.

52. For an approximately normal distribution with a minimum value Min = 11.4 ft and a maximum value Max = 57.6 ft, estimate the mean μ and the standard deviation σ.

E. The Honest- and Dishonest-Coin Principles

53. An honest coin is tossed $n = 3600$ times. Let the random variable Y denote the number of tails tossed.

(a) Find the mean and the standard deviation of the distribution of the random variable Y.

(b) What are the chances that Y will fall somewhere between 1770 and 1830?

(c) What are the chances that Y will fall somewhere between 1800 and 1830?

(d) What are the chances that Y will fall somewhere between 1830 and 1860?

54. An honest coin is tossed $n = 6400$ times. Let the random variable X denote the number of heads tossed.

    (a) Find the mean and the standard deviation of the distribution of the random variable X.

    (b) What are the chances that X will fall somewhere between 3120 and 3280?

    (c) What are the chances that X will fall somewhere between 3080 and 3200?

    (d) What are the chances that X will fall somewhere between 3240 and 3280?

55. Suppose a random sample of $n = 7056$ adults is to be chosen for a survey. Assume that the gender of each adult in the sample is equally likely to be male as it is female. Find the probability that the number of females in the sample is

    (a) between 3486 and 3570.

    (b) less than 3486.

    (c) less than 3570.

56. An honest die is rolled. If the roll comes out even $(2, 4, \text{or } 6)$, you will win \$1; if the roll comes out odd $(1, 3, \text{or } 5)$, you will lose \$1. Suppose that in one evening you play this game $n = 2500$ times in a row.

    (a) What is the probability that by the end of the evening you will not have lost any money?

    (b) What is the probability that the number of even rolls will fall between 1250 and 1300?

    (c) What is the probability that you will win \$100 or more?

    (d) What is the probability that you will win exactly \$101?

57. A dishonest coin with probability of heads $p = 0.4$ is tossed $n = 600$ times. Let the random variable X represent the number of times the coin comes up heads.

    (a) Find the mean and standard deviation for the distribution of X.

    (b) Find the first and third quartiles for the distribution of X.

    (c) Find the probability that the number of heads will fall somewhere between 216 and 264.

58. A dishonest coin with probability of heads $p = 3/4$ is tossed $n = 1200$ times. Let the random variable X represent the number of times the coin comes up heads.

    (a) Find the mean and standard deviation for the distribution of X.

    (b) Find the first and third quartiles for the distribution of X.

    (c) Find the probability that the number of heads will fall somewhere between 900 and 945.

59. Suppose that an honest die is rolled $n = 180$ times. Let the random variable X represent the number of times the number 6 is rolled.

    (a) Find the mean and standard deviation for the distribution of X.

    (b) Find the probability that a 6 will be rolled more than 40 times.

    (c) Find the probability that a 6 will be rolled somewhere between 30 and 35 times.

60. Suppose that one out of every ten cereal boxes has a prize. Out of a shipment of $n = 400$ cereal boxes, find the probability that there are

   (a) somewhere between 34 and 40 prizes.

   (b) somewhere between 40 and 52 prizes.

   (c) more than 52 prizes.

61. Each day a machine produces 1600 widgets. During 95 of the last 100 days, the machine has produced between 117 and 139 defective widgets. What are the chances that a randomly selected widget produced by this machine is defective?

62. At Tasmania State University, the probability that an entering freshman will graduate in four years is 0.89. What are the chances that of a class of 2000 freshmen, 1750 or more will graduate in four years?

JOGGING

Percentiles. *The **pth percentile** of a sorted data set is a number x_p such that p % of the data fall at or below x_p and $(100 - p)$% of the data fall at or above x_p. (For details, see Chapter 14, Sec. 14.3.) For normally distributed data sets, there are detailed statistical tables that give the location of the pth percentile for every possible p between 1 and 99. The following table is an abbreviated version giving the approximate location of some of the more frequently used percentiles in a normal distribution with mean μ and standard deviation σ. For approximately normal distributions, the table can be used to estimate these percentiles.*

Percentile	Approximate location	Percentile	Approximate location
99th	$\mu + 2.33\sigma$	1st	$\mu - 2.33\sigma$
95th	$\mu + 1.65\sigma$	5th	$\mu - 1.65\sigma$
90th	$\mu + 1.28\sigma$	10th	$\mu - 1.28\sigma$
80th	$\mu + 0.84\sigma$	20th	$\mu - 0.84\sigma$
75th	$\mu + 0.675\sigma$	25th	$\mu - 0.675\sigma$
70th	$\mu + 0.52\sigma$	30th	$\mu - 0.52\sigma$
60th	$\mu + 0.25\sigma$	40th	$\mu - 0.25\sigma$
50th	μ		

In Exercises 63 through 68, you should use the table to make your estimates.

63. The distribution of weights for 6-month-old baby boys is approximately normal with mean $\mu = 17.25$ pounds and standard deviation $\sigma = 2$ pounds.

   (a) Suppose that a 6-month-old baby boy weighs in the 95th percentile of his age group. Estimate his weight in pounds approximated to 2 decimal places.

   (b) Suppose that a 6-month-old baby boy weighs in the 40th percentile of his age group. Estimate his weight in pounds approximated to 2 decimal places.

64. Several thousand students took a college entrance exam. The scores on the exam have an approximately normal distribution with mean $\mu = 55$ points and standard deviation $\sigma = 12$ points.

(a) For a student that scored in the 99th percentile, estimate the student's score on the exam.

(b) For a student that scored in the 30th percentile, estimate the student's score on the exam.

65. Consider again the distribution of weights of 6-month-old baby boys discussed in Exercise 63.

(a) Jimmy is a 6-month-old-baby who weighs 17.75 lb. Estimate the percentile corresponding to Jimmy's weight.

(b) David is a 6-month-old baby who weighs 16.2 lb. Estimate the percentile corresponding to David's weight.

66. Consider again the college entrance exam discussed in Exercise 64.

(a) Mary scored 83 points on the exam. Estimate the percentile in which this score places her.

(b) Adam scored 45 points on the exam. Estimate the percentile in which this score places him.

(c) If there were 250 students that scored 35 points or less on the exam, estimate the total number of students that took the exam.

67. In 1998, 1,172,779 college-bound seniors took the SAT exam. The distribution of scores in the math section of the SAT was approximately normal with mean $\mu = 512$ and standard deviation $\sigma = 112$. (*Source: www.collegeboard.org.*)

(a) Estimate the 99th percentile score on the exam. Use the fact that SAT scores are given in multiples of 10. (See Example 16.7.)

(b) Estimate the 75th percentile score on the exam.

(c) Estimate the percentile corresponding to a test score of 540.

68. Consider a normal distribution with mean $\mu = 0$ and standard deviation $\sigma = 1$.

(a) Find the 90th percentile (rounded to 2 decimal places).

(b) Find the 10th percentile (rounded to 2 decimal places).

(c) Find the 80th percentile (rounded to 2 decimal places).

(d) Find the 20th percentile (rounded to 2 decimal places).

(e) Suppose that you are given that the 85th percentile is approximately 1.04. Find the 15th percentile.

69. An honest coin is tossed n times. Let the random variable X denote the number of heads tossed. Find the value of n so that there is a 95% chance that X will be between $n/2 - 10$ and $n/2 + 10$.

70. An honest coin is tossed n times. Let the random variable Y denote the number of tails tossed. Find the value of n so that there is a 16% chance that Y will be at least $n/2 + 10$.

71. Explain why when the dishonest-coin principle is applied with an honest coin, we get the honest-coin principle.

RUNNING

72. A dishonest coin with probability of heads $p = 0.1$ is tossed n times. Let the random variable X denote the number of heads tossed. Find the value of n so that there is a 95% chance that X will be between $n/10 - 30$ and $n/10 + 30$.

73. An honest pair of dice is rolled n times. Let the random variable Y denote the number of times a total of 7 is rolled. Find the value of n so that there is a 95% chance that Y will be between $n/6 - 20$ and $n/6 + 20$.

74. On an American roulette wheel, there are 18 red numbers, 18 black numbers, plus 2 green numbers (0 and 00). Thus, the probability of a red number coming up on a spin of the wheel is $p = 18/38 \approx 0.47$. Suppose that we go on a binge and bet \$1 on red 10,000 times in a row. (A \$1 bet on red wins \$1 if red comes up; otherwise, we lose the \$1.)

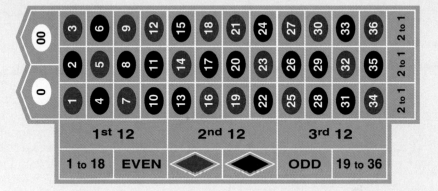

(a) Let Y represent the number of times we lose (i.e., the number of times that red does not come up). Use the dishonest-coin principle to describe the distribution of the random variable Y.

(b) Approximately what are the chances that we will lose 5300 times or more?

(c) Approximately what are the chances that we will lose somewhere between 5150 and 5450 times?

(d) Explain why the chances that we will break even or win in this situation are essentially zero.

75. After polling a random sample of 800 voters during the most recent gubernatorial race, the *Tasmania Gazzette* reports:

> *As the race for governor of Tasmania heads into its final days, our most recent poll shows Mrs. Butterworth ahead of the incumbent Mrs. Cubblson by 6 percentage points—53% to 47%. The results of the poll indicate with near certainty that if the election had been held at the time the poll was taken, Mrs. Butterworth would be the next governor of Tasmania.*

(a) Estimate the standard error for this poll.

(b) Compute a 95% confidence interval for this poll.

(c) Compute a 99.7% confidence interval for this poll.

PROJECTS AND PAPERS

A. Bernouilli Trials and Binomial Distributions

A random experiment with only two outcomes (which for convenience we call "success" and "failure") is called a **Bernouilli trial**. Typical examples of Bernouilli trials are tossing a coin (be it honest or dishonest) and shooting a free throw. One of the important questions discussed in this chapter is what happens when we independently repeat the same Bernouilli trial many times (see Examples 16.8, 16.10, and 16.11). The number of *successes* when we independently repeat a given Bernouilli trial n times is given by an important distribution called the **binomial distribution**. The rule that defines the binomial distribution is as follows:

> *In n independent Bernouilli trials each with probability of success p, the probability of having exactly r successes is given by $_nC_r p^r(1-p)^{n-r}$ [where $_nC_r = n! / r!(n-r)!$].*

In this project you are to explore binomial distributions and their connection with normal curves. In the first part of the project you should explain how binomial distributions and normal curves are related. In the second part of the project you are to illustrate how one can use some of the facts about normal curves we learned in this chapter to compute approximate values for the probability of any number of successes in n Bernouilli trials, regardless of how big n is. For the purposes of illustration choose a real-life example that involves a large number of repeated Bernouilli trials in a field that is of special interest to you (sports, the economy, politics, public health, etc.).

B. Confidence Intervals

The concept of a *confidence interval* was introduced in Example 16.13 in this chapter. The two most frequently used types of confidence intervals are 95% confidence intervals (sometimes described as intervals at a *95% confidence level*) and 99.7% confidence intervals (intervals at a 99.7% confidence level). Other confidence intervals, such as intervals at a 90% confidence level are sometimes used, and in principle, it is possible to define confidence intervals at any level.

In this project, you are to describe the process of constructing confidence intervals in general. Given a target confidence level of $x\%$, how do you construct the corresponding confidence interval? Conversely, given a specified interval, how do you find the confidence level that best fits that interval? Illustrate the relationship between confidence level and the size of the confidence interval using a real-life poll. Conclude with a discussion of some common misconceptions about confidence intervals. When no confidence interval is mentioned, how should we properly interpret the results of the poll?

C. Book Review: *The Bell Curve* and Rebuttals

In 1994, the late Richard J. Herrnstein of Harvard and Charles Murray of M.I.T. wrote a book entitled *The Bell Curve: Intelligence and Class Structure in American Life*. The book used social statistics (in particular, bell-shaped distributions of data on IQs and other tests of human intelligence) to justify the conclusion that intelligence levels vary across racial, ethnic, and economic groups. The book became one of the most discussed books of its time and raised a firestorm of controversy and many rebuttals.

In this project you should (1) summarize the main arguments presented by Herrnstein and Murray in their book and the justification given for these arguments, and (2) summarize some of the main rebuttal arguments to the premises of the book.

Suggested Readings: Much of what you need for this project you will find in
* *The Mismeasure of Man*, by Stephen Jay Gould, 1996 (reference 5).
* *Wringing the Bell Curve*, by B. Devlin, S. E. Fienberg, D. P. Resnick, and K. Roeder, in *Chance*, 8 (1995), 27–36.
* *The Bell Curve Debate: History, Documents, Opinions*, by R. Jacoby and N. Glauberman, 1995 (reference 7).

REFERENCES AND FURTHER READINGS

1. Clemons, T., and M. Pagano, "Are Babies Normal?" *The American Statistician*, 53:4 (1999), 298–302.
2. Converse, P. E., and M. W. Traugott, "Assessing the Accuracy of Polls and Surveys," *Science*, 234 (1986), 1094–1098.
3. Frankel, Max, "Margins of Error," *New York Times Magazine*, December 15, 1996, 34.
4. Freedman, D., R. Pisani, R. Purves, and A. Adhikari, *Statistics*, 2d ed. New York: W. W. Norton, 1991, chaps. 16 and 18.
5. Gould, Stephen Jay, *The Mismeasure of Man*. New York: W. W. Norton, 1996.
6. Herrnstein, R. J., and C. Murray, *The Bell Curve: Intelligence and Class Structure in American Life*. New York: Free Press, 1994.
7. Jacoby, Russell, and Naomi Glauberman (eds.), *The Bell Curve Debate: History, Documents, Opinions*. New York: Times Books, 1995.
8. Kerrich, John, *An Experimental Introduction to the Theory of Probability*. Witwatersrand, South Africa: University of Witwatersrand Press, 1964.
9. Larsen, J., and D. F. Stroup, *Statistics in the Real World*. New York: Macmillan Publishing Co., Inc., 1976.
10. Mosteller, F., W. Kruskal, et al., *Statistics by Example: Detecting Patterns*. Reading, MA: Addison-Wesley Publishing Co., Inc., 1973.
11. Mosteller, F., R. Rourke, and G. Thomas, *Probability and Statistics*. Reading, MA: Addison-Wesley Publishing Co., Inc., 1961.
12. Tanner, Martin, *Investigations for a Course in Statistics*. New York: Macmillan Publishing Co., Inc., 1990.

ANSWERS TO SELECTED PROBLEMS

Chapter 1

WALKING

A. Ballots and Preference Schedules

1. **(a)** 7

   (b) The Country Cookery: plurality.

   (c)

Number of voters	5	3	1	3
1st choice	A	C	B	C
2nd choice	B	B	D	B
3rd choice	C	A	C	D
4th choice	D	D	A	A

3. **(a)** 21   **(b)** 11   **(c)** A   **(d)** E   **(e)** C   **(f)** E

5. **(a)** B and E

   (b)

Number of voters	5	3	5	3	2	3
1st choice	A	A	C	D	D	A
2nd choice	C	D	D	C	C	C
3rd choice	D	C	A	A	A	D

   (c) A

7.

Number of voters	47	36	24	13	5
1st choice	B	A	B	E	C
2nd choice	E	B	A	B	E
3rd choice	A	D	D	C	A
4th choice	C	C	E	A	D
5th choice	D	E	C	D	B

B. Plurality Method

9. **(a)** It's a tie between B and D.

   (b) D

11. **(a)** 1st choice: C. 2nd choice: B. 3rd choice: A. 4th choice: D.

    (b) If Miss Insincere made B her first choice and C her second choice, then she would tip the scales in favor of B.

13. **(a)** 23 votes will guarantee A at least a tie for first; 24 votes guarantee A is the only winner. (With 23 of the remaining 30 votes, A has 49 votes. The only candidate with a chance to have that many votes is C. Even if C gets the other 7 remaining votes, it would not have enough votes to beat A.)

    (b) 11 votes will guarantee C at least a tie for first; 12 votes guarantee C is the only winner. (With 11 of the remaining 30 votes, C has 53 votes. The only candidate with a chance to have that many votes is D. Even if D gets the other 19 remaining votes, it would not have enough votes to beat C.)

15. **(a)** 82   **(b)** 83

C. Borda Count Method

17. (a) Prof. Chavez

(b)

Number of voters	5	3	5	5	3
1st choice	A	A	C	D	B
2nd choice	B	D	D	C	A
3rd choice	C	B	A	B	C
4th choice	D	C	B	A	D

A is now the winner.

19. (a) Borrelli

(b) Dante has a majority (13) of the first-place votes but does not win the election.

(c) Dante, having a majority of the first-place votes, is a Condorcet candidate but does not win the election.

21. (a) $40, 25, 20, 15$. Winner: B.

(b) $40N, 25N, 20N, 15N$. Winner: B.

(c) No. The number of points for each candidate is just multiplied by N.

23. (a) 200 points

(b) 50 points

25. (a) 10 points

(b) 1100 points

(c) 310 points

D. Plurality-with-Elimination Method

27. Prof. Argand

29. (a) Dante

(b) Dante has a majority of the first-place votes (13 first-place votes out of a total of 24 votes).

31. B

33. (a) Atlanta

(b) Chicago

(c) Chicago, which is the Condorcet candidate, fails to win the election under the plurality-with-elimination method.

E. Pairwise Comparisons Method

35. A

37. A

39. E

F. Ranking Methods

41. (a) Winner: A. Second place: C. Third place: D. Last place: B.

(b) Winner: A. Second place: C. Third place: D. Last place: B.

(c) Winner: C. Second place: A. Third place: D. Last place: B.

(d) Winner: D. Second place: C. Third place: A. Last place: B.

43. (a) Winner: A. Second place: C. Third place: D. Last place: B.

(b) Winner: A. Second place: C. Third place: D. Last place: B.

(c) Winner: C. Second place: D. Third place: A. Last place: B.

(d) Winner: D. Second place: C. Third place: A. Last place: B.

G. Miscellaneous

45. 125,250

47. 5,060,560

49. **(a)** 105
   (b) 1 hour and 45 minutes

51. **(a)** *A*
   (b) *B*
   (c) *A*
   (d) Condorcet criterion; independence-of-irrelevant-alternatives criterion. (*Note:* The majority criterion is also violated in this election.)

53. **(a)** Candidate *A* is a Condorcet candidate.
   (b) No
   (c) No
   (d) No

JOGGING

55. Suppose the two candidates are *A* and *B* and that *A* gets *a* first-place votes and *B* gets *b* first-place votes and suppose that $a > b$. Then *A* has a majority of the votes and the preference schedule is

Number of voters	*a*	*b*
1st choice	*A*	*B*
2nd choice	*B*	*A*

It is clear that candidate *A* wins the election under the plurality method, the plurality-with-elimination method, and the method of pairwise comparisons. Under the Borda count method, *A* gets $2a + b$ points while *B* gets $2b + a$ points. Since $a > b$, $2a + b > 2b + a$ and so again *A* wins the election.

57. **(a)** In this variation, each candidate gets 1 point less on each ballot. Thus, if there are *k* voters, each candidate gets a total of *k* fewer points, i.e., $q = p - k$.
   (b) Since each candidate's total is decreased by the same amount *k*, using this variation of the Borda count method the extended ranking of the candidates is the same. (If $a < b$ then $a - k < b - k$.)

59. One possible example is the following.

Number of voters	8	4	3	2
1st choice	*A*	*B*	*B*	*D*
2nd choice	*C*	*D*	*C*	*C*
3rd choice	*B*	*C*	*D*	*B*
4th choice	*D*	*A*	*A*	*A*

61. If *X* is the winner of an election using the plurality method and, in a reelection, the only changes in the ballots are changes that only favor *X*, then no candidate other than *X* can increase his/her first-place votes and so *X* is still the winner of the election.

63. **(a)** 1st place: 14 points. 2nd place: 9 points. 3rd place: 8 points. 4th place: 7 points. 5th place: 6 points. 6th place: 5 points. 7th place: 4 points. 8th place: 3 points. 9th place: 2 points. 10th place: 1 point.
   (b) Suppose player *A* gets 15 first-place votes, 11 second-place votes, and 2 last-place votes, giving him a total of 311 points. Likewise, suppose that player *B* gets 13 first-place votes and 15 second-place votes, giving him a total of 317 points. In this situation, player *A* has a majority of the first place votes, but player *B* wins the MVP award.

65. 126 points

Chapter 2

WALKING

A. Weighted Voting Systems

1. **(a)** 6   **(b)** 20   **(c)** 4   **(d)** 65%

3. **(a)** $[10:8,4,2,1]$ **(b)** $[11:8,4,2,1]$
   (c) $[12:8,4,2,1]$ **(d)** $[13:8,4,2,1]$

5. **(a)** 14   **(b)** 27

7. **(a)** There is no dictator: P_1 and P_2 have veto power; P_3 is a dummy.
   (b) P_1 is a dictator; P_2 and P_3 are dummies.
   (c) There is no dictator, no one has veto power, and no one is a dummy.

9. **(a)** There is no dictator; P_1 and P_2 have veto power; P_5 is a dummy.
   (b) P_1 is a dictator; P_2, P_3, P_4 are dummies.
   (c) P_1 and P_2 have veto power; P_3 and P_4 are dummies.
   (d) All 4 players have veto power.

B. Banzhaf Power

11. **(a)** 10
   (b) $\{P_1, P_2\}, \{P_1, P_3\}, \{P_1, P_2, P_3\}, \{P_1, P_2, P_4\}, \{P_1, P_3, P_4\}, \{P_2, P_3, P_4\}, \{P_1, P_2, P_3, P_4\}$
   (c) P_1 only
   (d) $P_1: \frac{5}{12}; P_2: \frac{3}{12}; P_3: \frac{3}{12}; P_4: \frac{1}{12}$

13. **(a)** $P_1: \frac{3}{5}; P_2: \frac{1}{5}; P_3: \frac{1}{5}$
   (b) $P_1: \frac{3}{5}; P_2: \frac{1}{5}; P_3: \frac{1}{5}$

15. **(a)** $P_1: \frac{4}{12}; P_2: \frac{3}{12}; P_3: \frac{2}{12}; P_4: \frac{2}{12}; P_5: \frac{1}{12}$
   (b) $P_1: \frac{7}{19}; P_2: \frac{5}{19}; P_3: \frac{3}{19}; P_4: \frac{3}{19}; P_5: \frac{1}{19}$

17. **(a)** $P_1: 1; P_2: 0; P_3: 0; P_4: 0$
   (b) $P_1: \frac{7}{10}; P_2: \frac{1}{10}; P_3: \frac{1}{10}; P_4: \frac{1}{10}$
   (c) $P_1: \frac{3}{5}; P_2: \frac{1}{5}; P_3: \frac{1}{5}; P_4: 0$
   (d) $P_1: \frac{1}{2}; P_2: \frac{1}{2}; P_3: 0; P_4: 0$
   (e) $P_1: \frac{1}{3}; P_2: \frac{1}{3}; P_3: \frac{1}{3}; P_4: 0$

19. $A: \frac{1}{3}; B: \frac{1}{3}; C: \frac{1}{3}; D: 0$

21. $P_1: \frac{1}{3}; P_2: \frac{1}{3}; P_3: \frac{1}{3}; P_4: 0; P_5: 0; P_6: 0$

C. Shapley-Shubik Power

23. **(a)** $\langle P_1, P_2, P_3 \rangle, \langle P_1, P_3, P_2 \rangle, \langle P_2, P_1, P_3 \rangle, \langle P_2, P_3, P_1 \rangle, \langle P_3, P_1, P_2 \rangle, \langle P_3, P_2, P_1 \rangle$
   (b) $\langle P_1, \underline{P_2}, P_3 \rangle, \langle P_1, \underline{P_3}, P_2 \rangle, \langle P_2, \underline{P_1}, P_3 \rangle, \langle P_2, P_3, \underline{P_1} \rangle, \langle P_3, \underline{P_1}, P_2 \rangle, \langle P_3, P_2, \underline{P_1} \rangle$
   (c) $P_1: \frac{4}{6}; P_2: \frac{1}{6}; P_3: \frac{1}{6}$

25. $P_1: \frac{7}{12}; P_2: \frac{3}{12}; P_3: \frac{1}{12}; P_4: \frac{1}{12}$

27. **(a)** $P_1: 1; P_2: 0; P_3: 0$
   (b) $P_1: \frac{4}{6}; P_2: \frac{1}{6}; P_3: \frac{1}{6}$
   (c) $P_1: \frac{4}{6}; P_2: \frac{1}{6}; P_3: \frac{1}{6}$
   (d) $P_1: \frac{1}{2}; P_2: \frac{1}{2}; P_3: 0$
   (e) $P_1: \frac{1}{3}; P_2: \frac{1}{3}; P_3: \frac{1}{3}$

29. **(a)** $P_1: 1; P_2: 0; P_3: 0$
   (b) $P_1: \frac{4}{6}; P_2: \frac{1}{6}; P_3: \frac{1}{6}$
   (c) $P_1: \frac{1}{2}; P_2: \frac{1}{2}; P_3: 0$
   (d) $P_1: \frac{1}{2}; P_2: \frac{1}{2}; P_3: 0$
   (e) $P_1: \frac{1}{3}; P_2: \frac{1}{3}; P_3: \frac{1}{3}$

31. **(a)** $P_1: \frac{5}{12}; P_2: \frac{3}{12}; P_3: \frac{3}{12}; P_4: \frac{1}{12}$
    (b) $P_1: \frac{5}{12}; P_2: \frac{3}{12}; P_3: \frac{3}{12}; P_4: \frac{1}{12}$
    (c) $P_1: \frac{5}{12}; P_2: \frac{3}{12}; P_3: \frac{3}{12}; P_4: \frac{1}{12}$

33. $A: \frac{1}{3}; B: \frac{1}{3}; C: \frac{1}{3}; D: 0$

D. Miscellaneous

35. **(a)** 6,227,020,800
    (b) 6.402374×10^{15}
    (c) 6.204484×10^{23}

37. **(a)** 3,628,800
    (b) 121,645,100,408,832,000
    (c) 100
    (d) 504

39. 9.33262×10^{157}

41. **(a)** 63
    (b) 2
    (c) 720

JOGGING

43. **(a)** If the coalition consisting of all players other than P is a losing coalition, then any other coalition without P must also be a losing coalition.
    (b) Since P is a critical member in the grand coalition, the coalition consisting of all players other than P is a losing coalition and hence P has veto power.

45. **(a)** 720
    (b) The player must be the last (sixth) player in the sequential coalition.
    (c) 120
    (d) $\frac{120}{720} = \frac{1}{6}$
    (e) $\frac{1}{6}$ (Each player is the last player in 120 of the 720 sequential coalitions.)
    (f) If the quota equals the sum of all the weights, then the only way a player can be pivotal is for the player to be the last player in the sequential coalition. Since every player will be the last player in the same number of sequential coalitions, all players must have the same Shapley-Shubik power index. It follows that each of the N players has Shapley-Shubik power index of $1/N$.

47. **(a)** $[4: 2, 1, 1, 1]$ and $[9: 5, 2, 2, 2]$ are among the possible answers.
    (b) $H: \frac{1}{2}; A_1: \frac{1}{6}; A_2: \frac{1}{6}; A_3: \frac{1}{6}$

49. **(a)** $7 \le q \le 13$
    (b) For $q = 7$ or $q = 8$, P_1 is a dictator.
    (c) For $q = 9$, only P_1 has veto power since P_2 and P_3 together have just 5 votes.
    (d) For $10 \le q \le 12$, both P_1 and P_2 have veto power since no motion can pass without both of their votes. For $q = 13$, all three players have veto power.
    (e) For $q = 7$ or $q = 8$, both P_2 and P_3 are dummies. For $10 \le q \le 12$, P_3 is a dummy since all winning coalitions contain $\{P_1, P_2\}$, which is itself a winning coalition.

51. **(a)** Both have Banzhaf power distribution $P_1: \frac{2}{5}; P_2: \frac{1}{5}; P_3: \frac{1}{5}; P_4: \frac{1}{5}$.
    (b) In the weighted voting system $[q: w_1, w_2, \ldots, w_N]$, P_k is critical in a coalition means that the sum of the weights of all the players in that coalition (including P_k) is at least q but the sum of the weights of all the players in the coalition except P_k is less than q. Consequently, if the weights of all the players are multiplied by $c > 0$ ($c \le 0$ would make no sense), then the sum of the weights of all the players in the coalition (including P_k) is at least cq but the sum of the weights of all the players in the coalition except P_k is less than cq. Therefore, P_k is critical in the same coalition in the weighted voting system $[cq: cw_1, cw_2, \ldots, cw_N]$. Since the critical players are the same in both weighted voting systems, the Banzhaf power distributions will be the same.

53. (a) If a player X has Banzhaf power index 0, then X is not critical in any coalition and so the addition or deletion of X to or from any coalition will never change the coalition from losing to winning or winning to losing. It follows that X can never be pivotal in any sequential coalition and so X must have Shapley-Shubik power index 0.

(b) If a player X has Shapley-Shubik power index 0, then X is not pivotal in any sequential coalition and so X can never be added to a losing coalition and turn it into a winning coalition. It follows that X can never be critical in any coalition and so X has Banzhaf power index 0.

55. You should buy your vote from P_1. The following table explains why.

Buying a vote from	Resulting weighted voting system	Resulting Banzhaf power distribution	Your power
P_1	$[6:3,2,2,2,2]$	$P_1:\frac{1}{5}; P_2:\frac{1}{5}; P_3:\frac{1}{5}; P_4:\frac{1}{5}; P_5:\frac{1}{5}$	$\frac{1}{5}$
P_2	$[6:4,1,2,2,2]$	$P_1:\frac{1}{2}; P_2:0; P_3:\frac{1}{6}; P_4:\frac{1}{6}; P_5:\frac{1}{6}$	$\frac{1}{6}$
P_3	$[6:4,2,1,2,2]$	$P_1:\frac{1}{2}; P_2:\frac{1}{6}; P_3:0; P_4:\frac{1}{6}; P_5:\frac{1}{6}$	$\frac{1}{6}$
P_4	$[6:4,2,2,1,2]$	$P_1:\frac{1}{2}; P_2:\frac{1}{6}; P_3:\frac{1}{6}; P_4:0; P_5:\frac{1}{6}$	$\frac{1}{6}$

57. (a) You should buy your vote from P_2. The following table explains why.

Buying a vote from	Resulting weighted voting system	Resulting Banzhaf power distribution	Your power
P_1	$[18:9,8,6,4,3]$	$P_1:\frac{4}{13}; P_2:\frac{3}{13}; P_3:\frac{3}{13}; P_4:\frac{2}{13}; P_5:\frac{1}{13}$	$\frac{1}{13}$
P_2	$[18:10,7,6,4,3]$	$P_1:\frac{9}{25}; P_2:\frac{1}{5}; P_3:\frac{1}{5}; P_4:\frac{3}{25}; P_5:\frac{3}{25}$	$\frac{3}{25}$
P_3	$[18:10,8,5,4,3]$	$P_1:\frac{5}{12}; P_2:\frac{1}{4}; P_3:\frac{1}{6}; P_4:\frac{1}{12}; P_5:\frac{1}{12}$	$\frac{1}{12}$
P_4	$[18:10,8,6,3,3]$	$P_1:\frac{5}{12}; P_2:\frac{1}{4}; P_3:\frac{1}{6}; P_4:\frac{1}{12}; P_5:\frac{1}{12}$	$\frac{1}{12}$

(b) You should buy 2 votes from P_2. The following table explains why.

Buying 2 votes from	Resulting weighted voting system	Resulting Banzhaf power distribution	Your power
P_1	$[18:8,8,6,4,4]$	$P_1:\frac{7}{27}; P_2:\frac{7}{27}; P_3:\frac{7}{27}; P_4:\frac{1}{9}; P_5:\frac{1}{9}$	$\frac{1}{9}$
P_2	$[18:10,6,6,4,4]$	$P_1:\frac{5}{13}; P_2:\frac{2}{13}; P_3:\frac{2}{13}; P_4:\frac{2}{13}; P_5:\frac{2}{13}$	$\frac{2}{13}$
P_3	$[18:10,8,4,4,4]$	$P_1:\frac{11}{25}; P_2:\frac{1}{5}; P_3:\frac{3}{25}; P_4:\frac{3}{25}; P_5:\frac{3}{25}$	$\frac{3}{25}$
P_4	$[18:10,8,6,2,4]$	$P_1:\frac{9}{25}; P_2:\frac{7}{25}; P_3:\frac{1}{5}; P_4:\frac{1}{25}; P_5:\frac{3}{25}$	$\frac{3}{25}$

(c) Buying a single vote from P_2 raises your power from $1/25 = 4\%$ to $3/25 = 12\%$. Buying a second vote from P_2 raises your power to $2/13 \approx 15.4\%$. The increase in power is less with the second vote, but if you value power over money, it might still be worth it to you to buy that second vote.

59. (a) The losing coalitions are $\{P_1\}$, $\{P_2\}$, and $\{P_3\}$. The complements of these coalitions are $\{P_2, P_3\}$, $\{P_1, P_3\}$, and $\{P_1, P_2\}$, respectively, all of which are winning coalitions.

(b) The losing coalitions are $\{P_1\}, \{P_2\}, \{P_3\}, \{P_4\}, \{P_2, P_3\}, \{P_2, P_4\}$, and $\{P_3, P_4\}$. The complements of these coalitions are $\{P_2, P_3, P_4\}, \{P_1, P_3, P_4\}, \{P_1, P_2, P_4\}, \{P_1, P_2, P_3\}, \{P_1, P_4\}, \{P_1, P_3\}$, and $\{P_1, P_2\}$, respectively, all of which are winning coalitions.

(c) If P is a dictator, the losing coalitions are all the coalitions without P; the winning coalitions are all the coalitions that include P. The complement of any coalition without P (losing) is a coalition with P (winning).

(d) Take the grand coalition out of the picture for a moment. Of the remaining $2^N - 2$ coalitions, half are losing coalitions and half are winning coalitions, since each losing coalition pairs up with a winning coalition (its complement). Half of $2^N - 2$ is $2^{N-1} - 1$. In addition, we have the grand coalition (always a winning coalition). Thus, the total number of winning coalitions is 2^{N-1}.

Chapter 3

WALKING

A. Fair Division Concepts

1. **(a)** $9.00   **(b)** $3.00   **(c)** $3.00

3. **(a)** $6.00   **(b)** $4.00   **(c)** $2.00
   (d) Piece 1: $1.00; piece 2: $1.50; piece 3: $2.00; piece 4: $2.50; piece 5: $3.00; piece 6: $2.00

5. **(a)** Ana: s_2, s_3
   (b) Ben: s_3
   (c) Cara: s_1, s_2, s_3

7. **(a)** Adams: s_1, s_4
   (b) Benson: s_1, s_2
   (c) Cagle: s_1, s_3
   (d) Duncan: s_4
   (e) Adams: s_1; Benson: s_2; Cagle: s_3; Duncan: s_4

B. The Divider-Chooser Method

9. **(a)** (iii) only
   (b) either piece

11. **(a)** Answers may vary. For example,

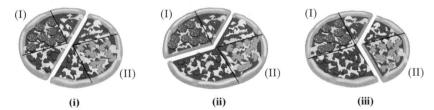

| (i) | (ii) | (iii) |

   (b) (i) either piece; (ii) II; (iii) I

13. **(a)** 50%
   (b) the left piece: 40%; the right piece: 60%
   (c) Jamie takes the right piece; Mo gets the left piece.

C. The Lone-Divider Method

15. **(a)** There are three possible answers, shown in the following table.

Chase	Chandra	Divine
s_2	s_1	s_3
s_2	s_3	s_1
s_3	s_1	s_2

   (b) There are two possible answers, shown in the following table.

Chase	Chandra	Divine
s_3	s_1	s_2
s_2	s_1	s_3

   (c) The only possible fair division is

Chase	Chandra	Divine
s_1	s_2	s_3

(d) Divine can pick between s_2 and s_3—let's say Divine picks s_2. Then s_1 and s_3 can be combined into a single property to be divided between Chase and Chandra using the divider-chooser method.

17. (a) One possible fair division of the cake is

Chooser 1	Chooser 2	Chooser 3	Divider
s_2	s_3	s_1	s_4

(b) Another possible fair division of the cake is

Chooser 1	Chooser 2	Chooser 3	Divider
s_3	s_1	s_2	s_4

(c) No. None of the choosers chose s_4, which can only be given to the divider.

19. (a) A fair division of the cake is

Chooser 1	Chooser 2	Chooser 3	Chooser 4	Divider
s_2	s_4	s_3	s_5	s_1

(b) Another fair division of the cake is

Chooser 1	Chooser 2	Chooser 3	Chooser 4	Divider
s_4	s_2	s_3	s_5	s_1

(c) No. None of the choosers chose s_1, which can only be given to the divider.

21. (a) A fair division of the cake is

Chooser 1	Chooser 2	Chooser 3	Chooser 4	Chooser 5	Divider
s_5	s_1	s_6	s_2	s_3	s_4

(b) Chooser 5 must get s_3, which forces chooser 4 to get s_2. This leaves only s_5 for chooser 1, which in turn leaves only s_6 for chooser 3. Consequently, only s_1 is left for chooser 2, and the divider must get s_4.

23. (a) Chooser 1: $\{s_3, s_4\}$: chooser 2: $\{s_1, s_3, s_4\}$; chooser 3: $\{s_3\}$
(b) The only possible fair division of the land is

Chooser 1	Chooser 2	Chooser 3	Divider
s_4	s_1	s_3	s_2

D. The Lone-Chooser Method

25. (a) One possible second division by Angela is

Angela's piece

(b) One possible second division by Boris is

Boris's piece

(c) One possible fair division is

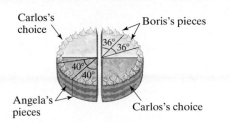

(d) The value of Angela's final share (in Angela's eyes) is $4.00; the value of Boris's final share (in Boris's eyes) is $4.00; the value of Carlos's final share (in Carlos's eyes) is $7.02.

27. (a) One possible first division is

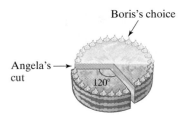

(b) One possible second division by Angela is

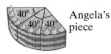

(c) One possible second division by Boris is

(d) One possible fair division is

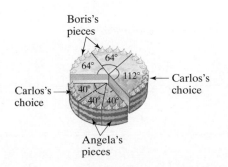

(e) Angela thinks her share is worth $4.00; Boris thinks his share is worth $7.11; Carlos thinks his share is worth $4.53.

29. (a) One possible first division is

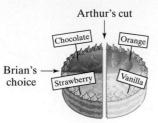

(b) One possible second division by Arthur is

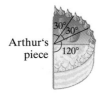

(c) One possible second division by Brian is

(d) One possible fair division is

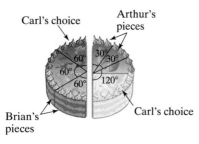

(e) Arthur thinks his share is worth $33\frac{1}{3}\%$; Brian thinks his share is worth $66\frac{2}{3}\%$; Carl thinks his share is worth $83\frac{1}{3}\%$.

31. (a) One possible first division is

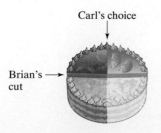

(b) One possible second division by Brian is

Brian's piece

(c) One possible second division by Carl is

Carl's piece

(d) One possible fair division is

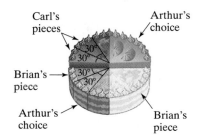

Carl's pieces Arthur's choice

Brian's piece

Arthur's choice Brian's piece

(e) Arthur thinks his share is worth $66\frac{2}{3}\%$; Brian thinks his share is worth $33\frac{1}{3}\%$; Carl thinks his share is worth $33\frac{1}{3}\%$.

E. The Last-Diminisher Method

33. (a) Yes. Although P_4 ends up with the piece at the end of round 1, it was diminished and so a piece of s went back to be a part of the R-piece to be divided in round 2 among several players including P_3.

(b) P_4 **(c)** P_1 **(d)** P_5

35. (a) P_9 **(b)** P_1 **(c)** P_{12} **(d)** P_5 **(e)** P_1 **(f)** P_2 **(g)** P_{12}

37. (a) P_3 **(b)** P_1 **(c)** P_2 **(d)** P_2 **(e)** P_4 **(f)** P_6

F. The Method of Sealed Bids

39. Ana gets the desk and receives $200 in cash; Belle gets the dresser and receives $80; Chloe gets the vanity and the tapestry and pays $280.

41. (a) Bob gets the partnership and pays $155,000.

(b) Jane gets $80,000 and Ann gets $75,000.

43. A ends up with items 1, 2, and 4 and must pay $170,666.66; B ends up with $90,333.33; C ends up with item 3 and $80,333.33.

45. A ends up with items 4 and 5 and pays $739; B ends up with $608; C ends up with items 1 and 3 and pays $261; D ends up with $632; E ends up with items 2 and 6 and pays $240.

G. The Method of Markers

47. (a) A gets items 10, 11, 12, 13; B gets items 1, 2, 3; C gets items 5, 6, 7.

(b) Items 4, 8, and 9 are left over.

49. (a) A gets items 1, 2; B gets items 10, 11, 12; C gets items 4, 5, 6, 7.

(b) Items 3, 8, and 9 are left over.

51. (a) A gets items 19, 20; B gets items 15, 16, 17; C gets items 1, 2, 3; D gets items 11, 12, 13; E gets items 5, 6, 7, 8.

(b) Items 4, 9, 10, 14, and 18 are left over.

53. (a) A gets items 4, 5; B gets item 10; C gets item 15; D gets items 1, 2.

(b) Items 3, 6, 7, 8, 9, 11, 12, 13, and 14 are left over.

JOGGING

55. Paul would choose the larger portion, worth $2.70 to him.

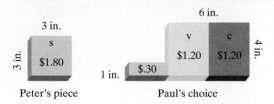

Peter's piece   Paul's choice

Since Paul likes all flavors the same, he would divide his piece into 3 shares of equal volume, each worth $.90 to him. Peter would also divide his piece into 3 shares of equal volume.

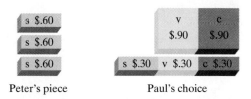

Peter's piece   Paul's choice

Mary would choose any one of Peter's pieces (each worth $.10 to her) and the vanilla piece from Paul (worth $1.35 to her).

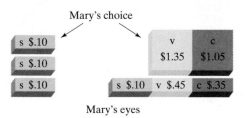

Mary's eyes

After all is said and done, Peter thinks his share is worth $1.20, Paul thinks his share is worth $1.80, and Mary thinks her share is worth $1.45.

57. (a) The total area is 30,000 m² and the area of C is only 8000 m². Since P_2 and P_3 value the land uniformly, each thinks that a fair share must have an area of at least 10,000 m².

(b) Since there are 22,000 m² left, any cut that divides the remaining property in parts of 11,000 m² will work. For example

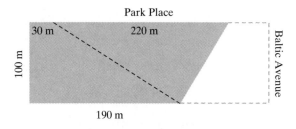

(c) The cut parallel to Baltic Avenue which divides the parcel in half is

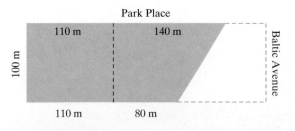

59. (a) A's original fair share is worth $x/2$; B's original fair share is worth $y/2$.

(b) $(y - x)/2$

(c) $(x + y)/4$

61. (a)

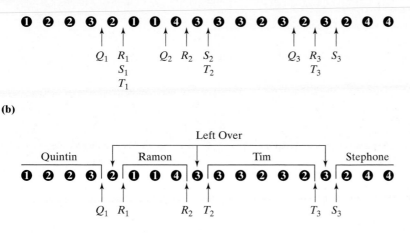

(b)

(c) These are discrete, indivisible items, so the only available methods are the method of markers and the sealed bids method. With only a few items to share, the sealed bids method is appropriate.

Chapter 4

WALKING

A. Standard Divisors and Quotas

1. (a) 50,000

(b) Apure: 66.2; Barinas: 53.4; Carabobo: 26.6; Dolores: 13.8

(c)

State	Apure	Barinas	Carabobo	Dolores
Upper quota	67	54	27	14
Lower quota	66	53	26	13

3. (a) The bus routes.

(b) 1000. The standard divisor represents the average number of passengers per bus per day.

(c) A: 45.3; B: 31.07; C: 20.49; D: 14.16; E: 10.26; F: 8.72

(d)

Bus route	A	B	C	D	E	F
Upper quota	46	32	21	15	11	9
Lower quota	45	31	20	14	10	8

5. (a) 119

(b) 200,000

(c) A: 8,100,000; B: 5,940,000; C: 4,730,000; D: 2,920,000; E: 2,110,000

B. Hamilton's Method

7. A: 66; B: 53; C: 27; D: 14

9. A: 45; B: 31; C: 21; D: 14; E: 10; F: 9

11. A: 40; B: 30; C: 24; D: 15; E: 10

13. (a) Bob: 0; Peter: 3; Ron: 8.

(b) Bob: 1; Peter: 2; Ron: 8.

(c) Yes. For studying an extra 2 minutes (an increase of 3.70%) Bob gets a piece of candy while Peter, who studies an extra 12 minutes (an increase of 4.94%) has to give up a piece. This is an example of the population paradox.

15. (a) Bob: 0; Peter: 3; Ron: 8

  (b) Bob: 1; Peter: 3; Ron: 7; Jim: 6

  (c) Ron loses a piece of candy to Bob. This is an example of the new-states paradox.

C. Jefferson's Method

17. A: 67; B: 54; C: 26; D: 13

19. A: 46; B: 31; C: 21; D: 14; E: 10; F: 8

21. A: 41; B: 30; C: 24; D: 14; E: 10

23. This illustrates that Jefferson's method can produce upper-quota violations.

D. Adams's Method

25. A: 66; B: 53; C: 27; D: 14

27. A: 45; B: 31; C: 20; D: 14; E: 11; F: 9

29. A: 40; B: 29; C: 24; D: 15; E: 11

31. This illustrates that Adams's method can produce lower-quota violations.

E. Webster's Method

33. A: 66; B: 53; C: 27; D: 14

35. A: 45; B: 31; C: 21; D: 14; E: 10; F: 9

37. A: 40; B: 30; C: 24; D: 15; E: 10

F. Miscellaneous

39. (a) 0.8%

  (b) A: 7.8; B: 32.7; C: 35.6; D: 48.9

  (c) A: 8; B: 33; C: 35; D: 49

41. (a) A: 7.67; B: 32.14; C: 34.99; D: 48.06

  (b) A: 8; B: 33; C: 35; D: 49

43.

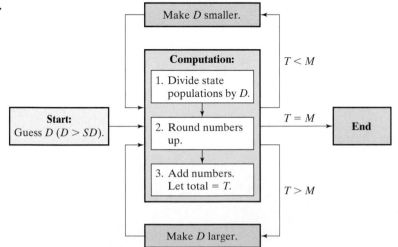

JOGGING

45. (a) Take, for example, $q_1 = 3.9$ and $q_2 = 10.1 \, (M = 14)$. Under both Hamilton's method and Lowndes's method A gets 4 seats and B gets 10 seats.

(b) Take, for example, $q_1 = 3.4$ and $q_3 = 10.6 \, (M = 14)$. Under Hamilton's method A gets 3 seats and B gets 11 seats. Under Lowndes's method A gets 4 seats and B gets 10 seats.

(c) Assume $f_1 > f_2$. Under Hamilton's method the surplus seat goes to A. Under Lowndes's method, the surplus seat would go to B if

$$\frac{f_2}{q_2 - f_2} > \frac{f_1}{q_1 - f_1},$$

which can be simplified to $\dfrac{q_1}{q_2} > \dfrac{f_1}{f_2}$.

47. (a) In Jefferson's method the modified quotas are larger than the standard quotas and so rounding downward will give each state at least the integer part of the standard quota for that state.

(b) In Adams's method the modified quotas are smaller than the standard quota and so rounding upward will give each state at most one more than the integer part of the standard quota for that state.

(c) If there are only two states, an upper quota violation for one state results in a lower quota violation for the other state (and vice versa). Since neither Jefferson's nor Adams's method can have both upper and lower violations of the quota rule, neither can violate the quota rule when there are only two states.

49. (a) 49,374,462

(b) 1,591,833

51. (a) $A: 5; \; B: 10; \; C: 15; \; D: 21$

(b) For $D = 100$, the modified quotas are $A: 5, \; B: 10, \; C: 15, \; D: 20$. For $D < 100$, each of the modified quotas above will increase and so rounding upward will give at least $A: 6, \; B: 11, \; C: 16, \; D: 21$ or a total of at least 54. For $D > 100$, each of the modified quotas above will decrease and so rounding upward will give at most $A: 5, \; B: 10, \; C: 15, \; D: 20$ or a total of at most 50.

(c) From part (b) we see that there is no divisor such that after rounding the modified quotas upward, the total is 51.

Chapter 5

WALKING

A. Graphs: Basic Concepts

1. (a) Vertices: A, B, C, D; Edges: AB, AC, AD, BD; $\deg(A) = 3, \deg(B) = 2, \deg(C) = 1, \deg(D) = 2$

(b) Vertices: A, B, C; Edges: none; $\deg(A) = 0, \deg(B) = 0, \deg(C) = 0$

(c) Vertices: V, W, X, Y, Z; Edges: $XX, XY, XZ, XV, XW, WY, YZ$; $\deg(V) = 1, \deg(W) = 2, \deg(X) = 6, \deg(Y) = 3, \deg(Z) = 2$

3. (a)

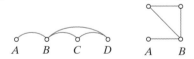

(b)

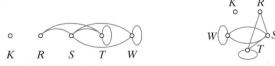

5. (a) Both graphs have four vertices $A, B, C,$ and D and (the same) edges AB, AC, AD, BD.

(b)

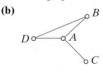

7. (a) 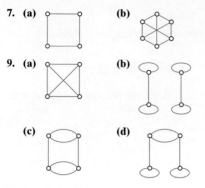 **(b)**

9. (a) **(b)**

(c) **(d)**

11. (a) *C, B, A, H, F*
(b) *C, B, D, A, H, F*
(c) *C, B, A, H, F*
(d) *C, D, B, A, H, G, G, F*
(e) 4 (*C, B, A*; *C, D, A*; *C, B, D, A*; *C, D, B, A*)
(f) 3 (*H, F*; *H, G, F*; *H, G, G, F*)
(g) 12 (Any one of the paths in [e] followed by *AH*, followed by any one of the paths in [f].)

13. (a) *D, C, B, D*
(b) *G, F, H, G, G*
(c) *HA* and *FE*

B. Graph Models

15.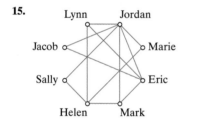

17.

19.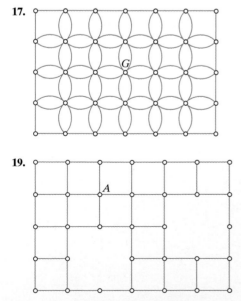

21.

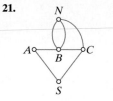

C. Euler's Theorems

23. **(a)** Has an Euler circuit because all vertices are even.

(b) Has no Euler circuit, but has an Euler path because there are exactly two odd vertices.

(c) Has neither an Euler circuit nor an Euler path because there are four odd vertices.

25. **(a)** Has an Euler circuit because all vertices are even.

(b) Has no Euler circuit, but has an Euler path because there are exactly two odd vertices.

(c) Has no Euler circuit, but has an Euler path because there are exactly two odd vertices.

27. **(a)** Has neither an Euler circuit nor an Euler path because there are eight odd vertices.

(b) Has no Euler circuit, but has an Euler path because there are exactly two odd vertices.

(c) Has no Euler circuit, but has an Euler path because there are exactly two odd vertices.

D. Finding Euler Circuits and Euler Paths

29.

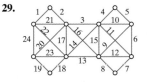

31. $A, B, C, D, E, F, G, A, C, E, G, B, D, F, A, D, G, C, F, B, E, A$

33.

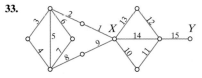

35.

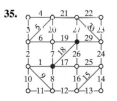

E. Unicursal Tracings

37. **(a)** Has neither because there are more than two odd vertices.

(b) Open unicursal tracing. For example,   **(c)** Open unicursal tracing. For example,

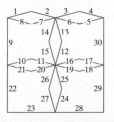
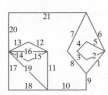

39. (a) Open unicursal tracing. For example,   **(b)** Open unicursal tracing. For example,

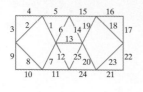 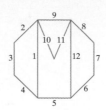

(c) Has neither because there are more than two odd vertices.

F. Eulerizations and Semi-eulerizations

41. (a)   **(b)**

43. (a)   **(b)**

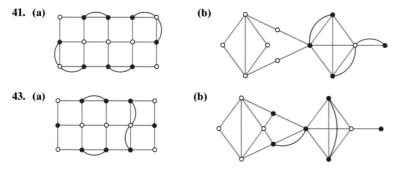

G. Miscellaneous

45. (a) None   **(b)**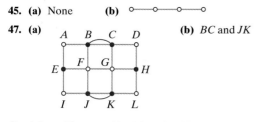

47. (a)   **(b)** *BC* and *JK*

49. 4 times. There are 10 odd vertices. Two can be used as the starting and ending vertices. The remaining 8 odd vertices can be paired so that each pair forces one lifting of the pencil.

51.

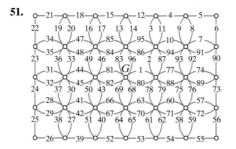

53.

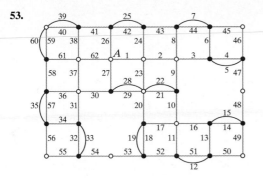

JOGGING

55. (a) An edge XY contributes 2 to the sum of the degrees of all the vertices (1 to the degree of X, and 1 to the degree of Y).

(b) If there were an odd number of odd vertices, the sum of the degrees of all the vertices would be odd.

57. (a) Eulerizing the graph shown in Fig.5–17(b) requires the addition of two edges so the cheapest walk will cost $9.00. One possible such walk is shown in the figure.

(b) Semi-eulerizing the graph shown in Fig. 5–17(b) requires the addition of one edge so the cheapest walk will cost $8.00. One possible such walk (starting at L and ending at A) is shown in the figure.

59.

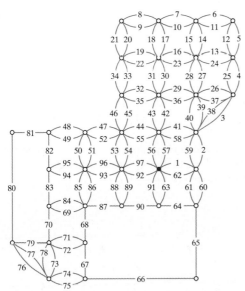

61. (a) 12

(b)

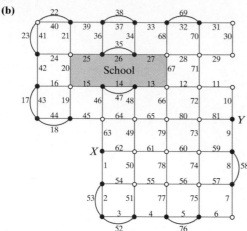

63. (a) The office complex can be represented by a graph (where each vertex represents a location and each edge a door).

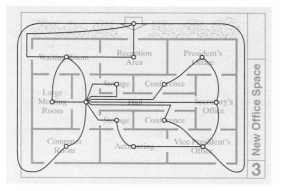

Since there are odd vertices, there is no Euler circuit.

(b) Since there are exactly 2 odd vertices (the secretary's office has degree 3 and the hall has degree 9), there is an Euler path starting at either the secretary's office or the hall and ending at the other.

(c) If the door from the secretary's office to the hall is removed (i.e., the edge between the secretary's office and the hall is removed), then every vertex will be even and so there will be an Euler circuit. Consequently, it would be possible to start at any location, walk through every door exactly once and end up at the starting location.

65. (a) The graph model has 4 odd vertices (B, D, L, R). To start and end at R, the photographer will have to recross at least two of the bridges, for example, the Kennedy bridge (BD) and the Jefferson bridge (LR). One of the many possible optimal routes is given by the following sequence of bridges: Hoover, Lincoln, Truman, Wilson, Monroe, Kennedy, Kennedy (second pass), Adams, Roosevelt, Grant, Washington, Jefferson, Jefferson (second pass).

(b) To start at the Adams bridge and end at the Grant bridge, the photographer should start and end the shoot on the Left Bank. One of the many possible optimal routes is given by the following sequences of bridges: Adams, Kennedy, Kennedy (second pass), Monroe, Wilson, Truman, Lincoln, Hoover, Jefferson, Jefferson (second pass), Washington, Roosevelt, Grant.

Chapter 6

WALKING

A. Hamilton Circuits and Hamilton Paths

1. (a) 1. A, B, D, C, E, F, G, A; 2. A, D, C, E, B, G, F, A; 3. A, D, B, E, C, F, G, A

(b) A, G, F, E, C, D, B

(c) D, A, G, B, C, E, F

3. 1. *A, B, C, D, E, F, G, A*   5. *A, G, F, E, D, C, B, A*
  2. *A, B, E, D, C, F, G, A*   6. *A, G, F, C, D, E, B, A*
  3. *A, F, C, D, E, B, G, A*   7. *A, G, B, E, D, C, F, A*
  4. *A, F, E, D, C, B, G, A*   8. *A, G, B, C, D, E, F, A*

5. (a) *A, F, B, C, G, D, E*
  (b) *A, F, B, C, G, D, E, A*
  (c) *A, F, B, E, D, G, C*
  (d) *F, A, B, E, D, C, G*

7. (a) 1. *A, B, C, D, E, F, A*   3. *A, F, E, D, C, B, A*
    2. *A, B, E, D, C, F, A*   4. *A, F, C, D, E, B, A*
  (b) 1. *D, E, F, A, B, C, D*   3. *D, C, B, A, F, E, D*
    2. *D, C, F, A, B, E, D*   4. *D, E, B, A, F, C, D*
  (c) The circuits in (b) are the same as the circuits in (a), just rewritten with a different starting vertex.

9. The degree of every vertex in a graph with a Hamilton circuit must be at least 2 since the circuit must "pass through" every vertex. A graph with a Hamilton path can have at most 2 vertices (the starting and ending vertices of the path) of degree 1 since the path must "pass through" the remaining vertices. This graph has 4 vertices of degree 1.

11. (a) 6   **(b)** 4
  (c) *A, B, C, D, E, A* (weight 32)   **(d)** *A, D, B, C, E, A* (weight 27)

13. (a) 11   **(b)** *A, B, C, F, E, D, A* (weight 37)
  (c) *A, D, F, E, B, C, A* (weight 41)

B. Factorials and Complete Graphs

15. (a) 6,227,020,800
  (b) 6,227,020,800
  (c) 6.204484×10^{23}

17. (a) 3,628,800
  (b) 362,880
  (c) 3,628,800

19. (a) 66
  (b) 39,916,800

21. (a) 6
  (b) 10
  (c) 201

C. Brute Force and Nearest-Neighbor Algorithms

23. (a) *A, C, B, D, A* (weight 62)
  (b) *A, D, C, B, A* (weight 80)
  (c) *B, D, C, A, B* (weight 74)
  (d) *C, D, B, A, C* (weight 74)

25. (a) *B, C, A, E, D, B* (121 + 119 + 133 + 199 + 150 = 722)
  (b) *C, A, E, D, B, C* (119 + 133 + 199 + 150 + 121 = 722)
  (c) *D, B, C, A, E, D* (150 + 121 + 119 + 133 + 199 = 722)
  (d) *E, C, A, D, B, E* (120 + 119 + 152 + 150 + 200 = 741)

27. (a) *E, C, I, G, M, T, E* (19.4 years)
  (b) *E, T, M, I, C, G, E* (18.9 years)

29. (a) Atlanta, Columbus, Kansas City, Tulsa, Minneapolis, Pierre, Atlanta (3887 miles)

(b) Atlanta, Kansas City, Tulsa, Minneapolis, Pierre, Columbus, Atlanta (3739 miles)

D. Repetitive Nearest-Neighbor Algorithm

31. *C, D, E, A, B, C* (weight 9.8)

33. *E, T, M, I, C, G, E* (18.9 years)

35. Atlanta, Columbus, Minneapolis, Pierre, Kansas City, Tulsa, Atlanta (3252 miles)

E. Cheapest-Link Algorithm

37. *A, B, E, D, C, A* (weight 9.9)

39. *E, C, I, G, M, T, E* (19.4 years)

41. Atlanta, Columbus, Pierre, Minneapolis, Kansas City, Tulsa, Atlanta (3465 miles)

F. Miscellaneous

43. (a)

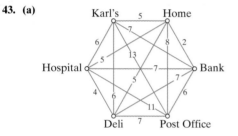

(b) Home, Bank, Post Office, Deli, Hospital, Karl's, Home. The total length of the trip is 30 miles.

45. (a)

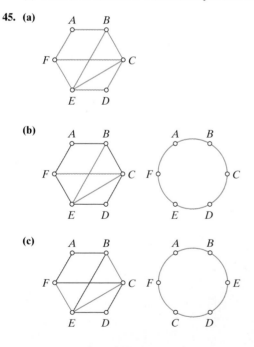

(b)

(c)

47. If we draw the graph describing the friendships among the guests (see figure) we can see that the graph does not have a Hamilton circuit, which means it is impossible to seat everyone around the table with friends on both sides.

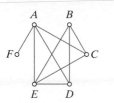

JOGGING

49.

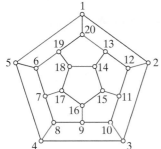

51. *A, B, C, D, J, I, F, G, E, H*

53. The 2-by-2 grid graph cannot have a Hamilton circuit because each of the 4 corner vertices as well as the interior vertex *I* must be preceded and followed by a boundary vertex. But there are only 4 boundary vertices—not enough to go around.

55. (a)

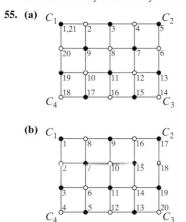

(b)

(c) Think of the vertices of the graph as being colored like a checker board with C_1 being a black vertex. Then each time we move from one vertex to the next, we must move from a black vertex to a white vertex or from a white vertex to a black vertex. Since there are 10 white vertices and 10 black vertices and we are starting with a black vertex, we must end at a white vertex. But C_2 is a black vertex. Therefore, no such Hamilton path is possible.

57. One possible example is as follows:

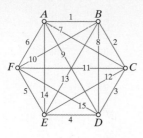

59. Each vertex is adjacent to each of the other vertices, so each vertex has degree $N - 1$. Since there are N vertices, the sum of the degrees of all the vertices is $N(N - 1)$. But the sum of the degrees of all the vertices in a graph is always equal to twice the number of edges. Therefore, the number of edges in a complete graph with N vertices is $\frac{N(N - 1)}{2}$.

61. Dallas, Houston, Memphis, Louisville, Columbus, Chicago, Kansas City, Denver, Atlanta, Buffalo, Boston, Dallas

Chapter 7

WALKING

A. Trees

1. **(a)** Is a tree.
 (b) Is not a tree (has a circuit and is not connected).
 (c) Is not a tree (has a circuit).
 (d) Is a tree.

3. **(a)** (II) A tree with 8 vertices must have 7 edges.
 (b) (II) A tree with 8 vertices must have 7 edges.
 (c) (III)

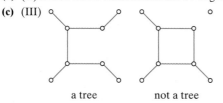

 a tree   not a tree

 (d) (II) A tree has no circuits.

5. **(a)** (I) If there is exactly one path joining any two vertices of a graph, the graph must be a tree.
 (b) (II) A tree with 8 vertices must have 7 edges and every edge must be a bridge.
 (c) (I) If every edge is a bridge, then the graph has no circuits. Since the graph is also connected, it must be a tree.

7. **(a)** (III)

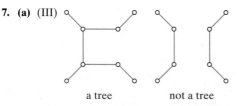

 a tree   not a tree

 (b) (II) A tree has no circuits.
 (c) (I) A graph with 8 vertices, 7 edges, and no circuits must also be connected and hence must be a tree.

9. (a) (II) Since the degree of each vertex is even, it must be at least 2. Thus, the sum of the degrees of all 8 vertices must be at least 16. But a tree with 8 vertices must have 7 edges, and the sum of the degrees of all the vertices would have to be 14.

(b) (III)

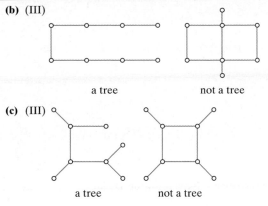

a tree   not a tree

(c) (III)

a tree   not a tree

B. Spanning Trees

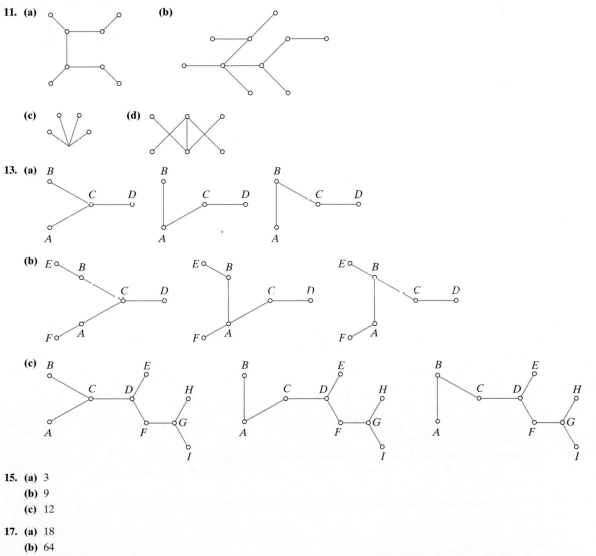

11. (a)   **(b)**

(c)   **(d)**

13. (a) B   C   D   A

(b) E   B   C   D   F   A

(c) B   C   E   D   H   A   F   G   I

15. (a) 3
(b) 9
(c) 12

17. (a) 18
(b) 64

C. Minimum Spanning Trees and Kruskal's Algorithm

19. Total weight is 855.

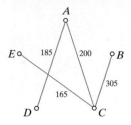

21. Total weight is 9.3.

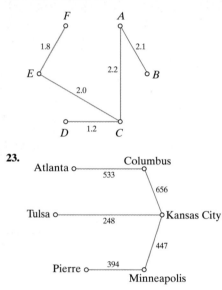

23.

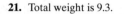

25. A spanning tree for the 20 vertices will have 19 edges and so the cost will be $(19/2)(\$40,000) = \$380,000$.

D. Steiner Points and Shortest Networks

27. (a) 580 miles

(b) 385 miles

29. 366 km

31. 334 km

33. Z. (The sum of the distances from Z to A, B, and C is 232 miles, the sum of the distances from X to A, B, and C is 240 miles, and the sum of the distances from Y to A, B, and C is 243 miles.)

35. Y. (If we call the Steiner point S, then angle $CSB = 120°$, and angles WCS and WBS are both 30°. It is clear from the picture that Z cannot be the Steiner point because the angles WCZ and WBZ are much smaller than 30°. Likewise, X cannot be the Steiner point because the angles WCX and WBX are much larger than 30°.)

37. (a) $CE + ED + EB$ is larger since $CD + DB$ is the shortest network connecting the cities C, D, and B.

(b) $CD + DB$ is the shortest network connecting the cities C, D, and B since angle CDB is 120° and so the shortest network is the same as the minimum spanning tree.

(c) $CE + EB$ is the shortest network connecting the cities C, E, and B since angle CEB is more than 120° and so the shortest network is the same as the minimum spanning tree.

E. Miscellaneous

39. (a) $BC = 10.2$ cm: $AC \approx 17.7$ cm

(b) $AB = 23.0$ cm: $AC \approx 19.9$ cm

(c) $BC \approx 12.1$ cm: $AB \approx 24.2$ cm

41. In a 30°-60°-90° triangle, the side opposite the 30° angle is $\frac{1}{2}$ the hypotenuse and the side opposite the 60° angle is $\frac{\sqrt{3}}{2}$ times the hypotenuse. Therefore, the distance from Alcie Springs to J is $\frac{\sqrt{3}}{2} \times 500 \approx 433$ miles and so the total length of the T-network (rounded to the nearest mile) is 433 miles + 500 miles = 933 miles.

JOGGING

43. In a tree there is one and only one path joining any two vertices. Consequently, the only path joining two adjacent vertices is the edge connecting them and so if that edge is removed, the graph will become disconnected.

45. (a)

(b)

(c)

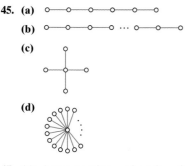

(d)

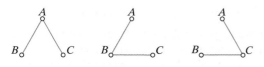

47. (a) A tree must have at least 2 vertices of degree 1. Exercise 45(b) shows that there are trees with N vertices having just 2 vertices of degree 1. To show that a tree cannot have fewer than 2 vertices of degree 1, let v be the number of vertices in the graph, e the number of edges, and k the number of vertices of degree 1. Recall that in a tree $v = e + 1$ and in any graph the sum of the degrees of all the vertices is $2e$. Now, since we are assuming there are exactly k vertices of degree 1, the remaining $v - k$ vertices must have degree at least 2. Therefore, the sum of the degrees of all the vertices must be at least $k + 2(v - k)$. Putting all this together, we have

$$2e \geq k + 2(v - k) = k + 2(e + 1 - k).$$
$$2e \geq k + 2e + 2 - 2k.$$
$$k \geq 2.$$

(b) For $N > 2$, a tree can have at most $N - 1$ vertices of degree 1. Exercise 45(d) shows that a tree with N vertices can have $N - 1$ vertices of degree 1, but if all N vertices had degree 1, the sum of the degrees of all the vertices would be N, which contradicts the fact that in a tree, the sum of the degrees of all the vertices is $2N - 2$ (for $N > 2$, $2N - 2 > N$).

(c) If all the vertices had the same degree, they would all have to have degree 1 by part **(a)**, which would contradict part **(b)**.

49. (a) According to Cayley's theorem, there are $3^{3-2} = 3$ spanning trees in a complete graph with 3 vertices, which is confirmed by the following figures.

Likewise, for the complete graph with 4 vertices, Cayley's theorem predicts $4^{4-2} = 16$ spanning trees, which is confirmed by the following figures.

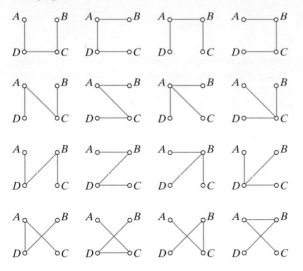

(b) The complete graph with N vertices has $(N-1)!$ Hamilton circuits and N^{N-2} spanning trees. For $N \geq 3$, $(N-1)! < N^{N-2}$ since we can write $(N-1)! = 2 \times 3 \times 4 \times \ldots \times (N-1)$, and $N^{N-2} = N \times N \times N \times \ldots \times N$. Both expressions have the same number of factors, with each factor in $(N-1)!$ smaller than the corresponding factor in N^{N-2}.

51. If some edge had weight more than the weight of e, deleting that edge would result in a spanning tree with total weight less than that of the minimum spanning tree.

53. The minimum cost network connecting the 4 cities has a 3-way junction point at A and has a total cost of 205 million dollars.

55. (a) If J is a Steiner point then $\angle BJC = 120°$. But since $\angle BJC > \angle BAC = 130°$, this is impossible. [See part (b) for a proof.]

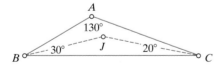

(b) In $\triangle ABC$, $\angle BAC + \angle ABC + \angle BCA = \angle BAC + u + v + r + s = 180°$ (see figure). In $\triangle BJC$, $\angle BJC + v + s = 180°$. Therefore, $\angle BAC + u + v + r + s = \angle BJC + v + s$ and so $\angle BAC + u + r = \angle BJC$. Consequently, $\angle BAC < \angle BJC$ and so if $\angle BAC > 120°$ then $\angle BJC > 120°$. Thus, J cannot be a Steiner point since $\angle BJC \neq 120°$.

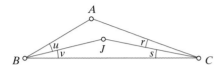

57. The length of the network is $4x$ (see figure). Since the diagonals of a square are perpendicular, by the Pythagorean theorem we have $x^2 + x^2 = 500^2$ which gives $2x^2 = 500^2$ or $x = \frac{500}{\sqrt{2}} = \frac{500\sqrt{2}}{2} = 250\sqrt{2}$. Thus $4x = 1000\sqrt{2} \approx 1414$.

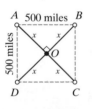

59. (a) The length of the network is $4x + (300 - x) = 3x + 300$, where $200^2 + \left(\frac{x}{2}\right)^2 = x^2$. (See figure.) Solving the equation gives $x = \frac{400\sqrt{3}}{3}$, and so the length of the network is $400\sqrt{3} + 300 \approx 993$.

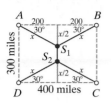

(b) Length of the network is $4x + (400 - x) = 3x + 400$, where $150^2 + \left(\frac{x}{2}\right)^2 = x^2$. (See figure.) Solving the equation gives $x = \frac{300\sqrt{3}}{3}$, and so the length of the network is $300\sqrt{3} + 400 \approx 919.6$.

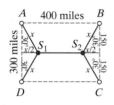

Chapter 8

WALKING

A. Directed Graphs

1. (a)

Vertex	Degree	Indegree	Outdegree	Vertex is incident to	Vertex is incident from
A	3	2	1	C	B, D
B	2	0	2	A, D	—
C	1	1	0	—	A
D	2	1	1	A	B

(b)

Vertex	Degree	Indegree	Outdegree	Vertex is incident to	Vertex is incident from
A	4	2	2	B, C	C, E
B	2	1	1	D	A
C	4	1	3	A, D, E	A
D	3	3	0	—	B, C, E
E	3	1	2	A, D	C

3. (a) Vertices: A, B, C, D, E, F.
Arcs: $AB, BD, CF, DE, EB, EC, EF$.

	Vertex	Indegree	Outdegree	Vertex is incident to E	Vertex is incident from E
(b)	A	0	1		
	B	2	1		✓
(c)	C	1	1		✓
(d)	D	1	1	✓	
	E	1	3		
(e)	F	2	0		✓

(f) EB, EC, EF

5. (a)

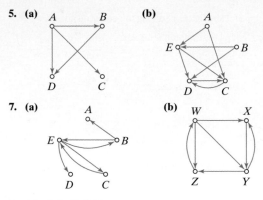

(b)

7. (a)

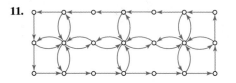

(b)

9. (a) *A, B, D, E, F*

  (b) *A, B, D, E, C, F*

  (c) *B, D, E, B*

  (d) The outdegree of *F* is 0.

  (e) The indegree of *A* is 0.

11.

13. (a) *B. B* is the only person that everyone respects.

   (b) *A. A* is the only person that no one respects.

15. (a)   **(b)**

B. Project Digraphs

17.

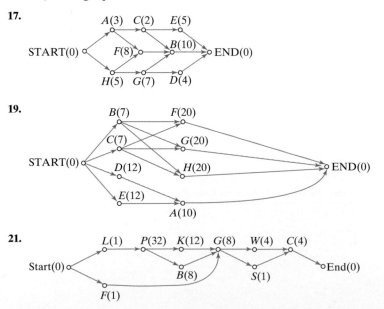

19.

21.

C. Schedules, Priority Lists, and the Decreasing-Time Algorithm

23. (a) 18 hours

(b) There is a total of 75 hours of work to be done. Three processors working without any idle time would take $75/3 = 25$ hours to complete the project.

25. There is a total of 75 hours of work to be done. Dividing the work equally between the six processors would require each processor to do $75/6 = 12.5$ hours of work. But there are no half-hour jobs, so the completion time could not be less than 13 hours.

27. Time: 0 1 2 3 4 5 6 7 8 9 10 11 12 13 14 15 16 17 18 19 20 21 22 23 24 25 26

P_1: C(9) | E(6) | G(2) | Idle
P_2: A(8) | B(5) | D(12) | F

Finishing time = 26

29. Time: 0 1 2 3 4 5 6 7 8 9 10 11 12 13 14 15 16 17 18 19 20 21 22 23 24 25 26

P_1: C | E | G | Idle | F
P_2: A | D | Idle
B | Idle

Finishing time = 21

31. Time: 0 1 2 3 4 5 6 7 8 9 10 11 12 13 14 15 16 17 18 19 20 21 22 23 24 25 26

P_1: C(9) | E(6) | G(2) | Idle
P_2: A(8) | B(5) | D(12) | F

Finishing time = 26

33. According to the precedence relations, G cannot be started until K is completed.

35. (a) Time: 0 1 2 3 4 5 6 7 8 9 10 11 12 13 14 15 16 17 18 19 20 21 22 23 24 25 26 27 28 29 30 31 32 33 34

F L | P ...

34 35 36 37 38 39 40 41 42 43 44 45 46 47 48 49 50 51 52 53 54 55 56 57 58 59 60 61 62 63 64 65 66 67 68 69 70 71

B | K | G | S | W | C

Finishing time = 71

(b) Time: 0 1 2 3 4 5 6 7 8 9 10 11 12 13 14 15 16 17 18 19 20 21 22 23 24 25 26 27 28 29 30 31 32 33

W_1: F | P ...
W_2: L | Idle

33 34 35 36 37 38 39 40 41 42 43 44 45 46 47 48 49 50 51 52 53 54 55 56 57 58 59 60 61

B | Idle | G | S | Idle | C
K | Idle | W | Idle

Finishing time = 61

37. Time: 0 1 2 3 4 5 6 7 8 9 10 11 12 13 14 15 16 17 18 19 20 21 22 23 24 25 26 27 28 29 30 31 32 33

W_1	F	P
W_2	L	Idle
W_3		Idle

...

33 34 35 36 37 38 39 40 41 42 43 44 45 46 47 48 49 50 51 52 53 54 55 56 57 58 59 60 61

K		G		W	C
B	Idle	Idle	S	Idle	
Idle					

Finishing time = 61

39. (a) Time: 0 2 4 6 8 10 12 14 16 18 20 22 24 26 28 30 32 34 36 38 40 42 44 46 48 50 52

P_1	M(12)	I(6)	H(5)	F(5)	D(4)	B(3)	Idle
P_2	L(7)	K(7)	J(6)	G(5)	E(5)	C(4)	A(3)

Finishing time = 37

(b) Time: 0 2 4 6 8 10 12 14 16 18 20 22 24 26 28 30 32 34 36 38 40 42 44 46 48 50 52

P_1	M(12)	I(6)	H(5)	F(5)	D(4)	C(4)
P_2	L(7)	K(7)	J(6)	G(5)	E(5)	B(3)	A(3)

Finishing time = 36

41. (a) Time: 0 1 2 3 4 5 6 7 8 9 10 11 12 13 14 15 16 17 18 19 20 21 22 23 24 25 26

P_1	M(12)	Idle
P_2	L(7)	D(4)	A(3)
P_3	K(7)	C(4)	Idle
P_4	J(6)	F(5)	Idle
P_5	I(6)	E(5)	Idle
P_6	H(5)	G(5)	B(3)	Idle

Finishing time = 14

(b) Time: 0 1 2 3 4 5 6 7 8 9 10 11 12 13 14 15 16 17 18 19 20 21 22 23 24 25 26

P_1	A(3)	C(4)	E(5)
P_2	B(3)	D(4)	F(5)
P_3	G(5)	K(7)
P_4	H(5)	L(7)
P_5	I(6)	J(6)
P_6	M(12)

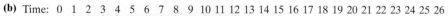

Finishing time = 12

(c) The completion time is 12 hours and one of the tasks takes 12 hours and so the job cannot be completed in less than 12 hours.

43. (a) Time: 0 1 2 3 4 5 6 7 8 9 10 11 12 13 14 15 16 17 18 19 20 21 22 23 24 25 26

P_1	M(12)
P_2	L(7) — Idle
P_3	K(7) — Idle
P_4	J(6) — B(3) — Idle
P_5	I(6) — A(3) — Idle
P_6	H(5) — E(5) — Idle
P_7	G(5) — D(4) — Idle
P_8	F(5) — C(4) — Idle

Finishing time = 12

(b) One of the tasks takes 12 hours.

(c) Any schedule with 12 hours completion time is optimal, so the schedule given in (a) is optimal. Also, the schedule given in 41(b) with 6 copiers is optimal letting copiers 7 and 8 be idle for the whole project.

45. Time: 0 2 4 6 8 10 12 14 16 18 20 22 24 26 28 30 32 34 36 38 40 42 44 46 48 50 52

P_1	AD — AP — IF — IW — ID — PU — EU — FW
P_2	AW — AF — Idle — PL — Idle — IP — HU — PD — Idle

Finishing time = 44

D. Critical Paths and the Critical-Path Algorithm

47. (a)

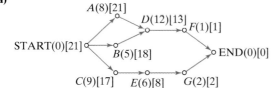

A(8)[21]

D(12)[13] F(1)[1]

START(0)[21]

B(5)[18]

END(0)[0]

C(9)[17] E(6)[8] G(2)[2]

(b) 21

(c) Time: 0 1 2 3 4 5 6 7 8 9 10 11 12 13 14 15 16 17 18 19 20 21 22 23 24 25 26

P_1	A(8) — D(12) — G(2)
P_2	B(5) — C(9) — E(6) — F — Idle

Finishing time = 22

(d) There is a total of 43 hours of work (assuming the numbers represent hours) and there is no task less than 1 hour long, so the shortest time the project can be completed in is 22 hours (43 ÷ 2 = 21.5 which rounded up to nearest hour is 22).

49. Time: 0 2 4 6 8 10 12 14 16 18 20 22 24 26 28 30 32 34 36 38 40 42 44 46 48 50 52

P_1	B — E — F — I — K
P_2	A — C — Idle — G — Idle
P_3	D — H — J — Idle

Finishing time = 49

51.

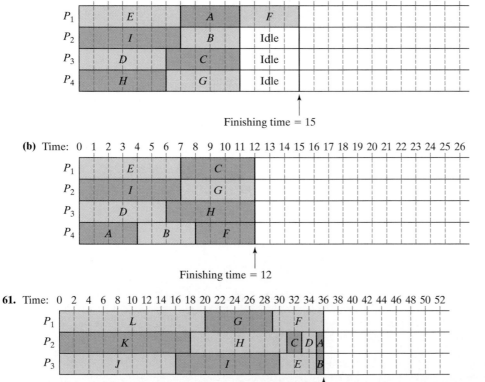

53. Time: 0 2 4 6 8 10 12 14 16 18 20 22 24 26 28 30 32 34 36 38 40 42 44 46 48 50 52

Finishing time = 36

JOGGING

55. Each arc of the graph contributes 1 to the sum of the indegrees and 1 to the sum of the outdegrees.

57. (a) True. If no processor is idle, then the total of the processing times of all the tasks is the same as the completion time of the schedule and so there can be no shorter schedule.

(b) False. Precedence relations may force idle time for one or more processors. (See Example 1 in the chapter.)

59. (a) Time: 0 1 2 3 4 5 6 7 8 9 10 11 12 13 14 15 16 17 18 19 20 21 22 23 24 25 26

Finishing time = 15

(b) Time: 0 1 2 3 4 5 6 7 8 9 10 11 12 13 14 15 16 17 18 19 20 21 22 23 24 25 26

Finishing time = 12

61. Time: 0 2 4 6 8 10 12 14 16 18 20 22 24 26 28 30 32 34 36 38 40 42 44 46 48 50 52

Finishing time = 36

This schedule is obviously optimal, since the processors are always busy.

63. (a) Time: 0 1 2 3 4 5 6 7 8 9 10 11 12 13 14 15 16 17 18 19 20 21 22 23 24 25 26

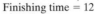

P_1	A(3)	C(4)	E(5)
P_2	B(3)	D(4)	F(5)
P_3	G(5)	K(7)	
P_4	H(5)	L(7)	
P_5	I(6)	J(6)	
P_6	M(12)		

Finishing time = 12

(b) Time: 0 1 2 3 4 5 6 7 8 9 10 11 12 13 14 15 16 17 18 19 20 21 22 23 24 25 26

P_1	A(3)	L(7)	M(12)
P_2	B(3)	K(7)	Idle
P_3	C(4)	J(6)	Idle
P_4	D(4)	I(6)	Idle
P_5	E(5)	H(5)	Idle
P_6	F(5)	G(5)	Idle

Finishing time = 22

65. (a) $4/15 \approx 0.267$; $5/18 \approx 0.278$; $6/21 \approx 0.286$; $7/24 \approx 0.292$; $8/27 \approx 0.296$; $9/30 = 0.3$

(b) $(N-1)/3N \le N/3N = 1/3$

67. (a) The finishing time of a project is always more than or equal to the sum of the processing times for all the tasks divided by the number of processors doing the work.

(b) The schedule is optimal with no idle time.

(c) The total idle time in the schedule.

Chapter 9

WALKING

A. Fibonacci Numbers

1. (a) $F_{10} = 55$

(b) $F_{10} + 2 = 55 + 2 = 57$

(c) $F_{10+2} = F_{12} = 144$

(d) $F_{10} - 8 = 55 - 8 = 47$

(e) $F_{10-8} = F_2 = 1$

(f) $3F_4 = 3 \times 3 = 9$

(g) $F_{3 \times 4} = F_{12} = 144$

3. (a) Add 1 to three times the Nth Fibonacci number.

(b) Three times the $(N + 1)$st Fibonacci number.

(c) Add 1 to the Fibonacci number in position $3N$.

(d) The Fibonacci number in position $(3N + 1)$.

5. (a) $F_{38} = F_{37} + F_{36} = 24,157,817 + 14,930,352 = 39,088,169$

(b) $F_{35} = F_{37} - F_{36} = 24,157,817 - 14,930,352 = 9,227,465$

7. I. $F_{N+2} = F_{N+1} + F_N$ is an equivalent way to express the fact that each term of the Fibonacci sequence is equal to the sum of the two preceding terms.

9. **(a)** $47 = 13 + 34$ **(b)** $48 = 1 + 13 + 34$
   (c) $207 = 8 + 55 + 144$ **(d)** $210 = 3 + 8 + 55 + 144$

11. **(a)** $1 + 2 + 5 + 13 + 34 + 89 = 144$
    (b) 22

13. **(a)** $(1 + 1 + 2 + 3 + 5 + 8) + 1 = 21; (1 + 1 + 2 + 3 + 5 + 8 + 13) + 1 = 34$
    (b) 10
    (c) $N + 2$

15. **(a)** $2(F_3) - F_4 = 2(2) - 3 = 1 = F_1$
    (b) $2(F_6) - F_7 = 2(8) - 13 = 3 = F_4$
    (c) $2F_{N+2} - F_{N+3} = F_N$

B. The Golden Ratio

17. **(a)** 46.97871
    (b) 46.97871
    (c) 21

19. **(a)** 21
    (b) 34
    (c) 13

21. **(a)** $38 + 17\sqrt{5}$
    (b) $161 + 72\sqrt{5}$

23. 1.394×10^{104}

25. **(a)** 169
    (b) $\dfrac{169}{70} \approx 2.41429$
    (c) $\dfrac{5741}{2378} \approx 2.41421$
    (d) 2.41421

C. Fibonacci Numbers and Quadratic Equations

27. **(a)** $x = 1 + \sqrt{2} \approx 2.41421, x = 1 - \sqrt{2} \approx -0.41421$
    (b) $1 + \sqrt{2}$ is, to five decimal places, the same as in 25(d).

29. $x = -1, x = \frac{8}{3} \approx 2.66667$

31. **(a)** $x = 1$
    (b) $x = \frac{34}{55} - 1 = -\frac{21}{55} \approx -0.38182$

33. **(a)** Putting $x = 1$ in the equation gives $F_N = F_{N-1} + F_{N-2}$, which is true since the F's are Fibonacci numbers.
    (b) The sum of the roots of the equation is $-\dfrac{-F_{N-1}}{F_N} = \dfrac{F_{N-1}}{F_N}$ and so the other root is $\dfrac{F_{N-1}}{F_N} - 1.$

D. Gnomons and Similarity

35. **(a)** 156 m
    (b) 2880 sq. m

37. $c = 24$

39. 20 by 30

41. $x = 4$

43. $x = 12, y = 10$

45. 10 by approximately 6.18

47. The ratio of the longer side to the shorter side must be $1 + \sqrt{2}$.

49. (a) III

    (b) II

    (c) I

    (d) I

JOGGING

51. $A_N = 5F_N$

53. (a) $T_1 = aF_2 + bF_1 = a + b$

    (b) $T_2 = aF_3 + bF_2 = 2a + b$

    (c) $T_N = aF_{N+1} + bF_N$

$$= a(F_N + F_{N-1}) + b(F_{N-1} + F_{N-2})$$
$$= (aF_N + bF_{N-1}) + (aF_{N-1} + bF_{N-2})$$
$$= T_{N-1} + T_{N-2}$$

55. $x = 6, y = 12, z = 10$

57. $x = 3, y = 5$

59. (a) Φ

    (b) Φ

61. We must have $\frac{b+y}{b} = \frac{h+x}{h}$ or, equivalently, $1 + \frac{y}{b} = 1 + \frac{x}{h}$. This gives $\frac{y}{b} = \frac{x}{h}$ or, equivalently, $\frac{y}{x} = \frac{b}{h}$.

63. (a) Since we are given that $AB = BC = 1$, we know that $\angle BAC = 72°$ and so $\angle BAD = 180° - 72° = 108°$. This makes $\angle ABD = 180° - 108° - 36° = 36°$ and so $\triangle ABD$ is isosceles with $AD = AB = 1$. Therefore, $AC = x - 1$. Using these facts and the similarity of $\triangle ABC$ and $\triangle BCD$, we have $\frac{x}{1} = \frac{1}{x-1}$ or, $x^2 = x + 1$ for which we know the solution is $x = \Phi$.

    (b) $36° - 36° - 108°$

    (c) $\dfrac{\text{longer side}}{\text{shorter side}} = \dfrac{x}{1} = x = \Phi$

65. (a) $\dfrac{10}{\Phi}$

    (b) $\dfrac{10r}{\Phi}$

Chapter 10

WALKING

A. Linear Growth and Arithmetic Sequences

1. (a) $P_1 = 205, P_2 = 330, P_3 = 455$

    (b) $P_{100} = 12.580$

    (c) $P_N = 80 + 125N$

3. (a) $P_{30} = 225$

    (b) 185

    (c) 186

5. (a) $d = 3$

    (b) $P_{50} = 158$

    (c) $P_N = 8 + 3N$

7. (a) $A_3 = -19$

    (b) $A_0 = 26$

    (c) None. The sequence is decreasing and starts at 26.

9. (a) $P_N = P_{N-1} + 5; P_0 = 3$

    (b) $P_N = 3 + 5N$

    (c) $P_{300} = 1503$

11. 24,950

13. 16,050

15. (a) 3.519.500

    (b) 3.482.550

17. (a) 213

    (b) $137 + 2N$

    (c) $7124

    (d) $2652

B. Exponential Growth and Geometric Sequences

19. $\approx \$4587.64$

21. $\approx \$1874.53$

23. (a) $\approx \$9083.48$

    (b) $\approx 12.6825\%$

25. The Great Bulldog Bank: 6%; The First Northern Bank: $\approx 5.9\%$; The Bank of Wonderland: $\approx 5.65\%$.

27. (a) $\approx \$10,834.71$

    (b) $\approx \$11,338.09$

    (c) $\approx \$10,736.64$

29. (a) $P_1 = 13.75$

    (b) $P_9 \approx 81.956$

    (c) $P_N = 11 \times 1.25^N$

31. (a) $P_{100} = 3 \times 2^{100}$

    (b) $P_N = 3 \times 2^N$

    (c) $3 \times (2^{101} - 1)$

    (d) $3 \times 2^{101} - 3 \times 2^{50} = 3 \times 2^{50} \times (2^{51} - 1)$

33. $\approx \$1133.56$

35. (a) $\approx \$6209.21$

    (b) $\approx \$6102.71$

    (c) $\approx \$6077.89$

C. Logistic Growth Model

37. (a) $p_1 = 0.1680$

    (b) $p_2 \approx 0.1118$

    (c) 7.945%

39. (a) $p_1 = 0.1680, p_2 \approx 0.1118, p_3 \approx 0.0795, p_4 \approx 0.0585, p_5 \approx 0.0441,$
        $p_6 \approx 0.0337, p_7 \approx 0.0261, p_8 \approx 0.0203, p_9 \approx 0.0159, p_{10} \approx 0.0125$

    (b) extinction

41. (a) $p_1 = 0.4320, p_2 \approx 0.4417, p_3 \approx 0.4439, p_4 \approx 0.4443, p_5 \approx 0.4444,$
        $p_6 \approx 0.4444, p_7 \approx 0.4444, p_8 \approx , 0.4444, p_9 \approx 0.4444, p_{10} \approx 0.4444$

    (b) The population becomes stable at 44.44% of the habitat's carrying capacity.

43. (a) $p_1 = 0.3570$, $p_2 \approx 0.6427$, $p_3 \approx 0.6429$, $p_4 \approx 0.6428$, $p_5 \approx 0.6429$,
$p_6 \approx 0.6428$, $p_7 \approx 0.6429$, $p_8 \approx 0.6428$, $p_9 \approx 0.6429$, $p_{10} \approx 0.6428$

(b) The population becomes stable at $\frac{9}{14} \approx 64.29\%$ of the habitat's carrying capacity.

45. (a) $p_1 = 0.5200$, $p_2 = 0.8112$, $p_3 = 0.4978$, $p_4 = 0.8125$, $p_5 = 0.4952$,
$p_6 = 0.8124$, $p_7 = 0.4953$, $p_8 = 0.8124$, $p_9 = 0.4953$, $p_{10} = 0.8124$

(b) The population settles into a two-period cycle alternating between a high-population period at 81.24% and a low-population period at 49.53% of the habitat's carrying capacity.

47. (a) exponential   **(b)** linear   **(c)** logistic   **(d)** exponential
    (e) logistic   **(f)** linear   **(g)** linear, exponential, or logistic (they all apply!)

D. Miscellaneous

49. (a) $P_2 = 22$

(b) $P_3 = 42$

(c) The sum and product of two even numbers is even.

51. (a) $40.50

(b) 40.5%

(c) 40.5%

53. 39.15%

JOGGING

55. 6%

57. The only right triangles having sides in an arithmetic sequence are $3, 4, 5$ triangles.

59. 100%

61. $10,737,418.24

63. $P_0 = 8$, $\quad r = \frac{1}{2}$

65. No. A constant population implies $p_0 = p_1$, i.e., $p_0 - 0.8(1 - p_0)p_0$. The only solutions to the preceding equation are $p_0 = 0$ and $p_0 = -0.25$, neither of which is possible.

67. $\approx 14,619$ snails

69. 6425

71. $(a + ar + ar^2 + \cdots + ar^{N-1})(r - 1) = ar + ar^2 + ar^3 + \cdots + ar^{N-1} + ar^N - a - ar - ar^2 - \cdots - ar^{N-1}$
$$= -a + ar^N$$
$$= a(r^N - 1)$$

So,

$$a + ar + ar^2 + \cdots + ar^{N-1} = \frac{a(r^N - 1)}{r - 1}$$

Chapter 11

WALKING

A. Reflections

1. (a) C

(b) F

(c) E

(d) B

3.

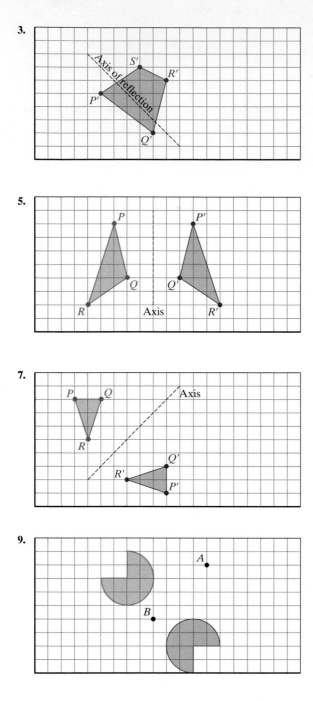

5.

7.

9.

B. Rotations

11. **(a)** *I*
  (b) *E*
  (c) *G*
  (d) *A*
  (e) *F*
  (f) *C*

13. **(a)** 110°

    (b) 350°

    (c) 10°

    (d) 81°

15.

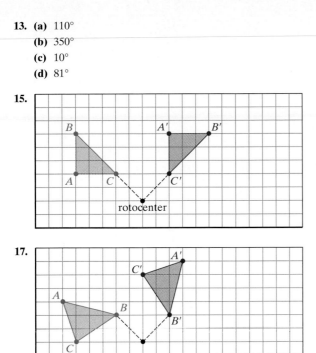

C. Translations

19. **(a)** *C*

    (b) *C*

    (c) *A*

    (d) *D*

21.

D. Glide Reflections

23.

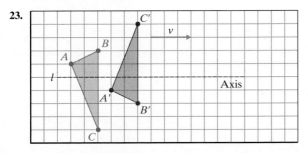

25.

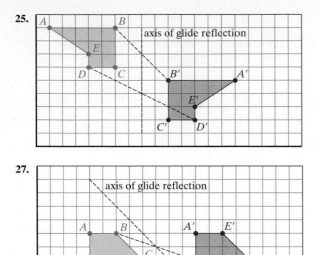

27.

E. Symmetries of Finite Shapes

29. (a) Reflection with axis going through the midpoints of *AB* and *DC*; reflection with axis going through the midpoints of *AD* and *BC*; rotations of 180° and 360° with rotocenter the center of the rectangle.

   (b) No reflections. Rotations of 180° and 360° with rotocenter the center of the parallelogram.

   (c) Reflection with axis going through the midpoints of *AB* and *DC*; rotation of 180° with rotocenter the center of the trapezoid.

31. (a) Reflections (three of them) with axis going through pairs of opposite vertices; reflections (three of them) with axis going through the midpoints of opposite sides of the hexagon; rotations of 60°, 120°, 180°, 240°, 300°, 360° with rotocenter the center of the hexagon.

   (b) Reflections with axis *AD, GJ, BE, HK, CF, IL*; rotations of 60°, 120°, 180°, 240°, 300°, 360° with rotocenter the center of the star.

33. (a) D_2 **(b)** Z_2 **(c)** D_1

35. (a) D_6 **(b)** D_6

37. (a) D_1 **(b)** D_1 **(c)** Z_1

   (d) Z_2 **(e)** Z_1

39. (a) J   **(b)** T

   (c) Z   **(d)** I

41. (a) Symmetry type D_5 is common among many types of flowers (daisies, geraniums, etc.). The only requirements are that the flower have 5 equal, evenly spaced petals and that the petals have a reflection symmetry along their long axis. In the animal world, symmetry type D_5 is less common, but it can be found among certain types of starfish, sand dollars, and in some single-celled organisms called diatoms.

(b) The Chrysler Corporation logo is a classic example of a shape with symmetry D_5. Symmetry type D_5 is also common in automobile wheels and hubcaps. One of the largest and most unusual buildings in Washington, D.C. has symmetry of type D_5.

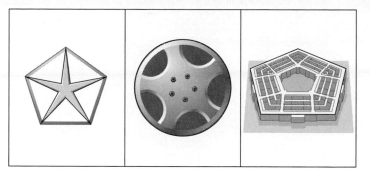

43. (a) Objects with symmetry type Z_1 are those whose only symmetry is the identity. Thus, any "irregular" shape fits the bill. Tree leaves seashells, plants, and rocks more often than not have symmetry type Z_1.

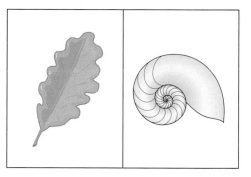

(b) Examples of man-made objects with symmetry of type Z_1 abound.

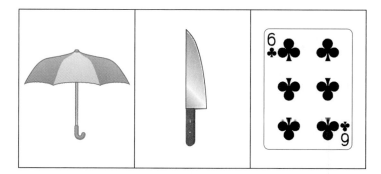

F. Symmetries of Border Patterns

45. (a) $m1$
  (b) $1m$
  (c) 12
  (d) 11

47. (a) mg
  (b) 12
  (c) $1g$
  (d) mg

G. Miscellaneous

49. Since every proper rigid motion is equivalent to either a rotation or a translation, and a translation has no fixed points, the specified rigid motion must be equivalent to a rotation.

51. **(a)** *D*
  (b) *D*
  (c) *B*
  (d) *E*
  (e) *G*

53. **(a)** improper
  (b) proper
  (c) improper
  (d) proper

55. The combination of two improper rigid motions is a proper rigid motion. Since *C* is a fixed point, the rigid motion must be a rotation with rotocenter *C*.

JOGGING

57. **(a)** The result of applying the reflection with axis l_1 followed by the reflection with axis l_2 is a clockwise rotation with center *C* and angle of rotation 2α.
  (b) The result of applying the reflection with axis l_2 followed by the reflection with axis l_1 is a counterclockwise rotation with center *C* and angle of rotation 2α.

59. **(a)**

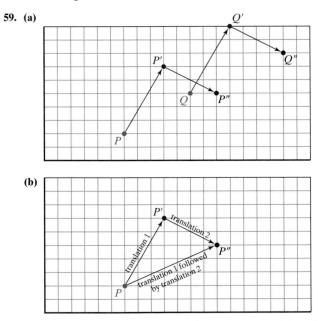

61. **(a)** By definition, a border pattern has translation symmetries in exactly one direction (let's assume the horizontal direction). If the pattern had a reflection symmetry along an axis forming 45° with the horizontal direction, there would have to be a second direction of translation symmetry (vertical).
  (b) If a pattern had a reflection symmetry along an axis forming an angle of $\alpha°$ with the horizontal direction, it would have to have translation symmetry in a direction that forms an angle of $2\alpha°$ with the horizontal. This could only happen for $\alpha = 90°$ or $\alpha = 180°$ (since the only allowable direction for translation symmetries is the horizontal).

63. **(a)** F⅃Ⴈ⅂FⴈႱႱ
  (b) FⴈF⅂Fⴈ
  (c) FⴈFⴈFⴈ
  (d) FFFFFF

65. A reflection is an improper rigid motion and hence reverses the left-right orientation. The propeller blades have a flat edge on the right side (facing a blade). After a reflection the flat edge will be on the left side.

67. Rotations and translations are proper rigid motions, and hence preserve clockwise-counterclockwise orientations. The given motion is an improper rigid motion (it reverses the clockwise-counterclockwise orientation). If the rigid motion was a reflection, then PP', RR', and QQ' would all be perpendicular to the axis of reflection and hence would all be parallel. By default, the rigid motion must be a glide reflection.

Chapter 12

WALKING

A. The Koch Snowflake and Variations

1.

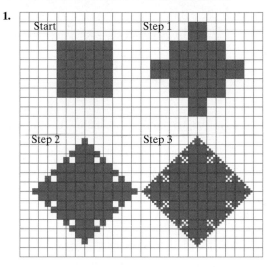

3. (a) $\frac{5}{3} \cdot 4a$; $\left(\frac{5}{3}\right)^2 \cdot 4a$; $\left(\frac{5}{3}\right)^3 \cdot 4a$

    (b) $\left(\frac{5}{3}\right)^N \cdot 4a$

    (c) infinite

5. (a)

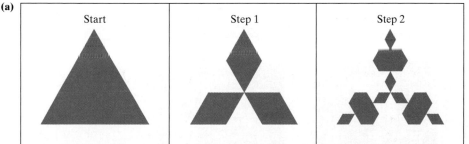

    (b) **Step 1:** $4 \times 3 = 12$

          Step 2: $4^2 \times 3 = 48$

          Step 3: $4^3 \times 3 = 192$

          Step 4: $4^4 \times 3 = 768$

    (c) $4^N \times 3$

7. (a) 3; $4 \times 3 = 12$; $4^2 \times 3 = 48$

    (b) $4^{N-1} \times 3$

9.

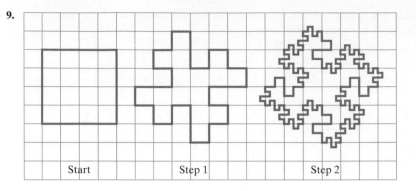

Start   Step 1   Step 2

11. (a) $2P$; $4P$; $8P$

   (b) $2^N P$

   (c) infinite

B. The Sierpinski Gasket and Variations

13. (a) 6

   (b) 1.5 (Triangle $M_1 A M_3$ is similar to triangle BAC, with scaling factor one-half. The area of $M_1 A M_3$ is one-fourth the area of BAC.)

   (c) 1.5

   (d) 4.5

15.

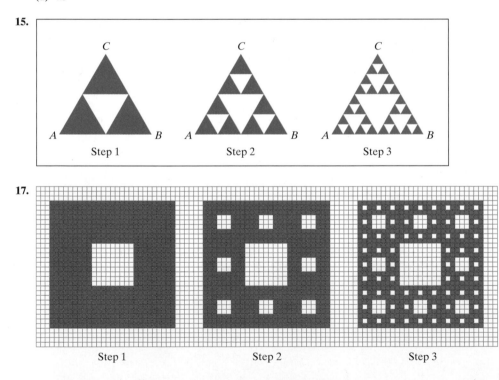

17.

Step 1   Step 2   Step 3

19. (a) Step 1: $4 + \frac{4}{3} = \frac{16}{3}$ [Outside boundary plus boundary of middle hole (with sides of length $\frac{1}{3}$).]

     Step 2: $\frac{16}{3} + 8 \cdot (\frac{4}{9}) = \frac{80}{9}$ [Previous boundary plus boundary of eight new holes (each with sides of length $\frac{1}{9}$).]

     Step 3: $\frac{80}{9} + 64 \cdot (\frac{4}{27}) = \frac{496}{27}$ [Previous boundary plus boundary of $8^2 = 64$ new holes (each with sides of length $\frac{1}{27}$).]

     Step 4: $\frac{496}{27} + 512 \cdot (\frac{4}{81}) = \frac{3536}{81}$ [Previous boundary plus boundary of $8^3 = 512$ new holes (each with sides of length $\frac{1}{81}$).]

   (b) $L + 8^N \left(\dfrac{4}{3^{N+1}} \right)$ [In step $(N + 1)$, 8^N new square holes are added to the figure in step N, each with sides of length $\dfrac{1}{3^{N+1}}$ for a total

increase in the boundary of $8^N \cdot \left(\dfrac{4}{3^{N+1}} \right)$.]

21.

	Start	Step 1	Step 2	Step 3	. . .	Step N
Number of solid triangles	1	6	$6^2 = 36$	$6^3 = 216$		6^N

23.

	Start	Step 1	Step 2	Step 3
Area of the figure	A	$\left(\frac{2}{3}\right)A$	$\left(\frac{2}{3}\right)^2 A = \frac{4}{9}A$	$\left(\frac{2}{3}\right)^3 A = \frac{8}{27}A$

C. The Chaos Game and Variations

25.

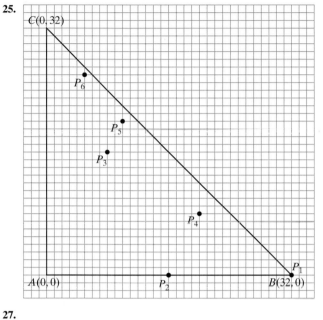

27.

Number rolled	3	1	2	3	5	5
Point	P_1	P_2	P_3	P_4	P_5	P_6
Coordinates	(32,0)	(16,0)	(8,0)	(20,0)	(10,16)	(5,24)

29. (a)   **(b)**

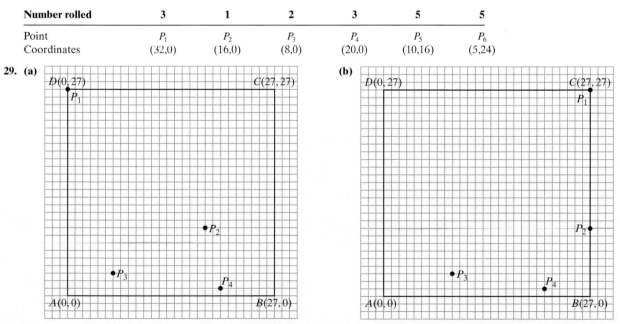

(c)

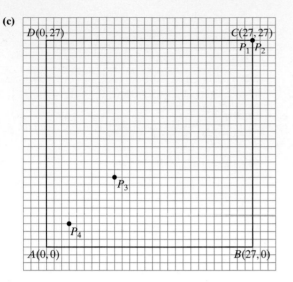

31. (a)

Number rolled	4	2	1	2
Point	P_1	P_2	P_3	P_4
Coordinates	(0,27)	(18,9)	(6,3)	(20,1)

(b)

Number rolled	3	1	1	3
Point	P_1	P_2	P_3	P_4
Coordinates	(27,27)	(9,9)	(3,3)	(19,19)

(c)

Number rolled	1	3	4	2
Point	P_1	P_2	P_3	P_4
Coordinates	(0,0)	(18,18)	(6,24)	(20,8)

D. Mandelbrot Sequences

33. (a) $2, 2, 2, 2, 2$

   (b) $s_{100} = 2$

   (c) Attracted to the number 2.

35. (a) $s_1 = -0.25$, $s_2 = -0.4375$, $s_3 = -0.3086$, $s_4 = -0.4048$, $s_5 = -0.3362$

   (b) -0.3360

   (c) Attracted to the number -0.3360 (rounded to four decimal places).

37. (a) $s_1 = \frac{3}{4}$, $s_2 = \frac{17}{16}$, $s_3 = \frac{417}{256}$

   (b) If $s_N > 1$, then $(s_N)^2 > s_N$ and thus $(s_N)^2 + \frac{1}{2} = s_{N+1} > s_N + \frac{1}{2} > s_N$.

   (c) The sequence is escaping. [From (b), each term is more than the preceding term plus $\frac{1}{2}$.]

39. (a) $s = 2$

   (b) $N = 1$

JOGGING

41. (a) $\left(\frac{3}{4}\right)^N$

   (b) $0.0563, 0.0032, 0.0000$

   (c) $\left(\frac{3}{4}\right)^N$ gets closer and closer to 0 as N gets bigger and bigger.

43. There will be infinitely many points left. For example, the 3 vertices of the original triangle will be left as well as the vertices of every black triangle that occurs at each step of the construction.

45.

	Start	Step 1	Step 2	Step 3	. . .	Step N
Number of cubes removed	0	7	20×7	$20^2 \times 7$		$20^{N-1} \times 7$

47. **(a)** Reflection with axis the line passing through A and perpendicular to BC; reflection with axis the line passing through B and perpendicular to AC; reflection with axis the line passing through C and perpendicular to AB.

(b) Rotations of $120°, 240°, 360°$ with rotocenter the center of the equilateral triangle ABC.

(c) D_3

49. D_4

51. The sequence is attracted to $x = -0.5$, one of the two solutions of the equation $x^2 - 0.75 = x$.

53. The first twenty steps: 0.3125, 0.34765625, 0.370864868, 0.38754075, 0.400187833, 0.410150302, 0.41822327, 0.424910704, 0.430549106, 0.435372533, 0.439549242, 0.443203536, 0.446429375, 0.449299187, 0.451869759, 0.454186279, 0.456285176, 0.458196162, 0.459943723, 0.461548228. Step 99: 0.49060422. step 100: 0.490692501. This sequence is attracted to 0.5, a solution of the equation $x^2 + 0.25 = x$.

55. Step 1: $2 + \sqrt{2}$, step 2: $6 + 5\sqrt{2}$, step 3: $86 + 61\sqrt{2}$. The sequence is escaping.

Chapter 13

WALKING

A. Surveys and Public Opinion Polls

1. The gumballs in the jar.

3. 64

5. **(a)** The 680 registered voters polled by telephone.

(b) Approximately 8.2% $(680/8325 \approx 0.082)$.

7. 3%

9. **(a)** All married people who read Dear Abby's column.

(b) Abby's target population appears to be all married people. However, she is only sampling from those married couples that read her column.

(c) sclf-selection

(d) 85% is a statistic, since it is based on data taken from a sample.

11. **(a)** 74.0%

(b) 81.8%

(c) Not very accurate. The sample was far from being representative of the entire population.

13. **(a)** The citizens of Cleansburg.

(b) The sampling frame is limited to that part of the target population that pass by a city street corner between 4 and 6 P.M.

15. **(a)** The choice of street corner could make a great deal of difference in the responses collected.

(b) D. (We are making the assumption that people who live or work downtown are much more likely to answer yes than people in other parts of town.)

(c) Yes, for two main reasons: (i) People out on the street between 4 P.M. and 6 P.M. are not representative of the population at large. For example, office and white collar workers are much more likely to be in the sample than homemakers and school teachers. (ii) The five street corners were chosen by the interviewers and the passersby are unlikely to represent a cross section of the city.

(d) No. No attempt was made to use quotas to get a representative cross section of the population.

17. **(a)** Assuming that the registrar has a complete list of the 15,000 undergraduates at TSU, the target population and the sampling frame both consist of all undergraduates at Tasmania State University.

(b) $N = 15,000$

19. (a) In simple random sampling, any two members of the population have as much chance of both being in the sample as any other two. But in this sample, two people with the same last name—say Len Euler and Linda Euler—have no chance of both being in the sample.

(b) Sampling variability. The students sampled appear to be a fair cross section of all TSU undergraduates that would attempt to enroll in Math 101.

21. (a) Stratified sampling

(b) Quota sampling

B. The Capture-Recapture Method

23. 2000

$$N = \frac{n_2}{k} \cdot n_1 = \frac{120}{30} \times 500 = 2000$$

25. 84 quarters. (*Note*: To estimate the number of quarters, we disregard the nickels and dimes—they are irrelevant. Thus, $n_1 = 12, n_2 = 28$, and $k = 4$.)

27. 124 dimes. (*Note*: We disregard quarters and nickels—they are irrelevant. Thus, $n_1 = 23, n_2 = 43$, and $k = 8$. This gives $N = 123.625$, which should be rounded to the nearest integer.)

C. Clinical Studies

29. (a) Anyone who could have a cold and would consider buying vitamin X (i.e., pretty much all adults).

(b) The sampling frame is only a small portion of the target population. It only consists of college students in the San Diego area who are suffering from colds.

(c) Yes. This sample would likely underrepresent older adults and those living in colder climates.

(d) No. There was no control group.

31. (i) Using college students. (College students are not a representative cross section of the population in terms of age and therefore in terms of how they would respond to the treatment.)

(ii) Using subjects only from the San Diego area.

(iii) Offering money as an incentive to participate.

(iv) Allowing self-reporting (the subjects themselves determine when their colds are over) is a very unreliable way to collect data and is especially bad when the subjects are paid volunteers.

33. Though the sampling frame consists only of male doctors, the target population appears to be anyone who could potentially suffer from arteriosclerosis (clogging of the arteries).

35. (a) There was a treatment group (the ones getting the beta-carotene pill) and there was a control group. The control group received a placebo pill. These two elements make it a controlled placebo experiment.

(b) The group that received the beta-carotene pills.

(c) Both the treatment and control groups were chosen by random selection.

37. All potential knee surgery patients.

39. (a) Yes, there was one control group receiving the sham surgery (placebo) and two treatment groups.

(b) The first treatment group consisted of those patients receiving arthroscopic debridement. The second treatment group consisted of those patients receiving arthroscopic lavage.

(c) Yes, since the 180 patients in the study were assigned to a treatment group or control group at random.

(d) This was a blind experiment. The doctors certainly knew which surgery they were performing on each patient.

41. The professor was conducting a clinical study because he was, after all, trying to establish the connection between a cause (10 milligrams of caffeine a day) and an effect (improved performance in college courses). Other than that, the experiment had little going for it: it was not controlled (no control group); not randomized (the subjects were chosen because of their poor grades); no placebo was used and consequently the study was not double-blind.

43. (a) It is likely that the study was blind but not double-blind since the professor knew who was in the study.

(b) (i) A regular visit to the professor's office could in itself be a boost to a student's self-confidence and help improve his or her grades.

(ii) The "individualized tutoring" that took place during the office meetings could also be the reason for improved performance.

(iii) The students selected for the study all got F's on their first midterm making them likely candidates to show some improvement.

JOGGING

45. (a) (i) the entire sky; (ii) all the coffee in the cup; (iii) the entire Math 101 class.

(b) In none of the three examples is the sample random.

(c) (i) In some situations one can have a good idea as to whether it will rain or not by seeing only a small section of the sky, but in many other situations rain clouds can be patchy and one might draw the wrong conclusions by just peeking out the window. (ii) If the coffee is burning hot on top, it is likely to be pretty hot throughout, so Betty's conclusion is likely to be valid. (iii) Since Carla used convenience sampling and those students sitting next to her are not likely to be a representative sample, her conclusion is likely to be invalid.

47. (a) The question was worded in a way that made it almost impossible to answer yes.

(b) "Will you support some form of tax increase if it can be proven to you that such a tax increase is justified?" is better, but still not neutral. "Do you support or oppose some form of tax increase?" is bland but probably as neutral as one can get.

49. (a) Under method 1, people whose phone numbers are unlisted are automatically ruled out from the sample. At the same time, method 1 is cheaper and easier to implement than method 2.

(b) For this particular situation, method 2 is likely to produce much more reliable data than method 1. The two main reasons are as follows: (i) People with unlisted phone numbers are very likely to be the same kind of people that would seriously consider buying a burglar alarm, and (ii) the listing bias is more likely to be significant in a place like New York City. (People with unlisted phone numbers make up a much higher percentage of the population in a large city such as New York than in a small town or rural area. Interestingly enough, the largest percentage of unlisted phone numbers for any American city is in Las Vegas, Nevada.)

51. (a) Both samples should be a representative cross section of the same population. In particular, it is essential that the first sample, after being released, be allowed to disperse evenly throughout the population, and that the population should not change between the time of the capture and the time of the recapture.

(b) It is possible (especially when dealing with elusive types of animals) that the very fact that the animals in the first sample allowed themselves to be captured makes such a sample biased (they could represent a slower, less cunning group). This type of bias is compounded with the animals that get captured the second time around. A second problem is the effect that the first capture can have on the captured animals. Sometimes the animal may be hurt (physically or emotionally), making it more (or less) likely to be captured the second time around. A third source of bias is the possibility that some of the tags will come off.

Chapter 14

WALKING

A. Frequency Tables, Bar Graphs, and Pie Charts

1.

Score	10	50	60	70	80	100
Frequency	1	3	7	6	5	2

3. (a)

Grade	A	B	C	D	F
Frequency	7	6	7	3	1

(b)

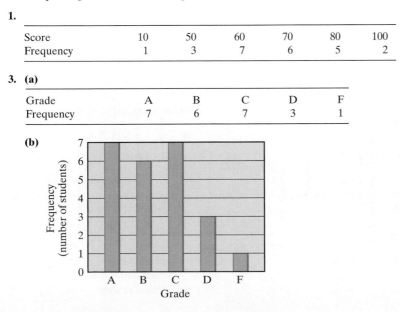

5. (a) 80   **(b)** 30%

(c)

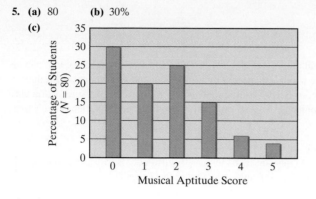

7.

Distance (miles) to school	0.0	0.5	1.0	1.5	2.0	2.5	3.0	5.0	8.5
Frequency	5	3	4	6	3	2	1	1	1

9. (a)

Class interval	Very close	Close	Nearby	Not too far	Far
Frequency	8	10	5	1	2

(b)

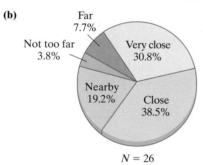

$N = 26$

11. (a) 30

(b) 0%

(c) approximately 56.7%

13. Asian: 40°; Hispanic: 68°; African American: 86°; Caucasian: 140°; Other: 25°.

15. (a) 132   **(b)** 9

17.

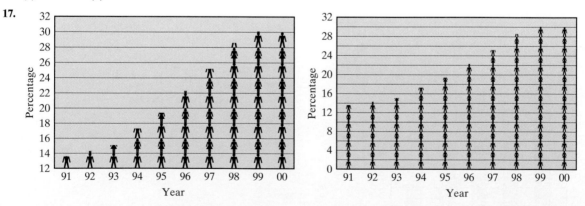

B. Histograms

19. (a) 12 ounces

(b) The third class interval: "more than 72 ounces and less than or equal to 84 ounces." Values that fall exactly on the boundary between two class intervals belong to the class interval to the left.

21.

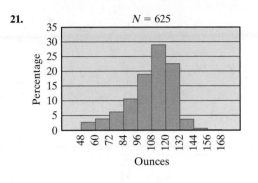

C. Averages and Medians

23. (a) 2

(b) 3

(c) average = 2, median = 2.5

25. (a) average = 4.5, median = 4.5

(b) average = 5, median = 5

(c) average = 5.5, median = 5.5

27. (a) 50

(b) 50

29. (a) 1.5875

(b) 1.5

31. (a) 6.77

(b) 7

D. Percentiles and Quartiles

33. (a) −3

(b) 7

(c) $Q_1 = -3, Q_3 = 7$

35. (a) 75th percentile = 75.5, 90th percentile = 90.5

(b) 75th percentile = 75, 90th percentile = 90

(c) 75th percentile = 75, 90th percentile = 90

(d) 75th percentile = 74, 90th percentile = 89

37. (a) 29

(b) 32

(c) 37

39. (a) The 586,390th score, $d_{586,390}$.

(b) The 293,195th score, $d_{293,195}$.

(c) The 879,585th score, $d_{879,585}$.

E. Box Plots and Five-Number Summaries

41. (a) Min $= -6, Q_1 = -3, M = 3, Q_3 = 7$, Max $= 8$

(b)

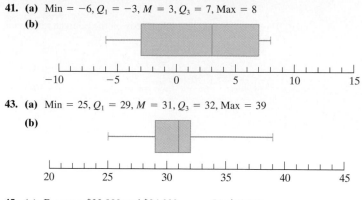

43. (a) Min $= 25, Q_1 = 29, M = 31, Q_3 = 32$, Max $= 39$

(b)

45. (a) Between \$33,000 and \$34,000   **(b)** \$40,000

(c) The line indicating the median salary in the engineering box plot is to the right of the box in the agriculture box plot.

F. Ranges and Interquartile Ranges

47. (a) 14

(b) 10

49. (a) \$41,000

(b) At least 171 homes

51. Outliers are 1 and 24.

53. There are no outliers.

G. Standard Deviations

55. (a) 0   **(b)** $\sqrt{\frac{50}{4}} = \frac{5\sqrt{2}}{2} \approx 3.5$ **(c)** $\sqrt{\frac{550}{4}} \approx 11.7$

57. (a) ≈ 2.87 **(b)** ≈ 2.87 **(c)** ≈ 2.87

H. Miscellaneous

59. 10

61. 4, 5, and 8

63. Caucasian

65. D, C, B, and A

JOGGING

67. 100

69. Ramon gets 85 out of 100 on each of the first four exams and 60 out of 100 on the fifth exam. Josh gets 80 out of 100 on all 5 of the exams.

71. (a) $\{1, 1, 1, 1, 6, 6, 6, 6, 6, 6\}$ (Average $= 4$; Median $= 6$)

(b) $\{1, 1, 1, 1, 1, 1, 6, 6, 6, 6\}$ (Average $= 3$; Median $= 1$)

(c) $\{1, 1, 6, 6, 6, 6, 6, 6, 6, 6\}$ (Average $= 5$; $Q_1 = 6$)

(d) $\{1, 1, 1, 1, 1, 1, 1, 1, 6, 6\}$ (Average $= 2$; $Q_3 = 1$)

73. (a) The five-number summary for the original scores was Min $= 1, Q_1 = 9, M = 11, Q_3 = 12$, and Max $= 24$. When 2 points are added to each test score, the five-number summary will also have 2 points added to each of its numbers (i.e., Min $= 3, Q_1 = 11$, $M = 13, Q_3 = 14$, and Max $= 26$).

(b) When 10% is added to each score (i.e., each score is multiplied by 1.1), then each number in the five-number summary will also be multiplied by 1.1 (i.e., Min $= 1.1, Q_1 = 9.9, M = 12.1, Q_3 = 13.2$, and Max $= 26.4$).

75. (a) 4   **(b)** 0.4   **(c)** 0.4

77. (a) Male: 10%, Female: 20%

    (b) Male: 80%, Female: 90%

    (c) The figures for both schools were combined. A total of 820 males were admitted out of a total of 1200 that applied—an admission rate for males of approximately 68.3%. Similarly, a total of 460 females were admitted out of a total of 900 that applied—an admission rate for females of approximately 51.1%.

    (d) In this example, females have a higher percentage ($\frac{100}{500} = 20\%$) than males ($\frac{20}{200} = 10\%$) for admissions to the School of Architecture and also a higher percentage ($\frac{360}{400} = 90\%$) than males ($\frac{800}{1000} = 80\%$) for admissions to the School of Engineering. When the numbers are combined, however, females have a lower percentage ($\frac{100 \ + \ 360}{500 \ + \ 400} \approx 51.1\%$) than males ($\frac{20 \ + \ 800}{200 \ + \ 1000} \approx 68.3\%$) in total admissions. The reason that this apparent paradox can occur is purely a matter of arithmetic: Just because $\frac{a_1}{a_2} > \frac{b_1}{b_2}$ and $\frac{c_1}{c_2} > \frac{d_1}{d_2}$, it does not necessarily follow that $\frac{a_1 + c_1}{a_2 + c_2} > \frac{b_1 + d_1}{b_2 + d_2}$.

79. (a) $1142/16 = 71.375$

    (b) Min $= 52, Q_1 = 67, M = 71.5, Q_3 = 75$, Max $= 90$

    (c) 75 exercises

    (d) $\sqrt{\dfrac{1245.75}{16}} \approx 8.8$ exercises

Chapter 15

WALKING

A. Random Experiments and Sample Spaces

1. (a) $\{HHH, HHT, HTH, THH, TTH, THT, HTT, TTT\}$

    (b) $\{0, 1, 2, 3\}$

    (c) $\{0, 1, 2, 3, 4, 5, 6, 7, 8, 9, 10\}$

3. (a) $\{ABCD, ABDC, ACBD, ACDB, ADBC, ADCB, BACD, BADC, BCAD, BCDA, BDAC, BDCA, CABD, CADB, CBAD, CBDA, CDAB,$ $CDBA, DABC, DACB, DBAC, DBCA, DCAB, DCBA\}$

    (b) $N = 24$

5. (a) Answers may vary. A typical outcome is a string of 10 letters each of which can be either an H or a T. An answer like $\{HHHHHHHHHH, \ldots, TTTTTTTTTT\}$ is not sufficiently descriptive. An answer like $\{ \ldots, HTTHHHTHTH, \ldots, TTHTHHTTHT,$ $\ldots, HHHTHTTHHT, \ldots \}$ is better. An answer like $\{X_1 \, X_2 \, X_3 \, X_4 \, X_5 \, X_6 \, X_7 \, X_8 \, X_9 \, X_{10}|$ each X_i is either H or $T \}$ is best.

    (b) $N = 1024$

7. (a) Answers will vary. An outcome is an ordered sequence of four numbers, each of which is an integer between 1 and 6. The best answer would be something like $\{(n_1, n_2, n_3, n_4): $ each n_i is $1, 2, 3, 4, 5,$ or $6\}$. An answer such as $\{(1,1,1,1), \ldots, (1,1,1,6), \ldots, (1,2,3,4), \ldots,$ $(3,2,6,2), \ldots, (4,3,1,5), \ldots, (6,6,6,6)\}$ showing a few typical outcomes is possible, but not as good. An answer like $\{(1,1,1,1), \ldots, (2,2,2,2), \ldots, (6,6,6,6)\}$ is not descriptive enough.

    (b) $N = 6^4 = 1296$

9. (a) Answers will vary. An outcome is an ordered sequence of four letters A through J with no repeated letters. The best answer would be something like $\{X_1X_2X_3X_4|$ each X_i is a different letter (A through J)$\}$. An answer such as $\{ABCD, \ldots, AGEB, \ldots, BDAC, \ldots,$ EDAH, $\ldots$, GHIJ, $\ldots$, JIHG$\}$ showing a few typical outcomes is acceptable. An answer like $\{ABCD, \ldots, JIHG\}$ is not descriptive enough.

    (b) $N = 10 \times 9 \times 8 \times 7 = 5040$

B. The Multiplication Rule

11. (a) $9 \times 26^3 \times 10^3 = 158,184,000$

    (b) $1 \times 26^3 \times 10^2 \times 1 = 1,757,600$

    (c) $9 \times 26 \times 25 \times 24 \times 9 \times 8 \times 7 = 70,761,600$

13. (a) $52^4 \times 10 = 73,116,160$

    (b) $52^3 \times 10 = 1,406,080$

    (c) $50 \times 52^3 \times 10 = 70,304,000$

    (d) $50^4 \times 10 = 62,500,000$

15. (a) $8! = 40,320$

    (b) $40,319$

17. (a) $35 \times 34 \times 33 = 39,270$

    (b) $15 \times 34 \times 33 = 16,830$

    (c) $35 \times 34 \times 33 - (15 \times 14 \times 13 + 20 \times 19 \times 18) = 29,700$

        The total number of all-female committees is $15 \times 14 \times 13 = 2730$.

        The total number of all-male committees is $20 \times 19 \times 18 = 6840$.

        The remaining 29,700 committees are mixed.

19. (a) $4,500,000$ **(b)** $1,800,000$ **(c)** $360,000$

C. Permutations and Combinations

21. (a) $_{15}P_4$ **(b)** $_{15}C_4$

23. (a) $_{20}C_2$ **(b)** $_{20}P_2$ **(c)** $_{20}C_5$

25. (a) $10 \times 9 = 90$ **(b)** $90/2 = 45$

    (c) $10 \times 9 \times 8 = 720$ **(d)** $720/6 = 120$

27. (a) 10 **(b)** $(10 \times 9)/2 = 45$

    (c) 100 **(d)** $(100 \times 99)/2 = 4950$

29. (a) $(20 \times 19)/2 = 190$ **(b)** 190

    (c) $(20 \times 19 \times 18)/(3 \times 2 \times 1) = 1140$

    (d) 1140

31. (a) $1 + 3 + 3 + 1 = 8$

    (b) $1 + 4 + 6 + 4 + 1 = 16$

    (c) $1 + 5 + 10 + 10 + 5 + 1 = 32$

    (d) 1024

33. (a) $\approx 6.7 \times 10^{11}$ **(b)** $\approx 1.26 \times 10^{14}$ **(c)** $\approx 1.26 \times 10^{14}$

    (d) $\approx 3.76 \times 10^{48}$ **(e)** Answer is too large.

D. General Probability Spaces

35. (a) 0.18 **(b)** 0.27

    (c) $\Pr(o_1) = 0.22, \Pr(o_2) = 0.24, \Pr(o_3) = 0.27, \Pr(o_4) = 0.17, \Pr(o_5) = 0.1$

37. $S = \{o_1, o_2, o_3, o_4, o_5, o_6, o_7\}$

    $\Pr(o_1) = \frac{2}{8}, \Pr(o_2) = \Pr(o_3) = \Pr(o_4) = \Pr(o_5) = \Pr(o_6) = \Pr(o_7) = \frac{1}{8}$

39. (a) 0.3

    (b) $\{$red, blue, white, green, yellow$\}$

    (c) $\Pr(\text{red}) = 0.3, \Pr(\text{blue}) = \Pr(\text{white}) = 0.2, \Pr(\text{green}) = \Pr(\text{yellow}) = 0.15$

E. Events

41. (a) $E_1 = \{HHT, HTH, THH\}$ **(b)** $E_2 = \{HHH, TTT\}$

    (c) $E_3 = \{\} = \varnothing$ **(d)** $E_4 = \{TTH, TTT\}$

43. (a) $E_1 = \{(1,1), (2,2), (3,3), (4,4), (5,5), (6,6)\}$

    (b) $E_2 = \{(1,1), (1,2), (2,1)\}$

    (c) $E_3 = \{(2,6), (3,5), (4,4), (5,3), (6,2), (3,6), (4,5), (5,4), (6,3), (4,6), (5,5), (6,4), (5,6), (6,5), (6,6)\}$

45. (a) $E_1 = \{HHHHHHHHH\}$

    (b) $E_2 = \{HHHHHHHHT, HHHHHHHTH, HHHHHHHTHH, HHHHHHTHHH, HHHHHTHHHH, HHHHTHHHHH,$
        $HHHTHHHHHH, HHTHHHHHHH, HTHHHHHHHH, THHHHHHHHH\}$

    (c) $E_3 = \{\} = \varnothing$

47. $\{\}, \{A\}, \{B\}, \{C\}, \{A, B\}, \{A, C\}, \{B, C\}, \{A, B, C\}$

F. Probability Spaces with Equally Likely Outcomes

49. (a) $\frac{3}{8} = 0.375$ **(b)** $\frac{2}{8} = 0.25$ **(c)** 0   **(d)** $\frac{2}{8} = 0.25$

51. (a) $\frac{6}{36}$ **(b)** $\frac{3}{36}$ **(c)** $\frac{15}{36}$

53. (a) $\frac{1}{1024} \approx 0.001$ **(b)** $\frac{10}{1024} \approx 0.01$ **(c)** 0

55. (a) $\frac{5}{36}$ **(b)** $\frac{31}{36}$ **(c)** $\frac{9}{36}$ **(d)** $\frac{15}{36}$

57. (a) $\frac{1}{10} = 0.1$ **(b)** $\frac{4}{10} = 0.4$ **(c)** $\frac{6}{10} = 0.6$ **(d)** $\frac{1}{5040} \approx 0.0002$

59. (a) $\frac{4}{15} \approx 0.267$ **(b)** $\frac{11}{15} \approx 0.733$ **(c)** $\dfrac{1}{1365}$

G. Odds

61. (a) 4 to 3   **(b)** 3 to 2

63. (a) $\Pr(E) = \frac{3}{8}$ **(b)** $\Pr(E) = \frac{15}{23}$ **(c)** $\Pr(E) = \frac{1}{2}$

65. (a) $\{YYXXX, YXYXXX, YXXYXX, YXXXYX, XYYXXX, XYXYXX, XYXXYX, XXYYXX, XXYXYX, XXXYYX\}$

   (b) $\{YYXXXX, YXYXXX, YXXYXX, YXXXYX, XYYXXX, XYXYXX, XYXXYX, XXYYXX, XXYXYX, XXXYYX, XXYYYY,$
   $XYXYYY, XYYXYY, XYYYXY, YXXYYY, YXYXYY, YXYYXY, YYXXYY, YYXYXY, YYYXXY\}$

67. (a) 3,628,800
   (b) 362.880
   (c) 28,800
   (d) 2880

69. 20

71. (a) $\dfrac{{}_{20}C_{10}}{2^{20}} \approx 0.1762$ **(b)** $\dfrac{{}_{20}C_{3}}{2^{20}} \approx 0.001$ **(c)** $1 - \left(\dfrac{1 + {}_{20}C_1 + {}_{20}C_2}{2^{20}} \right) \approx 0.9998$

73. $\frac{253}{4998} \approx 0.05$

75. $\frac{1}{2548} \approx 0.00039$

77. $1 - \left(\frac{5}{6}\right)^5 \approx 0.6$

79. (a) $(0.98)^{12} \approx 0.78$. We are assuming that the events are independent.
   (b) $(0.98)^{12} + 12(0.02)(0.98)^{11} \approx 0.98$

Chapter 16

WALKING

A. Normal Curves

1. (a) 83 lb   **(b)** 83 lb   **(c)** 7 lb

3. (a) 75 in.   **(b)** 3 in.   **(c)** $Q_1 \approx 73$ in., $Q_3 \approx 77$ in. (For $\sigma = 3$, $0.675 \times \sigma = 2.025 \approx 2$)

5. (a) $Q_1 \approx 72.8$ lb   **(b)** $Q_3 \approx 89.6$ lb

7. 20 in.

9. $\mu \neq M$

11. $\mu - Q_1 = 110$; $Q_3 - \mu = 100$

B. Standardizing Data

13. (a) 1   **(b)** 1.6   **(c)** -2 **(d)** -1.8

15. (a) 0.22   **(b)** -1.82 **(c)** 0

17. (a) 152.3 ft   **(b)** 199.1 ft
   (c) 111.74 ft   **(d)** 183.5 ft

19. 17

21. 5

23. $\mu = 52$, $\sigma = 16$

C. The 68-95-99.7 Rule

25. $\mu = 55, \sigma = 5$

27. $\mu = 92, \sigma = 3.4$

29. $\mu = 80, \sigma \approx 10$

31. **(a)** 97.5%   **(b)** 13.5%

33. $\sigma = \$15$

D. Approximately Normal Data Sets

35. **(a)** 52 points   **(b)** 50%
(c) 68%   **(d)** 16%

37. **(a)** $Q_1 \approx 44.6$ **(b)** $Q_3 \approx 59.4$ **(c)** IQR ≈ 14.8

39. **(a)** 1900   **(b)** 1630

41. **(a)** approximately the 3rd percentile
(b) the 16th percentile
(c) around the 22nd or 23rd percentile
(d) the 84th percentile
(e) the 99.85th percentile

43. **(a)** 95%   **(b)** 47.5%   **(c)** 97.5%

45. **(a)** 13   **(b)** 80
(c) 250   **(d)** 420
(e) 488   **(f)** 499

47. **(a)** 16th percentile   **(b)** 97.5th percentile   **(c)** 18.6 lb

49. **(a)** 97.5th percentile   **(b)** 99.85th percentile   **(c)** 8 lb

51. 1.35 standard deviations

E. The Honest- and Dishonest-Coin Principles

53. **(a)** $\mu = 1800, \sigma = 30$
(b) $\approx 68\%$ **(c)** $\approx 34\%$ **(d)** $\approx 13.5\%$

55. **(a)** $\approx 68\%$ **(b)** $\approx 16\%$ **(c)** $\approx 84\%$

57. **(a)** $\mu = 240, \sigma = 12$
(b) $Q_1 \approx 232, Q_3 \approx 248$
(c) ≈ 0.95

59. **(a)** $\mu = 30, \sigma = 5$
(b) ≈ 0.025
(c) ≈ 0.34

61. 8% chance

JOGGING

63. **(a)** 20.55 lb   **(b)** 16.75 lb

65. **(a)** 60th percentile   **(b)** 30th percentile

67. **(a)** 770 points
(b) 590 points
(c) 60th percentile

69. $n = 100$

71. For $p = \frac{1}{2}$, $\mu = np = \frac{n}{2}$, and $\sigma = \sqrt{np(1-p)} = \sqrt{n \cdot \frac{1}{2} \cdot \frac{1}{2}} = \frac{\sqrt{n}}{2}$.

INDEX

PHOTO CREDITS